oundations of Pulsed
ower Technology

IEEE Press
445 Hoes Lane
Piscataway, NJ 08854

Foundations of Pulsed Power Technology

Jane Lehr and Pralhad Ron

IEEE PRESS

WILEY

Published by John Wiley & Sons, Inc., Hoboken, New Jersey.
Published simultaneously in Canada.

Library of Congress Cataloging-in-Publication Data is available.

ISBN: 978-1-118-628393

Printed in the United States of America.

10 9 8 7 6 5 4 3 2 1

Contents

Preface

Pulsed power technology provides conditioned charging profiles for present and future applications requiring high peak power. This book is primarily concerned with providing the concepts, design information, and system techniques for optimizing this profile to the application. The field of pulsed power has been difficult to codify because each high-power application requires specific power sources. This customization, along with the innate interdisciplinary nature of the field, has inhibited comprehensive documentation. This book is meant to be useful for designers of pulsed power sources and the researchers who require them. The book will be valuable for graduate students, studying courses in pulsed power technology, plasma physics and applications, laser physics and technology, high-voltage insulation and power system engineering, measurement and diagnostics, high-power electromagnetics, particle beams, and electromagnetic interference and compatibility.

The authors have spent the bulk of their careers in government research laboratory settings in the design and construction of pulsed power technology and applications and have been heavily involved in the evolution of the field. Both authors have taught academic courses in pulsed power and recognize the deep need for a comprehensive book focused on the fundamental principles of the field. It is our deep conviction that a strong foundation in the fundamental principles – and the history of the field – will provide the future workforce with the necessary skills for emerging applications. This book, with its emphasis on engineering design and construction of pulsed power equipment, is intended for graduate students and practicing engineers with specialization in multiple disciplines, since it establishes a firm foundation in pulsed power components, systems, and measurements. Moreover, we include electromagnetic interference, compatibility, and topology concepts for the purpose of controlling noise and interference for modern designs of the pulsed power system. The book bridges the gap between a textbook for students and a monograph for research scientists. The length of the book is intended to provide in-depth insight into the theory, design, and construction of individual components of pulsed power equipment, but short enough to keep the attention anchored to overall system requirements. This book is illustrated with a large number of equations derived from fundamental concepts, figures, and solved design examples. *Foundations of Pulsed Power Science and Technology* complements the other books on the subjects of pulsed power: Gennady A. Mesyats, *Pulsed Power*, Kluwer Academic/Plenum Press, 2005; Paul W. Smith, *Transient Electronics: Pulsed Circuit Technology*, John Wiley & Sons, Inc., 2002; *JC Martin on Pulsed Power*, edited by T.H. Martin, A.H. Guenther, and M. Kristiansen, Plenum Press, 1996; S.-T. Pai and Qi Zhang, *Introduction to High Pulse Power Technology*, World Scientific, 1995; and W. James Sarjeant and R.E. Dollinger, *High Power Electronics*, TAB Books, 1989.

The entire subject of pulsed power technology is covered in 12 chapters. Each chapter contains a large number of references, to lead the researchers to greater depths in the field. This book is organized such that the Chapters 1–5 describe the "building blocks" of a pulsed power system. Chapters 6 and 7 describe considerations with examples of systems resulting from synergetic integration of individual components. The remaining five chapters describe the vital topics of electrical breakdown in insulators of interest (Chapters 8 and 9), pulsed voltage and current measurements (Chapter 10), and electromagnetic interference, compatibility, and topology for interference control (Chapters 11 and 12).

The following are the salient features of the various chapters:

- The design formulas, considerations, and examples of the widely used voltage-multiplying circuits based on the Marx generator are discussed in Chapter 1, Marx Generators and Marx-Like Circuits. A number of modified configurations of low-inductance Marx generators with capabilities for high-power delivery, fast erection with low jitter, and capability for repetitive pulse generation are discussed. Other circuits, such as the Maxwell Marx and Fitch circuit are introduced even though not widely used.
- Chapter 2, Pulse Transformers, introduces another popular means of voltage multiplication: the Tesla transformer. This chapter also includes a discussion of transmission line transformers for modifying high power pulses with minimum distortion, with its special application for impedance matching to the application device.
- Pulse forming lines, discussed in Chapter 3, receive input from a Marx generator or Tesla transformer, having capabilities to deliver gigawatts of power in a very short rise time and a flat-top. The various configurations of PFL, such as coaxial lines, striplines, Blumleins, stacked Blumleins, radial lines, helical lines, and spiral generators, are covered. The optimization of PFL design from the viewpoints of maximum charging voltage, maximum power delivery, choice of dielectric, and dielectric strength dependence on charging time are illustrated with solved design examples.
- The design considerations and performance parameters of self-triggered and externally triggered *spark gap switches* to transfer the energy to the load at

high efficiency with minimum distortion in waveform are covered in Chapter 4. A review of spark gap configurations, trigger geometries, trigger modes, and salient features of specialized spark gaps like krytrons, burst mode gaps, and radioisotope-aided gaps is included. Design examples are provided for calculation of inductive rise time, resistive rise time, and a rough estimation of number of spark channels.

- Chapter 5 describes a variety of *opening switches* that are a critical component of inductive energy storage systems. An exhaustive coverage of opening switches, their configurations, and performance are discussed at length. The subject covered in this chapter is taken largely from the NRC Report "Opening Switches in Pulsed Power Systems," Rep.TR-GD-007, by P.H. Ron & R.P. Gupta.
- Chapter 6 on multigigawatt pulsed power systems describes advanced systems with capabilities to deliver single or repetitive pulses at very high peak power levels. The major system categories are cascaded capacitor storage, cascaded inductor storage, magnetic pulse compression schemes, inductive cavity cells, and induction linacs. Some well-known machines comprised of fast Marx generators, fast pulse forming lines, and multichannel spark gaps are discussed at greater length, because of their historical importance and the relevance to the evolution of modern-day pulsed power systems.
- In Chapter 7, Energy Storage in Capacitor Banks, the theoretical, practical, and safety aspects involved in the design and construction of high current or energy storage capacitor banks are discussed. Capacitor banks are used for the delivery of large energy in the microsecond regime. The capacitor bank discharges have wide ranging applications in plasma heating, high magnetic field generation, and electromagnetic propulsion.
- Chapter 8, Electrical Breakdown in Gases, gives basic concepts of the kinetic theory of gases and ionization. The early experiments of Paschen and Townsend are described as well as insight into the fundamental mechanisms of electrical breakdown of gases. Pseudospark discharge and corona phenomena are also introduced. The techniques of optimum utilization of insulation by providing intermediate electrodes into electrically weak cylindrical and spherical geometries are discussed at depth. Practical hints are given for deriving maximum benefit out of SF_6 gas and its mixtures with other gases.
- Chapter 9 deals with the properties of electrical insulation and electrical breakdown in solids, liquids, and vacuum. The breakdown mechanisms in the above dielectrics and practical techniques to be adopted for enhancing insulation performance are discussed. The important topics of partial discharges and electrical trees, which govern long-term performance of solid dielectrics, behavior of liquids for PFL insulation, and vacuum in application

devices, and the theory of surface flashover and some mitigation techniques are described.

- The concepts and techniques for *pulsed voltage and current measurements*, which necessarily involves the accurate scaling down of parameters without distortion of pulse shape, are discussed at great depth in Chapter 10. The electro-optical and optoelectronic techniques, which possess high immunity to intense EMI, are introduced.
- The topic of Chapter 11 electromagnetic interference and compatibility, is important because of the intense radiated electromagnetic fields generated by the operation of the pulsed power system. This radiation may damage equipment or cause inaccurate measurements or even operation because of extraneous ground loops. After discussing the theory of fundamental mechanisms of capacitive coupling, inductive coupling, common impedance coupling, and radiative coupling responsible for electromagnetic interference, the practical methods of incorporating protection techniques such as shielded cables, power line and signal line filters, isolation transformers, effective earthing, and shielded enclosures are discussed at length.
- Techniques required for high frequencies are introduced in Chapter 12, EM Topology for Interference Control. Multiple, nested shields in complicated geometric shapes may be incorporated to enable the safe and reliable operation of electronic systems from high interfering levels in harsh environments. To maintain high shield integrity in the presence of numerous shield surface discontinuities requires specialized techniques in the various protection zones and high standards of connections between the penetrating devices and shield.

Jane Lehr
Pralhad Ron

About the Authors

Jane Lehr received the Bachelor of Engineering degree from Stevens Institute of Technology and the Ph.D. degree in Electrical Engineering from New York University in 1996 under the supervision of Professor Erich Kunhardt. Dr. Lehr joined the University of New Mexico's Electrical and Computer Engineering faculty in 2013 after spending 12 years at Sandia National Laboratories in the Pulsed Power Sciences Center. She served 5 years at the Air Force Research Laboratory's Directed Energy Directorate studying ultra-wideband high-power electromagnetics and repetitive pulsed power where she was awarded the USAF-wide Basic Research Award in 2001. Dr. Lehr's research interests are in all aspects of high-power electromagnetics, pulsed power, high-voltage engineering, and the physics and application of electrical breakdown in vacuum, gases, and liquids. Dr. Lehr is a Fellow of the IEEE.

Dr. Lehr served as President of the IEEE Nuclear and Plasma Sciences Society in 2007 and 2008. She has served the Society in a number of roles since then and currently serves as the Chair of the NPS Society Fellow Evaluation Committee. She has served on the Administrative Committees for both the Nuclear and Plasma Science and the IEEE Dielectric and Electrical Insulation Societies. She has served as an Associate Editor of the *IEEE Transactions on Dielectrics and Electrical Insulation*, Guest Editor of the *IEEE Transactions on Plasma Science*, and on the IEEE Technical Activities Board Publications Committee. Dr. Lehr received the IEEE Shea Distinguished Member Award in 2015 and the IEEE Region 6 Award for Leadership in 2001. She was named an Outstanding Woman of New Mexico and has been inducted into the New Mexico Hall of Fame for her technical accomplishments and volunteer activities serving her local community.

Pralhad Ron was born in 1939 in Dharwad, Karnataka, India. He obtained his Bachelor of Engineering degree from Pune Engineering College, India, in 1961. He was the winner of the Homi Bhabha Award for top performance in electronics engineering. He joined the Bhabha Atomic Research Centre

(BARC) in 1962. He obtained the M. Sc. (Engg.) degree from the University of Manchester, UK, in 1969 and a Ph.D. from the Indian Institute of Science, Bangalore, India, in 1984.

At BARC, Dr. Ron specialized in the design and development of high-voltage equipment for carrying out applications, such as (a) electron beam processing for melting and welding in vacuum., (b) industrial electron beams for radiation processing in atmosphere, (c) pulsed power technology for production of nanosecond multigigawattt electron beams, flash X-rays, and EMP testing, (d) pulsed high magnetic fields for magnetization, demagnetization, and magnetoforming, and (e) electromagnetic interference simulation and protection technology. Dr. Ron led a team of engineers, who successfully carried out garter spring repositioning in the nuclear power reactors at Narora and Kakrapar. Under his leadership, electron beam processing plants based on Cockcroft-Walton multipliers and RF Linacs were successfully developed for industrial applications.

Dr. Ron served as Head of the Accelerator and Pulsed Power Division of BARC from 1992–2001. He was a visiting scientist at Queen Mary College, London, UK in 1970 working on electrical breakdown and at the National Research Council in Ottawa, Canada in 1985 and McGill University, Montreal, Canada, in 1987 working on pulsed power technology. He taught a graduate course in pulsed power technology from 1996 to 2000 at the Devi Ahilya University in Indore, India. Dr. Ron was the Steering Committee Chairman of the Electron Beam Centre in Kharghar, New Bombay, to carry out radiation processing of polymers. He was Chair of the Atomic Energy Regulatory Board's Design Safety Review Committee for the construction of particle accelerators in India. He was Chairman of the Engineering Design Committee for the conversion of BARC's 5.5 MV van de Graff generator into a 7 MV folded tandem ion accelerator. He was a member of the following councils for the Department of Atomic Energy: (a) Trombay Council at BARC, (b) CAT Council, Center for Advanced Technology, Indore, Madhya Pradesh, India, and (c) Cyclotron Council, at Variable Energy Cyclotron Centre, Calcutta, West Bengal, India.

.cknowledgments

he authors are grateful to the large community in pulsed power whose works rm the basis for this book. The authors are indebted to Dr. D.V. Giri, ProTech, alifornia, and the great late Dr. Carl Baum, Air Force Research Laboratory, irected Energy Directorate, and the University of New Mexico for their ncouragement, valuable guidance, and constructive suggestions.

JML is grateful to Ian D. Smith of L-3 Communications for an in-depth review the breakdown scaling formulas; Keith Hodge and Zachariah Wallace of ndia National Laboratories; University of New Mexico colleagues Chuck ιeben for lessons and Edl Schamiloglu for advice; Jon Cameron Pouncey for mmentaries and encouragement; and Mark, for technical assistance and oral support, and to the memory of Lucius J. Lehr, my magnus opus, who is arly loved.

PHR is grateful to Dr. R. Chidambaram Scientific Advisor, Government of dia and former chairman, Department of Atomic Energy and Dr. Anil .kodkar, Homi Bhabha Chair and former chairman of the Department of omic Energy and Dr. S. Bhattacharjee, ex-Director, Bhabha Atomic Research nter (BARC), for their support; Dr. Raj Gupta for providing visiting fellowip grants; the following staff members of the Accelerator and Pulse Power vision (APPD) of BARC: Mr. S.H. Kamble, Mrs. Supriya Barje, Dr. G.V. Rao, rs. Geeta Patil, and Miss Poornalaxmi for the preparation of typescripts; his fe, Asha, for providing a loving home environment, thus enabling long ncentrated working hours on this book; and also Nitin and Sangeeta and esh and Monica for their cooperation.

ntroduction

ulsed power technology is an area of interest to physicists and engineers in elds requiring high voltages and large currents. Modern pulsed power runs the amut from its historical roots in flash radiography, X-ray generation, and the mulation of weapons effects, such as nuclear electromagnetic pulse (EMP), to ackaged pulsed power for directed energy weapons and biological and medical pplications. New applications and techniques continue to emerge.

Pulsed power has traditionally been described as the gradual accumulation of nergy over a relatively long timescale and the subsequent compression into ulses of high instantaneous power for delivery in the required form to a load. his process is illustrated in Figure I.1, and is discussed in Chapter 3. Depending the application, the slow accumulation of energy may be over minutes, such for charging large capacitive energy store, or milliseconds for systems perating in a repetitive burst mode. The fast discharge is usually less than ns of microseconds but may be measured in tens of picoseconds.

Pulsed power generally falls within the following range of parameters:

ergy per pulse	1–10^7 J
ak power	10^6–10^{14} W
ak voltage	10^3–10^7 V
ak current	10^3–10^8 A
se width	10^{-10}–10^{-5} s

This explanation, however, does not capture the two key elements of the field: e exploitation of the time dependence of electrical breakdown on insulating aterials and the specifics of the load requirements.

The observation that electrical breakdown of insulators – be it gas, liquid, or id – occurs at higher electric fields at shorter pulse durations was well known fore the "birth" of pulsed power in the United Kingdom in the late 1950s. finitive work on electrical discharges in gases and the mechanisms leading to electrical breakdown was performed in the first half of the twentieth century, starting with Townsend's experiments on current growth in low-pressure gases and its relation to ionization. The invention of the Marx generator, patented in 1923, allowed the generation of substantially higher voltages with shorter pulse durations than previously available. The streamer mechanism of electrical breakdown of gases was proposed in the 1940s and seminal work was performed by Loeb, Meek, Craggs, and Raether. At the same time, Llewellyn-Jones and Davies, as well as Raether, continued to advance the understanding of the Townsend breakdown mechanism. The focus, however, was on physics and scant attention was paid to the increased breakdown strength of insulators upon application of pulses of short duration. Even in the very thorough, modern text, *Gas Discharge Physics* by Yu. Raizer [1], the time dependence of electrical breakdown is rarely mentioned.

The integral nature of the time dependence of electrical breakdown to pulsed power is illustrated in an anecdote relayed by Goodman [2] and Martin [3]. The Atomic Weapons Research Establishment (AWRE) in the United Kingdom had acquired a used electron accelerator that was being used to radiograph explosive events. The radiographs were somewhat blurred and a better resolution was needed. The conventional option was to increase the accelerator beam current by three orders of magnitude – an enormously expensive option. Martin proposed an alternative concept of a high-voltage (~6 MV), high-current (~50 kA) accelerator lasting for 30–50 ns exploiting the time dependence of electrical breakdown. The second option was chosen on the basis of cost, and, with its success, pulsed power was born. Martin and his colleagues went on to develop a number of high peak power devices, and, along the way, an empirical relation for the time dependence of insulator breakdown that had the general form:

$$F \cdot t^a \cdot A^b = k$$

where F is the average electrical breakdown field, t is the charge time, A is the area, and a, b, and k are constants that depend on the insulating material. These empirical relations allowed electrical breakdown to be predicted and therefore exploited, and are discussed in detail in Chapter 3. The scaling relation clearly shows that higher breakdown electric fields are attained with shorter charge and discharge times. Thus, using the pulse compression scheme illustrated in Figure I.1 permits the generation of high peak power. An in-depth history of these early developments can be found in the article by Smith [4]. Much of the material presented here is derived from work done in the United States and the United Kingdom. Pulsed power was independently and simultaneously developed in the former Soviet Union and is well documented by Mesyats in the English translation of his incredibly detailed book *Pulsed Power* [5].

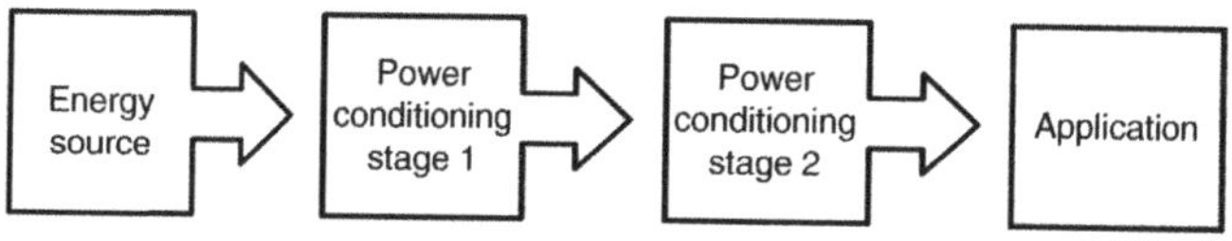

Figure I.1 Pulsed power may be represented schematically as a series of power conditioning stages that increases the peak power while decreasing the pulse width. This technique is known as pulse compression.

Pulsed power technology, simply stated, is the technique and equipment required to adapt the power characteristics of the prime power source to the electrical requirements of the load. Pulsed power in the context of power conditioning captures the quintessence since significant efforts must be made to optimize and specify the power demand. In the preface to the Pulsed Power Monograph series [6–8], Magne (Kris) Kristiansen and Art Guenther describe pulsed power as "special power conditioning for specific applications." This statement captures both the inextricable link between pulsed power and the application and their uniqueness. In applications requiring high peak power, pulsed power is a low-cost power conditioning technique. Electrical efficiency is increased by delivering the power in a specific optimal form – the "art" of pulsed power. Pulsed power provides unique solutions to certain physics applications.

Based on the above, the following succinct description of pulsed power is proposed:

> Pulsed power is a special power conditioning technique that transforms the characteristics of the prime energy source to the electrical requirements of the load. Energy from a primary source is accumulated over a relatively long time scale and compressed into pulses of high instantaneous power. Several stages may be needed to fully exploit the time dependence of breakdown of insulating materials to deliver energy with the required time dependence and amplitude for the application. The resulting peak power delivered to the load has a large ratio of instantaneous-to-average power.

Recognizing pulsed power as special power conditioning, the breadth of applications and the wide variety of implementation imply an overwhelming amount of knowledge is needed to participate. Certainly, this is true in part, but a large user-facility or a high-performance system is rarely designed by a single engineer. Our premise in choosing material for this book is that a strong foundation in fundamental principles – using realized systems as examples – provides a better perspective for the wide number of applications a pulsed power engineer should expect to encounter over a career. In modern times, it is rare that a pulsed power engineer spends an entire career on a single application. In general, an in-depth treatise on applications is avoided to highlight this and instead the focus is on the foundations of pulsed power technology on which most systems are built. Many of the references in this book are old, which is a reflection of fundamental nature of this book, but modern references are included where appropriate.

While traditional applications of pulsed power are still relevant and continue to incorporate advances in technology, a host of new applications with very different operational requirements are emerging. For this reason, we chose to avoid application-driven pulsed power and focus on the fundamentals to provide a strong technical foundation for the next generation, as well as to document the many innovations achieved thus far. It is our belief that once the basics are mastered, they can be combined in any number of ways to create the specified output. The dimensional scale of the equipment may be vastly different but a focus on the fundamentals allows the similarities to be seen. For example, the Marx bank was invented almost a century ago, but continues to play an integral part in many systems. The basic Marx architecture is very versatile: It has been used with solid-state switches to produce a few hundred volts and is used exclusively to produce tens of megavolts, and it stores energy ranging from joules to kilojoules. It is the basis for trigger generators as well as lightning simulators. Marx generators are used exclusively as the energy storage stage of multigigawatt pulsed power systems and voltages over 18 MV have been produced [9]. However, the fundamental circuit architecture used to produce a 10 J per pulse, 200 kV peak voltage Marx generator that is only 15 cm long [10] is also configured into a bank of Marx generators that yields 5 MV, requires a robust mechanical support, and powers a current source. While is it obvious that the application of these two voltage sources are vastly different, their basic operation is the same.

Sources of Information

The dissemination of information related to pulsed power has been a low priority for much of its history and this may be attributed not only to its initial rapid growth but also to its initial use for military applications. Thus, much of the early progress was preserved primarily in reports and internal memoranda and passed informally throughout the community. Copies of these early reports are increasingly difficult to find, with the exception of Carl Baum's Note Series. The Note Series was started in early 1964 by R.E. Partridge, a technical staff member at Los Alamos National Laboratory, to document the rapid progress in the simulation of the nuclear-generated effects. The Note Series recorded both simulation technology (largely pulsed power) and measurement techniques. The Note Series quickly passed into the care of Carl Baum and now stands at over 2000 documents, most of which are available electronically (The Note eries is maintained by the Summa Foundation and notes are available from ttp://www.ece.unm.edu/summa/notes/).

As an academic aid, Texas Tech University and the Air Force Weapons aboratory (now called the Air Force Research Laboratory Phillips Research te), under the leadership of Kristiansen, created the Pulsed Power Lecture eries. Forty-nine lectures were given and 35 of these leading researchers wrote self-consistent treatise on their specialty. Later, Kristiansen, along with Art uenther, began a book series in the category of *Advances in Pulsed Power echnology*. This series started with two books on high-power switching [6,7]. he third book was a collection of the internal memoranda and high-voltage ctures documenting the work of Charlie Martin's group at AWRE [8]. Many of e memoranda that comprise this book were also archived in the Carl Baum ote Series.

There are also a number of books that are widely used by some pulsed power chnologists. For instance, the two classic texts by Meek and Craggs [11,12] as ell as the textbooks by Raizer [1] and Fridman and Kennedy [13] give a orough treatment of gas discharge physics and electrical breakdown. obine [14] contains information on the practical implementation of gas scharges, such as the effects of humidity, which are difficult to find elsewhere. rover's book [15], which has been republished, is a thorough reference on ductance calculations, and Knoepfel has two books on high magnetic lds [16,17]. The MIT Radiation Series volume, *Pulse Generators* [18], conns essential information on a variety of pulsed lumped element circuit chniques, which are not included in this book. The volume by Lewis and ells [19] is another out-of-print book that contains a number of clever nsmission line configurations for pulsed circuits that are still relevant today moderate voltages. Martin credits Lewis and Wells as a source of considerle influence in the early days of pulsed power technology [3].

A primary source of information must be the biannual International Pulsed wer Conference and its proceedings. The first conference was held in bbock, Texas, in 1976 to aid the dissemination of information in the erging field. The next conference was held in 1979 and every other year ce. In 1995, the Pulsed Power Conference became sponsored by the IEEE and publications are now available worldwide through IEEE Explore. To encour- publication in archival, refereed articles, the *IEEE Transactions on Plasma ence* published the first *Special Issue on Pulsed Power Science and Technology* April 1997. Its success led to similar Special Issues with open call for papers olished in October of even years. Other conferences are the European-Asian sed Power Conference, held in even years, as well as a variety of other ferences specializing in either specific applications or subtopics. An example the former is the Electromagnetic Launch Conference; and the latter is the gagauss Conference, which focuses on the generation and use of high gnetic fields.

The evolution of pulsed power may also be viewed through the two issues of the *Proceedings of the IEEE*. The first appearance of pulsed power in the *Proceedings of the IEEE* was a portion of an issue, edited by Pace Van Devender, which included the classic paper by Martin [20]. Other review articles exist [21]. More recently, Edl Schamiloglu and Robert J. Barker devoted an issue of the *Proceedings of the IEEE* to applications of pulsed power [22]. Through this prism the evolution and expansion of pulsed power is apparent.

References

1. Yu. P. Raizer, *Gas Discharge Physics*, Springer, 1991.
2. M.J. Goodman, High Speed Pulsed Power Technology at Aldermaston, in *J.C. Martin on Pulsed Power*, in T.H. Martin, A.H. Guenther, and M. Kristiansen, eds., Plenum Press, New York, 1996.
3. J.C. Martin, Brief and Probably Not Very Accurate History of Pulsed Power at the Atomic Weapons Research Establishment Aldermaston, in *J.C. Martin on Pulsed Power*, in T.H. Martin, A.H. Guenther, and M. Kristiansen, eds., Plenum Press, New York, 1996.
4. I.D. Smith, The Early History of Western Pulsed Power. *IEEE Trans. Plasma Sci.*, Vol. 34, No. 5, pp. 1585–1609, 2006.
5. G.A. Mesyats, *Pulsed Power*, Kluwer Academic, 2005.
6. A. Guenther, M. Kristiansen, and T. Martin, eds., *Advances in Pulsed Power Technology: Vol. 1: Opening Switches*, Plenum Press, New York, 1987.
7. G. Schaefer, M. Kristiansen, and A. Guenther, eds., *Advances in Pulsed Power Technology: Vol. 2: Gas Discharge Closing Switches*, Plenum Press, New York, 1990.
8. T.H. Martin, A.H. Guenther, and M. Kristiansen, eds., *Advances in Pulsed Power Technology: Vol. 3: J.C. Martin on Pulsed Power*, Plenum Press, New York, 1996.
9. K.R. Prestwich and D.L. Johnson, Development of an 18 Megavolt Marx Generator. *IEEE Trans. Nucl. Sci.*, Vol. 16, Part II, No. 3, p. 64, 1969.
10. M.V. Fazio and H.C. Kirbie, Ultracompact Pulsed Power. *Proc. IEEE*, Vol. 92, No. 7, pp. 1197–1204, 2004.
11. J.M. Meek and J.D. Craggs, *Electrical Breakdown of Gases*, Clarendon Press, Oxford, 1953.
12. J.M. Meek and J.D. Craggs, eds., *Electrical Breakdown of Gases*, John Wiley & Sons, Inc., 1978.
13. A. Fridman and L.A. Kennedy, *Plasma Physics and Engineering*, Taylor & Francis Publishing, 2004.
14. J.D Cobine, *Gaseous Conductors: Theory and Engineering Applications*, Dover Publications, Mineola, NY, 1941 (and also 1958).

5 F.W. Grover, *Inductance Calculations Working Formulas and Tables*, Dover Publications, Mineola, NY, 1946 (and also 1973).
6 H.E. Knoepfel, *Magnetic Fields*, John Wiley & Sons, Inc., 2000.
7 H.E. Knoepfel, *Pulsed High Magnetic Fields*, North Holland Publishers, 1973.
8 G.N. Glasoe and J.V. Lebacqz, *Pulse Generators*, McGraw-Hill, 1948.
9 I.A.D. Lewis and F.H. Wells, *Millimicrosecond Pulse Techniques*, Pergamon Press, 1959.
10 J.C. Martin, Nanosecond Pulse Techniques. *Proc. IEEE*, Vol. 80, No. 6, pp. 934–945, 1992.
11 I.R. McNab, Developments in Pulsed Power Technology. *IEEE Trans. Magn.*, Vol. 37, No. 1, pp. 375–378, 2001.
12 E. Schamiloglu and R.J. Barker, "Special Issue on Pulsed Power: Technology and Applications," Proceedings of the IEEE, Volume: 92 Issue: 7, pp 1011–1013, 2004.

1

Marx Generators and Marx-Like Circuits

The simplest and most widely used high-voltage impulse generator is the device Erwin Marx introduced in 1925 for testing high-voltage components and equipment for the emerging power industry. The basic operation of a Marx generator is simple: Capacitors are charged in parallel through high impedances and discharged in series, multiplying the voltage. This simplicity, however, is somewhat misleading: The design of a Marx generator, when stray reactance is included and reliability and precise timing are needed, can be incredibly complex.

This chapter discusses the principles of operation and overall performance of Marx generators. For instruction, the design formulas for simple Marx generators based on their equivalent circuits are given in considerable detail. Some aspects are highlighted in the discussion of modified Marx configurations. The importance of overvoltages to Marx operation, as well as advanced triggering techniques, are reviewed. Various aspects of Marx generators such as electrical insulation, delay time and jitter, and the selection of components are discussed. A rigorous analysis is performed for pulse shaping using resistors, which is common in impulse generators.

1.1 Operational Principles of Simple Marxes

A Marx generator is a voltage-multiplying circuit that charges a number of capacitors in parallel and discharges them in series. The process of transforming from a parallel circuit to a series one is known as "erecting the Marx." In the common parlance, a "stage" is comprised of energy storage and switch. The energy storage elements are usually one or more capacitors, but pulse forming networks or transmission lines may also be used. The switches are almost always gas-insulated spark gaps, with varying sophistication, but other types of switches with low leakage current may be used. A careful evaluation of the role of overvoltage on the switches, however, is recommended.

Foundations of Pulsed Power Technology, First Edition. Jane Lehr and Pralhad Ron.

Figure 1.1 shows two simple ladder-type Marx generators, where a number N of capacitors with a capacitance value C_0 are charged in parallel through charging resistors R to a voltage V_0 and discharged in series through the spark gaps, producing an open-circuit voltage V_{OC}. The resistors play a dual role: During the charge cycle, the capacitors charge through the resistors on one side while the other completes the circuit to ground. During the discharge cycle, the resistors provide a high-impedance path, forcing the current through the spark gap. The resistance values are chosen sufficiently high to limit the current through the resistors and $R \sim$ few kΩ to a few $\sim$ MΩ is sufficient. The charge and discharge cycles and the erection of the Marx are treated separately. Inductors may also be used as isolation impedances.

The circuit of Figure 1.1a is an implementation that produces an output with the same polarity as the charge voltage. The advantage of this circuit is the elimination of a switch if the load can tolerate a modest DC voltage during the charge cycle. However, if the first switch is a triggered spark gap, as the Marx erects, a high-voltage transient is introduced into the trigger circuit. This issue is resolved in the circuit of Figure 1.1b. The first spark gap is at ground potential, making the choice of a trigatron as the triggered switch particularly attractive as

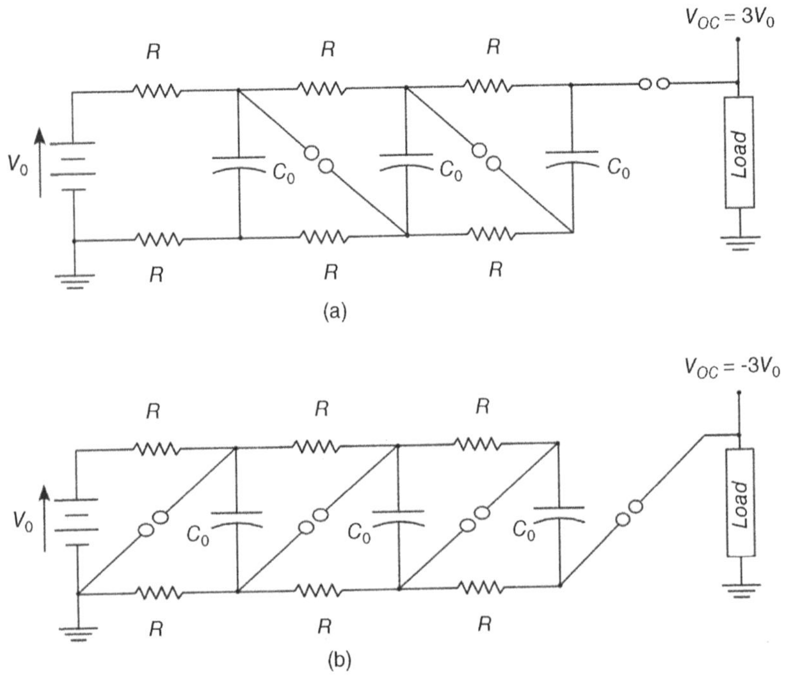

Figure 1.1 Ladder-type Marx generators may either (a) preserve the polarity of the charging voltage or (b) invert it.

the trigger pin may be embedded directly into the grounded electrode. The polarity of the output voltage is inverted from the charging polarity.

1.1.1 Marx Charge Cycle

During the charge cycle, the Marx charges a number of stages, N, each with capacitance C_0 to a voltage V_0, through a chain of charging resistors R, as shown in Figure 1.2.

The capacitors do not charge instantaneously, but do charge at different rates and sequentially. The time to charge the nth stage with a DC source is given approximately by Fitch [1] and validated by a rigorous analysis by Swift [2], as

$$\tau_{ch} = n^2 R C_0, \quad \text{where} \quad 1 \leq n \leq N \tag{1.1}$$

It is often advantageous to minimize the charge time T_M of Marx, which may be done by using a constant current-charging source. The Nth stage is the last to reach the full charge voltage, and the minimum charge time may be determined by the acceptable difference in charge voltage between the first and Nth (last) stage, illustrated in Figure 1.2, and given by [3]

$$\frac{V_{C,N} - V_{C,1}}{V_C} = \frac{N^2 R C_0}{T_M} \tag{1.2}$$

Given sufficient time, the last stage will charge to the full charge voltage V_0. However, the charge time determines the time of electrical stress on the insulation of the Marx, increasing the probability of unintentional insulation failure. Thus, minimizing the time to charge the Marx increases the reliability in highly stressed designs (Figure 1.3 1.3).

Moreover, during Marx erection, as the each stage is switched, the stored energy in that stage begins to discharge through the resistors R on each side with a time constant:

$$\tau_{disch} = \frac{RC_0}{2} \tag{1.3}$$

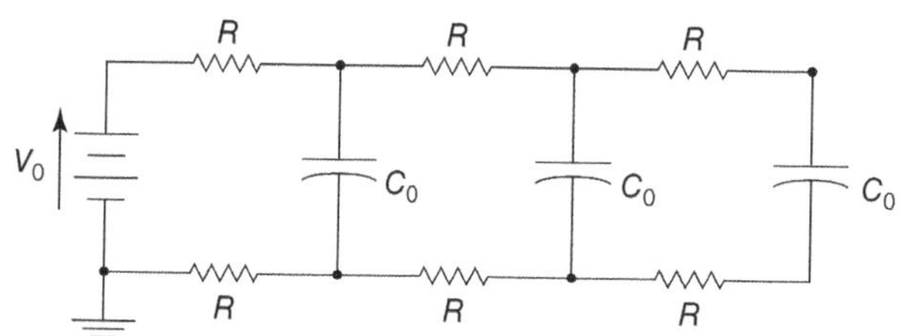

Figure 1.2 The Marx circuit during the charge cycle. Larger stage capacitances require longer charge times.

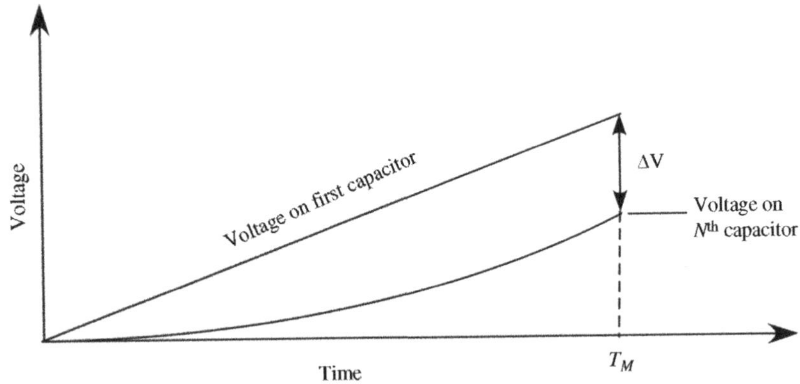

Figure 1.3 The difference in charge voltage ΔV between the first and Nth stage capacitor for a given Marx charge time TM charged with a constant charging current. The acceptable level of ΔV determines the minimum charge time

Energy dissipated in the resistor chain is energy lost to the load and contributes to inefficiency. The discharge time of the Marx is load dependent and should be kept short compared to τ_{disch} for maximum extraction of energy and generator efficiency.The maximum energy stored in the Marx is

$$E_{stored} = \frac{1}{2}(NC_0)V_0^2 \tag{1.4}$$

where (NC_0) is the parallel combination of the stage capacitance. The actual value is reduced from this maximum because of the reduced charge voltage on the upper stages and the energy lost to the resistors during the charge process.

1.1.2 Marx Erection

The Marx erection is the process of sequentially closing the switches to reconfigure the capacitors from the parallel charging circuit to the series discharge circuit. Marx erection is initiated when a spark gap fires resulting in an increased voltage across the remaining stages. The spark gap is said to be overvolted when the voltage exceeds its DC self-breakdown value. Sufficient spark gap overvoltages are critically important to reliable operation of the Marx.

Any switch in the Marx generator can initiate the erection process, but the maximum output voltage is ensured when the first spark gap initiates the discharge and fires each successive stage. Nonsimultaneity in the firing of spark gaps reduces the amplitude of the output voltage and distorts the waveform, as shown in Figure 1.4.

Each switch contributes its delay time and jitter to the Marx erection time and its overall jitter. Fast, reliable Marx erection requires that large overvoltages appear on each stage during discharge. The overvoltage as the Marx erects aids

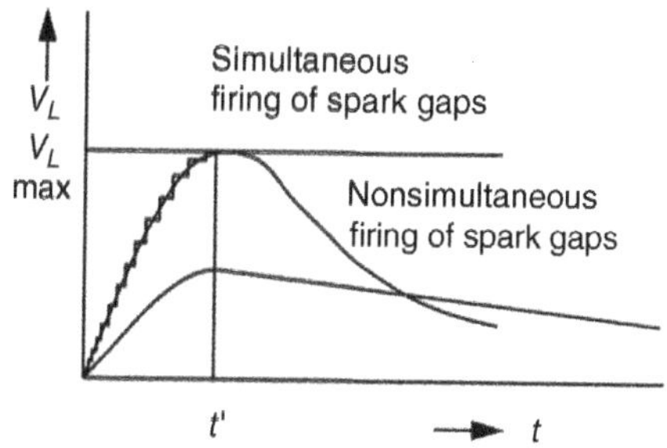

Figure 1.4 Sequential Marx erection minimizes pulse distortion and maximizes voltage.

in minimizing the switch jitter, but triggering methods may also be used. As Marx generators were used as primary energy storage for very high peak power applications, it was found the process of erecting a Marx is not necessarily straightforward as stray capacitances may limit the achievable stage overvoltage, the details of which are discussed in Section 1.3. Here, all stray impedances are neglected, and the erection of an ideal Marx is examined.

1.1.2.1 Switch Preionization by Ultraviolet Radiation

A convenient method of ensuring low jitter in simple Marx generators is to use the ultraviolet light generated by the firing of the first spark gap to trigger the next stage. The firing of the second stage generates ultraviolet light that aids the erection of the next stage and the Marx erects in a cascade fashion. The circuits of Figure 1.1 may be reconfigured as shown in Figure 1.5 so that the switches are arranged in line-of-sight. The spark gaps are often inserted into a pressurized gas column with the remaining components insulated with oil. This easily fabricated Marx can produce hundreds of kilovolts and erect reliably.

1.1.2.2 Switch Overvoltages in an Ideal Marx

Switch overvoltages during erection can be investigated with the aid of Figure 1.6 where, for simplicity, an ideal Marx has infinitely large charge resistors that draw no current. The remaining lossless circuit is a string of capacitors. Each stage capacitor C_0 is charged to V_0. Note that in this model, the voltages across each spark are equal and represented by V_g.

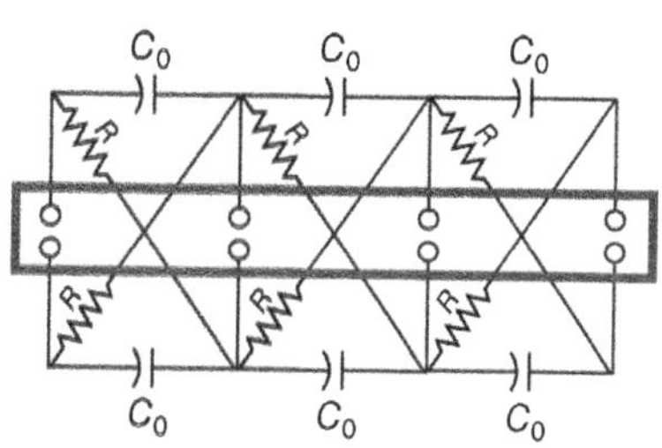

Figure 1.5 A Marx circuit arranged to allow the ultraviolet light generated by the closing of a spark gap to preionize the other spark gaps. The solid line represents a cylindrical support for the spark gaps and the resistors may be wound around the cylinder to connect the capacitors.

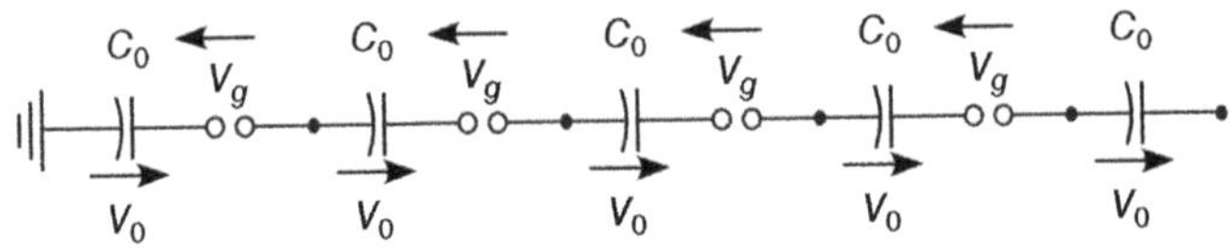

Figure 1.6 A circuit model for an ideal Marx shows how the erection sequence overvolts the gaps.

The unfired Marx draws no current and $V_{OC}=0$. Applying Kirchoff's voltage law (KVL) to each stage, the voltage across each spark gap is equal in magnitude and opposite in sense to the stage capacitor voltage. When the first spark gap fires, its voltage goes to zero and the voltage at its upper node becomes the stage voltage V_0. The Marx remains unfired and the voltage is redistributed across the remaining unfired spark gap changing from V_0 to a higher value, determined by the number of switches that have fired. Using KVL,

$$V_{OC} = 0 = \sum_{n=1}^{N} V_0 - \sum_{n=1}^{N-1} V_g = NV_0 - (N-1)V_g$$

$$V_g = \frac{N}{N-1} V_0 \tag{1.5}$$

The firing sequence is given by the following [4]:

Before firing, voltage on first gap V_{g1}	$NV_0 = NV_g$	$V_g = V_0$
First gap fires:	$NV_0 = (N-1)V_g$	$V_g = \frac{N}{N-1} V_0$
Second gap fires:	$NV_0 = (N-2)V_g$	$V_g = \frac{N}{N-2} V_0$
*n*th gaps fires:	$NV_0 = (N-2n)V_g$	$V_g = \frac{N}{N-n} V_0$
$(N-1)$th gap fires:	$NV_0 = V_g$	$V_g = NV_0 = V_{OC}$

As each spark gap fires, the voltage across each spark gap increases until the full open-circuit voltage is across the final spark gap, leading to the full erection of the Marx. Insufficient overvoltage results in late firing or nonfiring of some stages or energy following unintended paths. In practice, this model is too simplistic and implies significant overvolting that is not realized. In addition to losses in finite isolation resistors, the degree of overvolting is significantly affected by stray capacitance, which is explored further in Section 1.3.

1.1.3 Marx Discharge Cycle

The discharge of a Marx has two distinct phases, which will be treated separately. The first is the case where the final output gap of the Marx does not fire. The other is the case when the last gap discharges into a load.

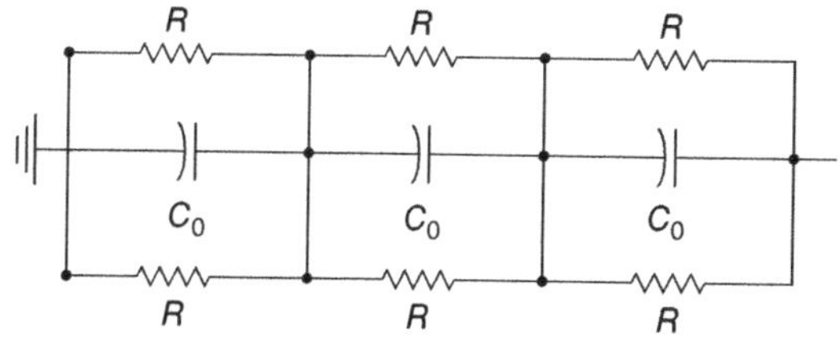

Figure 1.7 The equivalent circuit of the erected Marx. If the last gap in the Marx does not fire, the charged capacitors must discharge their energy through the resistor network.

1.1.3.1 No Fire

When the final output gap of the Marx does not fire, the circuit is represented by the equivalent circuit of Figure 1.7. The charge on the capacitors is discharged through the two charging resistors in parallel, with an effective time constant of $1/2RC_0$.

1.1.3.2 Equivalent Circuit Parameters During Discharge

Under proper Marx operation, where the last gap does fire, the Marx may be represented by the circuit in Figure 1.8. The output is essentially a capacitive discharge whose characteristics depend on the load. The Marx inductance, L_M, accounts for the inductance of the switches, capacitors, and connectors in the discharge circuit, and is given by

$$L_{Marx} = L_{Switches} + L_{Capacitors} + L_{connections}$$

Marx generators may be charged in a single polarity or differentially charged. Both configurations are common and may be treated identically if the Marx parameters C_M, L_M, and the energy are properly defined.

1.1.3.2.1 Single-Polarity Charging

If the resistor values are chosen so that little discharge current is drawn, the equivalent circuit is that given in Figure 1.8. The following are the parameters of the erected Marx:

Marx equivalent erected capacitance:

$$C_M = \frac{C_0}{N} \tag{1.6}$$

Figure 1.8 The equivalent circuit of an ideal Marx generator.

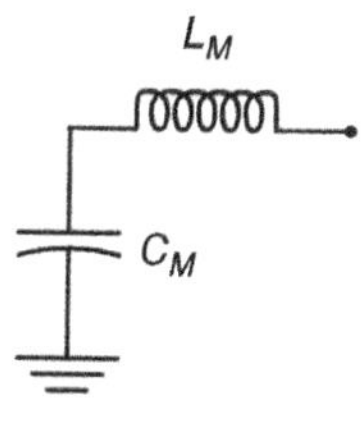

Marx erected impedance:

$$Z_M = \sqrt{\frac{L_M}{C_M}} \tag{1.7}$$

Marx intrinsic discharge time:

$$T_M = \sqrt{L_M C_M} \tag{1.8}$$

Erected Marx open-circuit voltage:

$$V_{OC} = -NV_0 \tag{1.9}$$

The Marx has an equivalent erected inductance, L_M, comprised of contributions from the capacitors, switches, and connections. When fully erected, the Marx is a capacitive discharge circuit whose output characteristics depend on the load. The peak output voltage under load conditions is designated V_M, to differentiate it from the open-circuit voltage V_{OC}. Fast Marx generators may be characterized by their intrinsic discharge times, with 500 ns being the state of the art, but designs for significantly smaller intrinsic discharge times exist [5–7].

The discharge cycle of the Marx is strongly dependent on the load characteristics. Of particular importance is to determine the discharge time so that it can be compared with Equation 1.3. This section calculates the discharge time for the cases where the Marx discharges into two important capacitive loads ($C_M \sim C_2$ and $C_M \gg C_2$) and a resistive load.

1.1.3.2.2 Dual-Polarity (Differential) Charging

A differentially charged, or bipolar, Marx, shown in Figure 1.9, uses two capacitors per stage charged to equal but opposite voltages. The differentially charged Marx has half the number of switches but at twice the stage voltage for a given output voltage. Two capacitors in series make up the stage capacitance.

In this arrangement, the Marx capacitance is given by

$$C_M = \frac{C_0}{2N} \tag{1.10}$$

The series Marx inductance, however, is lowered (compared to a single-polarity Marx with the same voltage) since N switches are used to switch $2N$ capacitors:

$$L_M = NL_S + 2NL_C \tag{1.11}$$

where L_S is the switch inductance and L_C is the capacitor inductance (Figure 1.9). The voltage per stage is $2V_0$, giving an open-circuit voltage of

$$V_{OC} = 2NV_0 \tag{1.12}$$

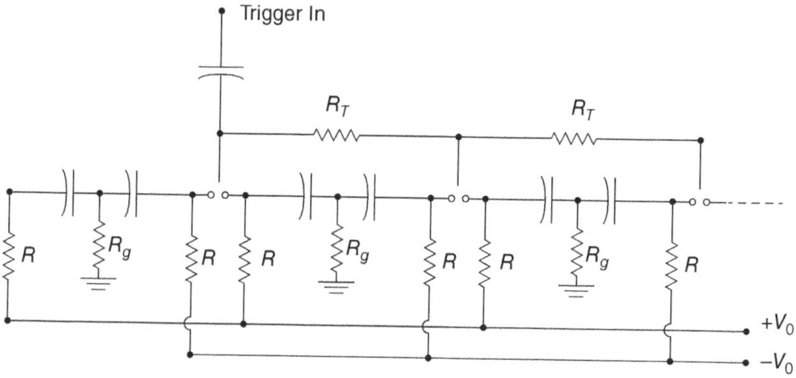

Figure 1.9 A differentially charged Marx generator uses a resistor to ground R_g to complete the circuit during the charge cycle. The first three switches are resistively coupled.

The energy stored in a bipolar Marx is

$$E_S = \frac{1}{2}(2NC_0)V_0^2 = NC_0V_0^2 \tag{1.13}$$

A high-impedance resistor to ground R_g is connected between the two capacitors of each stage. Generally, the value of R_g is chosen to be much greater than the charging resistors R_c to minimize the energy loss.

A differentially charged Marx generator is shown schematically in Figure 1.10, where the isolation resistor chain is replaced by an inductor chain to enable faster charging of the Marx. This allows higher pulse repetition rate capability and reduces the time of electric stress on the oil insulation during charging [8]. Bipolar charging, extensively used in very high energy storage generators, enables the efficient use of symmetrical three-electrode triggered

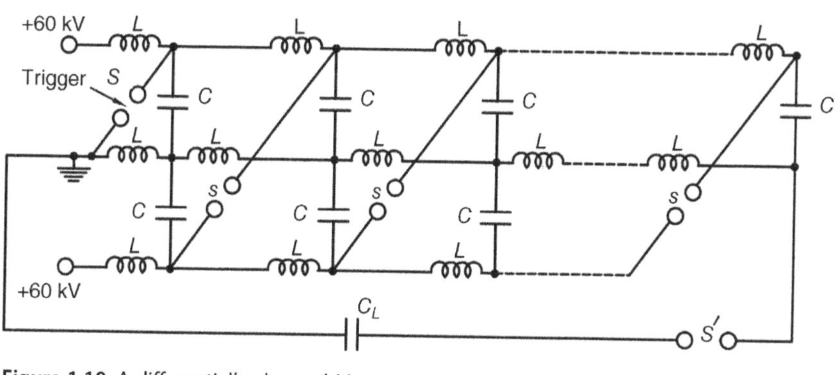

Figure 1.10 A differentially charged Marx using inductors as the charging and ground isolation impedances allows for higher repetition rates.

gaps. The center electrode, held to ground potential during charging, is the triggered electrode.

1.1.4 Load Effects on the Marx Discharge

A fully erected Marx generator is essentially a capacitive discharge. Thus, the load voltage depends not only on the characteristics of the Marx but also on the characteristics of the load. This is illustrated with the important cases when the load is a capacitor and when it is a resistor.

1.1.4.1 Capacitive Loads

The case of a Marx generator charging capacitive loads is of tremendous importance in pulsed power and forms the basis of many pulse compression schemes. A charged capacitor can transfer almost all of its energy to an uncharged capacitor if connected through an inductor. It is the basis for the intermediate storage capacitor architecture used in multigigawatt pulsed power machines. A Marx generator may also be used with a peaking switch to increase its rise time on an output load.

The energy transfer from a Marx generator to a capacitive load is considered by the equivalent circuit shown in Figure 1.11. The inductance includes the internal inductance of the Marx, L_M, as well as any additional inductance that may be added [9]. When the spark gap switch closes, the energy stored in the Marx capacitance C_M discharges through the inductor to charge the capacitor C_2.

Assume the erected Marx capacitance C_M has an initial charge of V_M, which may be different from the open-circuit value V_{OC}. Applying Kirchoff's law to the circuit of Figure 1.11,

$$V_1(t) - L\left(\frac{di}{dt}\right) = V_2(t) \tag{1.14}$$

where

$$V_1(t) - V_M = \frac{1}{C_M}\int i(t)dt \tag{1.15}$$

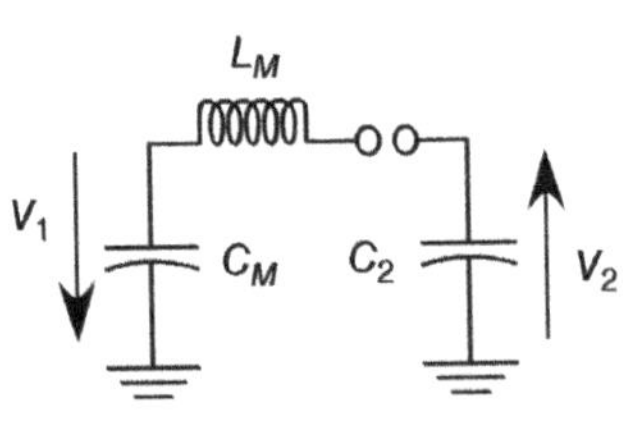

Figure 1.11 A Marx, with an equivalent capacitance C_M charging a capacitive load C_2.

and

$$V_2(t) = \frac{1}{C_2}\int i(t)dt \tag{1.16}$$

Differentiating (1.15) once, we obtain

$$L\left(\frac{d^2i(t)}{dt^2}\right) + i(t)\left(\frac{1}{C_M} + \frac{1}{C_2}\right) = 0 \tag{1.17}$$

With the initial conditions $i(0) = 0$ and $L(di(0)/dt) = V_M$. Defining

$$\omega = \sqrt{\frac{1}{L}\left(\frac{1}{C_M} + \frac{1}{C_2}\right)} = \sqrt{\frac{(C_M + C_2)}{LC_MC_2}} \tag{1.18}$$

The solution to (1.17) for the current in the circuit is

$$i(t) = \frac{\omega V_M \sin \omega t}{(1/C_M) + (1/C_2)} = \frac{V_M \sin \omega t}{\omega L} \tag{1.19}$$

The voltages on the two capacitors are

$$V_1(t) = V_M - \int \frac{i}{C_M}dt = V_M\left(1 - \int \frac{\sin \omega t}{\omega L C_M}dt\right) \tag{1.20}$$

which simplifies to

$$V_1(t) = V_M - \frac{V_M C_2}{(C_M + C_2)}(1 - \cos \omega t) \tag{1.21}$$

Similarly,

$$V_2 = \int \frac{i}{C_2}dt = V_M \int \frac{\sin \omega t}{\omega L C_M}dt \tag{1.22}$$

$$V_2(t) = \frac{V_M C_M}{(C_M + C_2)}(1 - \cos \omega t) \tag{1.23}$$

From Equation 1.23, the capacitive ringing gain can be defined as V_2/V_M and has a maximum value of

$$\left.\frac{V_2}{V_M}\right|_{Max} = \frac{2C_M}{C_M + C_2} \tag{1.24}$$

The ringing gain is easily measured and may be used to baseline a realized circuit against its design. When a Marx charges another capacitor through an inductor, the charging waveform has a (1 – cosine) waveshape.

Two cases of specific importance in pulsed power technology occur when (i) the Marx capacitance is approximately equal to the load capacitance $C_M \sim C_2$ and (ii) the Marx capacitance is much greater than the load capacitance, $C_M \gg C_2$.

1.1.4.1.1 Equal Marx and Load Capacitances

The case $C_M \sim C_2$ is the basis for the pulse compression schemes used in many pulsed power machines because the energy from the charged Marx generator can be transferred efficiently to the load. If the switch in Figure 1.11 is closed when the current is zero ($\omega t = \pi$), the energy is transferred from C_M to C_2,

$$V_1\left(t = \frac{\pi}{\omega}\right) = 0 \tag{1.25}$$

and

$$V_2\left(t = \frac{\pi}{\omega}\right) = \frac{V_M C_M}{(C_M + C_2)} \tag{1.26}$$

The voltage waveforms across the two capacitors are shown in Figure 1.12 for the case $C_M \approx C_2$. Since $C_M \sim C_2$ and $V_2 \sim V_M$, most of the energy initially stored in C_M has been transferred to C_2, and the energy transfer can be efficient. The capacitor C_2 is often an intermediate storage capacitor, as discussed in Chapter 3.

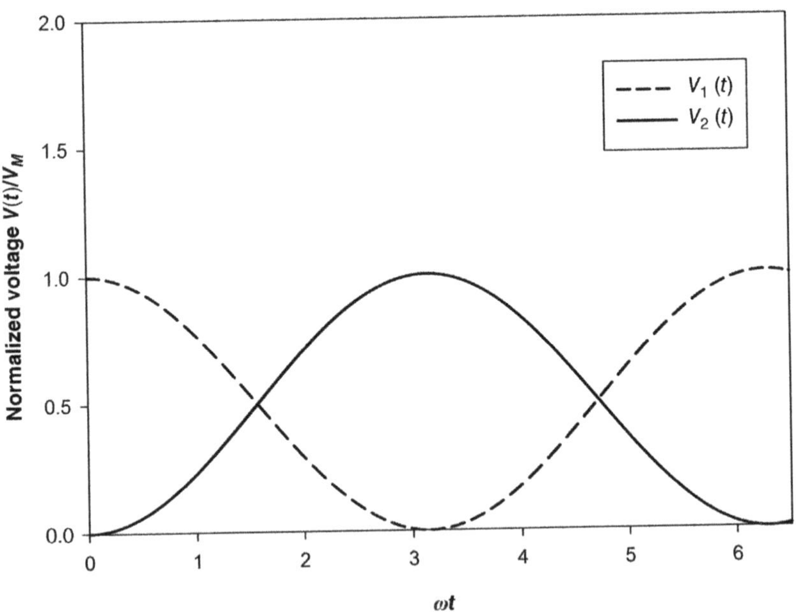

Figure 1.12 For the case when $C_2 \approx C_M$, the voltage $V_1(t)$ across the capacitor C_M charges the capacitor C_2 to a nearly identical voltage $V_2(t)$ and the energy transfer is very efficient. This is often used for the first stage of a pulse compression where the Marx charges an intermediate store capacitor in a time $t = \pi/\omega$ and the energy transfer is efficient.

1.1.4.1.2 The Peaking Circuit: $C_M \gg C_2$

The case $C_M \gg C_2$ is known as a peaking circuit. Here, energy transfer occurs when $(1 - \cos(\omega t)) = 2$, that is, when $(\omega t = \pi)$.

$$V_1\left(t = \frac{\pi}{\omega}\right) \approx V_M \tag{1.27}$$

and

$$V_2\left(t = \frac{\pi}{\omega}\right) \approx 2V_M \tag{1.28}$$

The voltage $V_2(t)$ across the capacitor C_2 is driven to nearly twice the Marx voltage, while the voltage $V_1(t)$ remains nearly the same, as shown in Figure 1.13. The energy transfer, however, is inefficient. It is usually sufficient to have a peaking capacitance such that $(C_M/C_2) \approx 10$.

1.1.4.1.3 A Peaking Circuit Driving a Resistive Load

A peaking capacitor ($C_2 = C_p$) can be used in conjunction with a Marx generator to sharpen the rise time of a Marx. The circuit is shown in Figure 1.14,

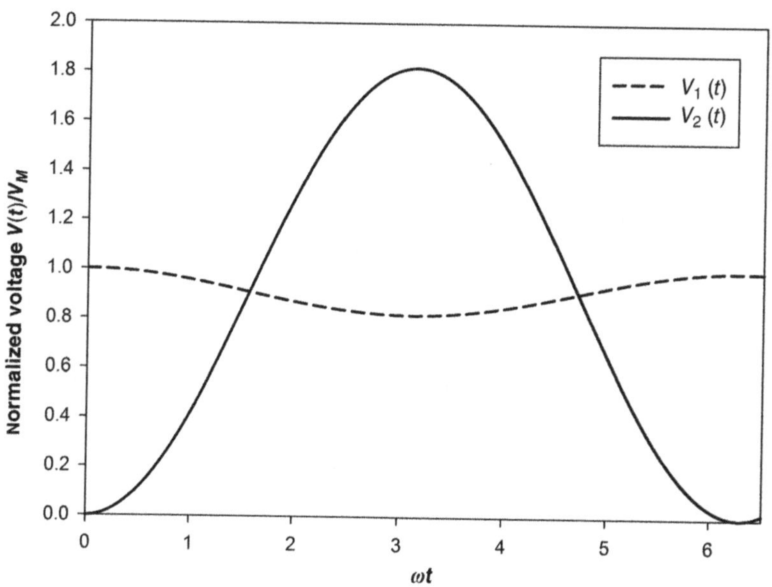

Figure 1.13 For the case when $C_2 = 0.1C_M$, the voltage $V_1(t)$ across the capacitor C_M varies little, while the voltage $V_2(t)$ on capacitor C_2 nearly doubles. The energy transfer is not efficient. Peaking capacitors are used to increase the pulse rise time and can be very difficult to implement.

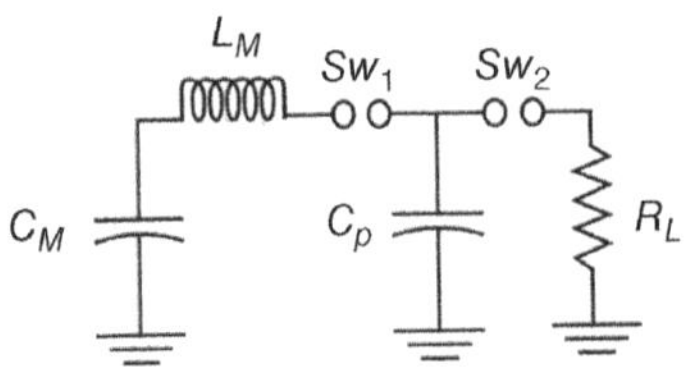

Figure 1.14 A peaking capacitor circuit may be used to sharpen the rise time of the Marx to a load.

The switch Sw_2 is called a peaking switch and is set to switch when the current is maximum [3]. An exponential waveform can be delivered to the resistive load if the value of the peaking capacitor C_p is chosen to be [10]

$$C_p = \frac{L_M C_M}{R_L^2 C_M + L_M} \tag{1.29}$$

And the switch Sw_2 is closed at a time t_p chosen:

$$t_p = \frac{1}{\omega} \cos^{-1}\left(\frac{-C_p}{C_M}\right) = \sqrt{\frac{L_M C_M C_p}{C_M + C_p}} \cos^{-1}\left(\frac{-C_p}{C_M}\right) \tag{1.30}$$

The frequency ω is usually quite high and care must be taken on the breakdown time and jitter of the peaking switch Sw_2, which manifest as variations of the load voltage. In practice, it may be difficult to make the low-value, high-voltage capacitor necessary for a peaking circuit.

1.1.4.2 A Marx Charging a Resistive Load

The case of a Marx generator charging a resistive load finds application where the current and voltage are in phase and proportional, such as for relativistic electron beam generation. If the Marx was comprised of a pure capacitance and charged a pure resistance, the voltage on the load resistor is

$$V_L(t) = V_M e^{-t/R_L C_M} \tag{1.31}$$

When the Marx fires at $t = 0$, the load voltage jumps instantaneously to the peak Marx voltage V_M and decays with a time constant equal to $R_L C_M$. This ideal case, however, is nonphysical since a Marx generator has a significant inductance. The effect of the equivalent series inductance of the Marx on a resistive load is analyzed using the circuit of Figure 1.13a.

$$V_1(t) - L\left(\frac{di}{dt}\right) = Ri(t) \tag{1.32}$$

and

$$i(t) = C_M \frac{dV_1(t)}{dt} \tag{1.33}$$

Differentiating (1.32), substituting (1.33), and simplifying,

$$\frac{d^2i(t)}{dt^2} + \frac{R}{L}\left(\frac{di}{dt}\right) - \frac{1}{LC_M}i(t) = 0 \tag{1.34}$$

With the initial conditions $i(0) = 0$ and $L(di(0)/dt) = V_M$ and setting

$$\gamma = \sqrt{\left(\frac{R}{L}\right)^2 - \frac{4}{LC_M}}$$

the solution is

$$i(t) = \frac{V_M}{\gamma L_M} e^{-(1/2)[(R/L_M)-\gamma]t}\left[1 - e^{-\gamma t}\right] \tag{1.35}$$

Equation 1.31 is a double exponential. The rise time is determined by the $(1 - e^{-\gamma t})$ term. As t gets large, this term is overtaken by the other decaying exponential term, as illustrated in Figure 1.15b. The series inductance of the Marx plays an important role in the performance of Marx generators because it increases the rise time of the output waveform in the case of a resistively loaded Marx generator and reduces the peak current capability.

1.2 Impulse Generators

Impulse generators are an important and long-used application of Marx generators. The shape of the pulse must be tailored for the specific testing requirements.

1.2.1 Exact Solutions

For impulse testing, the load is assumed to be capacitive. The equivalent circuit of the N stage Marx generator of Figure 1.16 is shown in Figure 1.17.

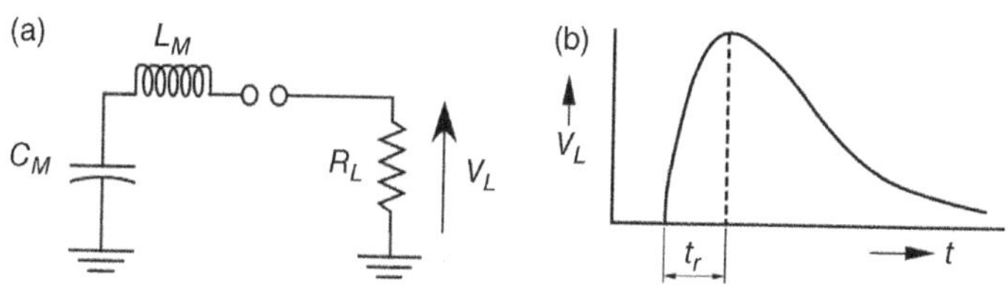

Figure 1.15 The equivalent circuit (a) of a conventional Marx generator with a resistive load and series inductor produces the voltage waveform (b) at the load.

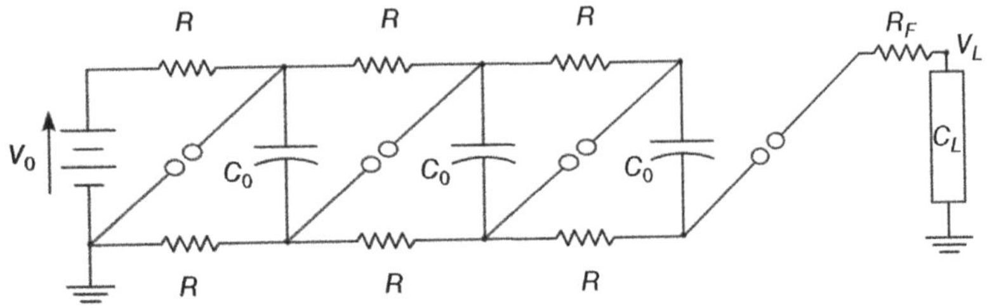

Figure 1.16 An *N*-stage Marx generator with a tailored waveshape for impulse testing.

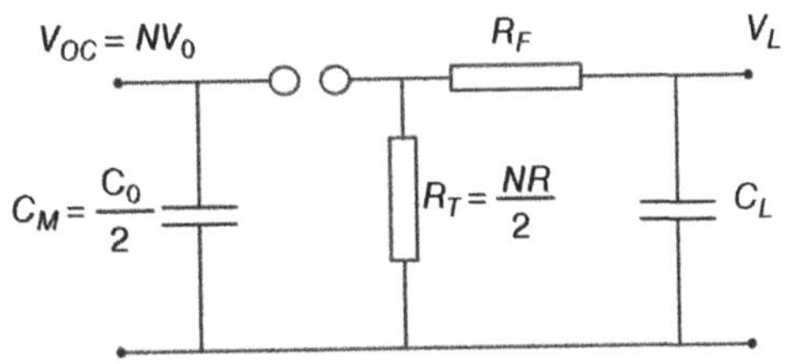

Figure 1.17 The equivalent circuit of the *N*-stage Marx generator of Figure 1.16.

The pulse duration can be adjusted by choosing proper values for the front resistor R_F and tail resistor R_T. The value of the tail resistor in the equivalent circuit of Figure 1.17 is

$$R_T = \frac{NR}{2} \tag{1.36}$$

In the Laplace domain, the impedance of the equivalent circuit is

$$\begin{aligned} Z(s) &= \frac{1}{C_M s} + \frac{R_T(R_F + (1/C_L s))}{R_T + R_F + (1/C_L s)} \\ &= \frac{R_T C_L s + R_F C_L s + 1 + C_M C_L R_F R_T s^2 + C_M R_T s}{C_M C_L R_T s^2 + C_M C_L R_F s^2 + C_M s} \end{aligned} \tag{1.37}$$

and the voltage on the load is

$$V_L(s) = \frac{V}{s}\frac{1}{Z(s)}\frac{R_T}{(R_T + R_F + (1/C_L s))}\frac{1}{C_L s} \tag{1.38}$$

Substituting $Z(s)$ from (1.33) into (1.34) leads to

$$\begin{aligned} V_L(s) &= \frac{V}{C_L R_F}\frac{1}{s^2 + s\left[(1/R_F C_M) + (1/R_T C_M) + (1/R_F C_L)\right] + (1/R_F R_T C_M C_L)} \\ &= \frac{V}{C_L R_F}\frac{1}{(s+\alpha)(s+\beta)} \end{aligned} \tag{1.39}$$

where α and β are roots of the following equation:

$$s^2 + s\left(\frac{1}{R_F C_M} + \frac{1}{R_T C_M} + \frac{1}{R_F C_L}\right) + \frac{1}{R_F R_T C_M C_L} = 0$$

The α and β are related to the circuit values by

$$\alpha + \beta = \frac{1}{R_F C_M} + \frac{1}{R_T C_M} + \frac{1}{R_F C_L} \tag{1.40}$$

$$\alpha\beta = \frac{1}{R_F R_T C_M C_L} \tag{1.41}$$

Converting Equation 1.39 into the time domain by taking the inverse Laplace transform, the Marx output voltage $V_L(t)$ is

$$V_L(t) = \frac{V_M}{R_F C_L(\beta - \alpha)}\left(e^{-|\alpha|t} - e^{-|\beta|t}\right) \tag{1.42}$$

A typical waveform representing Equation 1.42 is shown in Figure 1.18. The time t' when the output voltage $V_L(t)$ reaches the maximum value can be found by taking a time derivative of (1.42) and equating to 0, yielding

$$t' = \frac{\ln(\beta/\alpha)}{\beta - \alpha} \tag{1.43}$$

The maximum amplitude of the pulse voltage V_L^{Max} is obtained by substituting the value of t' from (1.43) into (1.42):

$$V_L^{Max} = V_L(t') = \frac{V_M}{C_L R_F (\beta - \alpha)}\left(e^{-|\alpha|t'} - e^{-|\beta|t'}\right) \tag{1.44}$$

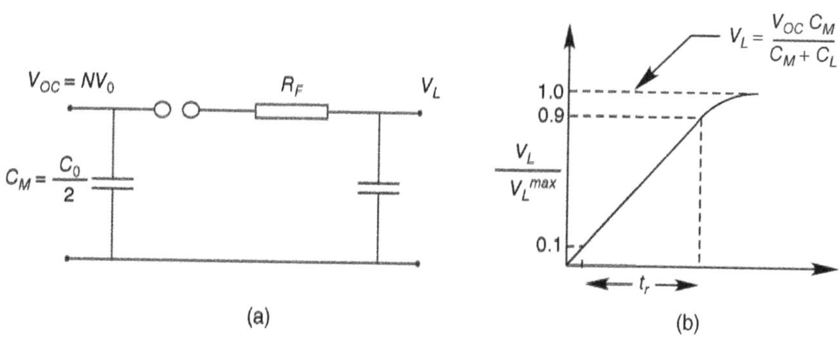

Figure 1.18 (a) The approximate Marx equivalent circuit for the front time calculation and (b) the output voltage during the front time.

1.2.2 Approximate Solutions

Equations 1.36–1.44 are the design formulas for the Marx generator output waveform. In most cases, the approximate formulas may suffice and are convenient for pulse shaping. If this is the case, Figure 1.18 can be used for the calculation of rise time t_r and front time t_f. The rise time t_r is defined as the time interval required for the voltage to rise from 0.1 V_L^{Max} to 0.9 V_L^{Max}. The front time t_f is defined as

$$t_f = 1.25 \cdot t_r \tag{1.45}$$

Figure 1.19 can be used for calculation of tail time t_t, the time for the falling edge of the output voltage $V_L(t)$ to fall to half its peak value. The values of t_r and t_t are indicated on the waveforms of Figures 1.18b and 1.19b.

The approximate solution for the output voltage $V_L(t)$ for the circuit shown in Figure 1.18 has the following form [11]:

$$V_L(t) \cong \frac{V_{OC}C_M}{C_M + C_L}(1 - e^{-t/R_F C_T}) \tag{1.46}$$

where

$$C_T = \frac{C_M C_L}{C_M + C_L} \tag{1.47}$$

The values of t_r and t_f deduced from Equation 1.46 are as follows:

$$t_r = 2.2 C_T R_F \tag{1.48}$$

$$t_f = 1.25 t_r = 2.75 C_T R_F \tag{1.49}$$

The approximate solution for the output voltage $V_L(t)$ for the circuit shown in Figure 1.17 has the following form [12]:

$$V_L(t) \cong \frac{V_{OC}C_M}{C_M + C_L} e^{-t/R_T(C_L + C_M)} \tag{1.50}$$

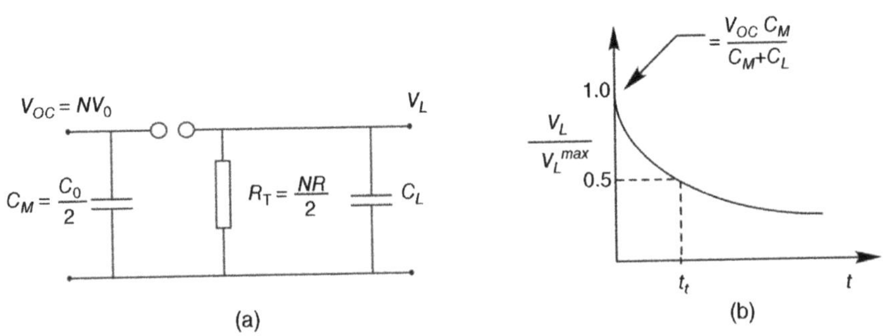

Figure 1.19 Approximate equivalent circuit for the tail time calculation and (b) the output voltage during the tail time.

The value of t_t deduced from Equation 1.50 is as follows:

$$t_t = (0.7) \cdot R_T (C_L + C_M) \tag{1.51}$$

Precise waveshaping can be attained with these techniques [13]. A voltage monitor is often added in parallel with the load to obtain a direct measurement. The impulse insulation testing of high-voltage insulators is a typical case of a capacitive load and therefore the above formulas apply. Impulse testing is usually notated in the form: (pulse rise time/FWHM fall time) and typical testing waveforms are 1.2/50 μs or 8/20 μs and are specified by specific standards. The effective capacitance of an erected Marx should be in the range of 4–10 times the load capacitance for effective voltage gain [14].

1.2.3 Distributed Front Resistors

One such modified configuration [12] is shown in Figure 1.20, where the externally connected front resistor R_F of Figure 1.16 is uniformly distributed throughout the individual stages of the Marx generator, forming an internal part of the Marx generator. The distributed front resistors need to only withstand a fraction of the total voltage. This design, however, increases the number of components in the Marx and makes varying the resistance value difficult.

1.3 Effects of Stray Capacitance on Marx Operation

The elegant Marx concept was used for many years with little change. Marx generators can be made to operate very simply when two-electrode spark gaps are triggered by overvoltages from switching previous stages and each switch contributes its delay time and jitter to the Marx erection time and the overall jitter. In the 1960s, interest in pulsed power spread rapidly as applications emerged in high-energy density physics. The increasingly specialized demands

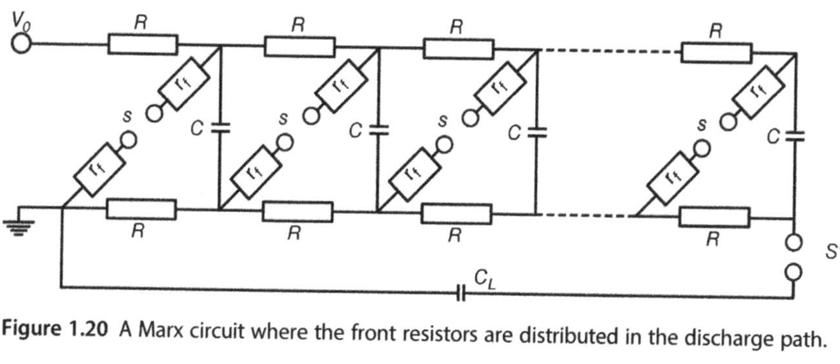

Figure 1.20 A Marx circuit where the front resistors are distributed in the discharge path.

revealed its limitations and sparked a period of careful study of the Marx circuit operation and construction.

These emerging capabilities had requirements that exceeded the state of the art in voltage, energy, and energy density. As facilities became large, efforts to reduce the Marx volume included insulating the Marx with transformer oil. The generation of very high voltages resulted in physically long Marx generators with many stages. These changing requirements resulted in increased stray capacitances that could no longer be neglected. Stray capacitance may limit the degree of spark gap overvoltage by establishing transient voltage dividers, resulting in increased erection time and jitter. Fast, reliable Marx erection requires that large overvoltages appear on each stage during discharge. The reduced overvoltage may have severe and unintended consequences: In developing a physically long, 42-stage, 18 MV Marx generator, Prestwich and Johnson [15,16] observed that after 23 stages fired, the switches began firing from the top, in reverse order from the output end, with the last 7 stages firing simultaneously from overvoltages, as discussed in Section 1.5.1.

There are three significant sources of stray capacitance. The first is the stray capacitance across the spark gaps. Prior to closing, a spark gap consists of two electrodes separated by an insulator, which also describes a capacitor. Thus, during the charge cycle, the spark gap may be represented by a capacitance and denoted C_g. The second source is the conducting connections of each stage of the Marx that are isolated from the system ground and are represented by the stray capacitance C_s. The third source is stray capacitance between stages. In many cases, the electrodes of the energy storage capacitors C_0 have only small separations between adjacent stages, leading to the stages being coupled by a capacitance C_c. In some designs, the stray capacitance between adjacent stages is minimized by physically separating the stages, in which case the dominant stray stage capacitance is between alternate stages and denoted C_R. The relative magnitudes of these stray capacitances determine the performance of the Marx.

This section is organized to first illustrate how stray capacitance can establish a voltage divider in the case when the nonnegligible stray capacitance consists of C_g and C_s and then show how the stray capacitance can be exploited. The wave erection Marx is a case when the energy is small, the component dimensions make for small inductance and coupling impedance, and the conventional Marx becomes capable of rapid erection. The next sections show the effect of the coupling capacitances C_c and C_R and how the Marx can be designed so that the stages with weak overvoltages are reinforced by capacitive or resistive coupling.

1.3.1 Voltage Division by Stray Capacitance

During the charge cycle, a spark gap consists of conductors separated by insulators, and is modeled as a stray capacitance C_g as shown. Each conductor

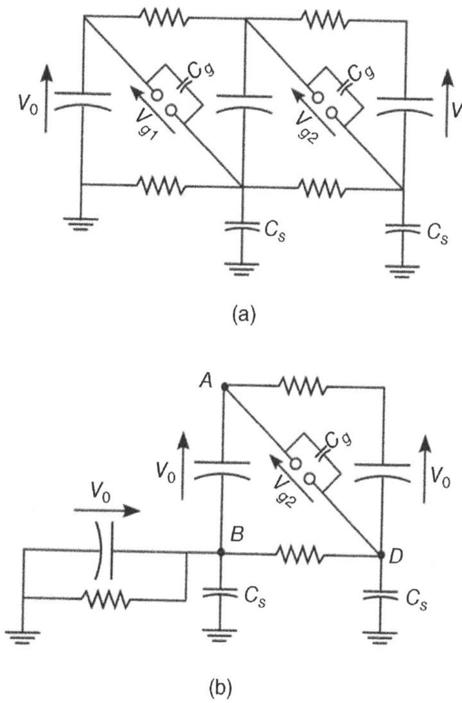

Figure 1.21 A Marx generator with some stray capacitance (a) during the charge cycle and (b) after the first stage has fired.

in the Marx also has capacitance to ground. The equivalent circuit of a Marx, including the stray capacitance of the gap and the conductors to ground, C_s, is shown in Figure 1.21.

Assume the first gap is triggered by an external source. The first stage erects, and the voltage at point B, V_B, equals V_0 because one terminal of the first-stage capacitance is connected directly to ground. The voltage at point A is given by $V_A - V_0 = V_B$. Invoking $V_B = V_0$ yields

$$V_A = 2V_0 \tag{1.52}$$

The voltage across gap 2 is $V_{g2} = V_A - V_D$. Recognizing the voltage at point D is the voltage across the stray capacitance C_s, $V_D = V_s$. The circuit stray capacitance across the spark gap, C_g, and the stray capacitance to ground, C_s, establish a voltage divider. The voltage across the spark gap can be calculated from Figure 1.21b and the voltage divider relation.

$$2V_0 - V_{g2} = V_s \tag{1.53}$$

$$V_s = 2V_0 \frac{C_g}{C_s + C_g} \tag{1.54}$$

Solving for V_{g2},

$$V_{g2} = \frac{2V_0}{1 + (C_g/C_s)} \tag{1.55}$$

The overvoltage on gap 2 is maximized from Equation 1.55 when $C_g \ll C_s$. The stray capacitance to ground, C_s, can be tailored somewhat by placing a ground plane next to the Marx, which is often accomplished by enclosing the Marx in a grounded cylinder.

Switch overvoltages can be increased by triggering several switches in a Marx. If, instead of triggering just the first gap, the first k gaps are triggered simultaneously, the overvoltage on the next $(k+1)$ gap is [3]

$$V_{g,k+1} = \frac{kV_0}{1 + \sqrt{1 + (4C_g/C_s)}} \tag{1.56}$$

Spark gap overvoltages are transient events. As the potential across C_s rises to V_0, the stray capacitance to ground of the next stage begins to charge to V_0, limiting the overvoltage on gap 2. Care must be taken in the design to ensure the spark gaps fire before an excessive reduction in its overvoltage happens. A fast erection time ensures that switching occurs at the maximum overvoltage leading to low jitter.

1.3.2 Exploiting Stray Capacitance: The Wave Erection Marx

The case where the coupling capacitance between stages, C_C, is small is typically one using ceramic capacitors with values of a few nanofarads. Platts [17] used this approach to develop a Marx in a compact geometry that generated a high peak voltage with low energy content. The circuit is arranged in a line so that the spark gaps are line-of-sight and the ultraviolet radiation produced in the breakdown acts as a preionization source for successive spark gaps. The Marx was encased in a grounded metal tube and insulated throughout with moderate pressure gas producing an open-circuit voltage of 200 kV. This sparked a renewed interest in Marx generators for new applications directly driving a load.

The stray capacitance to ground may be utilized to produce a Marx generator with a very fast rise time by designing the Marx to act like a cascading peak circuit [3,18]. The Marx is arranged so that $C_s \gg C_g$, and the circuit may be represented as shown in Figure 1.22a. As each stage of the Marx is switched, the peaking effect, shown in Figure 1.22b, becomes more marked, resulting in a very fast rise time.

The capacitance to ground, C_s, may be controlled by a suitably designed, grounded metal enclosure. The stage capacitance C_0 and the total number of stages N are selected to satisfy

$$\frac{C_0}{n} = C_n \gg C_s \quad n < N \tag{1.57}$$

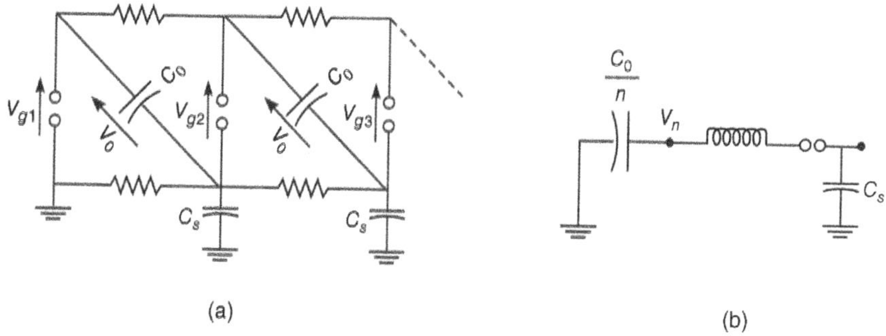

Figure 1.22 (a) The layout of a Marx circuit with a significant stray capacitance to ground and (b) the peaking equivalent circuit formed by the erecting Marx and its stray capacitance, C_S.

where C_n is the capacitance of the Marx as the nth stage has switched into the series discharge circuit. As the Marx erects, the Marx capacitance C_n and the stray capacitance to ground C_s produce a peaking circuit as shown in Figure 1.23b. Each Marx stage becomes a peaking circuit charging the next stage in an increasingly fast erection rate.

Impressive performance has been reported [19–23]. Kekez built a 600 kV, 1 ns output unit [19] and a 200 kV Petit Marx had an estimated 50 ps rise time into a 100 Ω load [20]. Mayes and his colleagues have also reported spectacular results using a wave erection scheme [21–23].

1.3.3 The Effects of Interstage Coupling Capacitance

For high energy storage applications, the open geometry yields a large device with significant inductance. For applications requiring high energies and relatively fast discharges, architectures become more compact to reduce the inductance but result in a significant increase in the coupling impedances. Typical high-energy density capacitors have capacitance values of a few

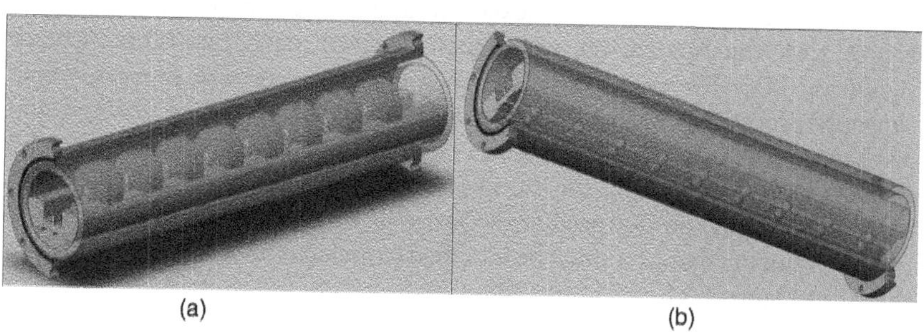

Figure 1.23 Two views of an enclosed compact Marx generator showing (a) the stage capacitance and (b) simple spark gap switches arranged to utilize UV preionization. (Reproduced with permission of Cockreham.)

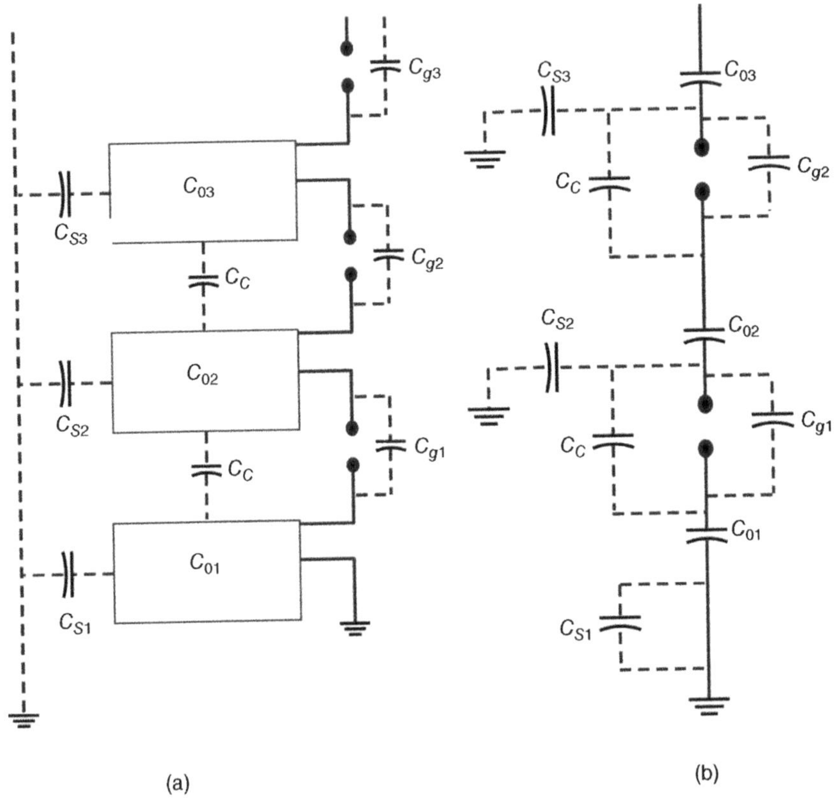

Figure 1.24 The (a) layout and (b) equivalent circuit of a Marx generator with significant stray coupling capacitances C_C between adjacent stages, as well as stray capacitance to ground and across the spark gap. The relative magnitudes determine the Marx operation.

microfarads and are arranged so that the metallic enclosure is one of the capacitors terminals. When stacked in a Marx generator, as shown in Figure 1.24, the capacitance established between stage capacitors may be significant [24]. The coupling capacitance slows the erection process, which can only proceed as fast as the interstage capacitances can be charged and discharged. Thus, the analysis of the erection process described in Section 1.1.2 predicts large overvoltages on the upper stages that do not materialize and, more importantly, do not predict the order in which the switches fire.

The circuit of Figure 1.1 must be modified to include the stray capacitances as shown in Figure 1.24.

The stray capacitances shown in Figure 1.24 are given as follows:

C_{gn}: The capacitance of the gap of the nth switch.
C_C: The coupling capacitance between adjacent capacitors.

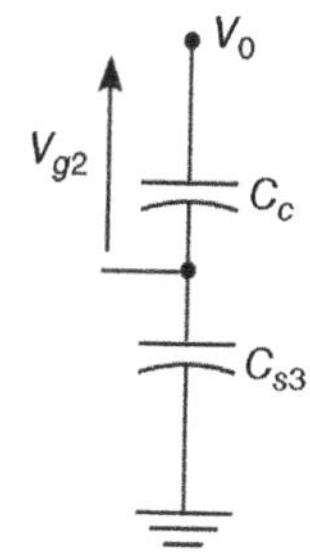

Figure 1.25 The voltage across the spark gap may be limited by the voltage divider set up by stray capacitance within the Marx.

C_R: The coupling capacitance between alternate capacitors.
C_{sn}: The stray capacitance of the nth capacitor to ground.

When the Marx is fully charged, the voltage across the second spark gap, V_{g2}, is the voltage on capacitor 2, V_0. From the simplified analysis of Section 1.1.2, the voltage across the second spark gap should increase to $2V_0$ when the first spark gap fires, but, instead, the transient overvoltage encounters the capacitive voltage divider formed from the stray capacities C_c and C_{s2}, reducing the overvoltage on the spark gap (Figure 1.25).

Only the voltage contribution from the erection of the first stage of the Marx sees the voltage divider. The charge voltage on stage 2 must be added to the divided contribution from stage 1, yielding the voltage on spark gap 2 to be

$$V_{g2} = V_0 + \frac{V_0 C_{s3}}{C_{s3} + C_c} \tag{1.58}$$

Morrison and Smith [4] did a thorough analysis of a 10-stage generator where the voltages in the Marx, including voltage division effects, were examined as the Marx erected. Their analysis showed that unexpected (and frightening) patterns can occur for various ratios of stray capacitances. Morrison and Smith use the nonsequential switch firing described by Prestwich and Johnson [15] to validate their analysis.

The effects of large interstage capacitance became apparent as Marx generator designs moved from the large, open constructions used in impulse testing to more compact configurations producing higher voltages. The high energy storage, fast discharge capacitors used in these generators typically use their outer case as one of the capacitor electrodes. When stacked in a Marx, as shown in Figure 1.24, the large area of the capacitor case results in significant interstage capacitance. Concerns about interstage capacitance are not only for high-energy generators. Ordinarily, in conventional Marx generators, the switch capacitance C_g is sufficiently low such that the stray stage to ground capacitance C_s is adequate to deliver a sufficient overvoltage to the next stage. In compact generators, C_g may be quite high because the switches may be set with small gaps to minimize the inductance in the discharge path. Discrete interstage coupling capacitors can be used to overcome this shortcoming [25].

1.4 Enhanced Triggering Techniques

Jitter assumes importance especially in cases where the Marx generator has to be synchronized to trigger other events in a given system or to other Marx generators. A Marx's erection time and jitter are closely linked to the degree of overvoltage on the spark gaps that may be adversely affected during the erection sequence by stray capacitance. The effects of stray capacitance can be mitigated by coupling key stages to control transients during the erection process. Interstage coupling is typically achieved with resistors or capacitors. Interstage capacitors may be lumped elements but are often attained by careful design of the layout.

1.4.1 Capacitive Back-Coupling

In the previous section, the effects of large interstage stray capacitance were shown to potentially severely affect the erection of the Marx. The interstage coupling capacitance can be reduced by arranging the Marx into columns, with each column containing alternating stages of the Marx, as illustrated in Figure 1.26.

The column arrangement changes the equivalent circuit, so the stray coupling capacitance between alternate stages, denoted C_R, is greater than between adjacent stages, C_C. This arrangement aids the erection of the Marx

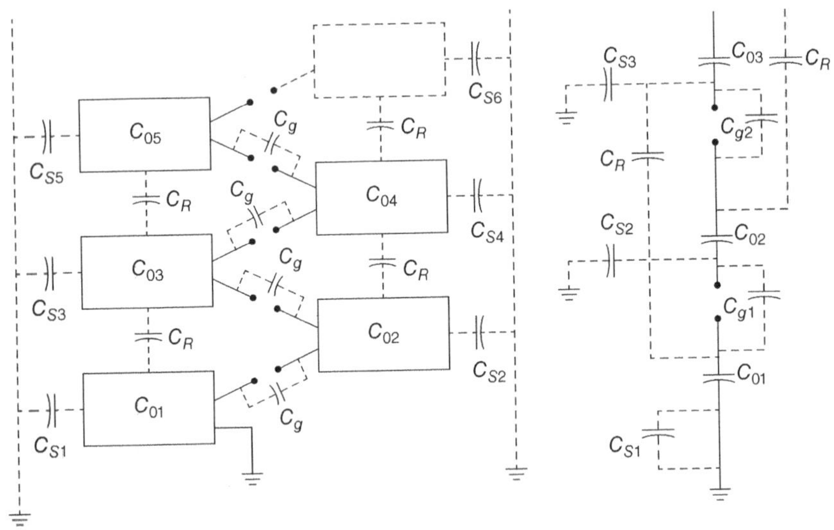

Figure 1.26 The (a) layout of a Marx arranged in two columns to promote capacitive back-coupling and (b) its equivalent circuit.

because C_R is in parallel to the gap capacitance C_g, increasing the overvoltage. Thus, capacitive coupling between alternate stages leads to enhanced triggering, and is sometimes referred to as "capacitive back-coupling" [1].

The resultant large overvoltages on the switches permit the Marx to be operated at a fraction of its self-breakdown voltage. By rearranging the Marx into two columns, an operating range approaching 2:1 can be achieved by triggering only 1 gap. In general, the Marx can be arranged in p columns with an operating range approaching p:1, but the first $(p-1)$ gaps must be triggered. In practice, due to stray impedances, the number of columns is limited to three. This analysis gave rise to a naming convention, not often used anymore, where the Marx is characterized by its interstage coupling with a designation of $n=p$. Thus, a $n=2$ Marx couples to every other stage. Another type of transient voltage appears in this type of construction. For a two-column Marx, if the conductors and switches were perfect, an overvoltage of $2V_0$ would appear between the two capacitors coupled by C_R. By including the inductance of the connections and switches, the actual voltage is [26]

$$V_2 = 2V_0 \cdot \left(1 - \cos\left(\frac{t}{\sqrt{LC_c}}\right)\right) \tag{1.59}$$

This high-frequency oscillation can reach a peak of $4V_0$, aiding triggering further, but is generally limited in practice to $3V_0$.

1.4.2 Resistive Back-Coupling

The use of one or more triggered gaps is common in many Marx generators in order to reduce jitter. The erection process can be further enhanced with strategically placed three-electrode triggered spark gaps [27–29]. This technique, pioneered by J.C. ("Charlie") Martin, uses more than one triggered spark gap combined with back-coupling to produce a highly reliable, low-jitter system and is sometimes referred to as a Martin Marx [1].

The Martin Marx is characterized by switches with trigger electrodes and an enhanced erection process. In addition to the triggered spark gaps to initiate the erection process from an external trigger generator, the Martin Marx places triggered spark gaps in upper stages and resistively couples the trigger electrode back to the earlier stages. The coupling impedance can be any combination of capacitors, inductors, or resistors but resistors are common. A common triggering scheme is shown in Figure 1.27 where every third gap is coupled and the first three stages are externally triggered.

In this circuit, the triggered spark gaps in the upper stages are used in a fundamentally different manner than in the first three stages. In the first three stages, the potential of the trigger electrode is changed, resulting in a large overvoltage and the spark gap fires. In the upper stages, the resistive coupling impedance

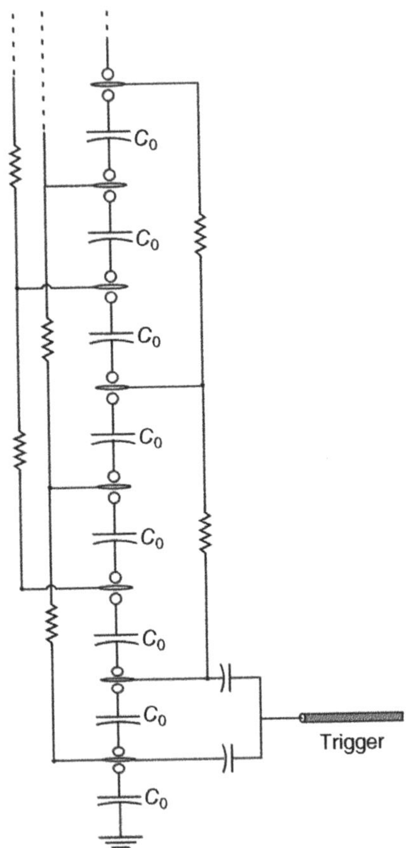

Figure 1.27 A Marx using resistive back-coupling to every third stage. The spark gaps are all symmetric three-electrode spark gaps [1].

holds the trigger electrode at a voltage determined by its lower stage coupling partner, while the voltage on the main electrodes responds to the erection process, resulting in large potential differences between the spark gap electrodes. The use of resistive back-coupling can make the design rather complex.

The main advantages of this Marx configuration are its control of prefires and its wide operating range, which in practice depends on the magnitude and distribution of the stray capacitance [29]. Voltages as low as 30% of the self-breakdown voltage can be achieved by varying the charging voltage with no adjustments to the spark gap spacing [15,16].

1.4.3 Capacitive and Resistively Coupled Marx

Coupling can be illustrated with the aid of Figure 1.28. The capacitors can be arranged so that the stray capacitance between alternate stages, C_R, is large. These stages are connected with charging resistors.

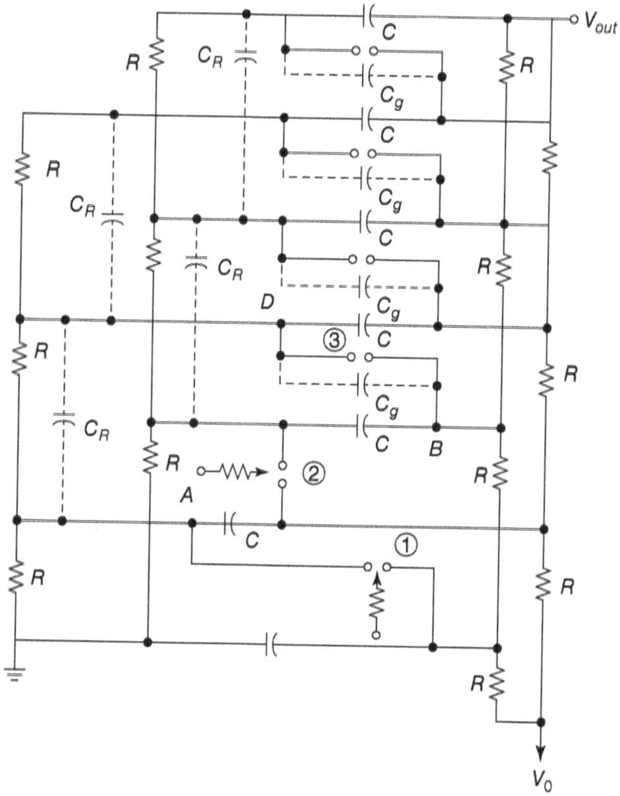

Figure 1.28 A Marx in which every other stage is coupled, called an $n = 2$Marx [15].

By triggering the first two spark gaps, point A is maintained at a voltage V_0 and point B at $3V_0$. The voltage at point D will initially be determined by the voltage division among stray capacitances. If $C_s \ll C_R$, as in the case of a long Marx generator, the voltage across the third spark gap, V_{g3}, approximately is

$$V_{g3} = \frac{2V_0C_R}{C_g + C_R} \tag{1.60}$$

The voltage across C_R decays with a time constant $\tau_{dis} = R(C_g + C_R)$ and the voltage across the spark gap approaches $(2V_0)$, making an $n = 2$ Marx. These relations can be generalized [16]. If the capacitors are coupled across every X capacitors, the voltage across the third spark gap, caused by the capacitive division, would immediately be

$$V_{g3} = X\frac{V_0C_R}{C_g + C_R} \tag{1.61}$$

The charge resistors, coupled across every Y stages, would discharge through C_g and C_R to give YV_0 across the spark gaps, making it an $n = Y$ Marx generator.

1.4.4 The Maxwell Marx

Very high energy storage Marx generators use physically large capacitors with large stray impedances. On the other hand, when the stored energy is small, the component dimensions have small coupling impedances and the low-energy Marx can erect rapidly. This rapid erection may be exploited to trigger a larger Marx, resulting in a precise, high-energy generator, with some added complexity. A small Marx generator, with the same basic characteristics of the large Marx, may be coupled, stage by stage, with the large Marx [1]. The small, low-energy fast Marx runs in parallel with the large Marx. The erection time is set by the small, fast Marx and can result in very precise operation of the large high-energy Marx and determines its erection time. This technique has been referred to as a Maxwell Marx [1].

Thus, for a long and large Marx system, the individual trigger generators can be arranged in ascending potential to match the potential of the erecting main Marx. An implementation of this design is shown in Figure 1.29, where the high energy storage Marx, with stage capacitance C_{OM}, is triggered by a small fast Marx with stage capacitance C_{Om}. The trigger Marx is arranged so that each of its stages acts as a trigger generator for a stage of the main Marx so that the erection rate of the large high-energy Marx is dictated by the small, low-energy trigger Marx.

Delays in erection between stages are caused by the channel formation time in the switches and the time to charge the stray capacitance associated with the next stage. Care must be taken to not to discharge the main Marx through the trigger Marx.

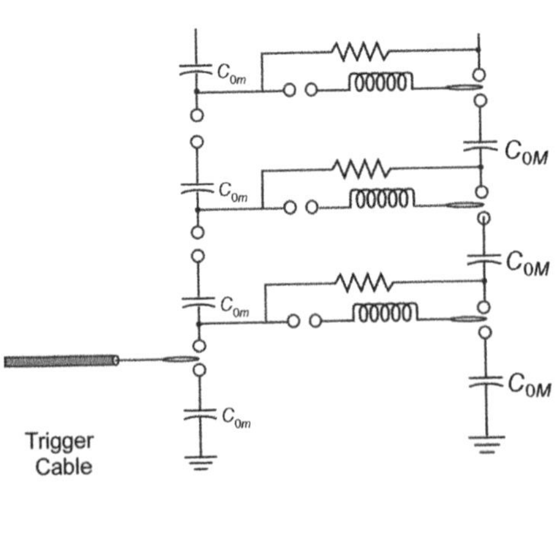

Figure 1.29 The Maxwell Marx uses a low-energy Marx to dictate the erection speed of a large high-energy Marx [1].

1.5 Examples of Complex Marx Generators

1.5.1 Hermes I and II

Using a differential charging scheme, where half the capacitors are charged to a positive voltage $+V_0$ and the other half to a negative voltage $-V_0$, Prestwich and Johnson [15] studied Marx generators with many stages for the HERMES program. By using a stage voltage of $2V_0$, half the number of switches is required to yield the same output voltage.

One of their generator designs is shown schematically in Figure 1.30. It is an $n=3$ Marx, using both resistive and capacitive coupling, with a measured inductance of 69 μH. It was demonstrated that with three gaps triggered, the Marx would fire down to 30% of its self-breakdown voltage. Triggering five gaps did not increase the firing range. The firing times of the individual spark gaps were measured and shown in Figure 1.31. It is interesting that after the first 23 stages erected, the Marx began firing from the high-voltage end down to the ground end, with the last seven gaps broke down simultaneously.

These studies yielded the design for the Marx generators for the Hermes I and Hermes II machines. Hermes I is a 100 kJ design consisting of six complete

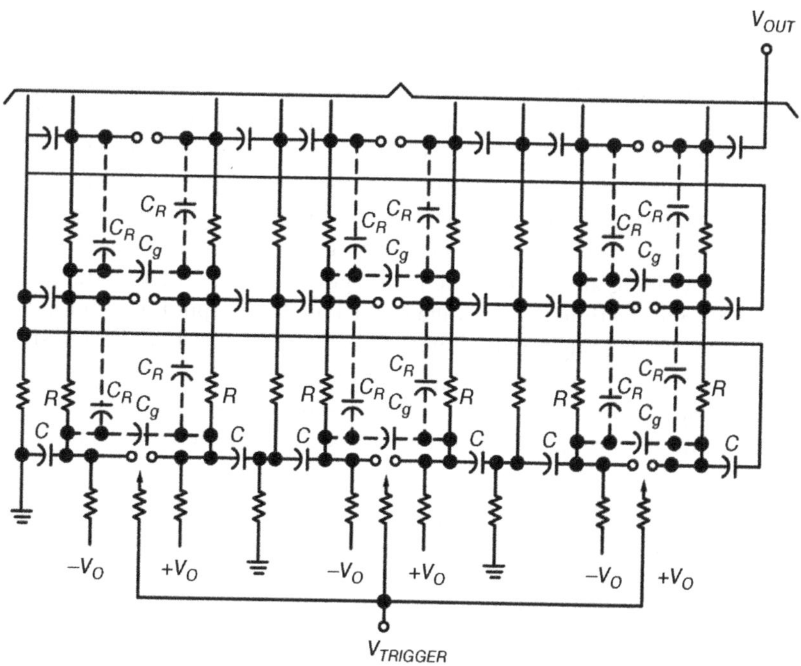

Figure 1.30 A prototype Marx developed as part of the HERMES program [15].

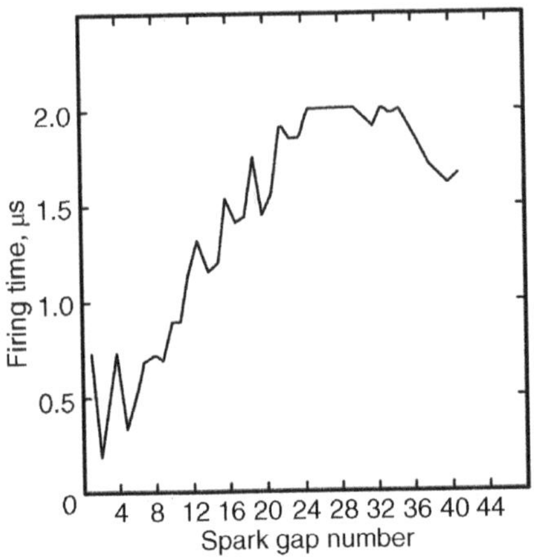

Figure 1.31 The firing times of individual gaps in the Marx of Figure 1.30. After erection was initiated, the generator began firing from the high-voltage end [15].

rows and 1 partial row to provide electrical grading at the high-voltage output end. The Hermes II Marx has 93 spark gaps (stages) arranged in 31 rows and stores 1 MJ when charged to ±103 kV. The parameters for these two generators are summarized in Table 1.1.

1.5.2 PBFA and Z

The innovation of triggered three-electrode switches allowed the development of advanced triggering for precision control of even large, high energy storage Marx generators. The triggering schemes can become quite complex. The Marx developed for the Particle Beam Fusion Accelerator programs at Sandia National Laboratories is a good example of the triggering complexities.

The PBFA II energy storage section consists of 36 bipolar Marx generators. Each Marx is composed of 60 capacitors charged to ±95 kV, storing 370 kJ of energy, and yielding an erected output voltage of 5.7 MV. The ringing gain is nearly unity and the time to peak voltage is 1.1 μs. The output for each Marx is connected through a single-pole, double-throw transfer switch to the first stage in the pulse compression scheme. The transfer switch is initially set to a liquid resistor dump load until the Marx is fully charged, and then the transfer switch is rotated into the firing position using a pneumatic actuator [30]. The Marx

Table 1.1 The equivalent circuit parameters for Hermes I and Hermes II.

	V_M	R_c	C_M	L_M	R_M	C_g	C_R	C_s
Hermes I	4 MV	1.2 kΩ	13.1 nF	22 μH	4 Ω	~45 pF	~90 pF	<20 pF
Hermes II	18 MV	1.5 kΩ	5.4 nF	80 μH	20 Ω	~45 pF	~190 pF	<10 pF

switches are three-electrode, SF6-filled spark gaps triggered through a midplane, field-enhanced electrode located at the midplane.

The reliability of low-jitter Marx generators was improved dramatically for PBFA II, and the basic design, with the exception of higher energy density capacitors, is still used on *Z*. Circuit modeling and design prototypes were studied extensively to determine the implications of stray charging paths. It was found that stray capacities within the Marx reduce the energy available to the spark gaps, and prevent cascade erection. Considerable effort was invested in the development of the trigger generator architecture. Trigger and ground resistors were installed to decouple detrimental strays and maintain sequential erection characteristics [31]. The resultant circuit layout is shown in Figure 1.30 [30,31].

The trigger system is arranged so that the operation of nine "MPUs" is initiated simultaneously with a 100 kV pulse with a 10 ns rise time. The MPU is also a bipolarly charged Marx generator providing a 540 kV pulse with an 80 ns rise time to the six gaps of row 1. The other gaps are sequentially triggered by the voltage pulses from the forward-feeding trigger resistors. Each MPU fires four Marx generators, resulting in all 36 modules of PBFA being triggered (Figures 1.32 and 1.33).

1.5.3 Aurora [9]

The Aurora Marx bank, with output parameters of 11 MV, 120 kA, and a stored energy of 5 MJ, was comprised of four individual Marx generators connected in parallel, as shown in Figure 1.34a. This design reduced the overall Marx inductance to an internal inductance of 12 μH. The Marx generators share a common 120 kV DC charging supply and are triggered by a common 600 kV Marx generator.

Each of the Marx generators has 95 stages with a stage capacitance of 1.85 μF charged to 120 kV. The basic schematic of a representative portion of the Aurora Marx generator [9] is shown in Figure 1.34b and uses both capacitive and resistive coupling. Weak overvoltages at particular spark gaps (S4, S6, S7, S9, S10) are resistively coupled to earlier stages, greatly increasing the reliability of the Marx erection upon command trigger, but the number of stages between the resistive coupling is not fixed. The Aurora Marx bank's erection time is 1 μs with an erection jitter of 10 ns. The enhanced coupling scheme allows operation down below 50% of its self-breakdown voltage with a prefire probability of less than 1%.

1.6 Marx Generator Variations

As a result of the large number of applications for Marx generators, a variety of modified versions of the conventional Marx generator have evolved, each of

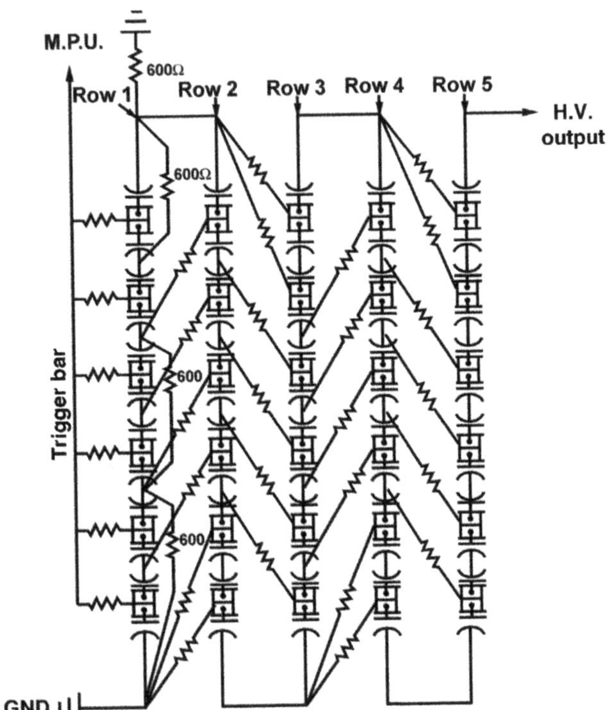

Figure 1.32 The circuit of the PBFA II Marx generator [30]. The six spark gaps of row 1 are triggered and the Marx generator uses both capacitive and resistive coupling to the upper stages to ensure a reliable erection process. The physical layout of the circuit is shown in Figure 1.33.

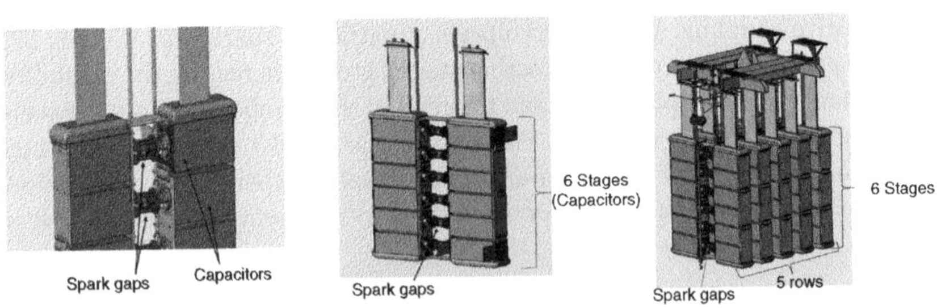

Figure 1.33 The implementation of Sandia Marx generator circuit is complex. The Marx is arranged in two columns of capacitors where each "stage" is connected by spark gaps and which constitutes a row. Five rows are assembled into the Marx with coupling connections as shown in Figure 1.32. The Marx banks are very sophisticated and have been shown to be astonishingly reliable. (Reproduced with permission of Sandia National Laboratories.)

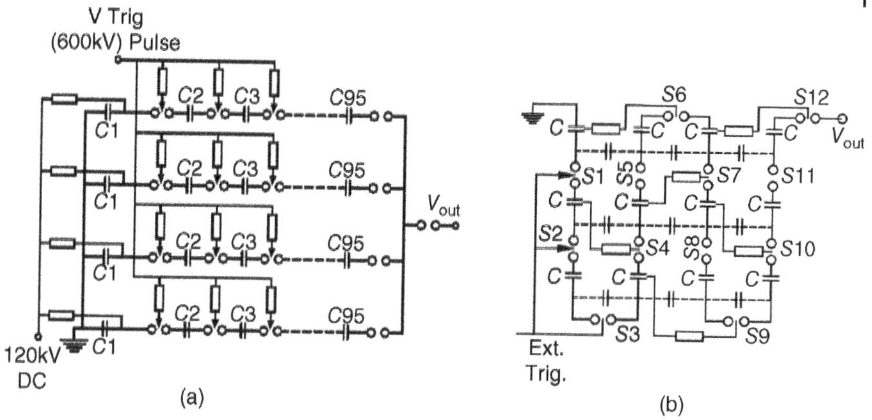

Figure 1.34 The Aurora Marx bank (a) is comprised of four Marx generators in parallel. The Aurora Marx generator design, shown in part (b), reinforces the weak overvoltage on various spark gaps with both capacitive and resistive back-coupling [9].

which is able to meet a specific application in a more efficient way. The salient features of a few of these modified Marx generators are described in this section.

1.6.1 Marx/PFN with Resistive Load

Conventional Marx designs do not produce rectangular pulse. The principles of Marx operation may be modified, however, to produce a flat-top pulse by making the Marx "capacitors" energy storage elements that produce a pulse with a flat-top pulse. Erection of the Marx adds the voltages of each stage while preserving the pulse shape. The circuit, shown in Figure 1.35, is known as a Marx-PFN.

Pulse forming networks (PFNs) made from lumped elements are very common and coaxial cables have also been used. Any type of PFN may be used, but E-type PFNs are common.

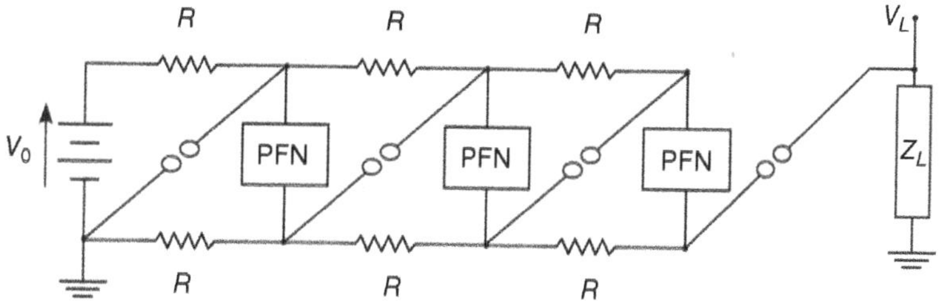

Figure 1.35 A PFN-Marx generator driving a resistive load.

From given requirements of pulse voltage, pulse current, and pulse duration, the equations for fixing the basic parameters assuming matched load conditions are given below:

Characteristic impedance (R'):

$$Z_{PFN} = \sqrt{L/C} \tag{1.62}$$

Pulse duration (T):

$$T = 2n\sqrt{LC} \tag{1.63}$$

Matched load voltage (V_L):

$$V_L = \frac{NV_0}{2} \tag{1.64}$$

Load resistance (R_L):

$$R_L = NZ_{PFN} \tag{1.65}$$

Maximum stored energy (E_S):

$$E_S = \frac{1}{2}N(nC)V_O^2 \tag{1.66}$$

where L and C are parameters of the PFN, n is the number of capacitors in the PFN, N is number of Marx stages, and V_0 is the charging voltage. The output waveform of a Marx-PFN into a matched load is shown in Figure 1.36. The rise time (t_r) in this case is governed by L'/R_L of the circuit and the main contribution to L' is from the spark gaps.

The energy content is reduced from its maximum value by inefficiencies, including the energy lost in the spark gaps and the resistors as well as charge remaining in the capacitors. Energy dissipated in the switches is reduced by low-gap spacing but necessitating high dielectric strength mediums. Both single insulator (gas) and hybrid insulation (gas/oil) schemes are common in Marx-PFNs. When a single insulating gas is used for insulation and switching, the contribution to the inductance from the connections is minimized, but the

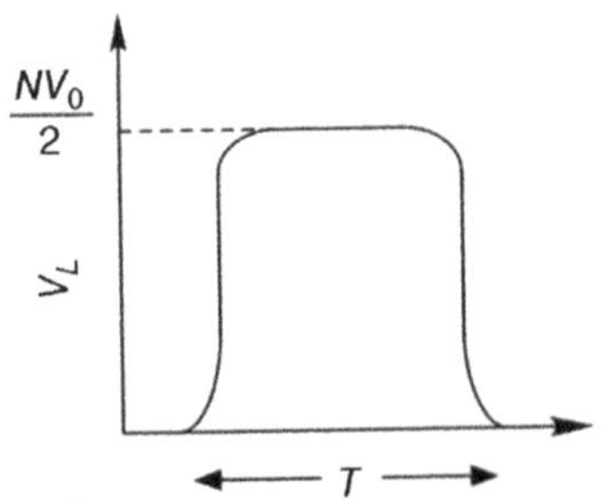

Figure 1.36 The output voltage of a Marx-PFN measured at a matched load.

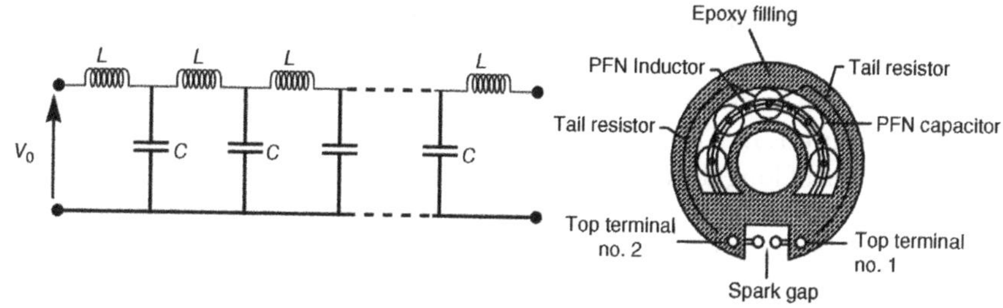

Figure 1.37 A single stage (a) circuit and (b) layout of a Marx/PFN using a Type E PFN.

switch inductance may increase because the gap length is set by the system pressure. At high voltages, transformer oil or epoxy may be used to insulate the Marx and pressurized gas is used in the switches. Figure 1.37 illustrates a single stage consisting of a PFN, resistors, and spark gap of a 20-stage Marx-PFN producing a 300 kV, 100 ns pulse [32,33]. The switches are aligned in a straight-line path so that the ultraviolet light generated when the first spark gap fires irradiates the rest of the spark gaps, providing initiatory electrons for the breakdown process, reducing the statistical lag time, and increasing reliability. The Marx-PFN uses pressurized N_2 gas for spark gaps and mineral oil for the rest of the components.

A Marx-PFN can be constructed using any type of PFN. Riepe [34] made the differentially charged Marx-PFN shown in Figure 1.38 with a type-C PFN to produce a 2.5 μs long pulse at 120 kA and 300 kV into a matched load with a reasonably flat top. The first spark gap is triggered.

More recently, Adler et al. [35] have developed a Marx-PFN with a novel charging supply to produce pulses with a duration greater than 1 μs with less than a 5% droop with high reliability. The Marx stages are type-C PFN network and generate 500 kV into a 50 Ω load with a rise time of less than 200 ns.

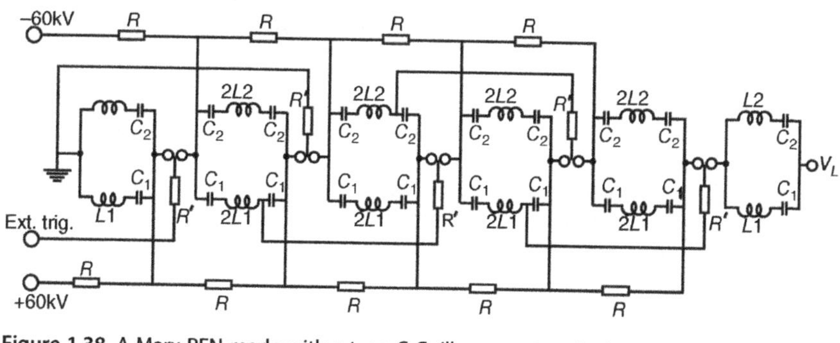

Figure 1.38 A Marx-PFN made with a type-C Guilleman network. (Reproduced with permission from Ref. [34]. Copyright 2008, AIP Publishing LLC.)

1.6.2 Helical Line Marx Generator

Helical lines can be used as energy storage elements in a Marx generator [36], as shown in Figure 1.39. The helical line Marx-PFN is useful for the generation of flat-top pulses of long duration (a few microseconds). The Marx capacitors are replaced by helical lines made by replacing the braided outer conductor of a typical low-inductance coaxial cable with a helical winding. The helical winding is a conductor of diameter d wound over a polyethylene insulator of diameter D with n turns per unit length.

The design formulas for a helical Marx are as follows:

Characteristic impedance:

$$Z_{helix} = \sqrt{L/C} \tag{1.67}$$

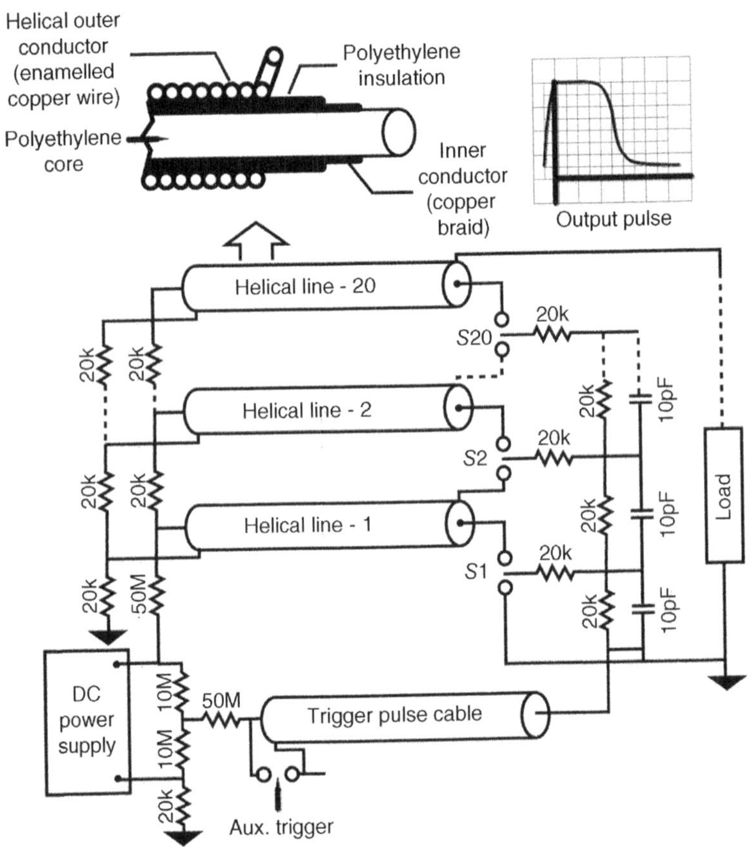

Figure 1.39 A Marx generator using helical lines as stage energy storage elements [36]. (Reprinted with permission from Ref. [36]. Copyright 2001, AIP Publishing LLC.)

Marx internal impedance:

$$Z_M = NZ_{helix} \tag{1.68}$$

Pulse duration:

$$T = 2\ell\sqrt{\tilde{L}_h\tilde{C}_h} \tag{1.69}$$

Matched load voltage:

$$V_L = NV_0 \tag{1.70}$$

Here

$\tilde{L}_h$	=	inductance per unit length of helical line
$\tilde{C}_h$	=	capacitance per unit length of helical line
ℓ	=	length of helical conductor $= \pi nD$ for $D \gg d$

The maximum load voltage can be greatly reduced by coupling between the cables of adjacent lines. A 20-stage Marx generator based on the above design using modified coaxial cables as helical line storage elements has delivered 400 kV and 20 A at a pulse duration of 1 μs into a matched load.

1.7 Other Design Considerations

This section deals with the salient features of Marx generators: (a) DC charging voltage and number of stages, (b) Marx capacitor selection, (c) pulse charging schemes, (d) Marx spark gap considerations, (e) Marx resistors, (f) delay time and jitter, and (g) triggering.

1.7.1 Charging Voltage and Number of Stages

The open-circuit output voltage of the Marx can be increased by increasing either the number of stages or the charging voltage per stage. Larger number of stages increases the number of switches, the inductance, and the cost. However, it also allows a common switching and insulating medium. The voltage per stage is set to be much lower than the self-breakdown voltage of the spark gaps to minimize the possibility of prefire. Prefires are often a significant risk in the operation of high-energy Marx banks and are often operated at 50–70% of their self-breakdown voltage. Sufficient overvoltages are ensured by carefully designed resistive triggering and capacitive coupling.

In advanced versions of fast Marx generators, a high DC charge voltage and fewer switches are preferred. Since the overall reliability of a Marx generator depends mainly on the precision of firing the spark gaps, any method that reduces the number of spark gaps increases the performance of the Marx generator. In practical systems, the maximum DC voltages are limited to approximately 100 kV, and such high DC voltages necessarily involve complicated insulation systems. This insulation requirement is one reason to consider bipolar charging since the insulation requirement per stage is similar to that of a single-polarity Marx, but the required number of stages (and switches) for a given output voltage is cut in half.

A Marx generator may be pulse charged, where the capacitors are charged in a very short time and immediately discharged. This was proposed to reduce the prefire rate and extend the useful life of the energy storage capacitors. Buttram and Clark built a pulse charged 1 MV, 10 kJ, 10 Hz Marx generator by interposing a triggered spark gap between the HVDC supply and the Marx generator [37]. Pulse charging greatly increases the complexity and necessitates sophisticated synchronization and a low-impedance HVDC supply.

1.7.2 Insulation System

Marx generators generally used by teaching institutions and high-voltage testing laboratories are usually of open type of construction utilizing atmospheric air insulation. Because of poor dielectric strength of air, these Marx generators tend to be very large in size and usually require high ceilings and large side clearances for the installation room. An open construction has the advantage that the components are accessible, easing maintenance and fault detection, as well as modification of the waveform. This is particularly important for impulse testing, where distinctly different waveforms are required or if the Marx is being used as a teaching aid. In an open construction, the insulation strength is easily affected by humidity levels and other environmental changes. The insulation systems that can be used in a Marx generator (other than switches) are (a) high-pressure gas, (b) insulating oil, and (c) pourable solids (potting).

Gas at elevated pressures, usually N_2, SF_6, or a mixture of the two, has good dielectric strength, considerably reducing the overall size of the Marx. The Marx containment vessel, however, becomes a pressure vessel. To ensure personnel safety, the pressure vessel should be designed according to an appropriate mechanical engineering standard, such as the ASME. Depending on the specific design, a single gas pressure may be used throughout the Marx or separate gas pressures may be used for the switches and the insulation. Oil insulation is widely used to insulate Marx generators in very high-voltage operation and have the considerable advantage of not requiring a pressure vessel. Oils are also suitable for repetitive Marx operation because it has a high heat capacity. However, contaminants and repeated stressing of the oil tend to decompose the oil and degrade its dielectric strength.

The most popular insulation system is one that uses gas in the spark gaps with the remainder of the Marx immersed in oil. The Marx output is easily variable by changing the pressure of the gas in the spark gaps, without affecting the basic insulation level of the overall Marx. By employing recirculation and filter system for this insulating oil, the Marx generator can be easily adapted for high repetitive pulsing.

1.7.3 Marx Capacitors

The Marx capacitors are generally fast-discharge, low-inductance energy storage capacitors with higher current capability. Considerable ringing may occur in Marx generators with capacitive and inductive loads and the voltage reversal capability of the capacitors should be considered. Resistive isolation resistors are preferable to inductive isolation since it damps the reversal.

For smaller Marx generators, ceramic capacitors such as those using a barium titanate dielectric can be used [38]. Experience indicates that with open Marx construction, barium titanate capacitors with epoxy enclosure are suitable but those with PVC and other plastic enclosures are susceptible to permeation of moisture on the dielectric surface, and subsequent failure by surface tracking.

The state of the art in capacitors for multimegavolt Marx generators is those made by General Atomics for Sandia National Laboratories' ZR project [39]. The new 2.6 μF, 100 kV capacitors doubled the energy density of the older 1.3 μF capacitors by doubling the capacitance in the same volume, while maintaining the low inductance (<30 nH) and high peak current (170 kA) capabilities. Testing predicts capacitor lifetimes of about 11,000–13,000 shots at 100 kV and 8000 shots at 110 kV. Several other manufacturers have comparable products.

1.7.4 Marx Spark Gaps

The first one or two spark gaps in a Marx generator are generally externally triggered and the remainder fire from the resultant overvoltage. Common electrode materials are brass [38], copper [40], and stainless steel [41]. Brass is often used in SF_6 insulated spark gaps because the chemical reaction with the SF_6 dissociation products stabilizes the self-breakdown level, greatly reducing the probability of prefire [42]. In three-electrode Marx spark gaps, the center plate may be made from stainless steel. For small or low-energy Marx, spherical electrodes are often used. In generators with many stages or high energies, long-lived commercial spark gaps are used. Triggered three-electrode spark gaps are used in combination with resistive or capacitive coupling to other stages to control overvoltages.

For a high-power repetitively pulsed Marx generator of 1 MV, 10 kJ, 10–100 Hz, Buttram and Clark employ a water-cooled spark gap with a high-velocity gas flow to remove the ionization products from the gap medium and reduce the erosion of the switch electrodes by not allowing the electrodes to heat up [37].

1.7.5 Marx Resistors

The purpose of the Marx resistors R in Figure 1.14 is to prevent the Marx capacitors from being short circuited when the Marx spark gaps fire, provide a dump for the energy stored in the Marx capacitors in case of emergency or malfunction, and shape the fall time of the output wave to specification. The maximum voltage drop across R corresponds to the charging voltage V_0, and it should be sized to prevent flashover across its surface. The resistor R_F should have a pulse insulation level for the full output voltage of the Marx. Resistors incorporated into Marx generators should be noninductive to avoid the introduction of oscillations in the erected Marx circuit.

Noninductive resistors may be fabricated by hand-winding on a fine-tooth comb, as shown in Figure 1.40 [43]. High-resistivity materials such as Nichrome, Manganin, and Kanthal can be conveniently used for this purpose. Manganin has the advantage of being easily soldered and has good stability with temperature. Potting in epoxy gives the resulting resistor mechanical rigidity and counteracts environmental influences such as humidity [44].

When designing for Marx-PFN, the value of R should be such as to make the R_C time constant much larger than the output pulse duration. Liquid resistors, particularly copper sulfate with copper electrodes or sodium thiosulfate with aluminum electrodes, can be used [29,45]. For copper sulfate solutions with resistivity between 60 and 1200 Ω cm and applied electric fields between 2 and 50 kV/cm, the resistance is linear [45,46]. Liquid resistors are low inductance and compact, are easily constructed, and have high energy dissipation but may be excessively capacitive in some applications.

1.7.6 Marx Initiation

A Marx generator can be self-fired, but this imposes severe limitations on its functionality: The generator must be operated at a large percentage of it's self-breakdown voltage, which results in difficulty in varying the output voltage and low overvoltages which leads to poor rise time performance. These drawbacks are overcome by externally triggering the first one or two gaps as shown in Figure 1.38, and Figure 1.39.

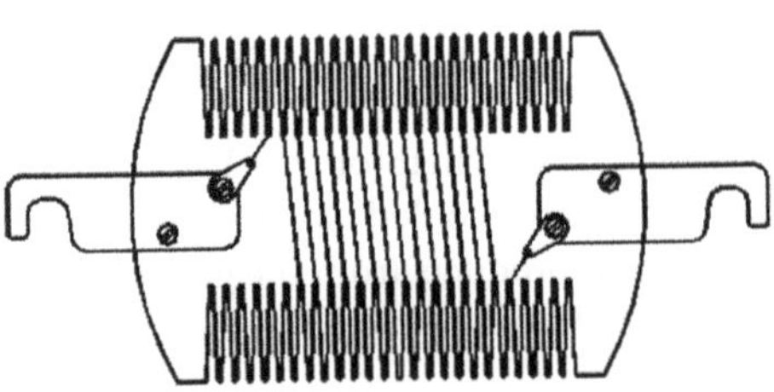

Figure 1.40 Hand-winding on a fine-tooth comb.

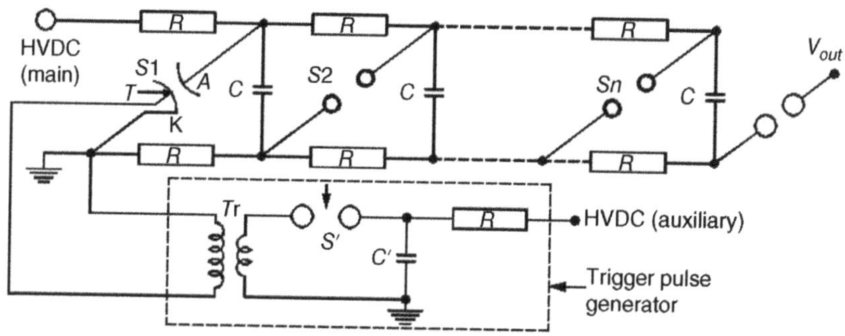

Figure 1.41 Triggering a Marx with a trigatron.

A trigatron, discussed extensively in Chapter 4, has two main electrodes, the anode A and the cathode K with an insulated trigger T embedded in the cathode. The trigger pulse is formed by the discharge of capacitor C' into the primary of a pulse transformer through switch S', as shown in the dotted box of Figure 1.41. The pulse transformer produces a high-voltage pulse of sufficient magnitude and duration to close the trigatron, leading to a command-triggered cascade erection of the generator.

Triggering along the midplane of a spark gap switch is a popular choice for Marx generators and is illustrated in Figure 1.39. For instance, in Kukhta's implementation [47], the spark gap consists of three electrodes: the cathode K, the anode A, and the trigger electrode T. The trigger electrode is located in the middle of the gap and held at potential ($V_0/2$) by a voltage divider comprised of R_1 and R_2. The trigger pulse is generated by the cable assembly, charged to a voltage $V_0/2$. The cable is short circuited when the spark gap S' fires, generating a pulse of magnitude ($-V_0/2$). When this pulse reaches the trigger electrode, it inverts its potential leading to breakdown of the spark gap S_1 (Figure 1.42).

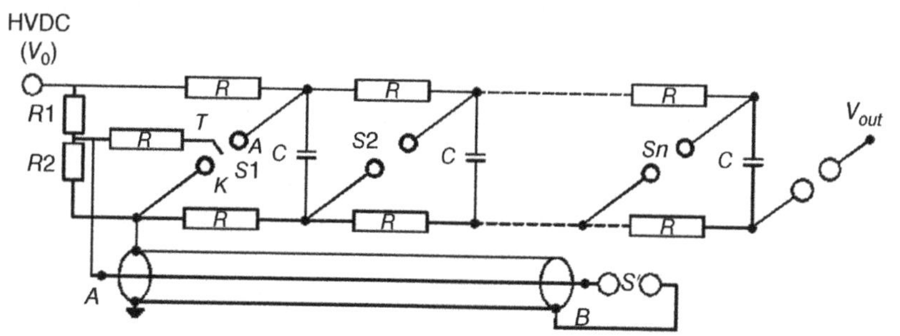

Figure 1.42 The operation of the Marx is initiated by triggering along the midplane of the spark gap in the first stage. The trigger generated is a simple transmission line.

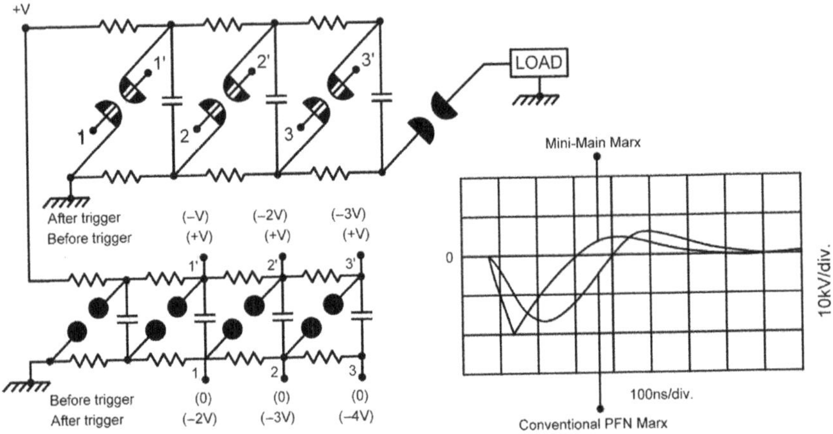

Figure 1.43 Triggering of a large Marx generator with a smaller Marx.

A Marx generator may be used as a trigger generator for another "main" Marx generator. This arrangement is used extensively on large high-energy banks to inject a high-energy, high-voltage pulse into a number of trigger points of the main Marx. Figure 1.43 shows a main Marx generator whose switches are all double-trigger mode, described in detail in Chapter 4, and fired from the trigger Marx [48]. The trigger Marx must have a fast rise time and firing reliability.

1.7.7 Repetitive Operation

The repetitive capability of Marx generators is severely limited relative to even conventional pulse transformers. For repetitive operation, the main advantage of Marx generators over other methods of generating high voltage is its versatility of packaging. The sequential erection of a Marx generator can be made to follow any number of paths because of the inherent flexibility in its architecture.

The pulse repetition rate is limited by the Marx cycle, which can be considered in three distinct parts: the time to charge the Marx, the high-voltage time, and the switch recovery time. Inductors can replace the isolation resistors for faster charging times. The Marx charging time has been investigated in both series and parallel charging using resistors, inductors, and resistive/inductor combinations as charging elements [49].

Pulse repetition rates of 600 Hz have been demonstrated with a series inductive charging system and spark gap switches insulated with 3 atm of an air/SF_6 mixture [50] (L. Veron, personal communication). Corona-stabilized switches, demonstrated to operate at 20 kHz, have been used in Marx

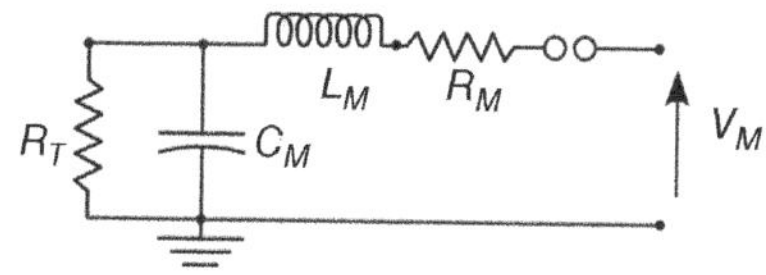

Figure 1.44 The equivalent circuit of a Marx generator used in a circuit simulation.

generators at pulse repetition frequencies of 400 Hz [51–53] and Marx generators using hydrogen-filled spark gaps have been demonstrated at 10 kHz in burst mode [54]. In practice, depending on the energy per pulse, the power supply may be the limiting factor.

1.7.8 Circuit Modeling

Circuit modeling is an important task of pulsed power design and both commercial and custom codes are used for this purpose. Circuit modeling is especially important for machines where currents are added, because even small deviations can manifest as large deviations from design when tens of modules are connected to the load. Custom codes may contain time-varying circuit elements to represent components unique to pulsed power, and commercial codes can usually be made to mimic them.

Marx banks can be quite intricate, but even a simple Marx generator would add unnecessary complexity to a circuit code. Instead, the Marx may be represented by the circuit of Figure 1.44. The capacitance C_M is the erected Marx capacitance with its discharge rate limited by the resistor R_T. The Marx inductance L_M accounts for the inductance of the switches, capacitors, and connectors in the discharge circuit. These inductances are difficult to calculate and, in practice, the Marx is charged to a reduced voltage and discharged into a short circuit. The equivalent circuit values of the Marx can be deduced from the ringing frequency and the decay time. These values are largely independent of voltage.

The equivalent Marx circuit can represent the energy storage stage quite effectively. Figure 1.45 is an overlay of a circuit code simulation and a measured current waveform from a pulse compression circuit showing excellent agreement. The dip that occurs at approximately 1700 ns is a reflection from the next capacitive energy storage stage.

1.8 Marx-Like Voltage-Multiplying Circuits

Other circuits may be used to multiply voltages that are not Marx generators. Like a Marx, these schemes feature a staged energy storage resulting in voltage multiplication. However, the fundamental principle of a Marx generator—namely, charging a circuit in parallel and discharging it in series—is missing.

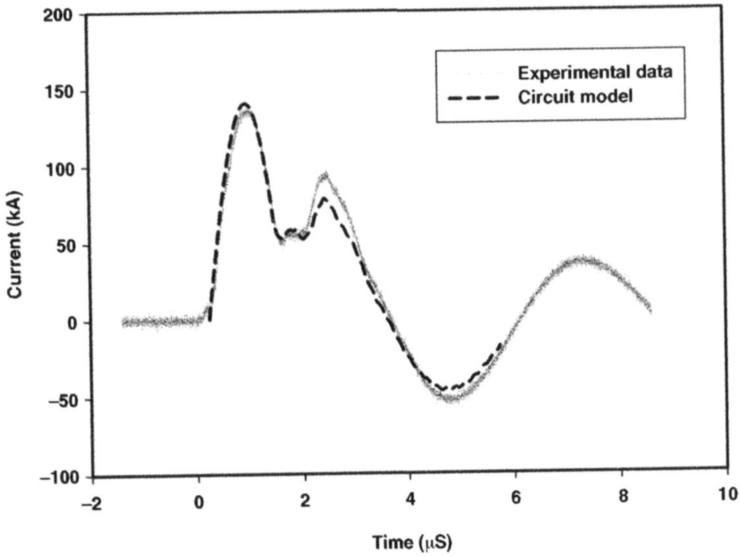

Figure 1.45 An overlay of a Marx current measurement and the output of a circuit simulation using the equivalent circuit of Figure 1.44 [55].

These circuits, particularly the *LC* inversion generator, are often mistakenly referred to as a Marx generator.

1.8.1 The Spiral Generator

A spiral generator [56,57] is a voltage multiplier constructed by placing another layer of insulation on a stripline and wrapping it around itself $(n-1)$ times in a spiral fashion, as shown in Figure 1.46. The resulting geometry, a three-electrode stripline, comprised of line 1 and line 2, is wound on an insulating form of diameter D in the form of a spiral of n turns with a width w and dielectric thickness h for each two-electrode line. In a spiral generator, voltage multiplication is achieved with a single switch—a big advantage—but the output waveshape is triangular, which limits its usefulness. The load is connected from the center of the generator to ground. The start switch S may be located as shown in Figure 1.46 or at the midpoint of the line length with the electrical lengths of the lines being equal.

The spiral generator operated on the manipulation of reflected waves, similar to the operation of a Blumlein. Line 1 is charged to V_0, and when the switch S is closed at $t=0$, a maximum voltage of $(-2nV_0)$ appears on the output terminal V_{out} at a time T after the switch closure. The time T corresponds to the sum of the electrical lengths of line 1 and line 2, which are equal. The status of the electric field vectors on line 1 and line 2 through the radial cross section of the

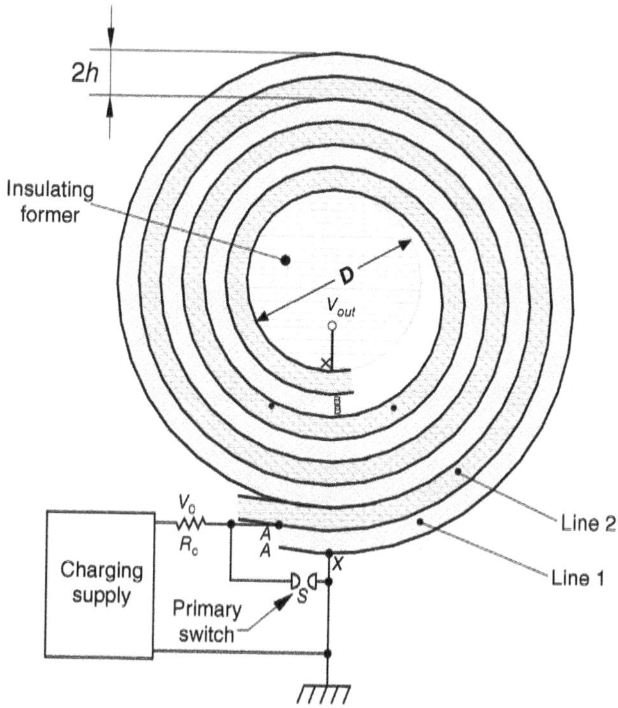

Figure 1.46 An ideal spiral generator.

spiral generator at times corresponding to $t=0$ and $t=T$ are shown in Figure 1.47. At $t=0$, just before the start switch closes, the *E*-field vectors in alternate sections of lines are in opposite directions and hence the net output voltage at the load is zero, as shown in Figure 1.47a. At $t=T$, the *E*-field vectors

Figure 1.47 A view of the status of the electric field on the cut *XX* in Figure 1.46. The electric field vectors on lines 1 and 2 of the spiral generator are initially opposed, but align to produce an output voltage V_{out}, with a maximum value of $-2nV_0$) at the output terminals at the time $t=T$.

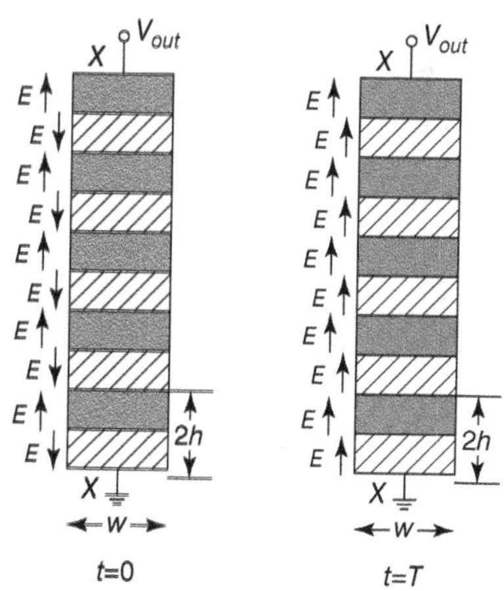

across the entire line 1 are reversed by the reflected voltage waveform, as shown in Figure 1.47b, producing an open-circuit output voltage V_{out} at the load.

If D represents the mean diameter of the spiral, under the assumptions that $D \gg 2nh$, the important parameters for a lossless spiral generator can be derived from its equivalent circuit [58]:

a) Time to reach maximum output voltage:

$$T = 2\pi Dn\sqrt{(\varepsilon_0 \varepsilon_r \mu_0)} \tag{1.71}$$

b) Output capacitance:

$$C = \frac{\varepsilon_0 \varepsilon_r \pi D w}{(2n)h} \tag{1.72}$$

c) Maximum electrostatic stored energy:

$$E = \frac{\varepsilon_0 \varepsilon_r \pi D w n V_0^2}{h} \tag{1.73}$$

The output voltage is triangular in shape, as shown in Figure 1.48, mainly due to the integrating effect of the inductive coupling of the spiral turns [3]. Fitch and Howell [57] provide excellent insight into losses in these types of generators. In particular, if the condition $D \gg 2nh$ is not satisfied, which is frequently the case, the voltage multiplication factor is reduced [3].

1.8.2 Time Isolation Line Voltage Multiplier

Lewis [59,60] conceived a high-voltage multiplying circuit, shown in Figure 1.49, that does not use switches. Instead, it uses long sections of transmission line, usually coaxial, driven by a low-voltage, low-impedance,

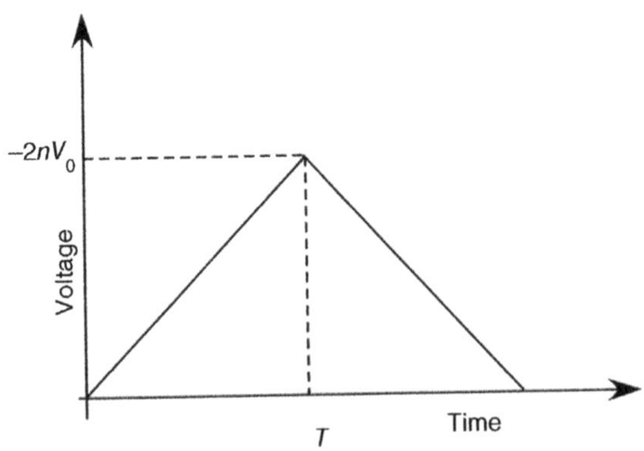

Figure 1.48 The output waveform of a spiral generator initially charged to a voltage V_0.

Figure 1.49 Schematic of a coaxial time isolation high-voltage generator showing connections.

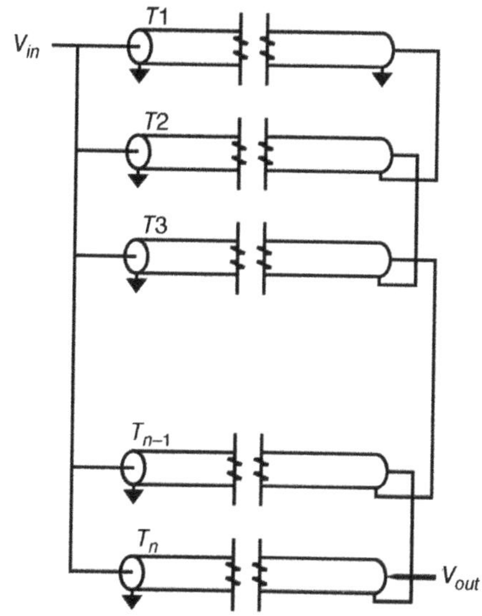

high-current source. It is termed a time isolation voltage multiplier because the long line propagation time T_T permits parallel excitation at the source and voltage multiplication by a series connection at the load.

The time isolation generator consists of N transmission lines connected in parallel at the input end and in series at the output end. When the input end is fed with a pulse voltage of duration, T_p, under the conditions of $T_p \ll T_T$, the output voltage is multiplied by a factor of ($2N$). The performance of this voltage multiplying is similar to the LC inversion generator described in the next section. Careful consideration must be shown regarding impedance matching at the input and output ends to avoid the waveform distortions from reflections. The effects of stray capacitances have been studied in considerable detail by Chodorow [61] who states a 700 kV, fast rise time Time Isolation generator was used as an EMP simulator. Voltage multiplication in coaxial Time Isolation generators was reported by Soto and Soto and Altamirano [62] and Carmel et al. [63]. The maximum theoretical output voltage is typically reduced, sometimes greatly, by coupling between the cables.

1.8.3 The LC Inversion Generator

The LC inversion generator is a voltage-multiplying circuit that was patented by Richard Fitch in 1968 [1,64]. Although similar in appearance to a Marx generator, the operation of an LC generator or inversion generator is not like a Marx generator but instead is related in its operation to the spiral generator, for which Fitch also has a patent [56], and a Blumlein generator.

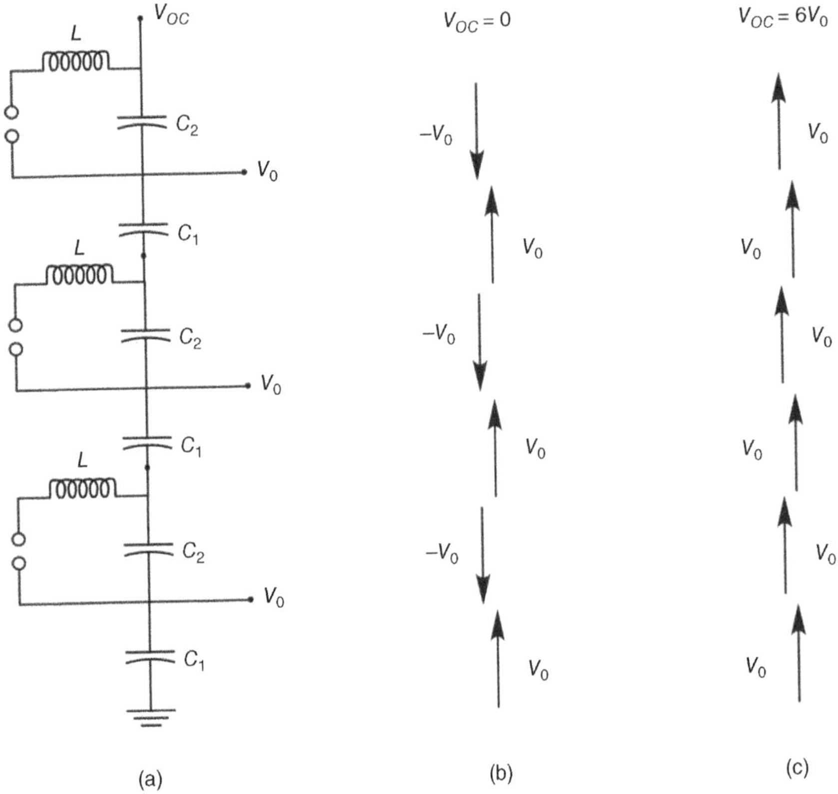

Figure 1.50 The three-stage *LC* inversion generator in part (a) has alternating potentials on the capacitor, as shown in part (b) for a net stage voltage of 0 when fully charged. The potentials add (c) when the spark gaps are simultaneously fired.

The LC generator operates on the transient inversion of alternate potentials in a series system of alternately opposed potentials, as shown in Figure 1.50. Each "stage" consists of two capacitors charged to equal and opposite potentials. During charge cycle, the net stage potential is zero and then switched at a precise time so that the potentials add.

Moderate power *LC* generators using diodes across the capacitors, often called a Fitch circuit, has found widespread use in industrial applications, such as lasers and pollution control circuits because the circuit can be made repetitive, highly reliable, and stable [65–67]. The principle of operation is most easily understood by looking at circuit where diodes or other rectifiers are used. The diodes in the circuit prevent the second half-cycle in the discharge cycle instead of relying purely precise switching times and relax requirements on the output pulse duration.

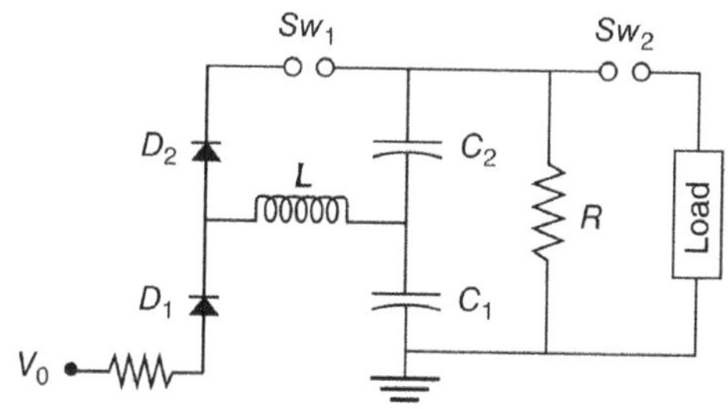

Figure 1.51 A Fitch circuit can be used to illustrate the operation of an *LC* inversion generator. The diodes prevent the second half of the discharge cycle of the capacitors and eliminate the need for precise timing.

A single-stage *LC* generator is shown in Figure 1.51. During the charge cycle, the capacitors C_1 and C_2 are charged through the diode D_1 and inductor L. The resistor R completes the charging circuit for C_2. D_2 is not involved in the charging cycle, but when more than one stage is used, it passes current through to subsequent capacitor pairs. Recognizing that when the capacitors are fully charged, no current flows in the circuit C_1–C_2–R, the application of Kirchoff's voltage law shows the voltages on capacitors C_1 and C_2 must be equal in magnitude and opposite in sign.

When the spark gap Sw_1 fires, the capacitor C_2 begins to discharge and current flows in the circuit C_2–L–D_2. The diode D_1 is reverse-biased and capacitor C_1 does not discharge but remains fully charged to the voltage V_0. The current in the circuit C_2–L–D_2 during this time, $I_2(t)$, is given by

$$I_2(t) = V_0\sqrt{\frac{C_2}{L}}\sin \omega t \tag{1.74}$$

where

$$\omega = \frac{1}{\sqrt{LC_2}} \tag{1.75}$$

As this resonant circuit reaches its zero crossing at a time

$$t_0 = \frac{\pi}{\omega} = \pi\sqrt{LC_2} \tag{1.76}$$

the sinusoidal current changes sign, the diode D_2 is now reverse-biased, and the spark gap extinguishes and opens. Capacitor C_2 is now inverted, and has a charge of $+V_0$. The capacitors C_1 and C_2 have the same polarity in the second period and the voltage at the output spark gap is $2V_0$. Switch Sw_2 may now be fired to discharge the circuit into the load. The switch Sw_2 may be omitted. The resistance R determines the decay time, in a manner similar to that derived for Marx generators. Switch timing and jitter are critical to the operation of the *LC* generator.

In general, an *LC* inversion generator stacks several of these circuits for voltage multiplication, yielding an open-circuit voltage:

$$V_{OC}(t) = NV_0\left[1 - e^{-\alpha t}\cos(\omega t)\right] \tag{1.77}$$

with

$$\alpha = \frac{R}{2L} \tag{1.78}$$

where N is the number of stages and R is the effective series resistance in the LC circuit. The $e^{-\alpha t}$ term accounts for losses in the discharge path including those introduced by component connections, and so on. The rise time of the output pulse is determined by the resonant portion of the circuit and is given by

$$t_r = \pi\sqrt{LC_2} \tag{1.79}$$

The pulse rise time can be recognized as the time of zero crossing of the LC_2 resonant circuit. For pulse generation with a fast rise time, a small value of inductance is required, but the chosen value also affects the pulse width. The realized inductor must also be designed to store the same energy as the capacitor C_2.

A high-power LC generator must achieve the same basic operation as described above without the benefit of diodes. Instead, the inversion is achieved by careful and precise firing of the switches. Figure 1.52 shows two variations of inversion generators with their charging impedances. With no diodes to prevent

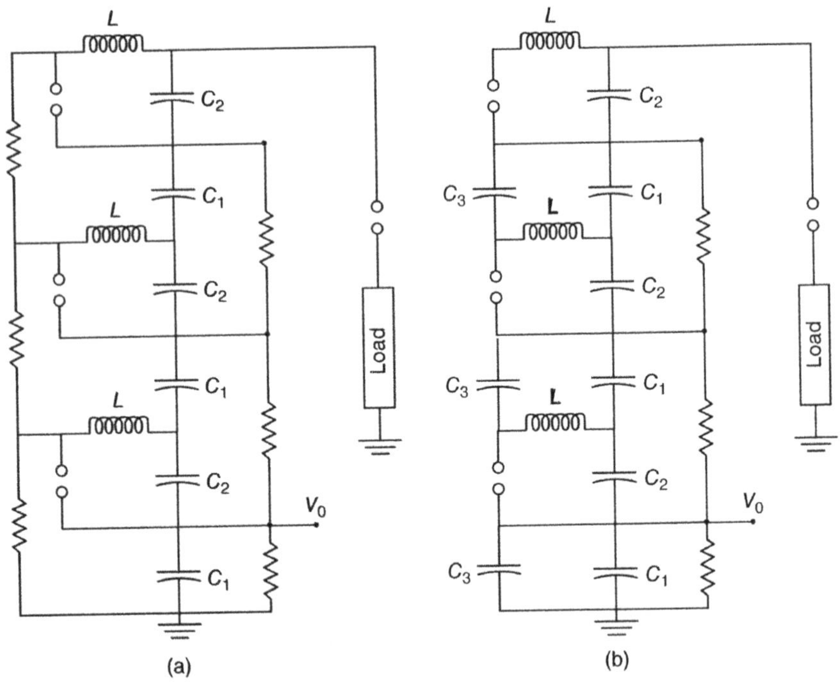

Figure 1.52 An *LC* generator circuit may use (a) resistors or (b) capacitors in its charging network.

the second half-cycle, the inversion time must be made two or three times longer than the required pulse duration.

In an inversion circuit, the switches are not part of the discharge circuit and the switch inductance and switch resistance do not contribute to inefficiency. The load currents are also independent of the switches and a low-inductance, fast current pulse can be produced. The output begins to appear as soon as the switches fire, with a rise time determined by the ringing portion of the circuit. Particularly for high-power operation, an output switch is typically included as part of the circuit.

The *LC* inversion generator has two very significant disadvantages: triggering and fault modes. The switches in an *LC* inversion generator must all be triggered simultaneously. Unlike a Marx, the discharge of the generator does not produce an overvoltage on the switches. In an inversion generator, each switch is independently fired and the additional cost of multiple trigger generators could be prohibitive. Trigger generators with multiple outputs may be used in lieu of independent triggers. Instead of independent switch triggers, nearby stages can be linked capacitively [1], as shown in Figure 1.53.

Protection against fault modes in an *LC* inversion generator is serious. If a switch fails to fire, the generator could develop large overvoltages or voltage reversals that can be dangerous to some capacitors. Switch prefires do not necessarily fire the remaining stages and prefire protection system is difficult to design. The complexity of protecting the circuit against its fault modes reduces the initial appeal of the *LC* inversion generator for some applications.

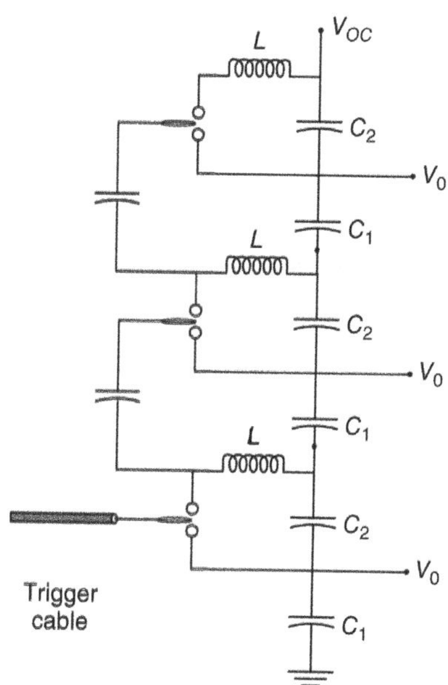

Figure 1.53 Coupling capacitors linking switch may be used so that each switch does not require a trigger generator.

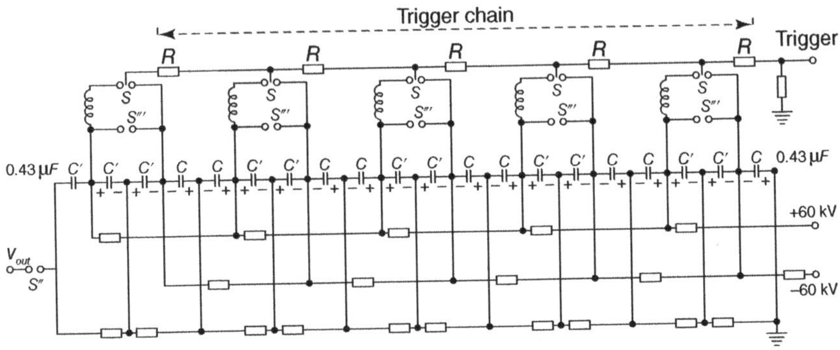

Figure 1.54 A dual-polarity vector inversion generator with a peak voltage of 1 MV and a protection circuit in the event a stage does not fire [68].

An *LC* inversion generator, developed by Harris and Milde [68], is shown in Figure 1.54. It is charged in dual polarity and produces an output voltage between 270 kV and 1 MV with a maximum stored energy of 15 kJ and a discharge time of 5 μs. When the capacitive voltages align, the load circuit is connected through the self-breaking transfer switch S''. Protective spark gaps S''' are incorporated into the design to prevent damage if one of the spark gaps fail to fire and puts a dangerously large overvoltage across its capacitor set.

1.9 Design Examples

Example 1.1

A four-stage Marx generator, having capacitors of 0.125 μF in each stage, discharges into a capacitive load of 200 pF. Using a simplified circuit approximation, calculate the values of stage resistors and front resistors to give (a) e-folding rise time of 50 ns, and (b) e-folding tail time of 50 μs. Calculate (c) overvoltage factor on the remaining spark gaps, when the first spark gap is triggered. Assume a charging voltage of 30 kV and self-breakdown voltage of the spark gap as 35 kV.

Solution

The equivalent circuit of this Marx generator is shown in Figure 1.2, with the following values:

$$C_M = \frac{C_0}{N} = \frac{0.125}{4}\,\mu\text{F} = 31.25\text{ nF}$$

$$C_L = 200\text{ pF}$$

a) The approximate circuit for rise time is shown in Figure 1.4, from which

$$\text{e-folding rise time } = 50\text{ ns} = \left[\frac{C_M \times C_L}{C_M} + C_L\right] \times R_F$$

$$\text{Front resistor } R_F = 250\,\Omega$$

b) The approximate circuit for tail time is shown in Figure 1.5, from which

$$\text{e-folding tail time } = 50\,\mu\text{s} = (C_M \times C_L) \times R_T$$

$$\text{Tail resistor } R_T = \frac{2R_T}{N} = 1.59\,\text{k}\Omega$$

$$\text{Stage resistor } R = \frac{2R_T}{N} = 795\,\Omega$$

c) When the first spark gap fires, additional voltage on the remaining spark gaps

$$V_g = \frac{N}{N-1} V_0 = \frac{4}{3}(30\text{ kV}) = 40\,kV$$

$$\text{Overvoltage factor} = \frac{V_{actual}}{V_{breakdown}} = \frac{40}{30} = 1.14$$

Example 1.2

A 10-stage Marx generator has the following parameters: (a) DC charging voltage =100 kV, (b) capacitance per stage = 0.25 μF, and (c) tail resistor per stage = 720 Ω. Calculate the magnitude of voltage reached on the output terminal at $t = 90\,\mu s$, when the Marx generator is fired on no load.

Solution

The equivalent circuit of a Marx generator with "no load" is similar to Figure 1.2, with C_L being open circuited. The values of C_M and R_T are as follows:

$$C_M = \frac{C_0}{N} = \frac{0.25\,\mu\text{F}}{10} = 25\text{ nF}$$

$$R_T = \frac{NR}{2} = \frac{10 \times 720}{2} = 3600\,\Omega$$

(continued)

(continued)

$$C_M R_T = 90$$

Output voltage $V_T = NV_0 e^{-t/(C_L R_T)}$

$$\therefore V_T(t = 90\ \mu\text{s}) = (10 \times 100)e^{-1} = 367.8\ \text{kV}$$

Example 1.3

A four-stage Marx generator discharging into an output capacitor (C_L) has the following parameters: (a) DC charging voltage = 100 kV, (b) capacitance per stage = 0.125 μF, (c) front resistor = 200 Ω, and (d) load capacitance = 200 pF.

Calculate the following: (a) output voltage at any time t, (b) amplitude of voltage at t = 0.01 μs, (c) time to reach maximum voltage, and (d) amplitude of peak voltage.

Solution

In Equation 1.6 and solving for the roots by Equations 1.4 and 1.5, we get the following corresponding values for the Marx equivalent circuit of Figure 1.2:

$$C_M = \frac{C_0}{N} = \frac{0.125}{4} = 31.25\ \text{nF}$$

$$R_T = \frac{NR}{2} = \frac{4 \times 1200}{2} = 2.4\ \text{k}\Omega$$

$$C_L = 200\ \text{pF}, \qquad R_F = 200\ \Omega$$

Substituting the above values:

$$\alpha = -0.0135\left(\frac{1}{\mu\text{s}}\right) \quad \text{and} \quad \beta = 25.159\left(\frac{1}{\mu\text{s}}\right)$$

a) Output voltage (V):

$$\begin{aligned} V_M(t) &= \frac{NV_0}{C_L R_F} \times \frac{1}{(\beta - \alpha)} \times \left(e^{-|\alpha|t} - e^{-\beta t}\right) \\ &= (0.994)(4)(100)\left(e^{-0.0135x10^6 t} - e^{-25.159x10^6 t}\right) \\ &= 397.6\left(e^{-0.0135x10^6 t} - e^{-25.159x10^6 t}\right)\text{kV} \end{aligned}$$

b) Amplitude of voltage at $t = 0.01\ \mu s$:

By substituting $t = 0.01 \times 10^{-6}$ in the relation (a) above, we get

$$V = 397.6 \times 0.222\ \text{kV} = 88.27\ \text{kV}$$

c) Time (t') to reach maximum voltage V_{max}:

By substituting the values of α and β in relation (3a), we get

$$t' = \ln\left[\frac{\beta/\alpha}{(\beta-\alpha)}\right] = 0.299\ \mu s$$

d) Amplitude of peak voltage (V_{max}):

By substituting the value of (t') in relation (a), we get

$$V_{\max} = (397.6)(0.995) = 395.6\ \text{kV}$$

References

1 R.A. Fitch, Marx- and Marx-Like High Voltage Generators. *IEEE Trans. Nucl. Sci.*, Vol. 18, No. 4, pp. 190–198, 1971.

2 G.W. Swift, Charging Time of a High Voltage Impulse Generator. *IEEE Electron. Lett.*, Vol. 5, No. 21, p. 534, 1969.

3 J.F. Francis, High Voltage Pulse Techniques. Report, AFOSR-74-2639-5, 1976.

4 R.W. Morrison and A.M. Smith, Overvoltage and Breakdown Patterns of Fast Marx Generators. *IEEE Trans. Nucl. Sci.*, Vol. 19, No. 4, p. 20, 1972.

5 I. Smith, V. Carboni, A.R. Miller, R. Crumley, P. Corcoran, P.W. Spence, W.H. Rix, W. Powell, and P.S. Sincerny, Fast Marx Generator Development for PRS Drivers, *Pulsed Power Plasma Science, 2001 (PPPS-2001). Digest of Technical Papers*, 2001.

6 P.S. Sincerny, S.K. Lam, R. Miller, T. Tucker, and L. Sanders, Fast Discharge Energy Storage Development for Improving X-Ray Simulators. *Proceedings of the IEEE International Pulsed Power Conference*, 2003.

7 I. Smith, P. Corcoran, A.R. Miller, V. Carboni, P.S. Sincerny, P. Spence, C. Gilbert, W. Rix, E. Waisman, L. Schlitt and D. Bell, Pulsed Power for Future and Past X-Ray Simulators. *IEEE Trans. Plasma Sci.*, Vol. 30, No. 5, pp. 1746–1754, 2002.

8 Y. Kubota et al. 2 MV Coaxial Marx Generator for Producing Intense Relativistic Electron Beams, in W.H. Bostick, V. Nardi and O.S.F. Zucker, eds., *Energy Storage, Compression and Switching*, Springer p. 63, 1976.

9 B. Bernstein and I. Smith, Aurora: An Electron Accelerator. *IEEE Trans. Nucl. Sci.*, Vol. NS-20, No. 3, p. 294, 1973.

10 J.L. Harrison, Solution of a Peaking Equation for Finite Storage Capacitor Size, Circuit and Electromagnetic System Design Note 32, January 1973.

11 F.W. Heilbronner, Firing and Voltage Shape of Multistage Impulse Generators. *IEEE Trans. Power Eng.*, Vol. PAS-90, No. 5, pp. 2233–2238, 1971.

12 T.W. Broadbent, High Voltage Impulse Generators. *Electrical Review*, April 14, 1961.

13 F.C. Creed and M.M.C. Collins, Shaping Circuits for High Voltage Impulses. *IEEE Trans. Power Apparatus Syst.*, Vol. PAS-90, No. 5, pp. 2239–2246, 1971.

14 Edwards et al., The Development and Design of High Voltage Impulse Generators. *Proc. IEE*, Vol. 98, Part I, p. 155, 1951.

15 K.R. Prestwich and D.L. Johnson, Development of an 18 Megavolt Marx Generator. *IEEE Trans. Nucl. Sci.*, 16, No. 3, p. 64, 1969.

16 T.H. Martin, K.R. Prestwich and D.L. Johnson, *Summary of the HERMES Flash X-Ray Program*, Radiation Production Note 3, PEP-3, October 1969. Available from http://www.ece.unm.edu/summa/notes.

17 D. Platts, A 10 Joule, 200 kV Mini-Marx, *Proceedings of the 5th International IEEE Pulsed Power Conference*, p. 834, 1981.

18 G.H.K. Simcox and J. Hipple, Pulse Generation Methods. *High Voltage Technology Seminar Handout Notes*, Ion Physics Corporation, 1969.

19 M.M. Kekez, J. LoVerti, A.S. Podgorski, J.G. Dunn, and G. Gibson, *Proceedings of the IEEE Pulsed Power Conference*, p. 123, 1989.

20 M.M. Kekez, Simple 50 ps Risetime High Voltage Generator. *Rev. Sci. Instrum.*, Vol. 62, No. 12, pp. 1991.

21 W.J. Carey and J.R. Mayes, Marx Generator Design and Performance. *IEEE Power Modulator Symposium*, 2002.

22 J.R. Mayes, M.B. Lara, M.G. Mayes, and C.W. Hatfield, A Compact MV Marx Generator. *IEEE Power Modulator Symposium*, 2004. Available at www.apelc.com (accessed June 19, 2010).

23 J.R. Mayes, M.B. Lara, M.G. Mayes, and C.W. Hatfield, An Enhanced MV Marx Generator for RF and Flash X-Ray Systems. *IEEE International Pulsed Power Conference*, 2005.

24 W.L. Willis, Pulse Voltage Circuits, in W.J. Sarjeant and R.E. Dollinger, eds., *High Power Electronics*, TAB Books, pp. 87–116, 1989.

25 D. Goerz, T. Ferriera, D. Nelson, R. Speer, and M. Wilson, An Ultra-Compact Marx-Type High Voltage Generator. *Conference Record of the Power Modulator Conference*, 2002.

26 R.A. Fitch, Marx-and Marx-Like-High-Voltage Generators, *IEEE Trans. Nucl. Sci.*, Vol. 18, No. 4, pp. 190–198, 1971.

27 J.C. Martin, Marx-Like Generators and Circuits, in T.H. Martin, A.H. Guenther, and M. Kristiansen, eds., *J.C. Martin on Pulsed Power*, Plenum Press, p. 107, 1996.

28 J.C. Martin, CESDN Note #3. Available at http://www.ece.unm.edu/summa/notes.

29 J.A. Nation, High Power Electron and Ion Beam Generation. *Part. Accelerators*, Vol. 10, p. 1, 1979.

30 B.N. Turman, T.H. Martin, E.L. Neau, D.R. Humphreys, D.D. Bloomquist, D.L. Cook, S. A. Goldstein, L.X. Schneider, D.H. McDaniel, J.M. Wilson, R.A. Hamil, G.W. Barr, and J.P. Van Devender, PBFA II, A 100 TW Pulsed Power Driver for the Inertial Confinement Fusion Program, in M.F. Rose and P.J. Turchi, eds., *Proceedings of the International Pulsed Power Conference*, pp. 155–161, 1985.

31 L.X. Schneider and G.J. Lockwood, Engineering High Reliabiltiy, Low Jitter Marx Generators, in M.F. Rose and P.J. Turchi, C eds., *Proceedings of the International Pulsed Power Conference*, 1985.

32 P. H. Ron, Configurations of Intense Pulse Power Systems for Generation of Intense Electromagnetic Pulses. *International Conference on Electromagnetic Interference and Compatibility (INCEMIC'99)*, New Delhi, India, p. 471, 1999.

33 P.H. Ron, A.K. Sinha, and V.K. Rohatgi, Flash X-Ray Generator for Dynamic Radiography, *Proceedings on High Energy Radiography Using Accelerators*, C Shar Centre, Sriharikota, India, 1981.

34 K.B. Riepe, High Voltage Microsecond Pulse Forming Network. *Rev. Sci. Instrum.*, Vol. 48, No. 8, p. 1028, 1977.

35 R. J. Adler, J.A. Gilbrech, and D.T. Price, *A Modular PFN Marx with a Unique Charging System and Feedthrough*. Available at http://www.appliedenergetics.com/highvoltage (accessed June 19, 2010).

36 V.P. Singhal, B.S. Narayan, K. Nanu, and P.H. Ron, Development of a Blumlein Based on Helical Line Energy Storage Element. *Rev. Sci. Instrum.*, Vol. 72, No. 3, p. 1862, 2001.

37 M.T. Buttram and R.S. Clark, Repetitively Pulsed Marx Generator Design Studies, Sandia Report 79–1011, p. 125, 1979.

38 Y. Kubota, S. Kawasaki, A. Miyahara, and H.M. Saad, Coaxial Marx Generators for Producing Intense Relativistic Electron Beams. *Jpn. J. Appl. Phys.*, Vol. 13, No. 2, p. 260, 1974.

39 R.A. Cooper, J.B. Ennis, and D.L. Smith, ZR Marx Capacitor Life Test and Production Statistics. *Proceedings of the IEEE International Pulsed Power Conference*, 2005.

40 G.J. Rohwein, Repetitively Pulsed Marx Generator Design Studies, Sandia Report 79-0002, p. 119, 1979.

41 J. Tulip et al. High Repetition Rate Short Pulse Gas Discharge, Los Alamos Scientific Laboratory Report, LA-UR, 78–714, 1978.

42 I.D. Smith, Liquid Dielectric Pulse Line Technology, in W.H. Bostick, V. Nardi, and O.S.F. Zucker, eds., *Energy Storage, Compression, and Switching*, Plenum Press, New York, p. 15, 1976.

43 W.G. Hawley, *Impulse Voltage Testing*, Chapman and Hall Ltd, London, 1959.
44 H. Aslin, Fast Marx Generator, in W.H. Bostick, V. Nardi, and O. Zucker, eds., *Proceedings of the International Conference on Energy Storage Compression and Switching*, Asti-Torino, Italy, p. 35, 1974.
45 R. Beverly, III, and R. Campbell, Aqueous-Electrolyte Resistors for Pulsed Power Applications. *Rev. Sci. Instrum.*, Vol. 66, pp. 5625–5629, 1995.
46 R. Beverly, III, Application Notes for Aqueous-Electrolyte Resistors. Available at www.reb3.com/pdf/r_app.pdf (accessed November 13, 2009).
47 V.R Kukhta, E.I. Logachev, V.V. Lopatin, G.E. Remnev, V.I. Tsvetkov, and V.P. Chernenko, Controlled Sparkgaps for High Voltage Pulse Generators. *Instrum. Exp. Tech.*, Vol. 19, No. 6, p. 1670, 1976.
48 A. Sharma, P.H. Ron and D.C. Pande, EMI Results from Pulse Excited Antennas. *Conference on Millimeter Wave and Microwave (ICOMM'90)*, Defense Electronics Applications Laboratory, Dehradun, India, 1990.
49 C.E. Baum and J.M. Lehr, Parallel Charging of Marx Generators for High Pulse Repetition Rate, in *Ultra-Wideband, Short-Pulse Electromagnetics 5*, Kluwer Academic/Plenum Publishers, pp. 415–422, 2002. ∗ (∗ Cited references such as Circuit and Electromagnetic System Design Notes, Switching Notes, Sensor and Simulation Notes and Interaction Notes, formerly edited by Dr. C.E. Baum, are available electronically at http://www.ece.unm.edu/Summa/ The Note Series is currently edited by Dr. D.V. Giri.)
50 L. Veron and J.C. Brion, Experimental Study of a Repetitive Marx Generator. *Proceedings of the IEEE International Pulsed Power Conference*, pp. 1054–1057, 2003.
51 S.M. Turnball, J.M. Koutsoubis, and S.J. MacGregor, Development of a High Voltage, High PRF, PFN Marx Generator. *Conference Record of the Power Modulator Conference*, pp. 213–216, 1998.
52 S.M. Turnball, S.J. MacGregor, and J.A. Harrower, A PFN Marx Generator Based on High Voltage Transmission Lines. *Meas. Sci. Technol.*, Vol. 11, pp. 51–55, 2000.
53 S.J. MacGregor, S.M. Turnball, F.A. Tuema, and O. Farish, Factors Affecting and Methods of Improving the Pulse Repetition Frequency of Pulse-Charged and DC-Charged High Pressure Gas Switches. *IEEE Trans. Plasma Sci.*, Vol. 25, No. 2, pp. 110–117, 1997.
54 M.G. Grothaus, S.L. Moran, and L.W. Hardesty, High Performance Pulse Generator, U.S. Patent 5,311,062, 1994.
55 J.M. Lehr, J.P. Corley, S.A. Drennan, D.W. Guthrie, H.C. Harjes, D.W. Bloomquist, D.L. Johnson, J.E. Maenchen, D.H. McDaniel, and K. Struve, SATPro: The System Assessment Test Program for ZR. *Proceedings of the IEEE International Power Modulator Symposium*, 2002.
56 R.A. Fitch, Pulse Generator, U.S. Patent 3,289,015, 1966.

57 R.A. Fitch and V.I.S. Howell, Novel Principles of Transient High Voltage Generation. *Proc. IEE,* Vol. III, No. 4, p. 849, 1964.

58 F. Ruhl and G. Herziger, Analysis of Spiral Generator. *Rev. Sci. Instrum.*, Vol. 51, p. 1541, 1980.

59 I.A.D. Lewis, Some Transmission Line Devices. *Elec. Eng.*, Vol. 27, p. 448, 1955.

60 I.A.D. Lewis and F.H. Wells, *Millimicrosecond Pulse Techniques,* Parmagon Press, Long England, 1959.

61 A.M. Chodorow, The Time Isolation High Voltage Impulse Generator, *Proc. IEEE,* 63, No. 7, pp. 1082–1084, 1975.

62 L. Soto and L. Altamirano, A Pulse Voltage Multiplier. *Rev. Sci. Instrum.*, Vol. 70, No. 3, pp. 1891–1892, 1999.

63 Y. Carmel, S. Eylon, and E. Shohet, A Repetitive 600 kV Stacked Line Transformer Pulse Generator. *J. Phys. E Sci. Instrum.*, Vol. 11, pp. 748–750, 1978.

64 R.A. Fitch, Electrical Pulse Generators, U.S. Patent 3,366,799, 1968.

65 E.M. Bazelyan and Yu. P. Raizer, *The Spark Discharge,* CRC Press, LLC, 1998.

66 A. Pokryvailo, Y. Yankelevich. M. Woldf, E. Abramzon, E. Shviro, S. Wald, A. Welleman, A 1 kW Pulsed Corona System for Pollution Control Applications. *IEEE Trans. Plasma Sci.*, 32,pp. 2042–2054, 2004.

67 Y. Sakai, S. Takahashi, T. Komatsu, I. Song, M. Watanabe, and E. Hotta, Highly Efficient Pulsed Power Supply System with a 2 Stage LC generator and a Step Up Transformer for Fast Capillary Discharge Soft X-Ray Laser at Shorter Wavelength. *Rev. Sci. Instrum.*, Vol. 81, p. 013303, 2010.

68 N.W. Harris and H. Milde, 100 GW Electron Beam Generator. *IEEE Trans. Nucl. Sci.*, Vol. NS-23, No. 5, p. 1470, 1976.

ulse Transformers

ılse transformers can be classified into induction transformers and transmis- on line transformers. Induction transformers may employ a ferromagnetic re or an air core. The primary and secondary windings can take the nfigurations of an autotransformer, independent windings, or cascaded ansformers [1]. This chapter deals with air core transformers and their havior under the conditions of double resonance, off resonance, and triple sonance.

1 Tesla Transformers

e Tesla transformer is one of the components of a pulsed power system own schematically in Figure 2.1. In contrast to a conventional transformer, ich uses iron or ferrite cores, the Tesla transformer employs an air core. The sence of a metallic core results in simpler electrical insulation and more icient energy transfer due to the absence of core loss. It can replace a Marx nerator when the output voltage is less than 2–3 MV and energy content is t more than a few kilojoules. Unlike a Marx generator, which produces a idirectional pulse, the Tesla transformer produces an oscillatory pulse, as own by the waveform at point A in Figure 2.1. Upon charging the pulse ming line (PFL) to a peak voltage V_2^{Max}, the start switch is triggered and rgy is transferred to the load in a unipolar pulse, as shown by the waveform point B, under matched conditions. Compared to Marx generators, the antages of Tesla transformers are inherent compactness and the ability to be etitively pulsed. Voltage multiplication factors on the order of 20 are routine, attempts to achieve higher gains have resulted in destructive breakdowns in secondary windings. Thus, voltage gains on the order of 25 are considered practical limit. Air core transformers are used extensively in short-pulse, h-peak power applications.

n contrast to a conventional transformer, which uses iron or ferrite cores, the la transformer employs an air core. The absence of a metallic core results in

dations of Pulsed Power Technology, First Edition. Jane Lehr and Pralhad Ron.

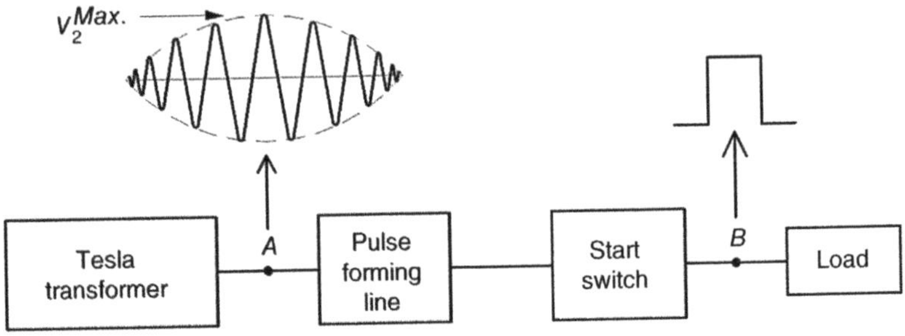

Figure 2.1 A typical pulsed power system employing a Tesla transformer.

simpler electrical insulation and more efficient energy transfer due to the absence of core loss.

2.1.1 Equivalent Circuit and Design Equations

The equivalent circuit of the Tesla transformer is shown in Figure 2.2. When the primary closing switch S_1 is triggered, the capacitor C_1 charged to a voltage V_0 discharges into the primary circuit L_1 of the Tesla transformer. The energy is coupled into the secondary circuit L_2C_2 through the mutual inductance M, where C_2 represents the equivalent capacitance of the PFL. S_2 is generally arranged to close when the voltage on the PFL reaches a peak value V_2^{Max}.

The Laplace loop equations for primary and secondary can be written in the following forms:

$$I_1(s) = \frac{V_0}{L_1} \frac{\omega_2^2 + s^2}{s^4(1-K^2) + s^2(\omega_1^2 + \omega_2^2) + \omega_1^2\omega_2^2} \tag{2.1}$$

$$I_2(s) = \frac{-V_0K}{\sqrt{L_1L_2}} \frac{s^{-2}}{s^{-4}(1-K^2) + s^2(\omega_1^2 + \omega_2^2) + \omega_1^2\omega_2^2} \tag{2.2}$$

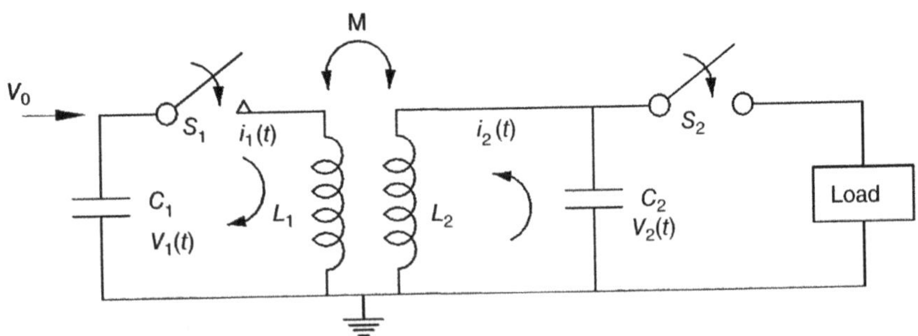

Figure 2.2 The equivalent circuit of a Tesla transformer.

where

$$M = K\sqrt{L_1 L_2}, \quad \omega_1^2 = \frac{1}{L_1 C_1}, \quad \omega_2^2 = \frac{1}{L_2 C_2} \tag{2.3}$$

The time domain solutions of the above equations for $\omega_1 = \omega_2 = \omega$ are as follows [2]:

$$i_1(t) = \frac{V_0}{2\omega L_1}\left[\frac{1}{\sqrt{1-K}}\sin\left(\frac{\omega t}{\sqrt{1-K}}\right) + \frac{1}{\sqrt{1+K}}\sin\left(\frac{\omega t}{\sqrt{1+K}}\right)\right] \tag{2.4}$$

$$i_2(t) = \frac{-V_0}{2\omega\sqrt{L_1 L_2}}\left[\frac{1}{\sqrt{1-K}}\sin\left(\frac{\omega t}{\sqrt{1-K}}\right) - \frac{1}{\sqrt{1+K}}\sin\left(\frac{\omega t}{\sqrt{1+K}}\right)\right] \tag{2.5}$$

$$V_1(t) = \frac{V_0}{2}\left[\cos\left(\frac{\omega t}{\sqrt{1-K}}\right) + \cos\left(\frac{\omega t}{\sqrt{1+K}}\right)\right] \tag{2.6}$$

$$V_2(t) = \frac{V_0}{2}\sqrt{\frac{L_2}{L_1}}\left[\cos\left(\frac{\omega t}{\sqrt{1-K}}\right) - \cos\left(\frac{\omega t}{\sqrt{1+K}}\right)\right] \tag{2.7}$$

2.1.2 Double Resonance and Waveforms

Double resonance occurs when the following conditions are simultaneously satisfied:

$$\omega_1 = \omega_2 = \omega \tag{2.8}$$

and

$$i_1(t_{2m}) = 0, \quad i_2(t_{2m}) = 0, \quad V_1(t_{2m}) = 0 \tag{2.9}$$

where t_{2m} is the time at V_2^{Max}.

The solution of Equations 2.4–2.7, with the conditions (2.8) and (2.9), yield the following discrete K values [3]:

$$K = 0.153,\ 0.18,\ 0.222,\ 0.28,\ 0.385,\ \text{and}\ 0.6 \tag{2.10}$$

The double-resonance waveforms are shown in Figure 2.3 for $K=0.6$, 0.385, and 0.153.

The following observations can be made from these waveforms:

- When the secondary voltage is maximum at $t = t_{2m}$ and the capacitor voltage at C_1, primary and secondary currents are zero, indicating that the entire energy stored in C_1 is transferred to C_2, resulting in a theoretical efficiency of 100% when the coupling coefficient K = 0 6 [4].

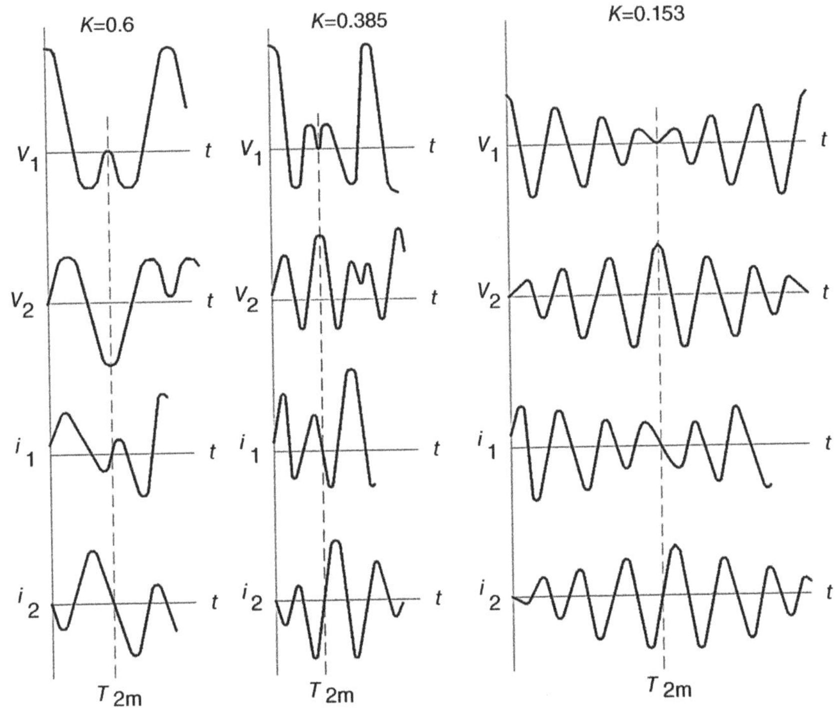

Figure 2.3 The double-resonance waveforms for $K=0.6$, $K=0.385$, and $K=0.153$.

- As the K value decreases, the time to peak secondary voltage increases. For example, as K reduces from 0.6 to 0.385 to 0.153, the secondary voltage peak shifts from the second half-wave to the third half-wave to the seventh half-wave. Because the voltage on the insulator is for a longer duration, statistically, it is more likely to fail from electrical breakdown.

This leads to the conclusion that the Tesla transformer should be operated at the highest possible K value. Additionally, in practice, the theoretical efficiency is lowered by parasitic capacitance losses in the windings and eddy current losses in the conductors.

2.1.3 Off Resonance and Waveforms

Although in general double-resonance operation is the most efficient from a maximum energy transfer viewpoint, "off resonance" operation, in spite of being energy inefficient, is preferred under certain conditions. For example, having $K = 0.8$ and $\omega_1 = \omega_2$ results in the peak secondary voltage V_2^{Max} being attained

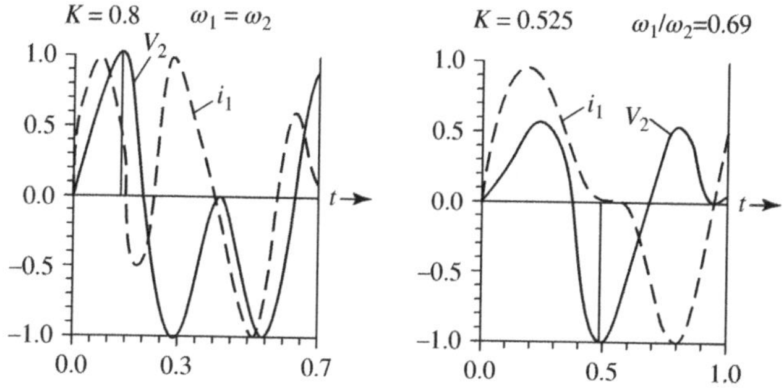

Figure 2.4 The secondary voltage and primary current for off-resonance transformers [2].

in the first half-wave itself. The higher dielectric strength available at smaller charge times allows optimum utilization of the dielectric provided in the transformer as well as the PFL. Similarly, having a Tesla transformer with $K = 0.525$ and $\omega_1/\omega_2 = 0.69$ results in nonreversal of primary current up to the time of secondary peak voltage. This has a specific advantage when a hydrogen thyratron is used in the primary circuit rather than a spark gap. Typical waveforms of secondary voltage and primary current for (a) $K = 0.8$ and $\omega_1 = \omega_2$ and (b) $K = 0.525$ and $\omega_1/\omega_2 = 0.69$ are shown in Figure 2.4. Hydrogen and deuterium thyratrons, having unidirectional conduction properties, cannot be employed in cases where the primary current reverses before V_2^{Max} is reached. The only alternative in such cases is the use of spark gaps, which have bidirectional conduction properties.

2.1.4 Triple Resonance and Waveforms

When the internal capacitance of the transformer C_3 is comparable to the PFL capacitance C_2, it has been shown [5] that the efficiency of energy transfer from C_1 to C_2 is significantly reduced. The suggested solution for improving the efficiency is to incorporate an additional tuning inductor L_T between the transformer secondary and the PFL capacitor C_2, as shown in Figure 2.5. The equivalent circuit, the improvement of performance resulting from addition of L_T, and the triple-resonance waveforms are shown in Figures 2.5a–c, respectively.

The conditions to be satisfied for triple resonance are

$$L_1C_1 = (L_2 + L_T)C_2 \tag{2.11}$$

$$\frac{M}{\sqrt{L_1(L_2 + L_T)}} = 0.6 \tag{2.12}$$

$$\omega_1 : \omega_2 : \omega_3 = 1 : 2 : 3 \tag{2.13}$$

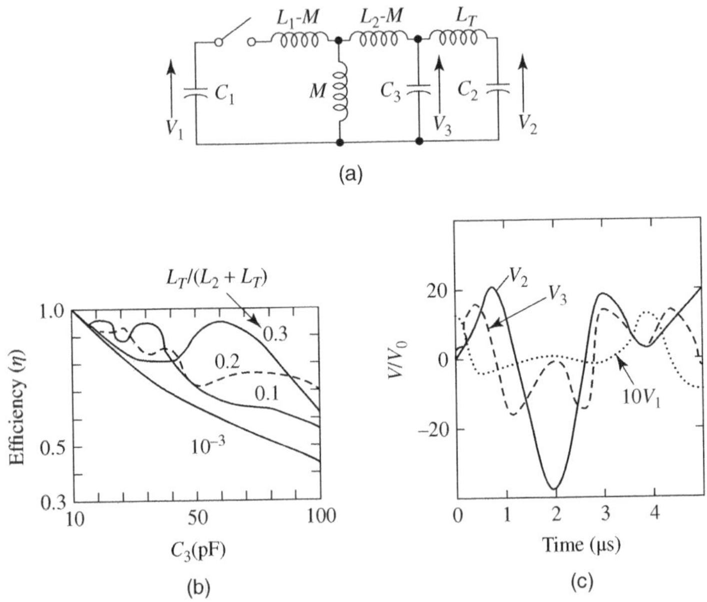

Figure 2.5 The performance of triple-resonance transformers. (a) The equivalent circuit, (b) The effect of internal capacitance and tuning inductor. (c) Typical waveforms. Reprinted with permission form [5]. copyright 1990 AIP Publishing LLC.

where

$$\omega_1^2 = L_1C_1, \quad \omega_2^2 = L_2C_3, \quad \text{and} \quad \omega_3^2 = L_TC_2 \tag{2.14}$$

A particular advantage of triple-resonance transformer is that the PFL can be charged to a much larger voltage than the peak voltage of the transformer, up to a ratio of 3.

2.1.5 No Load and Waveforms

If the spark gap switch S_2 in Figure 2.2 fails to trigger, the output waveforms will be periodic with a damping envelope of $e^{-t/\tau}$ [3]:

$$\tau = \frac{4}{\omega}\frac{Q_1Q_2}{Q_1 + Q_2}(1 - K^2) \tag{2.15}$$

The variables Q_1, Q_2, and K in Equation 2.15 represent the Q for the primary circuit, secondary circuit, and coupling coefficient, respectively. The unloaded waveforms of the Tesla transformer are shown in Figure 2.6.

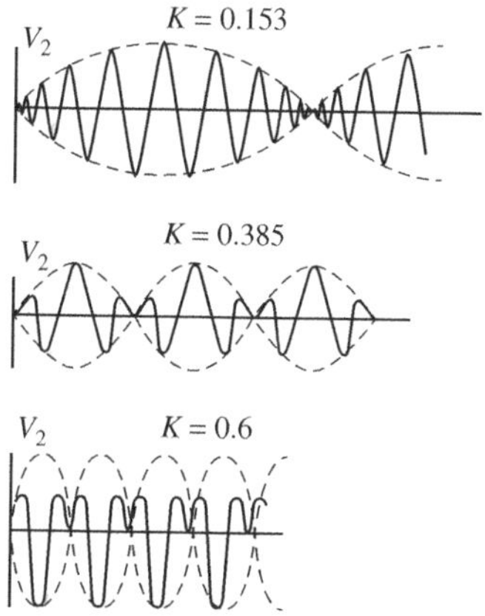

Figure 2.6 The unloaded waveforms of a Tesla transformer [3].

2.1.6 Construction and Configurations

The construction of Tesla transformers can take either of the two popular configurations shown in Figure 2.7: helical or radial. The advantages of the radial configuration over its helical counterpart are mainly in the superior capacitive grading of radial electric field lines along winding, resulting in better insulation utilization. Common insulation systems used in air core transformers are high-pressure gas, deionized water, oil, and copper sulfate solutions [6–17]. Solid sheet insulation with a liquid impregnate are also used and are chosen to have similar dielectric constants to allow uniform electric field grading along the radial dimension.

Enhanced performance and reliability in Tesla transformers are obtained [11,12] by (a) incorporating superior polyester film impregnated with oxidation-inhibiting oil for interturn insulation of high-voltage secondary winding; (b) shaping supports, shields, and corona guards for improving voltage grading and ensuring operation below corona inception voltage; (c) placing primary winding outside the container of secondary winding, thus allowing effective forced cooling of primary for efficient heat removal; (d) using partial ferrite or powdered iron core to improve the coupling K and voltage gain and achieve maximum secondary voltage on the first half-wave; (e) employing continuous recirculation of oil under vacuum to distribute corona-generated bubbles to the bulk volume of oil, thus preventing formation of bubble clouds, growth, and streamer initiation; and (f) designing with multiple primaries fed by independent energy sources but with single, high-voltage secondary for large voltage gain and reliability.

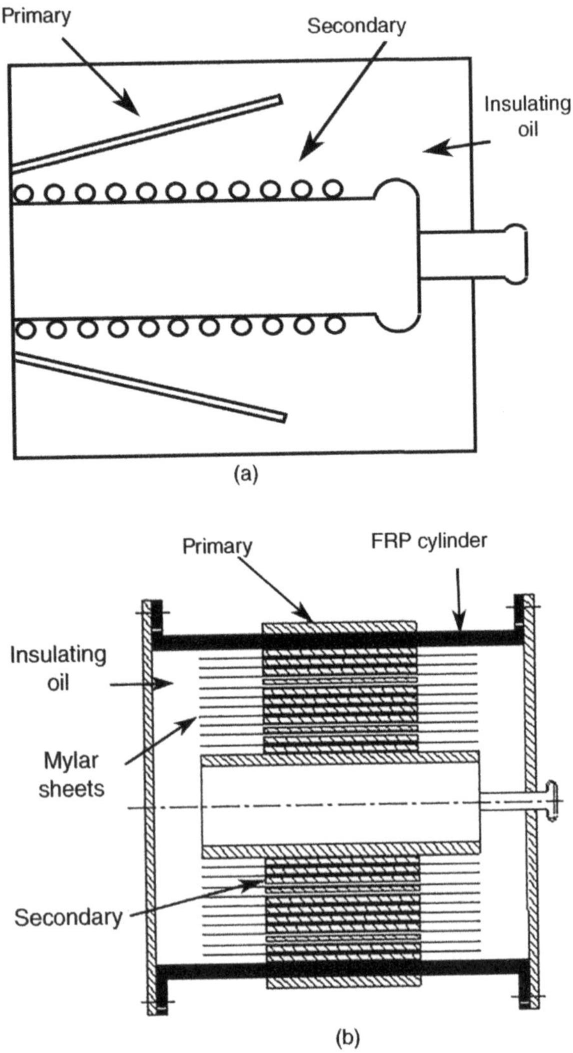

Figure 2.7 Tesla transformer configurations. (a) Helical. (b) Radial. Fiber-reinforced plastic (FRP) is used to house the transformer.

Some examples of efficient Tesla transformers are (a) PHOEBUS-I [8], a radial configuration storing 1.28 kJ with $V_2^{Max} = 1$ MV and $K = 0.76$; (b) RIUS-5 [9], a helical configuration employing $SF_6 + N_2$ insulation at 15 atmosphere, storing 2.5 kJ with V_2^{Max} of 7 MV and $K = 0.45$; and (c) 1.2 MV radial transformer [11] operated at 20 Hz and tested up to 100,000 shots.

2.2 Transmission Line Transformers

2.2.1 Tapered Transmission Line

Tapered transmission line transformers can be thought of as derivatives of the very well-known quarter wave transformer. The quarter wave transformer is a simple technique to match any real load impedance to a transmission line, at a fixed frequency, by inserting a resonant section of transmission line between them. For this section to be resonant, its length must be a multiple of one quarter of the wavelength of the frequency, with a characteristic impedance value given by the root mean square of the source and load impedances. Analysis quickly reveals that the source and load transmission lines are matched, with no reflections, at the frequency of interest. If more bandwidth is required, then a multisection transformer can be used. From each section, there is a reflection caused by the impedance mismatch, even if it is small. To further increase the bandwidth, the impedance of the transmission line is allowed to vary slowly, leading to the tapered transmission line transformer.

The tapered transmission line transformer has a characteristic impedance that varies continuously in a smooth fashion from the impedance of one line to that of the other. The impedance transition is typically achieved through a change in geometry, though advanced materials are being used at moderate peak power levels. Impedance tapers can be synthesized from any transmission line geometry and the impedance profile may be allowed to vary in any number of ways. In pulsed power applications, an exponential or linear profile are frequent impedance tapers, and common geometries are coaxial or planar transmission lines, as shown in Figure 2.8. When using a planar transmission line geometry with a homogeneous insulating media, the impedance taper can be attained by varying either the spacing between the plates, their width, or both.

2.2.1.1 Pulse Distortion

For frequencies such that the electrical length is greater than a quarter wavelength, the transmission line acts like a transformer. Above the minimum design frequency, the tapered section will be more than one quarter wavelength long and correct matching will take place, while below the design frequency the tapered section will be less than one quarter wavelength and it will not act as a transformer. For pulsed applications, only the frequency components comprising the pulse that meet this criteria experience gain. Thus, a critical determinant is the length of the tapered transmission line, or transit time, for a given characteristic impedance taper and input pulse shape. This restriction typically limits the application of tapered transmission lines to pulse widths on the order of 10 ns, for many realizable lengths.

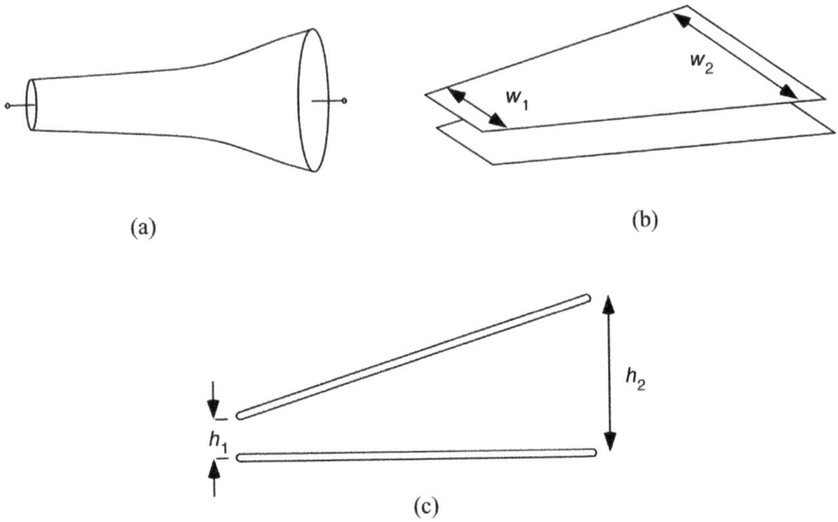

Figure 2.8 Types of tapered transmission line transformers commonly used in pulsed power applications are (a) the coaxial transmission lines and (b and c) planar transmission lines. The impedance taper is produced by a variation in the width of the line in part (b) and the height between the planes in part (c).

In practice, concern about pulse distortion can dominate the design of the tapered transmission line transformer. If insufficient length is chosen for the tapered transmission line transformer, the pulse will be distorted, as shown for the case of a rectangular input in Figure 2.9. Pulse distortion can also occur if the rate of taper is too large. These conditions come from an early derivation by Slater [16–18].

2.2.1.2 The Theory of Small Reflections

The relation between pulse distortion, length, and rate of taper are most easily understood by modeling the tapered transmission line as a succession of transmission line segments. This analysis is called the Theory of Small Reflections [18,19] and is presented here in detail. The analysis will start with a single-segment transmission line, proceed to a multisegment transmission line, and then to a tapered transmission line. It is instructive to become familiar with the Theory of Small Reflections for pulsed power applications because it aids not only in the understanding of pulse distortion in tapered transmission line transformers but also in the analysis of multisection transmission lines, sometimes used in high-current accelerators to optimize the peak current.

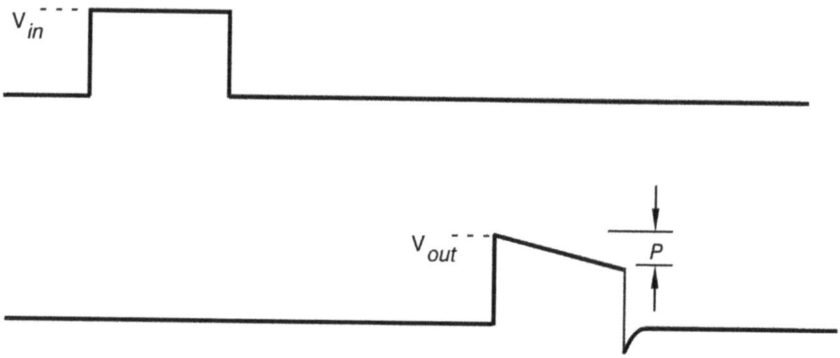

Figure 2.9 An illustration of pulse distortion through a tapered transmission line transformer. The rectangular input pulse will seem to fall linearly. The rate of taper and the pulse length employed must be such that the predicted distortion turns out to be small.

2.2.1.2.1 *Single-Section Transmission Line Transformers*

The analysis begins with the simple case of a segment of transmission line, shown in Figure 2.10, with characteristic impedance Z_2 and length ℓ, located between a semi-infinite transmission line of characteristic impedance Z_1 and a real load of impedance Z_L, which may also be a transmission line.

The partial reflection coefficients Γ_i and their associated transmission coefficients T_i at each juncture are combined into an overall reflection coefficient Γ (see Figure 2.11):

$$\Gamma_1 = \frac{Z_2 - Z_1}{Z_2 + Z_1} \tag{2.16}$$

$$\Gamma_2 = \frac{Z_1 - Z_2}{Z_1 + Z_2} = -\Gamma_1 \tag{2.17}$$

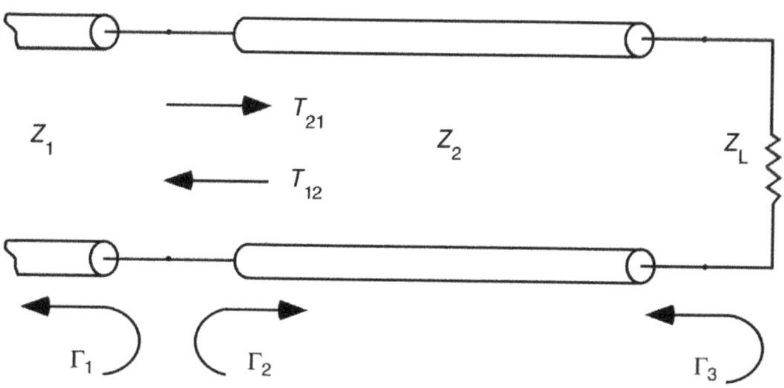

Figure 2.10 The single-segment transmission line transformer showing the partial reflection and transmission coefficients.

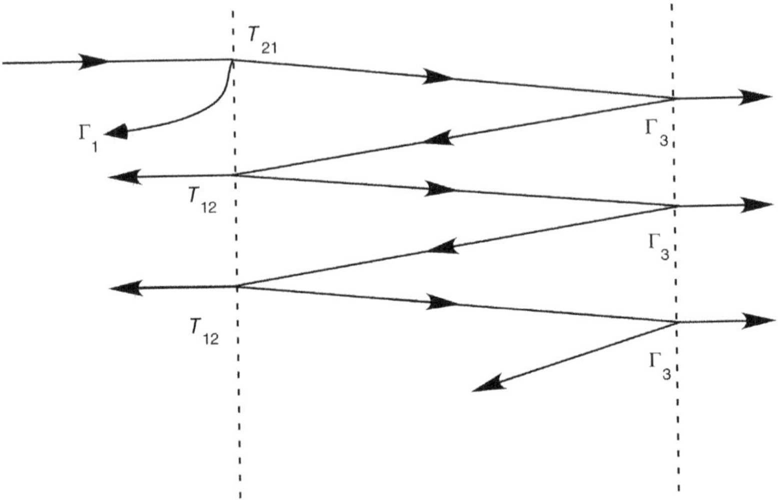

Figure 2.11 The partial reflection and transmission coefficients along the single-section transmission line.

$$\Gamma_3 = \frac{Z_\ell - Z_2}{Z_\ell + Z_2} \tag{2.18}$$

$$T_{21} = 1 + \Gamma_1 = \frac{2Z_2}{Z_1 + Z_2} \tag{2.19}$$

$$T_{12} = 1 + \Gamma_2 = 1 - \Gamma_1 = \frac{2Z_1}{Z_1 + Z_2} \tag{2.20}$$

Performing a multiple reflection analysis, using Figure 2.11, sometimes referred to as a "bounce diagram":

$$\Gamma = \Gamma_1 + T_{21}\Gamma_3 T_{12} e^{-2j\cdot(2\pi\ell/\lambda)} + \left(T_{21}\Gamma_3 T_{12} e^{-2j\cdot(2\pi\ell/\lambda)}\right)\Gamma_2\Gamma_3 e^{-2j\cdot(2\pi\ell/\lambda)} + \cdots \tag{2.21}$$

combining terms

$$\Gamma = \Gamma_1 + T_{21}T_{12}\Gamma_3 e^{-2j\cdot(2\pi\ell/\lambda)}(1 + \Gamma_2\Gamma_3 e^{-2j\cdot(2\pi\ell/\lambda)} + \Gamma_2^2\Gamma_3^2 e^{-4j\cdot(2\pi\ell/\lambda)} + \cdots \tag{2.22}$$

and summing

$$\Gamma = \Gamma_1 + T_{21}T_{12}\Gamma_3 e^{-2j\cdot(2\pi\ell/\lambda)}\left(\sum_{n=0}^{\infty}\Gamma_2^n\Gamma_3^n e^{-4jn\cdot(2\pi\ell/\lambda)}\right) \tag{2.23}$$

Using the geometric series

$$\sum_{n=0}^{\infty} x^n = \frac{1}{1-x}, \quad \text{for} \quad |x| < 1$$

the reflection coefficient can then be expressed in closed form as

$$\Gamma = \Gamma_1 + \frac{T_{21}\Gamma_3 T_{12} e^{-2j\cdot(2\pi\ell/\lambda)}}{1 - \Gamma_2\Gamma_3 e^{-2j\cdot(2\pi\ell/\lambda)}} \tag{2.24}$$

A further approximation can be made by using (2.17) and (2.19) along with

$$T_{21} = 1 - \Gamma_1$$

to express T_{21}, Γ_2, and T_{12} in terms of Γ_1 and applying $|\Gamma_1\Gamma_3|$ is small, so

$$\Gamma \cong \Gamma_1 + \Gamma_3 e^{-2j\cdot(2\pi\ell/\lambda)} \tag{2.25}$$

2.2.1.2.2 Multisection Transmission Line Transformers

The analysis of the previous section can be extended to multisection transformers. To simplify the analysis, as shown in Figure 2.12, several constraints will be imposed: the sections are all of the same length, the impedances change monotonically along the transformer; and the load impedance Z_L is real.

The partial reflection coefficients are defined at each junction as

$$\Gamma_0 = \frac{Z_1 - Z_0}{Z_1 + Z_0} \tag{2.26}$$

$$\Gamma_n = \frac{Z_{n+1} - Z_n}{Z_{n+1} + Z_n} \tag{2.27}$$

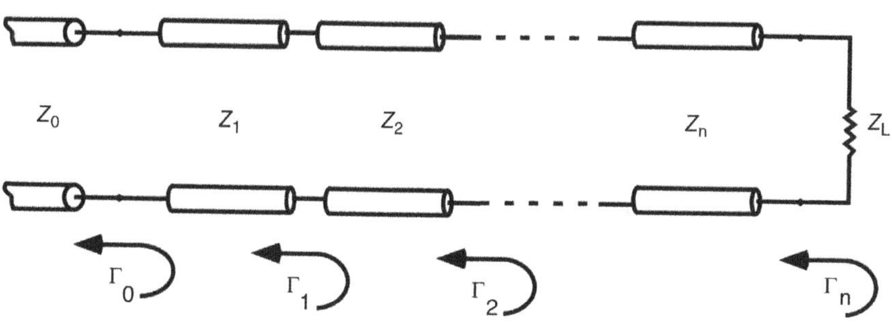

Figure 2.12 A multisection transmission line for analysis. The sections are all of the same length, with monotonically increasing impedances and real load impedance.

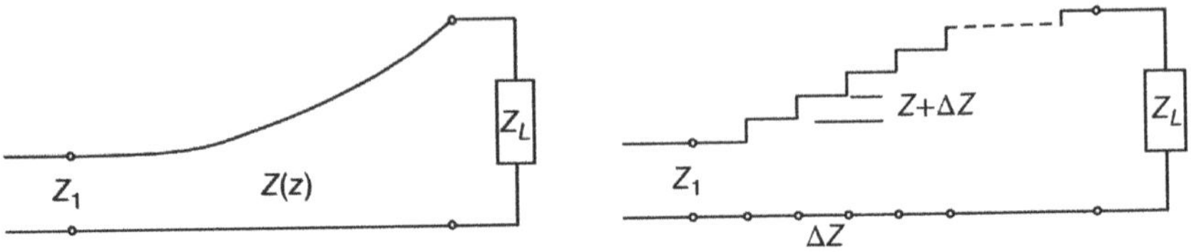

Figure 2.13 A tapered transmission line can be modeled as a multisection transmission line transformer with sections of small length.

and

$$\Gamma_N = \frac{Z_L - Z_N}{Z_L + Z_N} \tag{2.28}$$

Provided the reflections between segments are small, a similar analysis to that that resulted in Equation 2.24 can be performed on the multisection transmission line leading to

$$\Gamma\left(\frac{2\pi\ell}{\lambda}\right) = \Gamma_0 + \Gamma_1 e^{-2j\cdot(2\pi\ell/\lambda)} + \Gamma_2 e^{-4j\cdot(2\pi\ell/\lambda)} + \cdots + \Gamma_N e^{-2jN\cdot(2\pi\ell/\lambda)} \tag{2.29}$$

This important result illustrates that with a sufficient number of sections, almost any reflection coefficient can be attained as a function of frequency with appropriate choices of Γ_n. As the number of discrete sections increases and the impedance increment between adjacent sections decreases, the multisection transmission line becomes a tapered line, as illustrated in Figure 2.13.

2.2.1.2.3 Reflection Coefficient from a Tapered Transmission Line

Consider the impedance taper being made up of a number of small sections of length Δz with associated impedance changes $\Delta Z(z)$, as shown in Figure 2.13.

The incremental reflection coefficient at location z is

$$\Delta\Gamma = \frac{(Z+\Delta Z) - Z}{(Z+\Delta Z) + Z} \approx \frac{\Delta Z}{2Z} \tag{2.30}$$

In the limit as $\Delta z \to 0$, this becomes an exact differential with

$$d\Gamma = \frac{\Delta Z}{2Z} = \frac{1}{2}\frac{d(\ln Z/Z_0)}{dz}dz \tag{2.31}$$

Using the theory of small reflections, the total reflection coefficient at $z = 0$ can be found by summing up the partial reflections with their appropriate phase shifts:

$$\Gamma\left(\frac{4\pi}{\lambda}\cdot z\right) = \frac{1}{2}\int_0^\ell e^{-2j\cdot(2\pi/\lambda)z}\frac{d(\ln Z/Z_0)}{dz}dz \tag{2.32}$$

If $Z(z)$ is known, then the reflection coefficient, $\Gamma((4\pi/\lambda)\cdot z)$, can be found as a function of wavelength. It can be noted that the condition that the rate of taper should not be too large to minimize pulse distortion is related to the Theory of Small Reflections. If the impedance changes in adjacent sections are large, then the reflection coefficient is also large, resulting in pulse distortion.

2.2.1.3 Gain of a Tapered Transmission Line Transformer

When used in frequency domain circuits, tapered transmission lines are typically used in impedance-matching applications. In pulsed power, the tapered transmission line may be used for this purpose also, but a more common application is to increase the voltage or current on the load. For example, a pulse is injected from transmission line Z_1 and passed through a transformer into a transmission line with impedance Z_ℓ. The gain G is most precisely measured at early time because the high-frequency components are most efficiently transformed [20]. The early-time/high-frequency voltage gain G can be defined as

$$G = \frac{V_{out}}{V_{in}} \tag{2.33}$$

The maximum theoretical gain can be calculated from the power pulse, assuming no losses:

$$\frac{V_{in}^2(t)}{Z_1} = \frac{V_{out}^2}{Z_\ell} \tag{2.34}$$

yielding

$$G = \sqrt{\frac{Z_\ell}{Z_1}} \tag{2.35}$$

Since both the impedance and the maximum electric field are determined by material properties and physical dimensions, in practice, the attainable impedance ratios are limited by electrical breakdown considerations and voltage gains on the order of two can be realized. The gain is also limited by the length and the rate of impedance taper because of pulse distortion considerations.

2.2.1.4 The Exponential Tapered Transmission Line

In addition to being widely used, the exponential taper is an important example of the tapered transmission line transformer and one that has the advantage that exact solutions are known. Furthermore, it has been determined that the exponential taper is the geometry that allows for the shortest transmission line length for a given output pulse distortion [17].

The transformer has an impedance variation given by

$$Z(z) = Z_0 \cdot e^{az}, \quad \text{for} \quad 0 \leq z \leq \ell \tag{2.36}$$

with

$$Z(0) = Z_0 \quad \text{and} \quad Z(\ell) = Z_L$$

The inductance $L(z)$ and capacitance $C(z)$ of the tapered transmission line transformer vary as

$$L(z) = L_0 \cdot e^{az} \tag{2.37}$$

$$C(z) = C_0 \cdot e^{-az} \tag{2.38}$$

with the characteristic impedance given by

$$Z(z) = \sqrt{\frac{L(z)}{C(z)}} = \sqrt{\frac{L_0}{C_0}} \cdot e^{-az} \tag{2.39}$$

Knowing the desired impedances, the constant a can be determined to be

$$a = \frac{1}{\ell} \cdot \ln\left(\frac{Z_L}{Z_0}\right) \tag{2.40}$$

The time delay per unit length, τ_d, is constant:

$$\tau_d = \sqrt{LC} = \sqrt{L_0 C_0} \tag{2.41}$$

If the input pulse is rectangular of duration τ, the effect of pulse distortion in an exponential line can be predicted theoretically. An initially rectangular pulse will fall linearly by a total fractional amount. Pryce [17] developed a very useful design formula:

$$\frac{\tau_d}{\tau} = 50 \frac{(\ln G)^2}{P} \tag{2.42}$$

where τ_d is the total delay along the tapered transmission line, τ is the duration of the pulse, G is the voltage gain, and P is the percentage fall in amplitude at the end of the pulse.

Knowing $Z(z)$, the overall reflection coefficient for each wavelength can be calculated by substituting the expression for $Z(z)$ into the expression for the reflection coefficient:

$$\Gamma\left(\frac{4\pi z}{\lambda}\right) = \frac{1}{2} \int_0^{\ell} e^{-2j \cdot (2\pi/\lambda) z} \frac{d(\ln e^{az})}{dz} dz \tag{2.43}$$

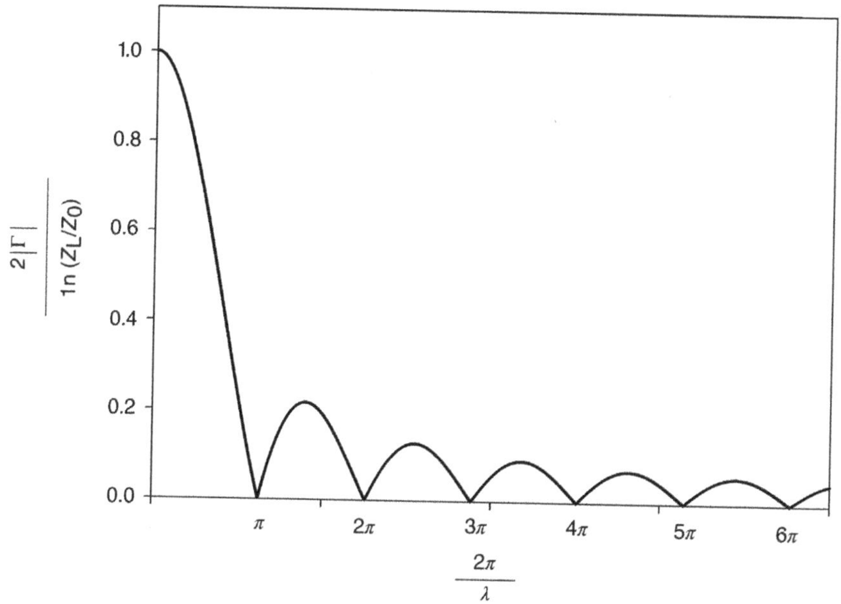

Figure 2.14 The magnitude of the overall reflection coefficient as a function of ℓ/λ.

Solving

$$\Gamma = \frac{\ln(Z_L/Z_0)}{2} e^{-j\cdot(2\pi\ell/\lambda)} \frac{\sin(2\pi\ell/\lambda)}{(2\pi\ell/\lambda)} \tag{2.44}$$

The magnitude of the reflection coefficient is shown in Figure 2.14. Note that the peaks in the magnitude of the input reflection coefficient, $|\Gamma|$, decrease with increasing length. When the length is greater than $\lambda/2$, the magnitude of the reflection coefficient drops to less than 0.25, minimizing the mismatch at low frequencies.

2.3 Magnetic Induction

The manufacture of large, laminated iron cores in the 1960s allowed the principle of magnetic induction to be developed for charged particle beam applications. These commercially available iron cores were constructed of wound layers of laminated ferromagnetic material, which could be constructed in the very large sizes required for high power. More importantly, the layered construction limited the eddy current losses to reasonable levels. Inductive voltage adders, linear induction accelerators, and linear transformer drivers all operate using the principle of magnetic induction.

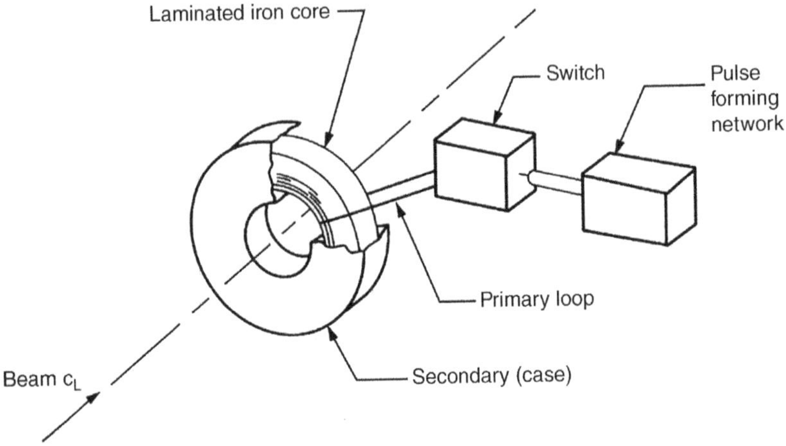

Figure 2.15 The principle of magnetic induction used in line pulse transformers uses a pulse forming line to drive a primary coil that couples to a beam load [21]. (Reproduced with permission from Ref. [21]. Copyright 2004, AIP Publishing LLC.)

The principle of magnetic induction is shown in Figure 2.15 where the current from a pulsed source is coupled through a ferromagnetic ring, known as a "core," and induces an axial electric field in the direction of the beam. Charged particles accelerate in this electric field. The energy from the pulse forming network couples to the beam by magnetic induction.

For high-power applications, pulsed current sources are used. As the pulse forming network discharges through the primary loop, the time-varying current $i(t)$ induces a magnetic field in the magnetic core and drives the core toward its saturated state. The time-dependent magnetic field $\vec{B}(t)$ produces a magnetic flux $\phi_B(t)$ in the core:

$$\phi_B(t) = \int_{A_m} \vec{B}(t) \cdot d\vec{S} \tag{2.45}$$

where A_m is the area of the cross section of the magnetic core. The changing magnetic flux $\phi_B(t)$ generates an inductive voltage V_0 according to

$$-\frac{\partial \phi_B}{\partial t} = V_0 \tag{2.46}$$

If the magnetic field in (2.45) associated with the pulsed current is sufficiently strong, the magnetic core is driven into its saturated state, where the magnetic permeability is small so that the duration of the induced voltage is determined by the saturation time of the magnetic core under operating conditions. Since the time for the magnetic core to saturate is determined by its size, it is the duration of the induced voltage. The relationship between the magnetic core size and saturation is discussed in Chapter 4.

2.3.1 Linear Pulse Transformers

The linear pulse transformer uses discrete pulse forming networks and magnetic induction to deliver power to a load by distributing the injection points. The linear pulse transformer consists of N primary windings with a common, single turn secondary, as shown in Figure 2.16 [22,23]. A capacitor or other energy storage element is discharged into each of the N primary windings simultaneously though a fast, triggered switch. Because the prime energy store is distributed, the energy delivered to the load may be very large. The linear pulse transformer has a low internal inductance since the secondary winding is a single turn and may be used for fast charging of high-energy stores. Often, energy is coupled to the load using magnetic cores.

Linear pulse transformers form the basis of such machines as SNOP-3 and HERMES III, the world's most powerful line pulse transformer, which is still in routine use as a radiation effects simulator.

2.3.2 Induction Cells

Induction cells allow pulsed power sources to be used in a cavity geometry by including a magnetic core. Humphries used the example of an electrostatic particle accelerator to illustrate the vital role the magnetic core plays in the induction cell [24].

A pulsed electrostatic gap, shown schematically in Figure 2.17, consists of a pulsed power source used to establish a voltage V_0 in the gap. The source is attached to the high potential plate through its inner conductor with its outer shield connected to the grounded extraction plate. Because the radial dimension is large, the pulsed source is connected to several points through transmission lines to distribute the voltage evenly around the particle source. Particles with an energy (qV_0) are extracted through an aperture in the grounded extraction plate. The current flows through the loop formed from the extraction plate, through the inner conductor of the transmission line, through the beam, and returns through the grounded extraction plate.

The high-voltage connections are difficult, so providing input power in the injector geometry of Figure 2.17 is challenging. A considerable advantage is gained if the source and extraction plates are at ground potential, which may be accomplished by electrically connecting the two plates through a conducting cylinder to form a toroidal cavity, as shown in Figure 2.18a. In this cavity, the source and extraction points are at the same (ground) potential, but now most of the current flows in the low-impedance outer loop. This outer loop current is an undesired leakage current I_{leak}, which results in a small potential in the accelerating gap. An adequate accelerating gap potential is attained by increasing the load current I_{load} relative to the leakage current I_{leak} by increasing the impedance of the outer loop by filling its volume with a magnetic material, as

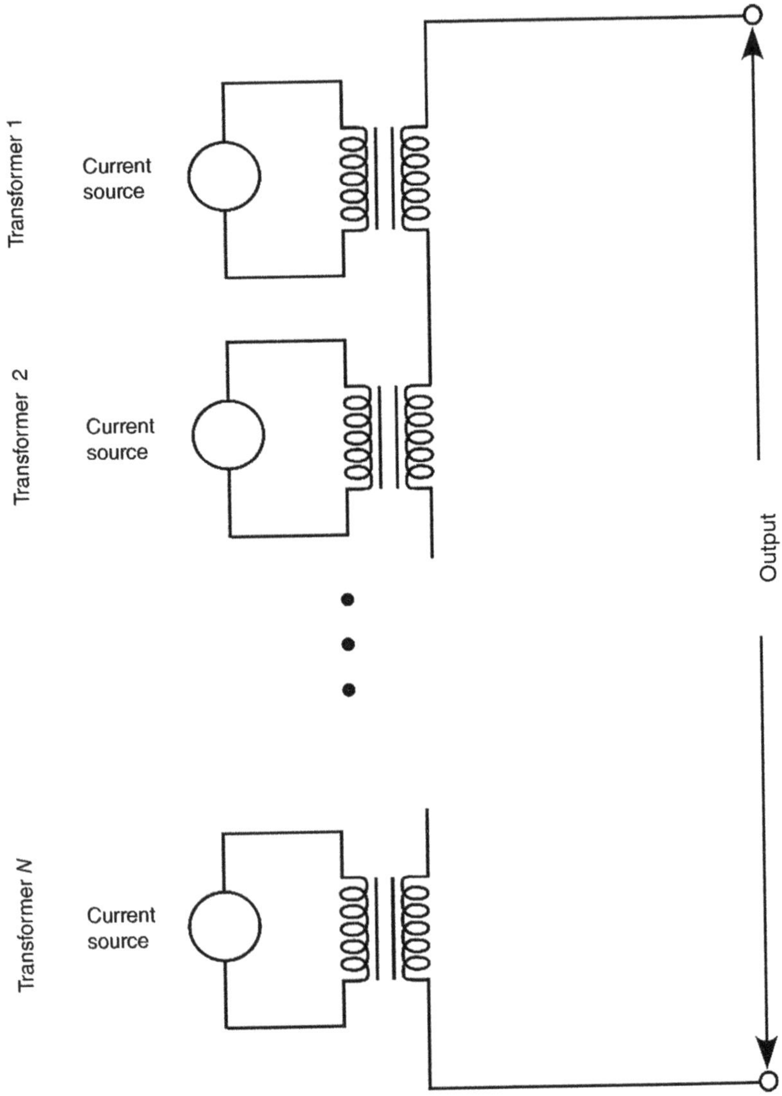

Figure 2.16 The line pulse transformer delivers power to a load with distributed current sources to a primary transformer with a common secondary.

shown in Figure 2.18b. The magnetic core increases the leakage inductance by a factor related to the μ_r relative magnetic permeability. The increase in leakage inductance as a result of the magnetic core results in a great reduction of the leakage current until the magnetic core becomes saturated. After saturation,

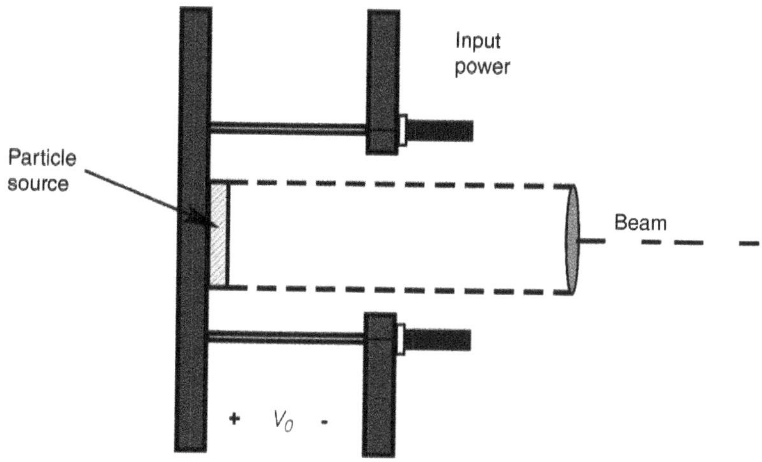

Figure 2.17 In an electrostatic injector, particles are accelerated by a voltage V_0 produced between two plates. The extraction plate is at ground potential. Current flows through the loop formed by the transmission line inner conductor and the charged particle beam.

the differential magnetic permeability approaches μ_0 and the cavity becomes a low-inductance load.

As the magnetic cores are driven into their saturated state, an inductive voltage appears in the gap. The inductive fields cancel electrostatic fields across most of the accelerator except for the beam line. The beam senses a large net accelerating field, while electrostatic potential differences in the accelerating structure are kept to manageable levels. This process is called inductive isolation and is the basis for linear induction cavities.

The induction accelerator is a series of induction cells where the voltage pulse is timed so that the accelerating electric field is present when beam particles pass through the core, illustrated in Figure 2.19. The pulsed voltage is applied through transmission lines to the primary turn surrounding the magnetic core. The secondary turn includes the induction gap. Induction accelerators are capable of producing very large beam currents—up to 10 kA—but generate lower voltages than RF accelerators.

2.3.3 Linear Transformer Drivers

The linear transformer driver (LTD) was introduced in 1997 as a fast, high-current circuit architecture inspired by linear pulse transformers [26,27]. LTDs use low-inductance, capacitive energy storage circuits (known as bricks) arranged in parallel to produce a high-peak current directly from the primary energy store. The bricks are contained within the enclosed volume of an induction cavity, shown schematically in Figure 2.20. Although LTDs have

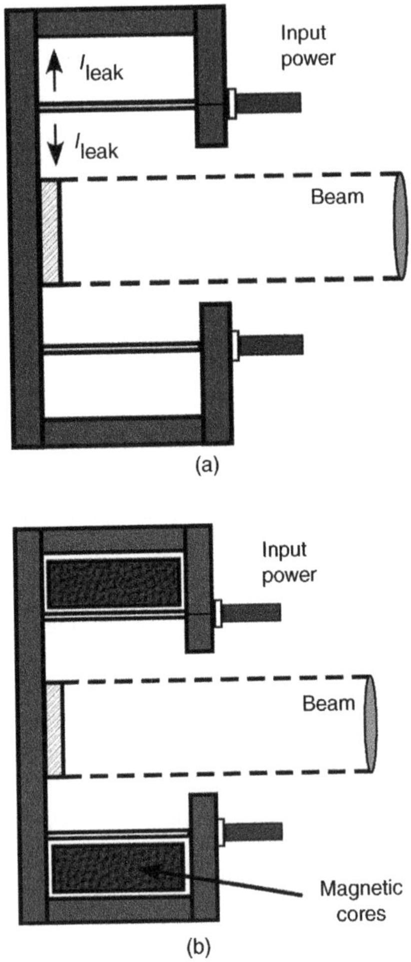

Figure 2.18 A short circuited electrostatic injector and (b) an injector with a high-inductance leakage path.

certain similarities to linear pulse transformers, the shift to increased modularity is a significant technological advance.

LTDs have a number of attractive features:

- High currents are generated by charging the capacitive storage elements to relatively low voltages (~100 kV).
- The current is coupled to the load by magnetic induction so that only the secondary turn has to be insulated from the high-output voltage.
- The circuit architecture is highly modular.

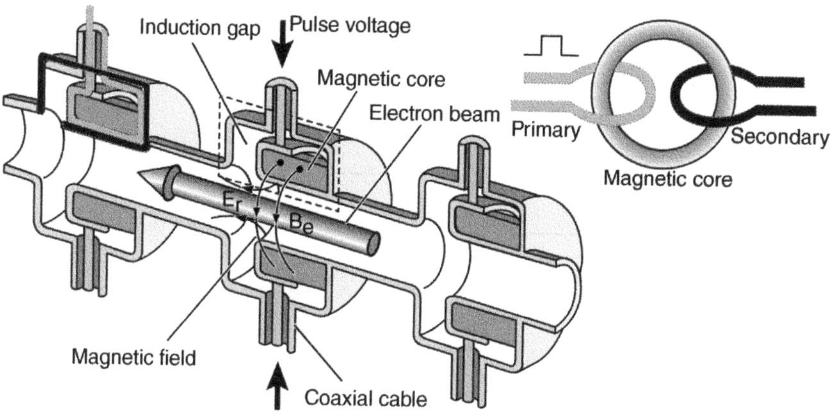

Figure 2.19 An induction accelerator is constructed from a series of induction cells. The pulsed voltage source supplies the primary that surrounds the magnetic core and the magnetic core acts as a 1:1 transformer. The primary and secondary loops are depicted on the cell on the left-hand side [25]. (Reproduced with permission of CERN Courier.)

- Fast, high-current pulses are achieved without multiple pulse forming stages at high voltages as in traditional pulse compression schemes.
- The current source is contained within the cavity itself.
- High potential for yielding compact designs for high-current applications.

2.3.3.1 Operating Principles

Linear transformer drivers are an arrangement of individual current sources encased in a cavity and coupled to a centrally located load by magnetic induction. A specific nomenclature has been developed to describe the various parts of the LTD.

The basic unit of the LTD architecture is a "cavity" or "cell" that is somewhat analogous to a "stage" in a Marx generator. An LTD cell is a toroidal cavity containing the primary energy storage capacitors, switches, resistors, and a magnetic core. The individual pulse forming circuits, known as "bricks," are arranged in a radial pattern surrounding the magnetic core. When the bricks are triggered into operation, current flow in the cavity establishes a time-dependent magnetic flux in the cores. As in an induction cavity, an inductive electric field is established to oppose the change in flux and a voltage appears across the load. The voltage duration depends on the time for the magnetic cores to be driven into their saturated state. The load for an LTD is often an accelerating gap or magnetically insulated transmission line.

The LTD was initially conceived as a linear pulse transformer with a parallel arrangement of capacitors and developed because of the limitations of very fast, very high-energy storage elements [28]. The typical approach for fast primary

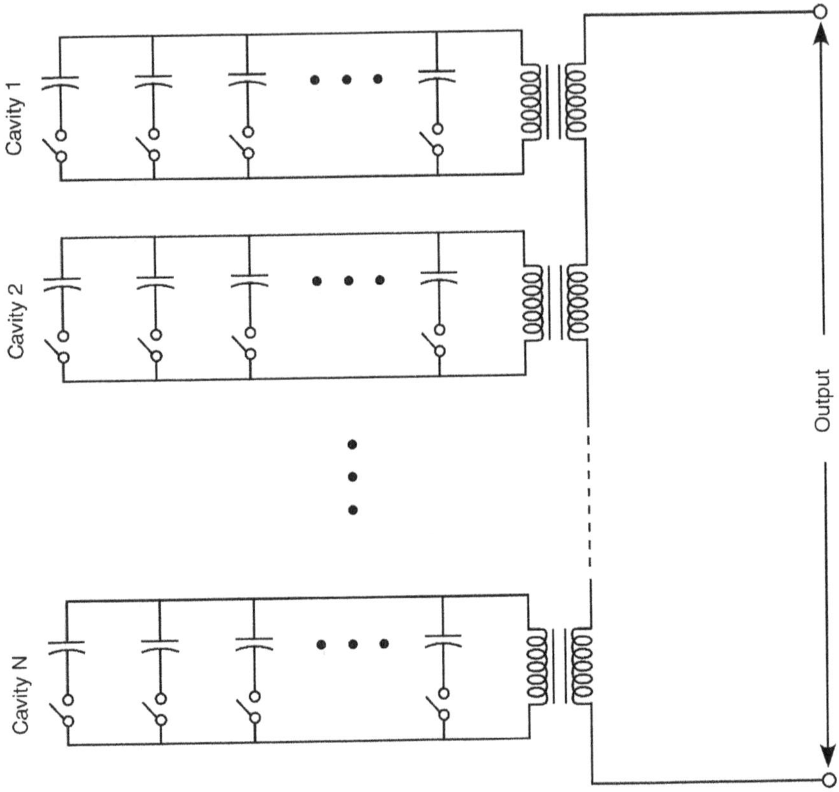

Figure 2.20 The equivalent circuit of an LTD is similar to the linear pulse transformer, but the topology may radically be different. The LTD typically uses multiple energy storage circuits, known as bricks, arranged in parallel within an induction cavity.

storage is to build an LC circuit from high-energy capacitors and reduce the circuit inductance to the lowest possible value to achieve the highest peak current. The rise time of the output pulse, t_r, is

$$t_r \sim \frac{\pi}{2}\sqrt{LC}$$

If a faster pulse is needed and the inductance is already minimal, the pulse rise time may only be reduced by lowering the capacitance. However, lowering the capacitance also lowers the stored energy. Fast delivery of a large energy requires a parallel arrangement of capacitors.

A schematic of the top half of the cross section of an LTD cell is shown in Figure 2.21 [27]. Here C is the storage capacitor, S is the switch, and Z_L is the load. The magnetic core serves the same purpose as in the induction cell and establishes a voltage between the two plates where the load is located. A feature

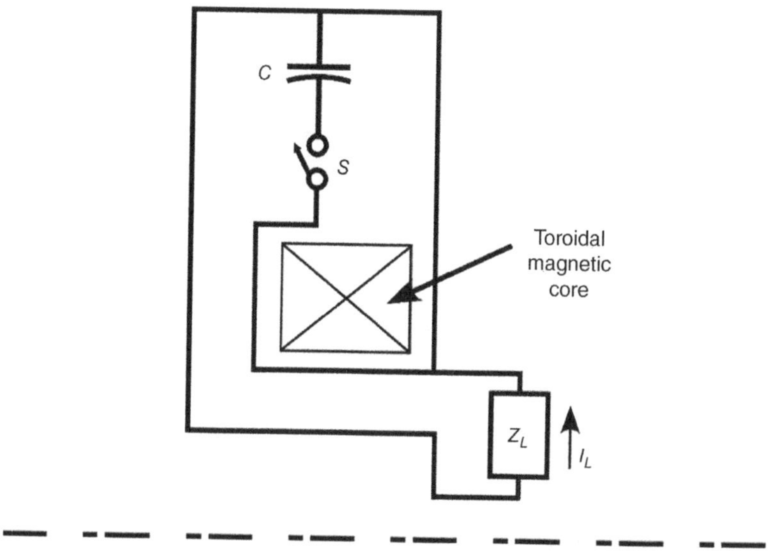

Figure 2.21 The upper half of the cross section of an LTD cavity showing the energy storage capacitor and a discharge switch for the brick and the magnetic core used for inductive isolation. The LTD establishes a voltage across the load Z_L.

of the LTD is that the energy stores and/or pulse forming lines are contained within the cavity that allows for compact, high-energy systems. The inductance of the secondary turn is defined by the geometry of the left part of the figure [27].

Typically, the brick is an R–L–C circuit formed by the primary energy storage capacitors, discharge switch, and the intrinsic inductance and resistance of the circuit. The circuit is usually designed to be underdamped with a load resistance of 1 Ω or less for high-current applications and higher load impedances may be considered. The circuit is made to have a fast discharge by minimizing the inductance of the circuit, which increases the peak current. Each brick is discharged with its own triggered switch so that high-performance switches are needed to simultaneously fire all the bricks in a cavity. In principle, a brick can be any pulsed current source and LTD designs using bricks with integrated pulse forming networks have been pursued [28,29].

LTDs impose severe requirements on the closing switches. The switches must be low inductance, yet trigger the DC capacitive discharge with jitter of less than ~10 ns and have a low prefire rate. In principle, each brick could be independently triggered but would be prohibitively expensive. Instead, LTD cells are triggered with a single trigger unit with the bricks connected with isolation resistors. Trigger pulses for the switches are injected through the cavity wall at numerous points along the outer circumference to minimize

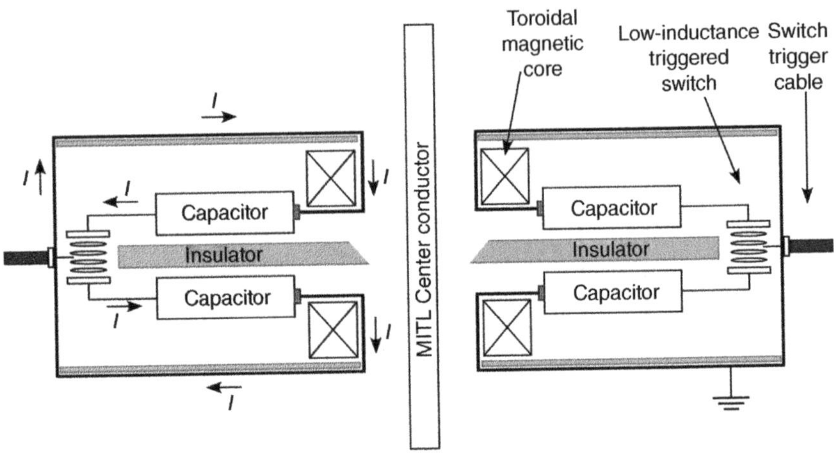

Figure 2.22 A side view of the cross section of an LTD cavity to charge magnetically insulated transmission line (MITL) center conductor. The capacitors are charged to opposite polarities and discharged with a triggered switch.

switching delays associated with transit time effects caused by the wave propagation of the trigger pulse from brick to brick. The magnetic cores represent major drawbacks for the LTD architecture because of cost and weight and must be reset—electrically driven to the initial state—for maximum flux swing.

2.3.3.2 Realized LTD Designs and Performance

In practice, LTD cells have coalesced to a symmetrical design such as that shown in Figure 2.22. Two half cavities are united to produce a voltage across a small gap. The two capacitors are charged in opposite polarities, one to $-V_0$ and the other to $+V_0$, so that the switch has a potential difference of $2V_0$ across its terminals. A single source is used to trigger the switches in a cell. The switches are electrically connected through isolating resistors.

Individual LTD cavities have been tested with 10 [30,31], 16 [32,33], 20 [34], and 40 [35,36] parallel bricks. Component lifetime testing of a single, high-current, 20-brick cavity, shown in Figure 2.23, has been performed and shown to be reliable to at least 13,500 shots [34]. LTDs have had very stable operation. An overlay of 200 output pulses from an LTD cavity is shown in Figure 2.24 [37]. The peak current is over 500 kA and has a rise time less than 100 ns.

LTDs are attractive not only for the low charging voltage of the brick but also because the current generators are completely enclosed in a grounded cavity cell. LTD cavities may be arranged in circuit topologies requiring their

Figure 2.23 A 20-capacitor LTD with the top cavity cover removed shows the 10 capacitors surrounding the magnetic core and connected together by liquid resistors. The other 10 capacitors are located in the bottom half of the cavity. The capacitors of the brick are connected together through spark gap switches so that current travels "into the page" in this view [34].

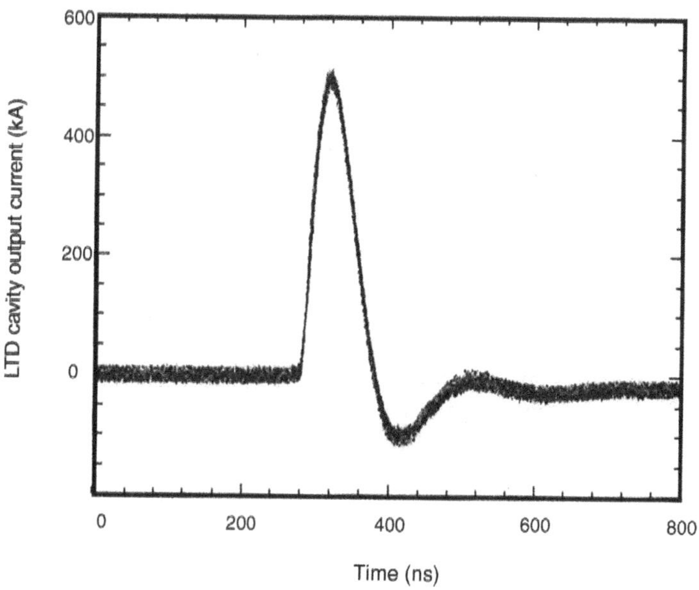

Figure 2.24 Overlay of 200 shots showing the LTD output has high shot-to-shot repeatability [37].

connection in series or parallel depending on the load requirements. High voltages may be attained by inductively adding the output voltage of LTD cavities in series as for an inductive voltage adder (IVA). A number of IVA designs using LTDs have been demonstrated. A high-current adder was tested at the Institute for High Current Electronics (HCEI) in Tomsk with 5 series cavities with 40 parallel bricks each [36]. An LTD-based IVA for radiography consisting of 7 series cavities containing 10 parallel bricks each [37–40] has been tested at Sandia National Laboratories in Albuquerque into an electron beam diode load. LTD-based next-generation drivers for z-pinch loads are envisioned [28]. The LTD concept is also under investigation using film capacitors and solid-state power MOSFETs for industrial applications [41,42].

2.4 Design Examples

Example 2.1

An air core resonant transformer is used to charge a 125 pF capacitor to 1 MV. Determine the (a) primary storage capacitor, (b) the secondary inductance, and (c) the charge time of the dual resonant transformer. Assume a primary charge voltage of 50 kV and a single-turn primary inductance of 0.45 μH.

Solution

a) Since the entire amount of energy is transferred from the primary energy storage capacitor C_1 to C_2, conservation of energy can be invoked:

$$\frac{1}{2}C_1V_1^2 = \frac{1}{2}C_2V_2^2 \quad \text{or} \quad C_1 = C_2\left(\frac{V_2}{V_1}\right)^2$$

For $C_2 = 125$ pF, $V_2 = 1$ MV or 1000 kV, and $V_1 = 50$ kV:

$$C_1 = 125\ \text{pF}\cdot\left(\frac{1000}{50}\right)^2 = 125\ \text{pF}\cdot 400 = 50\ \text{nF}$$

Note that the voltage step-up ratio $\frac{V_2}{V_1}$ can also be expressed as $\sqrt{\frac{C_1}{C_2}}$.

b) This transformer is double resonant. To minimize the time the secondary capacitor must store the charge, we will choose a coupling coefficient, $K = 0.6$. From Equation 2.8, $\omega_1 = \omega_2$ and from Equation 2.3 it follows that

$$L_1C_1 = L_2C_2$$

$$L_2 = \frac{L_1C_1}{C_2} = (0.45\ \text{mH})\cdot(20)^2 = 180\ \text{mH}$$

c) To calculate the charge time t_{2M}, the value of the trigonometric arguments in Equation 2.7 must be surmised. Since $i(t_{2M}) = 0$, so

$$\sin\left(\frac{\omega t_{2M}}{\sqrt{1-K}}\right) = \sin\left(\frac{\omega t_{2M}}{\sqrt{1+K}}\right) = 0$$

and both arguments must fall on multiples of π. Since $V_1(t_{2M}) = 0$, so from Equation 2.6,

$$\cos\left(\frac{\omega t_{2M}}{\sqrt{1-K}}\right) = -\cos\left(\frac{\omega t_{2M}}{\sqrt{1+K}}\right)$$

The arguments are both multiples of π, implying that

$$\left|\cos\left(\frac{\omega t_{2M}}{\sqrt{1-K}}\right)\right| = \left|\cos\left(\frac{\omega t_{2M}}{\sqrt{1+K}}\right)\right| = 1$$

The first extrema occurs when

$$\left(\frac{\omega t_{2M}}{\sqrt{1+K}}\right) = \pi \quad \text{and} \quad \left(\frac{\omega t_{2M}}{\sqrt{1-K}}\right) = 2\pi$$

Substituting $K = 0.6$ and $\omega = \left(\frac{1}{\sqrt{L_1 C_1}}\right)$ into either of the above equations yields $t_{2M} = 0.596\ \mu s$.

Example 2.2

Suppose a square pulse is to be transmitted from a 50 Ω source to a 200 Ω load. The exponentially tapered transmission line transformer has a physical length ℓ of 1 m long and is insulated with a new insulating oil that has a relative dielectric constant $\epsilon_r = 3$. What is the percent change that can be expected on the rectangular square pulse with a 1 ns pulse width?

Solution

From Equation 2.35,

$$G = \sqrt{\frac{Z_L}{Z_1}} = \sqrt{\frac{200\Omega}{50\Omega}} = 2$$

(continued)

(continued)

The delay per unit length for the exponentially tapered transmission line is constant along its length and its electrical length then is

$$T_{TL} = \frac{\epsilon \cdot \ell}{c} = \frac{(3) \times (1)}{3 \times 10^8} = 10 \text{ ns}$$

The percent change in the amplitude P can be calculated from Equation 2.42:

$$P = 50 \cdot (\ln G)^2 \frac{\tau}{T_{TL}} = 50 \cdot (\ln 2)^2 \cdot \frac{1}{10} = 2.4\%$$

Note that the pulse "droops" down in time because the exponential transformer is a high-pass filter, so the high frequencies in the rise time are preserved and the low frequencies comprising the pulse width are compromised.

Example 2.3

In the above example, the electrical length of the exponential transmission line transformer is chosen so that the electrical length of the transformer T_{TL} is 10 times the applied pulse width τ, resulting in an acceptable droop of 2.4%. Determine (a) how the length of the transmission line transformer, normalized to the pulse width, affects its gain with the percent change P as a parameter, and (b) the minimum relative electrical length for a voltage gain of 3 with an allowable droop is of 10%.

Solution

Rearranging Equation 2.42,

$$(\ln G)^2 = \frac{P}{50} \cdot \left(\frac{T_{TL}}{\tau}\right)$$

Its graph is a parabola restricted to the right-half plane $(T_{TL}/\tau) \geq 0$. The gain of the transmission line transformer, G, can be expressed more directly as

$$G = \exp\left[\frac{P}{50} \cdot \left(\frac{T_{TL}}{\tau}\right)\right]$$

which is plotted as a function of $(T_{TL}/\tau)$for values of P, the percentage change in Figure 2.25. The practical values of 2.5, 5, and 10% are chosen for the percentage change.

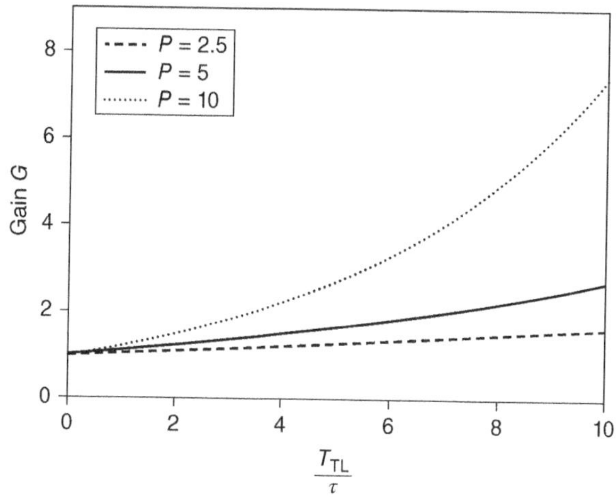

Figure 2.25 The gain of an exponentially tapered transmission line transformer depends on its electrical length relative to the transmitted pulse width. For a desired gain, the pulse distortion is minimized by increasing the physical length of the transmission line transformer.

For a gain of 3 and an allowable 10% change in the pulse amplitude, the length of the transformer is

$$\frac{T_{TL}}{\tau} = \frac{50}{P} \cdot (\ln G)^2 = \frac{50}{10} \cdot (\ln 3)^2 = 6$$

Thus, for a 10% allowable change in pulse amplitude, the transformer must be six times longer than the pulse width. However, if the gain is dropped to 2, then the transmission line needs to be only 2.4 times the pulse width.

References

1 G.A. Mesyats, *Pulsed Power*, Kluwer Academic/Plenum Publishers, New York, 2005.
2 E.G. Cook and L.L. Reginato, Off-Resonance Transformer Charging for 250 kV Water Blumlein. *IEEE Trans. Electron Devices*, Vol. 26, No. 10, p. 1512, 1979.
3 E.A. Abramyan, Transformer-Type Accelerators for Intense Electron Beams. *IEEE Trans. Nucl. Sci.* Vol. 18, No. 3, pp. 447–455, 1971.
4 D. Finkelstein et al., High Voltage Impulse System. *Rev. Sci. Instrum.*, Vol. 37, No. 2, p. 159, 1966.

5 F.M. Bieniosek, Triple Resonance Pulse Transformer Circuit. *Rev. Sci. Instrum.*, Vol. 61, No. 6, p. 1717, 1990.
6 K.V. Nagesh, Tesla Transformer for Microsecond Pulses. *Proceedings of Electrical Insulation and High Voltage Pulse Techniques, BARC, Bombay, India*, 1982.
7 I. Boscolo et al., ATRI: A Tesla Transformer-Type Electron Beam Accelerator. *J. Vac. Sci. Technol.*, Vol. 12, No. 6, p. 1194, 1975.
8 M. Akihiro et al., Relativistic Electron Beam Source with an Air Core Step–Up Transformer, Technical Report No. IPPJ–T–23, Institute of Plasma Physics, Nagoya, Japan, April, 1975.
9 E.A. Abramyan, High Voltage Pulse Generator of the Base of the Shock Transformer. *Proceedings of the 1st International Pulsed Power Conference*, 1976.
10 J.C. Martin et al., Notes on the Construction Methods of a Martin High Voltage Pulse Transformer, Technical Report No. CU–NRL/2, School of Electrical Engineering, Cornell University, New York, 1967.
11 G.J. Rohwein, 1.5 MV Resonant Transformer Pulser, Sandia National Laboratories Report, SAND80–0974, UC–21, p. 114, 1981.
12 G.J. Rohwein, Repetitive Pulsed Power and Reliability Studies, Sandia National Laboratories Report, SAND81–1459, UC–21, p. 118, 1982.
13 R.L. Cassel, A 200 kV, 100 ns, 4/1 Pulse Transformer. *Proceedings of the 20th Power Modulator Symposium*, pp. 234–236, 1992.
14 J.P. O'Loughlin, J.M. Lehr, G.J. Rohwein, and D.L. Loree, Air Core Transformer with a Coaxial Grading Shield, U.S. Patent 6,414,581, July 2002.
15 C.E. Baum, W.L. Baker, W.D. Prather, J.M. Lehr, J.P. O'Loughlin, D.V. Giri, I.D. Smith, R. Altes, J. Fockler, D. McLemore, M.D. Abdalla, and M.C. Skipper, JOLT: A Highly Directive, Very Intensive, Impulse Like Radiator. *Proc. IEEE*, Vol. 92, No. 7, pp. 1096–1109, 2004.
16 J.C. Slater, *Microwave Transmission*, McGraw-Hill, 1942.
17 I.A.D. Lewis and F.H. Wells, *Millimicrosecond Pulse Techniques*, 2nd ed., Pergamon Press, 1959.
18 P.W. Smith, *Transient Electronics*, John Wiley & Sons, Inc., New York, 2002.
19 R.E. Collin, *Foundations for Microwave Engineering*, McGraw-Hill, p. 372, 1992.
20 C.E. Baum and J.M. Lehr, Tapered Transmission Line Transformers for Fast High Voltage Transients. *IEEE Trans. Plasma Sci.*, Vol. 30, No. 5, pp. 1712–1721, 2002.
21 N.C. Christofilos, R.E. Hester, W.A.S. Lamb, D.D. Reagan, W.A. Sherwood, and R.E. Wright, High Current Linear Induction Accelerator for Electrons. *Rev. Sci. Instrum.*, Vol. 35, No. 7, p. 886, 1964.
22 G.A. Mesyats, Pulsed High Current Electron Technology. *Proceeding of the 2nd International Pulsed Power Conference*, pp. 9–16, 1979.
23 G.A. Mesyats, *Pulsed Power*, Chapters 1 and 3, Springer, 2005.

24 S. Humphries, *Principles of Charged Particle Acceleration*, Dover Publications Inc., 2012
25 CERN Courier, p. 22, April, 2005. Available at http://cerncourier.com/cws/article/cern/29307.
26 A.N. Bastrikov, A.A. Kim, B.M. Kovalchuk, E.V. Kumpjak, S.V. Loginov, B.I. Manylov, V.A. Visir, Y.P. Etlicher, L. Fresaline, J.F. Leon, P. Monjaux, F. Kovacs, D. Huet, and F. Bayol, Fast Primary Energy Storage Based on a Linear Transformer Scheme. *Proceedings of the 11th IEEE International Pulsed Power Conference*, pp. 489–497, 1997.
27 B.M. Kovalchuk, V.A. Vizir, A.A. Kim, E.V. Kumpjak, S.V. Loginov, A.N. Bastrikov, V.V. Chervjakov, N.V. Tsou, Ph. Monjaux, and D. Huet, Fast Primary Storage Based on Pulsed Linear Transformer. *Sov. Izw. Vuzov. Phys.*, Vol. 40, pp. 25–37, 1997.
28 A.N. Bastrikov, V.A. Vizir, V.A. Volkov, V.G. Durakov, A.M. Efremov, V.B. Zorin, A.A. Kim, B.M. Kovalchuk, E.V. Kumpjak, S.V. Loginov, V.A. Sinebryuhov, N.V. Tsou, V.V. Cervjakov, V.P. Yakovlev and G.A. Mesyats, Primary Energy Storages Based on Linear Transformer Stages. *Laser Part. Beams*, Vol. 21, pp. 295–299, 2003.
29 A.A. Kim, M. G. Mazarakis, V. A. Sinebryukhov, S.N. Volkov, S.S. Kondratiev, V.M. Alexeenko, F. Bayol, G. Demol and W.A. Stygar, Square Pulse Linear Transformer Driver. *Phys. Rev. Special Top. Accel. Beams*, Vol. 15, p. 040401, 2012.
30 A.A. Kim, A.N. Bastrikov, S.N. Volkov, V.G. Durakov, B.M. Kovalchuk, and V.A. Sinebryukhov, 1 MV Ultra-Fast LTD Generator. *Proceedings of the 14th IEEE International Pulsed Power Conference, Dallas, TX*, 2003, pp. 853–854.
31 J. Leckbee, S. Cordova, B. Oliver, D.L. Johnson, M. Toury, R. Rosol, and B. Bui, Load Line Evaluation of a 1-MV Linear Transformer Driver (LTD). *Proceedings of the IEEE International Power Modulator Conference, Las Vegas, NV*, 2008.
32 A.A. Kim, V. Sinebryukhov, B. Kovalchuk, A. Bastrikov, V.D.S. Volkov, S. Frolov, and V. Alexeenko, Superfast 75 ns LTD Stage. *Proceedings of the 16th IEEE International Pulsed Power Conference, Albuquerque, NM*, 2007, pp. 148–151.
33 M. Toury, C. Vermare, B. Etchessahar, L. Veron, M. Mouillet, F. Bayol, G. Avrillaud, and A.A. Kim, IDERIX: An 8 mv Flash X-Rays Machine Using a LTD Design. *Proceedings of the 16th IEEE International Pulsed Power Conference, Albuquerque, NM*, 2007, pp. 599–602.
34 S.T. Rogowski, W.E. Fowler, M. Mazarakis, C.L. Olson, D. McDaniel, K.W. Struve, and R.A. Sharpe, Operation and Performance of the First High Current LTD at Sandia National Laboratories. *Proceedings of the 15th IEEE International Pulsed Power Conference, Monterey, CA*, 2005, pp. 155–157.
35 A.A. Kim, A.N. Bastrikov, S.N. Volkov, V.G. Durakov, B.M. Kovalchuk, and V.A. Sinebryukhov, 100 GW Fast LTD Stage. *Proceedings of the 13th*

International Symposium on High Current Electronics (IHCE'04), Tomsk, Russia, 2004, pp. 141–144.

36 M.G. Mazarakis, W.E. Fowler, D.H. McDaniel, C.L. Olson, S.T. Rogowski, R.A. Sharpe, K.W. Struve, W.A. Stygar, A.A. Kim, V.A. Sinebryukhov, R.M. Gilgenbach, and M.R. Gomez, High Current Linear Transformer Driver (LTD) Experiments. *Proceedings of the 16th IEEE International Pulsed Power Conference, Albuquerque, NM*, 2007, pp. 222–225.

37 M.G. Mazarakis, S. Cordova, W. Fowler, K. Lechien, J.J. Leckbee, F.W. Long, M.K. Matzen, D.H. McDaniel, R.G. McKee, J.L. McKenny, B.V. Oliver, C.L. Olson, J.L. Porter, S.T. Rogowski, K.W. Struve, W. Stygar, J.R. Woodworth, A.A. Kim, Y.A. Sinebrynukhov, R.M. Gilgenbach, M.R. Gomez, D.M. French, Y.Y. Lau, J. Zier, D.M. VanDevalde, R.A. Sharpe, Linear Transformer Drive (LTD) Development at Sandia National Laboratories. *Proceedings of the IEEE International Pulsed Power Conference*, pp. 138–145, 2009.

38 D.V. Rose, D.R. Welch, B.V. Oliver, J.J. Leckbee, J.E. Maenchen, D.L. Johnson, A.A. Kim, B.M. Kovalchuk, and V.A. Sinebryukhov, Numerical Analysis of a Pulsed Compact LTD System for Electron Beam-Driven Radiography. *IEEE Trans. Plasma Sci.*, Vol. 34, No. 5, pp. 1879–1887, 2006.

39 J.J. Leckbee, J. Maenchen, S. Portillo, S. Cordova, I. Molina, D.L. Johnson, A.A. Kim, R. Chavez, and D. Ziska, Reliability Assessment of a 1 MV LTD. *Proceedings of the 15th IEEE International Pulsed Power Conference, Monterey, CA*, 2005, pp. 132–134.

40 J. Leckbee, S. Cordova, B. Oliver, D.L. Johnson, M. Toury, R. Rosol, and B. Bui, Load Line Evaluation of a 1-MV Linear Transformer Driver (LTD). *Proceedings of the IEEE International Power Modulator Conference, Las Vegas, NV*, 2008.

41 W. Jiang, Solid State LTD Module Using Power MOSFETs. *IEEE Trans. Plasma Sci.*, Vol. 38, No. 10, pp. 2730–2733, 2010.

42 W. Jiang and A. Tokuchi, Repetitive Linear Transformer Driver Using Power MOSFETs. *IEEE Trans. Plasma Sci.*, Vol. 40, No. 10, pp. 2625–2628, 2012.

3

Pulse Forming Lines

The discharge of primary energy storage systems such as Marx generators, transformers, and capacitor banks can produce the very high voltages required for many pulsed power applications but, because of their inherent inductance, their energy can generally only be extracted in relatively long times. Many applications, however, require energy to be delivered on faster timescales. In this case, the circuit may consist of a high-voltage generator and a pulse forming transmission line (PFL) controlled by a closing switch. The high-voltage generator, such as a Marx, charges the PFL in microseconds and the output switch discharges the PFL into the load.

For applications requiring very high peak power and very short durations, an additional capacitive storage section, usually called an intermediate storage capacitor, may be added between the high-voltage generator and the PFL. The intermediate storage allows the PFL to be charged faster than a Marx bank or transformer is capable of. The electrical breakdown strength of most dielectrics is greater for faster charge times, allowing for higher peak charge voltages. Intermediate storage capacitors can be lumped elements or transmission lines and numerous stages can be used to achieve the design objective. The design principles for intermediate stores and PFLs are the same.

This chapter gives an insight into mechanisms of transients in the PFLs as well as the equations required for design of various PFL configurations developed to meet specific requirements. The salient features of PFL performance parameters such as electrical breakdown, dielectric constant, self-discharge time constant, optimum impedance, charging source, and switching techniques are discussed.

3.1 Transmission Lines

Transmission lines are of tremendous importance in pulsed power technology. The treatment of waves traveling on transmission lines bridges the gap between electromagnetic field theory and analog circuit analysis, an underpinning of

Foundations of Pulsed Power Technology, First Edition. Jane Lehr and Pralhad Ron.

pulsed power design. Wave propagation on transmission lines can be approached from an extension of circuit theory appreciation of Maxwell's equations. How, then, is the appropriate analysis technique chosen? It is the relationship between the applied frequency and the physical dimensions of the circuit—its electrical size. Circuit analysis assumes that the physical dimensions of a network are much smaller than the electrical wavelength, while transmission lines may be a considerable fraction of a wavelength or many wavelengths in size. The fundamental frequency f_0 of a pulse with duration T_p is

$$f_0 = \frac{1}{T_p} \tag{3.1}$$

The remaining frequency components comprising the pulse are harmonics of this frequency with the highest-frequency components responsible for the rise time. A pulse with a fundamental frequency of ~100 MHz ($T_p = 10$ ns) has an electrical wavelength ~1 m, indicating the circuit should be treated as a transmission line. For $f \sim 1$ MHz ($T_p = 1$ μs), the wavelength is ~100 m, and the circuit can be treated as a lumped element. Thus, in most pulsed power applications, circuits are treated as transmission lines when the timescale is less than 1 ms because the physical dimensions of the circuit are on the order of the dominant wavelength. Transmission line theory may also apply when it is important to preserve the rise time.

A transmission line is a distributed parameter network where voltages and currents can vary in magnitude and phase over its length. A transmission line is usually represented by a two-wire line since the transverse electromagnetic mode (TEM) of propagation requires at least two conductors. If the lossless transmission line is divided into very short segments of length Δx, the transmission line can be modeled by its equivalent lumped element circuit, shown in Figure 3.1. The transmission line is characterized by its fundamental transmission parameters $\tilde{L}$ and $\tilde{C}$, where $\tilde{L}$, with units of Henry/meter, is the series self-inductance per unit length of the conductor pair, and $\tilde{C}$ is the shunt capacitance per unit length, in Farads per meter.

Transmission lines are generally analyzed on the principle of traveling waves. For waves propagating in the x-direction, the voltage at any point is

$$V(x,t) = V_I(x,t) + V_R(x,t) \tag{3.2}$$

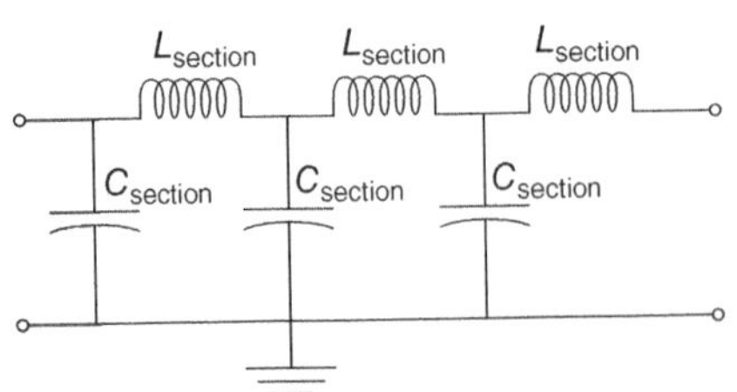

Figure 3.1 The equivalent circuit of a lossless transmission line along the x-direction with $L_{section} = \tilde{L}\Delta x$ and $C_{section} = \tilde{C}\Delta x$.

where $V_I(x,t)$, the incident (or the forward-going) wave, is assumed to be propagating in $+x$-direction and the reverse-going wave, $V_R(x,t)$, is assumed to propagate in the $-x$-direction wave. The voltage and current along any point on the transmission line are related by its characteristic impedance Z_0:

$$Z_0 = \frac{V(x,t)}{I(x,t)} \tag{3.3}$$

A wave traveling along an infinitely long transmission line will have only a forward-going wave unless it encounters a discontinuity. The discontinuity may be another transmission line, a lumped element reactance or an effect of a finite length such as an open- or short-circuit termination. The discontinuity introduces the reverse-going wave whose amplitude is a fraction of the incident wave. This fraction is known as the voltage reflection coefficient Γ. For a transmission line with characteristic impedance Z_0 terminated in a load impedance Z_L, the voltage reflection coefficient Γ is given by

$$\Gamma = \frac{Z_L - Z_0}{Z_L + Z_0} \tag{3.4}$$

The load impedance Z_L is not necessarily resistive and initial conditions must also be satisfied. The reverse-going wave, which is the fraction of the incident wave that is reflected from the load (or discontinuity), is given by

$$V_R(\ell,t) = \Gamma \cdot V_I(\ell,t) \tag{3.5}$$

The voltage at the discontinuity or load is given by

$$V(\ell,t) = V_I(\ell,t) + V_R(\ell,t) = (1+\Gamma)\cdot V_I(\ell,t) \tag{3.6}$$

The portion of the incident wave that is transmitted to the load (or through the discontinuity) is characterized by the voltage transmission coefficient T given by

$$T = 1 - \Gamma = \frac{2Z_0}{Z_L + Z_0} \tag{3.7}$$

In general, both Γ and T are complex numbers, and have both a magnitude and a phase. For a purely resistive load, the matched circuit condition occurs when $Z_L = Z_0$ and $\Gamma = 0$. In this case, the transmission coefficient $T = 1$ and voltage at the load is the forward-going wave.

3.1.1 General Transmission Line Relations

For a transmission line with a physical length ℓ, a number of quantities can be derived from the characteristic transmission line parameters $\tilde{L}$ and $\tilde{C}$:

Total capacitance of line with length ℓ:

$$C = \tilde{C}\cdot\ell \tag{3.8}$$

Total inductance of line with length ℓ:

$$L = \tilde{L} \cdot \ell \tag{3.9}$$

The characteristic impedance:

$$Z_0 = \sqrt{\frac{L}{C}} = \sqrt{\tilde{L}\tilde{C}} \tag{3.10}$$

Delay per unit length:

$$\delta = \sqrt{\tilde{L} \cdot \tilde{C}} \tag{3.11}$$

Lossless propagation velocity:

$$v_{pp} = {}^{1}\!/_{\delta} = \frac{1}{\sqrt{\tilde{L} \cdot \tilde{C}}} = \frac{c}{\sqrt{\mu_r \varepsilon_r}} \tag{3.12}$$

where $c = 3 \times 10^8$ m/s is the speed of light.
Electrical length:

$$T_T = \delta \cdot \ell = \sqrt{\tilde{L}\tilde{C}} \cdot \ell = \sqrt{LC} = \frac{\ell}{v_{pp}} \tag{3.13}$$

The electrical length is also known as the one-way propagation time and has units of time. The electrical length is particularly useful in pulsed power since it relates physical length with pulse length. Equations 3.8 and 3.13 can be combined to yield the useful manipulations:

Total capacitance of cable with length ℓ:

$$C = \frac{T_T}{Z_0} \tag{3.14}$$

Total inductance of line with length ℓ:

$$L = T_T \cdot Z_0 \tag{3.15}$$

Pulse width:

$$T_p = 2T_T = 2\sqrt{\tilde{L}\tilde{C}} \cdot \ell = 2\sqrt{LC} = \frac{2\ell}{v_{pp}} \tag{3.16}$$

These expressions are independent of the specific geometry and depend only on the characteristic parameters $\tilde{L}$ and $\tilde{C}$ and length ℓ of the transmission line.

For pulses with durations of ~1 μs and longer, the circuit of Figure 3.1 can be realized with lumped circuit elements and is known as an E-type PFN. Typically, five or more sections are needed to make a pulse with a reasonable flattop.

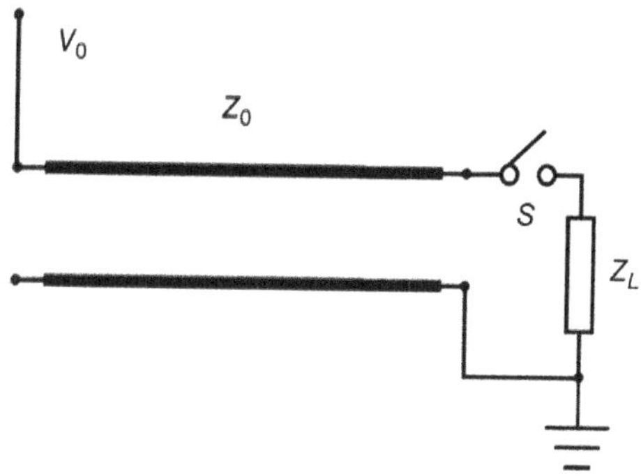

Figure 3.2 A transmission line charged to a voltage V_0 may be combined with a closing switch S to make a simple pulser to deliver a voltage pulse to a load Z_L. The two-conductor transmission line representation is used to illustrate that a variety of geometries may be realized.

3.1.2 The Transmission Line Pulser

A transmission line of any geometry of length ℓ and characteristic impedance Z_0 makes a pulse forming line (PFL), which when combined with a closing switch S makes the simple transmission line pulser shown in Figure 3.2.

Initially, switch S is open while a battery or power supply charges the transmission line or PFL through a charging resistor to a DC voltage V_0. From (3.2), the voltage along the line is composed of a forward-going wave V_I and a backward-going wave V_R. When the switch S closes, the incident wave V_I, with a peak voltage of $1/2V_0$, travels toward the load, while the reverse-going wave V_R, also with a peak voltage of $1/2V_0$, travels in the opposite direction. The incident wave V_I, then, supplies a voltage of $1/2V_0$ for a time determined by the electrical length of the transmission line T_T to the load. The reverse-going wave V_R travels along the transmission line for a duration T_T and then reflects from the high impedance of the voltage source, and becomes a forward-going wave traveling toward the load with peak voltage $1/2V_0$ and duration T_T. The two waves add at the load to produce a pulse of amplitude $1/2V_0$ and pulse duration $T_p = 2T_T$. In these circumstances, the load voltage may be expressed as

$$V_L(t) = \begin{cases} V_I = \dfrac{V_0}{2}, & t_1 < t < T_T \\ V_R = \dfrac{V_0}{2}, & T_T < t < 2T_T \end{cases} \tag{3.17}$$

For a transmission line initially charged to a voltage V_0, with a matched load $Z_L = Z_0$,

Peak power (W):

$$P = I \cdot V = \frac{1}{Z_0}\left(\frac{V_0}{2}\right)^2 = \frac{V_0^2}{4Z_0} \tag{3.18}$$

Stored energy (J):

$$E_S = 1/2CV_0^2 = \frac{V_0^2}{2Z_0} \cdot T_t \quad (3.19)$$

A transmission line pulser constructed with a length of commercially available coaxial cable restricts the maximum operating voltage, the impedance, and the stored energy per unit length. A wider range of parameters V_0, Z_0, E_s, and T_t may be attained by constructing a pulse forming line.

If the load is not matched ($Z_L \neq Z_0$) to the characteristic impedance of the pulse forming transmission line, the forward-going wave V_I will not be fully absorbed by the load and a reflected wave will travel along the transmission line in the opposite direction. A standing wave will be established until all the energy initially stored in the lossless transmission line is dissipated by the load.

3.2 Coaxial Pulse Forming Lines

A two-electrode coaxial PFL, shown in Figure 3.3, consists of two nested coaxial cylinders with radii R_1 and R_2 with the annular space filled with a dielectric material.

The coaxial pulse forming line can be used in a transmission line pulser similar to that described earlier. That is, the PFL is charged to a peak voltage V_0, a preset self-triggered or externally triggered fast spark gap S fires and delivers a pulse of $V_0/2$ into a load $Z_L = Z_0$. The maximum voltage V_0 to which the PFL can be charged depends on the dielectric strength E_{BD} of the dielectric and on the geometry of the PFL.

3.2.1 Basic Design Relations

In most cases, coaxial cables are made from nonmagnetic materials where $\mu = \mu_0$.

Inductance/unit length (H/m):

$$\tilde{L} = \frac{\mu}{2\pi} \ln\left(\frac{R_2}{R_1}\right) \quad (3.20)$$

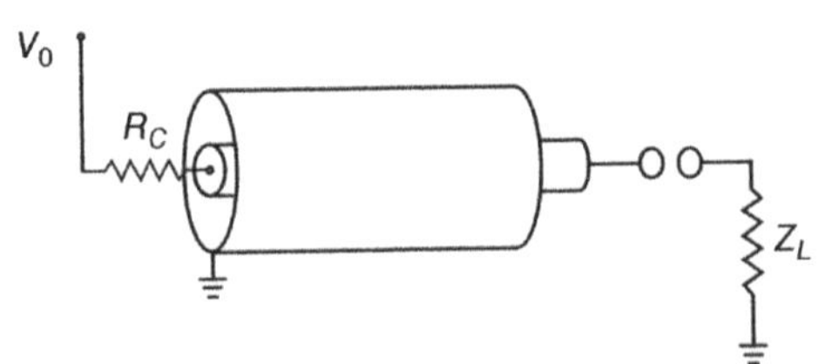

Figure 3.3 A coaxial pulse forming line may be made with concentric cylinders, filled with a high dielectric strength insulating media and configured into a pulser. The pulse width is twice the electrical length, which depends on the relative dielectric constant of the insulator and the physical length ℓ.

Since $\mu = \mu_0 \cdot \mu_r$ and $\mu_0 = 4\pi \times 10^{-7}\ H/m$, (3.20) reduces to

$$\tilde{L} = (0.2\mu_r) \cdot \ln\left(\frac{R_2}{R_1}\right)\ \mu\text{H/m} \tag{3.21}$$

Capacitance/unit length:

$$\tilde{C} = \frac{2\pi\varepsilon}{\ln(R_2/R_1)} \tag{3.22}$$

Since $\varepsilon = \varepsilon_r \cdot \varepsilon_0$ and $\varepsilon_0 = 8.85 \times 10^{-12}\ F/m$, (3.22) reduces to

$$\tilde{C} = \frac{(6.28)\varepsilon_r}{\ln(R_2/R_1)}\ \text{pF}/m \tag{3.23}$$

Characteristic impedance:

$$Z_0 = \frac{1}{2\pi}\sqrt{\frac{\mu}{\varepsilon}}\ln\left(\frac{R_2}{R_1}\right) \tag{3.24}$$

which reduces to

$$Z_0 = 60 \times \sqrt{\frac{\mu_r}{\varepsilon_r}} \cdot \ln\left(\frac{R_2}{R_1}\right)\Omega \tag{3.25}$$

The velocity of propagation along the transmission line depends only on its material properties:

$$v_{pp} = \frac{1}{T_t} = \frac{1}{\sqrt{\mu_o\mu_r\varepsilon_o\varepsilon_r}} = \frac{c}{\sqrt{\varepsilon_r\mu_r}} = \frac{30}{\sqrt{\varepsilon_r\mu_r}}\frac{cm}{ns} \tag{3.26}$$

Electric field at any radius:

$$E_r(r) = \frac{V_0}{r\ln(R_2/R_1)} \tag{3.27}$$

Maximum charging voltage:

$$V_0(\max) = E_{BD}R_1\ln\left(\frac{R_2}{R_1}\right) \tag{3.28}$$

where E_{BD} is the dielectric strength of the insulator.

Electric field on inner conductor:

$$E_M = \frac{V_0}{R_1 \ln (R_2/R_1)} \tag{3.29}$$

In the absence of an insulator with a polarity-dependent breakdown voltage, the electric field on the inner conductor is the maximum.
Stored energy/unit length:

$$E_s = \frac{\pi \varepsilon_0 \varepsilon_r V_0^2}{\ln (R_2/R_1)} \tag{3.30}$$

Dielectric radial resistance (Ω m):

$$R = \int_{R_1}^{R_2} \frac{\rho dr}{2\pi r} = \frac{\rho}{2\pi} \ln \left(\frac{R_2}{R_1}\right) \tag{3.31}$$

where ρ is the radial resistivity of dielectric medium (Ω m).
Self-discharge time constant (s):

$$T_d = RC = \rho\varepsilon \tag{3.32}$$

Large energy densities may be obtained by using insulating materials with high dielectric constants. Water, with a relative dielectric constant $\varepsilon_r \sim 81$, is often used to increase the energy density of the PFL. However, the resistivity of the water requires the water-insulated PFL to be pulse charged. The PFL charging time is chosen to be small compared to the self-discharge time constant $\rho\varepsilon$, where ε is the dielectric constant and ρ is the resistivity of the dielectric medium.

3.2.2 Optimum Impedance for Maximum Voltage

The electric field at any radius r within the annular gap in a two-electrode coaxial PFL, with inner conductor of radius R_1 charged to a voltage V_0 and outer conductor radius R_2 held at ground potential, is given by

$$E(r) = \frac{V_0}{r \ln (R_2/R_1)}, \quad \text{for} \quad R_1 \leq r \leq R_2 \tag{3.33}$$

From Equation 3.29, the maximum field E_M occurs at the inner electrode. Rearranging terms,

$$V_0 = E_M R_1 \ln \left(\frac{R_2}{R_1}\right)$$

The value of R_2/R_1 that optimizes the inner conductor voltage occurs when $dV_0/dR_1 = 0$, yielding

$$\ln\left(\frac{R_2}{R_1}\right) = 1, \quad \text{or} \quad \frac{R_2}{R_1} = e \tag{3.34}$$

and has an impedance of

$$Z_{opt}^{voltage} = \frac{1}{2\pi}\sqrt{\frac{\mu_0}{\varepsilon_0 \varepsilon_r}} = \frac{60}{\sqrt{\varepsilon_r}} \tag{3.35}$$

Substituting the values of $\varepsilon_r = 81$ for water and $\varepsilon_r = 2.4$ for oil in the above relation yields the following values of optimum impedances:

$$Z_{opt}^{water} = 6.7\,\Omega \quad \text{and} \quad Z_{opt}^{oil} = 38.7\,\Omega$$

These impedances do not account for the polarity dependence of water, which, when accounted for, may yield a different optimum impedance.

3.2.3 Optimum Impedance for Maximum Energy Store

For a two-electrode coaxial PFL, the energy stored per unit length is given by

$$E_s = 1/2CV_0^2$$

Using the capacitance per unit length given in Equation 3.22 and the expression for the maximum voltage from Equation 3.29,

$$E_s = \pi \varepsilon_0 \varepsilon_r E_M^2 R_1^2 \ln\left(\frac{R_2}{R_1}\right) \tag{3.36}$$

The ratio of radii that optimizes the stored energy per unit length is obtained from $dE_s/dR_1 = 0$,

$$\frac{dE_s}{dR_1} = \pi \varepsilon_0 \varepsilon_r E_M^2 \left[2R_1 \ln\left(\frac{R_2}{R_1}\right) - R_1\right] = 0$$

Solving, the radial ratio that optimizes the stored energy per unit length is

$$\frac{R_2}{R_1} = \sqrt{e} \tag{3.37}$$

3.3 Blumlein PFL

This pulse forming line geometry, named after its inventor A.D. Blumlein, enables the production of an output pulse into a matched load equal to the

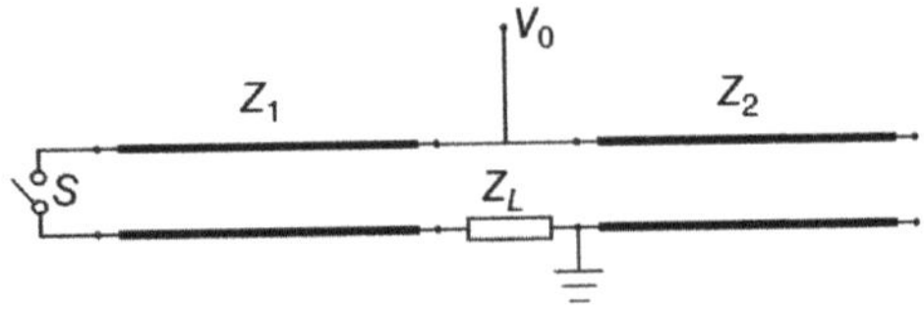

Figure 3.4 The two-conductor representation of a Blumlien circuit. The transmission lines with characteristic impedances Z_1 and Z_2 are each charged to a voltage V_0. The load is matched when $Z_L = Z_1 + Z_2$. Under matched conditions, when the switch S closes, the traveling waves combine to produce a voltage with magnitude V_0.

original charging voltage of the line. This is, of course, in contrast to the simple pulse line that delivers half of the charging voltage across a matched load. Blumleins are widely used in pulsed power technology and come in a variety of geometries. The most prevalent Blumlein geometry is coaxial cylinders because the distribution of internal electric fields is more easily managed due to the inherent symmetry but numerous examples of planar geometries may be found.

The Blumlein circuit is illustrated in Figure 3.4 using the generic two-conductor transmission line representation [1]. The two transmission lines are each charged to a common voltage V_0. During the charge cycle, both lines are terminated in open circuits. When the switch S is closed, traveling waves are established along both transmission lines and a voltage appears across the load impedance Z_L. A matched load is resistive and equal to the sum of the characteristic impedance of the two lines. The Blumlein circuit's main advantage is that it delivers a pulse with a peak amplitude of V_0 into a matched load. This advantage is illustrated in Figure 3.5, which compares the matched load waveform from a transmission line (a) and a Blumlein (b).

The Blumlein circuit lends itself to a variety of circuit topologies and both coaxial and planar geometries are common. A Blumlein circuit can be configured using two identical commercial transmission cables with characteristic impedance Z_0 and length ℓ, as shown in Figure 3.5b. The cables are connected to form three independent electrodes of the Blumlein. Electrode 1 is formed by the center conductors of the cables, electrode 2 is the braid of one cable, and electrode 3 is the braid of the other cable. When the closing switch S is triggered into conduction, the energy stored in the charged coaxial cables is discharged into the load Z_L, connected across electrodes 1 and 3. If each transmission line is charged to a voltage V_0, the Blumlein delivers a flat-top output pulse with amplitude of V_0 and duration of $(2\ell/v_{pp})$ to a matched resistive load, $Z_L = 2Z_0$. There is a time delay (ℓ/v_{pp}) between the appearance of output waveform and moment of closure of switch S. This is unlike in a two-electrode PFL, where there is no time difference between closure of switch and appearance of output waveform. For the same charging voltage, the output voltage in a Blumlein is twice that of a two-electrode PFL.

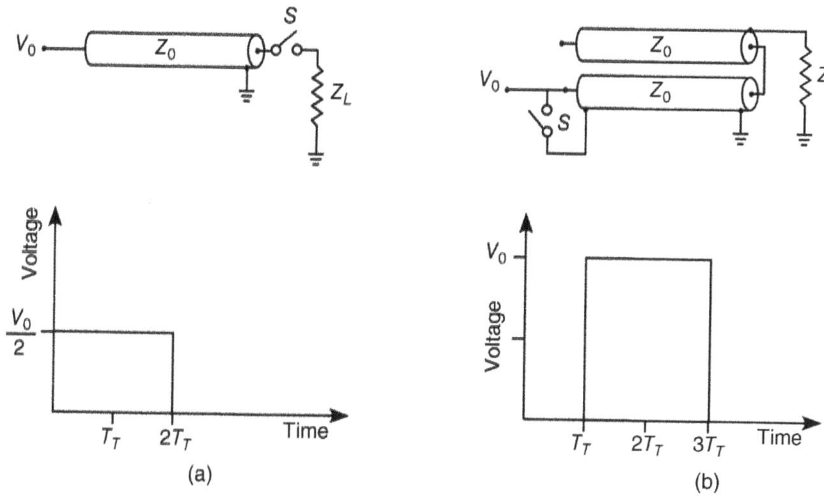

Figure 3.5 A transmission line pulser (a) produces a pulse with amplitude ½V_0 and duration of twice the electrical length T_T when triggered at $t = 0$ into a matched load. In contrast, a Blumlein circuit (b) constructed on coaxial cables produces a pulse with peak amplitude V_0 and duration of twice the electrical length T_T of a single cable when triggered at $t = 0$ into a matched load. Note the Blumlein is delayed by one electric length.

3.3.1 Transient Voltages and Output Waveforms

The transient voltages and output waveforms are analyzed in Figure 3.6 using traveling waves. When the switch S is closed, a voltage transient V_{I} of magnitude corresponding to V_0 starts traveling from terminals 1–2 toward the load end terminals 3–4, with a velocity corresponding to v_{pp} and the associated current transient I_I with magnitude V_0/Z_1, where Z_1 represents the characteristic impedance of line 1. This transient voltage is negative and negates the original line voltage as it travels. When this transient reaches the load end terminals 3–4 at a time corresponding to its electrical length $T_t (= \ell / v_{pp})$, it sees a terminating impedance of $(Z_2 + Z_L)$. If $(Z_2 + Z_L) > Z_1$, a portion of this voltage corresponding to V_R is reflected back into line 1. When the reflected voltage V_R reaches the closed switch end (terminals 1–2), it gets rereflected with the opposite polarity. At the terminals 3–4, a part of the transient voltage corresponding to V_T is transmitted into the terminating impedance comprised of Z_2 and Z_L. The part of V_T delivered into the terminals 5–6 sets up traveling transients into line 2, accompanied by rereflections at the terminals 7–8. The part of V_T delivered into the load terminals 3–5 results in the output voltage. The magnitudes of the various transients described above are as follows:

$$|I_I| = V_0/Z_1 \tag{3.38}$$

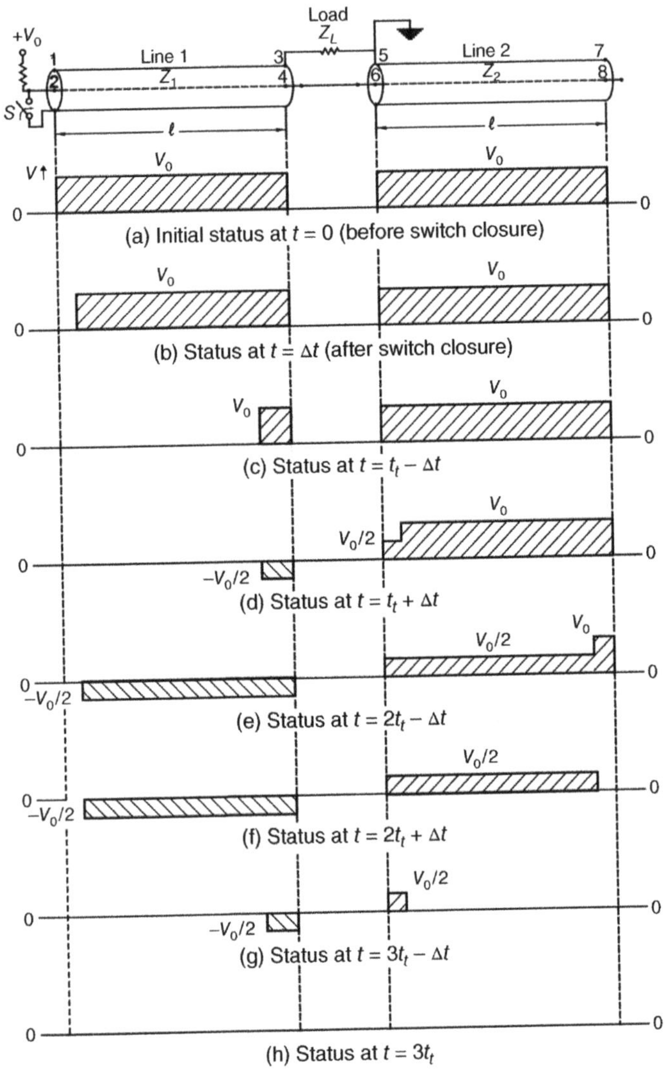

Figure 3.6 The relative positions of transient voltages along a Blumlein for the typical case where $Z_1 = Z_2 = Z_0$, and $Z_L = 2\ Z_0$.

$$|I_R| = \left[\frac{(Z_2 + Z_L) - Z_1}{(Z_2 + Z_L) + Z_1}\right] \cdot I_I \tag{3.39}$$

$$|I_T| = I_I - I_R \tag{3.40}$$

$$|V_R| = Z_1 \cdot I_R \tag{3.41}$$

$$|V_T| = (Z_2 + Z_L) \cdot I_T \tag{3.42}$$

The output voltage across the load Z_L:

$$V_L = V_T \frac{Z_L}{Z_L + Z_2} \tag{3.43}$$

For the usual conditions in a conventional Blumlein, when $Z_1 = Z_2 = Z_0$, and $Z_L = 2Z_0$, the above equations assume the following forms:

$$|I_R| = \frac{I_I}{2} \tag{3.44}$$

$$|I_T| = \frac{I_I}{2} \tag{3.45}$$

$$|V_R| = \frac{I_I}{2} \cdot Z_0 = \frac{V_0}{2} \tag{3.46}$$

$$|V_T| = \frac{I_I}{2} .(3Z_0) = \frac{3}{2} V_0 \tag{3.47}$$

Output voltage across the load,

$$V_\ell = \frac{3V_0}{2} \times \frac{2}{3} = V_0 \tag{3.48}$$

The voltage across Z_L appears between the intervals T_t and $3T_t$ with a pulse duration of $2T_t$.

3.3.2 Coaxial Blumleins

A coaxial Blumlein may be assembled by nesting cylinders, as shown in Figure 3.7. The intermediate cylinder is charged to a voltage V_0 (typically negative) determined by the breakdown stress of dielectric E_{BD}. In the most common implementation, the Blumlein switch, illustrated by a spark gap in Figure 3.7, is connected between the inner and intermediate cylinders. When the switch fires, an output pulse with a flat top is delivered to a matched load, connected between the center and outer cylinders. The load is matched when $Z_L = Z_1 = Z_2$.

The difficulty of making electrical contact with the center conductor is a complication of coaxial Blumleins. In order to charge the PFL formed between the intermediate and center conductors, an inductor L_i is connected between the center conductor and the grounded outer conductor. The inductor ideally acts as a short circuit during the charge cycle and an

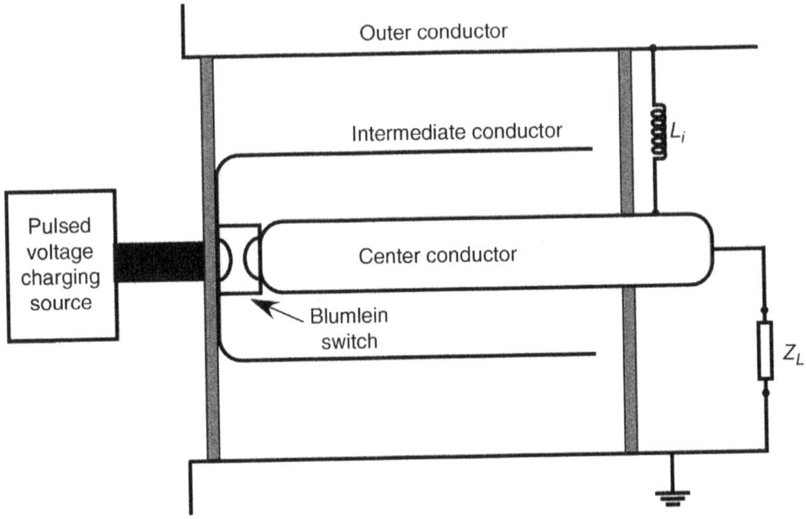

Figure 3.7 Blumlein constructed with three nested coaxial cylinders. Note the inductor L_i is needed to provide a ground path for charging the circuit. The load is matched when $Z_L = Z_1 + Z_2$.

open circuit during the discharge cycle. Thus, for the inductor to act like a short circuit during the charge cycle, the inductive impedance $\omega L_i \ll Z_0$, which is approximately

$$\frac{L_i}{T_{ch}} \ll Z_0 \tag{3.49}$$

During the much shorter discharge cycle, the inductance must act like an open circuit:

$$\frac{L_i}{T_p} \gg Z_0 \tag{3.50}$$

Because of the electrical access issues, the coaxial Blumlein must be pulse-charged. Typical values for the charge time $T_{ch} \sim 1\ \mu s$ and $T_p \sim 100$ ns, so that and inductance of a few microhenries is sufficient.

An alternative topology for the coaxial Blumlein is shown in Figure 3.8, where the switch is placed between the intermediate and outer conductors. Although most coaxial Blumleins are constructed as in Figure 3.7, the accessibility of the switch in the topology shown in Figure 3.8 has an advantage if multiple Blumleins are to be simultaneously fired.

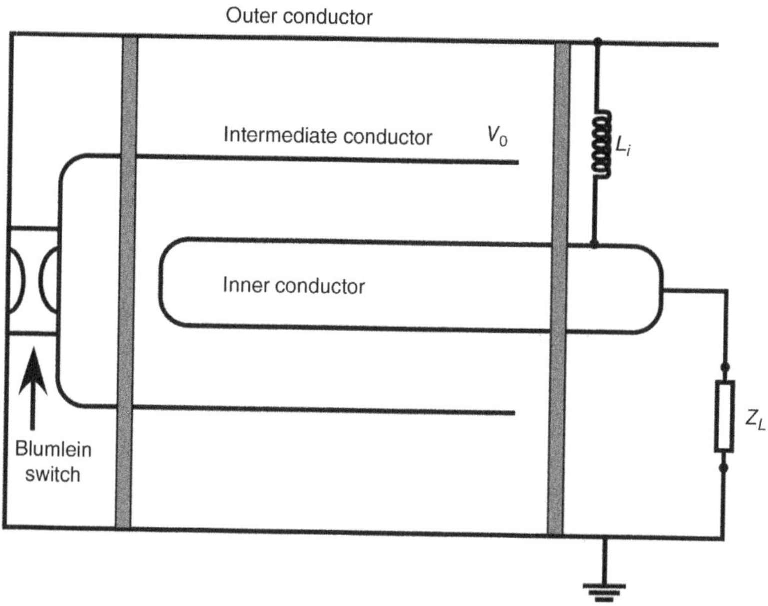

Figure 3.8 Blumlein constructed with three nested coaxial cylinders with the switch located between the intermediate conductor and the grounded outer conductor. This circuit is electrically identical to the Blumlein of Figure 3.7.

3.3.3 Stacked Blumlein

A configuration [2] of an assembly of (n) Blumleins, which are charged in parallel to a voltage V_0 at the input end and discharged in series and provide a pulsed voltage of (nV_0) on a terminating load impedance of impedance Z_ℓ ($=2nZ_0$), is shown in Figure 3.9. In practical systems, the output voltage will be $K\,(nV_0)$, where $K<1$ accounts for the efficiency factor. Similarly, the effective impedance in practical systems will be $K'(2nZ_0)$, where $K'>1$ accounts for contribution from extraneous impedances. Artificial Blumleins [3] comprised of lumped pulse forming networks or PFL-type Blumleins [4] can be used in stack configurations. Each Blumlein consists of an assembly of three metallic plates or strips having width w and length ℓ, separated by a dielectric of thickness h. The dielectric medium between metal strips is usually an assembly of multiple layers of insulating sheets such as Kapton, Mylar, or polyethylene, impregnated with mineral oil in the case of low-energy systems. For high-energy systems, robust dielectric mediums such as insulating oil or deionized water are used. Repetitive operation is determined by the switch [5]. The closing switches $S_1, S_2, \ldots, S_n$ could either be a single switch or separate switches made to fire simultaneously by an external trigger.

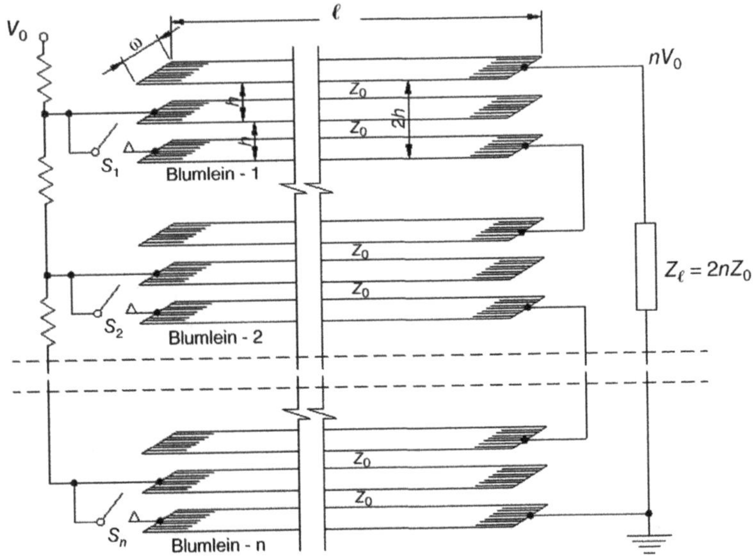

Figure 3.9 A stacked Blumlein and its associated parameters.

The values of characteristic impedance Z and pulse duration T_p of the stacked Blumlein are given by the following:

$$Z = 2n \cdot Z_0 = 2n \cdot \left(\frac{h}{w}\right)\sqrt{\frac{\mu}{\varepsilon}} \tag{3.51}$$

$$T_p = \frac{2\ell}{v_{pp}} = 2 \cdot \ell\sqrt{(\mu_0\varepsilon_0\varepsilon_r)} \tag{3.52}$$

The theoretical output voltage of a stacked Blumlein is (nV_0), as indicated in Figure 3.9, but coupling between the Blumleins reduces the output dramatically from this maximum.

A five-electrode Blumlein developed by Prestwich for driving a dedicated load of two symmetrical field emission diodes for generation of electron beams is shown in Figure 3.10 [6]. It comprises an assembly of five metal plates, 1.2 m wide and 2.4 m long, a plate separation of 10 cm, and immersed in mineral oil. The plates 2 and 4 are pulse charged to a voltage of $(-V_0)$ by a Marx generator. When the multichannel low-jitter oil dielectric rail-type spark gaps S_1 and S_2 are fired simultaneously, the Blumlein delivers twin pulses with parameters of $(V = V_0,\ T_p = 2\ell/v_{pp},\ I = V_0/2Z_0)$, which are carried by a triplate (A, B, C) transmission line and applied to a diode load, comprised of two cathode electrodes (K, K) and a central anode (A). Six such Blumlein assemblies, located

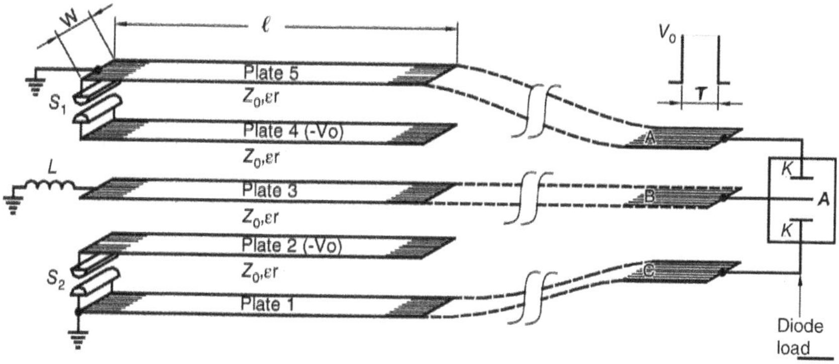

Figure 3.10 A five-electrode Blumlein charged in a single polarity.

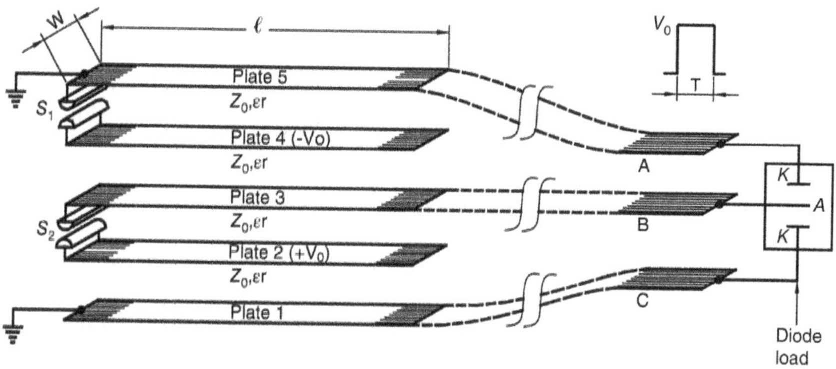

Figure 3.11 A five-electrode Blumlein with bipolar charging.

radially around a centrally placed diode, discharge in parallel to produce an electron beam with parameters of 3 MV, 800 kA, and 24 ns.

An improved version, employing a bipolar charge, is shown in Figure 3.11. The Blumlein is capacitively balanced, which obviates the need for connecting the inductor L between the central plate and ground. In this figure, a reflex diode configuration is shown. The upper diode produces a relativistic electron beam (REB) that enters into the lower diode through the anode foil A. In the lower diode, the REB is subjected to a decelerating force. As a result, the REB oscillates around the anode foil, producing microwave emission.

3.4 Radial Lines

Radial transmission lines are uncommon except in large pulsed power devices where energy is delivered to a centrally located load by multiple feeds distributed around the outer radius. Radial transmission lines also form the basis for coreless induction cavities.

Radial transmission lines are comprised of two disks separated in space at a relative angle. The geometry is shown in cross section in Figure 3.12. One terminal of the radial lines is located at the inner radius R_i and the other at the outer radius R_o so that when viewed from above, a radial transmission line looks like a disk. The direction of wave propagation is in the radial direction. Radial lines share the properties of other transmission lines: The voltage and current at any (radial) point is related to its characteristic impedance. That is, if a voltage is applied at the center—between the lower conductor (a disk) and the outer conical section, a wave will travel radially outward at a velocity determined by the medium of the line and remain constant in shape. If the line extends to infinity, the pulse never returns but if the line has a finite radius R_o (an open circuit), then the pulse is reflected and travels back toward the center.

Assume that one of the conductors is a disk located along the horizontal axis and the other conductor slopes at an angle θ relative to the first, as shown in Figure 3.12. The characteristic impedance Z_0 of this radial transmission line is

$$Z_0 = \frac{1}{2\pi}\sqrt{\frac{\mu}{\varepsilon}} \cdot \ln\left(\frac{1}{\sqrt{\tan\theta}}\right) \tag{3.53}$$

If the conductors are mildly diverging (the angle Θ is small), then (3.53) can be approximated by

$$Z_0 = \frac{1}{2\pi}\sqrt{\frac{\mu}{\varepsilon}} \cdot \tan\theta \tag{3.54}$$

Thus, for radial lines with mildly diverging angles, the characteristic impedance is constant for a given slope.

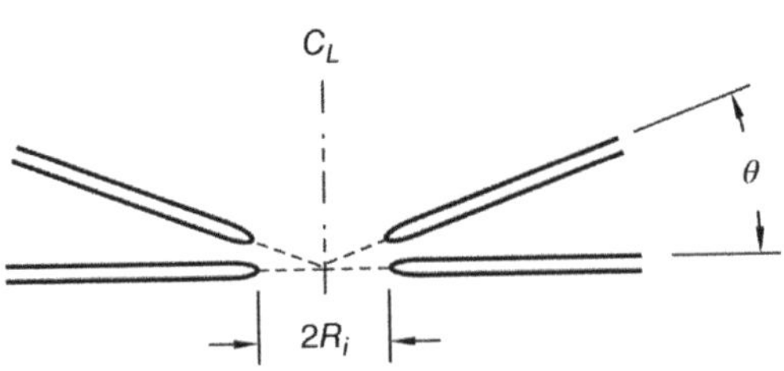

Figure 3.12 The geometry of a radial transmission line consists of two conductors with inner radius R_i and outer radius R_o. The characteristic impedance is determined by their separation angle θ.

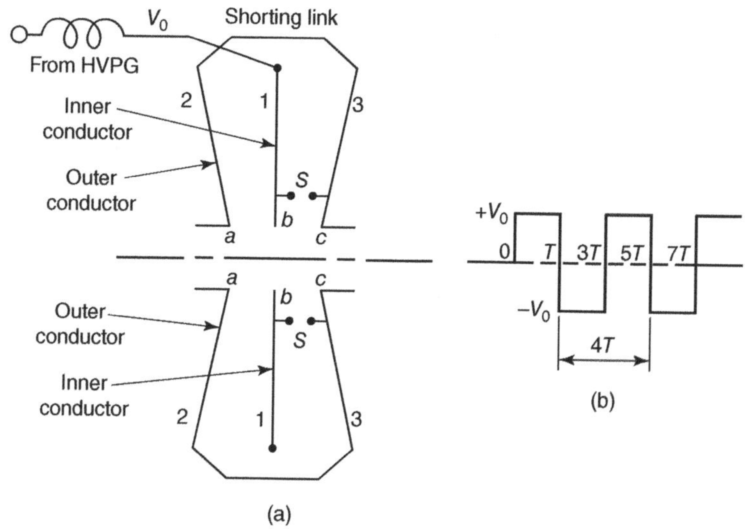

Figure 3.13 A radial line PFL in part (a) and the open-circuit voltage waveform (b) at terminal *ab*. Radial transmission lines are often used as part of a cavity.

Radial transmission lines are used extensively in coreless induction cavities where the cavities are formed by two radial transmission lines with a common conductor, each terminated in an open circuit, as shown in Figure 3.13. The outer conductor is shaped to provide a matched transition between the two waves. Thus, a wave traveling outward on one line propagates without reflection through the transition and travels radially inward, down the other line. A low-inductance, azimuthially symmetric switch is placed on one of the lines at the inner radius.

A radial line configuration is shown in Figure 3.13 [7]. It consists of an inner conductor 1 in the form of a flat circular disc with a central hole, surrounded on either side by slightly conical outer conductors 2 and 3 to form a radial tapered line of constant impedance [8]. When the inner conductor is pulse charged to a voltage V_0 and the spark gap S fires at a time $t = 0$, the transient voltage pulse of magnitude $(-V_0)$ generated at the short-circuited gap (bc) starts traveling toward the open-circuited gap (ab). If the transit time required for the transient voltage to travel from terminals (bc) to terminals (ab) is T, a symmetrical periodic square pulse of magnitude V_0 and period $4T$ appears at the output gap (ab) after a time delay of T subsequent to switch closure. Radial lines are compatible for generating electric fields for beam acceleration in a vacuum medium and are widely used in large linear induction accelerators [9–11]. Radial lines may be oil filled to obtain higher energy density, with the oil–vacuum interface located at $(a–b–c)$.

3.5 Helical Lines

For rectangular pulses with pulse durations on the order of a few microseconds, the physical lengths of the transmission lines become prohibitive – on the order of 1 km for a 10 μs pulse. The characteristic feature of a helical line storage element is that it provides a large PFL electrical length with a reasonable physical length [12]. The helical lines can be connected in parallel for enhanced energy storage or in series for enhanced output voltage. For a PFL based on coaxial cables, the helical line storage element can be realized by replacing the outer braid by a helically wound conductor and for one based on coaxial cylinders, the helical line storage element can be realized by helical winding of a conductor of circular or rectangular cross section over a metallic cylinder, as shown in Figure 3.14. The conductor helix and the inner cylinder should be appropriately insulated for withstanding the PFL charging voltage without breakdown either by insulation puncture through the insulator thickness $(R_2 - R_1)$ or by surface breakdown over the extended insulation length $(A + B)$. The spacing between the turns and hence the pitch of the helix should be sufficient to prevent electrical breakdown turn-to-turn. The values of Z_0 and T for the helical line storage element shown in Figure 3.10 are given by

$$Z_0 = \frac{(R_2 - R_1)}{d_h}\sqrt{\frac{\mu}{\varepsilon}} \tag{3.55}$$

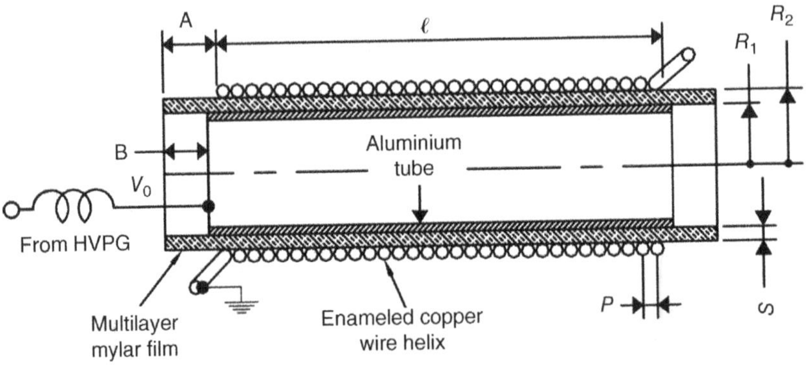

Figure 3.14 Schematic of helical line storage element charged to a voltage V_0 with a high-voltage pulse generator (HVPG). (Reproduced with permission from Ref. [12]. Copyright 2001, AIP Publishing LLC.)

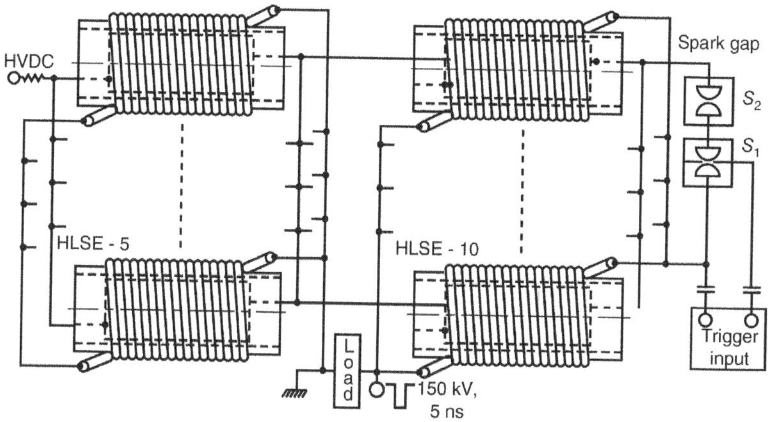

Figure 3.15 Configuration of a helical line storage element-based Blumlein generator [12]. (Reproduced with permission from Ref. [12]. Copyright 2001, AIP Publishing LLC.)

where $(R_2 - R_1)$ is the insulation thickness and d_h is the diameter of helical conductor in case of circular cross section. The parameter d_h is replaced by the width of helical conductor in case of rectangular cross section:

$$T_T = \frac{\ell'}{v_{pp}} = \frac{(2\pi R_{avg})n_h}{v_{pp}} \tag{3.56}$$

where ℓ' is the unfolded length of wire helix, R_{avg} is the average radius of helical winding, and n_h is the total number of turns in the conductor helix.

A schematic of a PFL using helical line storage elements in a paralleled Blumlein configuration with ratings of 150 kV, 200 A, and 5 μs and used for generation of an electron beam-controlled CO_2 laser is shown in Figure 3.15. This helical line storage element-based pulse generator employs a total of 10 helical line storage elements, which is equivalent to five Blumleins in parallel, each Blumlein employing two helical line storage elements. A hybrid insulation system comprised of Mylar and mineral oil is employed.

3.6 PFL Performance Parameters

A well-designed PFL should be able to provide optimum parameters of operating voltage, peak power ($P = IV$), and stored energy ($P{\cdot}T = I{\cdot}V{\cdot}T$). The PFL should be able to operate reliably for a large number of shots with minimum maintenance problems. The above objectives can be achieved by carrying out a proper design by considering electrical breakdown, dielectric constant, self-discharge time constant, optimum impedance, and charging source.

3.6.1 Electrical Breakdown

When charging voltage to a PFL is gradually increased, electrical breakdown will take place at a particular voltage, when the maximum electric field produced in the PFL exceeds the dielectric strength (or breakdown stress) E_{BD}, that is, $E_{max} > E_{BD}$. The role that local electric field enhancements play cannot be overstated. In cylindrical geometries, breakdown is more likely to occur at the inner electrode, where the local field is geometrically enhanced. In stripline or triplate PFLs, the edges of metal electrodes prove to have the highest field enhancements. However, electrical breakdown may initiate at either electrode, especially if the dielectric has a polarity effect.

The maximum voltage to which a PFL can be charged depends on the geometry and field distribution. It is prudent to note that dielectric strength values are statistical in nature, and thus the operating voltage should be lower than the maximum voltage. The maximum voltage is given by the following relations:

a) Cylindrical geometry:

$$V^{cyl}(\max) = E_{BD} \cdot R_1 \cdot \ln\left(\frac{R_1}{R_2}\right) \tag{3.57}$$

b) Stripline geometry:

$$V^{stripline}(\max) = E_{BD} \cdot d \tag{3.58}$$

c) Spherical geometry:

$$V^{sph}(\max) = E_{BD} \cdot R_1\left(1 - \frac{R_1}{R_2}\right) \tag{3.59}$$

In general, the lower the operating voltage relative to the maximum voltage, the higher the reliability of the PFL.

To prevent field enhancement at unintended locations, it is important to observe the following criteria are met during fabrication: (a) round off the edges in a metal strip or plate; (b) provide resistive grading at the interface between edges and dielectric (for this purpose, electrolytes like $CuSO_4$ or conducting polymers are generally used); (c) assemble in a dust-free environment; (d) impregnate solid insulating films with mineral oil under hot and vacuum conditions for effective degassing; (e) using degassed and filtered oil; and (f) use a large number of thin insulating sheets rather than a lesser number of thick insulating sheets, especially in case of stripline dielectrics. Various solid insulating materials such as polymethyl-methacrylate (PMMA), polystyrene, polyethylene, epoxy, or even ceramic can be used as insulator supports. The metallic parts should be noncorrosive and should not interact with the dielectric. The mechanical design of PFLs is very important because of the large mechanical forces induced by the high currents. Common electrode materials for PFLs are aluminum and stainless steel.

3.6.2 Dielectric Strength

3.6.2.1 Solid Dielectric

The PFLs based on striplines [13–16] make use of solid insulating sheets or foils as major dielectrics and are employed when there is requirement for high breakdown voltage and low impedance. A variety of materials such as polyethylene, polycarbonate, polypropylene, polyester, and so on are generally used. Careful fabrication and processing techniques are required to reduce field intensification caused at the sharp edges of metal electrodes, defects, and voids embedded in dielectric or gas bubbles trapped at the dielectric–electrode interface. Generally, the breakdown starts at these high field points and propagates into the interior of the dielectric. High performance can be achieved by assembling the PFLs in vacuum and immersing them in deionized water or copper sulfate ($CuSO_4$) solution or insulating oil for reduction in field enhancement at edges and voids. Special attention should be paid to the breakdown by surface flashover, whose value depends on the duration of voltage stress (t_f), area of electrodes (A), and the insulation environment.

The intrinsic breakdown strength for direct puncture through the thickness of a solid insulator is much higher than surface flashover strengths. With proper precautions, it is possible to achieve working intrinsic breakdown stresses of 300 MV/m. The working life of a PFL in terms of the number of discharges is dependent on the partial discharges inside voids in the insulation system.

3.6.2.2 Liquid Dielectric

Oil and water dielectrics are commonly used for PFLs when high charging voltages are required because of their high electrical breakdown strength and because they can provide large accurately shaped dielectric volumes by conforming to metal electrodes. Numerous studies have been conducted on liquid dielectrics to understand the process of electrical breakdown [17–24] and are discussed in Chapter 9. Others have detailed the application of liquids as insulators in pulsed high-voltage equipment [25–31]. Unfortunately, the mechanism of insulation failure is not as well understood in liquids as it is in gases.

Empirical expressions relating the average electrical breakdown field to the time of stress for a given area are extremely important in designing pulsed power technology. Much of the initial scaling was done by Martin's group at Aldermaston, England, in the 1960s, and several nuances of the formulas (such as the definition of the time of stress) can be traced to that early work. Many efforts have been made to increase the accuracy of the scaling laws for a variety of experimental conditions [32–43]. The following are the commonly accepted empirical formulas used in pulsed power design today.

The scaling relations take the following form:

$$F \cdot t^a \cdot A^b = k \tag{3.60}$$

where

t the time in μs; the pulse is above 63% of its peak voltage,
A the area in cm^2, which is stressed,
$F = \frac{V_p}{d}$ the average electric field at breakdown in MV/cm,
V_p the peak voltage across the electrodes,
d the distance between electrodes,

and

k is a constant that depends on the insulator.

In these expressions, the electric field F is the *average* electric field, which has a roughly 50% probability of producing breakdown. It is important that the proper units (MV/cm, cm^2, and μs) are used in order to get the correct predicted value. It is also important to note that these relations contain large uncertainties and deviations of a few percent are commonplace. The effect of field enhancement is accounted for by multiplying the F value by Martin's field enhancement factor α [13]:

$$\alpha = 1 + 0.12\sqrt{\frac{F_{Max}}{V_p/d} - 1} \tag{3.61}$$

where F_{Max} is the local electric field.

3.6.2.2.1 Transformer Oil

Transformer oil, with a relative permittivity $\varepsilon_r \sim 2.3$, is commonly used in high-impedance PFLs, where the characteristic impedance is between 10 and 100 Ω. Under pulsed charges, oil is considered as self-healing as water, since the decomposition products, other than bubbles, have a minor effect. Gaseous bubbles trapped in liquid dielectrics can dramatically lower its dielectric strength and hence it is useful to carry out some form of degassing during the first fill of the liquid dielectric. Transformer oil has an advantage that, unlike water, it is also an insulator under DC conditions.

From experimental data, under uniform field conditions, Smith [33] proposed for transformer oil

$$F \cdot t^{1/3} \cdot A^{0.075} = 0.48 \tag{3.62}$$

This expression can be used for large areas and additional insight can be found in Refs [34,35]. The breakdown in transformer oil is sometimes considered independent of polarity and sometimes dependent on polarity. The issue is poorly understood. In a uniform field, breakdown can be initiated from either polarity. For low-field enhancements, the effect of polarity is small, but for strong field enhancements (>1.5), the polarity effect is greater than 1.5 and less than 2. For strong field enhancements, a formula in use is [36]

$$F \cdot t^{1/3} \cdot A^{0.075} = 0.677 \quad (3.63)$$

The purity of transformer oils has a small effect on its pulsed breakdown strength. Oil in nearly continuous use has been used for decades in large facilities with only periodic filtering. Water is detrimental to its performance and mixing should be avoided. Typical pulsed power operation occurs at field levels between 100 and 350 kV/cm depending on the actual conditions.

3.6.2.2.2 Water

Water is an attractive insulator for PFLs because its high permittivity and good electrical breakdown strength allow for high-energy storage densities and low impedances. Coaxial PFLs with impedance of less than 10 Ω are exclusively made with deionized water. Pressurizing the water has also been found to improve the hold-off voltage as well as the time on stress, but this effect disappears at large areas [37]. Water has the well-known advantage of self-healing after accidental breakdown.

The breakdown strength of water shows a strong polarity effect of about a factor of 2 that can be used to great advantage. For a nonuniform field geometry in water, a negative voltage on the electrode in field region gives a higher dielectric strength than a positive voltage, which can be used to an advantage in designing PFLs. The overall diameter is minimized when the inner conductor is charged to a negative voltage.

The empirical scaling of water has been extensively studied and the best estimates have established validity regimes depending on the stressed areas:

1) For uniform fields, Smith and Champney [38–40] established

$$\text{Positive}\,polarity \quad F^{+} \cdot t^{1/3} \cdot A^{0.075} = 0.3 \quad (3.64)$$

for experimental data taken in the range of $0.02 \leq A \leq 2000\ \text{cm}^2$

$$\text{Negative}\,polarity \quad F^{-} \cdot t^{1/3} \cdot A^{0.075} = 0.6 \quad (3.65)$$

for experimental data taken in the range of $0.25 \leq A \leq 90\ \text{cm}^2$ and somewhat enhanced fields.

2) For larger areas, Eilbert and Lupton [41] re-examined the experimental data taken by Smith and colleagues at AWE, along with data from Shipman at

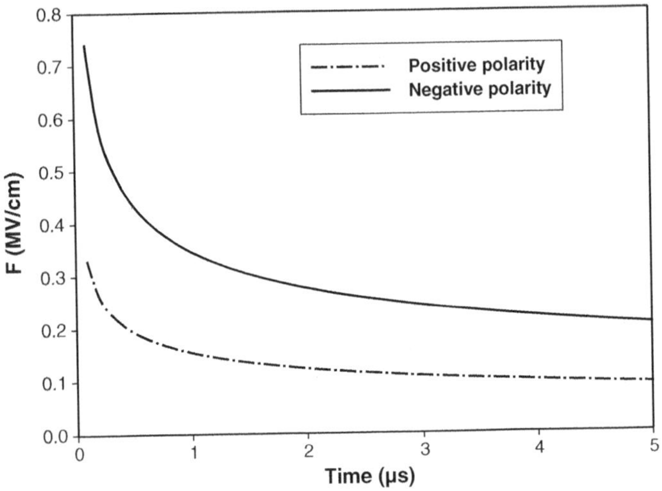

Figure 3.16 The Eilbert–Lupton formula for water breakdown, calculated at a fixed area of 1000 cm^2, shows a strong polarity effect of approximately 2.

NRL, and determined the best fit to data for water uniformly stressed over large areas in the range of $100 \le A \le 5000\ \text{cm}^2$ is

$$\text{Positive polarity} \quad F^+ \cdot t^{1/3} \cdot A^{0.058} = 0.230 \tag{3.66}$$

$$\text{Negative polarity} \quad F^- \cdot t^{1/3} \cdot A^{0.069} = 0.557 \tag{3.67}$$

The average breakdown fields, as given by the Eilbert–Lupton formulation, are plotted in Figure 3.16, where the polarity dependence is about a factor of 2, consistent with the Smith and Champney relations.

3) For nonuniform fields, where one electrode is highly enhanced and the other is either a large electrode or a plane, Martin [42] proposed the "point–plane" formula:

$$F \cdot t^{1/2} = 0.09 \tag{3.68}$$

The point–plane formula is independent of polarity with no area dependence. Recently, Stygar *et al.* [43] have examined a wide range of data and concluded that a more accurate point–plane result is

$$F \cdot t^{1/3} = 0.135 \pm 0.009 \tag{3.69}$$

and suggested that the prudent design criterion is

$$Ft^{1/3} \le 0.108, \quad \text{for} \quad A \ge 10^4\ \text{cm}^2 \tag{3.70}$$

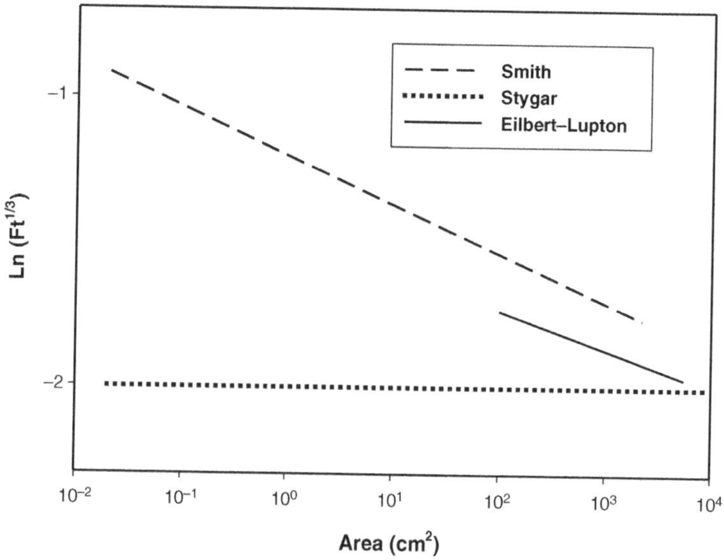

Figure 3.17 The Smith and Eilbert–Lupton empirical scaling laws for water breakdown are plotted within the ranges of the data. Stygar's point–plane relation is plotted for reference.

This criterion is attractive since it scales with the same time dependence as uniform fields. Smith (I.D. Smith, private communication) suggested plotting the empirical formulas in the format shown in Figure 3.17. On this plot, the Eilbert–Lupton scaling is seen to be conservative, consistent with the lack of reported failures due to water breakdown.

As many workers, including Stygar, correctly point out, the Eilbert-Lupton formula must cease to be applicable at some large area since it suggests as $A \to \infty, F \to 0$. Instead, the electrical breakdown field F must come to a limiting value. Figure 3.17 suggests that the limiting value may be the point–plane formula and recognizing for unenhanced electrodes breakdown may never occur at fields that low. The point–plane scaling formulas for water, plotted in Figure 3.18, indicate the Stygar relation yields a higher threshold field for a given effective time. This implication is important because operating large PFLs at higher electric fields can lead to considerable cost savings in large pulsed power machines.

3.6.2.2.3 *Gaseous Dielectric*

Gases may be used as the insulating medium in PFLs [42]. Common gaseous insulators are nitrogen, air, and SF_6. Table 3.1 shows the relative dielectric strengths of commonly used gaseous insulators. Note that mixtures of SF_6 tend

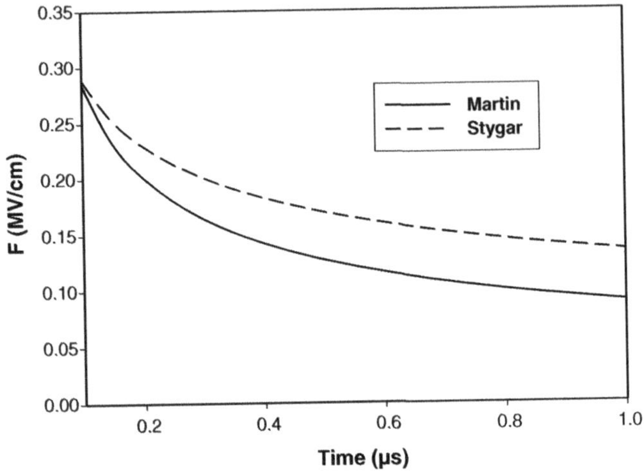

Figure 3.18 The Martin and Stygar point-plane formulas plotted over the time range 0.1–1 μs.

Table 3.1 Relative dielectric strengths.

Air	1
Nitrogen	1
Hydrogen	0.5
SF_6	2.5
30% SF_6 + 70% air (by volume)	2

to retain a large portion of its insulating strength even if diluted with another gas such as air or nitrogen.

The electric field distribution in a PFL may be considered as uniform in many cases such as parallel plate or even cylindrical geometries if the field enhancement factor is low. However, failure may be dictated by divergent electric fields introduced by local field enhancements, such as fasteners, during construction.

3.6.2.2.4 Uniform Fields

Dry atmospheric air has electrical breakdown strength E_{BR} of approximately 30 kV/cm for gaps in the centimeter range. This value rises dramatically for small gaps in the millimeter range. The electrical breakdown value is strongly dependent on the gas pressure, and the gas pressure is often a design parameter.

In uniform electric fields, the pressure dependence of air's electrical breakdown strength, valid up to gas pressures of 10 atm, is

$$F = 24.6p + 6.7\sqrt{\frac{p}{d}} \tag{3.71}$$

where

F is the average electrical breakdown field in kV/cm,
p is the gap pressure in atmospheres,

and

d is the gap spacing in centimeters.

The electrical breakdown strength is also a function of applied pulse duration. In air at 1 atm, the breakdown strength for a 10 ns pulse is a factor of ~2.3 greater than the DC value. For pulse durations in the microsecond regime, the DC values for electrical breakdown may be used since it differs by only a few percent.

Gaseous PFLs often use SF_6 for its excellent electrical breakdown strength of 90 kV/cm at atmospheric pressure. Electrical breakdown is pressure-dependent, and at pressures of 10 atm fields of 600 kV/cm can be achieved with proper electrode materials [42]. Various aspects of SF_6 breakdown and its dependence on gap spacing, gas pressure, and contaminating particles are discussed in Chapter 8.

3.6.2.2.5 Divergent Fields

For very divergent fields, such as those found around points or edges, the time dependence for the onset of breakdown is much more critical than in regions with uniform electric fields. An approximate relation is [42]

$$F^{\pm}(dt)^{1/6=k^{\pm}p^{n}} \tag{3.72}$$

where

F is the average electrical breakdown field in kV/cm,
p is the gap pressure in atmospheres,
d is the gap spacing in centimeters,
t is the pulsewidth in microseconds,

and n and $k^{\pm}$ are constants that depend on voltage polarity and gas type. The values of n and $k^{\pm}$ for air and SF_6 are shown in Table 3.2. The dependency on pressure is valid in the range 1–5 atm. For air, the time dependence disappears for pulses longer than a microsecond in the negative polarity and for pulses longer than hundreds of microseconds in the positive polarity [42].

Table 3.2 The empirical constants n and $k^{\pm}$ for air and SF_6 gas.

	Air	SF_6
k^+	22	44
k^-	22	72
n	0.6	0.4

3.6.3 Dielectric Constant

In general, a high relative dielectric constant ε_r results in a lower value of Z_0 and hence enhances the capability of PFL in terms of high output current and large peak power. It also increases the electrical length T_t and reduces the velocity of propagation v_{pp} resulting in a compact length of PFL for a given pulse duration. The wave transit times calculated from the above formulas for water and mineral oil are 30 and 5 ns/m, respectively. This explains why deionized water, having a very high relative dielectric constant, is the preferred dielectric in many PFLs. The relative dielectric constants for liquid and solid dielectrics commonly used in pulsed power technology are listed in Table 3.3.

Other common names for PMMA are Lucite and Acrylic. Ethylene glycol and water may be mixed in varying quantities to construct a liquid dielectric having a dielectric constant between 38 and 80.

3.6.4 Self-Discharge Time Constant

Equation 3.17 indicates that to have a high value of T_d, it is desirable to have high value of ρ for the dielectric. T_d of a liquid-filled PFL can be experimentally determined [40] by noting the time required for the initial voltage V_0 to fall to (V_0/e), after disconnecting the charging source. The value of T_d

Table 3.3 Relative dielectric constants ϵ_r.

Liquids		Solids	
Water	81	Polyethylene	2.25
Mineral oil	2.4	Mylar	2.92
Castor oil	4.7	Polystyrene	2.56
Glycerine	44	PTFE	2.1
Ethylene glycol	38	PMMA	2.5

decreases with increases in the temperature of the dielectric, since the ionic carrier density and its mobility increase rapidly with temperature. In case of water, the ionic carrier density contributed by impurities and thermal dissociation of liquid molecules need to be reduced by using resin column deionizers. A high discharge time T_d relaxes the fast charging requirements for the PFL and eliminates the need for expensive fast high-voltage power generators. An interesting finding [25] has been the substantial increase in resistivity (ρ) in chilled, deionized liquid dielectrics wherein the charging times of hundreds of microseconds can be employed even for water. In addition, the chilled liquid dielectrics also exhibit increased values of ε_r and E_{BD}, both of which are beneficial to PFL performance.

3.6.5 PFL Switching

A PFL's output switch requires careful selection and a number of switches have been developed to meet specific needs. In general, the output switch is required to hold off the full charging voltage of the PFL (twice the output voltage) as well as a sufficiently large current-carrying capability. It should have a fast rise time so that it can faithfully transmit the time profile of the pulse generated by the PFL. Connections between components can add a significant inductance to the system and must be considered when choosing a switch technology.

A small jitter allows the PFL to be discharged very close to its charge voltage, allowing maximum energy to be extracted. Moreover, with little energy left in the system, the lifetime of all components is extended. In many switch technologies, and spark gaps in particular, erosion of the switches is an important consideration. Moreover, the lifetime or maintenance schedule for the pulsed power system is typically determined by that of the switch. A small switching jitter enables synchronous switching when multiple PFLs are connected in parallel or when multiple, discrete spark gaps are operated in parallel.

Overall inductance is minimized when the PFL geometry is considered when choosing a discharge switch technology. For instance, multiple switch sites or rail gap switches (discussed in Chapter 4) are used in a planar geometry. In Figure 3.9, a parallel plate Blumlein is switched with a rail gap. The rail gap provides multiple arcs along with width of the PFL and contributes minimum additional inductance to the system. In high-voltage, high-peak-power systems with large coaxial PFLs, a rimfire laser-triggered gas switch, discussed in Chapter 4, is often used to provide low-inductance, low-jitter switching at very high voltages. Typical jitters are of a few nanoseconds, with switching voltages up to 6 MV with an expected lifetime of a few hundred shots. The lifetime is strongly dependent on the operating voltage.

3.7 Pulse Compression

A PFL charged directly from a DC source, as shown in Figure 3.19, has a limited energy storage density $W_1 = \varepsilon E^2$ because the maximum electric field is limited to the static breakdown value. Pulse charging of the PFL takes advantage of the increased electric field strength of insulators with decreased time of stress. This process of transferring stored energy from one stage to the next in a successively fast sequence is known as pulse compression and forms the basis for most pulsed power architectures. Even though energy is lost in the transfer, the decrease in charge time is sufficient to raise the peak power.

Typically, the first stage is a DC-charged Marx bank or HV transformer with a discharge time longer than a few microseconds. The energy storage stages may be capacitors, but for peak voltages above 100 kV, pulse forming lines are practical. Pulse compression allows the use of higher voltages and higher electric fields than could be achieved otherwise. Higher voltages allow more energy to be stored. For a given PFL output voltage, a high E_{BD} value enables reduction in the transverse dimensions of PFL, resulting in lower impedance and producing higher peak currents. The time dependence of insulation breakdown can be exploited to increase the energy density by charging the PFL quickly. Table 3.4 quantifies the increase in electric field for oil- and

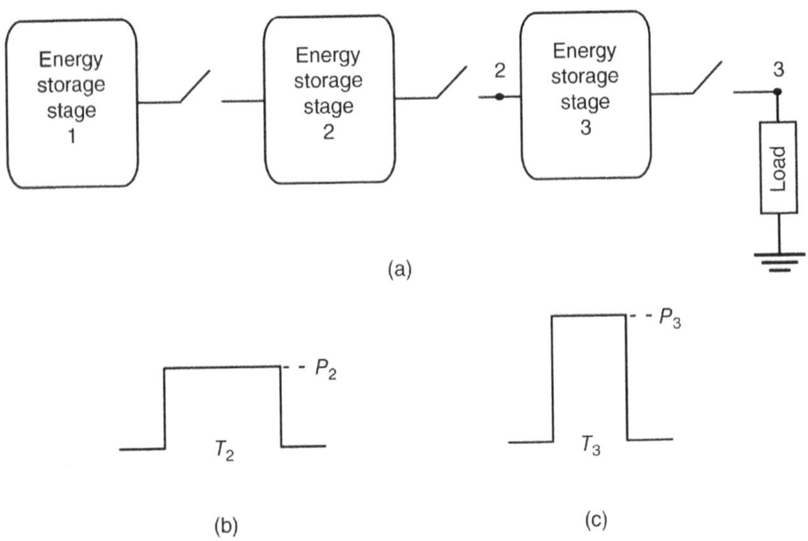

Figure 3.19 A generic pulse compression scheme is shown in part (a), where stage 1 is assumed to be charged from a DC source and discharges its energy into stage 2 when the switch is closed. Pulse compression produces pulses with higher peak powers and shorter durations, as shown in part (b) for node 2 and in part (c) for node 3.

Table 3.4 Time-dependent E_{BD} values for liquid dielectrics.

T_p	A (cm^2)	$t^{1/3}$	$A^{0.075}$	Breakdown field, F (kV/cm)			
				Water		Water	Oil
				$k^+ = 0.6$	$k^- = 0.6$	Point–plane	$k^+ = 0.6$
1 μs	1000	1	1.679	180	360	135	285
100 ns	1000	0.464	1.679	385	770	290	615
1 μs	10^4	1	1.995	150	300	135	240
100 ns	10^4	0.464	1.995	325	650	290	520

water-filled dielectrics, which can be achieved by reducing the charging time from 1 μs to 100 ns. The value of the electrical breakdown field increases by a factor of two.

3.7.1 Intermediate Storage Capacitance

The self-inductance of large energy storage Marx banks or transformers limits its discharge time to a few microseconds. The relatively long timescale limits the advantage of pulse charging on the breakdown field. Fast charging of a PFL is accomplished through the use of an intermediate storage capacitor (ISC). The ISC then is charged by the Marx in several microseconds and discharged through its output switch to the PFL. The PFL is charged by the ISC in a fraction of the discharge time of the Marx, allowing the PFL to be charged to a higher voltage for given dimensions than could be achieved by charging directly from the primary energy storage. The ISC may take any form suitable for the PFL and more than one ISC can be used. The majority of the system's pulse compression is achieved in the first two stages and power multiplication in successive stages is not dramatic.

3.7.2 Voltage Ramps and Double-Pulse Switching

Instead of the (1-cos) waveshape typical of one capacitor charging another, the output voltage of a PFL can approximate a linear ramp if the PFL is charged on a timescale that is comparable to its output pulse duration and then switched before the charge is complete. This allows the next stage of pulse compression to have greatly reduced time of stress on the dielectric. This interplay between the charge time and the length of the PFL can be extended to increase the peak power to the load without significantly increasing the voltage by using a technique known as "double-bounce switching" (DBS).

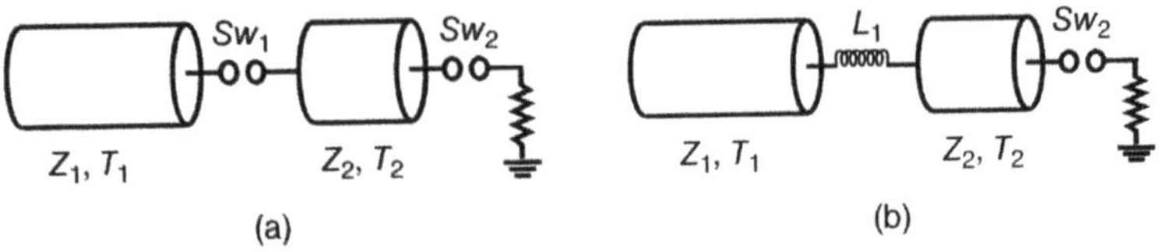

Figure 3.20 The circuit diagram for double bounce switching (a) before the output switch Sw_1 discharges the first PFL to charge the second PFL, and (b) after Sw_1 is fired. The switch Sw_1 is designed to have a significant inductance.

"Double-bounce switching" was first used on EAGLE and uses multiple reflections along the PFL transmission line to optimize the output power relative to voltage [44]. For a required peak power at the load, double-bounce switching provides the advantage of lower electrical stress on the ISC and PFL, and the output switch loss and jitter of the PFL are reduced. The essential element of DBS is to choose the length of the PFL so that the wave reflects from both the open switch at the output end and the switch inductance at the input end [45,46]. In Figure 3.20, an ISC has an electrical length T_1 and impedance Z_1 and transfers its energy to a PFL with an electrical length T_2 and impedance Z_2 when the switch Sw_1 is closed. The ISC output switch Sw_1 is designed to be sufficiently inductive so that the closed switch is represented by an inductor L_1.

The ISC charges the PFL through the inductor L_1 with a time constant τ_c given by

$$\tau_c = \frac{L_{12}}{(Z_1 + Z_2)} \tag{3.73}$$

If the electrical length of the PFL, T_2, is chosen so that

$$T_2 \sim \tau_c \text{ or } L_{12} = T_2(Z_1 + Z_2) \tag{3.74}$$

then the PFL acts as a transmission line. The charging waveform from the ISC [46] given by

$$V_{inc}(t) = \frac{V_0}{2}\left(1 - e^{-t/\tau_c}\right) \tag{3.75}$$

is reflected upon reaching S_{w2}, which is an open circuit. The wave travels back toward the ISC until it reaches S_{w1}, which is still conducting current and inductive. This backward-going wave is partially reflected from the inductance L_1 and travels back toward the PFL output switch S_{w2}. The partially reflected wave is given by

$$V_{ref}(t) = \frac{V_0}{2} \times \frac{t}{\tau_c} \times e^{-t/\tau_c} \tag{3.76}$$

The peak value of the reflected wave occurs at $t = \tau_c$ and is given by

$$V_{ref}^{max} = V_{ref}(\tau_c) = \frac{1}{e} \times \frac{V_0}{2} = (0.37) \times \frac{V_0}{2} \tag{3.77}$$

If switch S_{W2} were fired just as the reflected wave reached the length of the PFL, the voltage waveform would be given by

$$V(t) = \frac{V_0}{2} \cdot \left[1 - e^{-t/\tau_c} e^{-2T_2/\tau_c} + \frac{t}{\tau_c} e^{-t/\tau_c} \right] \tag{3.78}$$

The voltage waveform has a peak voltage 37% higher than would be achieved in conventional pulse compression. However, if S_{W2} is closed before the arrival of the wave reflected from L_1, the full charging waveform is transmitted to the load, and the voltage continues to rise. By switching the PFL before the arrival of the reflected waveform $V_{ref}(t)$, the voltage on the PFL and its output switch are minimized, without sacrificing peak power at the load. The maximum increase in power using double-bounce switching P_{DBS} is proportional to the square of the maximum increase in voltage, yielding

$$\frac{P_{DBS}}{P} = (1 + 0.37)^2 = 1.88 \tag{3.79}$$

The power delivered using double-bounced switching can be increased by 88% over conventional pulse compression techniques.

The value of L_1 is crucial because if it is too large, the ISC will not charge the PFL, and if it is too small, the reflected wave will not be partially rereflected back onto the PFL. The ratios of the electrical lengths T_1/T_2 and impedances Z_1/Z_2 are significant and have an optimal ratio between 1.5 and 2. Additionally, the transmission line geometry of the PFLs should end abruptly to minimize the stray capacitance, which would filter the reflected waves.

Double-bounce switching serves to increase the power to the load without a proportional increase of voltage on the PFL. The traveling wave does not increase the PFL voltage significantly, and allows the PFL and its output switch voltages to be reduced without reducing the load voltage.

3.7.3 Pulse Compression on Z

An example of multistage pulse compression is the module design for the Z Machine, used to study high energy density physics and inertial confinement fusion [47–49]. Sandia National Laboratories' Z consists of 36 modules combined in parallel to drive a current pulse with peak amplitude ~25 MA into a single, centrally located load [47]. The pulse compression circuit, shown in Figure 3.21, uses both voltage ramping and double-bounce switching.

Primary energy storage is an oil-insulated Marx generator arranged in 30 stages charged to a maximum voltage of ±100 kVDC and has the following

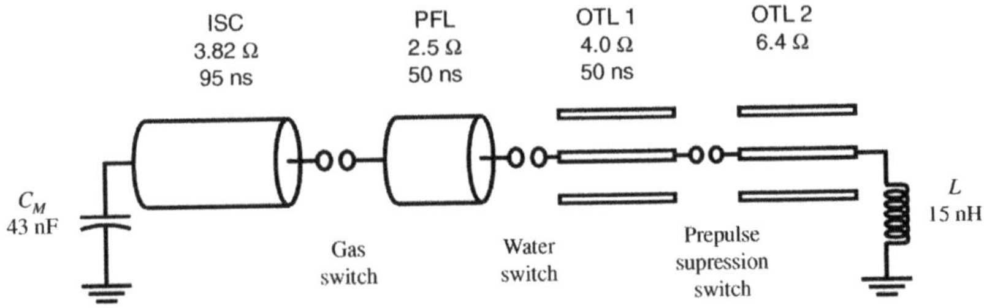

Figure 3.21 The circuit diagram for the *ZR* circuit.

erected parameters: $C_M = 43\,\text{nF}$, $L_M = 11\,\mu\text{H}$, an equivalent series resistance $R_M = 1.4\,\Omega$, and a peak current of 180 kA at ±90 kV charge. The Marx charges a 24.8 nF ISC with a time to peak voltage of 1.4 μs, but the ISC is normally switched before peak. The ISC is a coaxial, water insulated transmission line with an electrical length of 86 ns and an impedance of 3.46 Ω. Insulation external to the ISC is provided by immersing the ISC in the same oil tank as the Marx banks. A laser-triggered gas switch (LTGS) [50] transfers the energy from the ISC to a PFL with a 15 nF capacitance in approximately 250 ns. The PFL is also a water-insulated, coaxial transmission line, but with an electrical length of 43 ns and an impedance of 2.86 Ω. The negatively charged PFL is switched via water switches into a 4.2 Ω, 30 ns water-insulated triplate output transmission line (OTL). The OTL is terminated in an additional set of water switches whose primary purpose is to reduce the prepulse arising from the capacitive coupling across the main water switches. The 15 nH inductor in Figure 3.21 represents the initial inductance in vacuum for typical loads [49]. Final pulse shaping is achieved by synchronizing the modules with each other by triggering the laser triggered gas switch in each module. In doing this, complex waveshapes at high currents have been achieved [49]. The increase in instantaneous peak power from the pulse compression is shown in Figure 3.22. Note the peak power on the OTL is the same as on the PFL, but later in time.

The evolution of the voltage in the *Z* pulse compression circuit is shown in Figure 3.23. In the following experimental waveforms, the Marx charges the ISC in ~1.2 μs. The electrical length of the ISC is short compared to the charge time, and its output LTGS is triggered before peak, resulting in the ISC having a ramped output waveshape. Note the ISC voltage is higher than the Marx voltage because of the ringing gain, even though the ISC is switched before reaching its peak voltage.

The electrical length of the PFL is short compared to that of the ISC and double-bounce switching for current generators is utilized. In Figure 3.23, the voltage waveforms are those that are measured at the output. Evidence of double-bounce switching can be extracted by computing the forward-going voltage waveforms $V_{For}(t)$ from colocated current and voltage measurements,

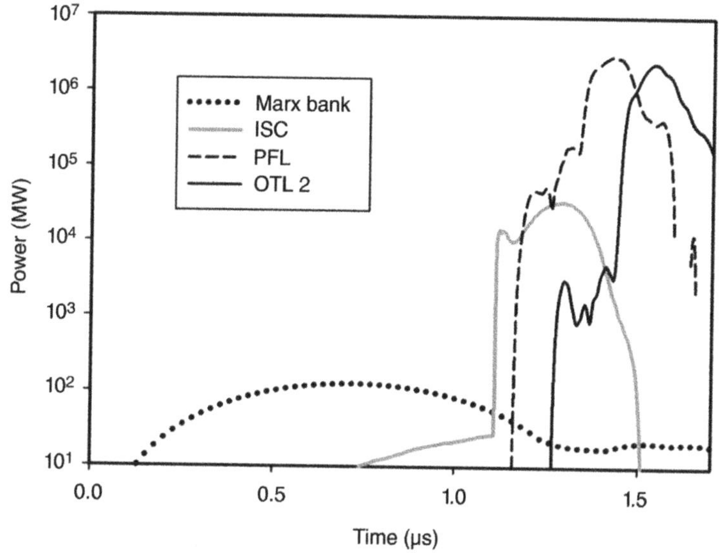

Figure 3.22 The increase in instantaneous peak power in the pulse compression circuit.

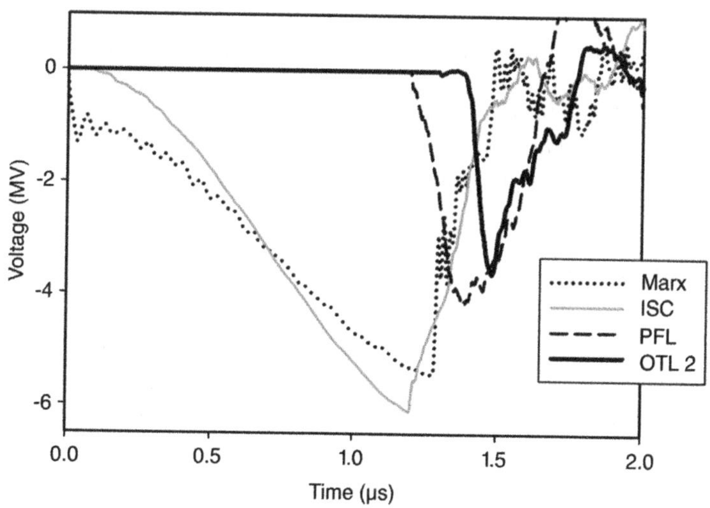

Figure 3.23 The measured voltage on each of the components.

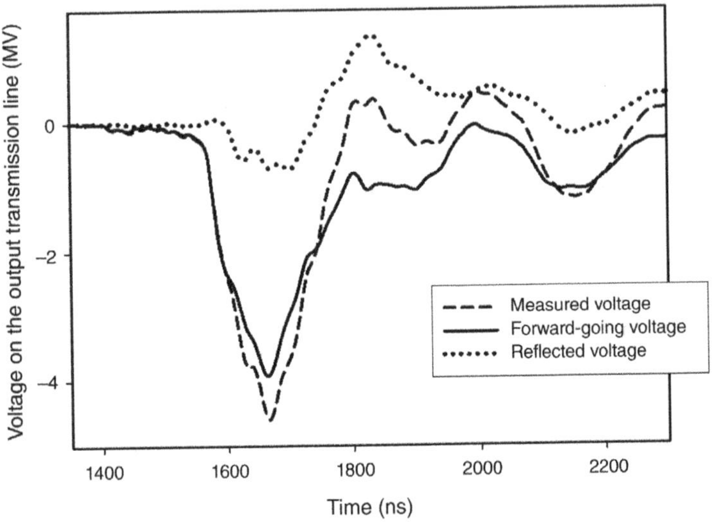

Figure 3.24 The forward-going voltage wave can be extracted from the measured waveforms and shows the "fast rise followed by a ramp" waveshape characteristic of double-bounce switching.

from

$$V_{For}(t) = \frac{V(t) + Z_{PFL} I(t)}{2} \tag{3.80}$$

where $V(t)$ and $I(t)$ are the measured waveforms at the output of the PFL transmission line and Z_{PFL} is the PFL impedance. The forward-going voltage has less amplitude than the measured voltage and shows the characteristic waveshape, suggested in Equation 3.78, of a fast-rising portion followed by a ramped voltage, as is shown in Figure 3.24.

3.8 Design Examples

Example 3.1

A coaxial pulse forming line using water dielectric ($\varepsilon_r = 80$) is 1.0 m long. The inner and outer conductor diameters are 150 and 300 mm, respectively. Calculate the following parameters: (i) impedance, (ii) pulse duration, and (iii) maximum charging voltage. Assume the charging time for the PFL to be 10 times the pulse duration.

Solution

Using Equations 3.7 and 3.8, we obtain the PFL basic parameters as follows:

$$\tilde{L} = \frac{\mu_0}{2\pi} \ln\left(\frac{D_2}{D_1}\right) = \frac{4\pi \times 10^{-7}}{2\pi} \ln\left(\frac{300}{150}\right) = 140\,\text{nH/m}$$

$$\tilde{C} = \frac{2\pi\varepsilon_0\varepsilon_r}{\ln(D_2/D_1)} = \frac{2\pi \times 8.85 \times 10^{-12} \times 80}{\ln(300/150)} = 6.4\,\text{nF/m}$$

The values of impedance and pulse durations are as follows:
Impedance of PFL:

$$Z_0 = \sqrt{\frac{L}{C}} = \sqrt{\frac{140}{6.4}} = 4.7\Omega$$

Pulse duration:

$$T = 2lT_t = 2l\sqrt{LC} = (2)(1)\sqrt{(140)(6.4)}\,\text{ns} = 60\,\text{ns}$$

To determine the electrical breakdown field F, the area is calculated from the inner conductor dimensions and converted to units of cm^2.

$$A = \pi D_1 \ell = \pi(15\,\text{cm})(100\,\text{cm}) = 4700\,\text{cm}^2$$

The Eilbert–Lupton formula is appropriate. The charge time t is 600 ns = 0.6 μs, yielding

$$F^+ = \frac{0.230}{(0.6)^{1/3}(4700)^{0.058}} = 0.1671/3\frac{\text{MV}}{\text{cm}} = 167\,\frac{\text{kV}}{\text{cm}}$$

$$F^- = \frac{0.557}{(0.6)^{1/3}(4700)^{0.069}} = 0.369\,\frac{\text{MV}}{\text{cm}} = 369\,\frac{\text{kV}}{\text{cm}}$$

Martin [2] notes that the empirical breakdown formulas are accurate to within ~10%. Prudence dictates that a safety factor be applied to ensure breakdown does not occur. Charging the PFL negative, let $E_{BD} = -185$ kV/cm; the operating voltage is calculated from Equation 3.13,

$$V_0(\max) = E_{BD}R_1 \ln\left(\frac{R_2}{R_1}\right) = \left(-185\frac{\text{kV}}{\text{cm}}\right)\cdot\left(\frac{15}{2}\text{cm}\right)\cdot\ln\left(\frac{300}{150}\right) = 960\,\text{kV}$$

Example 3.2

Determine the time to break down the PFL described in Example 1, when it is charged from a Marx bank with an erected capacitance of 50 nF and an equivalent inductance of 8 μH.

Solution

From Section 3.6.2.2, the charge time is defined as the time when the pulse is above 63% of its peak voltage. Since $C_M \sim C_2$, the PFL is charged by a (1−cos) waveform given in Chapter 1 by

$$V_2(t) = \frac{V_M C_M}{(C_M + C_2)}(1 - \cos \omega t)$$

The maximum voltage on the PFL occurs at a time t_M when $(1-\cos(\omega t_M)) = 2$, and $(\omega t_M) = \pi$ with a peak voltage of

$$V_2(t_M) = V_2^{peak} = \frac{2V_M C_M}{(C_M + C_2)}$$

The time at which the voltage reaches 63% of its peak, t_{63}, is determined by

$$V_2(t_{63}) = (0.63)\frac{2V_M C_M}{(C_M + C_2)} = \frac{V_M C_M}{(C_M + C_2)}(1 - \cos \omega t_{63})$$

$$(0.63) \times 2 = (1 - \cos \omega t_{63})$$

$\cos \omega t_{63} = 0.26$ and $\omega t_{63} = 1.3$.

The charge time t is

$$t = t_M - t_{63} = \frac{\pi}{\omega} - \frac{1.3}{\omega} = \frac{1.84}{\omega}$$

Using the values, $C_M = 8$ nF, $C_2 = 6.4$ nF, $L = 8$ μH,

$$\omega = \sqrt{\frac{(C_M + C_2)}{L C_M C_2}} = \sqrt{\frac{(8\text{nF} + 6.4\text{nF})}{8\text{mH} \times 8\text{nF} \times 6.4\text{nF}}} = 6\ \text{MHz}$$

The charge time t is

$$t = t_M - t_{63} = \frac{1.84}{\omega} = \frac{1.84}{6}\text{ms} = 300\ \text{ns}$$

Example 3.3

If the PFL of Example 1 is filled with oil dielectric ($\varepsilon_r = 2.5$), calculate the following parameters: (i) impedance, (ii) pulse duration, and (iii) maximum charging voltage.

Solution

The PFL inductance does not change, and $L = 140$ nH/m. Repeating the procedure of Example 1,

$$\tilde{C} = \frac{2\pi\varepsilon_0\varepsilon_r}{\ln(D_2/D_1)} = \frac{2\pi \times 8.85 \times 10^{-12} \times 2.4}{\ln(300/150)} = 0.192\,\frac{\text{nF}}{\text{m}}$$

Impedance of PFL:

$$Z_0 = \sqrt{\tilde{L}/\tilde{C}} = 27\,\Omega$$

Pulse duration:

$$T_p = 2\ell \times T_t = 2\ell \times \sqrt{\tilde{L}\tilde{C}} = (2)(1)\sqrt{(140) \times (0.192)}\text{ ns} = 10.4\text{ ns}$$

The electrical breakdown field F for transformer oil is calculated from

$$F = \frac{0.48}{(0.6)^{1/3}(4700)^{0.075}} = 300\,\frac{\text{kV}}{\text{cm}}$$

Using a factor of safety of 1.5,

$$V_0(max) = E_{BD}R_1 \ln\left(\frac{R_2}{R_1}\right) = \left(-200\,\frac{\text{kV}}{\text{cm}}\right) \cdot \left(\frac{15}{2}\text{cm}\right) \cdot \ln\left(\frac{300}{150}\right) \cong 1\text{ MV}$$

Example 3.4

For the PFL of Example 1, calculate the values of the impedance that optimizes voltage for the following dielectrics: (i) mineral oil ($\varepsilon_r = 2.5$); (ii) castor oil ($\varepsilon_r = 4.7$); (c) glycerine ($\varepsilon_r = 40$); and (d) water ($\varepsilon_r = 80$).

(continued)

(continued)

Solution

Optimum impedance is independent of the PFL dimensions:

$$Z_0^{optimum} = \frac{1}{2\pi}\sqrt{\frac{\mu_0}{\varepsilon_0\varepsilon_r}} = \frac{1}{2\pi} \times 238.0 = 38.0\,\Omega \text{ for mineral oil}$$

$$= \frac{1}{2\pi} \times 173.8 = 27.0\,\Omega \text{ for castor oil}$$

$$= \frac{1}{2\pi} \times 39.6 = 9.5\,\Omega \text{ for glycerine}$$

$$= \frac{1}{2\pi} \times 42.0 = 6.7\,\Omega \text{ for water}$$

Example 3.5

A coaxial Blumlein of 4.0 m length has (i) inner cylinder diameter of 800 mm, (ii) intermediate cylinder diameter of 1.0 m, and (iii) outer cylinder diameter of 1.2 m. The Blumlein employs a liquid dielectric having $\varepsilon_r = 4.7$ and $E_{BD} = 110$ kV/cm. Calculate the following parameters: (i) the impedance, (ii) the pulse duration, and (iii) the maximum charging voltage.

Solution

The basic parameters of L_{12} and C_{12} are as follows:

$$L_{12} = \frac{\mu_0}{2\pi}\ln(D_2/D_1) = \frac{4\pi \times 10^{-7}}{2\pi}\ln(1.0/0.8) = 0.446 \times 10^{-7}\ \text{H/m}$$

$$C_{12} = \frac{2\pi\varepsilon_0\varepsilon_r}{\ln(R_2/R_1)} = \frac{2\pi \times 8.85 \times 10^{-12} \times 4.7}{\ln(1.0/0.8)} = 1171.0 \times 0^{-12}\ \text{F/m}$$

Substituting the values L_{12} and C_{12} from (i) and (ii) into Equation 3.3, we get

$$Z_{12} = \sqrt{\frac{L_{12}}{C_{12}}} = \sqrt{\frac{0.446 \times 10^{-7}}{1171.0 \times 10^{-12}}} = 6.17\,\Omega$$

$$L_{23} = \frac{\mu_0}{2\pi}\ln(D_3/D_2) = \frac{4\pi \times 10^{-7}}{2\pi}\ln(1.2/1.0) = 36.5\ \text{nH/m}$$

$$C_{23} = \frac{2\pi\varepsilon_0\varepsilon_r}{\ln(R_3/R_2)} = \frac{2\pi \times 8.85 \times 10^{-12} \times 4.7}{\ln(1.2/1.0)} = 1.4\ \text{nF/m}$$

$$Z_{23} = \sqrt{\frac{L_{23}}{C_{23}}} = \sqrt{\frac{0.365 \times 10^{-7}}{1433.0 \times 10^{-12}}} = 5.04\,\Omega$$

Blumlein Impedance:

$$Z = Z_{12} + Z_{23} = 6.17 + 5.04 = 11.21\,\Omega$$

Pulse duration:

$$T_p = 2\ell T_t = 2\ell\sqrt{L_{12}C_{12}} = 2\cdot 4\times\sqrt{0.446\times 10^{-7}\times 1171.0\times 10^{-12}} = 57.8\text{ ns}$$

Maximum voltage for L_{12} can be obtained from Equation 3.13.

$$V_0(max)\text{ for }L_{12} = E_{bd}R_1\ln(R_2/R_1) = 110\frac{\text{kV}}{\text{cm}}\cdot\frac{80\text{ cm}}{2}\cdot\ln(1.0/0.8) = 982\text{ kV}$$

Similarly, maximum voltage for L_{23} is obtained as

$$V_0(max)\text{ for }L_{23} = E_{bd}R_2\ln\frac{R_3}{R_2} = 110\frac{\text{kV}}{\text{cm}}\cdot\frac{100\text{ cm}}{2}\cdot\ln(1.2/1.0) = 1\text{ MV}$$

Since the breakdown voltage for L_{12} is less than L_{23}, the maximum voltage to which the Blumlein can be charged is $V_0(max) = 982$ kV.

Example 3.6

A three-electrode strip Blumlein with plate dimensions of 1 m wide, 3 m long, with a separation distance of 10 cm between plates, employs an oil dielectric ($\varepsilon_r = 2.5$). Assuming a working stress for oil dielectric of 200 kV/cm, calculate the peak power and energy delivered into a matched load.

Solution

Inductance and capacitance per unit length per two-electrode line:

$$L = \frac{\mu_0\mu_r S}{W} = \frac{(4\pi\times 10^{-7})\times(1)\times(0.1\text{ m})}{(1\text{ m})} = 125.7nH/m$$

$$C = \frac{\varepsilon_0\varepsilon_r W}{S} = \frac{(8.85\times 10^{-12})\times(2.5)\times(1\text{ m})}{(0.1\text{ m})} = 221pF/m$$

Impedance of Blumlein:

$$Z = 2Z_0 = 2\sqrt{\frac{L}{C}} = 2\sqrt{\frac{125.7\times 10^{-9}}{221\times 10^{-12}}} = 47.7\,\Omega$$

(continued)

(*continued*)

Maximum voltage on load:

$$V_0^{max} = 200\frac{\text{kV}}{\text{cm}} \cdot 10\text{ cm} = 2\text{ MV}$$

Peak power on matched load:

$$P_p = \frac{V_0^2}{Z} = \frac{(2\times10^6)^2}{47.7} = 84\text{ GW}$$

Transit time per unit length:

$$T_t = \sqrt{LC} = \sqrt{(125.7\times10^{-9})\cdot(221\times10^{-12})} = 5.27\frac{\text{ns}}{\text{m}}$$

Pulse duration:

$$T = T_t \times 2\ell = \left(5.27\frac{\text{ns}}{\text{m}}\right)\cdot(2\times3\text{ m}) = 31.6\text{ ns}$$

Energy delivered to a matched load:

$$E = P_p \times T = (84\text{ GW})\cdot(31.6\text{ ns}) = 2.65\text{ kJ}$$

Example 3.7

In Chapter 1, it was stated that if $C_M \sim C_2$, the voltage transfer can be done efficiently. Suppose a pulse compression circuit consists of a coaxial PFL, with characteristic impedance Z_2 and one-way transit time T_{T2} charged with a Marx bank with an erected capacitance C_M and peak voltage V_M. Show that if $C_M \sim C_2$, the process is efficient.

Solution

The energy stored in the Marx W_{Marx} is the energy stored in its capacitors:

$$W_{Marx} = \frac{1}{2}C_M V_M^2$$

The PFL is a transmission line that will discharge with a voltage equal to half the charge voltage and pulse duration T_p equal to twice the electrical length.

The energy stored in the PFL W_{PFL} can be calculated from the output pulse parameters, the peak power P_k, and transit time T_T by

$$W_{PFL} = P_k \Delta T = \frac{1}{Z_2}\left(\frac{V_2}{2}\right)^2 (2T_T\ell)$$

Rearranging terms,

$$W_{PFL} = \frac{1}{2}V_2^2\left(\frac{T_T\ell}{Z_2}\right)$$

From Equation 3.3,

$$C_2 = C_{PFL}\ell = \frac{T_T}{Z_2}\ell$$

$$W_{PFL} = \frac{1}{2}C_2V_2^2$$

The ringing gain for the condition

$$C_M \sim C_2 \text{ is } \frac{V_2}{V_M} = \frac{2C_M}{C_M + C_2} \sim 1$$

Thus, for $C_M \sim C_2$, $V_M \sim V_2$ and

$$\frac{1}{2}C_2V_2^2 = W_{Marx} \sim W_{PFL} = \frac{1}{2}C_2V_2^2$$

the energy stored in the Marx is transferred to the PFL efficiently.

References

1 J.H. Crouch and W.S. Risk, A Compact High Speed Low Impedance Blumlein Line for High Voltage Pulse Shaping. *Rev. Sci. Instrum.*, Vol. 43, No. 4, p. 632, 1972.
2 J.C. Martin, Nanosecond Pulse Techniques, in T.H. Martin, A. H. Guenther, and M. Kristiansen, eds., *JCM on Pulsed Power*, Plenum Press, 1996.
3 J.D. Ivers and J.A. Nation, Compact 1 GW Pulse Power Source. *Rev. Sci. Instrum.*, Vol. 54, No. 11, p. 1509, 1983.

4 F. Davanloo, J.J. Coogan, T.S. Bowen, R.K. Krause, and C.B. Collins, Flash X-Ray Source Excited by Stacked Blumlein Generator. *Rev. Sci. Instrum.*, Vol. 59, No. 10, p. 226, 1988.
5 F. Davanloo, C.B. Collins, F.J. Agee, and L.E. Kingsley, Repetitively Pulsed High Power Stacked Blumlein Generators. *Nucl. Instrum. Methods Phys. Res.*, Vol. B99, p. 713, 1995.
6 K.R. Prestwich, HARP: A Short Pulse High Current Electron Beam Accelerator. *IEEE Trans. Nucl. Sci.*, Vol. 22, No. 3, p. 975, 1975.
7 A.I. Pavlovski et al., *Sov. Phys. Dokl.*, Vol. 25, p. 120, 1980.
8 D. Keefe, *Research on High Beam–Current Accelerators*, Report No. LBL–12210, Lawrence Berkeley Laboratory and Physical Dynamics Inc., CA, 1981.
9 A.I. Pavlovski et al., *Sov. Phys. Dokl.*, Vol. 20, p. 441, 1975.
10 R.B. Miller, K.R. Prestwich, J.W. Poukey, B.G. Epstein, J.R. Freeman, A.W. Sharpe, W.B. Tucker, and S.L. Slope, Multistage Linear Electron Acceleration Using Pulsed Transmission Lines. *J. Appl. Phys.*, Vol. 52, No. 3, p. 1184, 1981.
11 C.A. Kapetanakos and P. Sprangle, Ultra High Current Electron Induction Accelerators. *Phys. Today*, Vol. 38, p. 58, 1985.
12 V.P. Singhal, B.S. Narayan, K. Nanu, and P.H. Ron, Development of Blumlein Based on Helical Line Storage Elements. *Rev. Sci. Instrum.*, Vol. 72, No. 3, p. 1862, 2001.
13 A. B. Garasimov et al., High Precision Fast Acting Discharger with a Solid Dielectric, Instrument and Experimental Techniques. p. 446, 1970.
14 W.C. Nunnally et al., Simple, Solid Dielectric Start Switch, in W.H. Bostick, V. Nardi, and O.S.F. Zucker, eds., *Energy Storage, Compression, and Switching*, Plenum Press, New York, p. 429, 1976.
15 P. Dokopoulos, Fast 500 kV Energy Storage Unit with Water Insulation and Solid Multichannel Switching. *Proceedings of the Symposium on Fusion Technology*, Aachen, Germany, p. 393, 1970.
16 J.A. Nation High Power Electron and Ion Beam Generation. *Particle Accelerators*, Vol. 10, Gordon & Breach Science Publishers Inc., pp. 1–30, 1979.
17 D.B. Fenneman, Pulsed High Voltage Dielectric Properties of Ethylene Glycol/Water Mixtures. *J. Appl. Phys.*, Vol. 53, No. 12, p. 8961, 1982.
18 P.K. Watson, W.G. Chadband, and M. Sadeghzadeh-Araghi, The Role of Electrostatic and Hydrodynamic Forces in the Negative-Point Breakdown of Liquid Dielectrics. *IEEE Trans. Dielectr. Electr. Insul.*, Vol. 26, No. 4, pp. 543–559, 1991.
19 A. Beroual, M. Zahn, A. Badent, K. Kist, A.J. Schwabe, H. Yamashita, K. Yamazawa, M. Danikas, W. Chadband, and Y. Torshin, Propagation and Structure of Streamers in Liquid Dielectrics. *IEEE Electr. Insul. Mag.*, Vol. 14, No. 2, pp. 6–17, 1998.

20 E.A. Abramyan and K.A. Kornilov et al., The Development of Electrical Discharge in Water. *Sov. Phys. Dokl.*, Vol. 16, No. 11, p. 983, 1972.

21 J. Qian, R.P. Joshi, J. Kolb, K.H. Schoenbach, J. Dickens, A. Neuber, M. Butcher, M. Cevallos, H. Krompholz, E. Schamiloglu, and J. Gaudet, Microbubble-Based Model Analysis of Liquid Breakdown Initiation by a Submicrosecond Pulse. *J. Appl. Phys.*, Vol. 97, p. 113304, 2005.

22 V.M. Atrazhev, E.G. Dmitriev, and I.T. Iakubov, The Impact Ionization and Electrical Breakdown Strength for Atomic and Molecular Liquids. *IEEE Trans. Dielectr. Electr. Insul.*, Vol. 26, No. 4, pp. 586–591, 1991.

23 D.A. Wetz, J.J. Mankowski, J.C. Dickens, and M. Kristiansen, The Impact of Field Enhancements and Charge Injection on the Pulsed Breakdown Strength of Water. *IEEE Trans. Plasma Sci.*, Vol. 34, No. 5, pp. 1670–1679, 2006.

24 K.R. Prestwich, Electron and Ion Beam Accelerators, in M. Month and M. Diens., eds. *The Physics of Particle Accelerators*, AIP Conference Proceedings No. 249, Vol. 2, American Institute of Physics, p. 1725, 1992.

25 R.E. Tobazeon, Streamers in Liquids, in E.E. Kunhardt, L. Christophorou and L.L. Luessen, eds., *The Liquid State and Its Electrical Properties*, Plenum Press, 1988.

26 R.J. Gripshover and D.B. Fenneman, Water/Glycol Mixtures as Dielectric for Pulse Forming Lines in Pulse Power Modulators. *Conference Record of the Power Modulator Symposium*, p. 174, 1982.

27 I.D. Smith, Liquid Dielectric Pulse Line Technology, in W.H. Bostick, V. Nardi, and O.S.F. Zucker, eds., *Energy Storage, Compression, and Switching*, Plenum Press, New York, p. 15, 1976.

28 P.I. John, Design of Water Pulse Forming Line, *Proceedings of the Electrical Insulation and High Voltage Pulse Techniques*, BARC, Mumbai, India, p. 335, 1982.

29 R.S. Kalghatgi, Design of Castor Oil Dielectric Pulse Forming Line. *Proceedings of the Electrical Insulation and High Voltage Pulse Techniques*, BARC, Mumbai, India, p. 307, 1982.

30 A.P. Alkhimov, Megavolt Energy Generating Device. *Sov. Phys. Dokl.*, Vol. 15, p. 959, 1971.

31 I.D. Smith and H. Aslin, Pulsed Power for EMP Simulators. *IEEE Trans. Electromagn. Compatibility*, Vol. 20, No. 1, p. 53, 1978.

32 J.P. Van Devender and T.H. Martin, Untriggered Water Switching. *IEEE Trans. Nucl. Sci.*, Vol. 22, p. 979, 1975.

33 I.D. Smith, SSWA Notes "Breakdown of Transformer Oil," Aldermaston, Berkshire, England, AWFL Dielectric Strength Notes, Note 12, 1966.

34 I.D. Smith, D. Weidenheimer, D. Morton, and L. Schlitt, Estimating and Projecting Operating Fields for Liquid Dielectrics in KrF Laser Fusion Power Plants. *IEEE Trans. Plasma Sci.*, Vol. 30, No. 5, pp. 1967–1974, 2005.

35 I.D. Smith, D. Weidenheimer, D. Morton, L. Schlitt, and J. Sethian, Large Area, High Reliability Liquid Dielectric Systems: Provisional Design Criteria and Experimental Approaches to More Realistic Projections. *Proceedings of the IEEE International Pulsed Power and Plasma Science Conference*, p. 237, 2001.

36 R.J. Adler, *Pulsed Power Formulary*, C June 2002. C Available at www.highvoltageprobes.com/PDF/fromweb1.pdf (accessed April 19, 2010).

37 A.R. Miller, Sub-ohm Coaxial Pulse Generators, Blackjack 3, 4 and 5, *3rd International Pulsed Power Conference*, pp. 200–205, 1981.

38 I.D. Smith, SSWA Notes, "Impulse Breakdown of Deionized Water," Aldermaston, Berkshire, England, AWFL Dielectric Strength Notes, Note 4, 1965.

39 I.D. Smith, SSWA Notes, "Further Breakdown Data Concerning Water," Aldermaston, Berkshire, England, AWFL Dielectric Strength Notes, Note 13, 1966.

40 P.D.A. Champney, SSWA Notes, "Impulse Breakdown of Deionised Water with Asymmetric Fields," Aldermaston, Berkshire, England, AWFL Dielectric Strength Notes, Note 7, 1966.

41 R.A. Eilbert and W.H. Lupton, Extrapolating AWRE Water Breakdown Data, NRL Note (unpublished report), 1968.

42 J.C. Martin, Nanosecond Pulse Techniques. *Proc. IEEE*, Vol. 80, No. 6, 1992.

43 W.A. Stygar, T. Wagoner, H.C. Ives, Z. Wallace, V. Anaya, J. Corley, M. Cuneo, H.C. Harjes, J.A. Lott, G.R. Mowrer, E.A. Puetz, T.A. Thompson, S.E. Tripp, J.P. Van Devender, and J.R. Woodworth, Water-Dielectric-Breakdown Relation for the Design of Large Area Multi-Megavolt Pulsed Power Systems. *Phys. Rev. Special Top. Accel. Beams*, Vol. 9, 070401, 2006.

44 G.B. Frazier, S.R. Ashby, D.M. Barrett, M.S. Di Capua, L.J. Demeter, R. Huff, D.E. Soias, R. Ryan, P. Spence, D.F. Strachan, and T.S. Sullivan, Eagle: A 2 TW Pulsed Power Research Facility. *Proceedings of the International Pulsed Power Conference*, pp. 8–14, 1981.

45 G.B. Frazier and S.R. Ashby, Double Bounce Switching, *Proceedings of the International Pulsed Power Conference*, pp. 556–558, 1983.

46 A. Kuchler, Th. Dunz, J. Dams, and A.J. Schwab, Power and Voltage Gain of Pulse Forming Lines with Double Bounce Switching. *Proceedings of the International Pulsed Power Conference*, p. 559, 1983.

47 K.W. Struve, H.C. Harjes, and J.P. Corley, Circuit Model Predictions for the Performance of ZR, *Proceedings of the IEEE International Pulsed Power Conference*, 2007.

48 M.E. Savage et al., An Overview of Pulse Compression and Power Flow in the Upgraded Z Pulsed Power Driver. *Proceedings of the IEEE International Pulsed Power Conference*, pp. 979–984, 2007.

49 M. K. Matzen, M.A. Sweeney, J.R. Asay, G. Donovan, J. Porter, W. Stygar, K.W. Struve, E.A. Weinbrecht et al., Pulsed-Power-Driven High Energy Density Physics and Inertial Confinement Fusion Research. *Phys. Plasmas*, Vol. 12, p. 055503, 2005.

50 J.E. Maenchen, J.M. Lehr et al., Fundamental Science Investigations to Develop Triggered Gas Switches for ZR, First Annual Report, SAND2007-0217, Sandia National Laboratories, Albuquerque, NM, February 2007.

losing Switches

losing switches are a topic of premier interest to pulsed power practitioners nce the performance of the switch often determines the capability of the stem. Closing switches are required to withstand high voltages and then pidly enter a conducting state that will pass high currents with minimal losses, shown in Figure 4.1.

The most common high-voltage, high-current switches operate in a glow scharge mode (pseudosparks, thyratrons) or in an arc mode (spark gaps, nitrons), although solid-state switches are increasingly being used in oderate-voltage applications. Critical parameters for switches are as llows:

ocking Voltage: The voltage that causes the switch to close and commute current.

ak Current: The maximum current that can pass through the switch.

rward Voltage Drop: The change in switch terminal voltage during commutation.

'dt: The maximum rate at which current can rise to its peak value from the switch.

itch Recovery Time: The minimum time between successive pulses for the switch to recover its holdoff voltage capability.

lay Time: The time for significant current to flow after application of a voltage sufficient to induce switching.

er: The statistical variation of the delay time.

'his chapter describes spark gap closing switches in considerable detail :ause of their ubiquitous use in pulsed power technology. Other closing tches, including thyratrons, ignitrons, and pseudospark switches, as well as nmercially available solid-state switches are also described.

dations of Pulsed Power Technology, First Edition. Jane Lehr and Pralhad Ron.

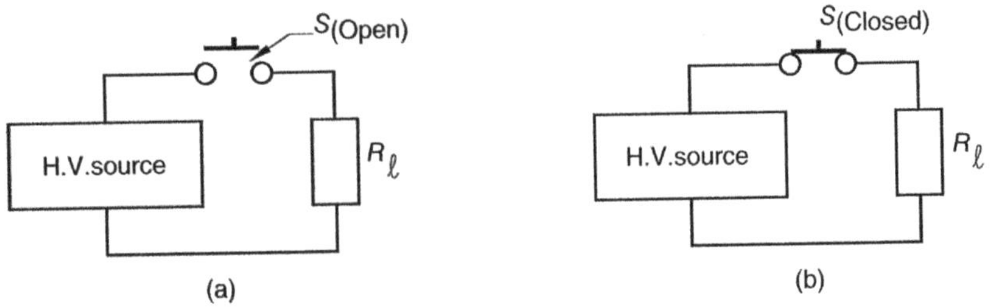

Figure 4.1 High-voltage pulsed power switch. (a) Spark gap switch before breakdown. (b) Spark gap switch after breakdown.

4.1 Spark Gap Switches

A spark gap is comprised of two conducting electrodes separated by an insulating medium, usually a gas, but liquids or vacuum are also used. A voltage applied to the electrodes does not produce current until a threshold value V_{BD} is reached. At V_{BD}, the spark gap is said to have "fired" and an arc forms through the insulating media and drastically reduces the resistance, allowing current to flow. The switch is said to be closed when the current through the switch is limited only by the external circuit, and the arc is self-sustaining. Two typical spark gap configurations are shown in Figure 4.2, where part (a) is supported with an insulating flange and part (b) is supported with a metal flange. For a voltage applied between the two main electrodes E_1 and E_2, separated by a gap spacing d, the sparking region is between the electrodes. The bright, luminous arc discharge between the electrodes of a DC-charged spark gap is shown in Figure 4.3. The switching media is 2 atm of dry air.

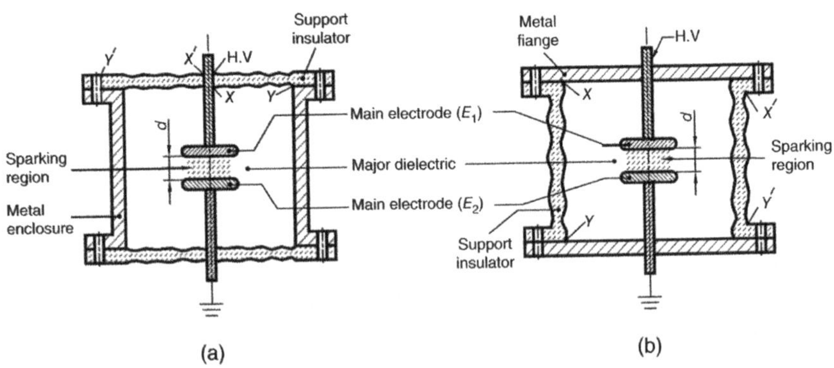

Figure 4.2 Typical configurations of a spark gap switch. (a) Typical spark gap configuration with insulator flange. (a) Typical spark gap configuration with metal flange.

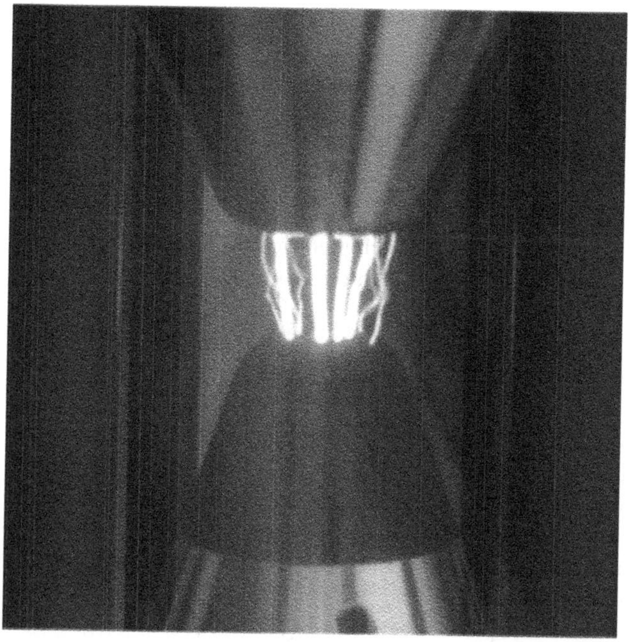

Figure 4.3 A time-integrated photo of an arc in a spark gap. The switch is insulated with 2 atm of dry air. The upper electrode has a hole in its center, resulting in a high field enhancement.

A spark gap that closes only under the action of a voltage applied across its electrodes is known as a self-breaking spark gap. In contrast, a triggered spark gap is one in which the voltage applied across the electrodes is below the self-breakdown value until switch closure is initiated by an external influence, which may take a number of forms. The initial conditions existing in the gap before command trigger are unfavorable to breakdown initiation, and the application of the trigger pulse to the spark gap modifies the conditions of the gap medium so as to initiate electrical breakdown and consequent closure of the switch. The trigger may be any method that initiates electrical breakdown in the spark gap such as an electron beam, a laser beam, or simply raises the electric field in a portion of the spark gap. Triggered spark gaps are used to control the time and precision of switching and to increase reliability, and are valued because of their wide voltage operating range.

It is assumed that the spark gap has been well designed, so the breakdown does not occur along the surface of the supporting insulators but through the main gap. This is usually achieved by increasing the breakdown strength at all other regions of the spark gap assembly, particularly across the surfaces *XY* and *X′Y′* indicated in Figure 4.2 of the support insulator. Surface flashover, where an arc is formed along the interface between the gaseous and solid insulators, is a common failure mode in spark gaps. In many cases, the electrical field required to produce a surface flashover is smaller than that required to fail the gas insulator and hence needs to be reinforced appropriately. One of the techniques of increasing the surface flashover voltage of an insulator in air is to arrange for a longer surface path for leakage current by providing corrugations on the surface, as shown in Figure 4.2b. In practice, with an AC or DC charge voltage, a rule of thumb is to design for an electric field of 10 kV/cm along the surface when using a dielectric like atmospheric air and 20 kV/cm in superior dielectrics such as oil or pressurized gas. Higher electric fields may be used when the switch is pulse charged since thresholds for surface flashover is also governed by a volt–time characteristic.

4.1.1 Electrode Geometries

Spark gap electrodes can be made in a wide variety of shapes and configurations. The spark gap is said to be symmetrical when both electrodes are the same. Both symmetrical and asymmetrical spark gap designs are common.

Uniform electric fields between the electrodes can be attained by using parallel plane electrodes, shown in Figure 4.4a. These electrodes are designed to have a flat center section with edges that are sufficiently curved so that the electric field distribution in the planar section is higher than at the edges. The shape of the parallel plane electrodes can be determined analytically, but is typically determined using electrostatic field software. Spheres are sometimes used as electrodes, but hemispheres are more common. Two variations of spark gaps with hemispherical electrodes are shown in Figure 4.4b and c. The hemispherical electrodes in Figure 4.4c are encased in a ground plane to control stray impedances. Open geometries, such as shown in Figure 4.4a and b, may have their effective capacitance value affected by the location of conductors.

The most common reason for using electrode shapes other than parallel plane is to use the geometry of the electrode to locally enhance the electric fields in the gap. The effect is quantified by the field enhancement factor *FEF*, defined

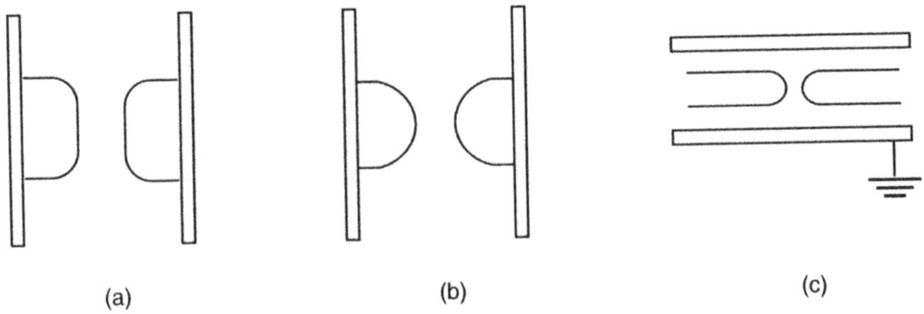

Figure 4.4 Common electrode shapes are (a) uniform field parallel plane and (b) divergent field parallel plane and (c) hemispheres in a transmission line structure.

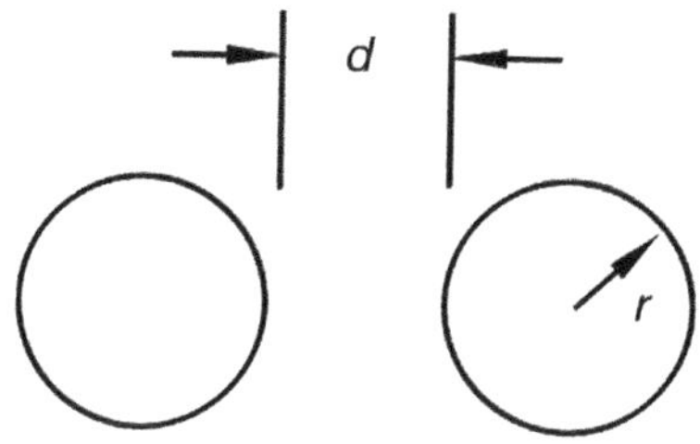

Figure 4.5 The FEF may be calculated for the sphere–sphere or cylinder–cylinder geometries of radius r and separated by a distance d.

as follows:

$$FEF = \frac{F_{\max}}{F_{avg}} = \frac{F_{\max}}{(V/d)} \tag{4.1}$$

where F_{max} is the maximum electric field in the gap, F_{avg} is the average electric field in the gap, and V is the voltage applied to the electrodes that are separated by a distance d. Analytic expressions for the FEF exist for certain geometries [1] relevant to spark gaps such as sphere–sphere and cylinder–cylinder. Figure 4.5 may represent either two identical spheres or two identical cylinders of radius r separated by a distance d.

The *FEF*s for these geometries are as follows:

Spheres with equal radii:

$$FEF = \frac{((d/r)+1)+\sqrt{((d/r)+1)^2+8}}{4} \tag{4.2}$$

$$\cong \frac{d}{2r}, \quad \text{for} \quad \frac{d}{r} \gg 1 \tag{4.3}$$

Parallel cylinders with equal radii:

$$FEF = \frac{\sqrt{(d/r)^2+4\cdot(d/r)}}{2\ln\left\{((d/2r)+1)+1/2\times\sqrt{(d/r)^2+4\cdot(d/r)}\right\}} \tag{4.4}$$

$$\cong \frac{d}{2r\cdot\ln(d/r)}, \quad \text{for} \quad \frac{d}{r} \gg 4 \tag{4.5}$$

The symmetry of the sphere–sphere gap can be used to obtain the FEF for a sphere over a plane. The FEF for a sphere located a distance d away from a plane is identical to a sphere–sphere gap spaced a distance $2d$ apart.

The FEF for a sphere or cylinder of radius r over a ground plane can be obtained by using the transformation

$$d \rightarrow 2d \tag{4.6}$$

and substituting it into Equation 4.2 or 4.3 for a sphere or Equation 4.4 or 4.5 for a cylinder. The transformation (4.6) can be used for other electrode geometries that have symmetry.

A large FEF can be attained by having a pin as at least one of the electrodes. A useful geometry is a pin to plane geometry, illustrated in Figure 4.6a. Its two-dimensional analogy is shown in Figure 4.6b and is called an edge–plane gap. The "edge" electrode may also be turned into a cylinder as shown in Figure 4.6c. The FEF can be further enhanced by using another edge in place of the plane, as shown in Figure 4.6d.

Spark gaps may also be made with more than two electrodes. When the electrodes are arranged so that the single large gap is replaced with a series of smaller gaps by inserting additional electrodes between the two main ones so that the gaps are in series, it is called a cascade arrangement. Large variations in the implementation of cascade electrodes are found, as shown in Figure 4.7.

The electrodes may float in the electric field and the voltage is then distributed by the equivalent capacitance of the cascade gaps. For equally spaced, identical electrodes, the voltage distributed across each gap is

$$V_{cas} = \frac{V}{m} \tag{4.7}$$

where V is the voltage across the switch and V_{cas} is the voltage across each gap or cascade section. This voltage division is a great advantage. For instance,

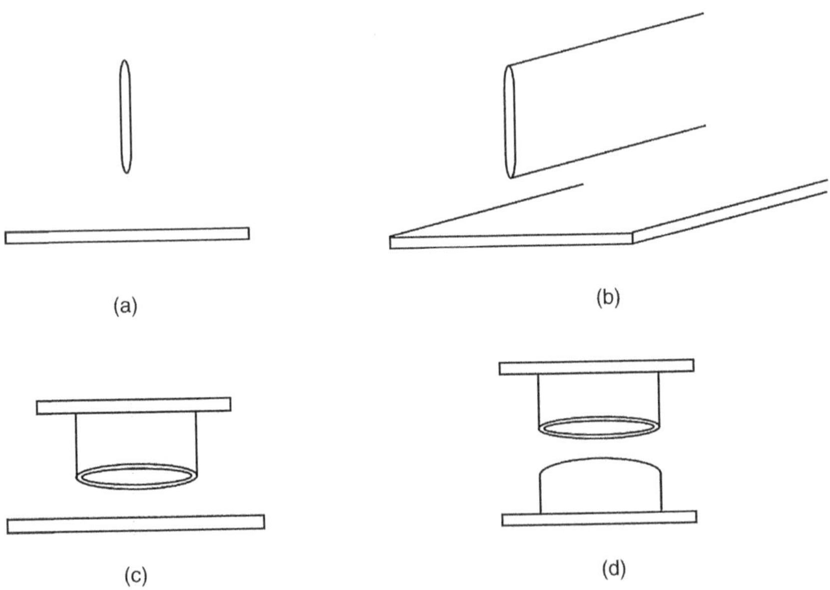

Figure 4.6 A pin-to-plane spark gap and its two-dimensional equivalent, an edge plane gap.

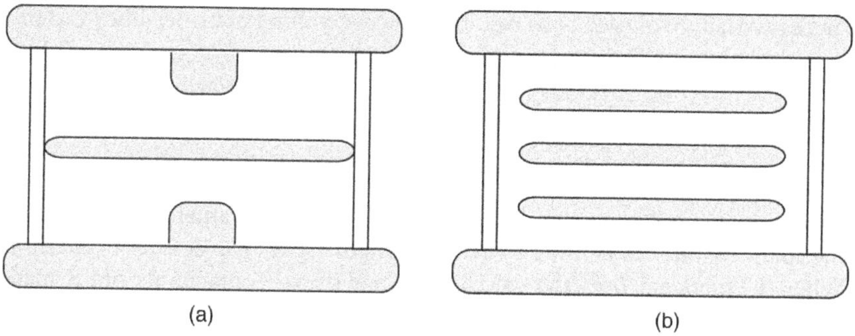

Figure 4.7 The electrodes in a cascade spark gap are not necessarily identical. A three-electrode spark gap (a) and a four-gap switch (b) use multiple electrodes to enhance the performance of the switch. A switch with m gaps has $(m+1)$ electrodes.

three-electrode spark gaps, as in Figure 4.7a, are used extensively in circuits charged in dual polarity, such as Marx banks. The middle electrode, located midway between the two charged electrodes, may be tied directly to ground or allowed to float. Cascade switches with many gaps, as in Figure 4.7b, may be used to switch very high voltages. A cascade spark gap from the 1970s, called a "Rope Switch," [2] shown in Figure 4.8, switched 1 MV with a length of less than 10 in. The electrodes are shaped as in Figure 4.7b to encourage the formation of multiple channels.

When the cascade electrodes are allowed to float, the voltage of each electrode is dictated by the capacitive distribution and determined largely by its position within the switch. The voltage distribution can be arbitrarily assigned by using a resistive divider network across the switch and explicitly ties the electrodes to a specific voltage. The resistor values should be very large and a corona discharge may be used to provide this resistance [3,4].

Cascade spark gaps may close very quickly if the capacitance to ground of each electrode exceeds the capacitance between adjacent gaps. A breakdown of any gap will cause the adjacent gaps to be overvolted, causing breakdown. The

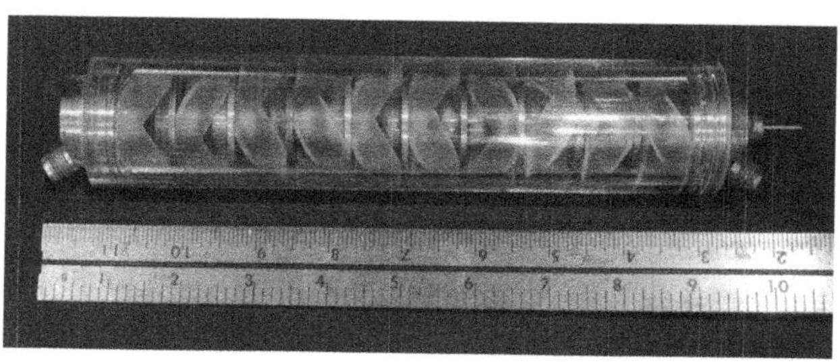

Figure 4.8 An early version of a cascade gap, named a Rope Switch, had 10 gaps and a hold-off voltage of 1 MV. (Reproduced with permission of Keith Hodge.)

successive failure of each gaps results in the overvolting of its neighbor gap in a cascade fashion, ultimately causing the switch to fire.

4.1.2 Equivalent Circuit of a Spark Gap

A spark gap can be represented by the equivalent circuit [5] shown in Figure 4.9, consisting of a self-capacitance C_g in parallel with the parameters representing the spark channel. When the switch is open, the spark gap acts as a capacitor. When the spark gap fires, the switch closes and the self-capacitance of the spark gap is shunted by the series combination of spark channel resistance $R_s(t)$ and spark channel inductance $L_s(t)$ representing the arc. The resistance of the fully formed arc channel is a fraction of a milliohm. During the breakdown process, R_S and L_S are time-dependent terms.

4.1.2.1 Capacitance of the Gap

An expression for the capacitance of a spark gap can be derived from its definition:

$$C_g = \frac{Q}{V} = \frac{\epsilon \oint \vec{E} \cdot d\vec{A}}{\int \bar{E} \cdot dl}$$

For parallel plate electrodes, the above expression simplifies to

$$C_g = \frac{\epsilon_r \epsilon_0 A}{d} \text{ F} \tag{4.8}$$

where

ϵ_r is the relative permittivity of the dielectric medium,
ϵ_0 is the permittivity of free space (=8.85 pF/m),
A is the area of the electrodes in m^2, and
d is the gap spacing in meter.

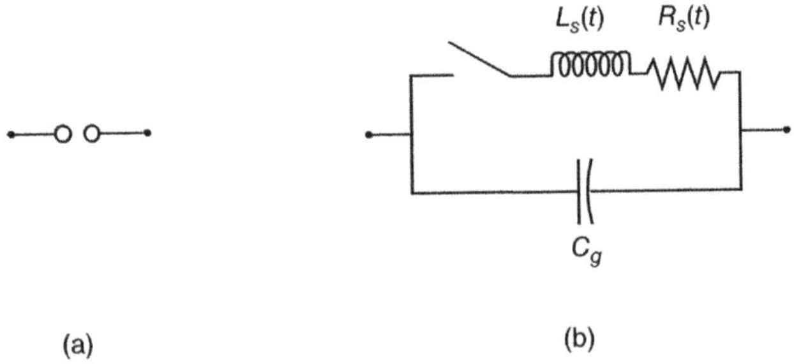

Figure 4.9 The equivalent circuit of a spark gap where the dynamic impedances $R_s(t)$ and $L_s(t)$ represent the arc channel.

The spark gap capacitance may be calculated with an electrostatic circuit code. For spark gaps that are physically long or housed in an open geometry, any charged structural components, such as field grading rings, should be included for accuracy. If the spark gap is pulse charged, the capacitance can be obtained directly from measurement with suitably calibrated monitors from

$$C = \frac{I_c(t)}{10^{-12} \times ((dV_c(t)/dt))} \text{ pF} \tag{4.9}$$

where the factor of 10^{-12} reduces the denominator to a reasonable numerical value, allowing the calculation to be done with electronic waveforms.

4.1.2.2 Resistance of the Arc Channel

When the spark gap thus breaks down, the switch is closed, and the energy is transmitted from the input to the load, as shown in Figure 4.9. For an ideal spark gap switch, the impedance changes from infinity to 0 Ω instantaneously when the breakdown takes place.

The values of R can be calculated from Toepler's empirical formula:

$$R = \frac{K_T d}{Q} \ \Omega \tag{4.10}$$

where K_T is the Toepler constant, d is the length of the spark gap, and Q is the charge transferred through spark gap in ampere-seconds. The Toepler constant depends on the gas species and is 0.8×10^{-3} for air at 1 kg/cm^2.

Rompe and Weizel [6] developed an expression for the spark channel resistance based on an energy balance given by

$$R = \frac{pd}{2\kappa} \left[\int_0^t I^2 dt \right]^{-1} \Omega \tag{4.11}$$

where κ is a constant that depends on the gas and p is the gas pressure. Values are $\kappa = (0.8 - 1)((\text{atm} \cdot \text{cm}^2)/\text{s} \cdot \text{V}^2)$ for air and nitrogen (N_2).

Braginskii developed a model for the time dependence of the spark channel expanding in the radial direction that included hydrodynamics [7,8]. The current flows through a fixed-conductivity gas and expands the channel at supersonic velocity, driving a cylindrical shock wave. The expansion of the channel reduces the resistance of the arc in the Braginskii model. The channel radius $r_c(t)$ is

$$r_c^2(t) = \left(\frac{4}{\pi^2 \rho_0 \xi \sigma} \right)^{1/3} \int_0^t I^{2/3}(\tau) d\tau \tag{4.12}$$

where ρ_0 is the undisturbed density of the gas, σ is the fixed conductivity, and ξ is a constant that depends on the gas properties and is given by [4]

$$\xi = K_P\left[1 + \frac{2}{3\cdot(\gamma-1)}\right] \tag{4.13}$$

where

$\gamma = \frac{C_P}{C_V}$, the ratio of the specific heat, and
$K_P = 0.9$, the coefficient of resistance constant.

The resistance of the arc channel, $R(t)$, is

$$R(t) = \frac{1}{\sigma}\cdot\frac{d}{\pi r_c^2(t)} \tag{4.14}$$

As the spark channel expands, the resistance falls from the initial high-value characteristics of an open switch to a much smaller value of a closed switch, as shown in Figure 4.10 and described by the Martin–Braginskii formulation [9].

4.1.2.3 Inductance of Arc Channel

The inductance of an arc channel is difficult to know precisely. It is generally thought to be negligibly small before the arc is fully formed and is modeled as a conducting wire once the arc has been established. A number of expressions have been used to estimate the inductance of the spark channel, L_s.

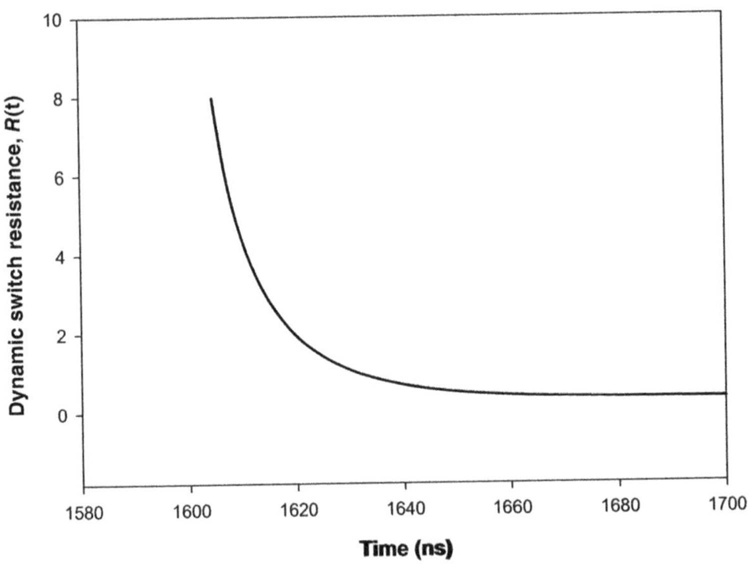

Figure 4.10 The resistance of a spark channel drops as the channel expands during the electrical breakdown process.

If the arc channel is far from the ground plane, it may be modeled as a circular single conductor. The self-inductance of a wire of length d is [10]

$$L_s = \frac{\mu_0 d}{2\pi}\left[\ln\left(\frac{2d}{r_c}\right) - 3/4\right] \tag{4.15}$$

where r_c is the radius of the arc channel. If both r_c and d are in centimeters, then the above equation can be written as

$$L_s = 2d \cdot \left[\ln\left(\frac{2d}{r_c}\right) - 3/4\right] \text{nH}$$

Martin [11] uses the model of a wire of radius r_c being fed by a disk of radius r, yielding an inductance of

$$L_s = 2d \cdot \ln\left(\frac{r}{r_c}\right) \text{nH} \approx 14d \text{ nH} \tag{4.16}$$

In the case of a cylindrically symmetric spark gap where the return conductor is close to the electrode radius, the inductance is estimated using the transmission line equations [12,13]:

$$L_s = d \cdot \frac{\mu_0}{2\pi} \cdot \ln\left(\frac{R}{r_c}\right) \tag{4.17}$$

The difference in Equations 4.15 and 4.17 is the location of the return conductor. The importance of considering the switch geometry is illustrated in the important case of a cylindrically symmetric, multichannel spark gap shown in Figure 4.11. The purpose is to induce a number of spark channels N, of radius r_c equally spaced equally distant on a circle of radius r. The electrode is encased in a cylindrical return conductor of radius R.

If N equidistant channels are produced in the spark gap, the inductance of the ensemble is [14]

$$L_s = \frac{\mu_0 d}{2\pi N}\left[\ln\left(\frac{R}{Nr_c}\right) + (N-1)\ln\left(\frac{R}{r}\right)\right] \tag{4.18}$$

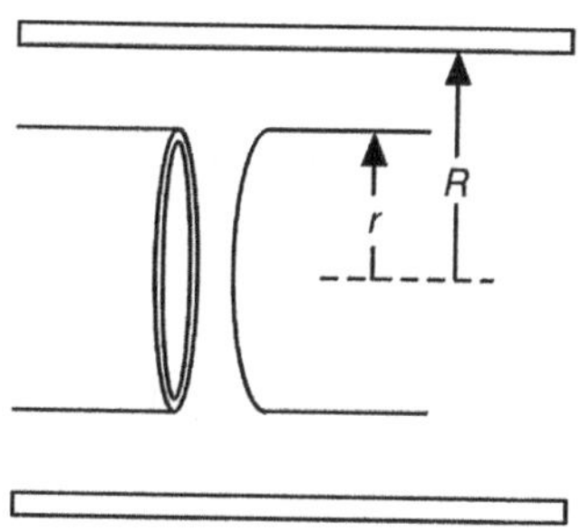

Figure 4.11 A cylindrically symmetric spark gap switch encourages multiple spark channels to be formed, resulting in a low-inductance geometry.

where d is the length of the current channels. This expression has only a weak dependence on the arc channel radius. The criterion for equidistant arc channels is important because the use of symmetry is used in the derivation of Equation 4.18. Additionally, the number of channels established should be greater than four for Equation 4.18 to be accurate.

The inductance of the arc channel is difficult to estimate because it generally requires knowledge about the arc radius. Precise information about the radius of the arc channel is difficult to measure, but values between 0.5 and 1.5 mm give reasonable values. Because inductance is derived from electrodynamic fields, it is influenced by the presence of conductors. A useful resource for estimating inductances in a wide variety of geometries can be found in Ref. [15].

4.1.3 Spark Gap Characteristics

Spark gaps have a number of important basic characteristics that apply to both triggered and untriggered switches. A brief description of these individual topics is presented in this section.

4.1.3.1 The Self-Breakdown Voltage and Probability Density Curves

Spark gaps will break down if the voltage applied across a given gap causes the electric field to exceed its threshold value for breakdown, including the effects of field enhancements. The self-breakdown voltage scales in the same manner as the Paschen curve; it is a function of fluid species and pressure, the gap spacing, and often the electrode material. It is also dependent on local electric field enhancements, imperfections in manufacture, and the charge time of the applied voltage pulse.

The static breakdown voltage is the voltage that produces an electrical arc in the insulating dielectric, without external triggering, when the spark gap is charged at rates slower than 1 kV/s. The static breakdown voltage depends on the properties of the dielectric, the electric field configuration, and, to a certain extent, on the properties of the electrodes such as material and surface condition.

The dynamic breakdown voltage is the value at which breakdown of the gap takes place when voltage is applied at a rate much faster than 1 kV/s. The dynamic breakdown voltage is larger than the static breakdown voltage and it is this time dependence that is exploited in most pulsed power systems to produce very high peak power pulses. The dynamic breakdown voltage is described in Chapter 3 by Martin's empirical formulas under uniform field conditions.

In describing realized spark gaps, the self-breakdown curve is an important characterization tool that is generated by measuring the breakdown voltage at discrete pressures under actual operating conditions. Thus, the self-breakdown voltage is a dynamic voltage measurement that accounts for local field enhancements, material effects, and manufacturing defects.

An example of a V–p curve for a DC-charged, air-insulated switch is shown in Figure 4.12. The data are plotted in discrete points to show the scatter in the breakdown voltage for a given pressure and a best-fit line is computed. Self-breakdown curves are generated for triggered spark gaps to determine the operating pressure, which is typically 70–85% of the self-breakdown value. Triggered spark gaps operated too close to the self-breakdown voltage may encounter the spread in the self-breakdown voltage, which manifests as switch prefires. The self-break curve is instructive as it is an indicator of the reliability of the spark gap, and large spreads in the self-breakdown voltage should be investigated.

The range in the self-breakdown voltage can be represented by a probability density function for a given pressure. The probability density function $P_V(V)$ may be generated by counting the number of breakdowns that occur within a small interval ΔV, normalized to the total number of shots. Thus, $P_V(V)$ is the probability that a spark gap breaks down in an interval between the voltages V and $V+\Delta V$. Figure 4.13a shows the probability distribution function $P_V(V)$ for an ideal spark gap, and Figure 4.13b shows a realized spark gap where there is a spread in the measured values of breakdown voltage.

The self-breakdown probability distribution function is greatly affected by the rate at which the spark gap is charged. Donaldson et al. [16] measured the self-breakdown voltage for charge rates differing by an order of magnitude and

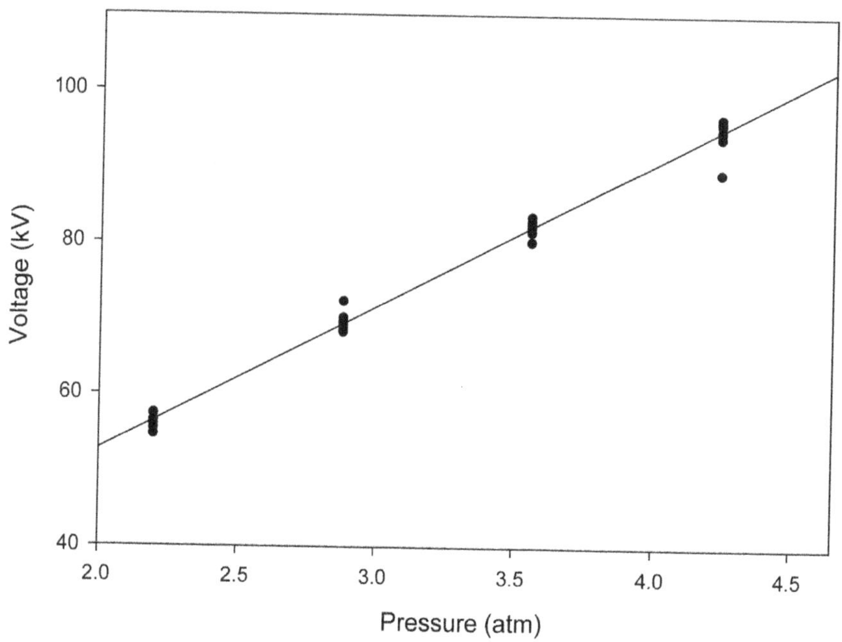

Figure 4.12 The V–p curve for an air-insulated switch with a gap spacing of 9 mm.

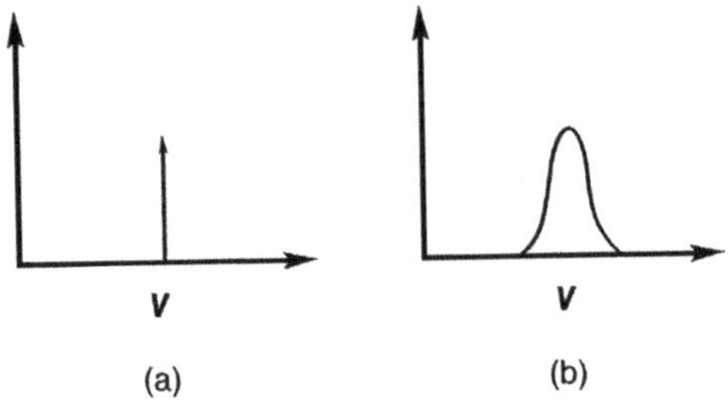

Figure 4.13 The probability density function $P_V(V)$ for (a) an ideal spark gap and (b) a spark gap with spread in its measured breakdown values.

summarized the data in the probability density distribution functions shown in Figure 4.14a. When the charge rate is 3 kV/s, the function $P_V(V)$ is narrow with a well-defined peak, indicating the breakdown of the switch is repeatable. The faster charge rate of 30 kV/s shows that higher dynamic breakdown values are achieved, but with a significant broadening of the probability distribution density function as well as skew toward lower breakdown values. It has been postulated [16] that the voltage is higher for the faster charge rate when the initiating electron appears, leading to a bigger scatter in breakdown voltage values.

Donaldson et al. also measured the effect of pressure on the probability density function of air with brass electrodes [16]. These data, shown in Figure 4.14b, indicate increasing the pressure results in higher breakdown voltages along with a tendency to prefire. This is attributed to the effects of the cathode microstructure. Brass electrodes can form protrusions of up to 500 μm in high-energy operation.

4.1.3.2 Delay Time

The delay time to breakdown, also called the time lag, is a measurement of the temporal development of the arc upon application of a voltage. The delay time is

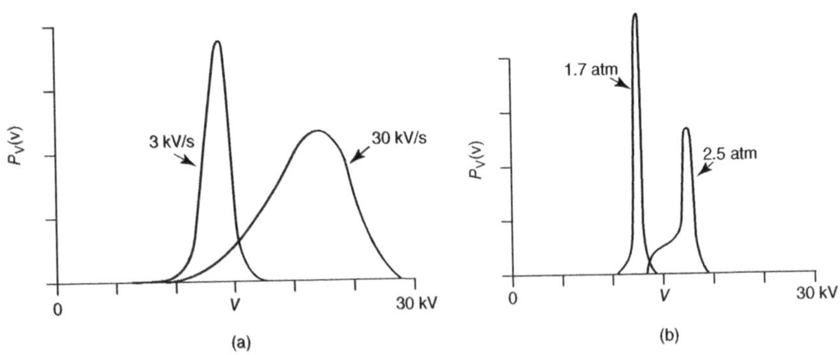

Figure 4.14 The effect of (a) charging rate and (b) pressure on the probability density distribution function $P_V(V)$ of an air-insulated spark gap.

a fundamental metric in the development and evaluation of spark gaps. The switch jitter is the standard deviation of the delay time.

The switching time or time lag t_d is generally considered to be separable into two components [17]: the statistical time lag t_s and the formative time lag t_f.

$$t_d = t_s + t_f \tag{4.19}$$

The switching time lag t_d of a self-breaking gap is obtained by examining the time-dependent voltage across the spark gap, as shown in Figure 4.15. The time lag is the time interval $(t_3 - t_1)$ measured from the moment of application of a voltage step across the gap, t_1, to the instant when the gap impedance starts to collapse at t_3. The statistical time lag t_s is the interval $(t_2 - t_1)$, with t_2 denoting the time at which an electron appears to initiate the breakdown process. The formative time lag t_f is the time interval $(t_3 - t_2)$ during which the ionization growth occurs and the channel formed. The time interval $(t_4 - t_3)$ is the time for the voltage across the spark gap to collapse, which corresponds to the rise time of the output pulse.

The statistical time is the time for electrons to appear in the high-field region of the gap and develop into an avalanche to initiate the breakdown process. The statistical time lag is negligible when electrons are provided, for example, with an ultraviolet (UV) source, which easily ionizes gases, or when strong electron emission from the electrodes is present.

In self-breaking spark gaps, the statistical time lag dominates the delay time. The influence of the UV trigger on the self-breakdown probability distribution density function $P_V(V)$ was measured [16] and is shown in Figure 4.16. The introduction of UV irradiation improved the static self-break characteristics of a spark gap to a statistical spread of just a few kilovolts. It is also interesting to note that the introduction of an UV source resulted in the self-breakdown voltage (without UV) collapsing to the lowest breakdown voltage of the distribution when the statistical time lag is eliminated.

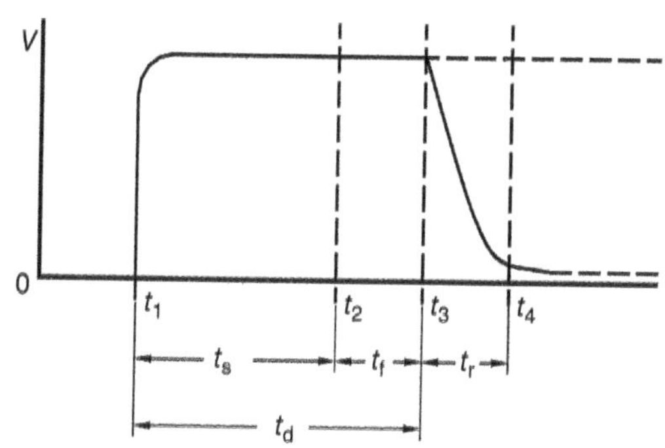

Figure 4.15 Switching time lag t_d of a self-breaking gap is comprised of a time t_s for initiatory electrons to appear and the time t_f for the current-carrying arc channel to be formed.

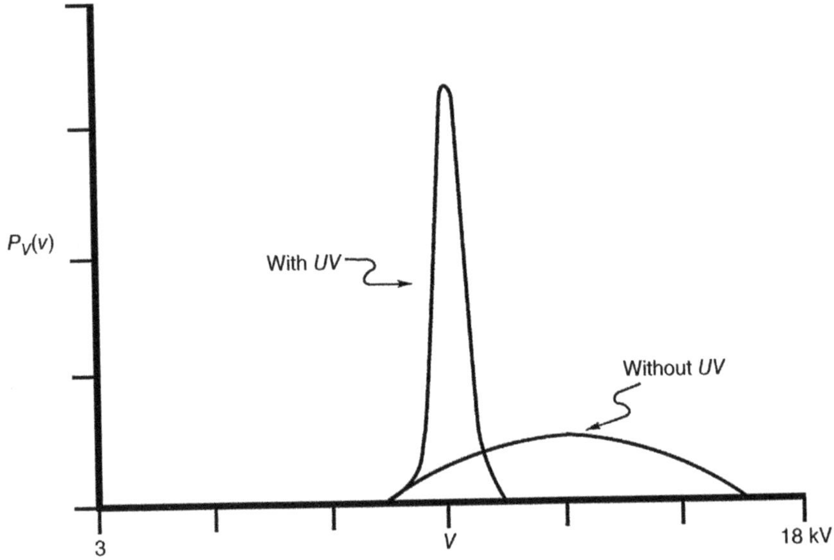

Figure 4.16 The influence of the statistical time lag is illustrated by Donaldson by examining the change in the probability density distribution function $P_V(V)$ of a self-breaking spark gap when a UV source is introduced [16].

The difficulty in measuring the delay time of self-breaking spark gaps is the determination of the initial time t_1, when the voltage is applied. In triggered spark gaps, t_1 is the time when the trigger mechanism is applied. Moreover, the statistical time lag for triggered spark gaps is usually considered to be negligible because the trigger supplies the initiatory electrons.

The formative time lag is the time for the discharge to form and establish a conducting channel across the gap. The formative time lag can be further broken down into the streamer formation time t_{sf} and the thermalization time t_{ch}. A streamer is formed when the head of an avalanche attains sufficient space charge that its self-generated electric field is equal to the applied electric field. The streamer channel propagates across the gap, but no current flows because the channel has a very high resistance—in the mega-ohm range. To conduct the current, the streamer channel must thermalize—become more conductive and expand to carry the current. The time for the streamer channel to thermalize is t_{ch} and is sometimes called the channel heating time.

Martin proposed an empirical model for the time lag to breakdown, given by [18]

$$\rho\, t_d = 97800\left(\frac{E}{\rho}\right)^{-3.44} \tag{4.20}$$

where ρ is the gas density in g/cm^3, E is the average electric field strength in kV/cm, and t_d is the delay time to breakdown in seconds.

4.1.3.3 Rise Time (t_r)

The rise time t_r of the spark gap is the time interval $(t_4 - t_3)$ in Figure 4.15 during which the impedance of the spark gap transitions from a high value at t_3 to a low value at t_4. The rise time of interest is the time for the current to rise between 10 and 90% of its peak value.

The predictions regarding t_R are based on two e-folding time constants: one related to the resistive phase of the developing spark channel, τ_R, and the other to its inductance, τ_L. These time constants predict the rise time of the pulse from the spark gap, t_r, according to the relation

$$t_r = 2.2\,\tau_{tot} \tag{4.21}$$

where

$$\tau_{tot} = \sqrt{\tau_R^2 + \tau_L^2} \tag{4.22}$$

4.1.3.3.1 Resistive Phase Time Constant (τ_R)

During this time, the weakly ionized gas at about 5000 K available at time t_3 gets converted to a plasma filament at 100,000 K at first and to a plasma column later, due to radial expansion of plasma filament. The initial reduction of impedance of the gap is due to the increase in the temperature of plasma from 5000 to 100,000 K. After this, further temperature increase becomes more difficult because the plasma column acts as a blackbody and radiates most of its energy. Further reduction of the gap impedance takes place by an increase in the cross section of the plasma expansion of the arc channel.

Martin [19] proposed the following widely accepted formula for the resistive-phase time contribution in a gas,

$$\tau_R = \frac{88}{Z^{1/3}E^{4/3}}\sqrt{\frac{\rho}{\rho_0}}\ \text{ns} \tag{4.23}$$

where

Z is the impedance of the source driving the channel in ohms,
E is the field in units of 10 kV/cm,
ρ is the density of the gas, and
ρ_0 is the density of air at STP.

It is derived from the collapse of the voltage at its maximum rate of change. Sorensen and Ristic [20] have proposed a slightly different formula:

$$\tau_R = \frac{44\sqrt{p_{\text{atm}}}}{E\cdot Z^{1/3}}\ \text{ns} \tag{4.24}$$

where p_{atm} is the gas pressure in atmospheres.

In liquids and solids, the resistive rise time is given by [8]

$$\tau_R = \frac{5\sqrt{\rho_d}}{Z^{1/3}E^{4/3}} \tag{4.25}$$

where ρ_d is the density in units of g/cm^3 and E is now in units of MV/cm.

4.1.3.3.2 *Inductive Phase Time Constant* (τ_L)

As the current builds up during the resistive phase discussed above, the resistance of the spark channel is gradually lowered. It is at this stage that inductance starts playing a dominant role. When the spark inductance is small, the electrode inductance $L_{electrode}$ must also be considered. The inductive rise time τ_L is given by

$$\tau_L = \frac{L}{Z} = \frac{L_a + L_{electrode}}{Z} \tag{4.26}$$

The units of the inductive rise time are nanoseconds when the inductance is in nanohenries and the impedance in ohms.

4.1.3.4 Burst-Mode Repetitively Pulsed Spark Gaps

Reliable repetitive spark gap switching occurs when the breakdown strength of the dielectric is restored to original value during the time interval between successive pulses. The repetition rate therefore is dependent on fast removal of ionized matter and heat in the spark channel. Low-molecular-weight gases like hydrogen (H_2) and deuterium (D_2) are optimum dielectrics for this purpose, because of their high mobility and large heat transfer capability.

4.1.3.4.1 *Low-Pressure Spark Gaps*

For many applications in pulsed power, it suffices to operate in a one-time burst mode with a group of few closely spaced pulses. Nagesh et al. [21–23] have developed a spark gap that operates on the left-hand side of the Paschen curve, using H_2 and D_2 at pressures between 1 and 34 Pa. It operates at the following parameters: (a) 160 kV at 8 kA peak in a 5-pulse burst at 10 Hz and (b) 35 kV at 5 kA peak in a 10-pulse burst at 100 Hz.

An interesting phenomenon of overrecovery takes place in these gaps, where the breakdown voltage for the succeeding pulse exceeds that of the preceding pulse. The salient features of this mechanism are illustrated in Figure 4.12, which shows switching of a group of five pulses with peak voltage of V_0. At the time t_1, when the first pulse appears at the switching gap, the breakdown takes place at Vb_1, corresponding to pressure P_1, on the Paschen curve. The formation of a hot plasma filament results in shock wave, which sweeps the

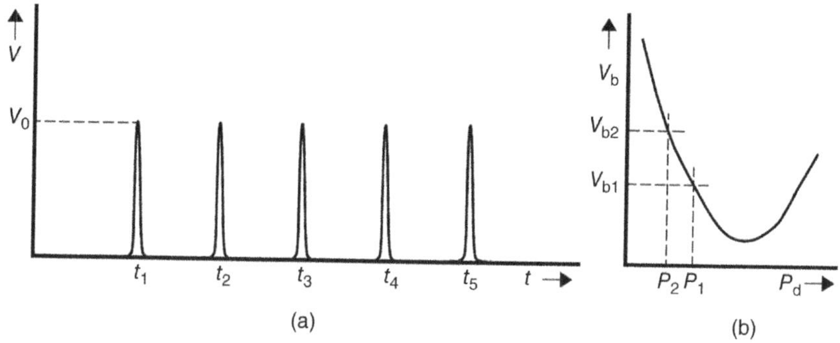

Figure 4.17 Overrecovery in low-pressure spark gap for burst-mode operation. (a) Five-pulse group. (b) Paschen curve.

gas away from the sparking region toward the walls. As a result of this, the dynamic pressure in the gap reduces to P_2 at time t_2, when the second pulse arrives at the switch. The corresponding breakdown voltage on the Paschen curve is Vb_2, which is greater than Vb_1 for the first pulse and may be termed overrecovery. This phenomenon is in contrast to the underrecovery that takes place in spark gaps using gas dielectric at high pressures, which operate on the right-hand side of the Paschen curve (Figure 4.17).

4.1.3.4.2 High-Pressure Spark Gaps

Without careful design, high-pressure spark gaps are limited in their repetitive capability because its switching media requires an amount of time to recover its insulating properties once an arc has occurred, resulting in successive pulses having a reduced amplitude. Sealed volume, gaseous spark gaps can achieve a pulse repetition rate (PRR) of ~100 Hz. Gas flow through the switch serves to sweep discharge by-products from the high electric field region and can result in much higher PRRs but at the expense of additional complexity. Transformer oils have been used in high PRR applications, but must be flowed at a considerable rate [24–26].

The PRR capability of spark gaps is determined by the minimum time between successive pulses for the insulator to recover its full voltage holdoff capability. This voltage recovery time has been studied extensively by Moran et al. [27–32] and MacGregor et al. [33–39] using a double-pulse generator with a variable time delay between the two pulses. The voltage recovery time cannot be directly applied to high-repetition-rate switches, but its study has uncovered the dominant operative physical processes and led to great increases in the repetitive capabilities of spark gaps.

There are three phases to voltage recovery. The channel must be deionized, which occurs via the atomic processes of recombination, de-excitation, and attachment. The channel reaches its background ionization level in ~10 μs, but remains hot [34,36]. During conduction, the gas in the arc column is heated to several thousand Kelvin, resulting in a severe drop in the local gas density. If the second voltage pulse is reapplied before the channel has cooled, the gap can break down at a much lower voltage. Voltage recovery can occur only when the gas is cooled, which usually takes ~1 ms. Once the channel is cooled, the static voltage breakdown is recovered.

The remaining voltage recovery affects the ability of the spark gap to be overvolted. The recovery of the pulse charge breakdown voltage may be delayed by as long as several hundred milliseconds and has been attributed to residual ions in the gap [33]. The residual ion population, albeit small, has been shown to significantly affect the breakdown probability function and results in the plateau observed in voltage recovery curves [34].

The timescales for these processes are as follows:

Channel charge neutralization:	tens of microseconds
Channel cooling:	few milliseconds
Plateau:	hundreds of milliseconds

Moran found that high pressures and small gap spacing led to enhanced switch recovery, with the rate of recovery directly proportional to the gap distance. The rate of recovery is also gas species dependent, with lighter gases like H_2, N_2, and air having higher PRR capabilities. These gases are extensively used in fast, high PRR switches such as those used for the generation of ultra-wideband high-power microwave (HPM).

Significant improvement in the PRR capability of spark gaps has been demonstrated. Moran operated a high-pressure H_2 trigatron at 50% of its static breakdown voltage while the channel was still hot and measured recovery in 100 μs [31]. The switch was operated at a voltage of 120 kV and peak current of 170 kA in a burst-mode PRR of 10 kHz with a few nanoseconds of jitter. MacGregor and coworkers, using a corona-stabilized switch, have demonstrated the performance of 20 kHz [37].

4.1.3.5 Shot Life

A spark gap can be said to have reached its end of life when its performance deteriorates beyond acceptable tolerance for a specific application. The performance of a spark gap changes with use because of (a) erosion of gap electrodes, (b) gas cleanup in case of sealed gas gaps, (c) gas evolution in case of sealed vacuum gaps, or (d) deposition of metal vapor on support insulators that ultimately leads to a flashover along the surface. The most important and noticeable changes in gap performance that occur over time before complete failure are a decrease in the voltage holdoff capability and an increase in jitter. The most serious factors that limit the shot life are integrally linked with erosion of material from the electrodes. The shot life of a spark gap may be expressed in terms of the number of shots or the charge transferred per shot.

4.1.3.6 Electrode Erosion

Electrode erosion occurs when material is permanently removed from the electrode surface, leaving it rough with "pits," as shown in Figure 4.18. Erosion will eventually cause the switch gap to widen, increasing the measured jitter, but the additional field enhancement of the roughened surface results in a decrease in the breakdown voltage. Experimental results on electrode erosion often appear to contradict one another largely due to the large number and interdependencies of affecting parameters. A basic understanding of the source of these contradictions in the literature has been reached by Donaldson et al. [40,41] building on the work of Belkin [42–44] and Watson [45]. Donaldson collected prodigious amounts of data over a large parameter space [41]. Despite these advances and the considerable engineering importance of electrode erosion, a thorough understanding of its mechanism remains elusive.

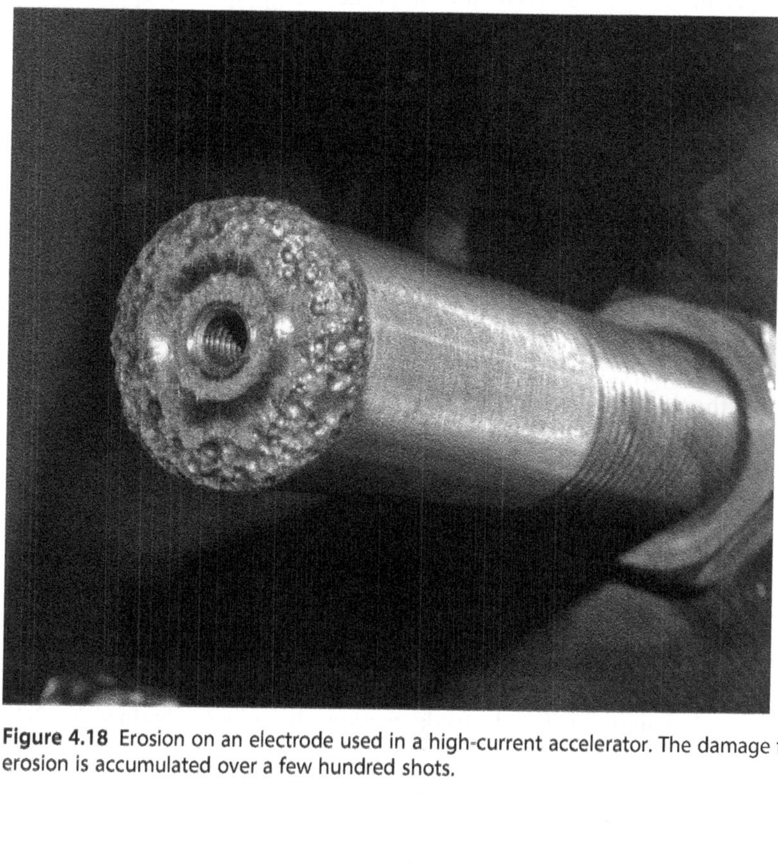

Figure 4.18 Erosion on an electrode used in a high-current accelerator. The damage from erosion is accumulated over a few hundred shots.

Permanent removal of the electrode surface is in one of three forms: particles, vapor, or liquid. The particles may be charged or neutral. The total volume eroded v_e can be expressed as

$$v_e = k_p v_p + k_m v_m + k_v v_v \tag{4.27}$$

where v_p is the charged particle volume, v_m is the volume melted, and v_v is the volume vaporized, and k_p, k_m, and k_v are the particle, molten, and vaporized metal ejection coefficients, respectively, which vary from 0 to 1. The material may be removed during or shortly after the applied current pulse. Thus, the erosion process can be thought of as being determined by two sets of mechanisms, namely, those thermal mechanisms that supply energy to the electrode surface and yield the various states of matter from the original solid electrode surface and those material removal mechanisms that determine how much material is actually removed in any given state.

The straightforward application of the fundamental equation of heat transfer to the problem of electrode erosion has proved to be quite successful in developing an understanding of the many of the observed features of electrode erosion [41,46–50]. The differential form of the energy equation applicable to general heat conduction problems is

$$\rho\, c_v \frac{\delta T}{\delta t} = \nabla \cdot k \nabla T + \frac{d\varsigma}{dt} \tag{4.28}$$

where ρ is the density of the material, c_v is the specific heat at constant volume, k is the thermal conductivity, T is the temperature, and $d\zeta/dt$ is the internal rate of heart generation within the material. The term $d\zeta/dt$ for an electrode is the Joule heating, given by J^2/σ where J is the magnitude of the current density and σ is the electrical conductivity. This term is generally assumed to be negligible in comparison to the surface (radiant) heat flux produced by the arc [45]. In a metallic conductor, $c_v \cong c_p$ and the material can be assumed to be isotropic with respect to the thermal conductivity. The applicable boundary condition for the surface receiving the heat flux from the arc is the Fourier rate equation for conduction:

$$k \cdot \nabla T = -\varsigma \cong -J \cdot V_f \tag{4.29}$$

where ζ, the surface heat flux per unit area from the arc, has been approximated by the product of the current density and the potential drop of the cathode fall of each electrode, V_f, which is on the order of 10 V for copper [50].

Solutions for the heat conduction equation when the electrodes can be assumed to be infinite in extent in the direction normal to the incident heat flux—a good assumption for electrode erosion—have been obtained by Belkin [42] and are represented in Figure 4.19.

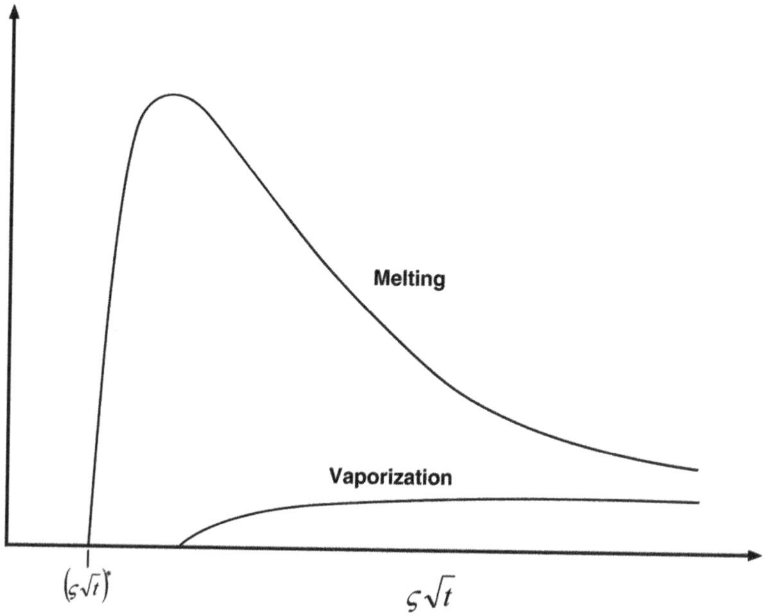

Figure 4.19 Solutions to the one-dimensional heat conduction equation.

Belkin derived a simple expression for the onset of melting of the electrode surface given by

$$\frac{\varsigma\sqrt{t}}{T_{mp}\sqrt{k_s\rho_s c_s}} = \sqrt{\frac{\pi}{4}} \tag{4.30}$$

where T_{mp} is the melting temperature of the material, k_s is the thermal conductivity of the solid phase of the material, ρ_s is the density of solid material, and c_s is the specific heat for the solid material. For material removal by vaporization, no simple expression for the onset condition has been found but has generally been found to be two to four times that for melting.

Electrode erosion may scale with any number of parameters related to the quantity of charge that passes through the electrode. Common parameters are peak current, I_p, the total charge transferred, $\int I(\tau)d\tau$, or the action integral, $\int I^2(\tau)d\tau$.

Erosion may be characterized by a generic curve, shown in Figure 4.19, which relates the volume eroded to any parameter of the pulse that has been found to affect the electrode erosion, such as peak current, effective charge transfer, action, and so on. Note that although the scaling of the erosion with a particular pulse parameter may be different than that for another, the generic shape of the curve will be the same regardless of the parameter chosen. Moreover, the phenomenology has been identified that dictates the transitions [46].

In Figure 4.20, curve A represents the generic electrode erosion curve that is typical of pure metals. Region 1 in Figure 4.20 corresponds to the region in Figure 4.19 for which the product $\varsigma\sqrt{t}$ is less than the threshold for bulk melting of the material. This means that the value of the current (as well as the heat flux ς and the charge transfer) is sufficiently low to keep the product $\varsigma\sqrt{t}$ below the threshold. There is a finite volume of material eroded in this region. While the overall combination of heat flux and pulse duration is too small to exceed the threshold value, the microscopic current density of the individual filament is quite high [51]—on the order of 10^{12} A/m^2—giving rise to a high local heat flux at the filament attachment site. Although the lifetime of a given filament is short (~10 ns), it is sufficient to produce local vaporization at the filament attachment site and results in a small, nonzero erosion rate.

As the pulse parameter is increased, the value of $\varsigma\sqrt{t}$ increases and eventually the threshold for bulk melting is reached, and the generic curve A transitions into region 2. Further increases in the pulse parameter result in large increases in the volume of material eroded. The steep slope of the melting curve in Figure 4.19 corresponds to the steep change in slope in region 2 of the generic erosion curve A of Figure 4.20. This region was extensively investigated by Watson et al. [45].

Increasing the pulse parameter even further eventually causes $\varsigma\sqrt{t}$ to reach the bulk vaporization threshold. When this threshold is reached, an appreciable fraction of the energy incident upon the electrode must now go into the latent heat of vaporization for the material, and the slope of the erosion curve decreases again, corresponding to region 3 in Figure 4.20. The exact value

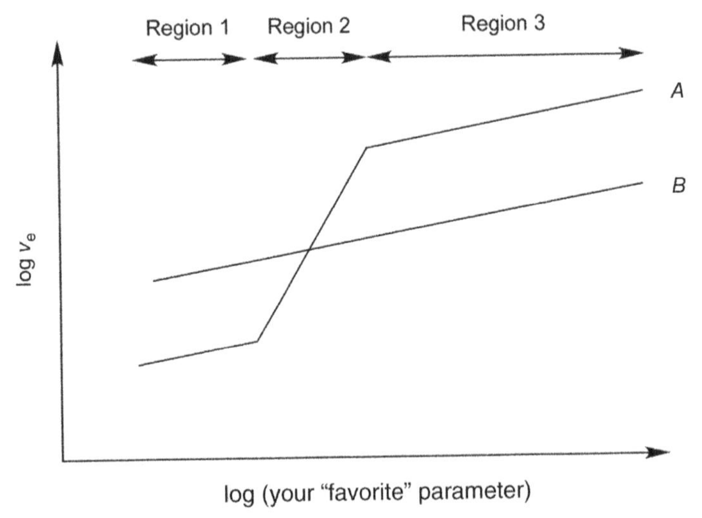

Figure 4.20 The generic curve for the volume erodes from an electrode as a function of one of the parameters related to erosion [52].

of a given pulse parameter at which this "break" in the erosion curve between regions 2 and 3 will occur for a particular material depends on the thermophysical properties of the material. The slope of the curve in region 3 will depend on these properties and the relative values of the molten and vaporized metal ejection coefficients.

A notable exception to curve *A* is curve *B* in Figure 4.20, which applies to materials that experience vaporization throughout the range, such as graphite. The generic curve *B* explains the contradictory experimental results that list graphite as a material with the lowest erosion rate [51,53,54] as well as the highest [49,55,56]. Care must be taken when using graphite, however, because unlike copper and molybdenum whose properties are fairly standardized for a given degree of purity, graphite properties vary widely (up to an order of magnitude in some cases) depending on the manufacturing process.

Actual calculations of the coefficients of material removal may never be realized but are known to increase with increasing current [43], gas flow rate [44], and decreasing gap distance [57] as well as being a function of several material properties. In small gaps, erosion can be worse than in larger gaps because of the formation of high-velocity vapor jets [56,57], which result in an electrode separation-dependent erosion rate.

Donaldson and Kristiansen [58] have shown the effect of various operating conditions on the generic arc erosion curve for pure metals, which is shown in Figure 4.21. The "transition region"—where bulk melting becomes operative—is a "collective" arc phenomenon.

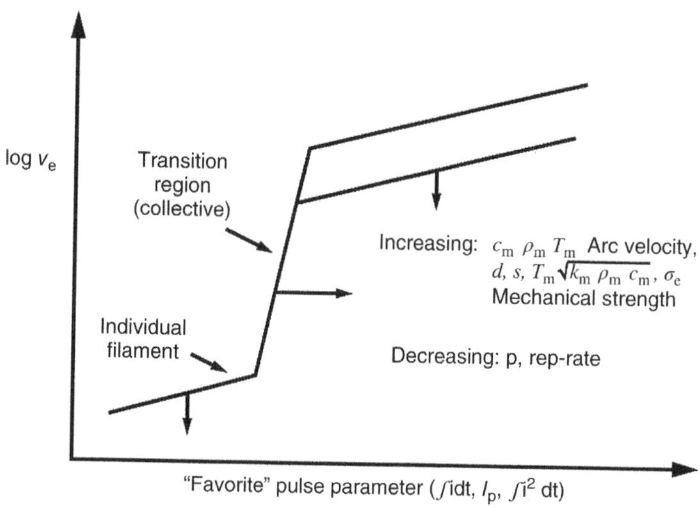

Figure 4.21 Electrode erosion has a characteristics shape that may scale with a variety of parameters such as charge transferred per shot, the peak current, or the action [58].

4.1.4 Current Sharing in Spark Gaps

In some instances, it is desirable to split the circuit current into a number of channels. This may be achieved by sharing the current by using several spark gaps switches in parallel or by designing a single switch to produce multiple channels. The latter type of switch is usually referred to as a multichannel switch.

There are three dominant reasons for provoking current sharing: (i) lifetime extension, (ii) inductance reduction, and (iii) uniform injection of energy from one energy storage stage to another. Current sharing extends the lifetime of electrodes and switches by reducing erosion and the associated switch contamination that scales roughly with current. The overall inductance of the switch is minimized because the contribution from the arc channel inductance is reduced due to the parallel configuration. The relation between the number of channels and the overall inductance depends on the specific geometry. Current sharing may be used to uniformly inject energy into a geometry of large dimensions and may be done with a single multichannel switch or with several switches in parallel.

Under normal circumstances, once a spark channel forms, it will prevent other channels from being formed because the voltage has collapsed to a very low voltage V_s determined by the resistance of the spark channel. The collapsed voltage wave rapidly travels across the electrode area with the velocity of propagation of an electromagnetic wave in a dielectric material to pull down the voltage across the entire gap to a low potential V_S. V_S, being very low, will not be able to support the development of other spark channels.

4.1.4.1 Parallel Operation

Operating spark gaps in parallel generally have the same requirements that encourage the generation of multiple channels in a single spark gap. The individual switches should have sufficient distance between them to encourage time isolation. The switches should have a small jitter. Triggered spark gaps should have a robust trigger pulse to ensure each spark gap fires within a short time window. A single trigger generator driving each of the current-carrying channels has the great advantage of eliminating the trigger generator contribution to the jitter.

Spark gaps operated in parallel may be triggered or untriggered and configured in any number of ways, including that shown in Figure 4.22 [59].

Untriggered spark gaps should be pulse charged and have significant time isolation. One way to enhance the time isolation is to use a dielectric impedance insulator between the switch sites. A magnetic material could be inserted between the switch sites [60]. A more common method is to use a high-electrical-permittivity material, such as water [61,62].

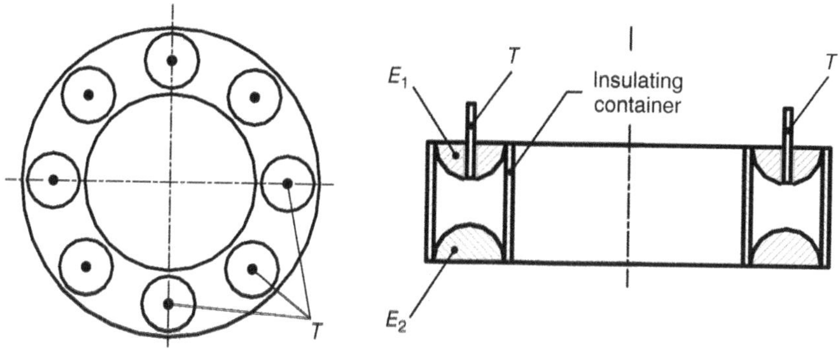

Figure 4.22 Paralleled high-voltage trigatron scheme with eight spark channels [59].

4.1.4.2 Multichanneling Operation

When an impulse voltage of sufficient amplitude is applied to a spark gap, a spark channel is formed at the weakest point and collapses the voltage at that point to a very low voltage V_s. Once formed, the spark channel prevents the formation of other channels across the area of the electrodes. When the first channel is formed, a voltage collapse wave ($V_0 - V_s$) travels rapidly across the electrode area, resulting in a low gap potential of V_s that is unable to support the development of additional spark channels.

It can be deduced from the mechanism of spark formation described above that the probability of multiple parallel channel development will be increased under the conditions of a reduced delay time. This implies a reduced statistical time lag t_s, indicating a large number of weak points are available simultaneously at many locations and a reduced formative time lag t_f, which means the growth of spark channel and its bridging across the electrodes are accomplished very rapidly. The methods to induce multiple channels are those that also reduce the jitter providing sharp edges to one of the electrodes in case of self-fired gaps or and increasing dV/dt on the gap. The principle in this is that the intensified field at a sharp point makes available a copious number of field-emitted electrons, thus enabling initiation of breakdown process at numerous locations simultaneously, reduction of τ_f by application of a large overvoltage $V_0 \gg V_{BD}$, where V_{BD} corresponds to the breakdown voltage, with a fast rise time dV/dt of more than 10^{13} V/s. This provides rapid heating of plasma and imparts enhanced velocity to rapidly bridge the gap between electrodes. Time isolation between adjacent channels can be aided by using a dielectric of high dielectric constant (like water with $\epsilon_r = 80$) to slow down the velocity of propagation of the collapse wave. This enables the spacing between adjacent spark channels to be reduced and increases the possibility of channel formation in a given electrode area.

4.1.4.2.1 Rise Time of a Multichannel Spark Gap

When the current is shared by creating multiple arc channels, the previous relation for the resistive rise time, Equations 4.16–4.18 hold with the transformation $Z \rightarrow (NZ)$, where N is the number of channels formed. For a multichannel spark gap, Equation 4.16 becomes [63]

$$\tau_R = \left(\frac{88}{E^{4/3} \cdot (NZ)^{1/3}} \right) \cdot \sqrt{\frac{\rho}{\rho_0}} \text{ ns} \tag{4.31}$$

The inductive time constant is

$$\tau_L = \frac{L_a/N + L_{electrodes}}{Z} \tag{4.32}$$

where

L_a is the inductance of one spark channel,
$L_{electrodes}$ is the inductance of the electrode/spark gap hardware,
N is the number of spark channels,
Z is the impedance of the driving source, and
E is the mean electric field (in units of 10 kV).

The total τ_{tot} and rise time τ_R are given by Equations 4.21 and 4.22.

An important observation is that as the number of channels increase, the inductive rise time τ_L decreases as N^{-1}. However, since the resistive rise time reduces only as $N^{-1/3}$, beyond a certain number of channels the incremental advantage gained in the reduction of τ_R is negligible. In most gases, with hydrogen being a notable exception, the resistive time constant is the dominant contribution to the rise time. Multichanneling assumes importance in high-voltage, low-impedance systems, since the large rise time resulting from low impedance may be offset by a large number of channels.

4.1.4.2.2 Optimizing the Number of Channels

The key to producing multiple channels in spark gaps is to ensure that each current-carrying channel is independently formed. For this to happen, the voltage across the spark gap must remain at the critical breakdown value during the time when channels are forming. Martin derived the expressions under which multichanneling may occur and performed the seminal work on its accuracy.

Martin used these arguments to examine multichanneling in terms of an elapsed time or a window of opportunity for two considerations that may begin to collapse the voltage across the spark gap. The first consideration is the time for the gap voltage to fall an amount that will no longer support channel formation. In his note on multichanneling, Martin states he initially assumed that time would be the e-folding rise time, τ_{tot}, which would decrease the voltage on the gap to approximately 63% of the charging voltage. Experiments indicated the elapsed time to be a fraction of this value, which Martin estimates to be about 0.1. This is an indication of the importance of maintaining voltage on the spark gap: Only a very small fraction of the voltage may decay and still result in channel formation.

When channels are formed, a time-varying current develops as the gap voltage falls. This current has an associated magnetic field that inductively couples one channel site to the next as the channels are formed. The inductive coupling has the effect of decreasing the voltage on the adjacent channel nucleation sites that are insufficiently established, and this wave is sometimes referred to as a "voltage collapse wave." For a number of channels N formed along a length ℓ of electrode edge, the average distance between channels is ℓ/N and the time to travel that distance, τ_{trans}, can be determined by the distance between channels and the speed of light in the dielectric medium. Because the distance between channels is not uniform, Martin derated the length between channels to factor of 0.8.

The window of opportunity ΔT for adjacent channels to be established is given by the semiempirical relation:

$$\Delta T = 0.1 \cdot \tau_{tot} + 0.8 \cdot \tau_{Trans} \tag{4.33}$$

The window of opportunity for a gas-insulated multichannel spark gap is obtained by inserting Equation 4.31 for τ_{tot} into Equation 4.33:

$$\Delta T = (0.1) \cdot \sqrt{\left(\frac{L}{NZ}\right)^2 + \left(\frac{88}{E^{4/3}(NZ)^{1/3}}\right)^2} + 0.8\left(\frac{\ell \cdot \sqrt{\epsilon_r}}{N \cdot c}\right) \tag{4.34}$$

where

N is the number of formed channels,
ℓ is the length along the edge,
ϵ_r is the dielectric constant of the dielectric medium used in the gap, and
c is the speed of light.

Martin [63] provided the definitive concepts and validation on spark formation in multiple channels. Martin's empirical formulas enable an estimate of the number of channels for a spark gap with given physical dimensions, insulating dielectric, and most importantly, the characteristics of the voltage applied to the main and trigger electrodes. The empirical formulas are based on the closure time characteristics of the voltage collapsing in the gap, after a few initial channels bridge the gap and the temporal characteristics of this collapsed wave as it sweeps across the electrodes, to collapse the voltage across the gap. Channels nucleated after this interval will either extinguish or have weak growth and be unable to carry substantial current.

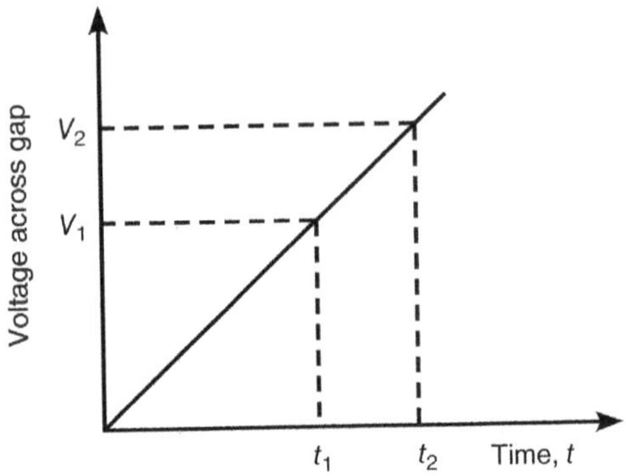

Figure 4.23 The voltage across the spark gap increases during the time $(t_2 - t_1)$ between the formation of adjacent channels.

The above explanation, however, assumes negligible switch timing jitter relative to the average delay time. Jitter may be quantified by examining the effect of jitter on the voltage. In spark gaps, multichanneling is encouraged either by pulse charging or by triggering the switch with a fast-rising pulse. Both of these methods result in the gap voltage changing rapidly with time [64], as shown in Figure 4.23.

The delay time is the average time it takes for channel to be fully formed and conduct the circuit-limited current and the jitter is the variance around that mean. A switch that commutes current at t_1 will be charged to a voltage of V_1 and one that commutes at t_2 will be charged to V_2. This is the shortest time for streamers to form. Thus, a distribution in time, $F_t(t)$, has an associated distribution in voltage, $G_V(V)$. The relation between the functions $F_t(t)$ and $G_V(V)$ could be determined by experimental conditions, but are rarely measured. Instead, in practice, only the deviations in time, $\sigma_T(T)$ and voltage $(\sigma_V(V))$ are measured.

Assume that $\sigma_V(V)$ and $\sigma_T(T)$ are given as percentages. The change is voltage, ΔV, and switching time, ΔT, are given by

$$\delta V = \sigma_V(V) \cdot V \quad \text{and} \quad \delta T = \sigma_T(T) \cdot T \tag{4.35}$$

$$\frac{dV}{dT} \approx \frac{\delta V}{\delta T} = \frac{\sigma_V(V) \cdot V}{\sigma_T(T) \cdot T} \tag{4.36}$$

Rearranging terms,

$$\Delta T = 2 \cdot \sigma_T(T) \cdot T = 2 \cdot \frac{\sigma_V(V) \cdot V}{dV/dT} \tag{4.37}$$

The factor of two is introduced to reflect the percentage change in voltage; as expressed, a percentage is actually $\pm\sigma_V(V)$.

Equation 4.37 represents the time window ΔT available for the growth of channels nucleated at various sites along the sharp electrode. The channels that are initiated at a later time will not lead to successful, current-carrying channels. The quantities $\sigma(V)$, V, and dV/dt are determinable in experiments.

For multiple channels to be formed, it is imperative that the voltage be maintained on the channel nucleation sites. From the above analysis, it is evident that the probability of multichanneling increases with high electric fields, high rates of voltage application (dV/dt), long electrode lengths, and large time isolation (high dielectric constant insulators). The parameters that produce multichanneling in spark gaps also result in low switching jitter.

4.1.5 Triggered Spark Gaps

Triggered switches are capable of producing subnanosecond jitter and can greatly affect the capability and reliability of a device. Electrical triggering methods dominate, but laser-triggered switches are also widespread. A good review of high-power switch technology is given by Burkes et al. [65]. A large number of other triggering methods have been investigated such as electron beam, X-ray flash, magnetic field, mechanical force, gas injection, metal vapor injection, and exploding wire. The salient features of these various trigger modes are described in this section.

There are three basic triggered spark gap types: the trigatron, the three-electrode field distortion spark gap, and the laser-triggered spark gap. The first two use an electrical pulse generator and the third uses a pulsed laser. To achieve a low overall command triggered performance, the trigger generator must also have a low jitter.

4.1.5.1 Operation of Triggered Spark Gaps

Triggered spark gaps are used to increase the control of the power conditioning system. Almost exclusively, increased precision in timing jitter is the objective. However, in some cases, the objective of triggering the switch is to provide pulses with a well-known amplitude or even to provide a timing signal for other components.

4.1.5.1.1 Switching Time of a Triggered Spark Gap

The delay time τ_D of a triggered spark gap is usually measured from the moment of application of trigger pulse until the impedance of the gap starts collapsing. Consider a spark gap that is closed by overvolting. Initially, the spark gap voltage is set below the self-breakdown voltage. The time for breakdown to develop is comprised of four parts: (i) the statistical time delay for the appearance of a free electron t_s, which may be reduced to zero with the application of the trigger pulse; (ii) the streamer formation time t_f, which is inversely proportional to the electric field; (iii) the channel heating time t_{ch}, which is also inversely proportional to the electric field; and (iv) the trigger pulse rise time.

The statistical portion of the time lag accounts for the generation of initiatory electrons. When a trigger source explicitly provides electrons, as is the case with a UV illumination, this assumption is clear. However, even in the case where a large, fast-rising voltage pulse is used to trigger the switch, the resulting large overvoltage across the switch will also result in a negligibly small contribution from the statistical time lag to the overall delay time, and the delay time may be considered to be wholly composed of the τ_f. The formative time lag is also reduced by using a robust trigger voltage to provide a substantial overvoltage to the switch since both the formative time lag and the channel heating time are inversely proportional to the applied electric field.

4.1.5.1.2 Triggered Switching Range

Untriggered spark gap switches operate at their self-break voltage. Triggered switches have a minimum operating voltage below which the switch does not fire upon application of a command trigger. The spread of voltages lying between the limits of minimum operating voltage V_{min} and the self-breakdown voltage V_{SB} is called its switching range, and is depicted by the hatched portion in Figure 4.24.

A design parameter of a triggered spark gap is the operating voltage V_{op}. For a given pressure, the operating range is $(V_{SB} - V_{min})$. V_{op} should be much lower than V_{SB}, line *BB* in Figure 4.24, to ensure that it does not prefire—that breakdown does not take place in the absence of a command trigger. The lower bound, V_{min}, denoted by the line *AA*, is more complicated to predict and

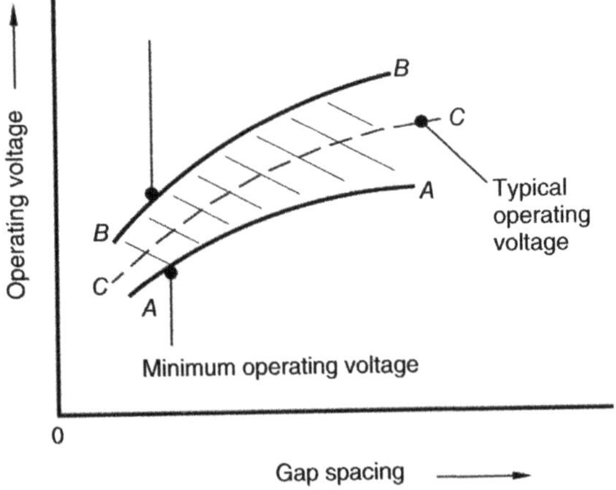

Figure 4.24 Switching range of a triggered spark gap for a fixed pressure.

depends on the trigger mode and the specifics of the trigger. For operating voltages V_{op} below V_{min}, the spark gap switch will not close even with command trigger. The spark gap switch will be most reliable for V_{op} located on the middle path *CC*, shown by dotted lines in Figure 4.24.

In practice, the gap spacing is determined by inductance considerations and the trigger is fixed *a priori*. The triggered spark gap switch is characterized by a *V–p* curve, as shown in Figure 4.25. The jitter may be low when operating close to V_{SB}, but the likelihood of a prefire increases greatly. The choice of operating voltage may be determined by jitter considerations since the timing jitter usually increases dramatically close to the V_{min}. Optimum V_{op} is typically 70–85% of self-breakdown voltage. As a compromise, Beverly and Campbell [66] use the following expression as a compromise:

$$V_{op} \approx 0.6\, V_{SB} + 0.2\, V_{\min} \tag{4.38}$$

Switches with a wide operating range are used in applications where the switch must first remain open for a duration when high voltage is applied and then close during a time of low voltage. A well-known example is a crowbar switch.

4.1.5.2 Types of Triggered Switches

Triggered spark gap switches can be designed with great variety in spark gap geometry and trigger methods. The specifics of the design include the switching media, the gap distances, and the trigger source. The trigger source, in particular, often comes with certain limitations. The most common triggered switches are initiated with an electrical pulse or a laser.

4.1.5.2.1 Electrically Triggered Switches

Electrically triggered switches generally fall into the specific category of a trigatron or the broad category of the directly overvolted spark gaps. The operation of electrically triggered switches is very closely linked to the

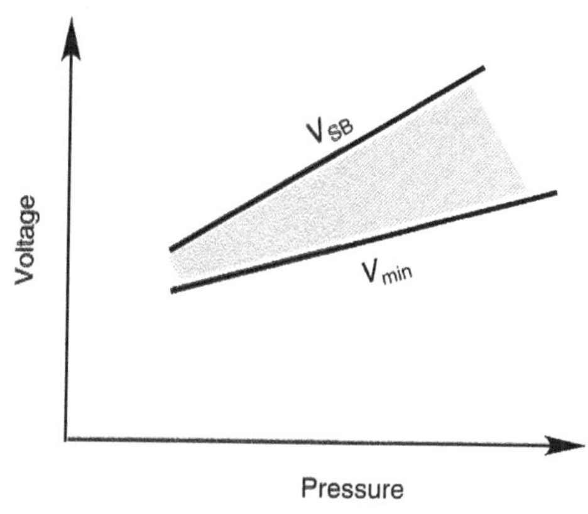

Figure 4.25 The operating range of a spark gap switch is measured by a *V–p* curve. The operating range may be given as a fraction of the self-breakdown voltage V_{SB}.

characteristics of the trigger source. A distinct difference is that the trigger voltage of a trigatron can be too high for optimum performance, but directly overvolted gaps perform more reliably with higher trigger voltages as long as it is applied fast enough.

Trigatrons Trigatrons, first described by Craggs et al. [67] and by Wilkinson [68], are widely used in pulsed power technology because of their wide voltage operating range and simplicity and are still actively studied. Trigatrons are generally comprised of uniform field electrodes with a third electrode, or trigger pin, embedded in one of the electrodes, as shown in Figure 4.26. The trigger pin is typically inserted into the load side, or earthed, electrode and is insulated by an annular gap that is filled with an insulator. This electrode is referred to as the adjacent electrode, and the electrode where the voltage is applied is the opposite electrode.

There are four polarity combinations of opposite electrode voltage and trigger voltage that are possible. When these voltages are of the same polarity, the trigatron is said to be homopolar, and when opposite, heteropolar. Triggering is almost universally observed to be better with the heteropolar configuration, where the trigger voltage increases the average electric field in the main gap [69,70]. Very slow discharge velocities have been observed in the homopolar trigatrons, consistent with the decrease in average electric field upon application of the trigger voltage.

Although the trigatron has been extensively used for over half a century, the mechanism of operation of the trigatron is not well established. Many researchers [71–73] erroneously assume that the initiation mechanism is a breakdown from the trigger pin to the adjacent electrode with a subsequent breakdown in the main gap aided by the ultraviolet emission from the trigger gap as well as the increased average electric field. This mode, "breakdown to the adjacent

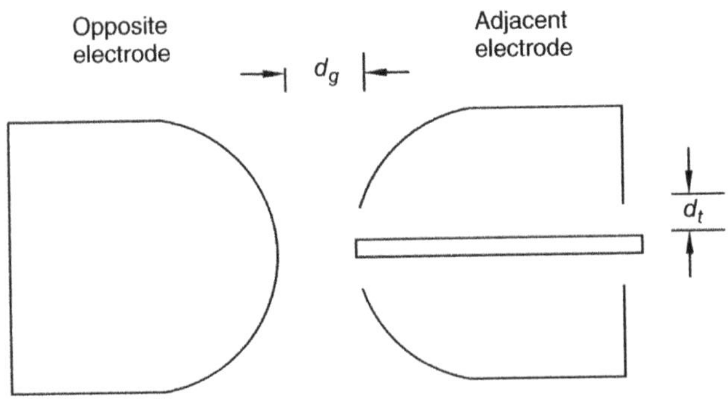

Figure 4.26 The basic geometry of a trigatron switch has a trigger pin embedded in one of the electrodes. The main voltage is usually applied to the opposite electrode.

electrode" (referred to as BAE by Beverly and Campbell [66]), is now generally accepted to result in long delay times and excessive jitter.

Another mode has been identified where the trigger pin breaks down to the opposite electrode—Beverly's BOE mode. The initiation mechanism—field distortion at the trigger pin resulting in streamer propagation to the opposite electrode—was first suggested by Craggs et al. [67] but generally discounted by subsequent researchers until Shkuropat's insightful work [74]. The field enhancement at the trigger electrode is estimated to be a factor of at least 2 over the field in the main gap [75]. Shkuropat suggested the initiating events occur before the collapse of the trigger gap voltage and that a streamer initiated directly from the enhanced field around the trigger pin tip was involved. This model required only that the streamer had to be initiated before the trigger gap broke down. This model was extended by Peterkin and Williams [76,77] to show that the enhanced field near the trigger pin tip launches a streamer that propagates across the gap, bridging it with a low-conductivity channel. This channel subsequently thermalizes to form the arc to carry current between the main electrodes and close the switch. Propagation of the streamer tip depends on the presence of an enhanced electric field in front of it. This model indicates it is not necessary for the streamer to be fully established in the gap, but simply to transverse a portion of the gap before the trigger gap breaks down. Mesyats [78] points out that this mode of operation can be viewed as a point-to-plane breakdown and the well-known polarity effect can be exploited. Additional insights into the physics of trigatron initiation are documented by MacGregor et al. [69] for homopolar as well as heteropolar configurations.

Both Wilkinson [68] and Wooton [79] noted that increased trigger voltage does not necessarily improve trigatron performance. Williams and Peterkin's model [76] predicts that increasing trigger voltage improves performance only until the trigger gap breaks down before the streamer transverses the gap. Thus, the design of an optimum triggering system involves balancing the requirements of maximizing the field enhancement near that trigger pin tip with delaying the breakdown of the trigger gap until the streamer at least partially transverses the gap.

In the design of a trigatron, the parameters to be chosen are the main gap distance d_g, the trigger gap distance d_t, the pin height, and the trigger voltage. Assuming heteropolar operation, the average electric field in the trigger gap E_t is established by the trigger voltage V_t across the annular gap d_t between the trigger pin and the adjacent electrode and is given by

$$E_t = \frac{V_t}{d_t} \tag{4.39}$$

In the main gap, the average electric field E_g is given by

$$E_g = \frac{V_g - V_t}{d_g + h} \tag{4.40}$$

where V_g is the charging voltage, d_g is the main gap distance, and h is the distance the pin is recessed into the adjacent electrode. The relative magnitudes of E_t and E_g determine the mode of operation of the trigatron:

$|E_t| > |E_g|$ breakdown occurs from the trigger pin to the adjacent electrode

$|E_t| < |E_g|$ breakdown occurs from the trigger pin to the opposite electrode

Optimal results with a megavolt trigatron occurred by setting $E_t = -E_g$, yielding a critical trigger voltage V_t^* given by

$$\frac{V_t^*}{V_g} = \frac{-d_t}{d_g + h - d_t} \tag{4.41}$$

Operation at the critical trigger voltage yielded simultaneous breakdowns between the trigger pin and both electrodes and with minimum jitter.

Beverly [66,80] has examined the performance of trigatrons based on their initiation modes and found that the mode of operation that minimizes the jitter (negatively charged opposite electrode with a positive trigger—denoted mode *A* in Figure 4.27) also yields the greatest voltage operating range. For optimal operation, the pin should be in level with the adjacent electrode surface. In Figure 4.27, for a trigatron operating in mode *B*, where the opposite electrode is

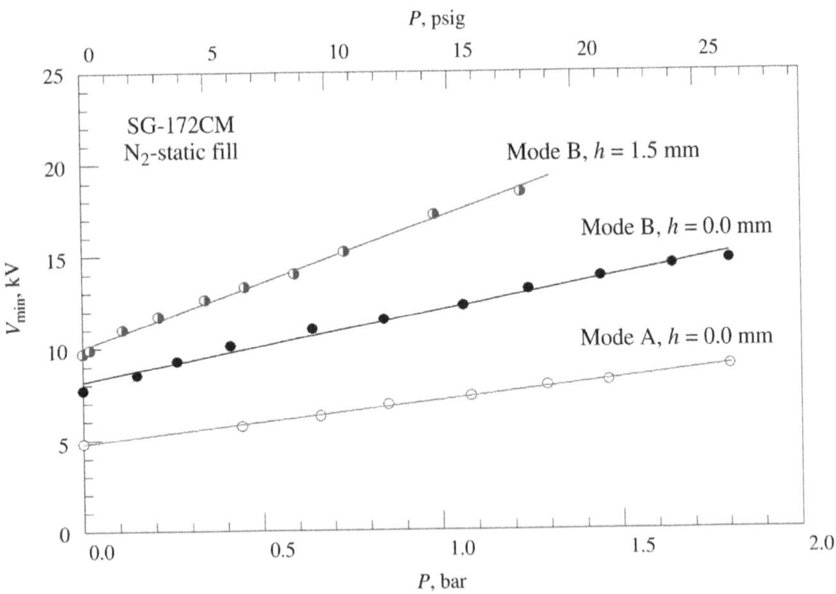

Figure 4.27 V_{min} is influenced by both the operating mode and the distance the pin is recessed. The widest voltage operating range is obtained for a negatively charged trigatron with a positive trigger voltage [63]. (Reproduced with permission of Robert E. Beverly.)

charged positive, a pin recessed by 1.5 mm into the adjacent electrode, the anode significantly affected the triggered voltage operating range.

Trigatrons can be used in a repetitive mode if care is taken in the electrical isolation between the trigger generator and the electrode. Electrical isolation is critical since the trigger pin can rise to the switching potential for a short time [81]. For repetitive systems, an inductor is ideal but suffers from a lack of access to the spark gap. In one trigatron operated at 1 kHz, the access and electrical isolation were provided by a ferrite-loaded, spiral-wound ignition wire [82]. The trigger pin is often recessed to address the erosion of the trigger pin that can be a significant issue. The material for the trigger pin affects the erosion and graphite has been successfully used as a trigger pin material.

The trigatron type of triggering has been developed using various switching media such as vacuum [83–91] and gas [92–96], as well as liquid dielectric [97–99]. An unconventional slot-type of vacuum gap based on trigatron triggering has been used by Kichaeva [83], which is shown in Figure 4.28a. Advanced vacuum trigatron gaps with graded electrodes [84,85,90,100] have also been used. An example of a graded electrode vacuum trigatron switch is shown in Figure 4.28b. Trigatrons have demonstrated subnanosecond jitter with high-performance triggers [101]. Investigations of the switching mechanism and optimization of trigatrons are ongoing [102].

Double-Mode Trigatron Superior results in the various parameters of operating range, delay time, rise time, and jitter have been obtained with a double-mode trigatron [103–105], shown in Figure 4.29. A double-mode trigatron has a trigger electrode embedded in each electrode: T_1 in E_1 and T_2 in E_2. The application of twin trigger pulses results in the simultaneous breakdown of both

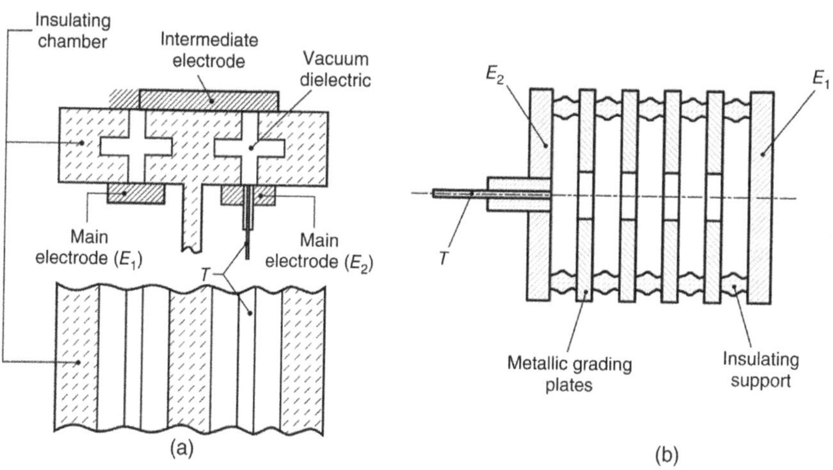

Figure 4.28 A vacuum trigatrons (a) of the slot type and (b) with graded electrodes.

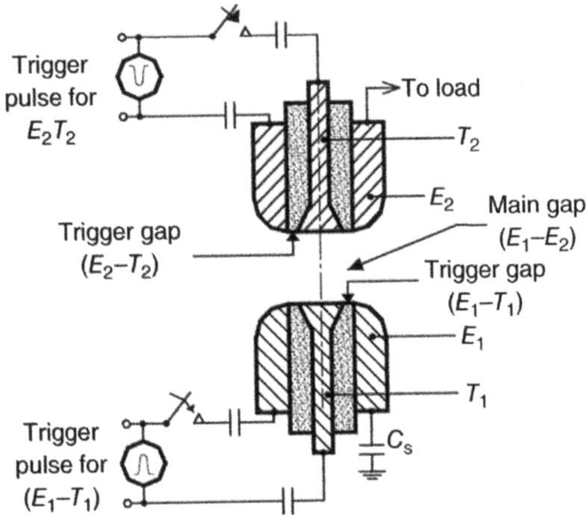

Figure 4.29 A double-mode trigatron uses a trigger pin in each electrode [103].

the auxiliary gaps. Minimum jitter in a double-mode trigatron is obtained when the polarity combinations are (E_1-, E_2+, T_1-, T_2+). The double-mode gas trigatrons are well suited for operation of many switches in parallel and have been successfully employed in an 85 kJ/20 kV capacitor bank for generation of pulsed high-magnetic fields [106]. The low jitter achieved with the double-mode trigatron has been used to study antenna arrays for radiating ultra-wideband pulses [105].

Field Distortion Spark Gaps Field distortion spark gaps, used extensively in applications requiring low jitter, are triggered by altering the electric fields between the electrodes by applying an external trigger pulse to a spark gap already charged to its operating voltage V_{op}. The increased voltage across the spark gap results in switch closure. Before triggering, the electric fields in the spark gap are fairly uniform so that the switch remains reliably open during the charge cycle and closes only upon application of the trigger pulse. The switch performance is largely determined by the characteristics of the trigger pulse. Precise switch operation with low jitter can be achieved with field distortion spark gaps by using a fast-rising trigger pulse (>10 kV/ns) with a recommended peak trigger voltage at least equal in magnitude to the charging voltage. Increased trigger peak voltage generally leads to better switch performance.

One way to distort the field in a spark gap is to directly overvolt the gap. The trigger pulse introduces an effective field enhancement of

$$\frac{V_T + V_{op}}{V_{op}} \tag{4.42}$$

where V_T is the change in voltage due to the incoming trigger pulse, with peak voltage V_t, and V_{op} is the operating voltage.

The simplest geometry is a two-electrode spark gap where the internal electric fields are distorted directly by the addition of a trigger voltage, as shown in Figure 4.30. This two-electrode gap is triggered by overvolting the main gap $E_1 - E_2$ with a fast-rising high-voltage pulse applied to the top electrode. The trigger pulse also drives a current through the load, which is usually undesirable, and reduces the amount of overvoltage on the main gap. This effect is minimized by increasing the impedance of the load circuit by inserting ferrite rings on the top electrode, which isolates the load circuit from the main gap during the trigger pulse period. The additional inductive loading is automatically removed when the energy storage capacitor C_s discharges, causing current flow and driving the ferrite into saturation.

The efficiency of the trigger pulse is enhanced by inserting a trigger electrode between the main electrodes of the spark gap. Applying the trigger voltage to the third electrode means the trigger pulse has only to overvolt one section of the total gap. Moreover, it permits the use of a high-impedance trigger circuit in a low-impedance switch.

Field distortion spark gaps can be operated in uniform or nonuniform electrode geometries. In nonuniform geometries, sharp edges are introduced to further distort the field upon application of a trigger pulse. In either case, very high trigger voltages are needed.

In the uniform field mode, triggering is achieved largely from overvolting one of the gaps by applying a trigger pulse. Gap closure is initiated by applying a pulsed voltage to the trigger electrode, which changes the local electric fields from its initial pretrigger distribution. This distortion of the electric field

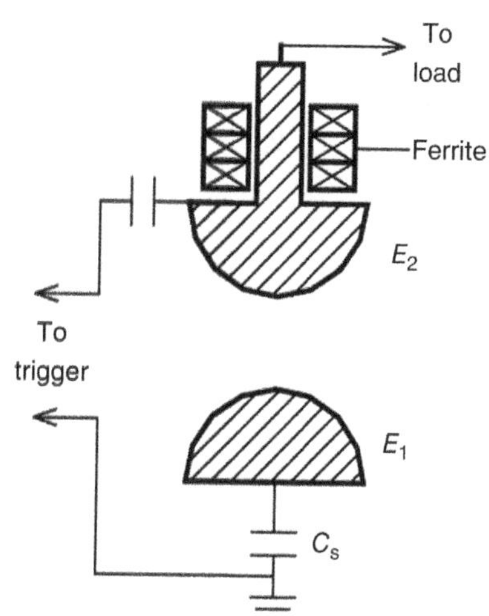

Figure 4.30 A two-electrode spark gap may be triggered by applying a trigger pulse directly to one of the electrodes to cause the switch to close.

distribution near the trigger electrode exceeds the breakdown threshold, causing a discharge from the trigger to main electrode. Nearly the full gap voltage then appears across the trigger electrode and the other main electrode, closing the switch. The polarity of the trigger pulse determines which gap closes first. Directly overvolted spark gaps usually consist of three-electrode gaps, but switches with many more gaps, arranged in a cascade arrangement, are increasingly being used [3,107].

Like a cascade gap, the voltage of the intermediate electrode before applying the trigger voltage is determined by the distribution of capacitances within the switch. This floating electrode arrangement can be problematic and it is common to bias the potential of the trigger electrode with a resistive network. Figure 4.31 shows a three-electrode gap with the trigger electrode located midway between the two main electrodes. Because the two gaps are equal, the values of the biasing resistors are also equal and hold the trigger electrode to a potential of $1/2V_{op}$.

Further enhancements may be achieved by preconditioning the switch with UV illumination before the application of the trigger pulse. A small spark gap can be inserted into the trigger electrode as a source of UV preionization, as shown in Figure 4.32, and is sometimes called a swinging cascade [108].

The distortion of the electric field may be further enhanced by adding some sharp points to the trigger electrode. A trigger pulse generates an enhancement of the average electric field first between the trigger electrode and the adjacent main electrode and, second, of the local field at the edge. In the first case—as in the uniform field case—the field enhancement is not large. In the second case, it is strongly dependent on the radius of curvature of the edge, which can be a factor of several hundreds. The trigger electrode may be a pin, as shown in Figure 4.33a, but is usually some type of plate (Figure 4.33b).

Before applying a trigger pulse, no additional field enhancement exists at the edge of the trigger electrode. When the trigger electrode has a strong field

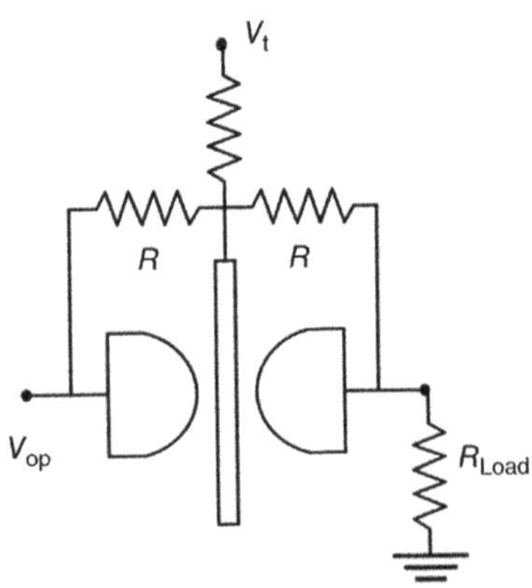

Figure 4.31 A three-electrode, uniform field triggered spark gap switch.

Figure 4.32 The swinging cascade switch uses a small spark gap inserted into the trigger electrode as a preionizing source.

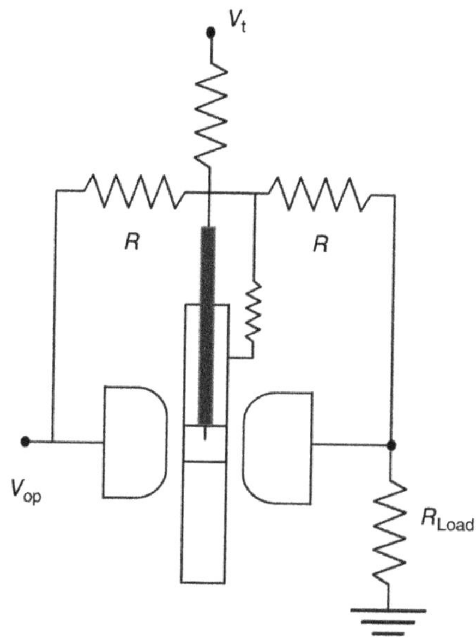

enhancement, it is held to a proportion of the voltage corresponding to its physical spacing with a biasing network so that the sharp edge does not appreciably disturb the equipotential lines from that of a uniform field distribution.

The trigger electrode may be positioned at any fraction of the gap, as shown in Figure 4.34, dividing the main gap into two gaps of spacing x and y, respectively. The ratio of the offset can vary from 1:2 to 1:15. For a balanced charge system, it is often convenient to place the trigger electrode midway between the main electrodes. However, in general, it has been found that field distortion gaps operate best when the trigger electrode is displaced toward one electrode and the polarity of the trigger is chosen such that the larger gap is triggered first.

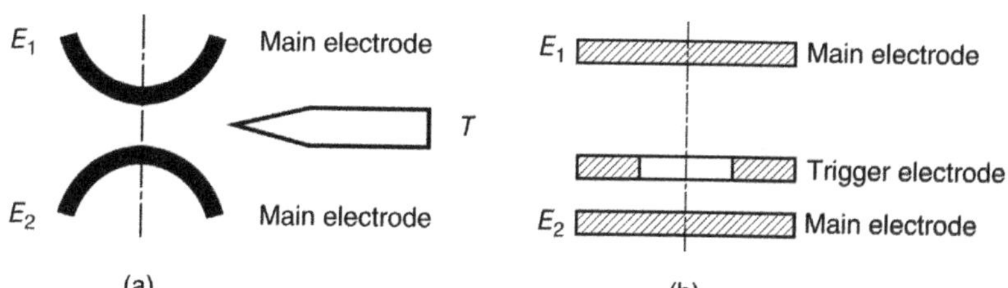

Figure 4.33 Field distortion triggered spark gaps can be implemented in many different ways. Local electric fields may be enhanced by incorporating sharp points on the trigger electrode.

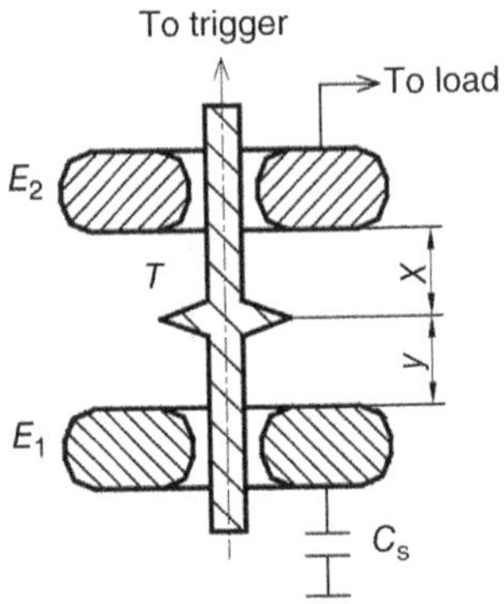

Figure 4.34 The trigger may be placed at any fraction of the gap length ($x+y$). Various insulating media and operating conditions have different optimal fractions.

By offsetting the trigger electrode, it is possible to arrange for both halves of the gap to break down almost simultaneously. Under these conditions, the gap may be triggered down to less than 40% of the gap self-breakdown voltage with a jitter of a few nanoseconds [109].

Two geometries of three-electrode field distortion spark gap switches are shown in Figure 4.35. An 85 kV, 200 kA switch, shown in Figure 4.35a, was developed by Post and Chen [110] and operated six in parallel to discharge a 220 kJ capacitor bank. The main electrode material is Kulite containing tungsten, nickel, and copper in ratio of 90:6:4 and the trigger electrode material is Elkonite containing tungsten and copper in ratio of 75:25. Falten et al.'s [111] coaxial geometry field distortion switch with tantalum inserts in the sparking region is shown in Figure 4.35b. The 220 kV, 40 kA switch operates at a pulse repetition rate of 1 kHz by flowing an SF_6–N_2 insulating gas mixture through the discharge region.

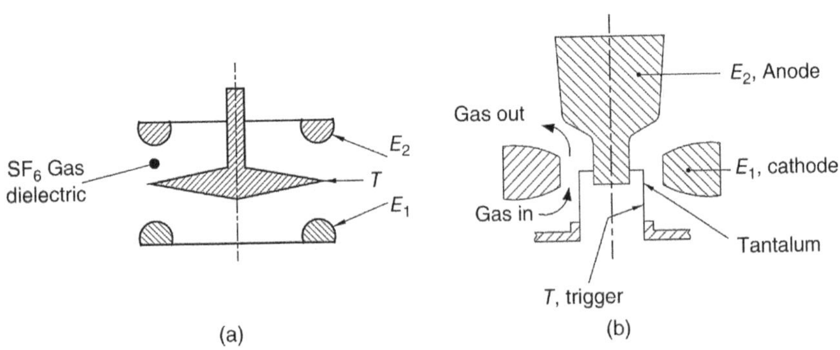

Figure 4.35 Field distortion-type triggered spark gap switches may be configured in a number of ways. In part (a) a cascade spark gap is common, but a coaxial geometry (b) may also be used.

Field distortion-triggered spark gaps may be used to switch any voltage level required with the proper choice of gas, pressure, and gap spacing. The field distortion triggering has been used in gaseous dielectrics [110–113], liquid dielectrics [114,115], and solids [116–118]. For liquids and gases, the field distortion and rail gaps are used. Generally, the preferred offset ratios for the trigger electrodes are 1:3 for gases and 1:7 for liquids. For switching voltages more than 5 MV, mineral oil is preferred to water dielectric.

4.1.5.2.2 Laser Triggering

With the development of Q-switched lasers of unprecedented power in 1963, optical breakdown of gases was discovered when the light of a ruby laser passed through a focusing lens, creating a spark in the focal region. Ionization of a gas by laser light occurs via multiphoton process—the simultaneous absorption of several photons. In the visible range, single-photon ionization is impossible because the photon energy is well below the atomic ionization potentials. While the ionization probability for multiphoton processes is low, it increases dramatically with light intensity and explains the importance of focusing optics. Experiments have shown that electron densities of $10^{13}\,cm^{-3}$ are produced [119]. This seed electron population acts as initiatory electrons that are quickly accelerated in the external field, leading to closure of the spark gap.

Laser triggering of spark gaps became viable when pulsed high-power lasers became commercially available. Early research on laser-triggered gaps used an air- or N_2-filled spark gap with a ruby laser focused on the electrode surface resulting in short time lag and nanosecond jitter [120,121]. The required laser power, however, was over 100 MW of peak power [120]. Later efforts used the fundamental wavelength of the KrF laser (1064 nm) with mixtures of argon (Ar) and achieved subnanosecond jitter with ~10 MW of power [121]. However, the low jitter of these laser-triggered switches required the spark gap use Ar gas mixtures. Ar, a gas with a breakdown voltage of ~5 kV/cm, led to interelectrode gap lengths of several centimeters for even moderate voltages. This triggering mechanism, when the laser is focused on the electrode, is not well understood but results in impressively short jitter and delay times. It has found few applications, however, because in practice the laser ablates the target electrode, coating the inside of the switch with conducting particles [121], severely limiting its lifetime.

Interest in using laser-triggered switches in large pulsed power machines was piqued when Rapoport used a harmonic of a KrF laser to trigger an SF_6-filled spark gap [122]. For switching very high voltages, lasers have a distinct advantage in that the required laser-triggered energy does not scale with voltage as field distortion gaps do, making laser triggering a relatively economical choice for very high-voltage spark gaps. The introduction of high-power, pulsed KrF lasers led to investigations on SF_6-filled spark gaps using the harmonics of the fundamental infrared wavelength and its harmonics (532 and 266 nm), as well as shifting the focal point to the ionization of the gas itself. Using this laser-induced optical breakdown mechanism, Woodworth triggered large megavolt-class spark gaps, with nanoseconds of jitter at a wavelength of 266 nm in the ultraviolet spectrum [123,124]. Later studies showed that when the focusing optics were arranged to produce a long spark between the electrodes, the switch jitter is minimized [125].

This work led to the development of laser-triggered spark gaps for large pulsed power machines. A single, high-power KrF laser beam was split to switch multiple pulsed power modules for simultaneous delivery to the load. The energy per pulse of modern, commercially available Nd:Yag infrared lasers now have a sufficiently high energy per pulse and low jitter that individually triggered lasers can be used to trigger multimodule pulsed power machines where the use of individual lasers allow the modules to be independently controlled.

Laser triggering does not affect the peak current or voltage holdoff capabilities of the switch that are dominated by the design of the electrode geometry. The limitations of the laser-triggered spark gap are the necessity for the focusing optics to be located in close proximity to the gap. This not only makes the system design more complex, but also allows the optics to be coated with discharge by-products, which eventually leads to degraded switch performance.

Laser triggering has been employed in switches using gases [126–129], solids [130], vacuum [131], and liquids [132], but high-pressure gases, especially SF_6, air, and N_2, are almost always preferred. Laser triggering can produce very reliable switching with nanoseconds of jitter.

4.1.5.2.3 Electron Beam Triggering

Gas-filled spark gaps can be triggered by inducing breakdown with an electron beam [133–136]. The electron beam may induce breakdown by forming either plasma in the gap from volume ionization of the insulator or a metal plasma from electrodes. An electron beam-triggered spark gap, shown in Figure 4.36, is capable of low inductance and short rise time. An electron beam-triggered spark gap can be operated without being self-sustained, meaning that conduction exists only as long as the electron beam is present.

4.1.5.2.4 X-Ray Flash Triggering

Triggering of spark gaps with X-rays was investigated with the experimental setup illustrated in Figure 4.37, and yielded encouraging results. Electrons are produced in the high-field region of the gap by ionization of the volume gas by the absorption of X-rays and by X-ray bombardment of the electrode. Unlike electron beams or lasers, X-ray flash triggering has the advantage that very little

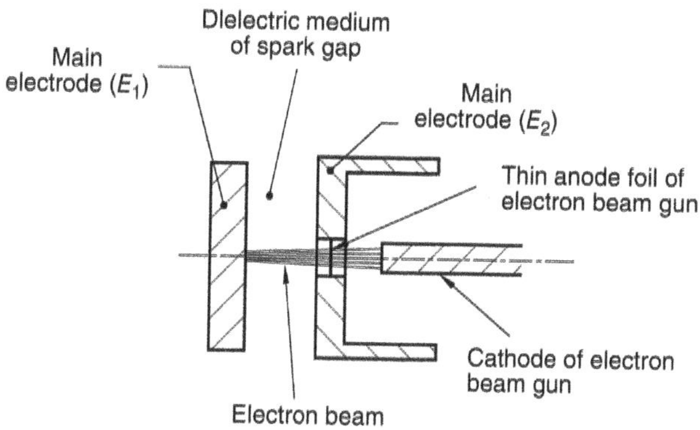

Figure 4.36 Electron beam triggering of a spark gap.

energy is deposited on the spark gap windows. Using this method, a voltage of 1 MV with a peak current of 144 kA switched with 10 ns of jitter [137].

4.1.5.2.5 Mechanical Force Triggering

These devices are also called metal-to-metal contact switches, wherein the switch closure is achieved by the physical movement of one electrode toward the other until contact is made. This results in an extremely low switch resistance of a few micro-ohms. Metal-to-metal contact switches are capable of switching millions of amperes with minimum loss of energy.

The mechanical force may be produced by a chemical explosive, an exploding wire shockwave [138,139], or a pulsed high magnetic

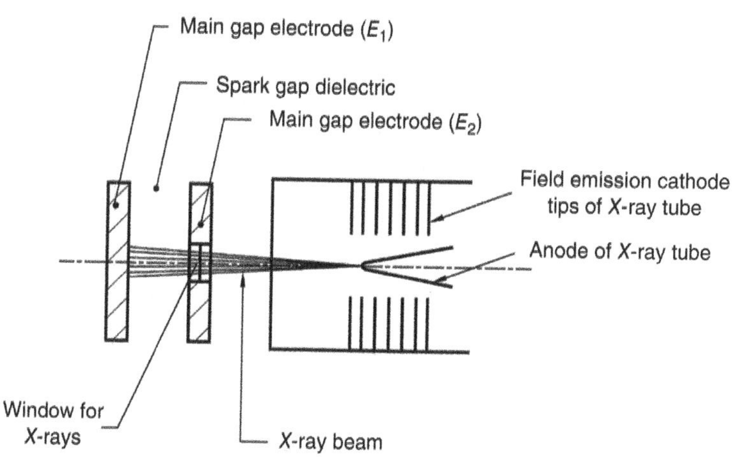

Figure 4.37 Triggering of a spark gap with X-rays.

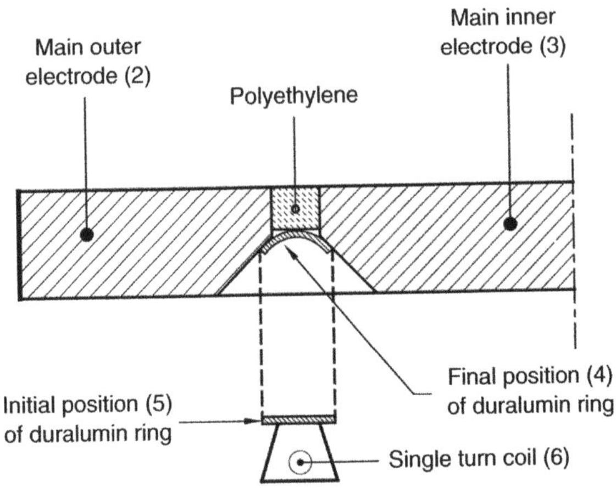

Figure 4.38 Configuration of a magnetically driven metal-to-metal switch [140]. (Reprinted with permission from Ref. [140]. Copyright 1978, AIP Publishing LLC.)

field [140,141]. A metal-to-metal switch driven by a pulsed magnetic field [140] is shown in Figure 4.38. When a pulsed current on the order of 70 kA is passed through the single-turn coil (6), the magnetic field produced drives the duralumin ring from its initial position (5) to the final position (4) to initiate the closure of the switch electrodes (2) and (3). The main drawbacks of mechanical force-triggered switches are their long delay time (on the order of few microseconds) and large jitter. Metal-to-metal switches are suited for some crowbar applications because of their small voltage drop across the switch electrodes.

4.1.5.2.6 Crossed Magnetic Field Triggering

Triggering of a spark gap by a crossed magnetic field, by Harvey [142], is shown in Figure 4.39. When the source grid *SG* is pulsed positive, plasma is formed between E_1 and *SG* because of the presence of crossed magnetic field in that region. This plasma is not normally drawn into the region between *CG* and E_2, because the control grid *CG* is biased to a negative potential. When the main gap, formed by electrodes E_1 and E_2, is to be triggered into conduction, a positive pulse is applied to the control grid *CG*, drawing the plasma toward E_2 and leading to breakdown. In earlier versions [143–146] of crossed magnetic field switch tubes, a magnetic field was required to be turned on and off for every trigger action, limiting the rise time and repetition rate capability. Even so, ratings of 40 kV, 40 kA, and 80 Hz were achieved [146]. This limitation has been overcome by the use of a fixed crossed magnetic field, as in the device shown in Figure 4.39. Experiments have been done with helium (He) gas at low pressures.

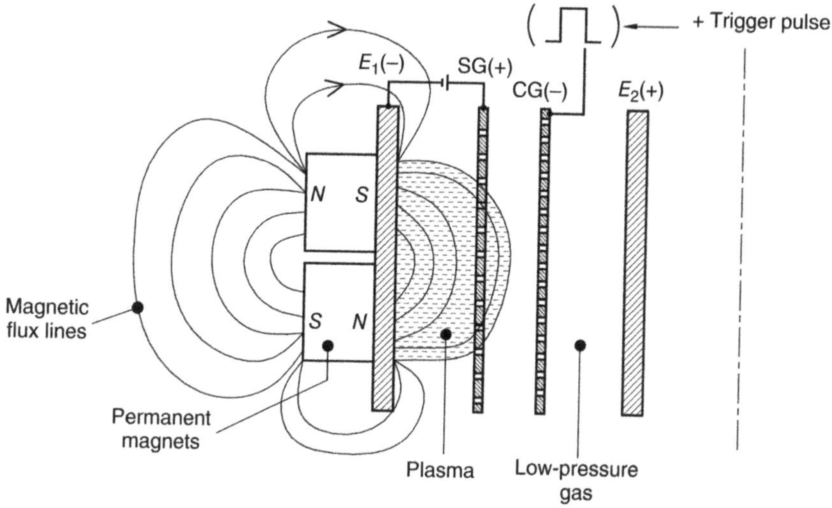

Figure 4.39 Crossed magnetic field triggering of a spark gap.

4.1.5.2.7 Gas Injection Triggering

A high-vacuum spark gap may be triggered by injecting a puff of gas into the high-field region that provides collision partners for the accelerating electrons, leading to the breakdown. Lafferty's [147] implementation of a gas injection trigger is shown in Figure 4.40. A command trigger pulse, applied between the electrodes E_2 and T, produces a flashover across the notch in the ceramic surface gap. The flashover produces an H_2 gas puff, introduced by the H_2-loaded titanium electrodes. The released H_2 expands into the vacuum gap,

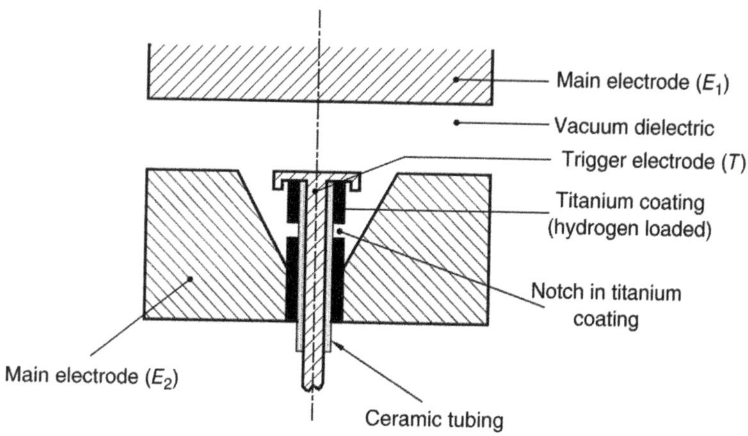

Figure 4.40 A vacuum gap triggered by a puff of gas [147].

leading to breakdown between the main gap electrodes E_1 and E_2. This switch has a high repetition rate capability.

4.1.5.2.8 Metal–Vapor Injection Triggering

A metal–vapor injection vacuum spark gap, used by Gilmour and Hope [148], is shown in Figure 4.41. This geometry is similar to a trigatron scheme, except that the exposed insulator surface between the trigger electrode T and main electrode E_2 is coated with a thin conducting film. When the trigger pulse is applied between E_2 and T, it vaporizes the thin metal film by joule heating. The metal vapor is accelerated in the electric field of the vacuum gap and leads to breakdown. The metal vapor injection vacuum gap has a high repetition rate capability and low trigger voltage requirement.

4.1.5.2.9 Exploding Wire Triggering

A vacuum gap may also be triggered by an exploding wire located along the axis in a two-cylinder coaxial vacuum gap [149,150], as shown in Figure 4.42a. The by-products of the exploding wire are injected into the main annular gap formed by the electrodes E_1 and E_2 through slots cut on inner electrode E_2, resulting in the switch being triggered into conduction. Depending on the energy deposited into the exploding wire, the by-products may be metal clumps, metal vapor, or metal plasma.

An interesting feature of the exploding wire trigger is the simultaneous formation of an opening switch formed by electrodes E_3 and E_4 and a closing switch formed by electrodes E_1 and E_2, as illustrated in Figure 4.42b. The exploding wire switch has potential for use in inductive energy storage systems

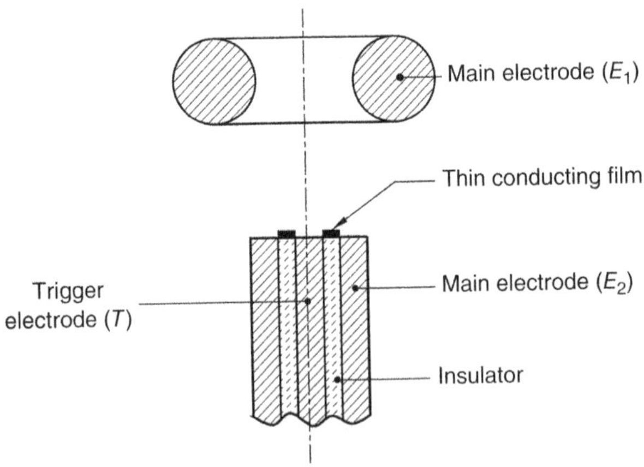

Figure 4.41 Vacuum gap triggered by a puff of metal vapor [148].

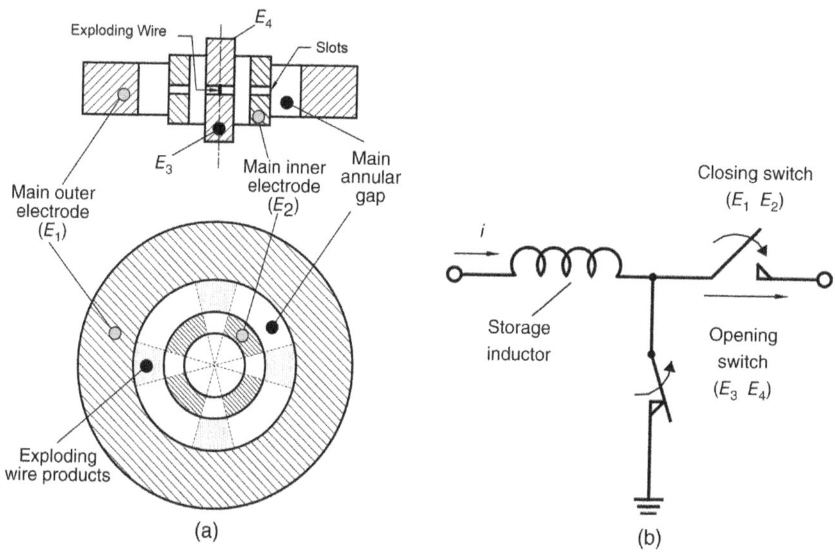

Figure 4.42 (a) An implementation of an exploding wire-triggered vacuum gap and (b) the equivalent circuit for an inductive storage application [149].

where a high degree of synchronism is required between the opening and closing switches for efficient energy transfer to the load. The behavior of the switching rise time t_r as a function of the energy E_e deposited in the exploding wire is illustrated in Figure 4.43. The regions *AB*, *BC*, and *CD* represent breakdown behavior corresponding to metal clumps, metal gas, and metal plasma, respectively. A reduction in the rise time has been observed when the density of the plasma injected through the annular gap is increased. This may be done by preventing the leakage of exploding wire plasma in the axial direction from the exploding wire chamber by inserting end plugs next near E_3 and E_4.

4.1.6 Specialized Spark Gap Geometries

4.1.6.1 Rail Gaps

A rail gap is a spark gap whose electrodes are made long to encourage multichanneling. The most common rail gap switch is a three-electrode field distortion switch whose typical configuration and mounting arrangement is shown in Figure 4.44. The anode and cathode are curved to provide a small field enhancement. The third electrode is highly enhanced—typically is a knife edge—and triggered with a fast-rising trigger pulse, (>5 kV/ns). The production of multiple channels requires the peak trigger voltage to be greater than the main switching voltage. As with other field distortion spark gaps, the distance between the anode and trigger rail electrodes and the trigger and cathode rails may be either equal or unequal but a 70:30 ratio is commonly used.

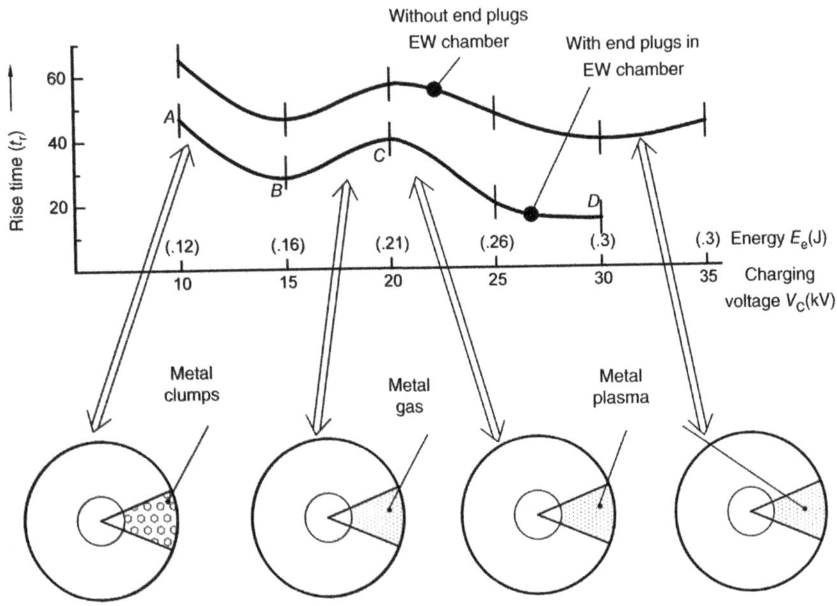

Figure 4.43 The rise time of the exploding triggered vacuum gap and form of the exploding wire by-products are a function of the applied energy and the containment.

Rail gaps generally make use of gaseous insulating mediums like SF_6, N_2, Ar, or mixtures of these at pressures of a few atmospheres. The number of spark channels increases with an increased rate of rise of trigger voltage and increased rail electrode length. The materials generally employed in the rail gaps are brass or stainless steel. Depending upon the peak current rating, the rail diameter up to 50 mm and lengths up to 600 mm are employed. Ratings up to 100 kV, 5 MA, and 10 C per pulse are available [151,152].

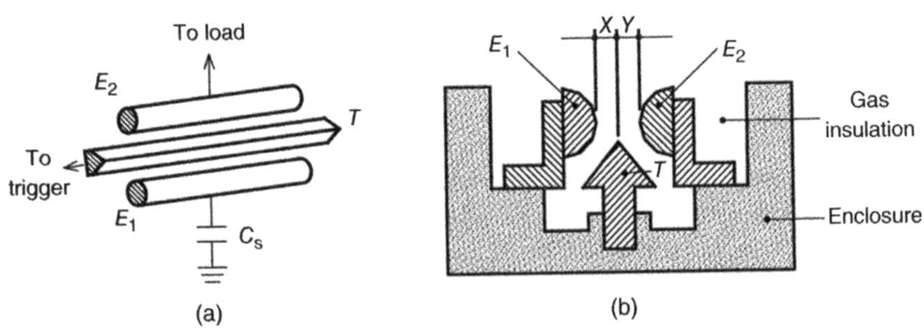

Figure 4.44 (a) A rail gap switch has long electrodes to encourage multichanneling. (b) A typical mounting arrangement for an electrically triggered rail gap.

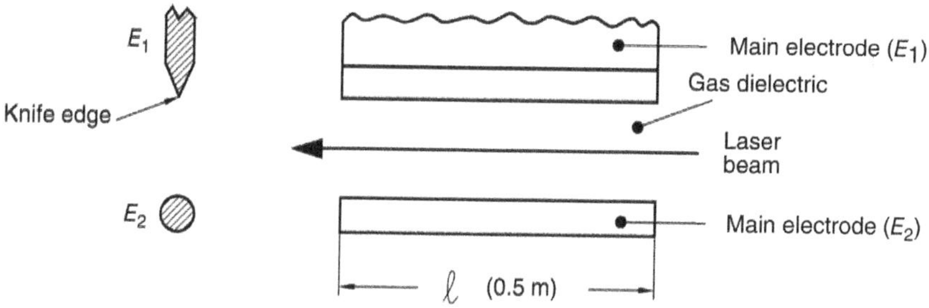

Figure 4.45 Laser-triggered knife-edge rail gap switch.

Two electrode rail gap switches have been triggered with a laser. A laser-triggered rail gap triggered by a laser aimed along the length of a 50 cm-long knife-edge trigger rail used by Taylor [153] and shown in Figure 4.45 operates with 100 spark channels per meter. Rail gap with jitter of 100 ps developed to operate over a length of 33 cm with an arc density of three channels per centimeter using a 1 mJ laser pulse at 248 nm and a pulse width of 500 fs [154].

An oil dielectric rail gap switch, employed by Prestwich [155] and illustrated in Figure 4.46, switched 2 MV with 10 spark channels with a fast trigger voltage of 1.5 MV. The electrodes are 122 cm long.

Rail gaps are used in Shiva Star 9 MJ capacitor bank where the rail gaps are mounted externally to the capacitors. They are operated under bipolar charging to ±90 kV and operated with a mixture of 30% Ar and 70% SF_6 to facilitate multichanneling [156,157]. High-energy operation has a failure mode that has been linked to the capacitor–switch interface.

4.1.6.2 Corona-Stabilized Switches

Moderate-pressure spark gap switch performance can be greatly enhanced by utilizing phenomena of corona stabilization, which occur in highly nonuniform field geometries in electronegative gases such as SF_6, oxygen, and their mixtures, including air.

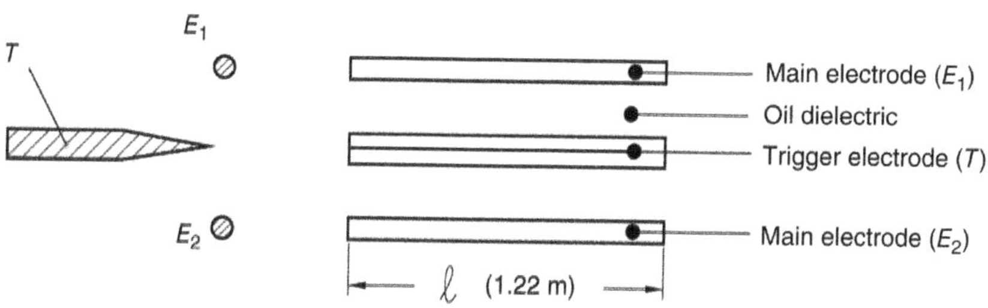

Figure 4.46 An oil dielectric rail gap switch, employed by Prestwich [155], switched 2 MV with 10 spark channels with a fast trigger voltage of 1.5 MV.

Under slow-rising and DC voltages, there is a range of pressures where breakdown is preceded by the formation of corona in the nonuniform electric fields established near the enhanced, high-voltage electrode. The breakdown characteristics for these conditions result in a highly nonlinear voltage–pressure (V–p) curve, illustrated in Figure 4.47. The full breakdown curve is the sparking criteria and the dashed curve is the voltage at which corona initiates for a given pressure. The two curves merge at the critical pressure P_c, where below P_c breakdown occurs in the presence of corona discharge, and above P_c breakdown occurs without a corona forming first. A corona-stabilized switch is operated in the parameter region between these two curves.

Corona-stabilized closing switches have been shown to have several advantages. For instance, the repetitive capability is greatly enhanced over conventional spark gaps where the corona activity produces space charge between the electrodes. This has the effect of redistributing the electric field such that the nonuniform electrode is effectively shielded from the rest of the gap, allowing the insulating properties of the gap to recover from a previous switch closure [158,159]. Recovery times corresponding to pulse repetition rates of 20 kHz have been demonstrated [37,160]. Under triggered conditions, the corona activity has been shown to significantly reduce the switch jitter [36].

Because the range of operation on the V–p curve is near the lower pressure, the operating voltage of corona-stabilized switches is relatively low and typically

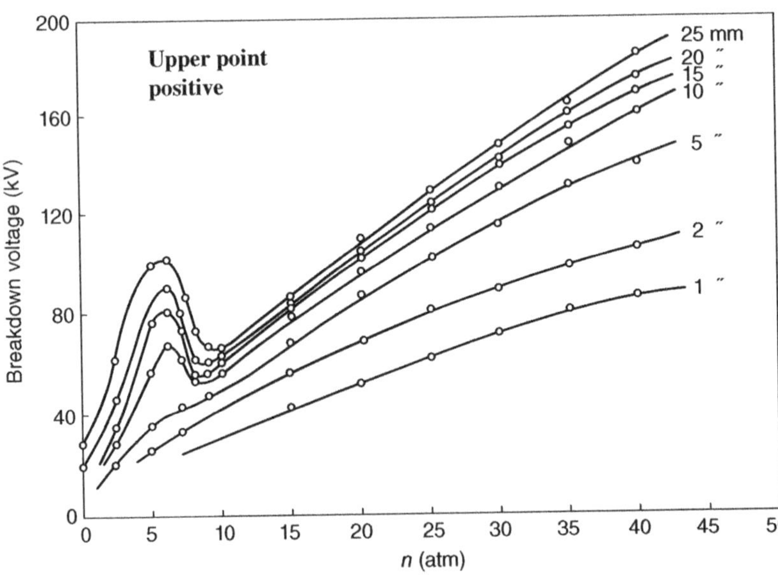

Figure 4.47 A typical V–p curve for an electronegative gas. The degree of electronegativity determines the degree of the "hump."

on the order of ~40 kV. The operating voltage can be increased by cascading several corona-stabilized spark gaps in series [161], and have been shown to operate with a switch jitter less than 2 ns [162]. A cascade corona-stabilized switch has the advantage of an inherent voltage grading between the electrodes because the corona is resistive, as indicated by the I_c–V characteristic. A commonly accepted form for the voltage dependence of the corona current I_c in a point–plane gaps is [163]

$$I_c(V) = \frac{2\mu\epsilon}{d_i}(V - V_0)^2 \tag{4.43}$$

where μ is the effective mobility for the charge carriers, ϵ is the permittivity of the insulator, d_i is the separation between the electrodes, and V_0 is a constant. The corona current, however, depends greatly on the field enhancement of the nonuniformity, which Given [107] has quantified and shown that the voltage distribution across the gaps can be controlled by the nonuniform field characteristics. There is, in effect, a feedback process that stabilizes each gap to a voltage that permits the average corona prebreakdown current to be constant through the switch. If each stage of the cascade is identical in design, and assuming the influence of the corona is the same for each stage of the switch, the voltage across each gap will be the same. This dependence of the gap voltage on the corona current provides a robust method of controlling the voltage distribution across the cascade, which may allow corona-stabilized cascade switches to be tuned to control the order in which gaps in the cascade break down, leading to improvements in switch jitter.

It should be noted that there are two design approaches to the use of corona stabilization in multielectrode switches. The first utilizes corona-active electrodes whereby the electrode that provides biasing for corona production is also the working electrode from which the main discharge occurs on closure of the switch. A second approach is to provide voltage biasing with a nonuniformity such as a needle protected within the body of the working electrode so that closure occurs across the more robustly profiled sections [164].

4.1.6.3 Ultra-Wideband Spark Gaps

Spark gaps have been used to produce impulse waveforms for many years [165]. The highest-frequency components comprise the rise time and the key to producing pulses with a wide bandwidth is to minimize the rise time of the pulse delivered to the load. For fast-rising pulses, the inductance of the spark gap, as well as the inductance of the spark channel itself, must be minimized. The spark gap may be made physically small to minimize the contribution to the rise time of the wave re-establishing itself after the spark gap is fully closed [166]. Ultrafast spark gaps may be embedded in transmission line geometries and tightly integrated to avoid dispersion and preserve the rise time. The inductance of the spark channel L_s is minimized with small gap spacing d, mandating high dielectric strength insulators and fast charging times. The inductance of the spark channel can be computed using the transmission line equations. Realizing that the impedance of the system is typically between 100 and 400 Ω of the antenna, and the logarithm is a slowly varying function, the inductance per unit length of the spark channel is approximately

$$L'_S = \frac{L_S}{d} \cong \frac{2\mu_0}{\pi} \tag{4.44}$$

which computes to approximately 1 nH/mm.

Moreover, the electric intensity in the far field of the transient antenna, to first order, is proportional to the rate of voltage rise, dV/dt, of the source, and estimates of the rate of voltage rise from spark gaps allow for estimates of system performance [167–169]. An estimation of the maximum rate of voltage rise can be derived [166] to be

$$\frac{dV_L}{dt} = \frac{Z_L E}{L'_S} \tag{4.45}$$

Thus, the rate of voltage rise from an ultrafast spark gap is determined by the electrical strength of the insulating media. To estimate the maximum rate of voltage rise, let the breakdown electric field E be 100 MV/m [170] for high-pressure H_2 and let the load impedance be 100 Ω. With these representative values, the maximum rate of voltage rise for a spark gap is on the order of $\sim 10^{16}$ V/s. To validate the approximation, a comparison of maximum dV/dt produced with spark gap pulsers with the literature appears in Table 4.1.

Table 4.1 Maximum rate of voltage rise measured with various pulsers.

Name	Output parameters	Spark gap media	$\frac{dV}{dt}$	Institute	Reference
RADAN	$V_{peak} \sim 150$ kV, $t_r = 150$ ps	N_2	$1 \times 10^{15} \frac{V}{s}$	Institute for Electrophysics	[171]
SNIPER	$V_{peak} \sim 140$ kV, $t_r = 50$ ps, gap = 0.5 mm	H_2 (2200 psi)	$2 \times 10^{15} \frac{V}{s}$	Sandia National Laboratories	[172]
SNIPER	$V_{peak} \sim 140$ kV, $t_r = 50$ ps, gap = 0.76 mm	N_2 (720 psi)	$1.4 \times 10^{15} \frac{V}{s}$	Sandia National Laboratories	[172]
JOLT	$V_{peak} = 1.1$ MV, $t_r = 130$ ps, gap = 0.76 mm	Flowing oil	$5 \times 10^{15} \frac{V}{s}$	Air Force Research Laboratory	[26]

Note: The maximum rate of rise is usually different from the measured rise time.

From the above discussion, it can be shown that a minimal inductance switch can be made by making the ratio of the inner and outer conductors of the spark gap a constant and compensating for the added inductance of the spark channel [166]. This spark gap was introduced as a "compensated" but when the gap spacing is small, the geometry is similar to that found in a bicone antenna and is commonly known as a bicone spark gap. This design has been realized [26].

4.1.7 Materials Used in Spark Gaps

A reliable spark gap must have adequate electrical performance, but it must also have mechanical reliability and a reasonable lifetime. A major failure mechanism of spark gap switches is surface flashover along either the inner housing or the outer housing.

4.1.7.1 Switching Media

Spark gaps have been made using solids, liquids, gases, and vacuum as the switching media. The breakdown mechanisms and predictive relations in gases are fairly well understood in comparison to other dielectrics. Gas-filled spark gaps are also very versatile and are the most widely used high-power switches today.

4.1.7.1.1 Gaseous Dielectric

Numerous gases are suitable for spark gaps, with dry air, N_2, H_2, and SF_6 being the most common. The gases are typically used at high pressures to increase the dielectric strength.

SF_6 has been widely used in spark gaps for its high electrical breakdown strength and its high electronegativity. For low-jitter applications, SF_6 should not be operated below 10 psig because of its erratic resistive phase [109]. The decomposition products of SF_6 react with certain metals and insulators, and therefore a careful selection of materials becomes necessary. Increasingly, spark gaps are using other gases, especially air, because of concerns for the environment as well as the cost—SF_6 is over 10 times the cost of other bottled, purified gases. Air is particularly attractive because it is also electronegative, although a much weaker effect than that in SF_6, and can be used in spark gaps exploiting the corona stabilization phenomena.

Gas mixtures of gases such as N_2/SF_6 and Ar have been studied for cost reduction as well as to tailor some properties. Gaseous mixtures of SF_6 have the advantage of retaining the dielectric strength of the SF_6 to a large extent. Common mixtures are SF_6/air, SF_6/N_2, and SF_6/Ar. Champney [109] has shown that the addition of Ar greatly reduces the probability of prefire, even when operating close to the self-breakdown voltage. It has also been observed that adding Ar to SF_6 improves the delay time, jitter, and the ability to multichannel [173]. A mixture of 70% SF_6 to 30% Ar is used in rail gaps for very high current operation.

Repetitive operation of even conventional spark gaps is strongly dependent on the gas species as well as on the pressure. Dry, bottled gases should be used since water vapor is detrimental to the dielectric strength of gases. Dessicants can be incorporated into switches intended for long-term use, such as those used in the power industry, to avoid degradation from water vapor. The gas handling system used for filling gases in spark gaps should ensure that the impurity contents of extraneous gases, hydrocarbon, and solid particles are reduced to acceptable levels. In SF_6-insulated systems, the purification process may be combined with a reclamation unit, allowing the gas to be used repeatedly.

4.1.7.1.2 Vacuum Dielectric

A significant advantage of a vacuum-triggered spark gap is the very wide switching range that can be achieved. The statistical delay time and jitter are generally quite large but very fast rise times can be achieved. Vacuum has been used in switches for closing as well as opening functions. The deionization efficiency is very high for vacuum and finds application in vacuum circuit breakers. Typical rates of rise of transient recovery voltages (RRRVs) for vacuum, SF_6, and air are 5, 1, and 0.5 kV/ms, respectively [174]. In continuously pumped vacuum switches, the initial SBV values are generally low but can be improved by spark conditioning, and further improvements may be gained by employing graded electrodes [175]. For nonuniform field gaps, the dielectric strength is polarity dependent with a positive high-field electrode yielding the higher value [65]. The vacuum employed should be better than 10^{-4} Torr. For sealed devices, the spark gap should be degassed at high temperatures before the sealing is done. Vacuum spark gaps are commercially available.

4.1.7.1.3 Solid Dielectric

Solid dielectrics, suitable when high V_{SB} and small inductance are required, usually employ thin films of polyester, polycarbonate, or polyethylene. The breakdown strength of solid dielectric is substantially lowered by voids or gas bubbles embedded within the film or trapped at the dielectric–electrode junction. Special care is typically needed in their assembly and solid dielectric switches are typically assembled in vacuum [116]. Breakdown due to edge effects may be mitigated by grading the field by immersion in deionized water, copper sulfate solution, or insulating oil [176]. Solid dielectric switches can be operated either in self-break mode or command triggered. A stabbed [63] dielectric switch is capable of multichannel sparking, yielding low inductance, short rise time, and high peak current [177]. A major disadvantage of solid dielectric switching is that the dielectric film is expended in a single shot and must be replaced or refreshed for every shot. A solid dielectric switch continued its reliable operation even after severe pitting and scarring [176].

4.1.7.1.4 Liquid Dielectrics

The commonly used liquid dielectrics are transformer oil and deionized water. The breakdown strength of liquids shows a polarity effect and this effect is enhanced with shorter and faster pulses. In water, this effect is much more pronounced than for oil. For nonuniform field geometry, negative voltage on the electrode in high-field region gives higher dielectric strength than positive voltage, which is in exact contrast [65] to vacuum dielectric. Liquid dielectrics are particularly suitable under fast pulsing conditions when their dynamic breakdown voltage is nearly three to four times [134] their static breakdown voltage value. The possible reason for this may be that the stress time is insufficient for the solid impurities to migrate and line up across electrodes. Pressurizing water has also been found to improve the holdoff voltage as well as the time-on stress [178]. Smith [179] reports usable field strengths of 200–300 kV/cm for oil in comparison to 100–150 kV/cm for water. Water has the advantage of self-healing after breakdown, whereas oil produces decomposition products. For both water and oil, a dynamic system of close-circuit recirculation and filtering is desirable to remove metallic particles arising out of electrode erosion. The gaseous bubbles trapped in liquid dielectrics can very much lower the dielectric strength and hence it is useful to vacuum degas during the first fill of liquid dielectric.

4.1.7.1.5 Hybrid Dielectric

Kitagawa and Hirano [180] have done some experiments with a spark gap comprised of two spark gaps in a series. One is a high-pressure gas spark gap, using air at a pressure of a few atmospheres as the dielectric. The other spark gap is also an air gap, but at low pressures. A high-pressure spark gap, which normally has neither a large switching range nor a low-voltage capability, now acquires both these characteristics when decoupled by a low-pressure gap. The combination gap also shows the superior reliability and low jitter, which is not normally available with a low-pressure gap alone. A spark gap with hybrid dielectric is therefore well suited for crowbar applications.

4.1.7.2 Electrode Materials

The various materials commonly used for electrodes of simpler spark gaps handling average power are aluminium, copper, brass, and stainless steel. For spark gaps handling high average power with less jitter, long life, and high repetition rate, the electrode materials used are inserts and liners of tungsten alloys, molybdenum and its alloys [162,181], and tantalum [182]. Copper tungsten is extensively used in high-peak-current spark gaps. Heat-treated beryllium–copper has been used by Skelton [183] in a metal-to-metal switch because of its high strength. The common electrode failures are cracking due to thermal stresses, delamination, and recrystallization. For low-cost spark gaps with average coulomb handling capacity, brass has been widely used because of its high resistance to thermal stress cracking [184]. Graphite has been shown to be a good material in some spark gap applications [185,186].

4.1.7.3 Housing Materials

The switch housing acts to support the electrodes and contains the switch's insulating medium. The insulator for the housing must be both resistant to surface flashover and have sufficient strength to contain the pressure of the insulating dielectric during switch operation. The integrity of the switch's mechanical design is particularly important when using a high-pressure gas as the switching media because of the possibility of catastrophic failure. A mechanical factor of safety of at least 4 is strongly recommended [187].

Acrylics are the most common switch housing material, but epoxy resins, polycarbonate, and Delrin are also common. Some plastics, such as polycarbonates, must be annealed after machining. Typically, the addition of particles, such as glass beads, makes the insulator mechanically stronger but electrically weaker, and should be avoided. Fiberglass [188,189] has good mechanical strength but the embedded fibers can be problematic, leading to either electrical failures or mechanical failures. A Teflon lining can be inserted to mitigate this problem [190]. Teflon has the interesting feature that it is not susceptible to damage from surface flashover, especially with an SF_6 gas [189]. The resistance to high-voltage tracking can be considerably improved by using grading [184] rings. Teflon and ceramic tubes are suitable for trigger gap insulation in case of trigatrons. Other materials with attractive properties are ULTEM, Torlon, and PEEK. These materials are strong with good resistance to electrical breakdowns, but are more expensive than the more common acrylics and polycarbonates. These materials are typically used in compact applications where the minimal length is attained carefully, contouring the inner surface of the spark gap to control the electric fields tangential to the surface [191].

4.2 Gas Discharge Switches

4.2.1 The Pseudospark Switch

The pseudospark [192–207] is a hollow cathode and hollow anode device that operates in a low-pressure discharge. The pseudospark switch exhibits remarkable switching properties, combining the advantages of thyratrons and spark gaps, specifically high dI/dt, reverse current, and charge transfer capabilities, along with long lifetime and low jitter. The voltage is limited to 30–40 kV, but may be extended by putting several in series. The pseudospark switch is particularly well suited for applications requiring high repetition rates. Operation has been demonstrated at 100 kHz [194].

A typical arrangement of a pseudospark device is shown in Figure 4.48. Each hollow electrode has a relatively small borehole in its center and arranged so that the boreholes are aligned. In a pseudospark switch, current is conducted by a diffuse discharge rather than an arc, and the mechanism of initiation is covered in Chapter 8. The pseudospark operates on the left-hand side of the Paschen curve and uses various gases such as H_2, D_2, He, Ar, and N_2 in the pressure range of 10–50 Pa. The anode and cathode are set close together, typically several millimeters, so that the (pd) product is sufficiently small so that breakdown does not occur here. Electrons emitted in the gap hit the anode without undergoing sufficient collisions and electron multiplication to become self-sustaining. Instead, the path length along the electric field lines through the aligned boreholes in the hollow electrodes is sufficiently long so that breakdown preferentially occurs there. The pseudospark is triggered by injecting electrons into the hollow cathode, initiating the diffuse discharge and closing the switch.

The diffuse nature of the discharge yields very low erosion rates. The size of the borehole is very important and determines the power capability of the switch. If the hole is too small, the discharge is likely to pinch and will not conduct the full current. Boreholes that are too large will degrade the voltage

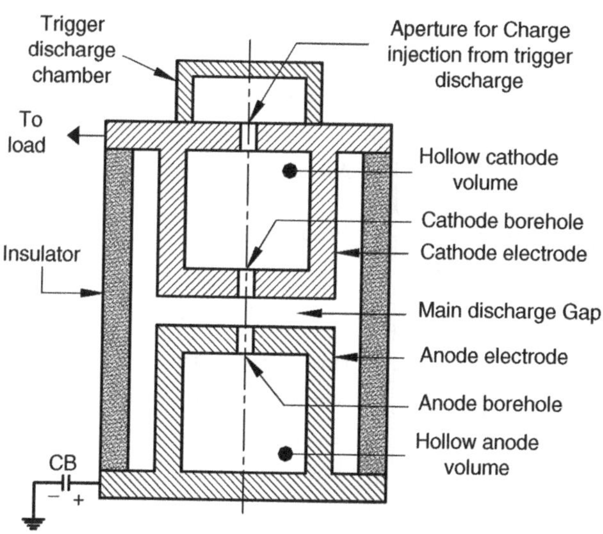

Figure 4.48 Typical arrangement of a single-channel coaxial pseudospark switch.

holdoff of the switch. Most research devices have been continuously pumped, but commercially available pseudospark switches have a sealed geometry with a gas reservoir and heater for pressure adjustment and gas depletion.

4.2.1.1 Trigger Discharge Techniques

The various methods of triggering hollow cathodes and pseudospark switches are (i) illumination of hollow cathode volume by UV flash, (ii) corona on a grid array, (iii) surface discharge across insulator, and (iv) glow discharge.

In the surface discharge technique, shown in Figure 4.49, on the application of a negative trigger pulse, V_T, a flashover takes place across the terminals P and Q on the surface of the insulator, resulting in the generation of surface discharge plasma. This plasma, being biased to a negative potential with respect to the cathode because of the resistor R acts as a source of electrons for the main gap. A semiconductor can also be used in place of insulator for the surface discharge, in which case there is no necessity for the resistor R and the trigger energy is efficiently utilized for plasma generation.

In the glow discharge technique shown in Figure 4.50, a trigger pulse, of either polarity but of sufficient magnitude, V_T, is applied to the trigger electrode E. As a result, a glow discharge is established between the electrode E and the hollow trigger, which diffuses through the charge injection aperture to the hollow cathode. The injection of electrons into the hollow cathode starts the conduction process in the pseudospark switch. The switch delay time can be substantially reduced by applying a preionizing DC voltage V_0 to the trigger electrode, with a positive trigger polarity resulting in shorter delays than a negative trigger.

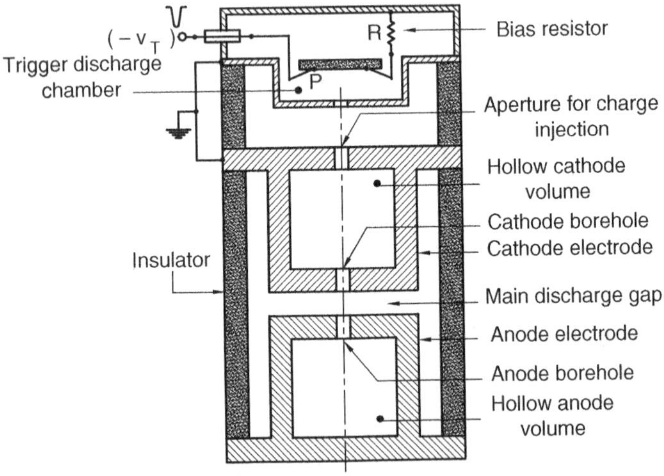

Figure 4.49 Surface discharge triggering of a pseudospark switch.

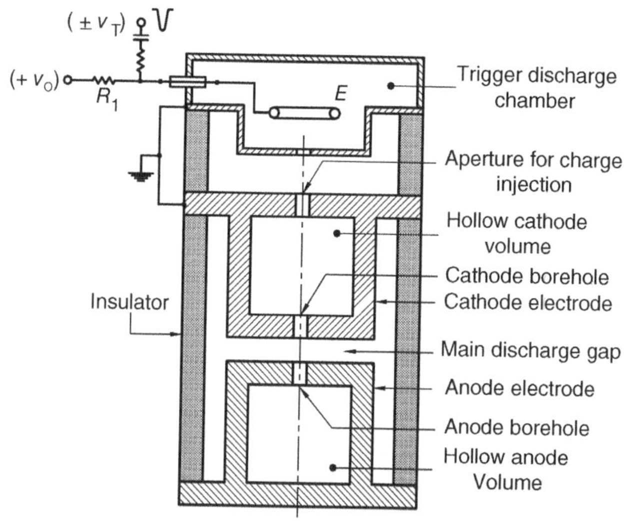

Figure 4.50 Glow discharge triggering of a pseudospark switch.

4.2.1.2 Pseudospark Switch Configurations

A variety of configurations of pseudospark switches have been successfully deployed to address specific capability, including increasing the peak switching current, increasing the holdoff voltage, and increasing the lifetime.

A number of geometries have used multichanneling to overcome the current-carrying limitation, including coaxial multichannel (Figure 4.51), linear multichannel (Figure 4.52), and radial multichannel (Figure 4.53). In a multichannel PSS, the total current I is shared by N channels, the current in each channel being roughly equal. Simultaneous triggering of the channels remains a challenge, albeit not a serious one. The voltage holdoff capability of the pseudospark switch is a serious limitation. To overcome it, a multigap configuration, shown in Figure 4.54, has been demonstrated, which increases the holdoff voltage in comparison to a single-gap arrangement, roughly in proportion to the number of gaps, but quickly increases the length to an unwieldy dimension. Although electrode erosion is not a serious concern in pseudosparks, as the peak current through the switch increases, the failure rate increases dramatically because of metal vapor deposition on the insulator separating the hollow electrodes causes a loss of surface resistance. The advantage of a radial PSS has an advantage over its longitudinal, coaxial cousin because it has no line-of-sight to the insulator and is protected from the deposition of metal vapor. Longitudinal pseudospark geometries are susceptible to reduced life in high-current applications, but the introduction of baffles, as shown in Figure 4.55, acts as a barrier for the metal vapor and protects the insulator.

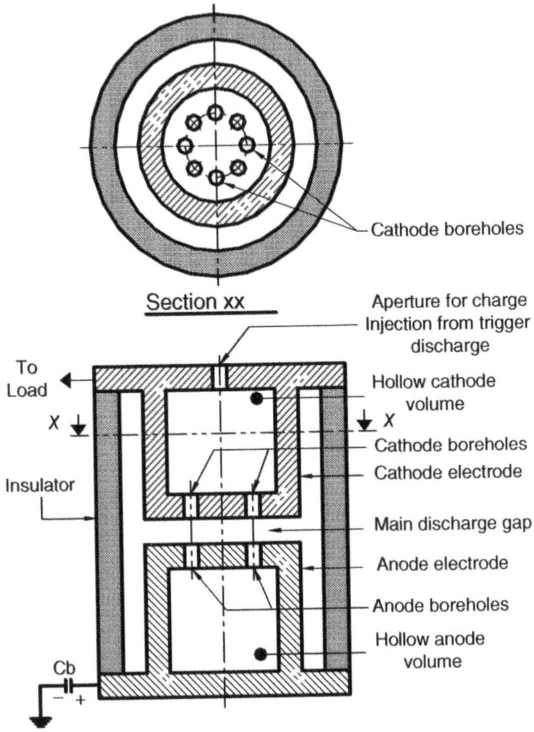

Figure 4.51 Typical arrangement of a multichannel coaxial pseudospark switch.

The pseudospark switches are commercially available for the following specifications: (i) stand-off voltage of tens of kilovolts, (ii) peak current of tens of kiloamperes, (iii) pulse repetition rate of several kilohertz, (iv) rate of rise of current up to 10^{12} A/s, (v) jitter of tens of nanoseconds, (vi) current reversal up to 100%, and (vii) lifetime of several hundreds of coulombs.

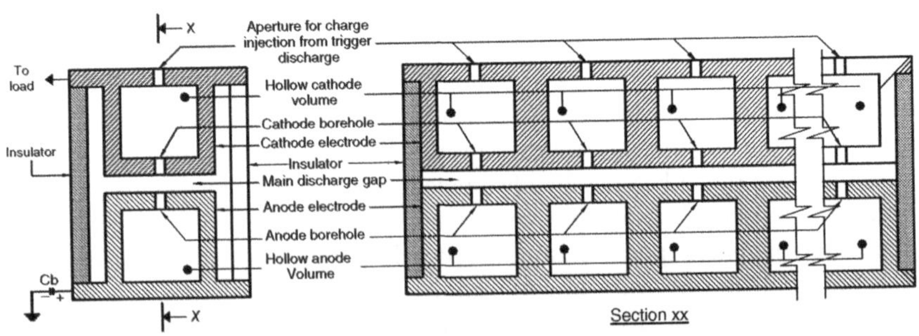

Figure 4.52 Typical arrangement of a linear multichannel pseudospark switch.

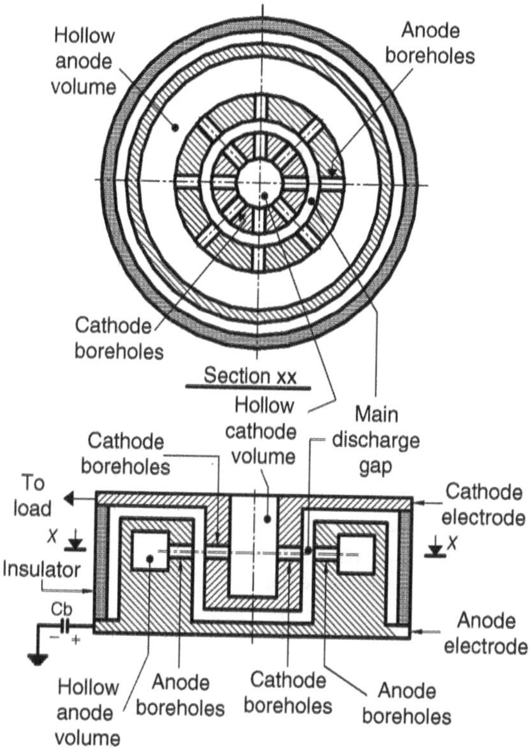

Figure 4.53 Typical arrangement of a radial multichannel pseudospark switch.

4.2.2 Thyratrons

Thyratrons are grid-controlled, low-pressure, hot-cathode tubes that are available in triode, tetrode, and pentode configurations [208,209]. A typical structure of a thyratron is shown in Figure 4.56. The presence of a large number of baffles aids in achieving high-voltage holdoff and short recovery times. When the anode voltage is positive, the transition from the off-state to the on-state is achieved by the application of a trigger voltage to the grid.

The characteristics of negative-grid thyratron and positive-grid thyratron are shown in Figure 4.57. The triggering in negative-grid thyratrons is achieved by suddenly reducing the negative bias on the grid, thus allowing the anode field to penetrate into the grid–cathode region. This accelerates the electrons to the anode, causing cumulative ionization. The positive grid thyratrons are triggered by the application of a positive voltage to the grid. Once the thyratron goes into conduction, the grid loses control because

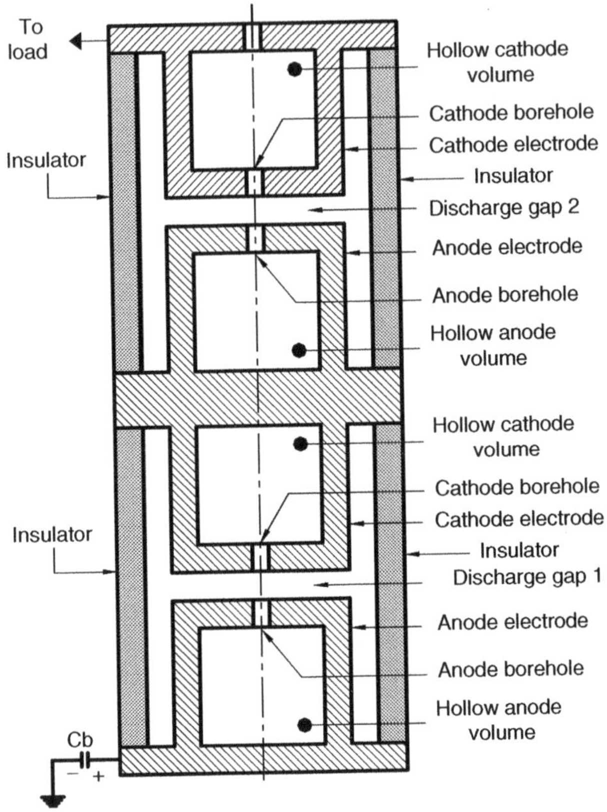

Figure 4.54 Typical arrangement of a coaxial multigap pseudospark switch.

of the formation of a positive ion sheath around the grid. The only way of reverting back to nonconducting state is by the removal or reversal of the anode voltage. In the case of a tetrode, grid 1 serves the purpose of ionizing the gas between the cathode and grid 1, whereas grid 2 does the gating function. A tetrode has a reduced firing time and less jitter compared to a triode. Some thyratrons, such as the pentode configuration and those employing hollow anode structures, enable current reversals up to 100%. High-power thyratrons at very high repetition rates generally require forced air or liquid immersion cooling. Other thyratrons operating at high voltage and high current ratings are fitted with internal X-ray shields to minimize X-ray emission to the outside. Thyratrons generally find wide use in trigger generators for firing of spark gaps and ignitrons. They are very widely used as pulse modulators in lasers, radar, and grid-controlled rectifiers.

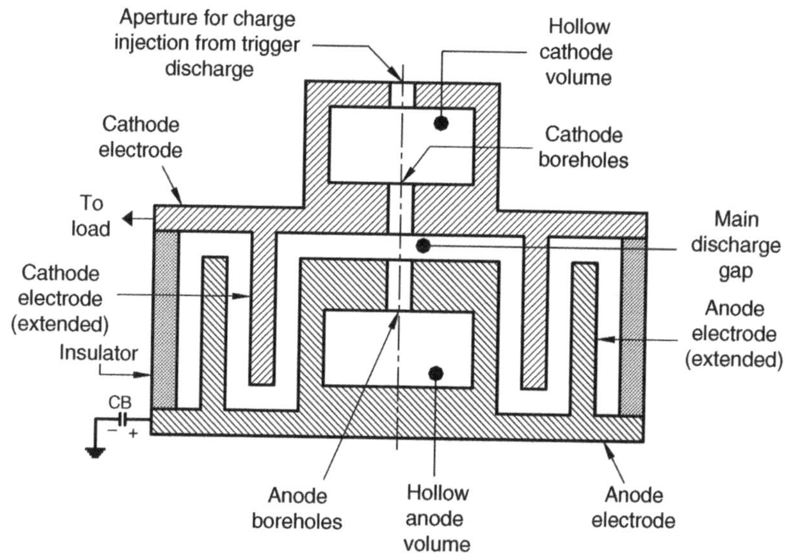

Figure 4.55 Modified arrangement of pseudospark electrodes for preventing metal vapor deposition on the insulator.

The fill gases used in thyratrons are Hg, H_2, D_2, Xe, or Ar. Mercury has a low average voltage drop of 10 V, but has the drawback that the vapor pressure is strongly dependent on temperature. Xe also has an average voltage drop of about 10 V. The pressure is not strongly dependent on temperature, but the absence of a gas reservoir poses problems for cleanup. Xe and Hg vapor

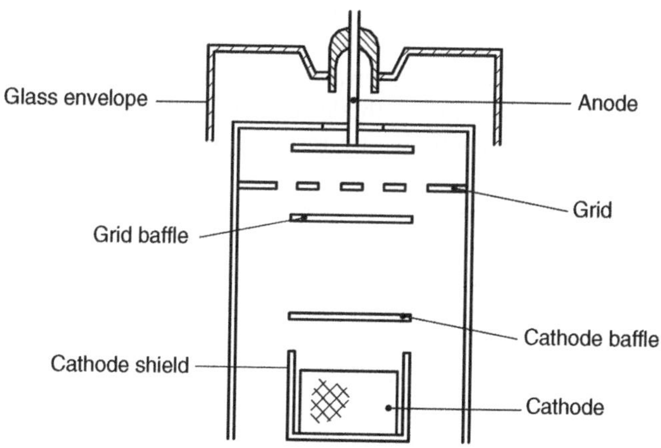

Figure 4.56 Typical structure of a thyratron.

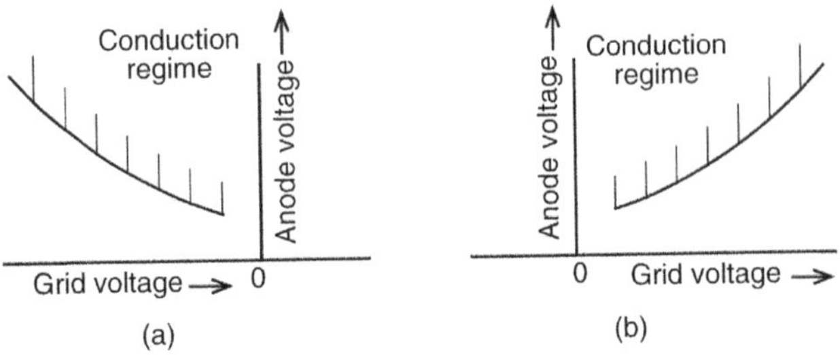

Figure 4.57 Characteristics of (a) a negative thyratron and (b) a positive grid thyratron.

thyratrons are generally used in low-frequency applications. H_2 thyratrons have high switching speeds, high peak current capability, high pulse repetition rates, and long life. These advantages may be attributed to the low mass and high mobility of the H_2 atom, which is also responsible for the high average voltage drop of 40 V due to its high ionization potential. D_2 has lower mobility [210] and hence high deionization time, which results in reduced pulse repetition rate capability but higher and more reliable holdoff voltage. To make up for the depleted H_2/D_2 gas due to physical and chemical adsorption, a gas reservoir of titanium hydride and an independent heater are usually provided. The H_2 pressure can be adjusted by varying the reservoir heater voltage. Low gas pressure results in higher holdoff voltage, whereas higher gas pressure gives better rate of rise of current.

Thyratron tetrodes can be operated in parallel or series. In parallel operation, the reliability is improved by adding a small inductance in series with each anode. The series operation, of course, requires the heater transformers to be insulated for high voltages. In series operation, only the thyratron near to the ground needs to be triggered externally and the rest can be made to fire automatically by suitable circuit arrangement [209]. The commercially available multigap thyratron in ceramic envelope is equivalent to a number of thyratrons in series, but with a single thermionic cathode. It contains equally spaced metallic annular rings along the length, with a resistive chain in parallel, for equal voltage distribution.

The rate of current rise in a thyratron typically has a maximum value of 300 kA/μs. Exceeding the specification results in enhanced dissipation and anode evaporation, which leads to further gas depletion. Thyratrons can usually handle maximum average currents up to 15 A. Pulse repetition rates up to 100 kHz are available. The maximum available ratings, although not simultaneously, for thyratrons in glass envelopes are (i) peak forward voltage ≈40 kV, (ii) peak anode current ≈5 kA, (iii) peak output power ≈20 MW, and (iv) anode heating factor $\approx 10^{10}$ (V A pps). The corresponding values for thyratrons in ceramic envelopes are 160 kV, 15 kA, 400 MW, and 5×10^{11} (V A pps).

4.2.3 Ignitrons

An ignitron (pronounced as in the word "ignite") is the gas discharge analog of a thyristor. It is a vacuum tube containing a cathode in the form of an Hg pool [211]. A semiconducting ignitor rod dips into the Hg pool to trigger the ignitron into conduction when the anode is positive. The voltage drop of an ignitron varies from 20 to 80 V depending upon the discharge current. Once the ignitron is triggered into conduction, the ignitor loses the control and the nonconduction state can be recovered only by reducing the anode voltage below the ionization potential of 10 V. The peak inverse voltage of the ignitor is about 5 V. An auxiliary anode helps to keep the arc alive at low discharge currents. The main discharge current is carried by the formation of an arc between the cathode and anode. In the absence of a trigger voltage, the voltage holdoff values for forward blocking and reverse blocking are similar, but may be lowered if the anode or anode seal is contaminated with Hg, in which case heating the anode and glass region to approximately 100 °C for 2 h will restore the voltage holdoff value. For reliable operation of the ignitron, the anode glass seal region should be at higher temperature than the rest of the ignitron, failing which the Hg may condense on the anode–glass seal region, leading to low reverse blocking voltage. Molybdenum, stainless steel, or graphite is typically used for the anode material and the current reversal capability decreases in that order. The maximum lifetime rating for an ignitron is 1000 operations but longer life is attainable by operating reduced ratings. For obtaining the best life, the rate of current rise should be limited to less than 10 kA/μs. Typically, operation at half the values of the specified maximum voltage or current increases the life by a factor of 10. The maximum specified ratings can be exceeded at the cost of reduced lifetime. For oscillatory currents, the calculation of the average current and ampere-second must involve the consideration of positive as well as negative cycles. Many ignitrons cannot tolerate significant current reversal.

The X-rays generated by the ignitron are generally attenuated to a safe level by the metallic casing of the ignitron. However, the need for additional shielding should be checked, particularly if the ratings are exceeded. Safe practices should be followed for the disposal of Hg in case of leakage.

When the anode is positive, the ignitron can be triggered into conduction by the discharge of a precharged capacitor into the ignitor. This capacitor may be a high capacitance value charged to a relatively low voltage (5 μF at 600 V, for example) or a low capacitance value charged to a higher voltage (0.25 μF at 3 kV, for example). The adaptation of the former results in a firing delay time of a few tens of microseconds, compared to less than 1 μs for the latter. For a large capacitor bank, where many ignitrons must be fired simultaneously in parallel, a low-capacitance, high-voltage trigger is employed since it gives smaller firing time jitter.

Ignitrons provide a rate of current rise limited to approximately 10 kA/μs. Exceeding this rating would give rise to increased dissipation and increased evaporation of anode material, leading to reduced life. Commercially available ignitrons, for capacitor bank service, have the following maximum ratings, although not all of them simultaneously.

Forward and reverse blocking voltage	50 kV
Peak current	200 kA
Ampere-second per pulse	500 As
Average anode current	10 A
Pulse duration 1	50 ms
Pulse repetition rate	1 pps

The coulomb requirement of an ignitron used in a capacitor bank discharge is given by the following relations:

For unidirectional current,

$$Q = CV_0 \, \text{A} - s \tag{4.46}$$

For oscillatory current,

$$Q = CV_0 \frac{1 + (I_1/I_{\text{max}})}{1 - (I_1/I_{\text{max}})} \text{ A-s} \tag{4.47}$$

where

I_{max} is the first current peak and I_1 the second current peak.

Ignitrons are widely used as crowbar switches because of their high coulomb capacity, and those with large coulomb rating are generally water cooled, employing either double walls or wound copper tubes. Ignitrons meant for power rectification service must have their arc extinguished near zero current to prevent conduction during the reverse cycle. This is achieved by incorporating deionizing baffles in the structure. Ignitrons must be mounted only in their vertical position.

A circuit [212] for switching 24 ignitrons, with one master trigger ignitron, is shown in Figure 4.58. When the master ignitron is triggered, the precharged capacitors C_1, C_2, . . . , C_{24} discharge through the ignitor in series with the 100 Ω resistors. This leads to triggering all the ignitrons, thus discharging the capacitors C_1, C_2, . . . , C_{24} into the load.

4.2.4 Krytrons

The krytron is a four-electrode switch employing low-pressure gas. A typical krytron switching circuit [213] is shown in Figure 4.59. A glow discharge with a

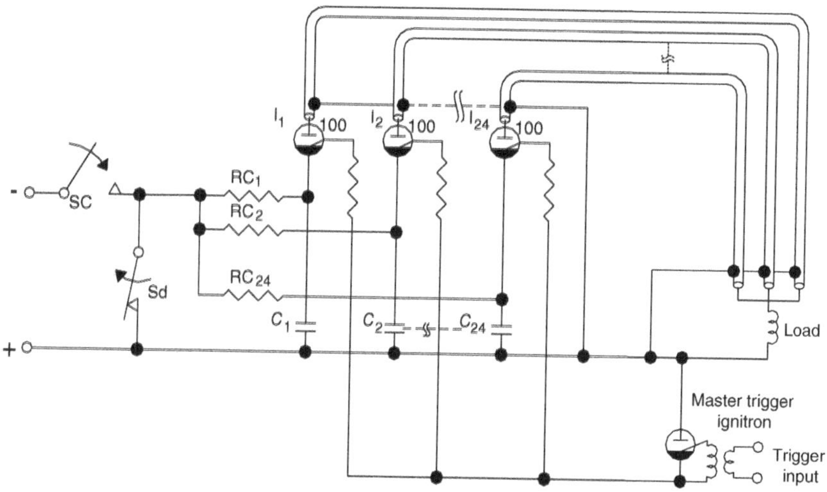

Figure 4.58 A circuit to trigger multiple ignitrons with a master ignitron switch.

current of a few microamperes is maintained between the cathode electrode K and corona electrode E_{CN}. This electrode is sometimes referred to as the "keep alive." The voltage for E_{CN} can be derived from the applied switching voltage itself or from an external power supply. The current-limiting resistor R ensures that the transition to an arc does not take place. A third electrode E_T interposed between cathode and anode is held at cathode potential in the absence of external trigger, through the secondary winding of the trigger transformer T. When a positive trigger pulse in the range of 500 V to 1 kV is applied to the trigger electrode E_T, the preformed plasma available between K and E_{CN} is drawn into the main gap region toward the anode A. The krytron's impedance collapses and the PFL discharges into the load R_ℓ. Krytrons have a voltage holdoff of a few kilovolts, peak discharge currents of a few kiloamperes, delay

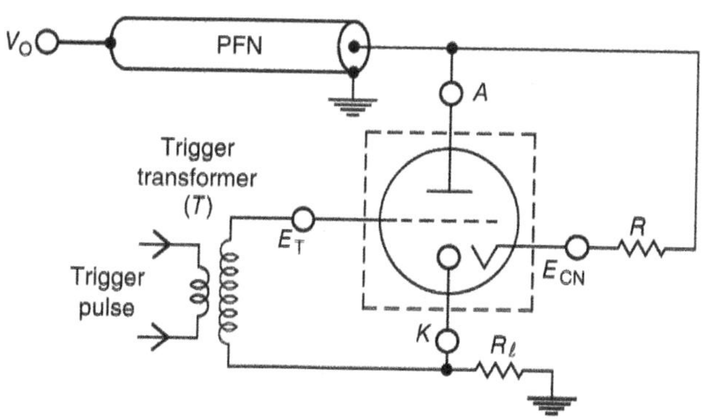

Figure 4.59 A typical krytron switching circuit.

times of tens of nanoseconds, and jitter of a few nanoseconds. Their performance may be further enhanced by incorporation of β-emitting radioisotopes such as ^{63}Ni.

The absence of a heated filament leads to high reliability and immunity to vibration and acceleration. These features, combined with small size and weight, make the krytron an attractive switch for military applications.

4.2.5 Radioisotope-Aided Miniature Spark Gap

A miniature self-triggered spark gap for applications such as high-speed surge suppressors for protection of electronic equipment against fast surges has been developed by Sharma et al. [214]. These spark gaps have operating range of 100 V to 5 kV with switching speeds of less than 10 ns. The statistical time lag is negligible because free electrons are available from ionization caused by emission from the radioisotope. Detailed studies have been conducted on the switch to minimize jitter and breakdown voltage. The gases used are Ar, neon, and SF_6 in the pressure range of 132 Pa to 20 kPa with gap spacing of 250, 500, 750, and 1000 μm. The radioisotopes investigated are ^{63}Ni (60 keV, β-source), ^{147}Pm (200 keV, β-source), and ^{241}Am (5 MeV, α-source), each at a dose of 1 μCi (3.7×10^4 Bq). The optimum low-jitter, high peak current performance was obtained from SF_6 gas with ^{63}Ni, oriented parallel to electric field. It has been observed that operation in certain pressure ranges gives rise to a high impedance discharge in a glow mode, whereas a low-impedance arc mode can be obtained in other pressure range.

In comparison to other surge suppressors like metal oxide varistors (MOVs) and transient absorbers (tranzorb), spark gaps have distinct advantages of high speed, low clamping voltage, minimum self-capacitance, and large energy absorption capability.

4.3 Solid Dielectric Switches

A solid dielectric switch, shown in Figure 4.60, consists of two main insulation sheets, a trigger insulation foil, and two metallic foils sandwiched between the main electrodes E_1 and E_2 by an external clamping force. When a trigger pulse is applied between the metallic foils, a surface discharge occurs at a small hole provided in the trigger insulation, leading to an explosion of the metal foil in the area around the hole. As a result of the interaction between trigger arc plasma and trigger discharge current, high magnetic fields are generated, leading to the puncture of the main insulation sheets by a shockwave, resulting in switch closure.

The electrode E_2 is slightly recessed to reduce the electrode erosion and an open hole in E_1 acts as a vent for the high-pressure gas generated during the

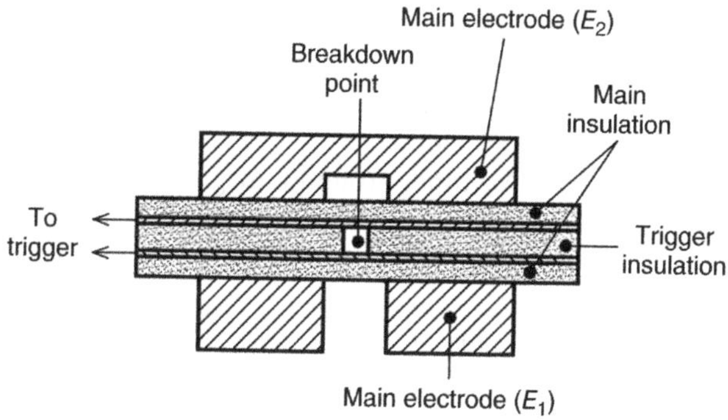

Figure 4.60 A solid dielectric switch.

closure of the solid dielectric switch. The insulation is destroyed during the shot and must be replaced. Electrode erosion can be controlled by shifting the mounting of the insulation to different areas on the main electrodes for subsequent shots. Alternative methods of triggering may be used such as exploding wires embedded in dielectric sheets or by detonating a charge of chemical explosives.

The solid dielectric switch has a wide voltage operating range and may be used down to a few percent of the self-breakdown voltage. This switch—with its ability to trigger at low voltage, inherent low inductance (few nanohenries) and discharge resistance (few milliohms), and high coulomb transfer capability (hundreds of coulombs)—finds wide application as a crowbar switch.

4.4 Magnetic Switches

Magnetic switches are a special type of saturable reactor—an inductor where the magnetic core can be deliberately saturated—and were first proposed as switches for radar applications by Melville [215]. However, whereas a saturable reactor generally uses the current in one winding to control the output of another winding, a magnetic switch uses both a single winding and the nonlinear properties of the magnetic material to delay the flow of current for a preset duration. Magnetic switches utilize the large change in inductance that occurs when its magnetic core is driven from its initial state to its saturated permeability. When the magnetic switch is closed at $t = 0$, it is designed such that the inductance is sufficiently large so that the inductive back voltage is opposite to the flux of the magnetic core material, preventing current flow. During the time before saturation, a small amount of leakage current passes to the load. When the magnetic core saturates, the relative permeability—and the inductance—plummets, and the voltage across the inductance of the magnetic switch drops to zero, resulting in circuit-limited current flow. Once the core is saturated, a large current may flow in the forward direction with the additional benefit that current is blocked in the reverse direction, so the magnetic switch may be considered to act as a temporary diode.

The inductance of a wire wrapped N_T times around a magnetic core is

$$L = \frac{\mu A_m N_T^2}{\ell_m} \tag{4.48}$$

where $\mu = \mu_0 \mu_r$ is the magnetic permeability, A_m is the cross-sectional area, N_T is the number of conductor turns about the magnetic material, and ℓ_m is the magnetic field path length. By changing the value of the magnetic permeability with time, the saturable inductor may be operated as a switch. A magnetic switch is usually represented in circuit diagrams by a variable inductor.

The magnetic field strength in the magnetic core material is controlled with the current I applied to the wire according to Ampere's law:

$$N_T I = \oint_{\ell_m} \vec{H} \cdot d\vec{\ell} \tag{4.49}$$

The magnetic field strength H, in turn, establishes a magnetic flux density B in the core according to the material's characteristic hysteresis (B–H) curve. Magnetic switching is achieved by using a time-dependent current $I(t)$ to induce a time-dependent magnetizing force $H(t)$ in the magnetic core material, which results in time-dependent magnetic flux density $B(t)$. The magnetic permeability μ, defined as the slope of the B–H *curve*,

$$\mu = \frac{dB}{dH} \tag{4.50}$$

is also time dependent. In fact, the time-dependent magnetic permeability $\mu(t)$ can be considered to be a function of the current applied to the saturable inductor winding.

$$\mu(t) = f(I(t)) \tag{4.51}$$

where the function depends on core material, geometry, and the number of windings. For applications in a magnetic switch, a material with a large swing in permeability is needed. The change from the initial high impedance state, where μ_r can be $\sim 10^6$, to a low impedance state, where $\mu_r \sim 1$, results in circuit-limited current flow switching.

4.4.1 The Hysteresis Curve

A hysteresis loop, often referred to as the B–H curve, is generated by measuring the magnetic flux density B of a magnetic material while the magnetizing force H is changed. It is a measure of the energy per unit volume of the material. The

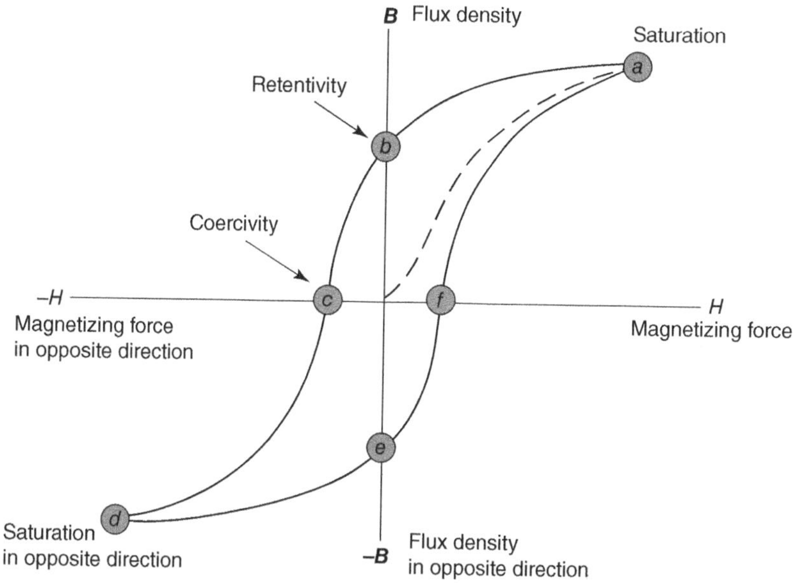

Figure 4.61 A hysteresis loop is a plot of the variation of magnetic flux density B with the change in magnetizing force H. When the magnetizing force returns to zero, there is a residual flux density B_r.

B–H curve for a magnetic material, shown in Figure 4.61, describes a number of primary magnetic properties.

A magnetic material that is not magnetized will follow the dashed line as H is increased from zero. Increasing the applied magnetizing force H results in an increasing magnetic flux density B until at point a where almost all of the magnetic domains are aligned. At point a, the cores are saturated and an additional increase in the magnetizing force H will produce very little increase in the magnetic flux density B. The material has been driven into saturation.

As H is reduced to zero from the saturation point, the curve will move from point a to point b. At point b, some residual magnetic flux density B_r remains in the material even though the magnetizing force H is zero and is referred to as retentivity. As the magnetizing force H is reversed, the curve moves to point c, the point of coercivity where the magnetic flux density B has been reduced to zero despite having a nonzero magnetizing force. At the coercive point, the reversed magnetizing force H_C has flipped enough of the domains so that the net flux within the material is zero. The coercive force removes the residual magnetism from the material and may be used to "reset" the magnetic material to its unmagnetized state.

As the magnetizing force is increased in the negative direction, the material will again become magnetically saturated but in the opposite direction (point d).

Reducing H to zero brings the curve to point e. It will have a level of residual magnetism equal to that achieved in the other direction. Increasing H back in the positive direction will return B to zero. Note the curve did not return to the origin of the graph because some force is required to remove the residual magnetism. The curve will take a different path from point f back to the saturation point where the loop is completed. The area enclosed by the *B–H curve* is the energy loss per unit volume.

4.4.2 Magnetic Core Size

Magnetic switches are passive devices whose switching (holdoff) voltage is determined by the time for the core to saturate. This implies that magnetic switches must be pulse charged and that their switching parameters are directly determined by the core size [216]. The applied voltage pulse $V(t)$ induces a time-varying magnetic flux ϕ_B in the material according to Faraday's law:

$$N_t \times \frac{\partial \phi_B(t)}{\partial t} = -V(t) \tag{4.52}$$

where N_t is the number of conductor turns about the magnetic material. Integrating from voltage application at $t = 0$ to the time it takes to saturate the core, τ_{sat},

$$\phi_B = -\frac{1}{N_T}\int_0^{\tau_{sat}} V(t)dt \tag{4.53}$$

Using the definition of magnetic flux and assuming uniformity,

$$\phi_B = \int_{A_m} \vec{B} \cdot d\vec{S} \approx \Delta B \cdot A_m \tag{4.54}$$

where ΔB is the change in the magnetic flux density in the material and A_m is the area of the core. Equating (4.53) and (4.54),

$$-\frac{1}{N_T}\int_0^{T_{sat}} V(t)dt = \Delta B \cdot A_m \tag{4.55}$$

which gives a relation between the switching properties and the magnetic core size.

Assume the magnetic switch is charged by a square pulse with amplitude V_0 and duration T_p. The core is sized so that the time for the core to saturate is the pulse duration, $T_{sat} = T_p$. Under these conditions, (4.55) reduces to

$$A_m = \frac{V_0 T_p}{N_T \times \Delta B} \tag{4.56}$$

and the area of the magnetic core may be determined.

In order to minimize the area of the magnetic core A_m, ΔB is usually made as large as possible. If the core is reset by biasing the magnetic material into reverse residual value $-B_r$ then $\Delta B = B_{sat} - B_r$. The required core area can then be determined by (4.56).

The magnetic core volume is efficiently utilized when the change in flux density is large for a relatively small change in magnetic force H. This occurs when the hysteresis curve is "square." Squareness is characterized by the ratio of the residual magnetic flux B_r to the saturation value B_s, and in a material with good squareness this ratio is $> 90\%$. At the same time, a small hysteresis loop area minimizes core loss and an optimum design may exist [217]. The area enclosed by the hysteresis loop is unrecoverable energy (loss). The area between the hysteresis loop and the vertical axis is recoverable stored energy.

Amorphous [218] magnetic core materials, such as Metglas®, offer large saturation flux density and reduce the necessary core size. For a Metglas® core, net flux swing from ΔB could be greater than 3 Tesla. For high-power applications, core materials often come in laminated to minimize losses due to eddy currents. The pulse rise time also limits the core cross section utilization in the case of low-resistivity cores because of large eddy current generation and nondiffusion of flux through its core. Ferrites offer the best resistivity and low loss, but also have a small flux density swing (<0.7 Tesla) that results in a larger core size.

4.5 Solid-State Switches

Unlike spark gap switches, where a single switch can block a megavolt or pass hundreds of kiloamperes, semiconductor switches need numerous switch elements to be connected in parallel and series arrangement for dynamic sharing of voltage and current. Moreover, semiconductor switches have outstanding high pulse repetition rate capability and can operate at the tens of kilovolt range. This introduces a new paradigm for pulse power design.

Rapid developments are taking place in this area to replace thyratrons and ignitrons and even exceed their limits of performance [219–221]. The utilization of semiconductor switches in pulsed power has tremendous potential with important applications in the field of pumping of lasers, high-frequency converters, industrial magneto-forming, and environment control. The emerging semiconductor switch technology relies on new material synthesis and triggering methods.

The behavior of semiconductor switches is governed by various device parameters such as resistivity modulation, charge storage, device capacitance, electrothermal interaction, and electrical breakdown. By an appropriate design, these chip structure parameters are selected to achieve the optimum desired switching performance.

Resistivity Modulation To achieve high blocking voltages, semiconductor switches incorporate thick and lightly doped semiconductor layers. The resistivity R of this region determines the voltage drop and power loss when the device is in its on-state. The resistivity is given by

$$R = \int_{x_1}^{x_2} \frac{dx}{qA\left(\mu_n n + \mu_p p\right)} \tag{4.57}$$

where

n and p are the electron and hole densities, respectively,
μ_n and μ_p are the charge mobilities, and
A is the cross-sectional area.

Charge Storage The transient behavior of the switch during the change from on-state to off-state or the reverse is governed by how quickly the stored charges can be removed from the lightly doped region or how fast the low-doped region is swamped by electrons and holes. The turn-on time and turn-off time are dependent on the temporal or spatial behavior of charge density $\rho(x,t)$, which are determined by the following equations:

Ambipolar Diffusion:

$$\frac{d\rho}{dt} = -\frac{\rho}{\tau} + D \cdot \frac{d^2\rho}{dx^2} \tag{4.58}$$

Transport Equation

$$I = \left(1 + \frac{\mu_p}{\mu_n}\right) \cdot \left(I_n - qAD \cdot \frac{d\rho}{dx}\right) \tag{4.59}$$

Charge Control Equation

$$\frac{dQ}{dt} = -\frac{Q}{\tau} + I_n(x_r) - I_n(x_l) \tag{4.60}$$

where

D is the diffusion coefficient,
τ is the charge carrier lifetime, and
x_r, x_l are the boundaries of the considered region.

Device Capacitance The device capacitances are formed by reverse biased junctions in bipolar devices or insulated gates in unipolar MOS structures. These capacitances result in positive or negative feedback and the dV/dt effect can result in unintended turn-on or turn-off or continued current conduction in forward or reverse direction even after the device is switched off.

Electrothermal Interaction The switching losses cause the device to heat up. The switch performance depends on the behavior of various device parameters on temperature.

Electrical Breakdown Breakdown in semiconductor devices may occur from avalanche, Zener, or punch-through mechanisms. Avalanche and Zener breakdown may be intentionally used as part of normal operation or in failure. Punch-through is destructive.

The semiconductor switches, broadly classified into the categories of thyristor-based switches and transistor-based switches, are briefly discussed in the following sections.

4.5.1 Thyristor-Based Switches

The thyristor-based switches have PNPN structure with three junctions. Varieties of switches with altered characteristics are obtained by varying the design parameters in the structure architecture such as (i) geometrical profile and thickness of n and p regions, (ii) carrier concentration and doping profile in different regions, (iii) introduction of buffer layers, (iv) geometry and profile of contact areas of electrodes with different regions, and (v) introduction of the metal–oxide–semiconductor field-effect transistor (MOSFET) structure within the thyristor structure.

The forerunner in the class of switches is the silicon-controlled rectifier (SCR). It still maintains its status as a workhorse in industrial electronics and its forward blocking voltage of 12 kV is yet to be exceeded by other switches. After a discussion of the SCR, other switch configurations are discussed.

4.5.1.1 Silicon-Controlled Rectifier

A typical shorted emitter structure of an SCR along with the two-transistor equivalent circuit is shown in Figure 4.62. In the absence of the gate signal, the SCR blocks in the forward as well as in the reverse direction. For the forward applied voltage, the transition from the off-state to the on-state is achieved by the application of a positive trigger signal to the gate terminal, so as to make the regenerative loop gain [222] $h_{FE1} \cdot h_{FE2} = 1$. From the direction of cathode–gate current in Figure 4.62, it is evident that the right-hand side of the emitter region, nearest to the gate, gets the highest forward bias initially and the cathode–anode current flow starts at that edge. The current subsequently spreads to the entire emitter region with a velocity typically 0.1 mm/μs.

The SCR can be turned on by the sudden application of the forward voltage—that is, by a dV/dt effect. This effect is reduced in a shorted emitter structure, but the shorted emitter structure also decreases the sensitivity of the gate, necessitating a larger gate injection current for turn-on. Typical characteristics of anode current versus anode voltage for various gate currents are shown in

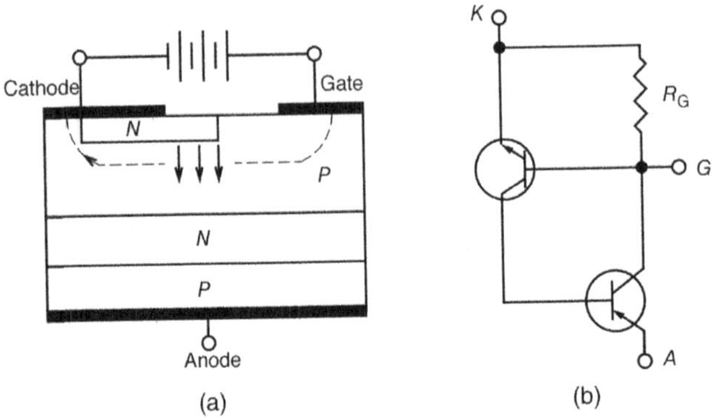

Figure 4.62 The structure (a) and equivalent circuit (b) of a shorted emitter SCR.

Figure 4.63. At high gate currents (>I_{G3}), the SCR behaves like a conventional PN diode. For reliable turn-on, a minimum holding current I_H is required before the gate signal is turned off.

The desirable parameters of an SCR for pulsed power applications are high forward and reverse blocking voltage, high peak current, high rate of rise of current, fast turn-on time, fast turn-off time, and long life.

High blocking voltage is obtained by increasing the *n*-base width and reducing the impurity carrier concentration, but this cannot be done without affecting the turn-off time and other thyristor properties. To take full advantage of the bulk breakdown of the semiconductor, the surface breakdown must be inhibited, which is usually done by beveling the surface.

Increased peak current and *di*/*dt* capability is generally achieved by increasing the initial turn-on area on the emitter, thus preventing the current concentration and destruction, with either the center gate–annular cathode or the center gate–double-annular cathode (also known as amplifying gate)

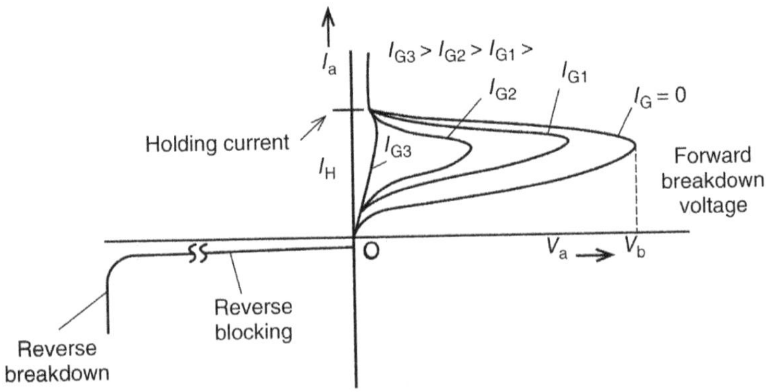

Figure 4.63 Voltage–current characteristics of an SCR.

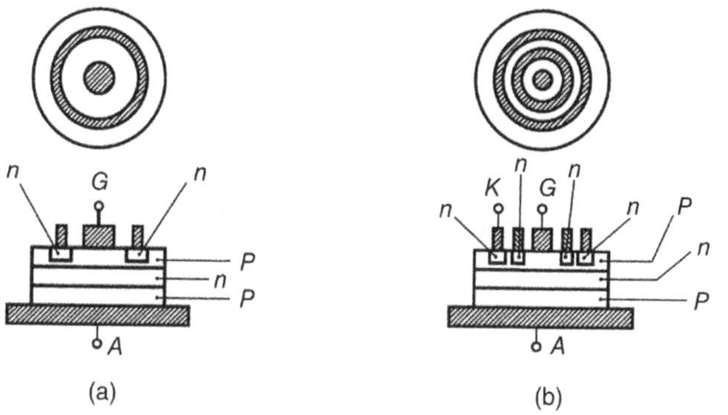

Figure 4.64 SCR structure to achieve a high peak current and a high rate of current rise using (a) a center gate and annular cathode and (b) an amplifying gate.

configurations, as shown in Figure 4.64a and b, respectively. These configurations result in the *di*/*dt* values [223] of 100 A/μs for the former and 1000 A/μs for the latter. Large periphery interdigited pattern [224] and involute contact pattern [225] are also used.

The turn-on time also depends on the buildup of charges in the central *p* and *n* base regions to support the holding current I_H, as well as on the plasma spreading velocity for the emitter coverage. The turn-on time can be reduced by reducing the *n*-base volume and increasing the *n*-base lifetime.

For high pulse repetition rates, a low turn-off time is required. The turn-off time depends on the charge stored in the *p* and *n* base regions. The turn-off time can be minimized by reducing either the *n*-base width or the carrier lifetime by introducing certain impurities and by circuit techniques like discharging a precharged capacitor through the SCR in the reverse direction.

Typical maximum parameters achieved in an SCR are peak voltage of ≈3 kV, peak currents of ≈150 kA, *di*/*dt* ≈ 800 A/μs, *dV*/*dt* ≈ 2 kV/μs, coulombs per pulse ≈250, life ≈2×10^4, voltage drop ≈10 V, and operating frequency ≈10 kHz. Maximum blocking voltages up to 12 kV are available at lower current ratings.

For higher voltage and current capabilities, SCRs can be connected in series and parallel configurations [226]. For uniform transient voltage distribution across the series string, the resistor–capacitor combination must be connected in parallel with each SCR. For an SCR series string, the external, synchronous triggering of many SCRs is complex. However, this can be simplified by suitable circuit techniques. For instance, instead of individually triggering each SCR, only the first SCR near the ground needs to be externally triggered. This SCR subsequently triggers the rest of the SCRs in a cascade [227] fashion. For SCRs connected in parallel, the inductors may be needed in individual SCRs to equally share the currents. A series–parallel combination of SCRs has been used to deliver [224] 4.7 kA at 15 kV.

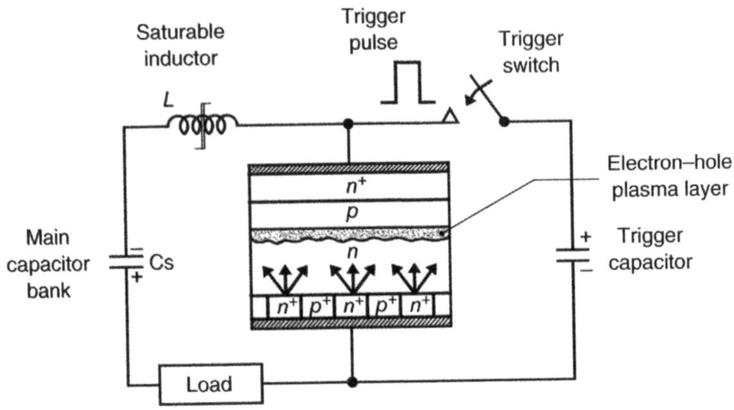

Figure 4.65 Structure and triggering of an RSD.

4.5.1.2 Reverse Switch-On Dynister

A technique of enhancing the peak current and *di/dt* capability, as used in a reverse switch-on dynister [224,228] by plasma triggering is illustrated in Figure 4.65. The triggering of RSD is achieved by applying a pulse of reverse polarity, which results in the breakdown of n^+p junction, injection of electrons from the n^+ region, and formation of a thin electron–hole plasma layer uniformly distributed on the collector junction plane. When the trigger pulse ends and the polarity reverts back to the initial state, the RSD turns on with the movement of the holes into the *p*-base and injection of electrons from the *n*-layer into the p^+ layer. The saturable inductor *L* helps to isolate the trigger pulse circuit from the main capacitor bank circuit. Once the discharge of the C_s starts, *L* saturates and hence has negligible effect on the load circuit. The main advantage of the RSD arises from the fact that the current is carried uniformly by the entire area of the device, rather than a restricted area near the gate electrode, as in an SCR.

Typical parameters in a single unit of RSD are as follows:

Repetitive peak off-state voltage	3 kV
Repetitive peak current	250 kA for 50 μs or 25 kA/10 ms
Rate of current rise	≈60 kA/μs
Off-state rate of voltage rise	0.8 kV/μs
Turn-off time	≈250 μs
Peak forward voltage drop	≈25 V
Operating frequency	≈100 kHz

4.5.1.3 Gate Turn-Off Thyristor

The basic structures [229–231] of conventional and emitter-shorted gate turn-off thyristors (GTOs) are shown in Figure 4.66. Similar to an SCR, a GTO can be

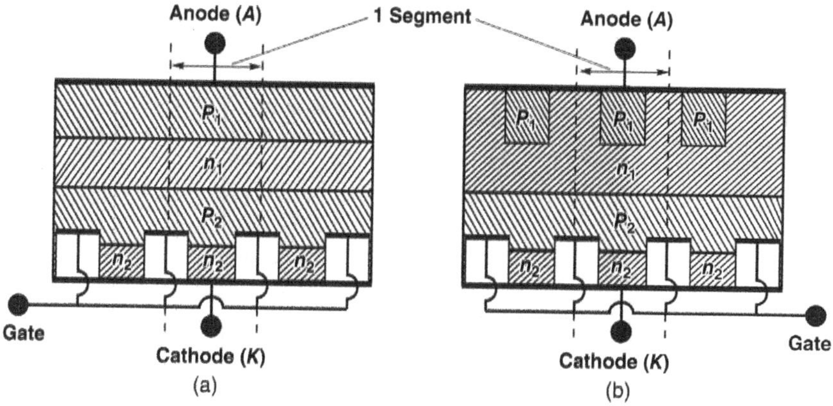

Figure 4.66 Basic GTO structures. (a) Conventional. (b) Emitter-shorted.

turned on by driving a positive gate current. However, it can also be turned off by driving a negative gate current. The turn-off gain of a GTO, which is the ratio of the anode current to the gate current for achieving a successful turn-off, is typically in the range of 3–5.

During the transition from the on-state to the off-state, the main current in the GTO is commuted from the cathode terminal to the gate terminal. To have low turn-off time and low off-state losses, a large surface area of contact between the gate electrode and P_2 region is necessary, which is achieved by having the gate electrode in the form of spiral fingers or strip contacts. An emitter-shorted GTO, where n_1 and P_1 are shorted to the anode electrode, has the advantages of low on-state voltage drop, less leakage current, and improved turn-off characteristics. However, its capability for reverse voltage blocking is compromised, because the entire reverse voltage has to be blocked by only one junction, the n_2–P_2 junction. This GTO needs to be protected by a snubber circuit or a diode connected across the GTO. In comparison to the turn-on time of an SCR (≈1 μs), the GTO has a longer turn-on time (≈4 μs) and it can be reduced by having a larger gate current. The di/dt rating is dependent on the rate of lateral spreading of current across the n_2–P_2 junction. If the di/dt rating of the device is exceeded, the resulting current concentration may destroy the device. The GTOs are available up to maximum ratings of 6 kV, 6 kA.

4.5.1.4 MOS Controlled Thyristor

The MOS controlled thyristor (MCT) was commercially introduced in 1988. The structure and circuit schematic of an MCT cell are shown in Figures 4.67 and 4.68, respectively. An overall MCT consists of a parallel combination of thousands of identical microcells on the same chip. A 500 V, 50 A MCT, for example [229], has 100,000 cells in parallel. The device is turned-on by the

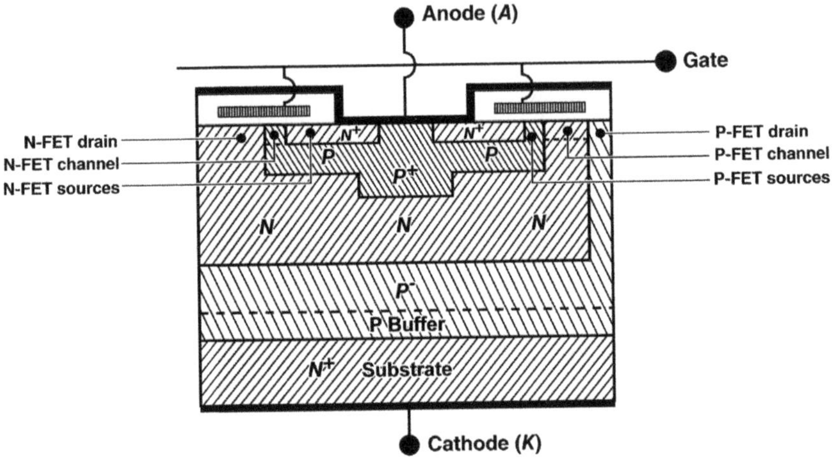

Figure 4.67 Structure of an MCT cell [229].

application of a pulse of negative voltage to the gate and turned off by a positive voltage pulse to the gate. It is seen from the equivalent circuit that a negative gate voltage turns on the *P*-FET, which results in driving a base current into the *n–p–n* transistor, thus initiating the regenerative feedback loop between the *n–p–n* and *p–n–p* transistors. When a positive voltage is applied to the gate, the *N*-FET is turned on, which short circuits the emitter base of the *p–n–p* transistor and breaks the regenerative loop. In comparison to a GTO, the turn-off current gain of an MCT is much larger.

4.5.1.5 MOS Turn-Off Thyristor

The structure and circuit representation of an MOS turn-off thyristor (MTO) [232] are shown in Figure 4.69. It is a four-terminal device with two

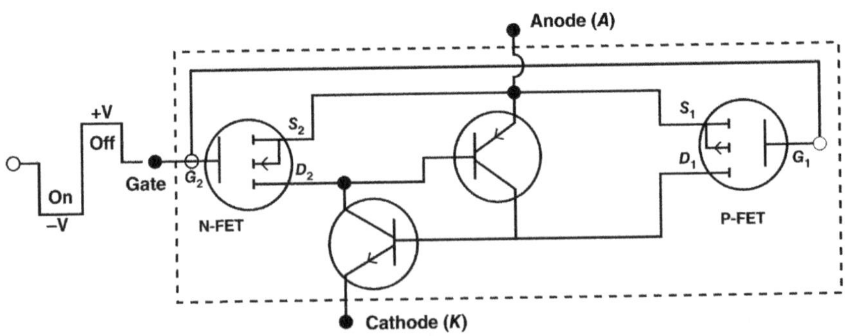

Figure 4.68 Equivalent circuit of an MCT.

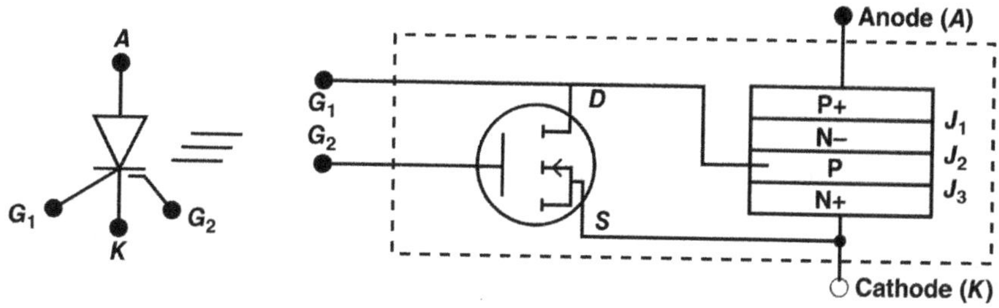

Figure 4.69 Structure and circuit schematic of an MTO [232].

gate terminals, G_1 and G_2. The device is turned on by the application of a positive voltage to the terminal G_1. It is turned off by the application of a positive voltage to the terminal G_2, which turns on the MOSFET. As a result, a short circuit (low-impedance path) is established in parallel to the junction J_3. The main MTO current is then commuted to the MOSFET. An MTO has the characteristics of a simplified gate drive, a short storage time, and an improved reverse blocking capability.

4.5.1.6 Emitter Turn-Off Thyristor

The structure and circuit schematic [233] of an emitter turn-off thyristor (ETO) are shown in Figure 4.70. In this structure, the power thyristor is supported by two MOSFETs, namely, NFET-1 and NFET-2. NFET-1 is connected to the gate terminal of the GTO and NFET-2 is in series with the GTO. The device is turned on when the gate drive is activated to result in NFET-1 becoming off and NFET-2 becoming on. The device gets turned off when NFET-1 is on and

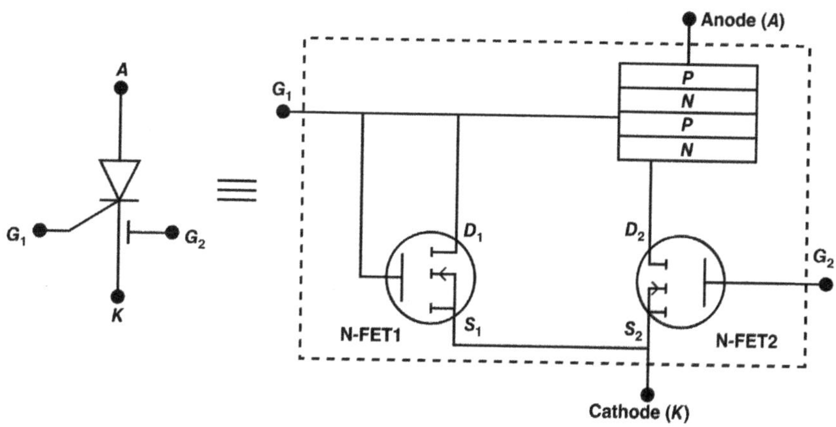

Figure 4.70 Structure and circuit schematic of an ETO [233].

NFET-2 is off. Under these conditions, the device turn-off mechanism is by commutation of the main current from NFET-2 to NFET-1. A quiescent off-state is reached when the ETO current becomes zero. The ETOs can be operated in series as well as in parallel arrangement to increase the forward blocking voltage and on-state current.

The beneficial features of ETO are the reduced dV/dt effects, minimum possibility of device failure by current filamentation, and reduced turn-off storage time. The ETOs are available up to maximum ratings of 6 kV, 6 kA.

4.5.1.7 Integrated Gate-Commuted Thyristor

The integrated gate-commuted thyristor (IGCT) is an improved version [234,235] of GTO, the changes being introduced in the structure and package. The changes in the structure are the addition of a buffer layer at the anode side, a reduction in the device thickness, and a diode incorporated in the wafer for reverse conduction. A significant change in the package from that of GTO is the provision of a low-inductance gate terminal in the form of a ring. The IGCT is available with a mechanically integrated hard drive circuit for the gate activation.

IGCTs are available in a variety of ratings such as 5.5 kV/2.3 kA, 4.5 kV/4 kA, and 6 kV/6 kA and may be connected in parallel to increase the current rating. The IGCT has a simplified and efficient low-inductance gate terminal and low losses, both in the on-state and the off-state.

4.5.2 Transistor-Based Switches

The characteristic feature of transistor-based switches, in contrast to thyristor-based switches, is that there is no regenerative feedback and hence latch-on is absent in the device. These switches are easily turned on and turned off by the application of appropriate signals to the gate terminal. The important versions are the MOSFET and the IGBT.

4.5.2.1 Insulated Gate Bipolar Transistor

The insulated gate bipolar transistor (IGBT) was commercially introduced in 1983. A typical structure [236–239] of a unit cell and equivalent circuit of an IGBT are shown in Figure 4.71a and b, respectively. Millions of such repetitive unit cells are internally connected in parallel on a chip to make up an IGBT power switch. When a positive voltage is applied to the gate with respect to the cathode, an n-channel is introduced in the P-region below the gate terminal, which forward biases the base-emitter junction (J_1) of the PNP transistor and the device gets turned on. When the positive gate voltage is removed by returning the gate voltage to zero, the n-channel is removed and the IGBT returns to the off-state. IGBTs can be connected in parallel and in series to make higher power modules if sufficient cooling is provided.

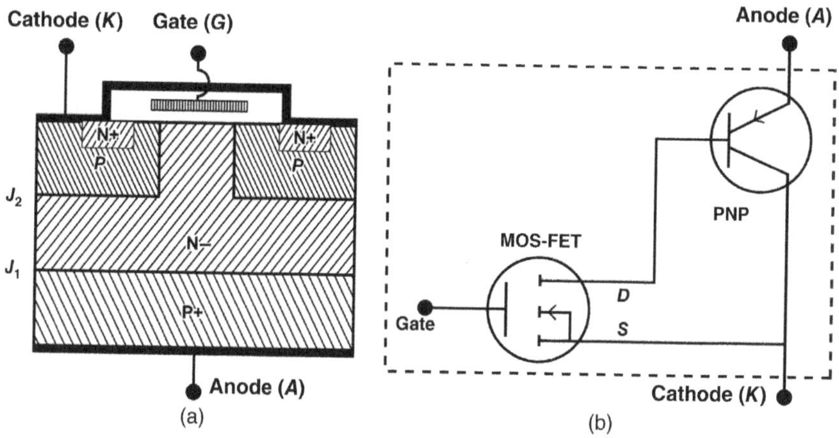

Figure 4.71 The (a) structure and (b) circuit representation of an IGBT.

The discrete IGBTs are available in various ratings such as 6.5 kV/600 A, 4.5 kV/1 kA, and power modules are available up to a maximum of 6 kV/2.5 kA.

4.5.2.2 Metal–Oxide–Semiconductor Field-Effect Transistor

The structure of a MOSFET is similar to that of an IGBT except that the *P*+ layer is absent. The MOSFET is turned on by the application of a positive voltage to the gate and turned off by returning the gate voltage to zero. MOSFETs are available in various ratings, such as 1 kV/100 A, 200 V/500 A, and 60 V/1 kA.

4.6 Design Examples

Example 4.1

A spark gap with a gap spacing of 5 cm and a nonuniform field geometry operates in air at a pressure of 300 kPa and breaks down at 100 kV. Calculate the rise time of the spark gap for driving source impedances of (i) 100 Ω, (ii) 10 Ω, and (iii) 1 Ω, respectively.

Solution

To compute the rise time, the contribution from the resistive and inductive components must be computed.

The inductive component is given by $\tau_L = L/Z$ and the inductance of the spark is estimated from Equation 4.9 to be $L = 14 \cdot \ell = 70$ nH for each of the three

(continued)

(continued)

feed impedances. The inductive rise time for the 100 Ω impedance is 0.7 ns, for the 10 Ω is 7 ns, and for the 1 Ω is 70 ns.

The resistive rise time for the 100 Ω feed impedance is given by Equation 4.16. The average electric field E is

$$E = \frac{V}{d} = \frac{100\ \text{kV}}{5\ \text{cm}} = 20\frac{\text{kV}}{\text{cm}}$$

Equation 4.16 lists the units of E in tens of kV/cm. The ratio of the gas density is approximately the ratio of the pressures, and realizing that 100 kPa is ~1 atm, the resistive rise time is computed for a feed impedance of 100 Ω as

$$\tau_R = \frac{88}{Z^{1/3}E^{4/3}} \cdot \sqrt{\frac{\rho}{\rho_0}} = \frac{88}{\sqrt[3]{(100)} \times 2^4 \sqrt{\frac{3}{1}} = 13\ \text{ns}}$$

Similarly, the resistive rise times for the other source impedances are $\tau_R = 28$ ns for $Z = 10\ \Omega$ and $\tau_R = 60$ ns for $Z = 1\ \Omega$.

The rise time t_r is given by Equation 4.15 and computes as follows:

For $Z = 100\ \Omega$,	$t_r = 2.2\sqrt{(13)^2 + (0.7)^2} = 13$ ns
For $Z = 10\ \Omega$,	$t_r = 2.2\sqrt{(28)^2 + (7)^2} = 29$ ns
For $Z = 1\ \Omega$,	$t_r = 2.2\sqrt{(60)^2 + (70)^2} = 92$ ns

Example 4.2

A solid dielectric spark gap breaks down at a voltage of 200 kV. The solid dielectric is a unit density polyethylene with a thickness of 3 mm. Calculate the total rise time if the source impedance is 100,10, and 1 Ω.

Solution

As in Example 1, to compute the rise time, the contribution from the resistive and inductive components must be computed. Since the insulation length is 3 mm, the inductance of the gap will decrease to

$$L = 14 \cdot \ell = 14 \cdot (0.3) = 4.2\ \text{nH}$$

The inductive rise time, then, is 0.042 ns for $Z = 100\ \Omega$, 0.42 ns for a $Z = 10\ \Omega$, and is 4.2 ns for $Z = 1\ \Omega$.

The resistive rise time is given by Equation 4.18 with the average electric field in units of MV/cm. The average electric field E is

$$E = \frac{V}{d} = \frac{200\ \text{kV}}{3\ \text{mm}} = 0.667\frac{\text{MV}}{\text{cm}}$$

$$\tau_R = \frac{5}{Z^{1/3}E^{4/3}} = \frac{5}{(100)^{1/3}(0.667)^{4/3}} = 1.9\ \text{ns}$$

Similarly, the resistive rise times for the other source impedances are $\tau_R = 4$ ns for $Z = 10\ \Omega$ and $\tau_R = 8.6$ ns for $Z = 1\ \Omega$.

The rise time t_r is computed from the resistive and inductive rise time using Equations 4.14 and 4.15 and computes as follows:

Source impedance	τ_L **(ns)**	τ_R **(ns)**	t_r **(ns)**
100 Ω	0.042	1.9	4.18
10 Ω	0.42	4	8.8
10 Ω	4.2	8.6	21

Example 4.3

Calculate the useful interval in which channels may be formed in an edge–plane spark gap charged to a voltage V_0 with a linear ramp voltage in 100 ns with a 0.3% deviation in voltage.

Solution

The linear ramp produces

$$\frac{dV}{dT} = \frac{V_0}{T_c} = \frac{V_0}{(100\ \text{ns})}$$

The useful interval for channel formation ΔT may be computed given that $\sigma_V(V) = 0.3\%$,

$$\Delta T = 2\sigma(t)T = 2 \cdot \frac{\sigma_V(V) \cdot V}{dV/dT} = \frac{2 \cdot (0.3/100) \cdot V_0}{V_0/100\ \text{ns}} = 0.6\ \text{ns}$$

Example 4.4

Typical gas spark gaps have deviations in voltage around 2%. Calculate the voltage rise time of the linear ramp pulse needed to give the same useful interval for channel formation ΔT calculated in Example 3.

Solution

Similar to Example 3, Equation 4.47 is used to calculate the pulse rise time to give a useful interval of 0.6 ns.

$$\Delta T = \frac{2\sigma(V)V_0}{V_0/T_C} = 2 \cdot \sigma(V) \cdot T_C \tag{4.61}$$

Rearranging terms,

$$T_C = \frac{\Delta T}{2 \cdot \sigma(V)} = \frac{0.6\text{ ns}}{2 \cdot (2/100)} = 15\text{ ns} \tag{4.62}$$

Thus, in rough gas sparks and good liquid gaps, the voltage must be applied in tens of nanoseconds in order to induce multichanneling.

Example 4.5

An edge–plane gap operating in air at 100 kPa has a gap spacing of 4.5 cm and the total length of the edge is 30 cm. It is driven by a PFL with an effective impedance of 3 Ω. The gap is charged to a voltage of 135 kV with a linear ramp in 100 ns with a 0.3% deviation in voltage. Assume the electrode inductance is negligible and the inductance per arc channel is 14 nH/cm. Calculate the probable number of spark channels in the gap.

Solution

The conditions are identical to those in Example 3, such that ΔT is 0.6 ns. The relationship between scatter in closure time ΔT, *e*-folding rise time t_{rT}, and transit time isolation t_{trans} is given by Equation 4.26.

$$\Delta T = 0.1 \cdot \tau_{TOT} + 0.8 \cdot \tau_{trans}$$

$$\tau_{trans} = \frac{\sqrt{\epsilon_r}\ell_{edge}}{Nc} = \frac{\left(\sqrt{1}\right) \cdot (30\text{ cm})}{N \cdot (30\text{ cm}/ns)} = \frac{1\text{ ns}}{N}$$

$$\tau_{TOT} = \sqrt{\tau_R^2 + \tau_L^2}$$

$$\tau_L = \frac{L}{NZ} = \frac{(14\,nH/\text{cm}) \cdot (4.5\,\text{cm})}{N \cdot (3\,\Omega)} = \frac{21}{N}\ \text{ns}$$

For air at a pressure of 100 kPa (~1 atm), $\sqrt{\rho/\rho_0}$~1 and noting that an average electric field value of 30 kV/cm yields a value of E in units of 10 kV/cm of 3,

$$\tau_R = \left(\frac{88}{E^{4/3} \cdot N^{1/3} \cdot Z^{1/3}}\right) = \left(\frac{88}{(3)^{4/3} \cdot N^{1/3} \cdot (3)^{1/3}}\right) = \frac{14\,\text{ns}}{N^{1/3}}$$

$$\tau_{TOT} = \sqrt{\left(\frac{14}{N^{1/3}}\right)^2 + \left(\frac{21}{N}\right)^2}\ \text{ns} = \sqrt{\frac{196}{N^{2/3}} + \frac{441}{N^2}}\ \text{ns} = \frac{1}{N}\sqrt{\frac{196}{N^{1/3}} + 441}\ \text{ns}$$

$$\Delta T = 0.1 \cdot \tau_{TOT} + 0.8 \cdot \tau_{trans} = \left[\frac{0.1}{N}\sqrt{\frac{196}{N^{1/3}} + 441} + 0.8 \cdot \frac{1}{N}\right]$$

$$\text{ns} = \frac{0.1}{N}\left[\sqrt{\frac{196}{N^{1/3}} + 441} + 8\right]\ \text{ns}$$

For $\Delta T = 0.6$ ns, the solution obtained from iteration is $N \sim 5$ channels.

References

1 L.L. Alston, *High Voltage Technology*, Oxford University Press, London, 1968.
2 C.H. Jones, W. Crewson, J.T. Naff, and J. Granados, *An Analytical Model for the High Voltage Rope Switch*, Pulsar Associates, Switching Note 19, May 1973.
3 B. M. Kovalchuk, Multi Gap Spark Switches. *Proceedings of the IEEE International Pulsed Power Conference*, pp. 59–67, 1997.
4 G.A. Mesyats, *Pulsed Power*, Kluwer Academic/Plenum Publishers, 2005.
5 F. Frungel, *High Speed Pulse Technology*, Vol. 1, Academic Press, 1965.
6 R. Rompe and W. Weizel, On the Toepler Spark Law. *Z. Phys.*, Vol. 122, pp. 636–639, 1944.
7 S.I. Braginskii, Theory of the Development of a Spark Channel. *Sov. Phys. JETP*, Vol. 34, No. 7, pp. 1068–1074, 1958.
8 T.H. Martin, J.F. Seaman, and D.O. Jobe, Energy Losses in Switches. *Proceedings of the 9th International Pulsed Power Conference*, pp. 463–470, 1993.

9 L.K. Warne, R. Jorgenson, and J. Lehr, *Resistance of a Water Spark*, SAND 2005–6994, Sandia National Laboratories, Albuquerque, NM, 2005.

10 H. Knoepfel, *Pulsed High Magnetic Fields*, Elsevier, New York, 1970.

11 J.C. Martin, Nanosecond Pulse Techniques. *Proc. IEEE*, Vol. 80, No. 6, pp. 934–945, 1992.

12 S. Levinson, E.E. Kunhardt, and M. Kristiansen, Simulation of Inductive and Electromagnetic Effects Associated with Single and Multichannel Triggered Spark Gaps. *Proceedings of the 2nd Pulsed Power Conference*, pp. 433–436, 1979.

13 W.C. Nunnally and A.L. Donaldson, Self Breakdown Gaps, in G. Schaefer, M. Kristiansen, and A. Guenther, eds., *Gas Discharge Closing Switches*, Plenum Press, 1990.

14 D.M. Weidenheimer, N.R. Pereira, and D.C. Judy, The Aurora Synchronization Project. *Proceedings of the International Pulsed Power Conference*, 1991.

15 F.W. Grover, *Inductance Calculations*, D. Van Nostrand, 1947.

16 A.L. Donaldson, R. Ness, M. Hagler, and M. Kristiansen, Modeling of Spark Gap Performance. *Proceedings of the International Pulsed Power Conference*, pp. 525–529, 1983.

17 J.M. Meek and J.D. Craggs, *Electrical Breakdown of Gases*, Oxford University Press, 1953.

18 T.H. Martin, An Empirical Formula for Gas Switch Breakdown Delay. *Proceedings of the IEEE Pulsed Power Conference*, pp. 73–79, 1985.

19 J.C. Martin, Duration of the Resistive Phase and Inductance of Spark Channels, in T.H. Martin, A.H. Guenther, and M. Kristiansen, eds., *JCM on Pulsed Power*, Plenum Press, 1996.

20 T.P. Sorensen and V.M. Ristic, Rise Time and Time Dependent Spark Gap Resistance in Nitrogen and Helium. *J. Appl. Phys.*, Vol. 48, No. 1, p. 114, 1977.

21 K.V. Nagesh, P.H. Ron, G.R. Nagabhushana, and R.S. Nema, Recovery Times of Low Pressure Spark Gaps Under Burst-Mode. *Fourth Workshop and Conference on EHV Technology*, Indian Institute of Science, Bangalore, India, 1998.

22 K.V. Nagesh, P.H. Ron, G.R. Nagabhushana, and R.S. Nema, Over Recovery in Low Pressure Spark Gaps. *IEEE Trans. Plasma Sci.*, Vol. 27, No. 1, p. 199, 1999.

23 K.V. Nagesh, P.H. Ron, G.R. Nagabhushana, and R.S. Nema, Over-Recovery in Low Pressure Spark Gaps: Confirmation of Pressure Reduction in Over Recovery. *IEEE Trans. Plasma Sci.*, Vol. 27, No. 6, p. 1559, 1999.

24 R. Curry, P. D'A. Champney, C. Eichenberger, J. Fockler, D. Morton, R. Sears, I. Smith, and R. Conrad, The Development and Testing of Sub-Nanosecond Rise, Kilohertz Oil Switches for the Generation of High-Frequency Impulses. *IEEE Trans. Plasma Sci.*, Vol. 20, No. 3, pp. 383–392, 1992.

25 P. Norgard and R.D. Curry, An In-Depth Investigation of the Effect of Oil Pressure on the Complete Statistical Performance of a High Pressure Flowing Oil Switch. *IEEE Trans. Plasma Sci.*, Vol. 38, No. 10, pp. 2539–2547, 2010.

26 C.E. Baum, W.L. Baker, W.D. Prather, J.M. Lehr, J. O'Loughlin, D.V. Giri, I.D. Smith, R. Altes, J. Fockler, M. Abdalla, and M. Skipper, JOLT: A Highly Directive, Very Intensive, Impulse-Like Radiator. *Proc. IEEE*, Vol. 92, No. 7, pp. 1096–1109, 2004.

27 S.L. Moran, L.W. Hardesty, and M.G. Grothaus, Hydrogen Spark Gap for High Repetition Rates. *Proceedings of the IEEE Pulsed Power Conference*, pp. 473–476, 1985.

28 R. Hutcherson, S.L. Moran, and E. Ball, Triggered Spark Gap Recovery. *Proceedings of the IEEE Pulsed Power Conference*, p. 221, 1987.

29 S.L. Moran and R. Hutcherson, *High PRF High Current Switch*. U.S. Patent No. 4,912,369 March 27, 1990.

30 S.L. Moran and L.W. Hardesty, High-Repetition-Rate Hydrogen Spark Gap. *IEEE Trans. Electron Devices*, Vol. 38, No. 4, pp. 726–730, 1991.

31 M.G. Grothaus, S.L. Moran, and L.W. Hardesty, Recovery Characteristics of Hydrogen Spark Gap Switches. *Proceedings of the IEEE Pulsed Power Conference*, pp. 475–478, 1993.

32 M.W. Ingram, S.L. Moran, and M.G. Grothaus, High Average Power Hydrogen Spark Gap Experiments. *Conference Record of the Power Modulator Symposium*, pp. 309–311, 1994.

33 S.J. MacGregor, F.A. Tuema, S.M. Turnbull, and O. Farish, The Operation of Repetitive High Pressure Spark Gap Switches. *J. Phys. D Appl. Phys.*, Vol. 26, pp. 954–958, 1993.

34 S.J. MacGregor, S.M. Turnbull, F.A. Tuema, and O. Farish, Enhanced Spark Gap Switch Recovery Using Nonlinear *V*/*p* Curves. *IEEE Trans. Plasma Sci.*, Vol. 23, No. 4, pp. 798–804, 1995.

35 S.J. MacGregor, S.M. Turnbull, F.A. Tuema, and A.D.R. Phelps, Methods of Improving the Pulse Repetition Frequency of High Pressure Gas Switches. *Tenth IEEE International Pulsed Power Conference, Digest of Technical Papers*, Vol. 1, pp. 249–254, 1995.

36 S.J. MacGregor, S.M. Turnbull, F.A. Tuema, and O. Farish, Factors Affecting and Methods of Improving the Pulse Repetition Frequency of Pulse-Charged and DC-Charged High-Pressure Gas Switches. *IEEE Trans. Plasma Sci.*, Vol. 25, No. 2, pp. 110–117, 1997.

37 F.A. Tuema, S.J. MacGregor, J.A. Harrower, J.M. Koutsoubis, and O. Farish, Corona Stabilization for High Repetition Rate Plasma Closing Switches. *11th International Symposium on High Voltage Engineering*, Vol. 3, pp. 280–284, 1999.

38 J.M. Koutsoubis and S.J. MacGregor, Electrode Erosion and Lifetime Performance of a High Repetition Rate, Triggered, Corona Stabilized Switch in Air. *J. Phys. D Appl. Phys.*, Vol. 33, pp. 1093–1103, 2000.

39 J.M. Koutsoubis and S.J. MacGregor, Effect of Gas Type on High Repetition Rate Performance of a Triggered, Corona Stabilised Switch. *IEEE Trans. Dielectr. Electr. Insul.*, Vol. 10, No. 2, pp. 245–255, 2003.

40 A.L. Donaldson, M. Kristiansen, A. Watson, K. Zinsmeyer, E. Kristiansen, and R. Dethlefsen, Electrode Erosion in High Current, High Energy Transient Arcs. *IEEE Trans. Magn.*, Vol. MAG-22, No. 6, pp. 1441–1447, 1986.

41 A.L. Donaldson, Electrode Erosion Resulting from High Current, High Energy Transient Arcing, Ph.D. dissertation, Texas Tech University, August 1990.

42 G.S. Belkin, Dependence of Electrode Erosion on Heat Flux and Duration of Current Flow. *Soviet Phys. Tech. Phys.*, Vol. 15, No. 7, pp. 1167–1170, 1971.

43 G.S. Belkin, Method for the Approximate Calculation of Electrode Erosion in Discharges for Computation of Large Pulse Flows. *Fizika Khim. Obrabot. Mat.*, No. 1, pp. 33–38, 1974.

44 G.S. Belkin, Methodology of Calculating Erosion of High Current Contacts During Action of an Electrical Arc. *Trans. Electrichestro*, Vol. 1, FSTC-HT-0787-85, pp. 61–64, 1972.

45 A. Watson, A.L. Donaldson, K. Ikuta, and M. Kristiansen, Mechanism of Electrode Surface Damage and Material Removal in High Current Discharges. *IEEE Trans. Magn.*, Vol. 22, No. 6, pp. 1799–1803, 1986.

46 F.M. Lehr, B.D. Smith, A.L. Donaldson, and M. Kristiansen, The Influence of Arc Motion on Electrode Erosion in High Current, High Energy Switches. *Proceedings of the International Pulsed Power Conference*, pp. 529–553, 1986.

47 A.L. Donaldson, F.M. Lehr, and M. Kristiansen, Performance of *In-Situ* Copper Alloys as Electrodes in High Current, High Energy Switches, *Proc. SPIE*, Vol. 0871, doi:10.1117/12.943673, 1988.

48 A.L. Donaldson, Electrode Erosion Measurements in a High Energy Spark Gap, Master's thesis, Texas Tech University, 1982.

49 F.M. Lehr, Electrode Erosion from High Current, High Energy Moving Arcs, Master's thesis, Texas Tech University, 1988.

50 R. Holmes, Electrode Phenomena, in J.M. Meek and J.D. Craggs, eds., *Electrical Breakdown in Gases*, John Wiley & Sons, Inc., New York, NY, pp. 839–867, 1978.

51 D. Affinito, E. Bar-Avraham, and A. Fischer, Design and Structure of an Extended Life High Current Sparkgap. *IEEE Trans. Plasma Sci.*, Vol. 7, pp. 162–163, 1979.

52 A.L. Donaldson, G. Engel, and M. Kristiansen, State of the Art Insulator and Electrode Materials for High Current Switching Applications. *IEEE Trans. Magn.*, Vol. 25, pp. 138–141, 1989.

53 G. Marchesi and A. Maschio, Influence of Electrode Material on Arc Voltage Waveforms in Pressurized Field Distortion Spark Gaps. *Proceedings of the International Conference on Gas Discharges*, Liverpool, UK, p. 145148, September 1978.

54 K.J. Bickford, K.W. Hanks, and W.L. Willis, Spark Erosion Characteristics of Graphite and CO Gas. *Proceedings of the International Power Modulator Symposium*, pp. 89–92, 1982.
55 A.L. Donaldson, M.O. Hagler, M. Kristiansen, G. Jackson, and L.L. Hatfield, Electrode Erosion Phenomena in a High Energy Pulsed Discharge. *IEEE Trans. Plasma Sci.*, Vol. 12, No. 1, pp. 28–38, 1984.
56 A.L. Donaldson, M. Kristiansen, A. Watson, K. Zinsmeyer, E. Kristiansen, and R. Dethlefsen, Electrode Erosion in High Current, High Energy Transient Arcs. *IEEE Trans. Magn.*, Vol. MAG-22, No. 6, pp. 1441–1447, 1986.
57 A.L. Donaldson, D. Garcia, M. Kristiansen, and A. Watson, A Gap Distance Threshold in Electrode Erosion in High Current, High Energy Spark Gaps. *Proceedings of the Power Modulator Symposium*, pp. 146–148, 1986.
58 A.L. Donaldson and M. Kristiansen, Electrode Erosion as a Function of Electrode Materials in High Current, High Energy Transient Arcs. *Proceedings of the International Pulsed Power Conference*, p. 83, 1989.
59 V.G. Emellyanov et al. Multi-Spark High Voltage Trigatron. *Instrum. Exp. Tech.*, Vol. 18, p. 1114, 1975.
60 C.E. Baum and J.M. Lehr, *Some Considerations for Multichannel Switching*, Switching Note 30, January 2002. Available at www.ece.unm.edu/summa/notes/.
61 J.M. Lehr, et al. SATPro: The System Assessment Test Program for ZR. *Conference Record of the IEEE Power Modulator Symposium*, pp. 106–110, 2004.
62 J.M. Lehr, J.E. Maenchen, J.R. Woodworth, W.A. Johnson, R.S. Coates, L.K. Warne, L.P. Mix, D.L. Johnson, I.D. Smith, J.P. Corley, S.A. Drennan, K.C. Hodge, D.W. Guthrie, J.M. Navarro, and G.S. Sarkisov, Multi-Megavolt Water Breakdown Experiments. *Proceedings of the IEEE International Pulsed Power Conference*, pp. 609–614, 2003.
63 J.C. Martin, Multichannel Gaps, in T.H. Martin, A.H. Guenther, and M. Kristiansen, eds., *JCM on Pulsed Power*, Plenum Press, 1996.
64 T.H. Martin, Pulsed Charged Gas Breakdown. *Proceedings of the International Pulsed Power Conference*, pp. 74–83, 1985.
65 T.R. Burkes et al., A Review of High Power Switch Technology. *IEEE Trans. Electron Devices*, Vol. 26, No. 10, p. 1401, 1979.
66 R.E. Beverly and R.N. Campbell, Transverse Flow 50 kV Trigatron Switch for 100 pps Burst-Mode Operation. *Rev. Sci. Instrum.*, Vol. 67, No. 4, pp. 1593, 1996.
67 J.D. Craggs, M.E. Haine, and J.M. Meek, The Development of Triggered Spark Gaps for High Power Modulators. *J. Inst. Electr. Eng.*, Vol. 93, No. Part III A, p. 963, 1946.
68 K.J.R. Wilkinson, Some Developments in High Power Modulators for Radar. *J. Inst. Electr. Eng.*, Vol. 93A, p. 1090, 1946.
69 S.J. MacGregor, F.A. Tuema, S.M. Turnball, and O. Farish, The Influence of Polarity on Trigatron Switching Performance. *IEEE Trans. Plasma Sci.*, Vol. 25, No. 2, pp. 118–123, 1997.
70 A.S. Husbands and J.B. Higham, The Controlled Tripping of High Voltage Impulse Generators. *J. Sci. Instrum.*, Vol. 28, No. 8, p. 242, 1951.

71 A.M. Sletten and T.J. Lewis, Characteristics of the Trigatron Spark Gap. *Proc. IEE*, Vol. 104, p. 54, 1957.

72 T.E. Broadbent, The Breakdown Mechanism of Certain Triggered Spark Gaps. *Br. J. Appl. Phys.*, Vol. 8, p. 37, 1957.

73 T.E. Broadbent and A.H.A. Shlash, The Development of the Discharge in the Trigatron Spark Gap at Very High Voltages. *Br. J. Appl. Phys.*, Vol. 14, No. 10, p. 687, 1963.

74 P.I. Shkuropat, The Investigation of Controlled Triggered Spark Gap. *Zh. Tekh. Fiz.*, Vol. 30, p. 954, 1960; *Sov. Phys. Tech. Phys.*, Vol. 5, p. 895, 1961.

75 B. Pashaie, G. Schaefer, K.H. Schoenbach, and P.F. Williams, Field Enhancement Calculations for Field Distortion Triggered Spark Gap. *J. Appl. Phys.*, Vol. 61, No. 2, pp. 790–792, 1987.

76 P.F. Williams and F.E. Peterkin, Triggering in Trigatron Spark Gaps: A Fundamental Study. *J. Appl. Phys.*, Vol. 66, No. 9, pp. 4163–4175, 1989.

77 F.E. Peterkin and P.F. Williams, Physical Mechanism of Triggering in Trigatron Spark Gaps. *Appl. Phys. Lett.*, Vol. 53, No. 3, pp. 182–184, 1988.

78 G.A. Mesyats, Pulsed High Current Electron Technology. *Proceedings of the International Pulsed Power Conference*, p. 9, 1979.

79 R.E. Wooton, Triggering of Spark Gaps at Minimum Voltage. *Proceedings of the International Pulsed Power Conference*, p. 258, 1985.

80 R.E. Beverly, III, Application Note 104. Available at http://www.reb3.com/ (accessed March, 2011).

81 P.F. Williams and F.E. Peterkin, Trigatron Spark Gaps, in G. Shaefer, M. Kristiansen, and A. Guenther, eds., *Gas Discharge Closing Switches*, Plenum Press, New York, pp. 63–84, 1990.

82 J.M. Lehr, M.D. Abdalla, F. Gruner, B. Cockreham, M.G. Skipper, and W.D. Prather, A Hermetically Sealed Trigatron for High Pulse Repetition Rate Operation. *IEEE Trans. Plasma Sci.*, Vol. 28, No. 5, pp. 1469–1475, 2000.

83 G.S. Kichaeva, A Controlled High Voltage Spark Gap of Low Inductance. *Instrum. Exp. Tech.*, Vol. 18, p. 122, 1975.

84 D.C. Hagerman and A.H. Williams, High Power Vacuum Spark Gap. *Rev. Sci. Instrum.*, Vol. 30, No. 3, p. 182, 1959.

85 W.L. Baker, High Voltage, Low Inductance Switch for Megaampere Pulse Currents. *Rev. Sci. Instrum.*, Vol. 30, No. 8, p. 700, 1959.

86 N. Vidyardhi and R.S.N. Rau, A Simple Triggered Vacuum Gap. *J. Phys. E Sci. Instrum.*, Vol. 6, p. 33, 1973.

87 G.A. Farall, Low Voltage Firing Characteristics of a Triggered Vacuum Gap. *IEEE Trans. Electron Devices*, Vol. 13, p. 432, 1966.

88 G.R. Govindaraju et al., Breakdown Mechanisms and Electrical Properties of Triggered Vacuum Gaps. *J. Appl. Phys.*, Vol. 47, p. 1310, 1976.

89 S. Kamakshaiah and R.S.N. Rau, Delay Characteristics of a Simple Triggered Vacuum Gap. *J. Phys. D Appl. Phys.*, Vol. 8, p. 1426, 1975.

90 A.M. Andrianov et al., Pulse Generator Producing a High Power Current. *Instrum. Exp. Tech.*, Vol. 14, p. 124, 1971.
91 J.E. Thompson et al., Design of a Triggered Vacuum Gap for Crowbar Operation. *Conference Record of the Power Modulator Conference*, p. 85, 1980.
92 V.V. Baraboshkin, Plasma Trigatron Gap. *Instrum. Exp. Tech.*, Vol. 20, p. 472, 1977.
93 I.I. Kalyatskii et al., Parallel Operation of Controlled Discharge Gaps in a Current Pulse Generator. *Instrum. Exp. Tech.*, Vol. 21, p. 973, 1978.
94 G. Baldo and M. Rea, Long Time Lag Discharge in a Triggered Spark Gap in Air. *Br. J. Appl. Phys.*, Vol. 1, p. 1501, 1968.
95 P.I. Shkuropat, Electrical Characteristics of Controlled High Current Triggered Air Spark Gaps. *Sov. Phys. Tech. Phys.*, Vol. 11, p. 779, 1966.
96 A.I. Gerasimov et al., Starting Characteristics of a 100 kV Trigatron Filled with SF_6. *Instrum. Exp. Tech.*, Vol. 18, p. 1455, 1975.
97 V. Ya. Ushakov, Megavolt Water-Filled Trigatron Type Spark Gaps. *Instrum. Exp. Tech.*, Vol. 20, p. 1364, 1977.
98 N.K. Kapishnikov et al., High Voltage Discharge Gaps Filled with Transformer Oil. *Instrum. Exp. Tech.*, Vol. 21, pp. 975–979, 1978.
99 J.L. Maksiejewski and J.H. Calderwood, Time Lag Characteristics of a Liquid Trigatron. *J. Phys. D Appl. Phys.*, Vol. 9, p. 1195, 1976.
100 D.N. Dashuk and G.S. Kichaeva, Controlled Vacuum Spark Gaps Rated at a Voltage of 50 kV for Multiple Commutation of Mega-Ampere Currents. *Instrum. Exp. Tech.*, Vol. 18, p. 463, 1975.
101 S.N. Ivanov and S.A. Shunailov, Low Jitter Generators of High Voltage Pulses with a Subnanosecond Rise Time and an Increased Actuation Accuracy. *Instrum. Exp. Tech.*, Vol. 43, No. 3, pp. 350–353, 2000.
102 H.H. Zhu, L. Huang, Z.H. Cheng, J. Yin, and H.Q. Tang, Effect of the Gap on Discharge Characteristics of the Three Electrode Trigatron Switch. *J. Appl. Phys.*, Vol. 105, p. 113303, 2009.
103 P.H. Ron, K. Nanu, S.T. Iyengar, and V.K. Rohatgi, Single and Double Mode Gas Trigatron: Main Gap and Triggered Gap Interactions. *J. Phys. D Appl. Phys.*, Vol. 21, p. 1738, 1988.
104 K. Nanu, P.H. Ron, S.T. Iyengar, and V.K. Rohatgi, Switching Characteristics of a Gas Trigatron Operated in Single and Double Trigger Modes. *IEEE Trans. Electr. Insul.*, Vol. 25, No. 2, p. 381, 1990.
105 R.J. Focia and C.A. Frost, *A Compact, Low Jitter, Fast Risetime, Gas-Switched Pulse Generator System with High Pulse Rate Capability.* Available at www.pulsedpwr.com (accessed June 19, 2009).
106 P.H. Ron, K. Nanu, S.T. Iyengar, K.V. Nagesh, R.K. Rajawat, and V.R. Jujaray, An 85 kJ High Performance Capacitor Bank with Double Mode Trigatrons. *Rev. Sci. Instrum*, Vol. 63, No. 1, p. 37, 1992.

107 M.J. Given, I.V. Timoshkin, M.P. Wilson, S.J. MacGregor, and J.M. Lehr, A Novel Design for a Multistage Corona Stabilised Closing Switch. *IEEE Trans. Dielectr. Electr. Insul.*, Vol. 18, No. 4, 2011.

108 R.A. Fitch and N.R. McCormick, Low Inductance Switching Using Parallel Spark Gaps. *Proc. IEE*, Vol. 106A, p. 117, 1959.

109 P.D.'A. Champney, Some Recent Advances in Three Electrode Field Enhanced Triggered Gas Switches, in W.H. Bostick, V. Nardi, and O.S.F. Zucker, eds., *Energy Storage, Compression, and Switching*, Plenum Press, New York, p. 463, 1976.

110 R.S. Post and Y.G. Chen, A 100 kV Fast High Energy Non-Uniform Field Distortion Switch. *Rev. Sci. Instrum.*, Vol. 43, No. 4, p. 622, 1972.

111 A. Faltens et al., High Repetition Rate Burst-Mode Spark Gap. *Conference Record of the 13th Power Modulator Conference*, p. 98, 1978.

112 V.R. Kukhta et al., Controlled Spark Gaps for High Voltage Pulse Generators. *Instrum. Exp. Tech.*, Vol. 19, p. 1670, 1976.

113 H.B. McFarlane and R. Kihara, The FXR: One Nanosecond Jitter Switch. *Conference Record of the Power Modulator Symposium*, p. 9, 1980.

114 X. Zou, R. Liu, N. Zeng, M. Han, J.Yuan, X. Wang, and G. Zhang, A Pulsed Power Generator for X-Pinch Experiments. *Laser Part. Beams*, Vol. 24.4, pp. 503–509, 2006.

115 K.R. Prestwich, 2 MV, Multi-Channel Oil Dielectric Triggered Spark Gap, in W.H. Bostick, V. Nardi, and O.S.F. Zucker, eds., *Energy Storage, Compression, and Switching*, Plenum Press, New York, p. 451, 1976.

116 A.B. Gerasimov, High-Precision Fast-Acting Discharger with a Solid Dielectric. *Instrum. Exp. Tech.*, Vol. 2, p. 446, 1970.

117 P. Dokopoulos, Fast 500 kV Energy Storage Unit with Water Insulation and Solid Multi-Channel Switching. *Proceedings of the 6th Symposium on Fusion Technology*, Aachen, Germany, p. 393, 1970.

118 P. Dokopoulos, A Fast Solid Dielectric Multi-Channel Crowbar Switch. *Proceedings of the 6th Symposium on Fusion Technology*, Aachen, Germany, p. 379, 1970.

119 Y.P. Raizer, *Gas Discharge Physics*, Springer, p. 151, 1991.

120 A.H. Guenther and J.R. Bettis, Laser Triggered Megavolt Switches. *IEEE J. Quantum Electron.*, Vol. 3, No. 6, pp. 265, 1967.

121 A.H. Guenther and J.R. Bettis, The Laser Triggering of High Voltage Switches. *J. Phys. D Appl. Phys.*, Vol. 11, p. 1577, 1978.

122 W.R. Rappoport, KrF Laser Triggered SF6 Spark Gap for Low Jitter Timing. *IEEE Trans. Plasma Sci.*, Vol. 8, pp. 167–170, 1980.

123 J.R. Woodworth, P.J. Hargis, L.C. Pitchford, and R.A. Hamil, Laser Triggering of a 500 kV Gas-Filled Spark Gap: A Parametric Study. *J. Appl. Phys.*, Vol. 56, p. 1382, 1984.

124 J.R. Woodworth et al., UV-Triggering of 2.8 Megavolt Gas Switches. *IEEE Trans. Plasma Sci.*, Vol. 10, No. 4, pp. 257–261, 1982.

125 W.T. Clark, Analysis of a Laser Induced Plasma in High Pressure SF6 Gas for High Voltage, High-Current Switching. Master's thesis, University of New Mexico, 2004.
126 R.S. Taylor et al., Laser Triggered Rail Gaps. *Conference Record of the Power Modulator Symposium*, p. 32, 1980.
127 R.J. Dewhurst et al., Picosecond Triggering of a Laser-Triggered Sparkgap. *J. Phys. D Appl. Phys.*, Vol. 5, p. 97, 1972.
128 Y.A. Kurbatov and V.F. Tarasenko, Time Characteristics of Spark Gaps Initiated by a Laser Beam. *Instrum. Exp. Tech.*, Vol. 16, p. 169, 1973.
129 E.A. Lenberg et al., Communication of Spark Gaps by Means of a Pulsed Gas Laser Operating in the Ultraviolet Range. *Instrum. Exp. Tech.*, Vol. 16, p. 165, 1973.
130 A.H. Guenther et al., Low Jitter, Low Inductance Solid Dielectric Switches. *Rev. Sci. Instrum.*, Vol. 50, p. 1487, 1979.
131 V.S. Bulygin et al., Laser Triggered Vacuum Switch. *Sov. Phys. Tech. Phys.*, Vol. 20, p. 561, 1976.
132 A.H. Guenther et al., Laser Triggered Switching of a Pulsed Charged Oil Filled Spark Gap. *Rev. Sci. Instrum.*, Vol. 46, No. 7, p. 914, 1975.
133 K. McDonald et al., An Electron Beam Triggered Spark Gap. *Proceedings of the 2nd International Pulsed Power Conference*, p. 437, 1979.
134 A.S. El'chaninov et al., Nanosecond Triggering of Mega-Volt Switches. *Sov. Phys. Tech. Phys.*, Vol. 20, p. 51, 1975.
135 K. McDonald et al., An Electron Beam Triggered Spark Gap. *IEEE Trans. Plasma Sci.*, Vol. 8, No. 3, p. 181, 1980.
136 L.A. Miles et al., Design of Large Area E-Beam Controlled Switch. *Conference Record of the Power Modulator Symposium*, 1980.
137 E.L. Neau, X-Ray Triggered Switching in SF_6 Insulated Spark Gaps. *Conference Record of the Power Modulator Symposium*, p. 42, 1980.
138 P. Dokopoulos, Fast Metal-to-Metal Switch with 0.1 Microsecond Jitter Time. *Rev. Sci. Instrum.*, Vol. 39, No. 5, p. 697, 1968.
139 D.E. Skelton et al., Development Aspects of Fast Metal Contact Solid Dielectric Switches. *Proceedings of the 6th Symposium on Fusion Technology*, Aachen, Germany, p. 365, 1970.
140 C. Boissady and F. Rioux-Damidav, Coaxial Fast Metal-to-Metal Switch for High Current. *Rev. Sci. Instrum.*, Vol. 49, No. 11, p. 1537, 1978.
141 C.A. Bleyset et al., A Simple Fast Closing Metal Contact Switch for High Voltage and Current. *Rev. Sci. Instrum.*, Vol. 46, p. 180, 1975.
142 R.J. Harvey, The Crossatron Switch: A Cold Cathode Discharge Device with Grid Control. *Conference Record of the Power Modulator Symposium*, p. 77, 1980.
143 M.A. Lutz et al., Feasibility of a High Average Power Crossed Field Closing Switch. *IEEE Trans. Plasma Sci.*, Vol. 4, No. 2, pp. 118–128, 1976.
144 M.A. Lutz, Gridded Crossed Field Tube. *IEEE Trans. Plasma Sci.*, Vol. 5, No. 4, p. 24, 1977.

145 R.J. Harvey, R.W. Holly, and J.E. Creedon, High Average Power Tests of a Crossed Field Closing Switch. *Proceedings of the 2nd International Pulsed Power Conference*, 1976.

146 R.J. Harvey et al., Operating Characteristics of the Crossed Field Closing Switch. *IEEE Trans. Electron Devices*, Vol. 26, No. 10, p. 1472, 1979.

147 J.M. Lafferty, Triggered Vacuum Gaps. *Proc. IEEE*, Vol. 54, p. 23, 1966.

148 A.S. Gilmour, Jr., and R.F. Hope, III, 10 kHz Vacuum Arc Switch Ignition. *Conference Record of the Power Modulator Symposium*, p. 80, 1980.

149 P.H. Ron, V.K. Rohatgi, and R.S.N. Rau, Rise Time of a Vacuum Gap Triggered by an Exploding Wire. *IEEE Trans. Plasma Sci.*, Vol. 11, No. 4, p. 274, 1983.

150 P.H. Ron, V.K. Rohatgi, and R.S.N. Rau, Delay Time of a Vacuum Gap Triggered by an Exploding Wire. *J. Phys. D Appl. Phys.*, Vol. 17, p. 1369, 1984.

151 Maxwell Engineering Bulletin Operating Instructions for Maxwell High Voltage SparkGap Switches, EB No. 1007, 1978.

152 Zeonics , Catalog on Rail Gaps, Zdd-Oct 1999, Zeonics, India.

153 R.S. Taylor et al., Laser Triggered Rail Gaps. *Conference Record of the Power Modulator Symposium*, p. 32, 1980.

154 G. Kovacs, S. Smatmari, and F.P. Schafer, Low Jitter Rail Gap Switch Triggered by Sub-Picosecond KrF Laser Pulses. *Meas. Sci. Technol.*, Vol. 3, pp. 112–119, 1992.

155 K.R. Prestwich, 2 MV, Multi-Channel Oil Dielectric Triggered Spark Gap, in W.H. Bostick, V. Nardi, and O.S.F. Zucker, eds., *Energy Storage, Compression, and Switching*, Plenum Press, New York, p. 451, 1976.

156 R.E. Reinovsky, W.L. Baker, Y.G. Chen, J. Holmes, and E.A. Lopez, Shiva Star Inductive Pulse Compression System. *Proceedings of the International Pulsed Power Conference*, pp. 196–201, 1983.

157 M.C. Scott, D.L. Ralph, J.D. Graham, Y.G. Chen, and R. Crumley, Low Inductance 180 kV Switch Development for Fast Cap Technology. *Proceedings of the IEEE International Pulsed Power Conference*, 1991.

158 J.M. Koutsoubis and S.J. MacGregor, Effect of Gas Type on High Repetition Rate Performance of a Triggered Corona Stabilised Switch. *IEEE Trans. Dielectr. Electr. Insul.*, Vol. 10, No. 2, pp. 245–255, 2003.

159 S.J. MacGregor, S.M. Turnball, F.A. Tuema, and O. Farish, The Application of Corona Stabilised Breakdown to Repetitive Switching. *Proceedings IEE Colloquium on Pulsed Power*, 1996.

160 J.A. Harrower, The Development and Characterization of Corona Stabilised Repetitive Closing Switches. Ph.D. thesis, University of Strathclyde, 2001.

161 S.J. MacGregor, Cascade High Voltage Switch. University of Strathclyde Internal Report, 1999.

162 J.R. Beveridge, S.J. MacGregor, M.J. Given, I.V. Timoshkin, and J. Lehr, A Corona Stabilised Plasma Closing Switch. *IEEE Trans. Dielectr. Electr. Insul.*, Vol. 16, pp. 948–955, 2009.

163 R.S. Sigmond, Simple Approximate Treatment of Unipolar Space Charge Dominated Coronas: The Warbur Law and the Saturation Current. *J. Appl. Phys.*, Vol. 53, No. 2, pp. 891–898, 1982.

164 B.M. Kovalchuk, A.A. Kim, E.V Kumpjak, N.V Zoi, J.P. Corley, K.W. Struve, and D.L. Johnson, Multi Gap Switch for Marx Generators. *Proceedings of the IEEE International Pulsed Power Conference*, pp. 1739–1742, 2001.

165 R.C. Fletcher, Impulse Breakdown in the 10-9 Second Range at Atmospheric Pressure. *Phys. Rev.*, Vol. 76, No. 10, pp. 1501–1511, 1949.

166 J.M. Lehr and C.E. Baum, Fundamental Physical Considerations for Ultrafast Spark Gap Switching, in E. Heyman, B. Mandelbaum, and J. Shiloh, eds., *Ultra-Wideband Short-Pulse Electromagnetics 4*, Kluwer Academic/Plenum Publishers, p. 11, 1999.

167 C.E. Baum, Radiation of Impulse-Like Transient Fields, Sensor and Simulation Note 321, November 1989.

168 D.V. Giri, *High Power Electromagnetic Radiators, Nonlethal Weapons and Other Applications*, Harvard University Press, Cambridge, MA, 2004.

169 D.V. Giri, J.M. Lehr, W.D. Prather, C.E. Baum, and R.J. Torres, Intermediate and Far Fields of a Reflector Antenna Energized by a Hydrogen Spark-Gap Switched Pulser. *IEEE Trans. Plasma Sci.*, Vol. 28, No. 5, pp. 1631–1636, 2000.

170 E.A. Avilov and N.V. Belkin, Electrical Strength of Nitrogen and Hydrogen at High Pressures. *Sov. Phys. Tech. Phys.*, Vol. 19, No. 12, 1975.

171 G.A. Mesyats, S.N. Rukin, V.G. Shpak, and M.I. Yalandin, Generation of High Power Subnanosecond Pulses, in E. Heyman, B. Mandelbaum, and J. Shiloh, eds., *Ultra-Wideband Short-Pulse Electromagnetics 4*, Kluwer Academic/Plenum Publishers, p. 1, pp. 1–10, 1999.

172 C.A Frost, T.H. Martin, P.E. Patterson, L.F. Rinehart, G.J. Rohwein, L.D. Roose, J.F. Aurand, and M.T. Buttram, Ultrafast Gas Switching Experiments. *Proceedings of the 9th IEEE International Pulsed Power Conference*, p. 491, 1993.

173 G.A. Mesyats, Pulsed High Current Electron Beam Technology. *Proceedings of the 2nd IEEE International Pulsed Power Conference*, p. 9, 1979.

174 R.B. McCann et al., Inductive Energy Storage Using High Voltage Vacuum Circuit Breakers, in W.H. Bostick, V. Nardi, and O.S.F. Zucker, eds., *Energy Storage, Compression, and Switching*, Plenum Press, New York, p. 491, 1976.

175 J.W. Mather and A.H. Williams, Some Properties of a Graded Vacuum Spark Gap. *Rev. Sci. Instrum.*, Vol. 31, p. 297, 1960.

176 W.C. Nunnally et al., Simple Dielectric Start Switch, in W.H. Bostick, V. Nardi, and O.S.F. Zucker, eds., *Energy Storage, Compression, and Switching*, Plenum Press, New York, p. 429, 1976.

177 P. Dokopoulos, A Fast Solid Dielectric Multi-Channel Crowabar Switch. *Proceedings of the 6th Symposium on Fusion Technology*, Aachen, Germany, p. 379, 1970.

178 A.P. Alkhimov et al., The Development of Electric Discharge in Water. *Sov. Phys. Dokl.*, Vol. 15, No. 10, p. 959, 1971.

179 I.D. Smith, Liquid Dielectric Pulse Line Technology, in W.H. Bostick, V. Nardi, and O.S.F. Zucker, eds., *Energy Storage, Compression, and Switching*, Plenum Press, New York, p. 15, 1976.

180 S. Kitagawa and K.-I. Hirano, Fast Air gap Crowbar Switch Decoupled by a Low Pressure Gap. *Rev. Sci. Instrum.*, Vol. 46, No. 6, p. 729, 1975.

181 J.G. Melton et al., Development of the Switching Components for ZT-40. *Proceedings of the 7th Symposium on Engineering Problems of Fusion Research*, Vol. II, p. 1076, 1977.

182 Faltens , High Rep Rate Burst Mode Spark Gap. *IEEE Trans. Electron Devices*, Vol. 26, No. 10, p. 1411, 1979.

183 D.E. Skelton, et al., Development Aspects of Fast Metal Contact Solid Dielectric Switches. *Proceedings of the 6th Symposium on Fusion Technology*, Aachen, Germany, p. 365, 1970.

184 S. Mercer and I. Smith, A Compact Multiple Channel 3 MV Gas Switch, in W.H. Bostick, V. Nardi, and O.S.F. Zucker, eds., *Energy Storage, Compression, and Switching*, Plenum Press, New York, p. 429, 1976.

185 M.E. Savage, Final Results from the High Current, High Action Closing Switch Test Program at Sandia National Laboratories. *IEEE Trans. Plasma Sci.*, Vol. 28, No. 5, pp. 1451–1455, 2000.

186 L. Li, C. Zhang, B. Yan, L. Zhang, and X. Li, Analysis on Useful Lifetime of High Power Closing Switch with Graphite Electrodes. *IEEE Trans. Plasma Sci.*, Vol. 39, No. 2, pp. 737–743, 2011.

187 American Society of Mechanical Engineers , Section VIII, Division I Type Vessels, 2013.

188 G.R. Neil and R.S. Post, Multi-Channel High Energy Rail Gap Switch. *Rev. Sci. Instrum.*, Vol. 49, No. 3, p. 401, 1978.

189 H.F.A. Verhaart and A.J.L Verhage, Insulator Flashover in SF_6 Gas. *Kema Sci. Tech. Rep.*, Vol. 6, No. 9, pp. 179–228, 1988.

190 J.M. Lehr, M.D. Abdalla, B. Cockreham, F. Gruner, M.C. Skipper, and W.D. Prather, Development of a Hermetically Sealed, High Energy Trigatron Switch for High Repetition Rate Operation. *Proceedings of the IEEE International Pulsed Power Conference*, p. 146, 1999.

191 J.M. Lehr, M.D. Abdalla, J.W. Burger, J.M. Elizondo, J. Fockler, F. Gruner, M.C. Skipper, I.D. Smith, and W.D. Prather, Design and Development of a 1MV, Compact Self Break Switch for High Repetition Rate Operation. *Proceedings of the IEEE International Pulsed Power Conference*, p. 1199, 1999.

192 W. Hartmann and M.A. Gundersen, Origin of Anomalous Emission in a Superdense Glow Discharge. *Phys. Rev. Lett.*, Vol. 60, No. 23, 1988.

193 A. Anders, S. Anders, and M.A. Gundersen, Model for Explosive Electron Emission in a Pseudospark Superdense Glow. *Phys. Rev. Lett.*, Vol. 71, No. 3, pp. 364–367, 1993.

194 E. Boggasch and H. Riege, The Triggering of High Current Pseudospark Switches. *XVII International Conference on Phenomena in Ionized Gases,*, 1985.

195 T. Mehr, J. Christiansen, K. Frank, A. Gortler, M. Stetter, and R. Tkotz, Investigations About Triggering of Coaxial Multichannel Pseudospark Switches. *IEEE. Trans. Plasma Sci.*, Vol. 22, No. 1, p. 78, 1994.

196 R. Tkotz et al., Pseudospark Switches: Technological Aspects and Applications. *IEEE Trans. Plasma Sci.*, Vol. 23, No. 3, p. 309, 1995.

197 A. Gortler, K. Frank, J. Insam, U. Prucker, A. Schwandner, R. Tkotz, J. Christiansen, and D.H.H. Hoffmann, The Plasma in High Current Pseudospark Switches. *IEEE Trans. Plasma Sci.*, Vol. 24, No. 1, p. 51, 1996.

198 R. Tkotz, M. Schlaug, J. Christiansen, K. Frank, A. Gortler, and A. Schwandner, Triggering of Radial Multi-Channel Pseudospark Switches by a Pulsed Hollow Cathode Discharge. *IEEE Trans. Plasma Sci.*, Vol. 24, No. 1, p. 53, 1996.

199 K. Frank et al., Scientific and Technological Progress of Pseudospark Devices. *IEEE Trans. Plasma Sci.*, Vol. 27, No. 4, p. 1008, 1999.

200 Y.D. Korolev and K. Frank, Discharge Formation Processes and Glow to Arc Transition in Pseudospark Switch. *IEEE Trans. Plasma Sci.*, Vol. 27, No. 5, p. 1525, 1999.

201 K. Frank et al., Pseudospark Switches for High Repetition Rates and High Current Applications. *Proc. IEEE*, Vol. 80, No. 6, p. 958, 1992.

202 V.F. Puchkarev and M.A. Gundersen, Spatial and Temporal Distribution of Potential in the Pseudospark Switch. *IEEE Trans. Plasma Sci.*, Vol. 23, No. 3, p. 318, 1995.

203 M. Legentil, C. Postel, J.C. Thomas, Jr., and V. Puech, Corona Plasma Triggered Pseudospark Discharges. *IEEE Trans. Plasma Sci.*, Vol. 23, No. 3, p. 330, 1995.

204 M. Stetter, P. Elsner, J. Christiansen, K. Frank, A. Gortler, G. Hintz, T. Mehr, R. Stark, and R. Tkotz, Investigation of Different Discharge Mechanisms in Pseudospark Discharges. *IEEE Trans. Plasma Sci.*, Vol. 23, No. 3, p. 283, 1995.

205 V.I. Bochkov, V.M. Djagilev, V.G. Ushich, O.B. Frants, Y.D. Korolev, I.A. Shemyakin, and K. Frank, Sealed Off Pseudospark Switches for Pulsed Power Applications. *IEEE Trans. Plasma Sci.*, Vol. 29, No. 5, pp. 802–808, 2001.

206 T. Mehr et al., Trigger Devices for Pseudospark Switches. *IEEE Trans. Plasma Sci.*, Vol. 23, No. 3, p. 324, 1995.

207 G. Lins, J. Stroh, and W. Hartmann, The Thermal Behaviour of Tantalum Carbide Cathodes in Pseudospark Switches. *IEEE Trans. Plasma Sci.*, Vol. 23, No. 3, p. 375, 1995.

208 G. Susskind, ed., *The Encyclopaedia of Electronics*, Reinhold Publ. Corp., 1962.

209 Hydrogen Thyratrons Preamble, E2V Technologies, 2002. Available at http://www.tayloredge.com/reference/Electronics/VacuumTube/thyratron_preamble.pdf

210 K.G. Cooks and G.G. Issacs, *Br. J. Appl. Phys.*, Vol. 9, p. 497, 1958.

211 English Electric Valve Company, *Product Data Manual on Ignitrons,*, 1980.

212 E.L. Kemp, *Principal Features in Large Capacitor Banks*, Pulsed Power Lecture Series, Department of Electrical Engineering, Texas Tech University, Lubbock, Texas.

213 Frank Frungel , *High Speed Pulse Technology*, Vol. III, Academic Press, New York, 1976.

214 A. Sharma, P.H. Ron, and M.S. Naidu, Radioisotope Aided Switching Speed Enhancement of Low Pressure Miniature Spark Gaps. *Proceedings of the International Conference on Applications of Radioisotopes and Radiation in Industrial Development*, p. 469, 1994.

215 W. S. Melville, The Use of Saturable Reactors as Discharge Devices for Pulse Generators. *J. Inst. Electr. Eng.*, Vol. 1951, No., 6, pp. 179–181, 1951.

216 D.L. Birx, E.J. Lauer, L.L. Reginato, J. Schmidt, M. Smith, Basic Principles Governing the Design of Magnetic Switches, LLNL Report, UCID 18831, November, 1980.

217 D.D. Kumar, S. Mitra, K. Senthil, D.K. Sharma, R.N. Rajan, A. Sharma, K.V. Nagesh, and D.P. Chakravarthy, A Design Approach for Systems Based on Magnetic Pulse Compression. *Rev. Sci. Instrum.*, Vol. 79, No. 4, p. 045104, 2008.

218 A. Sharma, S. Acharya, T. Vijayan, P. Roychoudhury, and P.H. Ron, High Frequency Characterization of an Amorphous Magnetic Material and Its Use in Induction Adder Configurations. *IEEE Trans. Magn.*, Vol. 39, No. 2, pp. 1040–1045, 2003.

219 H. Singh, J.L. Carter, and J. Creeden, Comparison of Switching Technologies for a Tactical EML Application. *IEEE Trans. Magn.*, Vol. 33, No. 1, p. 513, 1997.

220 I. Grekhov, Novel Semiconductor Devices and Pulsers for the Range of Power 10^4–10^{10} W and Pulse Durations 10^{-10}–10^{-4} Seconds. *Proceedings of the IEEE International Pulsed Power Conference*, 1997.

221 W. Jiang, K. Yatsui, K. Takayama, M. Akemoto, E. Nakamura, N. Shimizu, A. Tokuchi, S. Rukin, V. Tarasenko, and A. Panchenko, Compact Solid State Switched Pulsed Power and Its Applications. *Proc. IEEE*, Vol. 92, No. 7, p. 1180, 2004.

222 D.R. Graham and J.C. Hey, *SCR Manual*, 5th edition, General Electric, 1964.

223 J. Seymour, *Electronic Devices and Components*, Pitman Publishers, 1981.

224 W.M. Portnoy, *Thyristors as Switching Elements for Pulsed Power*, Pulsed Power Lecture Series, Department of Electrical Engineering, Texas Tech University, Lubbock, TX.

225 H.F. Strom and J.G. St. Clair, An Involute Gate Emitter Configuration for Thyristors. *IEEE Trans. Electron Devices*, Vol. 21, No. 8, p. 520, 1974.

226 R.M. Davis, *Power Diode and Thyristor Circuits*, Peter Peregrinus, 1976.

227 J.D. Campbell and J.V.V. Kasper, A Microsecond Pulse Generator Employing Series Connected SCRs. *Rev. Sci. Instrum.*, Vol. 43, No. 4, p. 619, 1972.

228 A.V. Gorbatyuk, I.V. Grekhov, and A.V. Nalivkin, Theory of Quasi-Diode Operation of Reversely Switching Dynistors. *Solid State Electron.*, Vol. 31, No. 10, p. 1483, 1988.
229 B.K. Bose, Recent Advances in Power Electronics. *IEEE Trans. Power Electron.*, Vol. 7, No. 1, p. 2, 1992.
230 E. Ohno, *Introduction to Power Electronics*, Oxford University Press, 1988.
231 H. Bluhm, *Pulsed Power Systems*, Springer, 2006.
232 B.J. Cardose and T.A. Lipo, Applications of MTO Thyristors in Current-Stiff Converters with Resonant Snubbers. *IEEE Trans. Ind. Appl.*, Vol. 37, No. 2, p. 566, 2001.
233 L.I. Yuxin, A. Huang, and K. Motto, Experimental and Numerical Study of the Emitter Turn-Off Thyristor (ETO). *IEEE Trans. Power Electron.*, Vol. 15, No. 3, p. 561, 2000.
234 V.A.K. Temple, MOS-Controlled Thyristors: A New Class of Power Devices. *IEEE Trans. Electron Devices*, Vol. 33, No. 10, 1986.
235 W. Wang, A.Q. Huang, and F. Wang, Development of Scalable Semiconductor Switches. *IEEE Trans. Power Electron.*, Vol. 22, No. 2, p. 364, 2007.
236 V.K. Khanna, *IGBT: Theory and Design*, Wiley-Interscience, 2003.
237 S. Bernet, Recent Developments of High Power Converters for Industry and Applications. *IEEE Trans. Power Electron.*, Vol. 15, No. 6, p. 1102, 2000.
238 F. Bauer, L. Meysene, and A. Piazzesi, Suitability and Optimization of High Voltage IGBTs for Series Connection with Active Voltage Clamping. *IEEE Trans. Power Electron.*, Vol. 20, No. 6, p. 1244, 2005.
239 N. McNeill, K. Sheng, B.W. Williams, and S.J. Finney, Assessment of Off-State Negative Gate Voltage Requirements for IGBTs. *IEEE Trans. Power Electron.*, Vol. 13, No. 3, pp. 436–440, 1988.

)pening Switches

n opening switch is a device that conducts current in a low-impedance state ntil a command trigger turns it to a high-impedance state with no current onduction. Opening switches are critical components for inductive storage stems and also find applications in pulse compression and power distribution stems. Inductive storage systems are very attractive because the stored energy ensity is orders of magnitude larger than can be stored in capacitors. These rcuits, however, inherently cause the opening switch to dissipate a larger action of the circuits' total energy in comparison to a closing switch [1]. ealizing the potential of inductive energy storage would enable smaller, less ostly pulsed power systems, especially for very high energy density applica- ons. Inductive energy systems are capable of producing machine parameters of illions of volts and millions of amperes in nanoseconds and potential plications include electromagnetic launch technology, intense radiation urces, dynamic materials effects, inertial confinement fusion research, and host of applications requiring compact pulsed power sources. A detailed eatise on this topic can be found in *Opening Switches* [2].

1 Typical Circuits

typical schematic of an inductive energy storage device is shown in Figure 5.1. itially, opening switch S_1 is closed and closing switches S_2 and S_3 are open, ich allows the power supply to charge the energy storage inductor L_s to the ximum current $i_{s,max}$. Just before discharge, S_3 is closed to protect the power pply from high-voltage surges. This is quickly followed by the simultaneous sing of S_2 and the opening of S_1. The inductor current is then commuted to load inductor L_ℓ. The waveforms of i_{S_1}, i_{L_ℓ}, and V_{L_ℓ} are shown in Figure 5.2. ulsed power system using capacitive storage and an opening switch for pulse npression is shown in Figure 5.3. The voltage waveforms of the pulsed power tem with and without an opening switch are qualitatively indicated in this ure. The waveform parameters of V_1, T_1 are transformed to V_2, T_2 by the

ndations of Pulsed Power Technology, First Edition. Jane Lehr and Pralhad Ron.

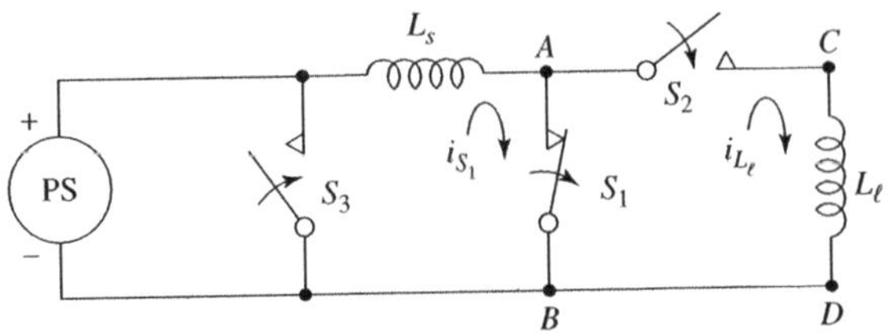

Figure 5.1 Typical circuit for inductive energy storage.

inclusion of an opening switch. The opening switch thus provides voltage amplification of V_2/V_1 and pulse compression of T_2/T_1.

A circuit using an opening switch to protect semiconductor components against overcurrents is shown in Figure 5.4. The power supply *PS* delivers power to the load through a semiconductor component *SC*. In the absence of an opening switch, if the load is short circuited, excessive current would flow through the semiconductor, damaging it. With the opening switch S_1 in place, once the short circuit is sensed, the current flow to S_1 is interrupted and the current from the power supply is commuted to a shunt circuit by the closing switch S_2.

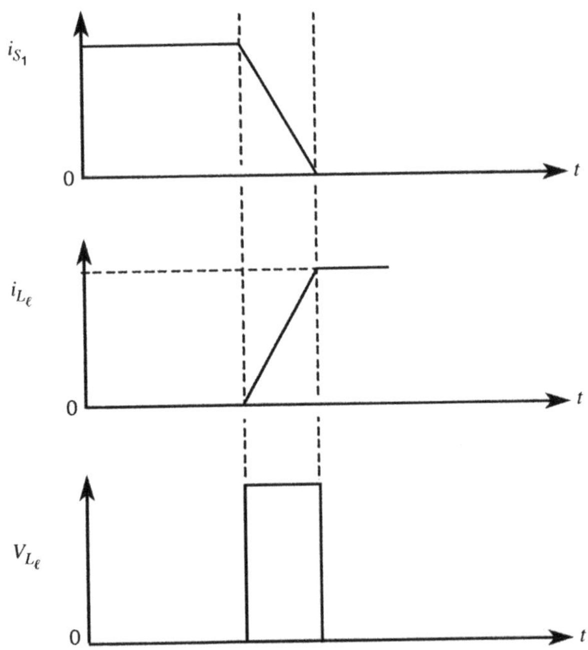

Figure 5.2 Typical current and voltage waveforms for the circuit shown in Figure 5.1.

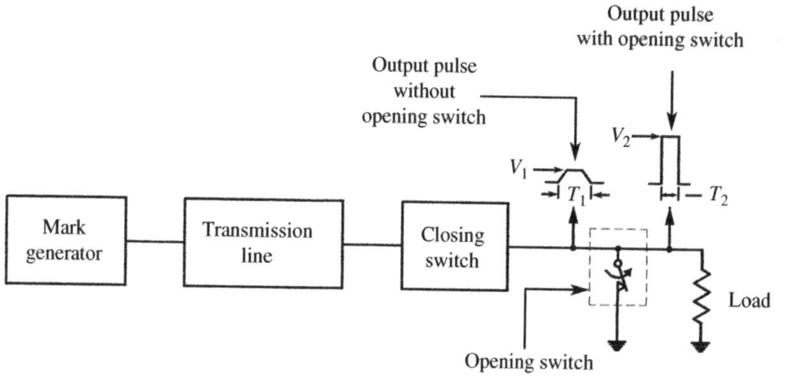

Figure 5.3 A pulsed power system based on capacitive energy storage with an opening switch as the final element.

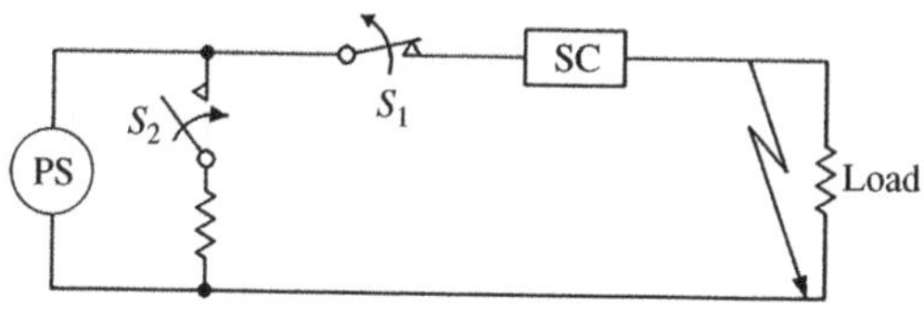

Figure 5.4 An opening switch used for overcurrent protection.

5.2 Equivalent Circuit

The equivalent circuit of an opening switch, shown in Figure 5.5, is a time-varying impedance, wherein the impedance increases with time. The ideal impedance profile is shown in Figure 5.6a, where, when closed, the switch impedance is zero but switches instantaneously at time t'. A practical opening switch impedance profile is shown in Figure 5.6b. When the switch is opened at

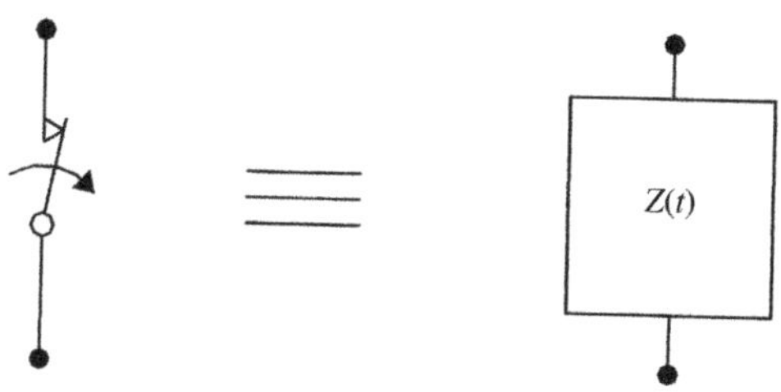

Figure 5.5 The equivalent circuit of an opening switch.

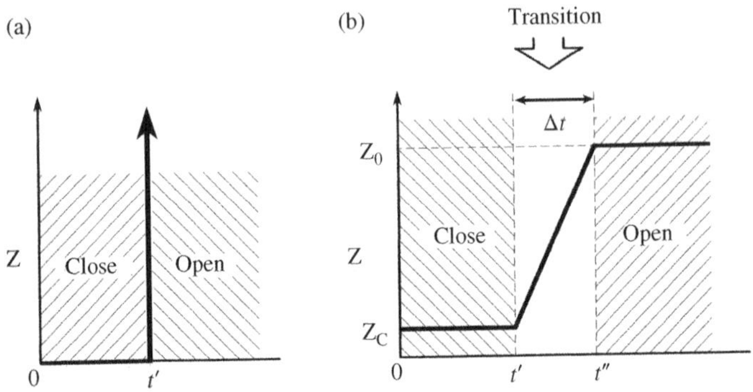

Figure 5.6 The opening switch impedance profile. (a) Idealized. (b) In practice.

time t', the impedance rises to Z_0 in a time $\Delta t = t'' - t'$. The time Δt is called the "opening time" of the switch. The impedance *Z(t)*, shown in Figure 5.5, could be resistance, inductance, or capacitance. Letting I_0 denote the current through the switch before it opens, the voltage that appears across the opening switch is given in Table 5.1.

5.3 Opening Switch Parameters

The overall efficiency of an opening switch in an inductive energy storage system is determined by conduction time and opening time of the switch, the trigger sources for opening and closing the switch, and the rate at which the dielectric recovers its electrical strength.

Table 5.1 Voltage across opening switch.

Switch impedance	Voltage	Assumption
Resistive:		
$R_o \to R_f$	$I_o R_f$	$R_f \gg R_o;\ t \ll L_s/R$
Inductive:		
$L_o \to L_f$	$\frac{d}{dt}(Li) = I_o \frac{dL_f}{dt}$	$L_f \gg L_o;\quad t \ll \frac{L_f + L_s}{R}$
Capacitive:		
$C_o \to C_f$	$\int \frac{idt}{C} = I_o \int \frac{dt}{C_f}$	$C_f \ll C_o;\quad t \ll \frac{\pi\sqrt{L_s C_f}}{2}$

5.3.1 Conduction Time

In some sense, the name "opening switch" is a misnomer because, before "opening," the switch must first "close" efficiently to carry the charging current to the storage inductor. The switch must carry the current for a sufficiently long time so that the current reaches its maximum value in the storage inductor. In this initial closed state, the switch is resistive with the energy dissipated during conduction, W_{sc}, given by

$$W_{SC} = V_{S_1} i_{S_1} t = i_{S_1}^2 R_{S_1} t \tag{5.1}$$

where

V_{S_1}	=	voltage across switch during conduction,
i_{S_1}	=	average conduction current,

and

R_{S_1}	=	resistance of the switch during conduction,
t	=	conduction time.

For maximum energy storage in the inductor, i_{S_1} and t should be as large as possible. To minimize the energy loss, the voltage drop across the switch and the switch resistance should be as small as possible.

5.3.2 Trigger Source for Closure

In some switches, external trigger sources such as electron beams, plasmas, or magnetic fields are used for initiating the closure of the switch. In these switches, the gap spacing of the electrodes is such that no breakdown takes place at the voltage level of the charging supply in the absence of external trigger. The conduction mechanism is by multiplication of charge carriers and electrical breakdown in the switch dielectric. Some switch configurations, such as exploding fuses, which rest permanently in a closed position, do not need external triggering but the initiation of inductor charging is executed by an external closing switch. Opening switches, such as conventional circuit breakers, employ the brute-force method of physically moving one electrode against the other, for the conduction process is initiated by the physical contact between the two electrodes. For efficient energy transfer, the trigger source energy should be minimized.

5.3.3 Trigger Source for Opening

For devices operating in a diffuse discharge mode, the switch can be made to open by removing the trigger source. Removal of the trigger source eliminates the source of charge carriers. These charge carriers, created when the switch was closed, are removed by the atomic processes of electron-attachment and electron–ion recombination. For spark gaps, the opening process of achieving an effective net zero current is initiated by passing a reverse current from an external capacitor bank. For exploding fuse switches, a high-impedance state is initiated by a sudden deposition of energy sufficient to vaporize the foil or wire.

5.3.4 Opening Time

The opening time is the interval during which the impedance increases from its initial low value to a final high value R_f. This important parameter determines the rate of voltage rise of the switch and is intrinsically related to the rate of dielectric strength recovery. The opening time of a switch depends on its impedance during closure. If the closure impedance is low, the high charge carrier density requires a larger opening time to neutralize the current. In the case of inductive switches, the opening time determines the magnitude of voltage ($I_0\, dL/dt$) generated across the switch, and hence decides the power amplification factor. The opening time of a switch can be reduced by a properly designed commutation circuit that rapidly transfers the current from the switch to the load.

5.3.5 Dielectric Strength Recovery Rate

As a switch begins to open, its current decreases, but the voltage across the switch is increasing. Voltages across opening switches typically become very high, ranging from tens of kilovolts to millions of volts. Therefore, the danger of electrical breakdown in the opening switch must be carefully avoided. As the switch opens, the dielectric strength of the switch medium must recover at a faster rate than the voltage across the switch rises, as shown in Figure 5.7. If the dielectric recovery rate falls in the hatched portion of Figure 5.7, a restrike will take place and the switch will close again. The dielectric recovery rate depends upon the characteristics of dielectric medium, the current and conduction time before opening, and the external means of removing heat from the medium.

5.4 Opening Switch Configurations

The key to unlocking the potential of inductive energy storage is the opening switch, and numerous opening switch concepts have been proposed for single-shot operation. Repetitive operation of opening switches would impose additional

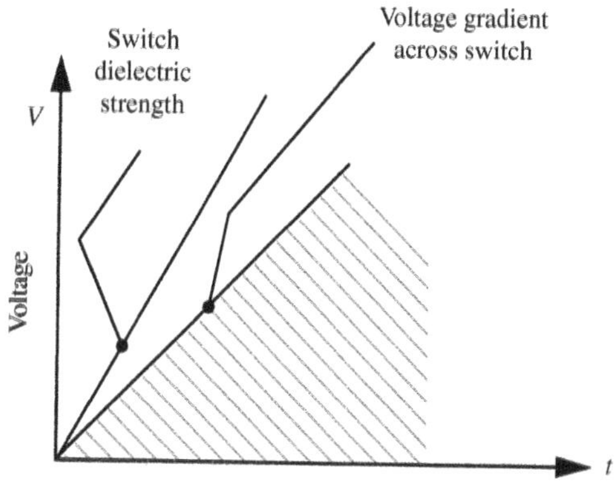

Figure 5.7 The rates of rise of dielectric strength and voltage gradient for successful current interruption.

design requirements for efficient heat removal. The obstacles to opening switch technology, especially for interrupting very high currents, is formidable. Not only does nature fight the disruption of current, but also the associated magnetic pressures mandate excellent mechanical integrity of the components.

The basic concepts and principal features of various opening switch configurations are discussed here. To date, the most successful opening switch has been the exploding fuse, even with the major drawback that the fuse assembly is destroyed after each shot. Fuse switches and electron beam-controlled switches have been analyzed in detail.

5.4.1 Exploding Fuse

An exploding fuse switch interrupts the circuit current from the rapid joule heating of the conducting wire or foil, resulting in the vaporization of the foil and sudden jump in resistance. Typical waveforms of current i_f, resistance R_f, and voltage V_f, for an exploding fuse in air are shown in Figure 5.8 [3].

Heating of the solid wire manifests in an increased resistance in the interval t_0–t_s. The large rate of resistance rise during the interval t_s to t_ℓ is due to the high resistivity of the liquid column. From t_ℓ up to t_v, superheating of the liquid column and reduction in cross section causes the resistance to increase. Note the maximum rate of resistance rise occurs during the vaporization stage. Beyond t_v, the resistance decreases due to thermal ionization and expansion of the vapor and the formation of a conducting plasma. However, if ionization of the vapor is prevented, the resistance will continue to rise, but at a slower rate. If too much energy is deposited, the temperature of the vapor column could reach 10,000 K, at which temperature the metal vapor is fully ionized and highly conductive. For an efficient exploding fuse switch, the plasma state must be

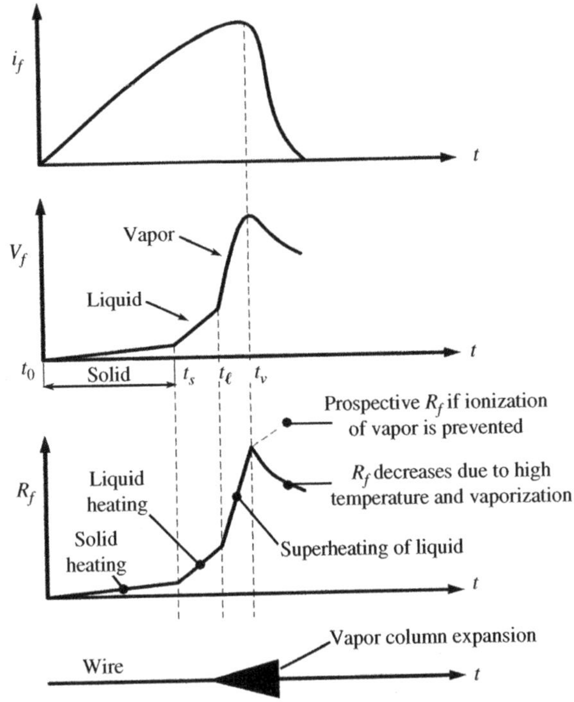

Figure 5.8 The current, voltage, resistance, and streak record for an exploding wire.

avoided by restricting the arc to small dimensions. Excellent references are available for further details [4–10].

5.4.1.1 Exploding Conductor Phenomenon

When sufficient energy is supplied to a wire on a timescale of a microsecond or less, the capillary forces dominate the gravitational forces, and the wire remains physically intact but goes through the four states of matter: solid to liquid to vapor to plasma. This process is known as the exploding conductor phenomenon (ECP), and is usually accompanied by brilliant flash of light and loud noise. The history of exploding conductors (ECs) started with Nairne [11] in 1774, but little was understood about the phenomena until 1950 with the advent of microsecond pulsed power sources and sophisticated diagnostic equipment such as fast oscilloscopes and high frame rate photography.

5.4.1.1.1 Exploding Conductor Waveforms

A typical circuit configuration for dumping energy into an EC is shown in Figure 5.9, where the basic element in the opening switch S_1 is an EC. If the

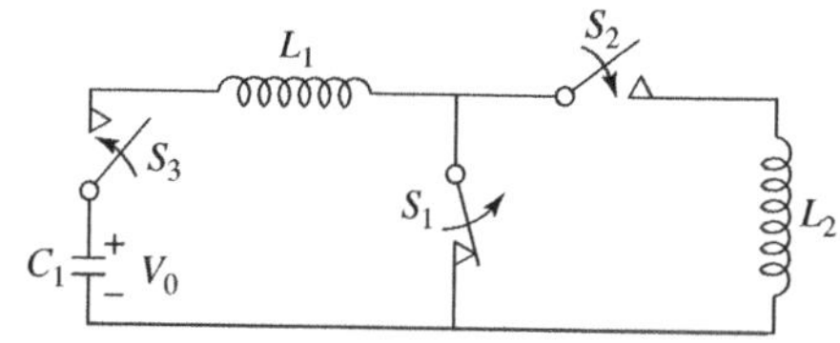

Figure 5.9 An inductive energy storage circuit with an inductive load.

energy invested in the EC is insufficient, the conductor does not explode and the current waveform will correspond to damped oscillations of conventional L–C–R circuit, with the degree of damping depending upon the value of R, the resistance of exploding conductor. If the energy invested in the conductor is sufficient, the conductor will expand, dropping in resistance. When the diameter is three or four times its original size, it can no longer conduct, and the current drops to zero [12]. The inductance of the circuit then gives rise to a voltage spike, known as the "time of burst."

The voltage appearing across the EC at the time of burst along with the EC environment decides the further course of the waveform. If the conditions are such that no peripheral arc takes place, the metal vapor starts expanding and at some time when the mean free path in the metal vapor is favorable, the breakdown [13] occurs in the metal vapor, resulting in a restrike, when the current increases again. The time interval between the current dip and the current spike is the "dwell phase"—where negligible current flows through the EC. Generally, the dwell phase is observed for the EC in vacuum. If the voltage across the EC at the time of burst is sufficient to cause breakdown in the gas around the EC or in its own vapor [14], a peripheral arc takes place and the restrike occurs immediately, almost coinciding with the time of burst. In such a case, the current dip is difficult to observe on the waveform. For many applications of the EC, the presence of the dwell phase may not matter. However, for opening switches the dwell phase should be strictly eliminated, since it results in the reclosing of an opened switch. This is generally achieved by suitable design of the EC surrounding, which should enable rapid and intense heat conduction, so as to quench the plasma.

5.4.1.1.2 Exploding Conductor Circuits

For efficiently exploding the conductor, the energy deposited into the conductor must be injected in the shortest possible time. This is typically achieved by a capacitor bank discharge. To satisfy this condition, the favorable factors are small circuit inductance and a fast discharge capacitor with a small ESR. A simple capacitor bank, because of limited output voltage, gives rate of energy deposition resulting in poor explosion. A Marx bank, used by Trolan, is an improvement [15]. A still better scheme is a coaxial cable energy storage used by

Tucker and Neilson [16]. An advanced scheme for achieving highly efficient EC performance is energy stored in a water capacitor, used by Scherrer [17].

5.4.1.1.3 Exploding Conductor Matching Criterion

For maximum energy deposition in the EC, Bennett [18] gives the criterion as follows:

$$1.1\sqrt{\frac{L}{C}} \leq R_{opt} \leq 1.3\sqrt{\frac{L}{C}} \tag{5.2}$$

where

R_{opt} is the resistance of wire at the time of burst.

A wire resistance smaller than the minimum for R_{opt} results in explosion when the current is rising, giving poor energy transfer. A wire resistance larger than R_{opt} leads to an explosion very late in time resulting in poor efficiency.

5.4.1.1.4 Exploding Conductor Classification

Conductor explosions can be classified into slow, fast, or superfast explosions, depending on the energy available in the first pulse and the corresponding *dE/dt* value. Slow explosions take place when the energy input in the first pulse is less than the value required for complete vaporization and when *di/dt* is small. Macroscopic instabilities take place, resulting in the breakup of the conductor and causing striations. Fast explosions are characterized by a large *di/dt* and a large, pulse energy. The inertial confinement force and kinetic pressure developed within the conductor are such that the melting starts at the surface and considerable superheating beyond the boiling point of the metal takes place at the axis of the metal. In superfast explosions, the duration of the applied pulse is on the order of current diffusion time and the skin effect predominates. During the fall time of the current, the inverse skin effect, as predicted by Haines [19], comes into play and an explosion takes place because of the tremendous radial outward directed force.

5.4.1.2 Switch Energy Dissipation in the Switch

Conditions for efficient interruption are derived for the circuit shown in Figure 5.9, where S_1 represents the exploding fuse switch, and the following relations for the circuit are valid.

$$I_1L_1 = I_2(L_1 + L_2) \tag{5.3}$$

$$W_0 = \frac{1}{2}C_1V_0^2 = \frac{1}{2}L_1I_1^2 \tag{5.4}$$

$$f = \frac{1}{T} = \frac{1}{2\pi\sqrt{L_1 C_1}} = \frac{\omega}{2\pi} \tag{5.5}$$

where

I_1 is the maximum current in L_1 before switch S_1 is opened;

and

I_2 is the maximum current in L_1 and L_2 after switch S_1 is opened.

The energy dissipated in the switch is given by

$$W_{S_1} = W_0 - W_1 - W_2 = W_0 \frac{L_2}{L_1 + L_2} \tag{5.6}$$

where

W_1	=	final energy stored in L_1

and

W_2	=	final energy stored in L_2

From Equation 5.6 it can be seen that when $L_1 = L_2$, the energy dissipated in the fuse is 50% of the initial energy stored in L_1. High energy efficiency with minimum switch loss is attained when $L_1 \ll L_2$.

5.4.1.3 Time for Vaporization

During the conduction in S_1, the maximum current is reached in the storage inductor L_1 at $t = T/4$. If the fuse vaporization takes place at this moment, it ensures maximum energy storage in L_1 and also maximum power amplification when the switch opens. For full vaporization to take place at $t = T/4$, the following relation is be satisfied:

$$\int_0^{T/4} i^2 R_f(t)\, dt = \int_{e_0}^{e_{max}} m \cdot de \tag{5.7}$$

where

m is the mass of the fuse;

and

e is the internal energy per unit mass.

Equation 5.7 can be rewritten as

$$\int_0^{T/4} [I_1 \sin(\omega t)]^2 \cdot \frac{\rho \ell}{A} \cdot dt = \int_{e_0}^{e_{T/4}} \gamma \cdot A\ell \cdot de$$

where

ρ is the resistivity,
ℓ is the length of fuse,
A is the fuse cross section,

and

γ is the mass density of fuse.

The above equation can be simplified [20] to

$$\frac{W_0^{3/2}}{V_0 L_1^{1/2} A^2} = \frac{\sqrt{2} \cdot \gamma}{\pi} \int_{e_0}^{e_{max}} \rho^{-1} de \tag{5.8}$$

For fast heating, rather than the slow adiabatic heating considered above, Equation 5.8 is modified [21] to the following,

$$\frac{W_0^{3/2}}{V_0 \sqrt{L_1} A^2} = \frac{K_1 \sqrt{2} \cdot \gamma}{\pi} \int_{e_0}^{e_{T/4}} \rho^{-1} de = K_1 a \tag{5.9}$$

where $1 \leq K_1 \leq 3$ accounts for pulse heating.

Equation 5.9 shows that once the circuit values and fuse material are finalized, the cross section A of the fuse can be calculated.

5.4.1.4 Energy for Vaporization

The energy dissipated in the switch (from the circuit point of view) must equal the energy required for vaporization (from the fuse material point of view). This can be expressed as

$$W_{S_1} = W_0 \cdot \frac{L_2}{L_1 + L_2} = \gamma \cdot (A\ell) e_v \cdot K_2$$

or

$$(A\ell / W_0)\left(\frac{L_1 + L_2}{L_2}\right) = \frac{1}{\gamma e_v K_2} = \frac{b}{K_2} \tag{5.10}$$

where

e_v is the latent heat of vaporization under normal conditions;
$1 \leq K_2 \leq 3$ accounts for pulse heating [21];

and

$$b = (\gamma\ e_v)^{-1}$$

Equation 5.10 enables the evaluation of ℓ, once the circuit parameters and fuse material are fixed, since A is already known from Equation 5.9.

5.4.1.5 Optimum Fuse Length

The effects of fuse dimensions on the opening time and fuse voltage have been discussed in the literature [22]. For a given cross section and energy, there is an optimum length that produces a maximum voltage. As the length is increased, the voltage starts increasing up to a certain value and then it starts decreasing. The initial increase is due to increase in the fuse resistance (since $V_f = I_0 R_f$). However, beyond the optimum length, even though the resistance increases, the energy of the storage bank is not sufficient to cause vaporization of the foil, whereas the possibility of restrike along the length of the fuse is greater at shorter lengths.

5.4.1.6 Fuse Assembly Construction

For handling large amounts of energy, the foils are generally preferred to wires because of their low inductance and other advantages such as compactness, large thermal capacity, and ease of clamping. A typical construction of an exploding foil fuse switch is shown in Figure 5.10 and consists of a folded copper foil (6.5 cm wide × 7 cm long) insulated by a combination of Mylar and fiberglass [23]. The self-inductance is less than 1 nH. Driven by a 10 kJ, 20 kV capacitor bank, this opening switch delivered a peak current of 400 kA to a 0.004 Ω, 1 nH dummy load with a di/dt value of 10^{13} A/s. In order to contain the pressure when the foil explodes, the terminal plates are clamped to a pressure of 350 kg/cm^2. It has been found that if fiberglass is not used, the pressure generated by the explosion is excessive and the fuse voltage is also substantially reduced [22].

An exploding fuse switch with folded aluminum foil of thickness 0.001 in., employing a fuse medium combination of Mylar, fine-grained quartz sand, and

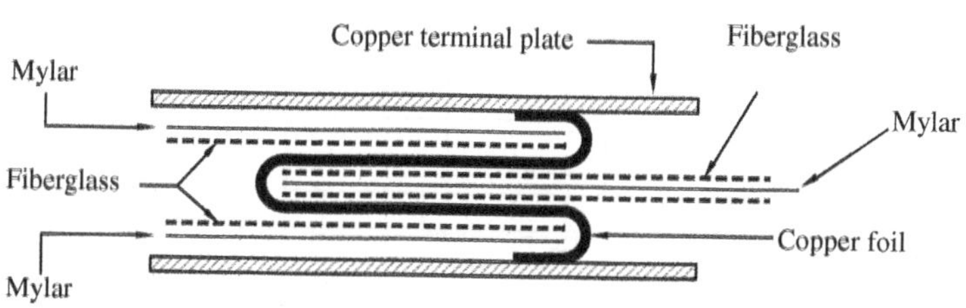

Figure 5.10 A typical low-inductance exploding fuse assembly. (Reproduced with permission from Ref. [23]. Copyright, 1965, AIP Publishing LLC.)

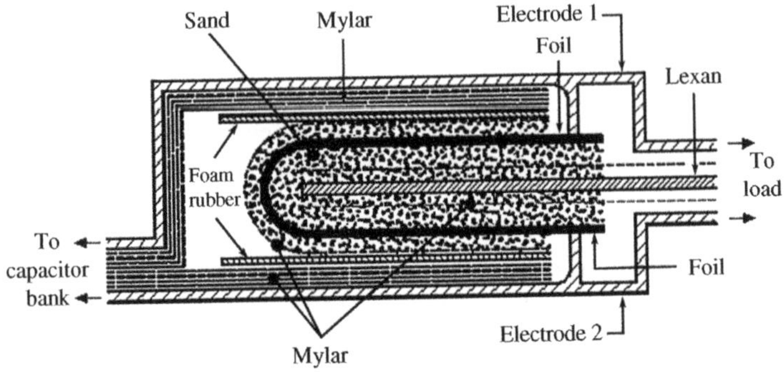

Figure 5.11 An exploding fuse assembly with Mylar, quartz sand, and foam rubber [24].

foam rubber, used by Reinovsky [24], is shown in Figure 5.11. The quartz sand serves two purposes: It helps rapid heat transfer from the exploded fuse column to the surrounding material, and it does not allow free expansion of the vapor column. Both factors help to maintain high resistivity of the column and inhibit the transformation of metal vapor to plasma. Driven by a capacitor bank with 300 kJ when charged to 90 kV, this switch has interrupted 3.9 MA with an interrupt time of 320 ns.

Water, oil, air, and vacuum have also been used as environments for fuse switches [25–28]. High fuse resistance and hence high fuse voltage, both characteristics of a good opening switch, have been obtained without any filler by simply enclosing the fuse in a slit of small dimensions, as shown in Figure 5.12 [3]. The enclosed narrow slit helps to restrict the cross section

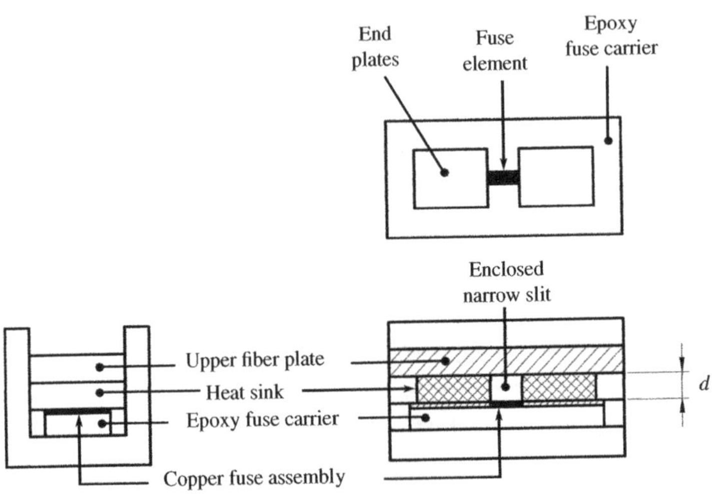

Figure 5.12 Exploding fuse assembly with an enclosed narrow slit [3].

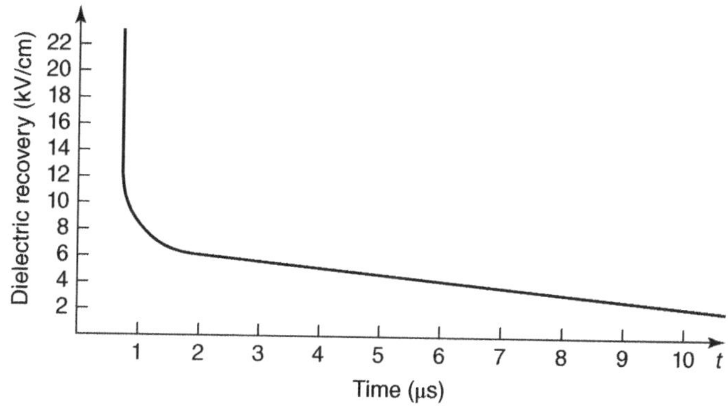

Figure 5.13 Recovery of the dielectric strength of a fuse exploded in air.

of the fuse arc, and provides very effective heat contact between the arc and the surrounding solid matter.

The various insulating mediums described above show different characteristics of recovery strength (kV/cm) versus recovery time. The characteristic recovery for a fuse exploded in air is shown in Figure 5.13 [26]. Initially, the recovery strength is high because the vapor is nonconducting. It then starts decreasing partly due to thermal ionization of vapor and partly due to drop in the (*pd*) value resulting from expansion of the vapor column cross section. It may be observed [29] from Figure 5.14 that after substantial time from the time of burst, the recovery strength increases again due to cooling, recombination, and diffusion of the fuse debris. The recovery in air is faster than in the sand, probably due to free expansion and faster disappearance of the vapor debris.

5.4.1.7 Multistage Switching

In a single-stage configuration, a fuse has to carry out the dual functions of efficiently conducting the charging current of the storage inductor before

Figure 5.14 The recovery of the dielectric strength of a fuse in (1) air and (2) sand.

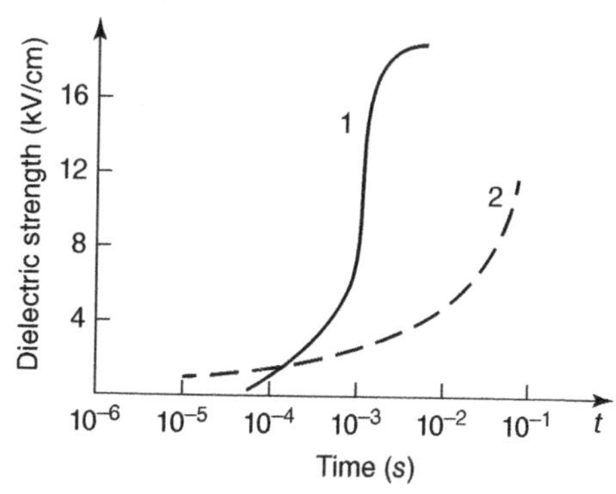

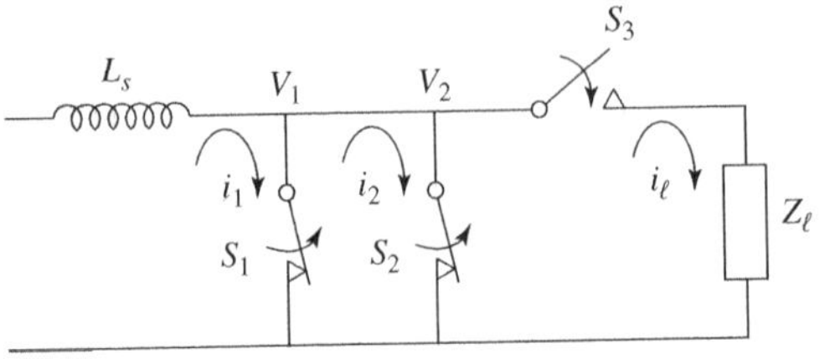

Figure 5.15 Inductive storage with multistage switching.

opening and interrupting the peak current in the shortest possible time. When the conduction period of the charging inductor is small (<100 μs), as in the case of a medium energy capacitor bank, a single-switch configuration is simple, reliable, and efficient and thus widely used. However, when high-impedance sources such as a DC supply, homopolar generator, or a motor-alternator are used for charging a storage inductor to high energy, the conduction periods involved are quite large (>100 ms), and multistage switching, shown in Figure 5.15, is usually employed.

The multistage switching is based on the observation that a switch that can carry out the function of conduction efficiently is typically inefficient in interrupting the current. The first-stage switch S_1 is designed for high conduction current, long conduction time, and low conduction loss. It need not have a small opening time. The second-stage switch S_2 must have a high conduction current and small opening time, but need not have a long conduction time. As a result of the optimum properties for switches S_1 and S_2, the overall system acquires the capabilities of high voltage generation, high power amplification, and large pulse compression. Typical temporal waveforms are shown in Figure 5.16.

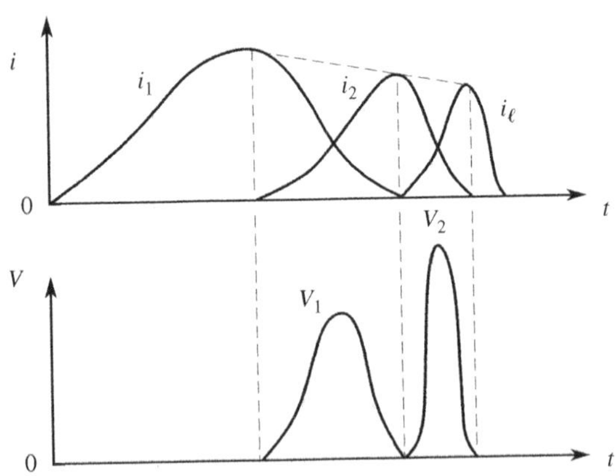

Figure 5.16 Current and voltage waveforms for multistage switching.

In multistage switch configuration, the moment of opening of a switch depends upon the recovery time of the preceding switch. In a typical configuration, the second-stage switch could be a fuse and the first stage may employ a moving contact circuit breaker [25], an exploding foil [30], or an explosive switch [30]. If a switch opens before the previous stage has recovered to its full dielectric strength, the restrike will take place, thereby inhibiting the energy transfer to the load. Multiple stages up to a maximum of three have been used [31].

5.4.1.8 Performances of Fuse Switches

Table 5.2 summarizes the performance parameters achieved using fuse switches.

5.4.2 Electron Beam-Controlled Switch

A typical inductive storage circuit employing an electron beam-controlled opening switch (EBCS) is shown in Figure 5.17. The EBCS consists of a high-vacuum chamber C_1 and a high-pressure chamber C_2, the two being separated by a titanium or Mylar foil window F_1, and mounted on the anode A. The high-pressure chamber C_2 contains the switch electrodes: a meshed electrode E_1 and a solid electrode E_2. The voltage applied between the electrodes is below the self-breakdown voltage and therefore the incident electron-beam (*e*-beam) does not lead to a self-sustained avalanche ionization. The foil F_1 is generally sandwiched between two metal meshes to give it mechanical strength to withstand the pressure differential. The *e*-beam is generated in the vacuum chamber C_1 by the application of an impulse voltage from the negative output Marx generator PS_2 to the diode terminals K (cathode) and A (anode), as shown in Figure 5.17. The cathode in the *e*-beam chamber could be a cold cathode [36,37] or one that operates by thermionic emission [38]. The *e*-beam enters the high-pressure chamber C_2, through F_1, only slightly attenuated. The *e*-beam collides with neutral molecules in the high-pressure chamber and causes ionization in the switch volume. The electrons liberated in the ionization process contribute to the conduction current between the electrodes in the gaseous dielectric and the switch S_1 turns on. Opening of the switch occurs when the *e*-beam is turned off, ionization ceases, and the residual charged particles in the switch volume are removed by recombination and attachment processes. The *e*-beam can be turned off by turning off the Marx voltage or by applying a negative voltage to a grid electrode placed next to the cathode.

Synchronization of the elements in this inductive storage circuit is important. The Marx generator PS_1 starts charging the storage inductor L_s as soon as the opening switch S_1 is turned on by the *e*-beam. When the inductor current reaches its peak, the *e*-beam is turned off, opening S_1, and the current is commuted to the load through the closing switch S_2.

Table 5.2 Performance parameters of fuse switches.

Inductor charging source	Switch configuration	Output parameters
Capacitor bank [20] (50 μF, 33 kV, 25 kJ) $L_s = 36$ nH, $L_\ell = 27$ nH	Single-stage exploding foil	$\Delta t = 0.4$ μs $V_{max} = 38$ kV
Capacitor bank [25] (20 nF, 720 kV) $L_s = 170$ μH	Single-stage exploding wire	$V_{max} = 1.7$ MV $I_\ell = 5$ kA
DC power supply [25] (2 kV, 500 A), $L_s = 1$ H	Two-stage switching Stage 1: AC circuit breaker Stage 2: Exploding wire	$V_{max} = 140$ kV
Capacitor bank [31] (266 μF, 50 kV, 388 kJ) $L_s = 2.5$ μH, $I_o = 500$ kA	Three-stage switching Stage 1: Explosive switch Stage 2: Exploding foil Stage 3: Exploding wire	$V_{max} = 700$ kV $I_{max} = 410$ kA $t_r = 150$ ns
Capacitor bank [30] $L_s = 80$ μH, $I_s = 35$ kA	Two-stage switching Stage 1: Explosive switch Stage 2: Exploding foil in water	$V_{max} = 300$ kV $T = 8$ μs
Capacitor bank [30] $L_s = 225$ μH, $I_s = 8$ kA	Two-stage switching Stage 1: Exploding foil in water Stage 2: Exploding wire in water	$V_{max} = 1$ MV $T = 0.25–2$ μs
B field compression [32] (energized by a 400 kJ capacitor bank/ detonated by chemical explosives)	Single-stage switching with an exploding foil	$L_\ell = 60$ nH $I_\ell = 2$ MA $t_r = 6.5$ μs
Homopolar generator [33] 200 kA, 3.8 MJ $L_s = 185$ μH 20 turns of Cu coil from 2 m (L) × 7.6 cm (W) 1 cm (t)	Two-stage switching Stage 1: Explosive switch [Al tube with 20 gaps; ℓ (total) = 1.25 m] Stage 2: Exploding foil in water	$V_{max} = 1$ MV t_c = (explosive 500) = 190 ms t_o = (explosive 500) = 70 μs
Capacitor bank [34] 240 μF, 9 kV, $L_s = 7$ μH	Two-stage switching Stage 1: Exploding wire Stage 2: Exploding wire	180 kV, 200 ns
Capacitor bank [35] 3.5 μF, 25 kV, $L_s = 0.5$ μH	Two-stage switching Stage 1: Exploding wire in water [Copper, 0.1 mm × 10 cm, NOS] Stage 2: Plasma erosion switch	230 kV, 35 kA Opening times: Fuse = 250 ns PES = 10 ns

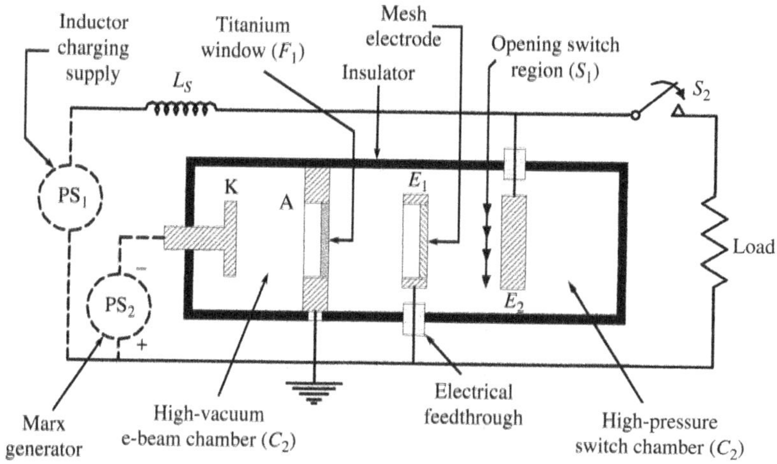

Figure 5.17 An inductive energy storage circuit using an *e*-beam-controlled opening switch.

5.4.2.1 Electron Number Density (n_e)

The rate of change of the electron number density n_e is governed by the continuity equation

$$\frac{dn_e}{dt} = \Psi_1 - \Psi_2 \tag{5.11}$$

where Ψ_1 is the source that increases n_e and Ψ_2 is the sink that decreases n_e. Note that the rate of change of electron number density can be either positive, denoting growth, or negative. The contributions to the source term Ψ_1 are the secondary electrons created by the ionization of gas molecules by the primary *e*-beam on its passage through the switch dielectric and further ionization from secondary electrons accelerated by the applied electric field *E*. Generally, the latter is smaller compared to the former and hence usually neglected [39,40]. Ψ_1 can be written in terms of ionization cross section [41] or mass stopping power [42] for electrons, as

$$\Psi_1 = \frac{j_b\tau}{e}\sigma_i N = \frac{j_b\tau \cdot \rho'}{e\epsilon_i} \cdot \left(\frac{\partial E}{\partial m}\right) \tag{5.12}$$

where

j_b is the *e*-beam current density,
σ_i is the ionization cross section,
N is the neutral gas density,
τ is the duration of pulsed *e*-beam,
$\delta E/\delta m$ is the mass electron stopping power of the gas,
ρ' is the mass density of gas,

and

ε_i is the effective ionization potential.

The e-beam density averaged over 1 s (j_b) is represented by $j_b\tau$. If the e-beam is pulsed at a repetition rate of γ per sec and the duration of each pulse is τ, then $J_b = j_b\gamma\tau$.

The rate of loss of electrons per cm^3, Ψ_2, due to the removal of electrons by the processes of attachment and recombination can be expressed as

$$\Psi_2 = \eta n_e N_a + \beta n_e n_+ \tag{5.13}$$

where

η is the attachment coefficient (cm^{-3} s^{-1}),
β is the two-body recombination coefficient (cm^{-3} s^{-1}),
N_a is the number density of attaching gas,

and

n_+ is the number density of positive ions.

For weakly ionized gases, that is, $N_a \gg n_e$, and for $n_e = n_+$, the above equation can be written as

$$\Psi_2 = \eta \cdot n_e + \beta \cdot n_e^2 \tag{5.14}$$

Examples of typical attachment processes are given as follows:

$$e + O_2 \rightarrow O + O^- \rightarrow \text{two-body attachments}$$
$$e + O_2 + N_2 \rightarrow O_2^- + N_2 \rightarrow \text{three-body attachments}$$

Examples of a typical two-body recombination is given as

$$e + A_2^+ \rightarrow A + A \rightarrow \text{two-body recombination}$$

Detailed mechanisms of attachment and recombination are described in the literature [42,43].

From Equations 5.11–5.13, the electron continuity equation can now be rewritten as

$$\frac{dn_e}{dt} = \frac{j_b \sigma_i N}{e} - \eta \cdot n_e - \beta \cdot n_e^2 \tag{5.15}$$

$$\frac{dn_e}{dt} = \frac{j_b(\partial E/\partial m)\rho'}{eE_i} - \eta n_e - \beta n_e^2 \tag{5.16}$$

During the closing phase of EBCS, ηn_e can be neglected in comparison to βn_e^2 because $\eta < \beta$, since E is small during closure, and attachment becomes effective at large E, and $n_e^2 \gg n_e$, since n_e is quite large during closure. Thus, during the closing phase of EBCS, the electron-loss mechanism is mainly governed by the recombination process.

For a steady state during the closing phase, that is, $dn_e/dt = 0$, Equation 5.16 can be written as

$$n_e = \sqrt{\frac{j_b \tau (\partial E/\partial m) \rho'}{eE_i \beta}} \tag{5.17}$$

5.4.2.2 Discharge Resistivity (ρ)

The discharge resistivity ρ or the discharge conductivity ($\sigma = 1/\rho$) is an important parameter for an EBCS that determines the power amplification and the rate and efficiency of energy transfer from an inductive storage circuit.

The discharge resistance R is related to resistivity by the following:

$$R = \frac{\rho \ell}{A} \tag{5.18}$$

where ℓ is the length and A the cross-sectional area of the EBCS. An efficient EBCS should have a low discharge resistance when closed and high discharge resistance when opened, with a minimum transition time. The resistivity ρ can be written as

$$\rho = \frac{E}{j_d} \tag{5.19}$$

where E is the electric field across the switch and j_d is the discharge current density, given by

$$j_d = en_e v_d = en_e \mu \cdot E \tag{5.20}$$

and v_d is the drift velocity and μ is the electron mobility.

Substituting Equation 5.20 into Equation 5.19, we get

$$\rho = \frac{E}{en_e v_d} = \frac{1}{en_e \mu} \tag{5.21}$$

Now substituting n_e from Equation 5.17 into Equation 5.21, we get

$$\rho = \frac{E}{ev_d \left((j_b \tau (\partial E/\partial m) \rho') / eE_i \beta \right)^{1/2}} = \frac{1}{e\mu \left((j_b \tau (\partial E/\partial m) \rho') / eE_i \beta \right)^{1/2}} \tag{5.22}$$

Equation 5.22 is an important relation for ρ, because it can be used to estimate the resistivity or conductivity of an EBCS for the given dielectric gas and e-beam parameters.

5.4.2.3 Switching Time Behavior

Since an EBCS is an opening switch, the closure time during the conduction phase and subsequently the opening time during the opening phase are of prime importance [39,41].

5.4.2.3.1 Switch Closure Time (t_c)

The switch closure time t_c may be defined as the time required for the discharge current j_d^- to rise to 90% of the maximum amplitude. An approximate dependence of t_c on various factors can be examined from Equation 5.15 with a few simplified assumptions. Neglecting attachment and recombination, Equation 5.15 can be written as

$$\frac{dn_e}{dt} = \frac{j_b \sigma_i N}{e} \tag{5.23}$$

If we assume that σ_i N does not vary with time, then the integration of Equation 5.23 leads to

$$n_e = \frac{j_b \sigma_i N}{e} t \tag{5.24}$$

It may be observed from Equation 5.24 that n_e increases linearly as a function of time, as shown in Figure 5.18.

$$\text{Since} \quad j_d = en'_e v_d, \qquad n'_e = \frac{j_d}{ev_d} = \frac{j_d}{e\mu E} \tag{5.25}$$

Equation 5.25 gives the electron number density n'_e required to carry the discharge current j_d. The time t_c required to reach electron number density n'_e after the switch-on of e-beam can be found from Equation 5.24 as

$$t_c = \frac{0.9 n'_e e}{N \sigma_i j_b} \tag{5.26}$$

Substituting the value of n'_e from Equation 5.25 into Equation 5.26, we get

$$t_c = 0.9 \left(\frac{j_d}{j_b} \right) \frac{1}{\sigma_i N v_d} = 0.9 \left(\frac{j_d}{j_b} \right) \frac{1}{\sigma_i N \mu E} \tag{5.27}$$

Sets of characteristics of n_e versus t are shown in Figure 5.18. Depending upon the properties of switch gas and its pressure, $\sigma_i N v_d$ varies. As the value of $\sigma_i N v_d$ increases, the characteristics shift from right to left (i.e., 5 to 1) and the closure time t_c for a given discharge current decreases. In practice, however, the presence of L and C in the circuit does not allow j_d to rise linearly, in which case t_c is defined as the time required for current to rise from 10 to 90% value. To get short switch closure time, Equation 5.27 indicates the requirements of large ionization cross section σ_i, high gas pressure (large N), and high drift velocity (v_d).

The assumptions made in the deduction of Equation 5.27 are not strictly valid, because σ_i and v_d are time-dependent as the electric field E across the EBCS continuously decreases with time during the closure time. Though crude, Equation 5.27 is valuable, since it highlights the dependence of t_c on various factors.

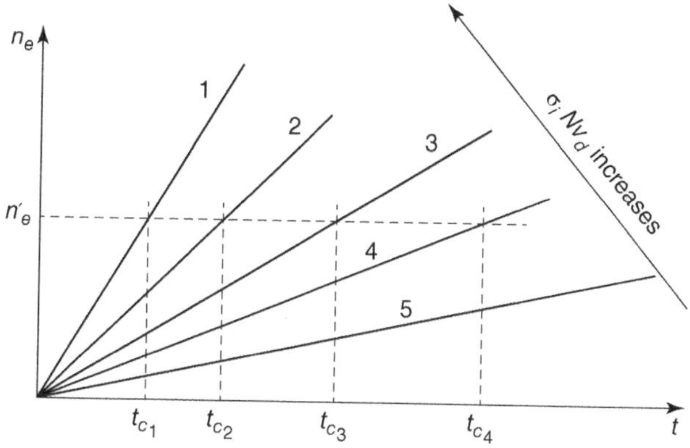

Figure 5.18 Rate characteristics of the electron number density n_e during switch closure.

5.4.2.3.2 Switch Opening Time (t_o)

When the *e*-beam is cut off, the source term $j_b \sigma_i N/e$ in Equation 5.15 becomes zero and the processes of attachment and recombination start removing the electrons. The opening time t_o may be defined as the time required for the discharge current to fall to 10% of its initial value starting from the time of switch-off of the beam. This time can be found from Equation 5.15. During the opening phase, Equation 5.15 can be written as

$$\frac{dn_e}{dt} = -\eta n_e - \beta n_e^2 \tag{5.28}$$

Depending on whether attachment or recombination dominates, the following two cases can be considered:

Case 1: Attachment regime with the two-body recombination coefficient $\beta = 0$: Equation 5.28 becomes

$$\frac{dn_e}{dt} = -\eta n_e \tag{5.29}$$

Integration of this equation, with the condition $n_e = n'_e$ at $t = 0$, gives

$$n_e = n'_e e^{-\eta t} \tag{5.30}$$

In Equation 5.30, n'_e is the electron number density present in the gas just before the *e*-beam is cut off. Using the chosen definition of opening time, namely, when the current drops to 10% of its initial value time measured from when the *e*-beam is cut off, we can estimate that

$$\eta t_o = \ln\left(\frac{n'_e}{n_e}\right) \approx \ln(10)$$

Therefore,

$$t_o = \frac{\ln 10}{\eta} = \frac{2.3}{\eta} \tag{5.31}$$

Case 2: Recombination regime with $\eta = 0$:
Equation 5.28 becomes

$$\frac{dn_e}{dt} = -\beta n_e^2 \tag{5.32}$$

Integration of this equation, with the condition $n_e = n_e'$ at $t = 0$, gives

$$n_e = \frac{n_e'}{1 + n_e'\beta t} \tag{5.33}$$

To find t_o, put $n_e = n_e'/10$. This leads to $1 + n_e'\beta t_o = 10$ or

$$t_O = \frac{9}{n_e'\beta} \tag{5.34}$$

It may be observed from Equations 5.31 and 5.34 that to obtain short opening time the coefficients of attachment and recombination should be as large as possible. Typical sets of opening time behavior are shown in Figure 5.19. The characteristics shift from right to left (i.e., 5 to 1) with increasing values of η and β.

5.4.2.4 Efficiency of EBCS

For efficiency calculations, the switch should not be considered in isolation from the total system. The efficiency depends on the *e*-beam current as well as

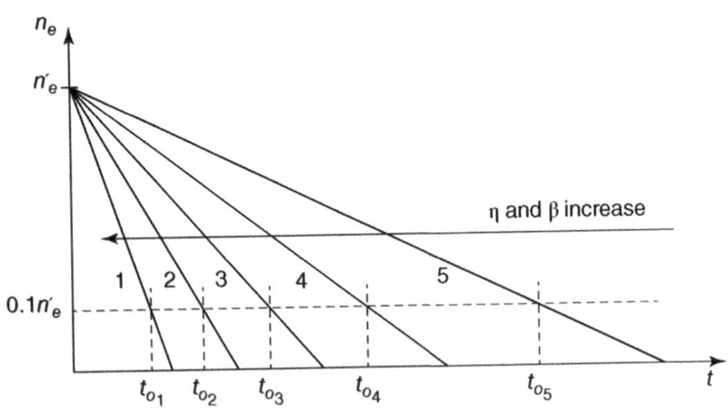

Figure 5.19 The opening time characteristics as functions of *h* and *b*.

the discharge current. Efficiency is discussed in terms of power efficiency [43] and energy efficiency [39], and they are briefly discussed in the following sections.

5.4.2.4.1 Power Efficiency (η_p)

Power efficiency (η_p) may be defined as the ratio of power delivered to the load and the power supplied to the circuit. The input power is the sum of that delivered to the switch, the e-beam and the load:

$$\eta_p = \frac{IV_\ell}{IV_c + V_b I_b + IV_\ell} = \frac{1}{1 + (R_c/R_o) + (V_b I_b / I^2 R_o)} \tag{5.35}$$

where

R_c is the resistance of the switch in the closed condition,
I is the maximum current,
R_o is the resistance of the switch in the opened condition,
V_ℓ is the highest voltage across the load,
V_b is the e-beam voltage,

and

I_b is the e-beam current.

Therefore, to achieve high power efficiency, the requirements to be satisfied in Equation 5.35 are (i) a large ratio of opened switch resistance to closed switch resistance, that is, R_o/R_c, (ii) a small e-beam power, that is., V_b I_b, (iii) a large conduction current, that is, I, and (iv) a high opening switch resistance R_o.

5.4.2.4.2 Energy Efficiency (η_E)

Energy efficiency (η_E) may be defined as the ratio of energy delivered to the load to the energy input into the system:

$$\eta_E = \frac{V_\ell I \tau_\ell}{V_\ell I \tau_\ell + I^2 R_c \tau_c + V_b I_b \tau_c + \left(V_\ell^2 \tau_o / R_o\right)} \tag{5.36}$$

where

τ_ℓ is the load pulse duration,
$I^2 R_c \tau_c$ is the energy losses in the switch during conduction,
V_b I_b τ_c is the energy carried by the electron beam,

and

$V_\ell^2 \tau_o / R_o$ is the energy losses in the switch during opening.

Equation 5.36 may be rewritten as

$$\eta_E = 1 \bigg/ \left(1 + \frac{R_c}{R_\ell}\frac{\tau_c}{\tau_\ell} + \frac{R_\ell}{R_O}\frac{\tau_o}{\tau_\ell} + \frac{V_b I_b}{V_\ell I}\frac{\tau_c}{\tau_\ell}\right) \tag{5.37}$$

It may be observed from Equation 5.37 that the following conditions should be satisfied to obtain high energy efficiency with an EBCS: (i) R_c/R_ℓ should be small, which means the switch should have a low closed resistance R_c in comparison to the load resistance R_ℓ. (ii) τ_c/τ_ℓ should be small, which requires that the conduction period should be small in comparison to load pulse duration. (iii) R_ℓ/R_o should be small, that is, the opened switch resistance should be large compared to load resistance. (iv) τ_o/τ_ℓ should be small, which requires that the opening time of the switch should be much smaller than the load pulse duration. (v) $V_b I_b/V_\ell I$ should be small, which is satisfied if the beam power is small in comparison to the load power.

Comparing the expressions for power and energy efficiencies reveals that the conditions for achieving an efficient system are the same, regardless of the metric.

5.4.2.5 Discharge Instabilities

An EBCS is presumed to operate in a diffuse discharge mode. Under certain conditions, the externally sustained volume discharge may transform into a spark discharge either during the *e*-beam-sustained conducting stage or the opening stage. The various instabilities causing the transformation from a diffuse discharge to a filamentary channel are discussed.

5.4.2.5.1 Cathode Instabilities

As in a glow discharge, a diffuse discharge has plasma sheaths near its anode and cathode resulting in steep voltage gradients across the sheath. As the *e*-beam current is increased, the cathode sheath thickness decreases, enhancing the *E* field on the electrode microprojections. This local electric field can be on the order of 10^8 V/cm [44], resulting in the explosive emission of electrons. The resulting cathode hot spot can lead to self-sustained breakdown.

5.4.2.5.2 Attachment Instabilities

This instability generally happens during the opening stage when the voltage across the switch is rising with time. Due to electron attachment, the n_e continuously decreases, thereby increasing the resistivity of the gas [45]. Therefore, in the regions of strong attachment, the *E* field increases, which further increases attachment and subsequent enhanced depletion of n_e. This interdependent process continues until the resulting high *E* field leads to local electrical breakdown and further to a channel discharge between the switch electrodes.

5.4.2.5.3 Injection Instabilities

The injection instability is mainly due to regions of nonuniform ionization. The injected *e*-beam is not generally monoenergetic but instead contains electrons having a wide spectrum of energy. Whereas the high-energy electrons are able to traverse the entire gap, the low-energy electrons are more easily stopped within the gap. Therefore, the region near the cathode has more intense ionization compared to the anode region, leading to a strong electric field near the anode where channel development may start. This instability may be prevented with the use of a high-energy *e*-beam (>200 keV), resulting in uniform ionization [46].

5.4.2.5.4 Flow and Acoustic Instabilities

If for some reason the temperature of the gas increases locally or in the entire volume, then the resulting increase in the gas pressure leads to the flow of gas from that region to the surrounding region. The reduced gas density, and hence higher E/N value, may favor the development of self-sustained avalanche discharge. In the case of a repetitively pulsed EBCS, the series of sound waves originating and propagating one after the other may also lead to hydrodynamic instability and fluctuating density regions [42].

5.4.2.5.5 Beam Pinching

Due to the presence of high magnetic fields at high current densities of either the *e*-beam or discharge current, the *e*-beam may pinch to a smaller diameter, causing filamentary discharge with intense ionization and high-temperature plasma. This type of instability can be avoided by guiding the *e*-beam in an external magnetic field [42].

5.4.2.6 Switch Dielectric

The switch dielectric employed in EBCS is usually a gas or a mixture of gases at high pressure. For high-power efficiency and high-energy efficiency, the switch has to satisfy certain conditions during the conduction phase and the opening phase.

5.4.2.6.1 Conduction Phase

During the conduction phase, the switch should have low resistivity and high conduction current. Both of these factors can be achieved by having high electron density factor n_e and large drift velocity. During conduction, the voltage drop across the switch is usually low, resulting in low E/N. It may, therefore, be generalized that the dielectric should possess large electron mobility and low attachment coefficient at small E/N.

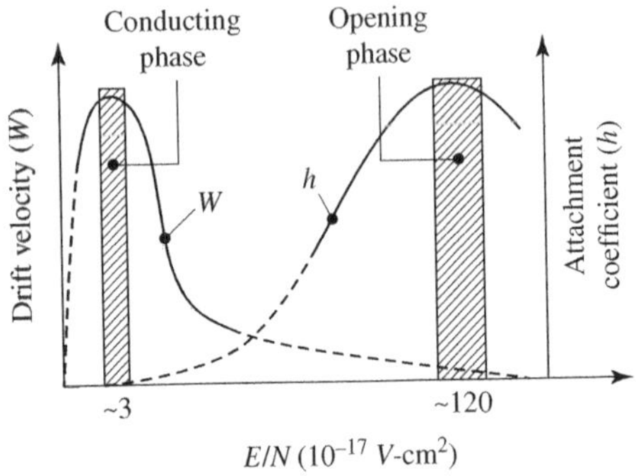

Figure 5.20 The desirable characteristics of the drift velocity and attachment coefficient for a gas in EBCS.

5.4.2.6.2 *Opening Phase*

During the opening phase, the switch should rapidly increase in resistivity by the fast removal of electrons, by choosing a gas with a high attachment coefficient η and low mobility μ. During the opening phase, the voltage drop is large and results in a high *E/N* in comparison to the closing phase. It may, therefore, be generalized that the dielectric should possess low drift velocity and high attachment coefficient at large *E/N*.

The desired overall characteristics for a good opening switch are illustrated in Figure 5.20. These characteristics can be achieved in buffer gas of argon or CF_4 by the addition of electron-attaching gases such as $(CF_3)_2S$, (1,2-$C_2Cl_2F_4$), (l,l,2-$C_2Cl_3H_3$), (l,l-$C_2Cl_2H_4$), or nC_3F_8 [47]. The characteristics can be tailored to peak at the desired *E/N* values by adjusting the percentage of individual components of the gas mixture. Attaching gases N_2O, SO_2, CO_2, and O_2 have also been used in a buffer gas of N_2 [38,48].

5.4.2.7 Switch Dimensions

The important switch dimensions, as indicated in Figure 5.21, are the cross section (*ab*) from where the *e*-beam is injected and the length ℓ between the electrodes E_1 and E_2.

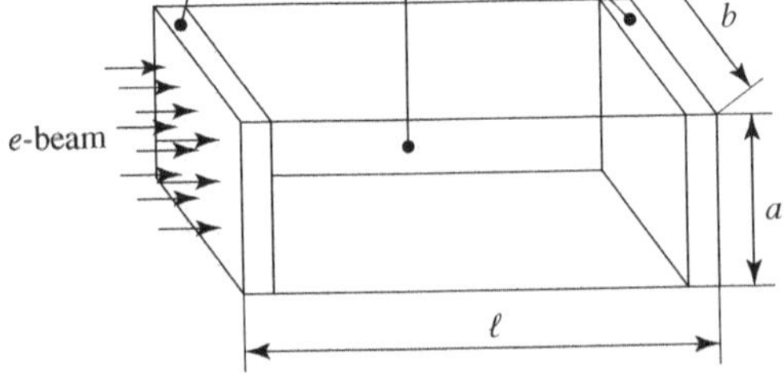

Figure 5.21 The dimensions of EBCS.

5.4.2.7.1 *Cross Section (ab)*

The switch cross section is usually determined by heat dissipation considerations [42]. The temperature rise in the gas should be limited to approximately 500 K to prevent the volume discharge from contracting to a spark channel. The energy dissipated W_d is given by

$$E_d = I^2 R\tau = I^2 \frac{\rho \ell}{ab} \tau \tag{5.38}$$

where

τ is the current pulse duration

and

ρ is the resistivity of the gas.

Assuming all the energy is utilized in raising the temperature of the gas to T,

$$W_d = NRT = \frac{(ab\ell)\rho'}{M} RT \tag{5.39}$$

where

N is the total number of molecules,
ρ' is the gas density,
M is the molecular weight,

and

R is the gas constant.

Relations 5.38 and 5.39 can be combined such that $I^2 \rho \ell \tau / ab = (\rho' \cdot ab \cdot \ell \cdot RT)/M$, or

$$ab = I \left(\frac{M\rho\tau}{RT\rho'} \right)^{1/2} \tag{5.40}$$

Knowing the parameters of the gas and e-beam, ρ can be calculated from Equation 5.22. Invoking that T should not exceed 500 °K, the cross section (ab) can then be calculated.

5.4.2.7.2 *Length*

The length of the switch ℓ can be calculated from the following relation:

$$V_\ell = K \cdot E_b \cdot \ell \tag{5.41}$$

where V_ℓ is the maximum voltage across the switch; E_b is the breakdown strength at the given gas pressure; and K is the factor of safety (<1).

If the dielectric strength is E_{b0} at ρ'_0, then Equation 5.41 can be written as

$$V_\ell = K\left(\frac{E_{b0}\rho'}{\rho'_0}\right)\ell \tag{5.42}$$

where ρ' is the gas density at the given pressure and ρ'_0 is the gas density at STP.

5.4.3 Vacuum Arc Switch

5.4.3.1 Mechanical Breaker

The basic device based on vacuum arc is a conventional mechanical breaker (MB) in which the switch closure is done by mechanically bringing the two electrodes together to make physical contact. The opening is achieved by physically separating the two electrodes. The arcing, which ensues upon separation, continues until the natural current zero (in the case of AC schemes) is achieved, which leads to the arc extinction. A typical configuration of a mechanical vacuum interrupter is shown in Figure 5.22. The bellows enable the movement of the electrode while maintaining the vacuum seal. The shield prevents the metal vapor from depositing on the ceramic wall. The desired performance parameters of a vacuum-arc opening switch are large current, high dI/dt, high transient voltage, and high dV/dt. For a given switch, if any of these factors are absent, successful interruption may not take place.

5.4.3.1.1 *Counterpulsing for Forced Current Zero*

The mechanical breaker, when used in DC circuits or inductive storage, cannot interrupt the current because of the absence of natural current zero. In such cases, a counterpulse circuit, shown in Figure 5.23, may be used to force a

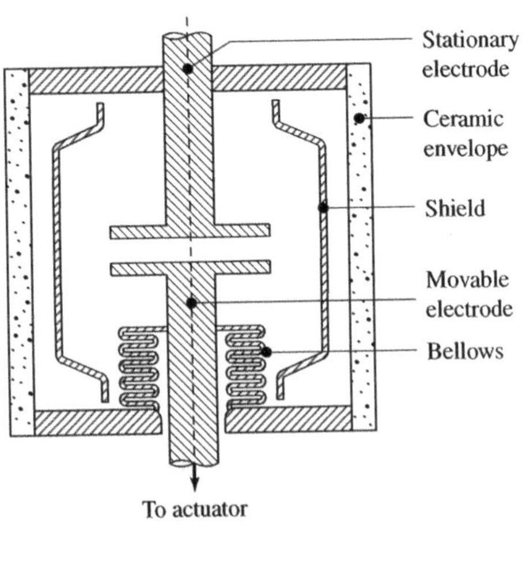

Figure 5.22 A mechanical vacuum interrupter.

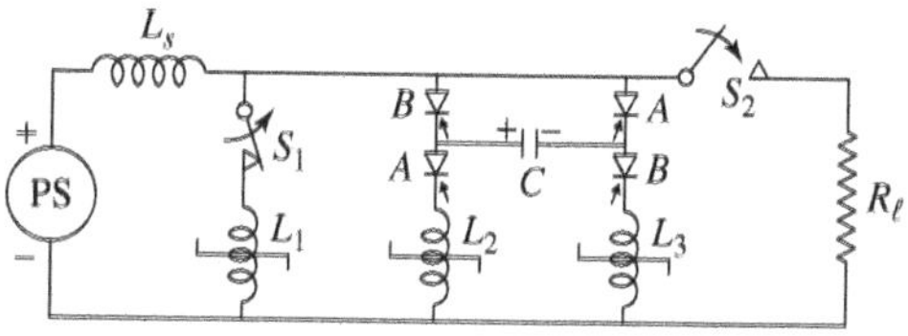

Figure 5.23 A counterpulse circuit for the repetitive operation of an inductive storage [49].

reverse current and cause "current zero" [49]. The circuit breaker S_1 opens after charging the storage inductor L_s to a maximum current I_0. To pass a reverse current through the circuit breaker, the silicon-controlled rectifiers (SCRs) A are triggered, so that the precharged capacitor C discharges through S_1 and the saturable inductors L_1 and L_2. At current zero, when S_1 interrupts the current, I_0 in L_s is redirected to charge C through the SCRs. When C charges to a voltage exceeding the self-breakdown voltage of S_2, then S_2 closes and the current transfers to the load R_ℓ. The SCRs A can recover because of the reverse polarity on C and the next cycle begins reclosing S_1, followed by opening of S_1 and subsequent counterpulsing by triggering the thyristors B.

The purpose of the saturable reactor is to slow down the dI/dt of the circuit breaker near the current zero and thus prolong the duration of low current flow, as shown in Figure 5.24.

The principle of operation of a saturable reactor is explained as follows: At t_1, the circuit breaker starts opening, but the current I_0 continues to flow through the arcing between the electrodes. At t_2, the counterpulse circuit initiates the reverse current in S_1 and, in the absence of saturable reactor, the current zero is produced at t_3. In the presence of saturable reactor, the current zero is delayed to t_4. The slowing down of current decay is due to the shifting of the operating point of the saturable reactor from X at I_0 to Y at I'_0. The substantial increase in the inductance of the saturable reactor caused by shifting the operating regime from saturation point X to high-permeability point Y results in an increased L/R

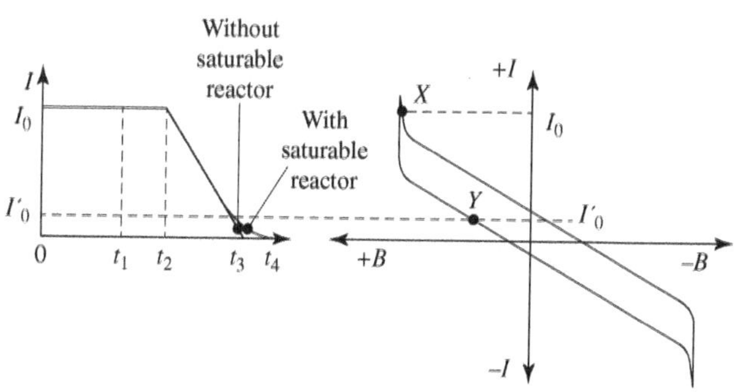

Figure 5.24 Current levels in the breaker and operating point for the saturable reactor.

of the circuit and hence the slowing down of current decay. The slower decay of current through the breaker ensures that there is sufficient time for the plasma to decay and thereby build up sufficient dielectric strength for a successful interruption.

5.4.3.1.2 *Early Counterpulsing*

In the conventional counterpulsing scheme discussed above, the breaker electrodes are separated at full current, and after a delay of few milliseconds the counterpulsing is initiated. The disadvantages of prolonged full current arcing are the formation of hot spots on the electrodes and re-ignition of the arc, and increased erosion of the electrodes due to continued arcing. To overcome these drawbacks, early counterpulsing can be employed [50], where the counterpulse is applied first and then the breaker is opened so that the arcing takes place at much lower currents. A typical waveform of a breaker current with early counterpulse is shown in Figure 5.25. The disadvantages of the early counterpulse scheme are the requirements of (i) a large capacitor bank for counterpulse driver circuit, (ii) a large saturable reactor, (iii) a fast actuator to enable rapid separation in the short time interval between t_0 and t_1, and (iv) a rugged interrupter to withstand the powerful mechanical impulse at the bellows.

5.4.3.2 Magnetic Vacuum Breaker

The main disadvantages of the mechanical breaker are long opening times and poor repetition rate capability (<50 pps). To overcome these drawbacks, a magnetic vacuum breaker (MVB) was developed where the electrodes are stationary [51]. The MVB, as shown in Figure 5.26, consists of an annular anode fitted with a magnetic coil for producing the axial magnetic field. On the axis and below the anode level is located a trigger gap assembly consisting of cathode and trigger electrode, which are bridged by an insulator with a thin metallic coating. The closing of the switch is initiated by applying a voltage pulse between cathode and trigger electrodes causing current flow and evaporation of the metal coating on the insulator. The metal vapor/plasma bridges the gap between the cathode and anode and causes the closure of the switch. The opening of the switch is achieved by energizing the magnet coil. The axial magnetic field thus produced exerts a force on the electrons and makes them

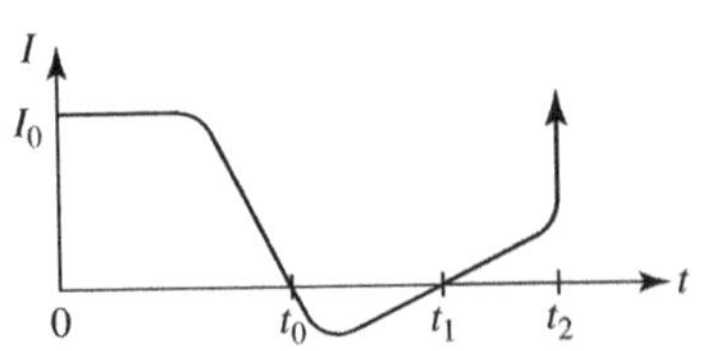

Figure 5.25 Typical breaker current waveform for early counterpulsing [50].

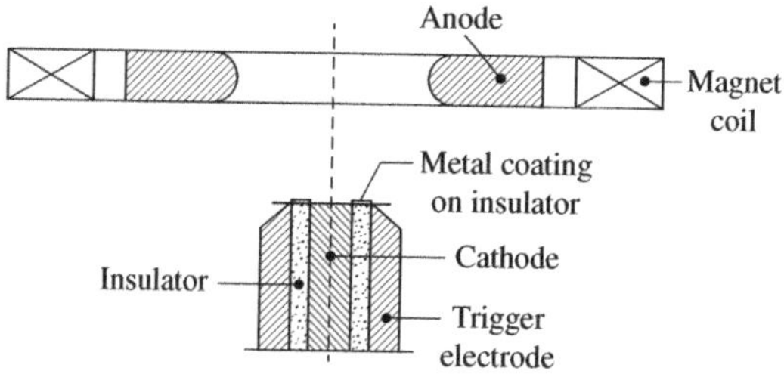

Figure 5.26 The configuration of a magnetic vacuum interrupter [51].

traverse a longer path to reach the anode. This, in effect, is equivalent to an increased cathode–anode gap, which increases the voltage drop and reduces the current. The arc current and the magnetic field are related by the following expression:

$$\frac{I_m}{I_o} = \frac{1}{1 + KB_z^2} \tag{5.43}$$

where I_m is the arc current with magnetic field, I_0 is the arc current without magnetic field, B_z is the magnetic flux density, and K is the constant.

Representative behavior of arc current is illustrated in Figures 5.27 and 5.28. Arc extinction can be achieved by applying the required magnetic field and holding it for a time that is longer than the average arc life at that particular current.

5.4.3.3 Mechanical Magnetic Vacuum Breaker

A mechanical breaker has the advantages of high conduction current and long conduction time, but has a poor repetition rate capability. A mechanical vacuum breaker has a good repetition rate capability but cannot conduct large current for long periods. Therefore, the advantages of both are combined in a mechanical magnetic vacuum breaker (MMVB) [52,53]. The trigger electrode

Figure 5.27 Behavior of arc current versus magnetic flux density.

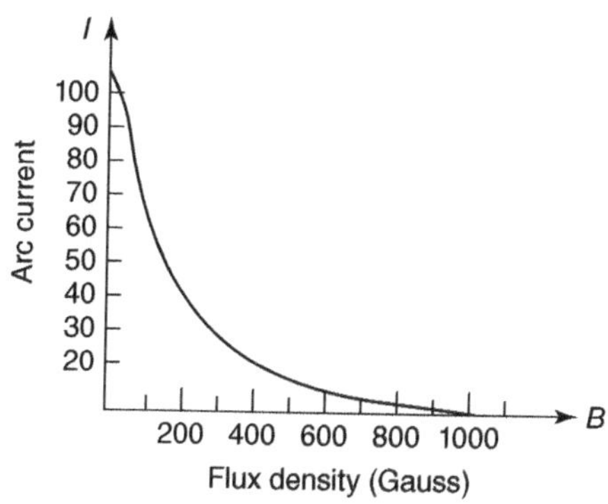

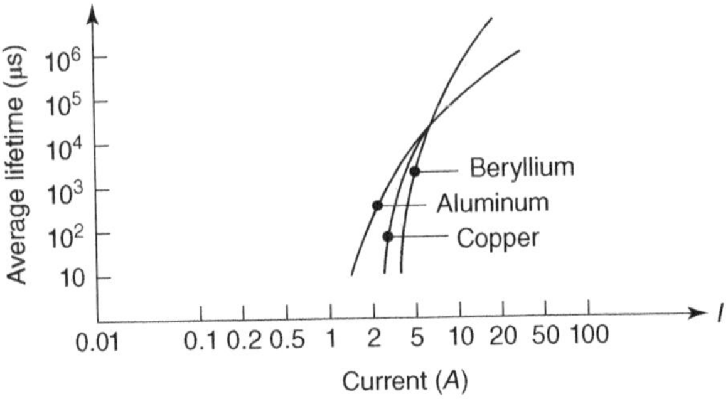

Figure 5.28 Average arc lifetime for different electrode materials [51].

for plasma source generation is incorporated either in the stationary electrode or in the shield, as shown in Figure 5.29. The charging of the storage inductor is done by bringing the movable electrode toward the stationary electrode for a physical contact and holding it for a sufficient time until the current reaches the maximum value of I_o^{max} in the inductor. The contacts are then separated and interruption obtained by forced current zero. The pulse to the trigger gap followed by the pulse to the magnet coil is applied repeatedly to deliver a train of output pulses at a high repetition rate. This sequence of operations is illustrated in Figure 5.30.

5.4.4 Explosive Switch

A typical configuration [54] of an explosive switch (ES) is shown in Figure 5.31. When closed, as shown in Figure 5.31a, current is carried by the aluminum cylinder. Large conduction currents with long conduction times are possible

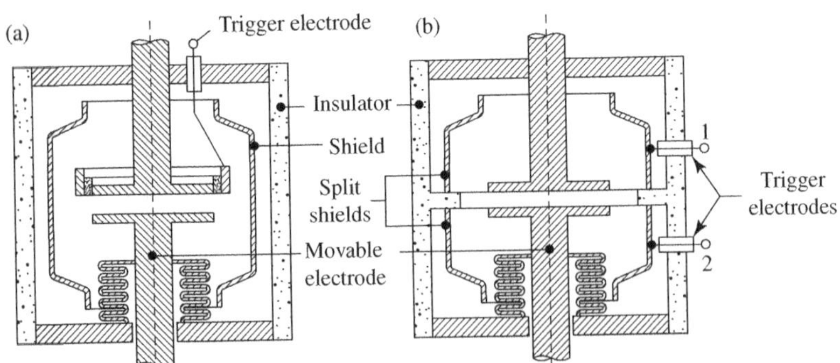

Figure 5.29 Trigger configurations in a mechanical magnetic vacuum breaker. (a) Trigger in stationary electrode. (b) Trigger in split shield [52].

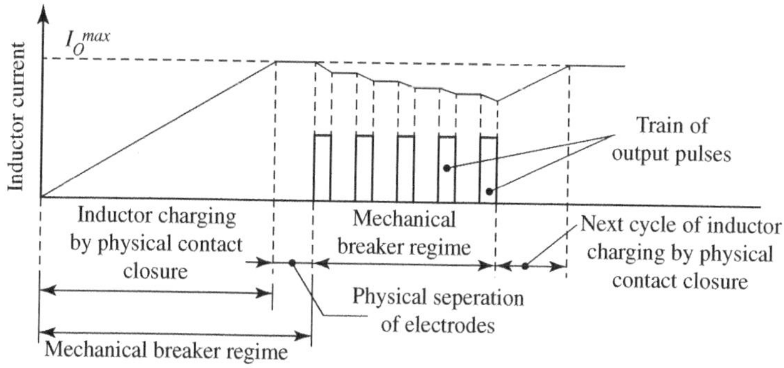

Figure 5.30 The sequence of operations in a mechanical magnetic breaker.

because the thermal mass of the cylinder is large. Opening of the explosive switch is initiated by detonating the explosive. The outward radial pressure cuts the cylinder at the edges of the cutting rings and folds it back onto the bending rings, as shown in Figure 5.31b. As a result, the current continues to flow by electrical arcs formed between the rings. The paraffin, which extrudes radially outward, acts as a heat sink, quenches the arc, and interrupts the current. The arc voltage is usually 0.5–1 kV/gap, which helps to commutate the current to the load or the succeeding switching stage.

As with MVBs, the voltage recovery rate is not fast when interrupted currents are large. To increase the recovery rate at high interrupting currents, the ES is made into two sections, as shown in Figure 5.32. The detonations in the two sections take place with a predetermined time delay. When the first section of the ES is detonated, arcing takes place only in that section, but part of the current is already commutated to the next stage. Section 2, detonated later, has to interrupt a much smaller current. This technique has been used to interrupt 400 kA at 600 kV with a delay of 40 μs with a two-section switch with 16 and 27 gaps [54]. Unlike an exploding fuse switch, the ES does not consume energy

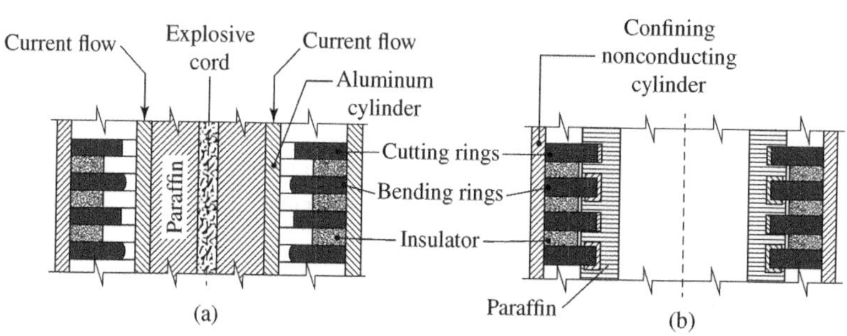

Figure 5.31 Configuration of an explosive switch. (a) Configuration during conduction. (b) Configuration after opening.

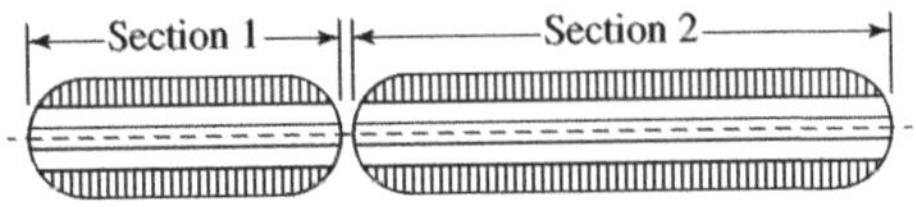

Figure 5.32 Explosive switch with two sections.

from the inductive store with the required energy being supplied by the explosive.

5.4.5 Explosive Plasma Switch

In an explosive plasma switch (EPS), the performance parameters of a fuse are enhanced appreciably by compressing the fuse plasma using the explosive force of a chemical explosive [55–57]. A typical explosive plasma switch is shown in Figure 5.33, where the switch current is carried by the outer tube, inner tube, and interconnected fuse foil. At the maximum current, the foil is converted into a plasma channel, which continues to carry the current. When the explosive is detonated along the inner surface, the explosive pressure, acting radially outward, compresses the plasma channel simultaneously along the entire length. The resistance of the plasma channel increases and cools due to the mixing of the explosive products, enabling interruption of the current. The EPS has interrupted 7 MA at 300 kV with a resistance ratio greater than 100 [55].

5.4.6 Plasma Erosion Switch

The plasma erosion switch (PES), shown in Figure 5.34, is a vacuum device in which the plasma is sprayed from outside through the anode A [58,59]. Figure 5.34a shows that when the plasma has bridged the AK gap, the pulse generator PG is energized and starts charging the inductor. As the current starts increasing, a double layer is formed near the cathode K, X–X being the interface between the plasma region and vacuum region, as shown in Figure 5.34b. The high electric field in the double layer, on the order of 10^7 V/cm, gives rise to

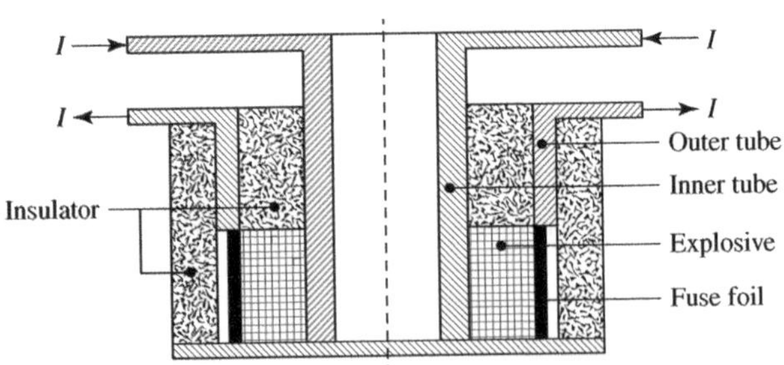

Figure 5.33 Schematic of an explosive plasma switch.

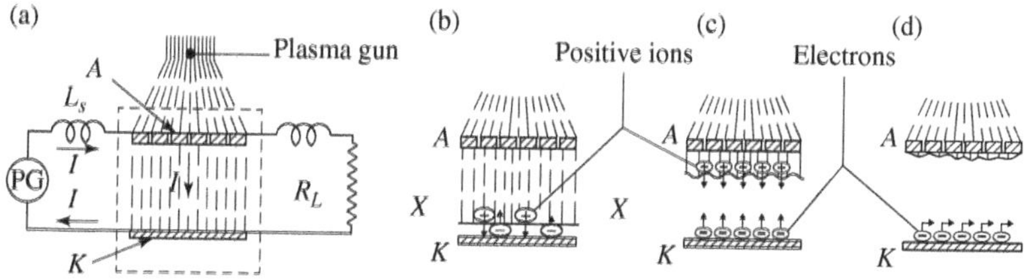

Figure 5.34 Phenomenon of conduction and opening in a plasma erosion switch. (a) Plasma bridges *A*–*K* gap. (b) Double-layer formation. (c) Plasma erosion growth. (d) Magnetic insulation.

explosive electron emission from the cathode. The positive ions are drawn from the plasma and accelerated toward the cathode. Much more current can be drawn from a PES than in a conventional bipolar diode because of space charge neutralization in the region between *X*–*X* and *K*.

At very high current densities, the ion flux drawn from the plasma is insufficient to sustain that current, leading to a large flux rate of plasma ions, making replacement impossible. The plasma is eroded, resulting in an increase of the separation d of the virtual *AK* gap, shown in Figure 5.34c. As a result, the impedance of the switch increases proportional to d^2. At very large current levels, the magnetic field produced by the discharge current deflects the electrons, leading to the magnetic insulation of electrons, as shown in Figure 5.34d. This magnetic insulation further enhances the rate of rise of the switch impedance. Sheath opening velocities of 10^8 cm/s are possible. Typical maximum operating parameters obtained in PES are 5 MA at 1 MV with conduction time of 1 μs and opening time of 100 ns [60]. Plasma erosion switches have also been used in repetitive pulsed power systems with typical parameters of 10–30 kA at 2–3 MV at repetition rates of 1–4 Hz [61]. The use of the PES, as a final stage in a fast pulse power system for particle beam generation, located next to the diode, will result in a large enhancement of power amplification and pulse compression. Sources used for plasma generation are surface discharges across flash boards, coaxial guns, or lasers impinging on a target [62].

5.4.7 Dense Plasma Focus

A typical configuration of a dense plasma focus (DPF) device is shown in Figure 5.35. Two coaxial cylinders are separated by an insulator and act as the anode and cathode. The space between the electrodes is filled with a suitable gas, such as argon, hydrogen, or a mixture of the two, at a pressure between $\sim 10^{-3}$ and 1 Torr. The evolution of plasma focus, shown in Figure 5.36, occurs as follows. The application of a pulsed voltage between the anode and cathode electrodes results in a surface discharge across the insulator surface (Figure

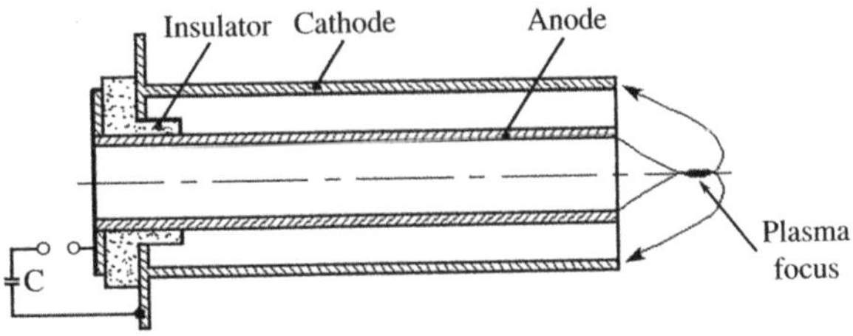

Figure 5.35 Configuration of a dense plasma focus device.

5.36a). The $\vec{j} \times \vec{B}$ force from the radial current density j_r and the azimuth magnetic field B_θ results in the propagation of the plasma front along the anode to the open end of the electrodes, illustrated in Figure 5.36b and c. As the plasma sheaths move beyond the end of the electrodes, as shown in Figure 5.36d, the sheath collapses onto itself and forms a plasma focus. Formation of an efficient plasma focus depends upon a proper discharge initiation: a high ionization rate and the formation of a uniform of discharge along the insulator surface. The ionization rate is governed by the (pd) value, where p is the pressure of the gas and d is the insulator length. The insulator discharge is governed by the energy range of electrons emitted from the cathode and accelerated across the cathode sheath. For a given gas, an optimum pressure can be found where these conditions are fulfilled [63]. If the gas pressure is higher than the optimum pressure, the ionization will be high but the insulator coverage will be lacking because of a reduced range of electrons. At the time of plasma pinching and focus formation, the impedance rises rapidly and interrupts the current [64]. The plasma resistance has been found to increase from 0.1 Ω to 1.0 Ω in about 10 ns [65]. A typical waveform of DPF discharge current showing the rapid decrease of current is shown in Figure 5.37 [66]. The current is commutated to the load with power amplification in the period Δt, time resolution of the current. During the interrupting period in a DPF it was found that the switch exhibits a series of openings and closings with a repetition rate of 10^5 [67]. An inductive storage system based on DPF has delivered 90 kA at 25 kV [67].

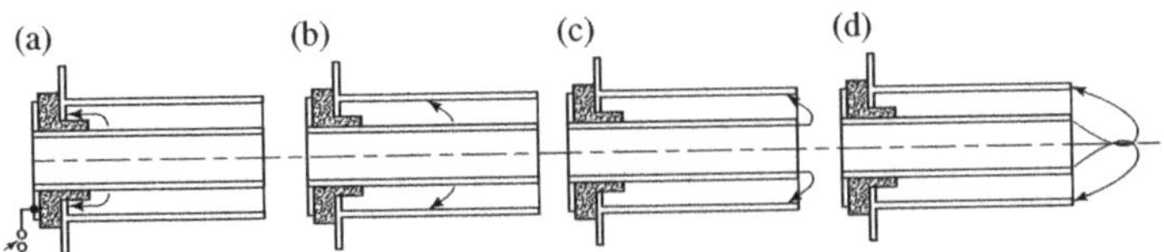

Figure 5.36 Stages of events in a dense plasma focus device leading to plasma focus. (a) Flashover across insulator. (b) Propagating current sheet. (c) Current sheet at end of electrodes. (d) Pinch forms plasma focus.

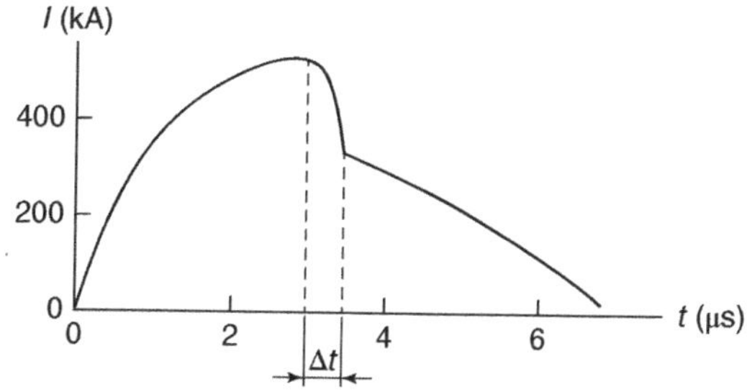

Figure 5.37 Plasma focus discharge current waveform.

5.4.8 Plasma Implosion Switch

A plasma implosion switch, shown in Figure 5.38, can be considered a modified form of a DPF, employing vacuum as the switch medium and a foil or array of wires for the plasma generation [68]. The current I from the pulse generator PG causes the foil or wires to explode, resulting in the formation of a circular plasma sheet. For good energy transfer efficiency, the design should ensure that the explosion takes place at the peak current. As in a DPF, the $\vec{j}_r \times \vec{B}_\theta$ force drives the plasma sheet in the axial direction. When the plasma reaches the end of the electrodes, a radial force $\vec{j}_z \times \vec{B}_\theta$ becomes effective and implodes the plasma on axis. The implosion results in a sudden jump in the switch inductance due to elongation and reduction in diameter of the plasma sheet and a thinning of the plasma sheet wall. This serves to increase the switch resistance R interrupting the current and opening the switch. In vacuum inductive storage applications, the plasma implosion switch does not need a dielectric interface, implying the potential for very high current and voltage capability. Switch opening times of 300 ns have been obtained at 10 MA at 90 kV [69]. Repetitive switching capability can be achieved by replacing the plasma-producing foil with a gas puff [68].

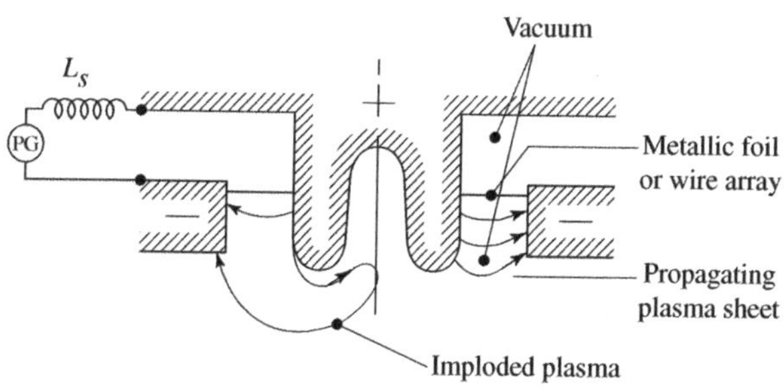

Figure 5.38 Configuration of a plasma implosion switch [69].

5.4.9 Reflex Switch

A reflex switch consists of a primary cathode K_1, a thin foil anode A, and a floating cathode K_2, as shown in Figure 5.39, insulated with vacuum [70,71].

Generally, the AK_2 chamber is at a slightly higher pressure (~10^{-4} Torr) compared to AK_1 chamber (~10^{-5} Torr). When the switch S_1 closes, A is positive with respect to K_1, and electrons are emitted from K_1 and accelerated to the anode. These energetic electrons pass through the anode foil into the AK_2 chamber to charge K_2 to a negative potential. The electrons traveling toward K_2 are repelled and flow toward K_1, which in turn repels them toward K_2. The electrons, therefore, start reflexing around A and depositing their energy in a number of repeated passes, and plasma is formed around A. The positive ions from the plasma source of A accelerate toward K_1 and K_2. The presence of positive ions in the region AK_1 and the high density of reflexing electrons around A modify the potential distribution, as shown in Figure 5.40.

The distribution for a conventional Langmuir bipolar flow is shown by curve (a), and a reflex mode distribution is shown by (b). In the reflex mode, most of the potential drop is within a very short distance of the anode A—the equivalent of shifting the cathode from position K_1 to K_1'. The bipolar current density is determined by the Langmuir–Child's law, so the reflex triode possesses higher current capability and low impedance [72]. This is evidenced by the I–V characteristics shown in Figure 5.41. The low-impedance operating point A

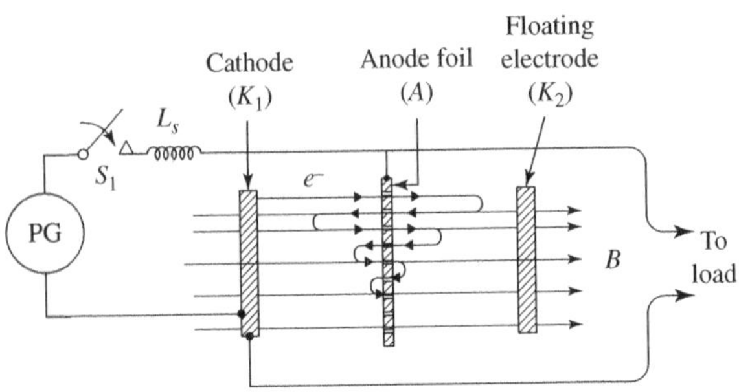

Figure 5.39 Schematic of a reflex switch.

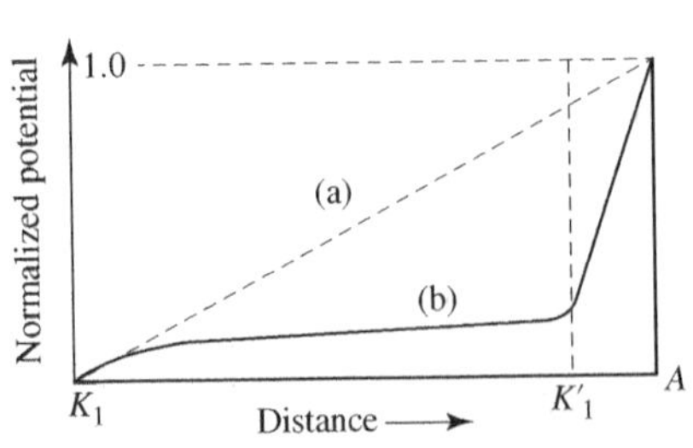

Figure 5.40 Potential distribution in (a) a bipolar diode and (b) a reflex triode.

Figure 5.41 *I–V* characteristics of a bipolar diode and reflex triode.

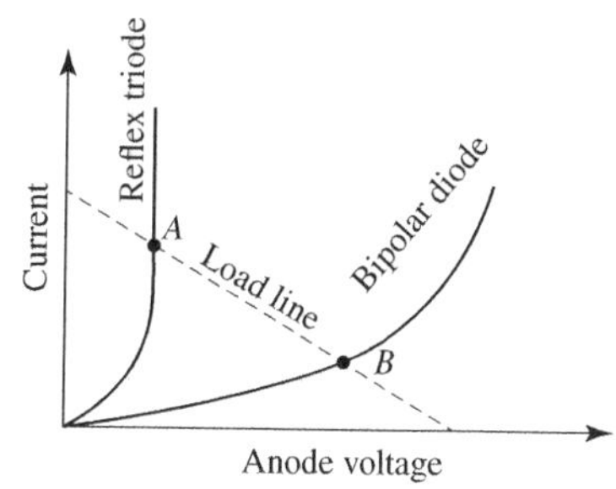

represents the closed switch. When the switch is opened, the operating point shifts to a high-impedance point *B*. This can be achieved by several methods, such as shorting K_2 to *A*, or by injecting a gas puff into the AK_2 chamber, effectively removing the reflexing triode mode and establishing a bipolar diode mode. An axial magnetic field is generally applied to prevent the loss of reflexing electrons. The reflex switch can also be configured in a tetrode mode [73]. The reflex switch has potential for use in vacuum inductive storage because of its compatibility with vacuum transmission lines. The reflex switch has been operated at 1.8 MV, 160 kA with an impedance jump from 0.8 to 15.3 Ω [71].

5.4.10 Crossed Field Tube

A crossed field tube (XFT) consists of a coaxial cathode and anode contained in a vacuum vessel and surrounded by a coaxial magnet coil, as shown in Figure 5.42. The space between the electrodes is filled with a chemically inert gas at low pressure (10^{-2}–10^{-1} Torr). The inert gas is chosen to minimize the cleanup of gas by chemisorption with the sputtered metal from the cathode [74].

When energized by a capacitor bank, the magnet coil produces a pulsed magnetic field in the axial direction. The highest operating voltage is set such

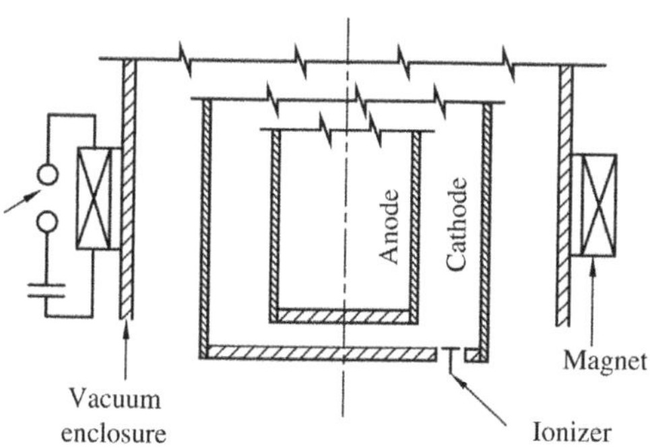

Figure 5.42 Typical configuration of a crossed field tube.

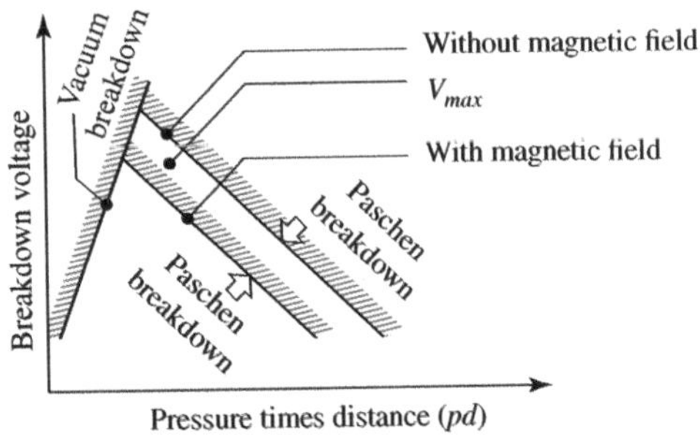

Figure 5.43 Pressure times distance (*pd*) regime of a crossed field tube.

that, in the absence of a magnetic field, the *pd* value is below both the Paschen breakdown and vacuum breakdown values, as shown in Figure 5.43 [74]. However, when a suitable magnetic field is applied, the electrons, which were earlier accelerated radially toward the anode, now experience an axial force of $\vec{j}_r \times \vec{B}_\theta$. The electrons become axially confined and in this process cause enhanced ionization of the gas molecules. This has the effect of shifting the Paschen curve to the left, as shown in Figure 5.44, making the applied voltage V_{max} higher than the modified breakdown voltage, which leads to breakdown and initiation of conduction. The current density is usually limited to less than 5 A/cm^2 to prevent the glow-to-arc transition [75]. Reducing the magnetic field below a critical value makes the radial electric force much stronger in comparison to the axially confining force $\vec{j}_r \times \vec{B}_\theta$. Thus, the electrons are rapidly collected by the anode, without significant ionization, resulting in the opening of the XFT switch. The critical magnetic field B_{cr} is given by the following relation [76]:

$$B_{cr} = \frac{1}{d}\sqrt{\frac{2mV}{e}} \tag{5.44}$$

where d is the electrode spacing between cathode and anode, V is the voltage across the electrodes, and e and m are the charge and mass of the electrons, respectively. Equation 5.44 is valid for conduction as well as interruption of the switch current.

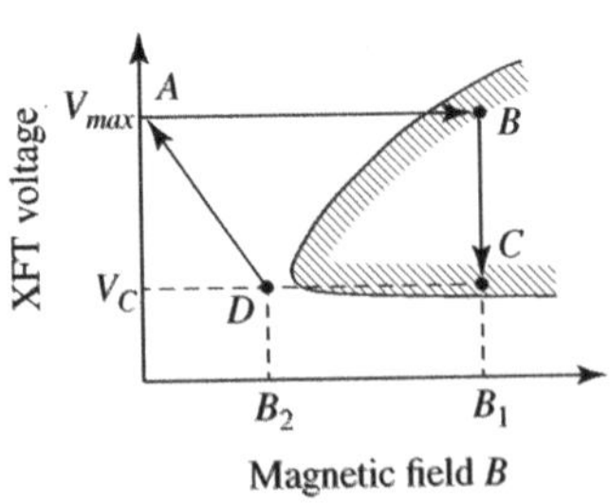

Figure 5.44 Voltage versus magnetic field characteristic for a crossed field tube.

The variation of the critical magnetic field with voltage for an XFT with fixed d is shown in Figure 5.44 [77]. Point A represents the hold-off stage, when the magnetic field is zero and the XFT is able to sustain the full applied voltage (V_{max}) without conduction. The operating point B represents the conduction phase, which takes place when the magnetic field is increased to B_1. When the conduction starts, the voltage across the tube collapses to a lower voltage V_c and conduction continues at the point C. Operating point D represents the interruption stage for which the magnetic field is reduced to B_2, with $B_2 < B_{cr}$. After the XFT recovers completely, the operating point jumps to A and continues to hold off the applied voltage. XFTs have achieved parameters of 100 kV at 10 kA under single-pulse conditions [75] and operated at pulse repetition rates of 120 Hz at 50 kV and 1.3 kA [78]. A triggered XFT has been demonstrated by inserting a grid adjacent to cathode [77].

A grid may be introduced to a cross field tube to control the operation and is known as a "Crossatron" [79–82]. The preformed plasma is already present but confined to the volume between the cathode and screen grid. Conduction in the device is achieved by pulsing the control grid, to a positive polarity, increasing its potential relative to the screen grid, and thereby allowing plasma flow to the anode electrode. Interruption of the current is achieved by pulsing the control grid negative, which erodes the plasma between the control grid and the anode. The opening time is governed by the time for diffusion of plasma to the boundary walls. These tubes generally employ hydrogen as the filling gas in the pressure region of 0.1–1 Torr. Typically, a Crossatron can carry peak discharge current of 2.5 kA at 50 kV and can interrupt a 1 kA current in about 0.55 μs. Average current up to a few amperes can be carried. Crossatrons have achieved peak repetition frequency (PRF) of 40 kHz in continuous operation and 1 MHz in burst mode. The lifetime of Crossatron is limited by the degradation of internal structures due to ion sputtering of the cathode during conduction and the control grid during interruption. By employing thermionic hollow cathode discharge in xenon at low pressure, low forward voltage drops of 10–20 V have been achieved. A planar configuration, called a "Hollotron," has demonstrated voltage standoff of up to 20 kV, switching time ≤300 ns, and pulse repetition frequencies of 20 kHz at 50% duty [80]. A high-power vacuum triode with grid-controlled plasma cathode, which does not need crossed magnetic field, has also been developed [81].

5.4.11 Miscellaneous

Many other opening switch concepts have been proposed for a variety of applications. Some of these are superconducting switches [82,83], saturable capacitor and inductor switches [84], Hall effect device [85], controlled plasma instability [86], optically controlled switch [87], spoiled electrostatic confinement [88], thyratrons [52], and thyristors [49].

5.5 Design Example

The two-stage inductive storage system shown in Figure 5.15 has a 0.5 μH storage inductor L_s, which is charged by a 3.5 μF capacitor charged to a voltage of 25 kV. The opening time T_1 of switch S_1 is 250 ns and the opening time T_2 of switch S_2 is 10 ns. Assuming 10% loss of energy in each opening switch, calculate (i) voltage amplification and pulse time compression produced by the first stage and (ii) voltage amplification and pulse time compression produced by the second stage.

Solution

The energy E_c stored in the capacitor bank is given by

$$E_c = 1/2CV^2 = 1/2\left(3.5 \times 10^{-6}\right)\left(25 \times 10^3\right)^2 = 1.09\ \text{kJ}$$

Since the energy loss in the first-stage switch S_1 is 10% of E_c, that is, 0.109 kJ, the energy E_L transferred to the inductor L_s can be written as

$$E_L = (1.094) - (0.109) = 0.985\ \text{kJ}$$

The maximum current i_1^{max} in the inductor is determined by the maximum stored energy: $E_L = 1/2L_s\left(i_1^{max}\right)^2$.

Rearranging terms,

$$i_1^{max} = \sqrt{\frac{2E_L}{L_s}}$$

Substituting the values for $E_L = 0.985$ kJ and $L_s = 0.5$ μH into the above expression, $i_1^{max} = 62.8$ kA. The current in the inductor reaches this value at a time $T_{ch} = T/4$where T is given by

$$T = 2\pi\sqrt{L_s C}$$

$$T = 2\pi\sqrt{\left(0.5 \times 10^{-6}\right)\left(3.5 \times 10^{-6}\right)} = 8.3\ \mu\text{s}$$

The inductor charging time $T_{ch} = T/4 = 8.3/4 = 2$ μs. When S_1 opens, the voltage V_1 generated across the inductor is given by

$$V_1 = L_S\left(\frac{i_1^{max}}{T_1}\right)$$

$$T = \left(0.5 \times 10^{-6}\right)\left(\frac{62.8 \times 10^3}{250 \times 10^{-9}}\right) = 126\ \text{kV}$$

Because of this large voltage, a current with high *di/dt* is commuted to the second-stage opening switch S_2 causing S_2 to open very fast—in 10 ns. The S_2 has a further energy loss of 10%, that is, 0.098 kJ. Hence, the maximum energy stored in the inductor at the time of opening of the second-stage switch, E'_L, is

$$E'_L = 0.985) - (0.098) = 0.887\ \text{kJ}$$

The maximum current in the inductor at the time the second-stage switch opens is given by $E'_L = 1/2L_s\left(i_2^{max}\right)^2$with

$$i_2^{max} = \sqrt{\frac{2E'}{L_s}} = \sqrt{\frac{2\times\left(0.887\times10^3\right)}{\left(0.5\times10^{-6}\right)}} = 59\ \text{kA}$$

When the switch S_2 opens, the voltage V_2 generated across the inductor is given by

$$V_2 = L_S\left(\frac{i_2^{max}}{T_2}\right) = \left(0.5\times10^{-6}H\right)\cdot\frac{\left(59\times10^3A\right)}{\left(0.5\times10^{-6}s\right)} = 2.95\ \text{MV}$$

From the above calculations, the required results are as follows:

a) First stage performance
   i) Voltage amplification:

$$\frac{V_1}{V_i} = \frac{126\ \text{kV}}{25\ \text{kV}} = 5$$

   ii) Pulse time compression:

$$\frac{T_1}{T_i} = \frac{250\ \text{ns}}{2\ \mu\text{s}} = 0.125$$

b) Second stage performance
   i) Voltage amplification:

$$\frac{V_2}{V_1} = \frac{2.95\ \text{MV}}{126\ \text{kV}} = 23$$

   ii) Pulse time compression:

$$\frac{T_2}{T_1} = \frac{10\ \text{ns}}{250\ \text{ns}} = 0.04$$

References

1 E.M. Honig, Repetitive Energy Transfers from an Inductive Energy Store. Ph. D. dissertation, Texas Tech University, and LANL Tech Report, LA-10238-T, 1985.

2 A. Guenther, M. Kristiansen, and T. Martin, *Opening Switches*, Plenum Press, New York, 1987.

3 L. Vermij, The Voltage Across a Fuse During the Current Interruption Process. *IEEE Trans. Plasma Sci.*, Vol. 8, No. 4, p. 460, 1980.

4 W.G. Chace and H.K. Moore, *Exploding Wires*, Vols. 1–4, Plenum Press, New York, 1968.

5 F.D. Bennett, High Temperature Exploding Wires, in C.A. Ronn, C ed., *Progress in High Temperature Physics and Chemistry*, Pergamon Press, 1968.
6 V.A. Burtsev, V.N. Litunovskii, and V. Prokopenko, Electrical Explosion of Foils-I. *Sov. Phys. Tech. Phys.*, Vol. 22, No. 8, p. 950, 1977.
7 V.A. Burtsev, V. Litunovskii, and V. Prokopenko, Electrical Explosion of Foils-II. *Sov. Phys. Tech. Phys.*, Vol. 22, No. 8, p. 957, 1977.
8 I.F. Kvarkhtsava, A.A. Pliutto, A.A. Chernow, and U.V. Bondarenko, Electrical Explosion of Metal Wires. *Sov. Phys. JETP*, Vol. 3, No. 1, p. 40, 1956.
9 P.H. Ron, Rise and Delay Time of a Vacuum Gap Triggered by an Exploding Wire. Ph.D. dissertation, Indian Institute of Science, Bangalore, India, 1984.
10 S.U. Lebedev, Explosion of a Metal by an Electric Current. *Sov. Phys. JETP*, Vol. 5, No. 2, p. 243, 1957.
11 E. Nairne, *Philos. Trans. R. Soc. Lond.*, p. 79, 1774.
12 F.D. Bennett, *Sci. Am.*, Vol. 106, p. 103, 1962.
13 W.G. Chace, *Phys. Today*, Vol. 17, p. 19, 1964.
14 D.P. Ross and O.H. Zinke, in W.G. Chace and H. K. Moore, eds., *Exploding Wires*, Vol. 4, Plenum Press, New York, p. 147, 1968.
15 J.K. Trolan et al., in W.G. Chace and H. K. Moore, eds., *Exploding Wires*, Vol. 3, Plenum Press, New York, p. 361, 1964.
16 T.J. Tucker and F.W. Neilson, in W.G. Chace and H.K. Moore, eds., *Exploding Wires*, Plenum Press, New York, p. 73, 1959.
17 V.E. Scherrer, in W.G. Chace and H. K. Moore, eds., *Exploding Wires*, Vol. 1, Plenum Press, New York, p. 118, 1959.
18 F.D. Bennett, High Temperature Exploding Wires, in Rouse, ed., *Progress in High Temperature Physics and Chemistry*, Vol. 2, Pergamon Press, 1968.
19 M.G. Haines, *Proc. Phys. Soc. Lond.*, Vol. 74, p. 576, 1959.
20 Ch. Maisonnier, J.G. Linhart, and C. Gourlan, Rapid Transfer of Magnetic Energy by Means of Exploding Foils. *Rev. Sci. Instrum.*, Vol. 37, No. 10, p. 1380, 1966.
21 F.H. Webb, Jr., H.H. Hilton, P.H. Levine, and A.V. Tollestrup, The Electrical and Optical Properties of Electrically Exploded Wires, in W.G. Chase and H. K. Moore, eds., *Exploding Wires*, Plenum Press, New York, Vol. 3, p. 37, 1962.
22 J.N. Dimarco and L.C. Burkhardt, Characteristics of a Magnetic Energy Storage System Using Exploding Foils. *J. Appl. Phys.*, Vol. 41, No. 9, p. 3894, 1970.
23 H.C. Early and F.J. Martin, Method of Producing a Fast Current Rise from Energy Storage Capacitors. *Rev. Sci. Instrum.*, Vol. 36, No. 7, p. 1000, 1965.
24 R.E. Reinovsky, D.L. Smith, W.L. Baker, J.H. Degnan, R.P. Henderson, R.J. Kohn, D.A. Kloc, and N.F. Roderick, Inductive Store Pulse Compression System for Driving High Speed Plasma Implosions. *IEEE Trans. Plasma Sci.*, 10, No. 2, p. 73, 1982.
25 J. Salge, U. Braunsberger, and U. Schwarz, Circuit Breaking by Exploding Wires in Magnetic Energy Storage Systems, in W.H. Bostick, V. Nardi, and

O.S.F. Zucker, eds., *Energy Storage, Compression, and Switching*, Plenum Press, New York, p. 477, 1976.

26 I.M. Vitkovitsky, Fuses and Repetitive Current Interruption, in M. Kristiansen and K.H. Schoenbach, eds., *Proceedings of the ARO Workshop on Repetitive Opening Switches*, Tamarron, CO, DTIC # AD-A110770, 1981.

27 J. Benford, H. Calvin, I. Smith, and H. Aslin, High-Power Pulse Generation Using Exploding Fuses, in W.H. Bostick, V. Nardi, and O.S.F. Zucker, eds., *Energy Storage, Compression, and Switching*, Plenum Press, New York, p. 39, 1976.

28 L. Vermij, Short Fuse Elements Enclosed in a Small Slit. *International Symposium on Switching Arc Phenomena*, Lodz-Poland, p. 247, 1970.

29 R.K. Borisov, V.L. Budovich, and I.P. Kuzhekin, Restoration of the Dielectric Strength Following Explosion of a Wire. *Sov. Tech. Phys. Lett.*, Vol. 3, No. 12, p. 516, 1977.

30 D. Conte, R.D. Ford, W.H. Lupton, and I.M. Vitkovitsky, Two Stage Opening Switch Techniques for Generation of High Inductive Voltages. *Proceedings of the 7 Symposium on Engineering Problems of Fusion Research*, Vol. 2, p. 1066, 1977.

31 D. Conte, R.D. Ford, W.H. Lupton, and I.M. Vitkovitsky, TRIDENT: A Megavolt Pulse Generator Using Inductive Energy Storage. *Proceedings of the 2nd International Pulsed Power Conference*, p. 276, 1979.

32 B. Antoni, Y. Landure, and C. Nazet, The Commutation of the Energy Produced by a Helical Explosive Generator Using Exploding Foils, in W.H. Bostick, V. Nardi, and O.S.F. Zucker, eds., *Energy Storage, Compression, and Switching*, Plenum Press, New York, p. 481, 1976.

33 W.H. Lupton, D. Conte, R.D. Ford, P.J. Turchi, and I.M. Vitkovitsky, Application of Homopolar Generators for High Voltage Plasma Experiments. *Proceedings of the 7th Symposium on Engineering Problems of Fusion Research*, p. 430, 1977.

34 R.J. Commisso, R.F. Fernsler, V.E. Scherrer, and I.M. Vitkovitsky, Application of Electron-Beam Controlled Diffuse Discharges to Fast Switching. *Proceedings of the 4th International Pulsed Power Conference*, p. 87, 1983.

35 N. Shimomura, H. Akiyama, and S. Maeda, Compact Pulse Power Generator by an Inductive Energy Storage System with Two Staged Opening Switches. *IEEE Trans. Plasma Sci.*, Vol. 19, No. 6, p. 1220, 1991.

36 B.M. Kovalchuk and G. A. Mesyats, Rapid Cutoff of a High Current in a Electron-Beam-Excited Discharge. *Sov. Tech. Phys. Lett.*, Vol. 2, No. 7, p. 252, 1976.

37 R.O. Hunter, Electron Beam Controlled Switching. *Proceedings of the International Pulsed Power Conference*, p. IC8-1, 1976.

38 H.C. Harjes, K.H. Schoenbach, G. Schaefer, H. Krompholz, and M. Kristiansen, E-Beam Triode for Multiple Sub-Microsecond Pulse Operation. *Proceedings of the 4th Pulsed Power Conference*, p. 87, 1983.

39 R.F. Fernsler, D. Conte, and I.M. Vitkovitsky, Repetitive Electron-Beam Controlled Switching. *IEEE Trans. Plasma Sci.*, 8, No. 3, p. 176, 1980.

40 R.J. Commisso, R.F. Fernsler, V.E. Scherrer, and I.M. Vitkovitsky, Electron-Beam Controlled Discharges. *IEEE Trans. Plasma Sci.*, 10, No. 4, p. 241, 1982.

41 B.M. Kovalchuk, Yu. D. Korolev, V.V. Kremnev, and G.A. Mesyats, The Injection Thyratron: A Completely Controlled Ion Device. *Sov. Radio Eng. Electron. Phys.*, Vol. 21, No. 7, p. 112, 1976.

42 D.H. Douglas-Hamilton, Diffuse Discharge Production (Electron Beam Discharge Switch), in M. Kristiansen and K.H. Schoenbach, eds., *Proceedings of the ARO Workshop on Repetitive Opening Switches*, Tamarron, CO, DTIC # AD-A110770, 1981.

43 L.E. Kline, Performance Prediction for Electron-Beam Controlled On/Off Switches. *IEEE Trans. Plasma Sci.*, 10, No. 4, p. 224, 1982.

44 G.A. Mesyats, Electric Field Instabilities of a Volume Gas Discharge Excited by an Electron Beam. *Sov. Tech. Phys. Lett.*, Vol. 1, No. 7, p. 292, 1975.

45 D.H. Douglas-Hamilton and S.A. Mani, Attachment Instability in an Externally Ionized Discharge. *J. Appl. Phys.*, Vol. 45, No. 10, p. 4406, 1974.

46 S.A. Genkin, Yu. D. Kosolev, G.A. Mesyats, and V.B. Ponomarev, Criterion for the Injection Instability of a Volume Gas Discharge with External Ionization by an Electron Beam. *Sov. Tech. Phys. Lett.*, Vol. 8, No. 6, p. 279, 1983.

47 L.G. Christophorou, S.R. Hunter, J.G. Carter, S.M. Spyrou, and V.K. Lakdawala, Basic Studies of Gases for Diffuse–Discharge Switching Applications. *Proceedings of the 4th International Pulsed Power Conference*, p. 702, 1983.

48 G. Schaefer, K.H. Schoenbach, M. Kristiansen, and M. Krompholz, An Electron-Beam Controlled Diffuse Discharge Switch. *Proceedings of the XVII International Conference on Phenomena in Ionized Gases*, Budapest, Hungary, p. 626, 1985.

49 W.M. Parsons, A Comparison Between an SCR and a Vacuum Interrupter System for Repetitive Opening, in M. Kristiansen and K.H. Schoenbach, eds., *Proceedings of the ARO Workshop on Repetitive Opening Switches*, Tamarron, CO, DTIC # AD-A110770, 1981.

50 R.W. Warren, The Early Counterpulse Technique Applied to Vacuum Interrupters. *Proceedings of the 2nd International Pulsed Power Conference*, p. 198, 1979.

51 A.S. Gilmour, Jr., The Present Status and Projected Capabilities of Vacuum Arc Opening Switches. *Proceedings of the International Pulsed Power Conference*, p. IC1-1, 1976.

52 E.M. Honig, Vacuum Interrupters and Thyratrons as Opening Switches, in M. Kristiansen and K.H. Schoenbach, eds., *Proceedings of the ARO Workshop on Repetitive Opening Switches*, Tamarron, CO, DTIC # AD-A110770, 1981.

53 K. Watanabe, E. Kaneko, and S. Yanabu, Technological Progress of Axial Magnetic Field Vacuum Interrupters. *IEEE Trans. Plasma Sci.*, Vol. 25, No. 4, p. 609, 1997.
54 I.M. Vitkovitsky, D. Conte, R.D. Ford, and W.H. Lupton, Current Interruption in Inductive Storage System Using Inertial Current Source. *2nd International Conference on Energy Storage, Compression and Switching*, Venice, Italy, p. 953, 1978.
55 A.I. Pavlovskii, V.A. Vasyukov, and A.S. Russkov, Magnetoimplosive Generators for Rapid-Risetime Megaampere Pulses. *Sov. Tech. Phys. Lett.*, Vol. 3, No. 8, p. 320, 1977.
56 B.N. Turman and T.J. Tucker, Experiments with an Explosively Opened Plasma Switch, in M. Kristiansen and K.H. Schoenbach, C eds., *Proceedings of the ARO Workshop on Repetitive Opening Switches*, Tamarron, CO, DTIC # AD-A110770, 1981.
57 J.H. Goforth and R.S. Caird, Experimental Investigation of Explosive Driven Plasma Compression Opening Switches. *Proceedings of the 4th International Pulsed Power Conference*, p. 786, 1983.
58 R.A. Meger, J.R. Boller, D. Colombant, R.J. Commisso, G. Cooperstein, S.A. Goldstein, R. Kulsrud, J.M. Neri, W.F. Oliphant, P.F. Ottinger, T.J. Renk, J.D. Shipman, Jr., S.J. Stephanakis, F.C. Young, and B.V. Weber, Application of Plasma Erosion Opening Switches to High Power Accelerators for Pulse Compression and Power Multiplication. *Proceedings of the 4th International Pulsed Power Conference*, p. 335, 1983.
59 C.W. Mendel, Jr., S.A. Goldstein, and P.A. Miller, The Plasma Erosion Switch. *Proceedings of the of the International Pulsed Power Conference*, p. IC2-1, 1976.
60 W. Rix, P. Coleman, J.R. Thompson, D. Husovsky, P. Melcher, and R.J. Commisso, Scaling Microsecond-Conduction-Time Plasma Opening Switch Operation from 2 to 5 MA. *IEEE Trans. Plasma Sci.*, 25, No. 2, p. 169, 1997.
61 G.I. Dolgachev, L.R. Zukatov, and A.G. Ushakov, Study of Repetitive Plasma Opening Switch Generator Technology. *IEEE Trans. Plasma Sci.*, 26, No. 5, p. 1410, 1998.
62 H. Akiyama et al., Repetitively Operated Plasma Opening Switch Using Laser Produced Plasma. *Rev. Sci. Instrum.*, Vol. 68, No. 6, p. 2378, 1997.
63 D.C. Gates, Studies of a 60 kV Plasma Focus. *2nd International Conference on Energy Storage, Compression and Switching*, Venice, Italy, p. 329, December 5–8, 1978.
64 J.W. Mather, Dense Plasma Focus, in R.H. Lovberg and H.S. Griem, eds., *Methods of Experimental Physics, Volume 9–Part B: Plasma Physics*, Academic Press, Inc., p. 187, 1977.
65 F. Venneri, J. Mandrekas, and G. Gerdin, Preliminary Studies of the Plasma Focus as an Opening Switch. *Proceedings of the 4th International Pulsed Power Conference*, p. 350, 1983.

66 G.M. Molen, H.C. Kirbie, and J.L. Cox, A Plasma Focus Interrupting Switch for Use with Inductive Energy Storage, in M. Kristiansen and K.H. Schoenbach, eds., *Proceedings of the ARO Workshop on Repetitive Opening Switches,* Tamarron, CO, DTIC # AD-A110770, 1981.

67 J. Salge, Problems of Repetitive Opening Switches Demonstrated on Repetitive Operation of a Dense Plasma Focus, in M. Kristiansen and K.H. Schoenbach, eds., *Proceedings of the ARO Workshop on Repetitive Opening Switches,* Tamarron, CO, DTIC # AD-A110770, 1981.

68 V.L. Bailey, J.M. Creedon, L. Demeter, and D. Sloan, Opening Switches for Vacuum Inductive Storage, in M. Kristiansen and K.H. Schoenbach, eds., *Proceedings of the ARO Workshop on Repetitive Opening Switches*, Tamarron, CO, DTIC # AD-A110770, 1981.

69 R.L. Bowers et al., Initiation and Assembly of the Plasma in a Plasma Flow Switch. *IEEE Trans. Plasma Sci.*, 24, No. 2, p. 510, 1996.

70 J.M. Creedon, B.A. Lippmann, and V.L. Bailey, The Reflex Switch: An Opening Switch for Use with Magnetic Energy Storage, in W.H. Bostick, V. Nardi, and O.S.F. Zucker, eds., *Energy Storage, Compression, and Switching,* Plenum Press, New York, p. 1007, 1983.

71 B. Ecker, J. Creedon, L. Demeter, S. Glidden, and G. Proulx, The Reflex Switch: A High Current, Fast-Opening Vacuum Switch. *Proceedings of the International Pulsed Power Conference,* p. 354, 1983.

72 R.K. Parker and C.A. Kapetanakos, in M. Kristiansen and A.H. Guenther, eds., *The Generation of Intense Electron and Ion Beams.* Pulsed Power Lecture Series, Lecture No. 6, Texas Tech University, Lubbock, TX, 1981.

73 J.A. Pasour, R.A. Mahaffey, J. Golden, and C.A. Kapetanakos. Reflex Tetrode with Uni-Directional Ion Flow. *Phys. Rev. Lett.*, Vol. 40, No. 7, p. 448, 1978.

74 M.A. Lutz and R.J. Harvey, Feasibility of a High Average Power Crossed Field Closing Switch. *IEEE Trans. Plasma Sci.*, 4, No. 2, p. 118, 1976.

75 R.J. Harvey, M. Lutz, and H. Gallagher, Current Interruption at Powers Up to 1 GW with Crossed Field Tubes. *IEEE Trans. Plasma Sci.*, 6, No. 3, p. 248, 1978.

76 R.J. Harvey and M.A. Lutz, The Crossed Field Switch Tube: A Re-Usable Fuse. *Proceedings of the 5th Symposium on Engineering Problems of Fusion Research,* p. 670, 1973.

77 M.A. Lutz, Gridded Cross Field Tube. *IEEE Trans. Plasma Sci.*, 5, No. 4, p. 273, 1977.

78 R.J. Harvey and M.A. Lutz, High Power On–Off Switching with Crossed Field Tubes. *IEEE Trans. Plasma Sci.*, 4, No. 4, p. 210, 1976.

79 D.M. Goebel, Cold Cathode Pulsed Power Plasma Discharge Switch. *Rev. Sci. Instrum.*, Vol. 67, No. 9, p. 3137, 1996.

80 D.M. Goebel, R.C. Preschel, and R. Schumacher, Low Voltage Drop Plasma Switch for Inverter and Modulator Applications. *Rev. Sci. Instrum.*, Vol. 64, No. 8, p. 2312, 1993.

81 S. Humphries, M. Savage, S. Coffey, and D.M. Woodall, High Power Opening–Closing Switches Using Grid Controlled Plasmas. *IEEE Trans. Plasma Sci.*, 13, No. 4, p. 177, 1985.

82 K.H. Schoenbach, M. Kristiansen, and G. Schaefer, A Review of Opening Switch Technology for Inductive Energy Storage. *Proc. IEEE*, Vol. 72, No. 8, p. 1019, 1984.

83 H. Laquer, Superconductivity, Energy Storage and Switching, in W.H. Bostick, V. Nardi, and O.S.F. Zucker, eds., *Energy Storage, Compression, and Switching*, Plenum Press, New York, p. 279, 1976.

84 O.S.F. Zucker and W.H. Bostick, Theoretical and Practical Aspects of Energy Storage and Compression, in W.H. Bostick, V. Nardi, and O.S.F. Zucker, eds., *Energy Storage, Compression, and Switching* Plenum Press, New York, p. 71, 1976.

85 P.J. Turchi, Magnetoplasmadynamic and Hall Effect Switching for Repetitive Interruption of Inductive Circuits, in M. Kristiansen and K.H. Schoenbach, eds., *Proceedings of the ARO Workshop on Repetitive Opening Switches*, Tamarron, CO, DTIC # AD-A110770, 1981.

86 K.H. Schoenbach, M. Kristiansen, E.E. Kunhardt, L.L. Hatfield, and A.H. Guenther, Exploratory Concepts of Opening Switches, in M. Kristiansen and K.H. Schoenbach, eds., *Proceedings of the ARO Workshop on Repetitive Opening Switches*, Tamarron, CO, DTIC # AD-A110770, 1981.

87 A.H. Guenther, Optically Controlled Discharges, in M. Kristiansen and K.H. Schoenbach, eds., *Proceedings of the ARO Workshop on Repetitive Opening Switches*, Tamarron, CO, DTIC # AD-A110770, 1981.

88 I. Alexeff, Opening Switch Using Spoiled Electrostatic Confinement, in M. Kristiansen and K.H. Schoenbach, eds., *Proceedings of the ARO Workshop on Repetitive Opening Switches*, Tamarron, CO, DTIC # AD-A110770, 1981.

Iultigigawatt to Multiterawatt Systems

Iultigigawatt pulsed power technology concerns the techniques for generation electrical pulses of short duration and high peak power. The energy stored in primary medium is released on a short timescale and converted to produce the esired electrical pulse parameters by the techniques of pulse compression and ower amplification. The various primary mediums [1] in which the energy can stored are electric (½CV^2), magnetic (½LI^2), mechanical (½$m\omega^2$, where ω is gular momentum), and chemical ($V_{VOL}{\cdot}\rho{\cdot}\zeta$, where V_{VOL} is the volume, ρ is e mass density, and ζ is the mass energy density). The energy density of the ediums, the basic schemes of energy conversion, efficiency of energy conver- on, and the range of electrical pulse parameters are illustrated in Figure 6.1. ulsed power techniques are useful for the generation of powerful bursts of tense relativistic electron beams (IREB), light ion beams (LIB), lasers, free ectron lasers (FEL), high-power microwaves (HPM), X-ray flash, and neutron ams. These generators may be used to study high-temperature, high-density asma physics, inertial/magnetic confinement fusion schemes, advanced accel- ators, particle beam weapons, material processing, and numerous scientific, edical, defense, and industrial applications. Gigawatt pulsed power technol- y is also useful for the creation of intense transient electric, magnetic, and ctromagnetic fields and hence is widely used as a source of intense electro- gnetic interference, especially in the simulation of nuclear electromagnetic lse (EMP), system-generated EMP (SGEMP), and directed high power crowave (HPM) HPM beams.

Gigawatt technology can be adapted for the generation of intense power ses in single-pulse mode, burst mode, and repetitively pulsed mode. The veforms required for the generation of beams, fields, and plasma are nerally a unidirectional flat-top, double-exponential, and periodic damped ve. The unidirectional flat-top pulses with specifications of nanosecond/ gavolt/gigawatt are usually generated by cascaded stages of pulse compres- n and power amplification using capacitive storage, inductive storage, or a nbination of them. The most suitable means of generation of double-

dations of *Pulsed Power Technology*, First Edition. Jane Lehr and Pralhad Ron.

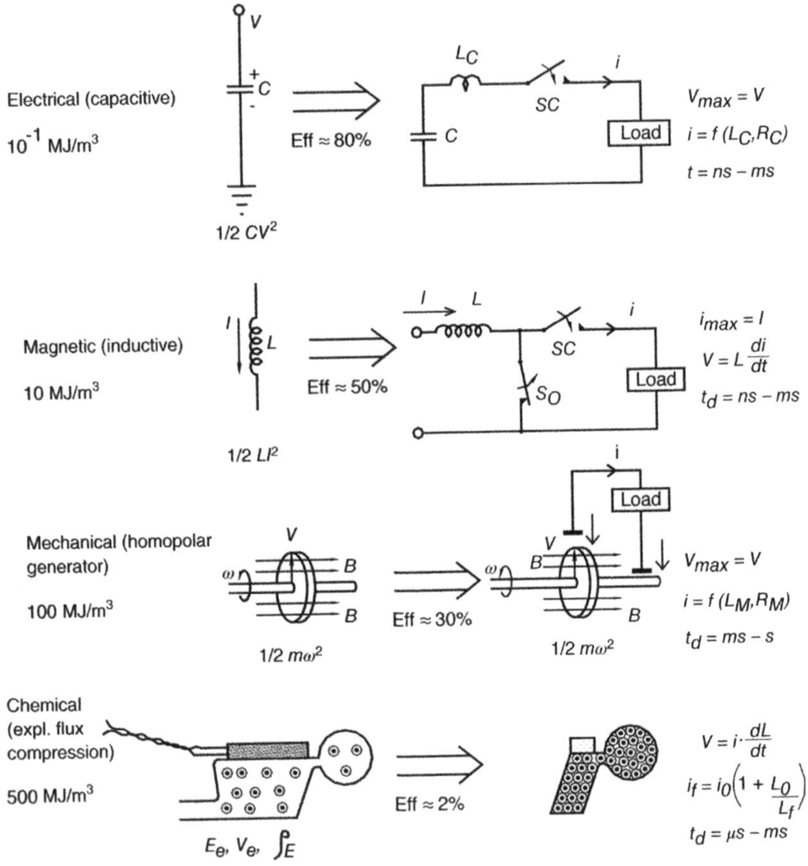

Figure 6.1 Energy storage media and electrical pulse delivery.

exponential pulse is Marx generator. Very high energy, up to many tens of megaelectronvolts (MeV), can be imparted to high-current electron beams by successively accelerating them through multiple gaps in induction linacs.

The design and performance of Marx generators, Tesla transformers, pulse forming lines (PFLs), spark gap switches, and opening switches, which are the basic building blocks of advanced pulse power systems, have been discussed in the earlier chapters. This chapter deals with the concepts and techniques involved in the design and construction of intense pulse power sources' primary capacitor storage, primary–intermediate capacitor storage, primary–intermediate–fast capacitor storage, primary inductor storage, cascaded inductor storage, magnetic pulse compression, inductive cavity cell multiplier, and induction linac.

6.1 Capacitive Storage

6.1.1 Primary Capacitor Storage

This is the simplest of the schemes in which a high-voltage pulse generator (HVPG) directly discharges into the load. The HVPG may take the form of a Marx generator, in which a number of capacitors are charged in parallel and discharged in series via the spark gap switches, as described in Chapter 1.

The generation of a flat-top pulse can be made directly with a Marx generator if a pulse forming network (PFN) is used in place of the conventional stage capacitor. The rise time of the output pulse is dependent on the overall *L/R* of the resultant equivalent circuit. It can be improved by reducing the physical size of the Marx generator and reducing the interelectrode spacing in the spark gaps, both of which reduce the overall Marx inductance. A typical schematic of a Marx-PFN generator, using a high-dielectric-strength oil and high-pressure gas as insulators, is shown in Figure 6.2. Transformer oil insulates the components like the PFNs and tail resistors allowing them to be placed closer together. The high-pressure gas is used as the dielectric medium for the spark gaps. Typical waveforms of a Marx-PFN generator are shown in Figure 6.3.

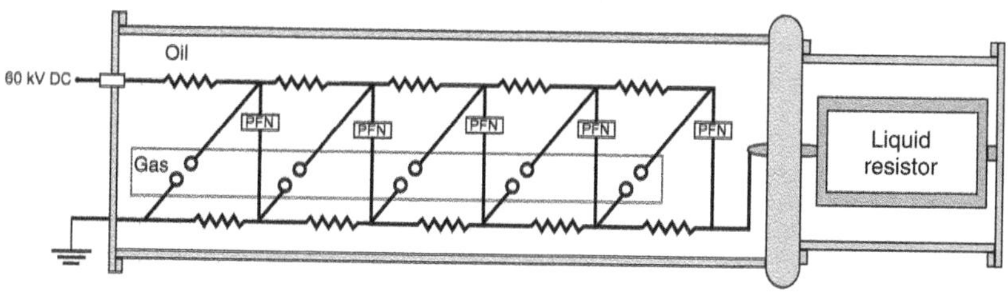

Figure 6.2 A general layout of a Marx generator with hybrid insulation of gas and oil.

Figure 6.3 Typical waveforms of a Marx-PFN generator.

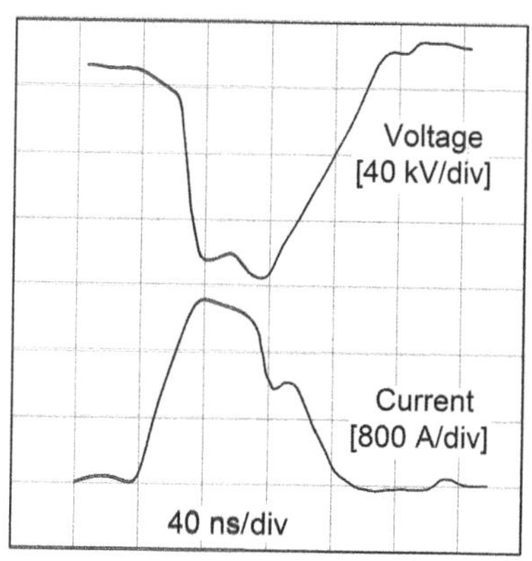

6.1.2 Primary–Intermediate Capacitor Storage

In this design [2], the primary capacitor pulse charges an intermediate capacitor, which then discharges into the load via a spark gap-based fast closing switch, shown schematically in Figure 6.4. The output pulse parameters of this system are determined entirely by the characteristics of the intermediate capacitor store. By designing the intermediate capacitor in the form of a coaxial PFL and choosing a good dielectric, it is possible to achieve good flat-top and current amplification. The intermediate capacitor store could take the form of a two-electrode coaxial PFL with oil or water dielectric, as shown in Figures 6.5 and 6.6 respectively. The intermediate store is in the form of a three-electrode

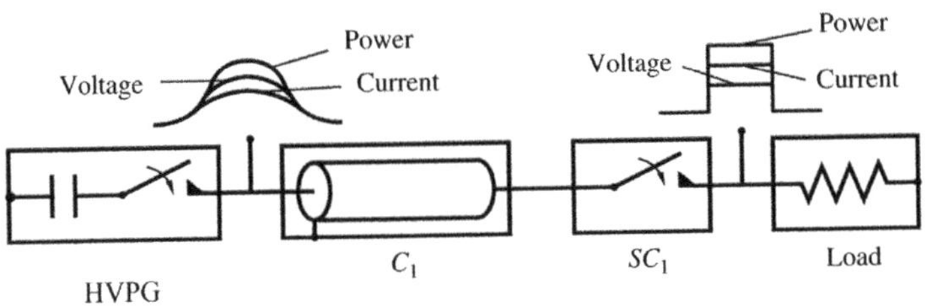

Figure 6.4 A primary energy store switches energy into an intermediate capacitor before being switched to the load.

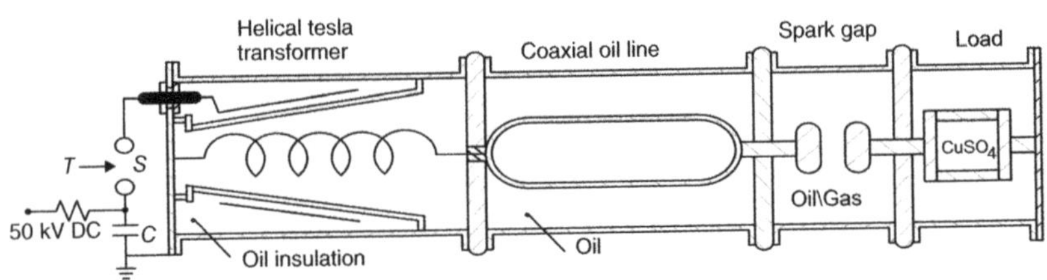

Figure 6.5 A high-voltage pulser based on a helical Tesla transformer as primary storage and a coaxial PFL as an intermediate energy storage.

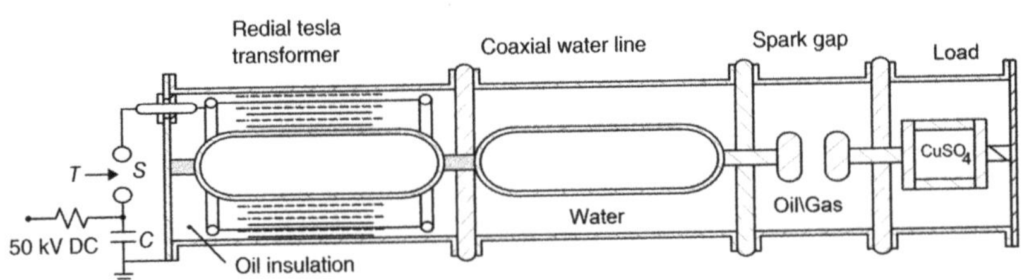

Figure 6.6 A high-voltage pulser based on a radial Tesla transformer as primary storage and a coaxial PFL as an intermediate energy storage.

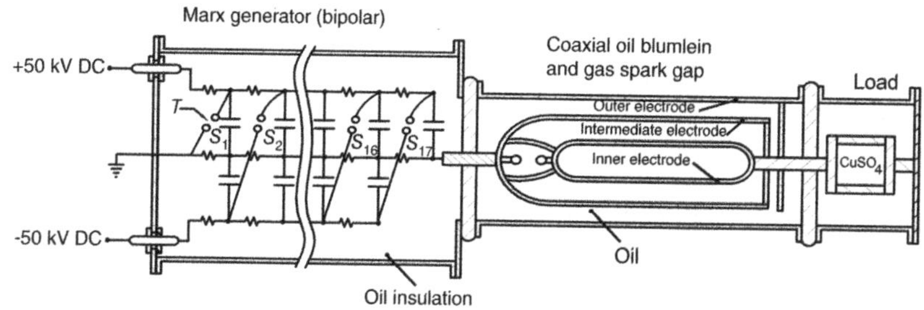

Figure 6.7 A high-voltage pulser based on a Marx generator as primary storage and a coaxial Blumlein as intermediate energy storage.

coaxial Blumlein using oil dielectric, as shown in Figure 6.7. The primary energy stores used are helical Tesla transformers (Figure 6.5), radial Tesla transformers (Figure 6.6), and Marx generators (Figure 6.7).

6.1.3 Primary–Intermediate–Fast Capacitor Storage

In the circuit topology in Figure 6.8, the intermediate capacitor store consisting of a coaxial capacitor C_1 and a closing switch SC_1 charges another coaxial capacitor C_2, which in turn discharges into the load via a fast-closing switch SC_2. Because C_2 is charged on a submicrosecond timescale, voltage appears across C_2 and SC_2 for only a very short time. Thus, higher electric fields can be used for C_2, with the actual value depending on the characteristics of the specific dielectric used in C_2. Using high breakdown strength insulators enables the reduction of the physical cross section of C_2, which in turn lowers its output impedance $\sqrt{L/C}$, and simultaneously enhances the current output V/Z. Deionized water, with a dielectric constant of ~80 and a high electrical breakdown strength, is a popular choice as a dielectric for the fast capacitor, especially when a very low impedance is needed. The ideal waveforms for pulsed voltage, current, and power at the various points are indicated in Figure 6.8.

The individual components in a pulsed power system based on primary–intermediate–fast capacitor storage have evolved substantially leading to peak powers greater than 1 TW at pulse durations less than 50 ns. The requirements

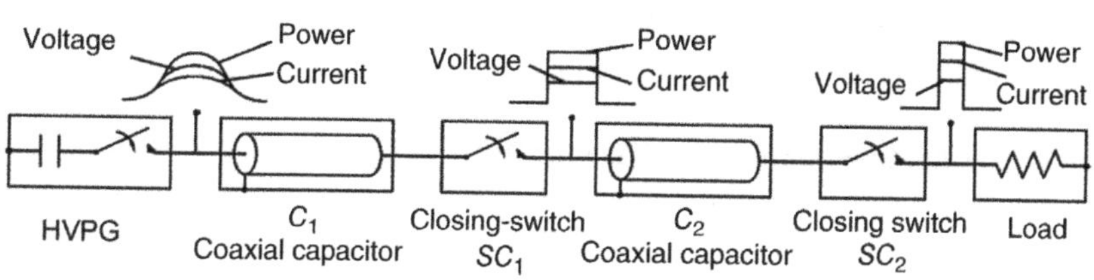

Figure 6.8 Primary–intermediate–fast capacitor storage schematic.

of individual components like Marx generators [3], pulse forming lines [4], and spark gaps [5] for an optimum performance from the overall system are discussed below.

6.1.3.1 Fast Marx Generator

In order for the Marx generator to charge the intermediate PFL as quickly as possible, the overall inductance of the Marx generator should be minimized. The overall inductance in a Marx generator equals to the summation of inductances contributed by individual components such as energy storage capacitors, spark gaps, and the connecting leads.

6.1.3.1.1 Reduction of Capacitor Inductance

The inductance per stage contributed by the energy storage capacitors can be reduced by (i) using capacitors possessing low self-inductance and low-inductance output lead configurations like coaxial or strip line, and (ii) connecting smaller capacitors in parallel rather than a single large capacitor.

6.1.3.1.2 Reduction of Spark Gap Inductance

The inductance of a spark gap is typically dominated by the inductance of the arc channel. Because there is no control over the diameter of the arc channel, the spark gap inductance can be reduced by reducing the length of the arc channel. This can be achieved by using a dielectric medium of high dielectric strength, such as a liquid or sulfur hexaflouride (SF6) SF_6 gas at high pressures. The inductance due to the spark channel can also be reduced by introducing multiple arc channels in parallel, as discussed in Chapter 4.

6.1.3.1.3 Reduction of Connecting Lead Inductance

The inductance contributed by the connecting leads of the Marx generator can be minimized by reducing the number of stages, and hence the overall length, but this mandates increasing the charging voltage per stage. Bipolar charging of the Marx may be considered, as well as using a high-dielectric-strength insulator to reduce the spacing between adjacent components or multiple folding of the Marx mounting configuration to increase the width at the expense of length. This contributes to lowered overall inductance due to currents flowing in opposite directions in the system.

6.1.4 Parallel Operation of Marx Generators

After incorporating the above measures, the overall Marx inductance can be further reduced by connecting more Marx generators in parallel. This, of course, needs careful design to obtain low jitter in the firing of the multiple

Marx generators acceptable for efficient synchronous operation. The following are the general performance [5] parameters achieved in an optimally designed fast Marx generator:

- 60–100 nH/stage in smaller Marx and 350–400 nH/stage in larger Marx generators.
- Erection time of less than 50 ns/MV.
- Erection jitter of 5 ns.
- Triggering range of up to 40% of self-breakdown voltage.
- Overall voltage gradients of 1.5 MV/m along the physical length of the Marx bank.

6.1.5 Pulse Forming Line Requirements for Optimum Performance

For its optimum performance, the PFL has to meet the requirements of high peak power delivery, low impedance, and large pulse time compression (PTC). These parameters are determined by the physical dimensions of the PFL and choice of an appropriate dielectric.

6.1.5.1 Peak Power Delivery into a Matched Load

For a two-electrode coaxial PFL, the voltage delivered to the load is one-half the charge voltage of the PFL. Therefore, the peak power (P) delivered to the load is given by

$$P = \frac{V_L^2}{Z_0} = \frac{V_0^2}{4Z_0} \tag{6.1}$$

where V_0 is the maximum voltage to which the PFL is charged, Z_0 is the characteristic impedance of the PFL, and Z_ℓ (=Z_0) is the matched load impedance.

The maximum voltage to which the PFL can be charged is given by

$$V_0 = E_{BD} \cdot R_1 \cdot \ln \frac{R_2}{R_1} \tag{6.2}$$

where E_{BD} is the maximum breakdown stress in the dielectric, R_1 is the radius of inner cylinder of the PFL, and R_2 is the radius of outer cylinder of the PFL.

The impedance of the PFL is given by

$$Z_0 = \frac{1}{2\pi} \cdot \left(\sqrt{\frac{\mu}{\varepsilon}}\right) \cdot \ln \frac{R_2}{R_1} \tag{6.3}$$

Assuming $\mu_r = 1$, (6.3) reduces to

$$Z_0 \approx \frac{60}{\sqrt{\varepsilon_r}} \cdot \ln \left(\frac{R_2}{R_1}\right) \tag{6.4}$$

where ε_r is the relative permittivity of the dielectric.

Substituting (6.3) and (6.2) into (6.1),

$$P = \left(\frac{\pi}{2} \cdot \sqrt{\frac{\varepsilon}{\mu}}\right) \cdot E_{BD}^2 \cdot R_1^2 \cdot \ln\left(\frac{R_2}{R_1}\right) \tag{6.5}$$

Simplifying,

$$P = \frac{\pi}{2} \cdot \frac{E_{BD}^2 \cdot \sqrt{\varepsilon_r} \cdot R_1^2}{377} \cdot \ln\left(\frac{R_2}{R_1}\right) \tag{6.6}$$

Hence, from Equation 6.6, the requirements to be satisfied for high peak power delivery are (i) large inner diameter of the PFL, (ii) large relative permittivity of the dielectric used in the PFL, (iii) large breakdown strength of the dielectric used in the PFL, and (iv) optimum ratio of R_2/R_1, which then determines the annular gap $(R_2 - R_1)$.

A good choice of dielectric would therefore be water having a relative permittivity of $\varepsilon_{r=}80$. High breakdown strength E_{BD} can be obtained by charging the PFL in the shortest possible time, as determined by the breakdown characteristics of the dielectric.

Since mineral oil possesses much larger E_{BD} for all charging times and shows polarity independence, it is used for very large voltages in preference to water. For very high powers, the PFLs in the final stage can be connected in parallel configurations. However, this would require the use of specially designed multichannel spark gaps as their output switches with the low jitter for synchronous operation of the PFLs.

6.1.5.2 Low-Impedance PFLs

Low-impedance PFLs can be made by choosing a dielectric having large ε_r and reducing the ratio R_2/R_1 that also tends to reduce the annual gap distance $R_2 - R_1$. Electrical breakdown considerations must be evaluated for a trade-off between low Z_0 and high V_0. Fast charging, however, yields a high value of E_{BD} and allows considerable reduction in the ratio of R_2/R_1. Fast charging can also be attained by inserting an intermediate energy storage stage between the primary energy storage (e.g., a Marx generator) and the final energy storage stage (fast PFL). This intermediate stage enables rapid charging of the PFL and hence high E_{BD}.

In cascaded capacitor storage, progressive increase of pulsed power amplification along the stages is mainly due to gradual impedance reduction along the advancing stages of the cascades; hence, Z_0 is an important design parameter.

6.1.5.3 Pulse Time Compression

The output pulse duration T_p of a PFL is dependent only on the relative dielectric constant ε_r of the insulating media and the length ℓ of the PFL:

$$T_p = \frac{2\ell}{v_{pp}} = \frac{2 \cdot \ell \cdot \sqrt{\varepsilon_r}}{c} = 6.67 \cdot \ell \cdot \sqrt{\varepsilon_r}\,\frac{ns}{m} \tag{6.7}$$

The *PTC* for a PFL is given by

$$PTC = \frac{\text{Charging}\ time}{\text{Output}\ pulse\ duration} \tag{6.8}$$

6.2 Inductive Storage Systems

6.2.1 Primary Inductor Storage

In the approach shown in Figure 6.9, a high-current pulse generator (HCPG) charges an inductor L to a peak current I_m in a time t_m through an initially conducting opening switch SO. At the moment t_m, switch SO opens and the energy stored in the inductor can be transferred to the load via a closing switch SC.

In any inductive storage system, the opening switch has to carry out the dual and conflicting functions of efficiently conducting the charging current of the storage inductor over a relatively long period of time before opening and then efficiently interrupting the peak current in the shortest possible time during the opening phase. A number of opening switch configurations [3] can be used, with the exploding foil switch, electron beam-controlled switch, and plasma erosion switch being the dominant technologies. An exploding wire that usually works as an opening switch can also be used as a closing switch [4]. It is possible to develop a single device [5] capable of carrying out the dual function of the opening and closing switches.

6.2.2 Cascaded Inductor Storage

The primary inductor storage consisting of L_1, SO_1, and SC_1 charges another inductor L_2 through an initially conducting opening switch SO_2, shown in Figure 6.10 [3]. The current is commuted to the load via the closing switch SC_2 when the SO_2 opens. The advantage of a cascaded inductive energy store accrues from the fact that an opening switch that can carry out the function of

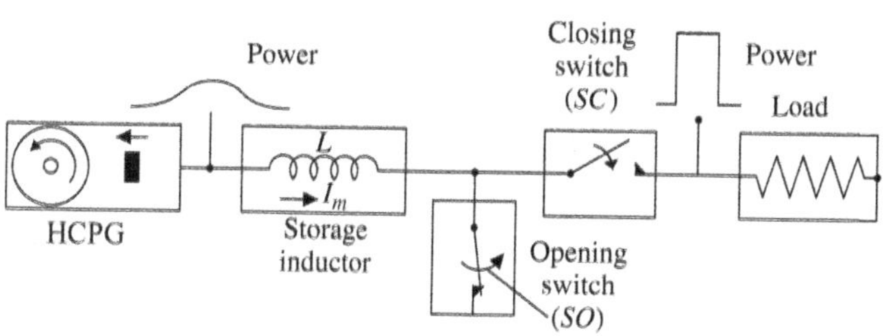

Figure 6.9 Primary inductor storage schematic.

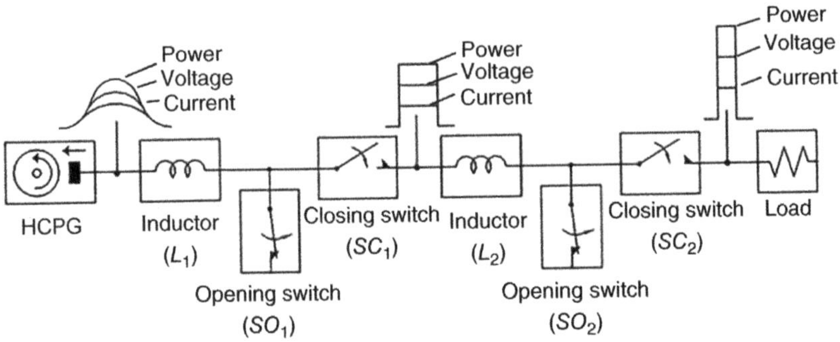

Figure 6.10 Cascaded inductive storage schematic.

conduction efficiently is necessarily inefficient in interrupting the current. The first-stage switch SO_1 is designed for high conduction current, long conduction time, and low conduction loss. It need not have a small opening time. The second-stage switch SO_2 has the features of high-conduction current and small opening time, but it need not have a long conduction time. As a result of this combination of switches SO_1 and SO_2, the overall system acquires the capabilities of high-voltage generation, high-power amplification, and large pulse compression. In a typical configuration, the first-stage switch SO_1 could be an exploding foil switch and the second-stage switch SO_2 a plasma erosion switch. The waveforms at various points in a cascaded inductive energy store are illustrated in Figure 6.10.

A single-stage capacitive energy store cannot deliver an output voltage larger than its charge voltage but can deliver a current much larger than its input charging current. Inductive energy store can produce significant gains in output voltage but cannot deliver a current larger than its charging current. Substantial peak power gain with pulse compression is possible by using a combination of capacitive and inductive energy storage. A capacitor-inductor storage system is shown in Figure 6.11, where the capacitor storage, consisting of a high-voltage

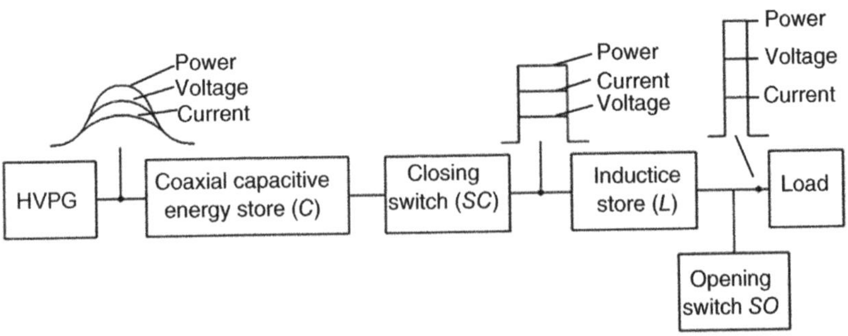

Figure 6.11 Capacitive-inductive storage schematic.

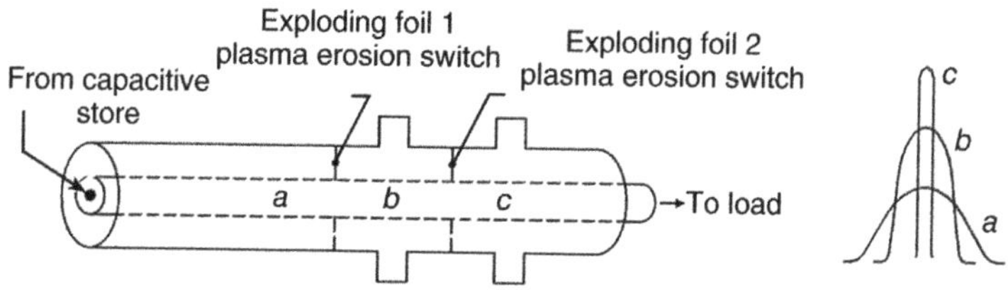

Figure 6.12 Schematic of a capacitor-cascaded inductor storage.

pulse generator (HVPG), coaxial capacitor *C*, and a closing switch *SC*, delivers the energy to an inductive storage unit composed of an inductor *L* and an opening switch *SO*. The inductive store delivers the energy to the load at an optimum time when the *SO* opens. The ideal waveforms of voltage, current, and power at various points are illustrated in the figure. A practical implementation of this scheme is shown in Figure 6.12.

6.3 Magnetic Pulse Compression

A schematic of a two-stage magnetic pulse compression circuit (MPC) and the resulting waveforms are shown in Figures 6.13 and 6.14, respectively. In typical design implementations, $C_0 = C_1 = C_2 = C_3 = C$. When the switch S is closed, the long-duration input pulse on C_1, V_{C1} of duration T_1, is compressed to V_{C2} with duration T_2 by the first compression stage comprised of C_1L_1 and further compressed to Vc_3 with duration T_3 by the second compression stage C_2L_2. The pulsed power amplification occurs as a result of successive enhanced currents i_1, i_2, and i_3. The sequential process [6] of stage-to-stage transfer of energy and resulting pulse compression and power amplification are depicted in Figure 6.14. For efficient pulse compression and pulsed power amplification, the following criteria should be satisfied: (a) The nonlinear saturable inductors should reach saturation at input volt-seconds corresponding to $\int_o^T V_c\,dt$, where T is the time at which V_c reaches the maximum voltage V_0. (b) The saturated inductance values of nonlinear inductors L_1 and L_2 should satisfy $L_{2s} \ll L_{1s}$.

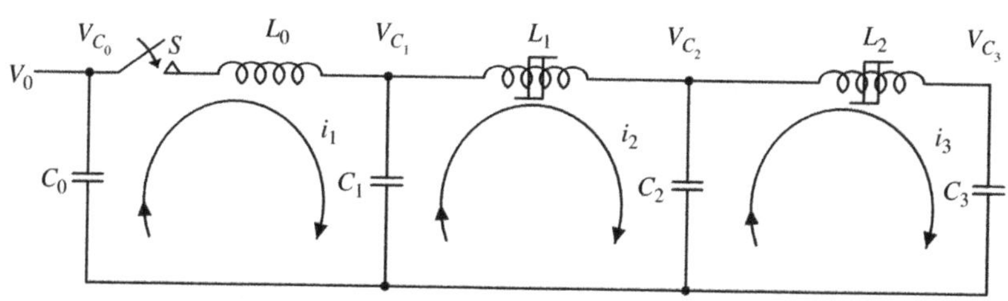

Figure 6.13 Schematic of a two-stage magnetic pulse compressor.

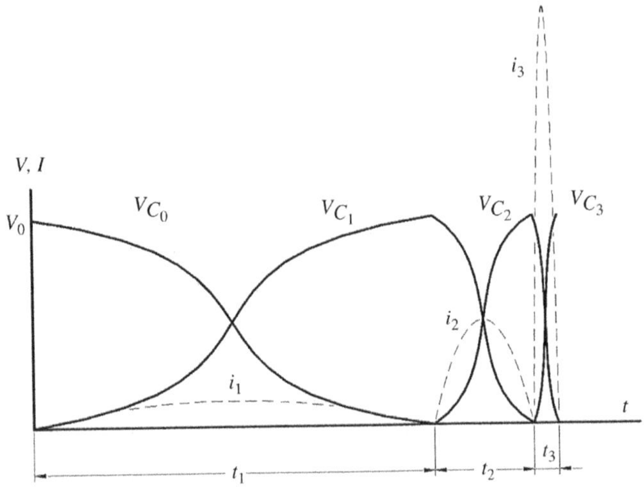

Figure 6.14 Waveforms for a two-stage magnetic pulse compression circuit.

(c) The unsaturated inductance values of nonlinear inductors L_1 and L_2 should be high enough to provide sufficient isolation between stages and thus limit the prepulse currents in the load.

The relations for V_c and T are given below [7,8]:

$$V_{c_1} = \frac{V_0}{2}(1 - \cos \omega_1 t) \tag{6.9}$$

where $\omega_1 = \dfrac{1}{\sqrt{L_0(C/2)}}$; $\quad T_1 = \dfrac{\pi}{\omega_1}$;

$$V_{c_2} = \frac{V_0}{2}(1 - \cos \omega_2 t) \tag{6.10}$$

where $\omega_2 = \dfrac{1}{\sqrt{L_{1s}(C/2)}}$; $\quad T_2 = \dfrac{\pi}{\omega_2}$; and

$$V_{c_3} = \frac{V_0}{2}(1 - \cos \omega_3 t) \tag{6.11}$$

where $\omega_3 = \dfrac{1}{\sqrt{L_{2s}(C/2)}}$; $\quad T_3 = \dfrac{\pi}{\omega_3}$.

The schematic of a typical MPC for generating a flat-top pulse is shown in Figure 6.15. Depending on the magnitudes of voltage amplification and pulse compression, the required numbers of pulse transformers and MPC stages are chosen. The first stage is generally a command resonant charging system comprised of a pulsed power source PPS, a charging inductor L_0, and an intermediate storage capacitor C_1. This stage gives a voltage gain of n ($\leq$2), thus charging C_1 to a voltage of nV_0, where V_0 is the magnitude of peak voltage of the

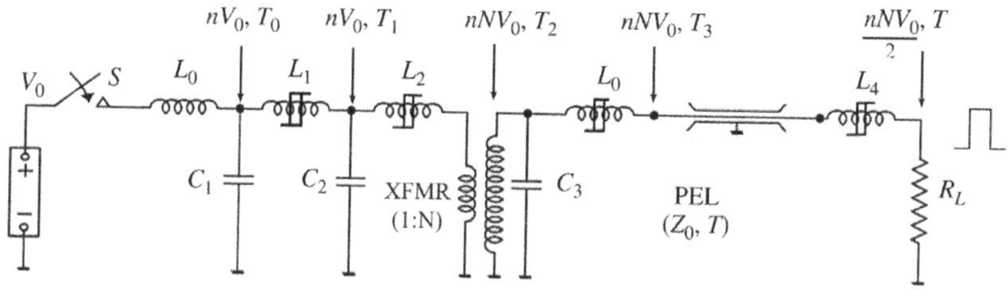

Figure 6.15 Typical magnetic pulse compression circuit for generation of rectangular pulses.

PPS. The three stages of MPC comprised of C_1L_1, C_2L_2, and C_3L_3 reduce the input pulse duration from T_0 to T_3 and the pulse transformer increases the magnitude of voltage from nV_0 to NnV_0. The output stage comprised of a PFL and a magnetic switch L_4 deliver a flat-top pulse of magnitude $NnV_0/2$ of duration T into the load.

The driver pulsed power source can be a Marx generator, Tesla transformer, or a solid-state voltage generator. The switches used in these may be spark gaps, thyratrons, thyristors, or insulated gate bipolar transistors (IGBTs). The magnetic switches employ a variety of core materials such as ferrites, FeNi, and FeZn and amorphous magnetic alloys such as Metglas® that have wide and square B–H loop characteristics. For efficient performance, the material should have a high unsaturated permeability, low saturated permeability, high saturated flux density, low hysteresis loss, and low eddy current loss. The magnetic core is usually in the form of a toroid wound with thin tapes of magnetic alloy, insulated between the overlapping turns. To minimize the size and cost of the magnetic core, it is important to choose an optimum number of MPC stages.

A realized pulse power system with output parameters of 200 kV, 5 kA, 50 ns, and 10–100 pps is shown in Figure 6.16 [9]. An input pulse voltage with parameters of 1 kV, 10 μs generated by a solid-state driver is amplified in magnitude and compressed in duration to ultimate parameters of 200 kV, 5 kA, 50 ns through a pulsed power system comprised of three magnetic compression devices, two pulse transformers, intermediate water dielectric capacitors, PFLs with water dielectric, magnetic compression switch, and an induction cavity multiplier.

6.4 Inductive Voltage Adder

An inductive cavity adder is shown in Figure 6.17. The pulsed output voltage from a system comprised of a HVPG, an intermediate coaxial capacitor store, and a fast-closing switch is carried by N coaxial cables and fed to N coaxial inductive cavity cells. The cavity cells, connected in series in an adder

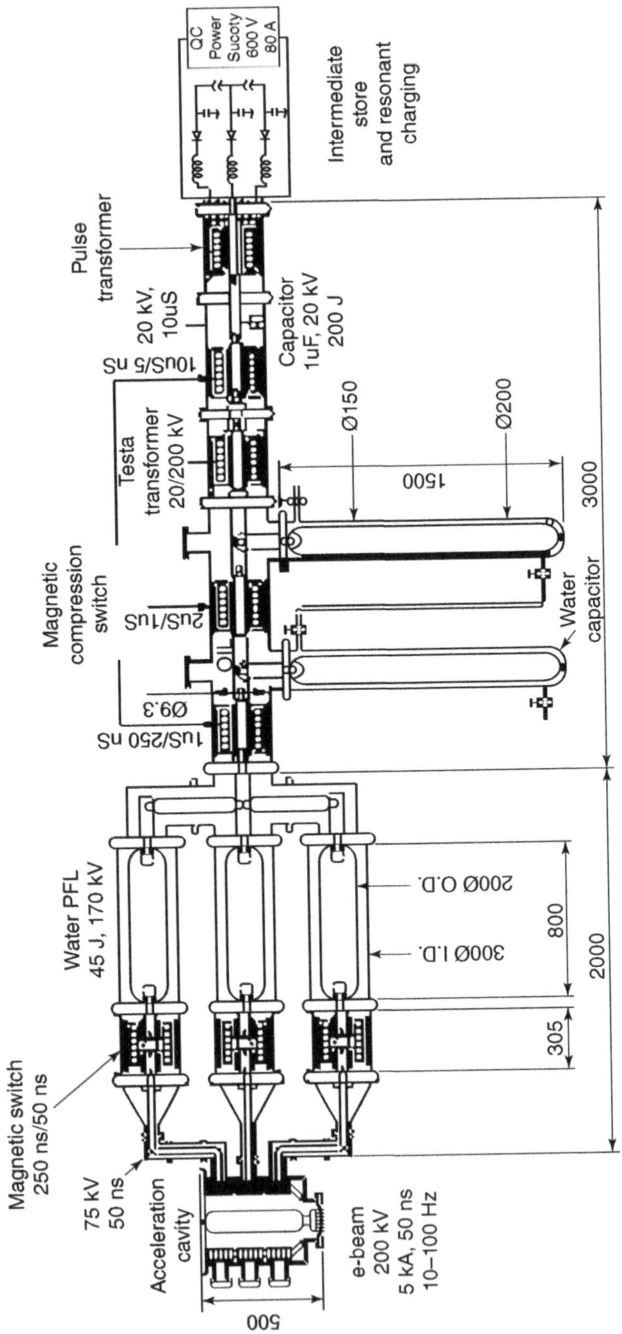

Figure 6.16 A pulsed power system based on magnetic compression [9].

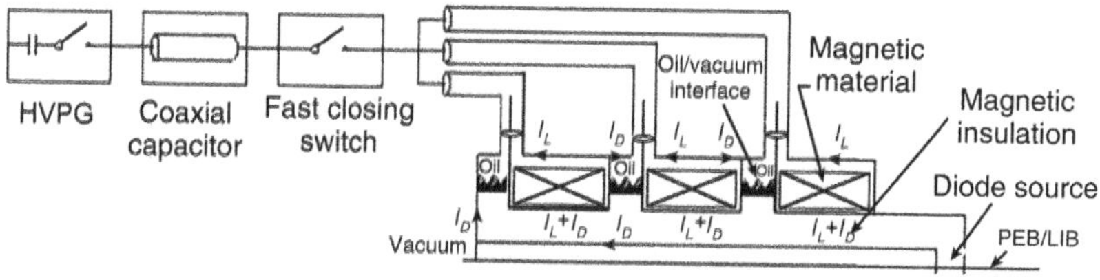

Figure 6.17 The circuit schematic of an inductive voltage adder.

configuration, are loaded with soft magnetic alloy cores having high permeability to provide good inductive isolation [10,11]. For a symmetrical feed, more than one cable may be fed into a cavity cell. Under the conditions of proper impedance matching between the driver and the cavity cells, the net output voltage of the cavity cells, which is ideally NV_0, is delivered to the load, which in Figure 6.17 is a relativistic electron beam diode. Figure 6.17 shows the three-cavity cells and the distribution of load current I_D and inductor current I_L in the various parts of the circuit and the oil–vacuum insulating interface. Under high load current, the azimuthal magnetic field produced by the coaxial central conductor will considerably enhance the vacuum insulation, thus resulting in a very compact magnetically insulated vacuum transmission line.

6.5 Induction Linac Techniques

In induction linacs, which are basically multigap accelerating structures, the high-current charged particles gain incremental energy while passing through each successive gap. The output beam energy ξ_0 is given by $\xi_0 = \xi_i + N\xi_g$, where ξ_i, N, and ξ_g represent the injection energy, number of accelerating gaps, and energy gain per gap, respectively. By proper phasing of the accelerating field in relation to the charged particle velocity, energies up to many tens of megaelectronvolts can be achieved. Induction accelerators—also called linacs—can be classified into (i) magnetic core induction [12–16], (ii) pulse line induction [15–20], and (iii) autoaccelerator [13,14,19,21]. Compact high-current induction accelerators where the electron beam makes multiple passes [22] through accelerating gaps have been developed. A tailored magnetic field is used to bend the electron beams for re-entry into the gap. The output beam energy ξ_0 for multipass accelerators is given by $\xi_0 = \xi_i + N'\xi_g$, where N' represents the number of passes per gap.

6.5.1 Magnetic Core Induction Linacs

Figure 6.18 illustrates the principle of electron beam acceleration by magnetic core induction linac.

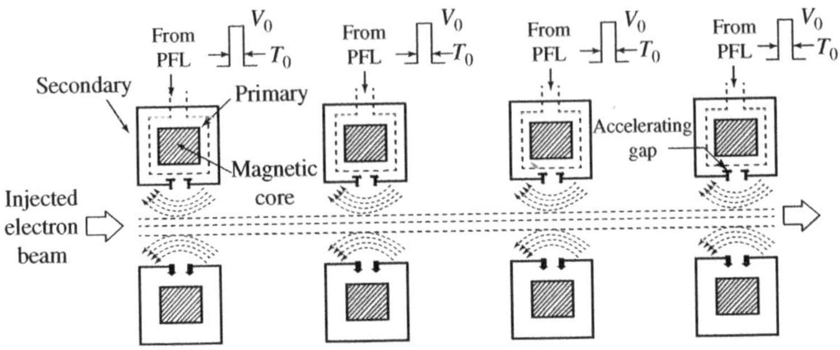

Figure 6.18 Principle of magnetic core induction linac.

A typical cross section of an induction accelerator cell is shown in Figure 6.19. The PFL discharges into a single-turn primary encircling a toroidal core. The PFL output pulse with parameters of V_{out} and T_0 produces a linearly rising current in the primary, which is given by $I = V_{out}/L$, where L represents the inductance of the primary. If the magnetic core is biased to a flux density of $+B_{sat}$ before the PFL is energized, the time-varying flux density produced in the core will change from $+B_{sat}$ to $-B_{sat}$ if the core area is chosen to be $V_{out}T_0/2B_{sat}$. Since the secondary completely encircles the core, the voltage induced in the secondary will be again V_{out}, with a pulse duration of T_0. This secondary voltage appears across the gap, which causes acceleration of the injected beam.

Typical characteristics of flux swing, induced gap voltage, and B–H characteristics are shown in Figure 6.20. By arranging a proper coordination between the phase of the accelerating field and the velocity of the particle, the particle

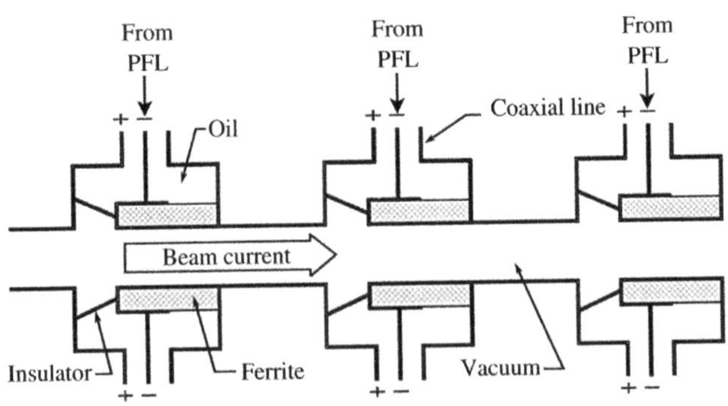

Figure 6.19 Typical cross section of accelerator cells.

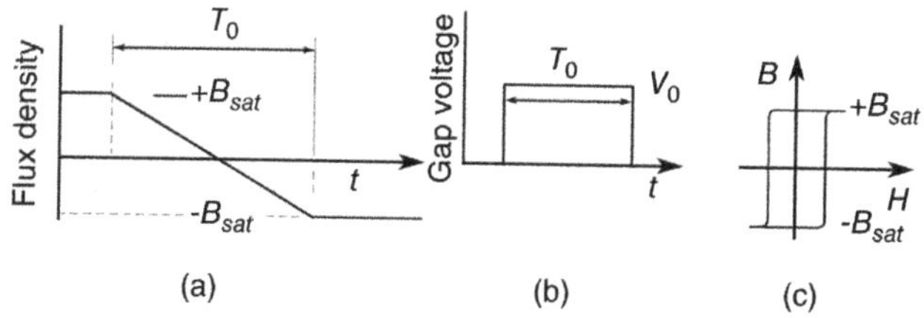

Figure 6.20 Characteristics of flux swing (a), induced gap voltage (b), and *B–H* characteristics (c).

gains energy at every gap. Ferrite having the properties of a relative permeability $\mu_r = 1000$, $\varepsilon_r = 10$ is generally used for pulse durations shorter than 50 ns, whereas ferromagnetic materials like silicon–iron, Metglas, and supermalloy are used for longer pulse durations up to 1 µs. Surface field up to 300 kV/cm can be achieved for accelerating gaps for an electropolished metal surface. Enhanced surface flashover strength for the oil–vacuum interface insulator can be obtained by inclining the insulator surface by 45° to the *E*-field lines, as shown in Figure 6.20. The large induction linac, the Advanced Test Accelerator (ATA), produces an electron beam of 50 MeV at a current of 10 kA with a pulse duration of 50 ns. With an injector stage of 2.5 MeV produced by a mini-induction linac of 10 gaps each with 250 kV, ATA incorporates 190 output stages each of 250 kV. The PFL for ATA is a radial line oil Blumlein energized by a Tesla transformer.

6.5.2 Pulsed Line Induction Linacs

The pulsed line induction linacs, also known as air core induction linacs, are employed for the generation of pulsed electron beams with durations less than 50 ns. In comparison to magnetic core linacs, much larger currents can be obtained. A four-stage accelerator based on radial lines is shown in Figure 6.21.

The center disk conductor is pulse charged to a voltage V_0. During the charging time, the axial components of electric fields in each of the two half-gaps, g_1 and g_1', of the accelerating gap are opposite in nature and present a nonaccelerating mode. When the spark gap switch *S* is triggered, the *E*-field in g_1'goes to zero, whereas the *E*-field in g_1 varies, as shown in Figure 6.22a, for the matched condition $Z_1 = Z_2 = Z_0$ and the configuration is transformed into an acceleration mode. The time period $2T$ in Figure 6.22a represents the transit time of the voltage step pulse to travel from g_1' to g_1 via the transition region III. The period $2T$ equals $2\ell/V_{pp}$, where ℓ represents the length of a single PFL and v_{pp} the wave propagation velocity. The waveform of Figure 6.22a can be readily synthesized by taking into consideration the incident pulses, reflected pulses, and their interaction with the originally existing voltage V_0. Any one of the

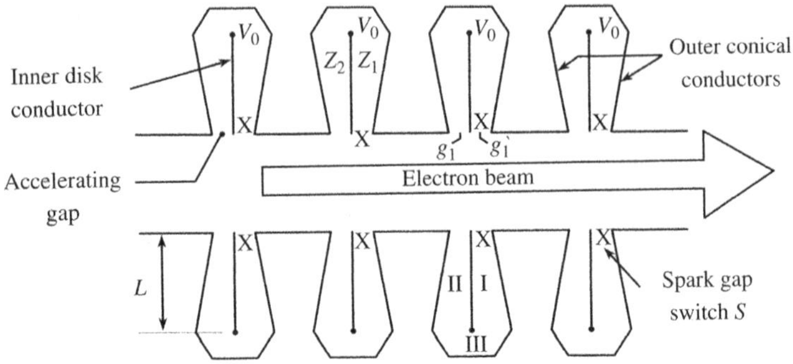

Figure 6.21 Schematic of a four-stage pulsed line induction LINAC.

pulses shown in Figure 6.22a can be used for acceleration purposes. However, because of larger pulse duration, the second pulse enables full extraction of the stored energy. By correct matching of the e-beam impedance V_g/I_{eb} to the PFL impedance, that is, $Z_1 = Z_2 = Z_0 = V_g/I_{eb}$, the accelerator voltage gets reduced to $V_0/2$, as shown in Figure 6.22b. Enhanced acceleration voltage for a given PFL charging voltage can be obtained in asymmetric lines and by properly designing for the transition region III between the two PFLs.

A typical asymmetric radial line and corresponding waveforms are shown in Figure 6.23. Some other examples of pulse lines adaptable for pulsed line induction linacs are shown in Figure 6.24 with (i) cylindrical coaxial geometry, (ii) radial line with a modified position of switch, and (iii) radial line with a modified transition zone III [18].

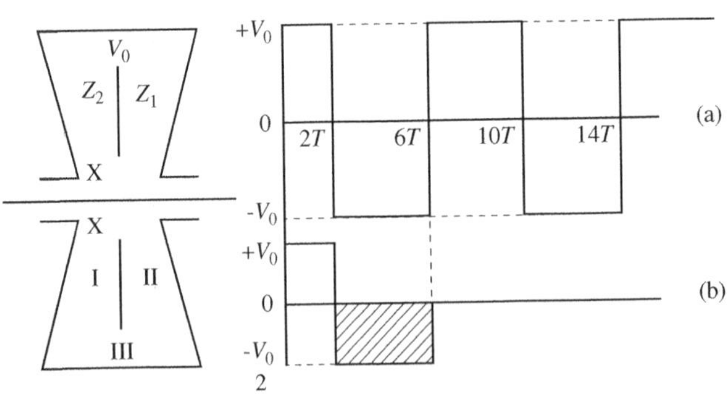

Figure 6.22 Symmetrical radial line PFL and waveforms for $Z_1 = Z_2 = Z_0$. Open circuit (a) with an e-beam (b). [18]

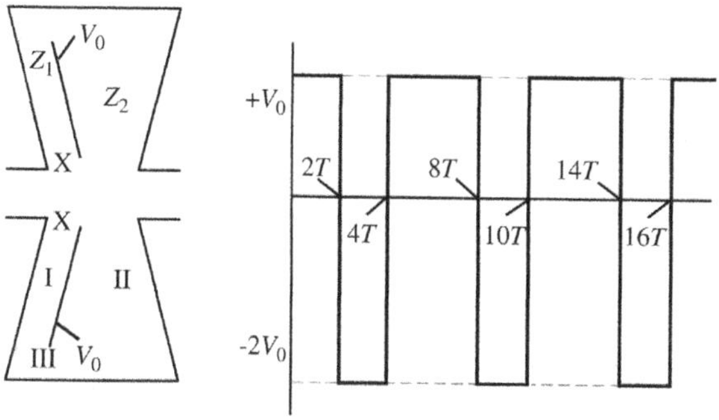

Figure 6.23 Asymmetric radial line PFL and waveforms for $Z_1 = Z_0$ and $Z_2 = 3Z_0$. [18]

An electron beam with parameters of 13 MeV, 50 kA, 20/40 ns has been obtained from LIU-10 with 52 cavities, water dielectric, and an initial charge voltage of 500 kV. The electron beam obtained from RADLAC with four oil-filled cavities has the parameters of 9.8 MeV, 28 kA, 12 ns.

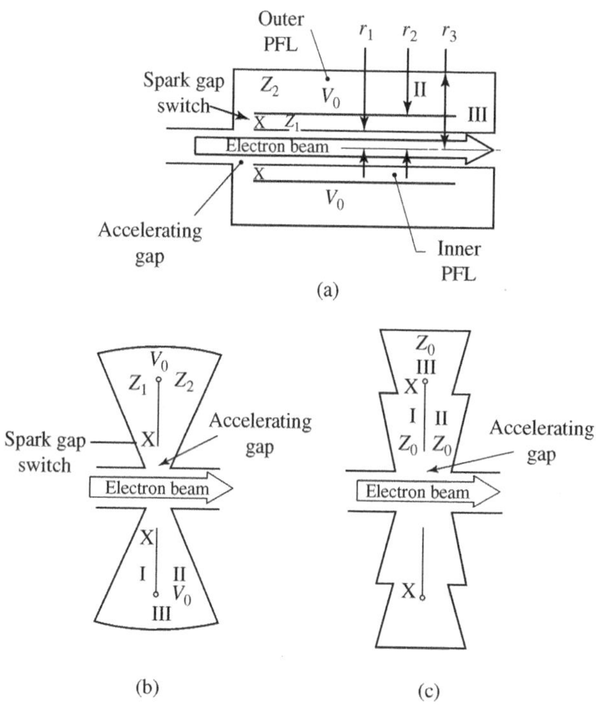

Figure 6.24 Pulse line configurations. (a) Cylindrical coaxial. (b) Radial line with modified switch position. (c) Radial line with modified transition zone III. [18]

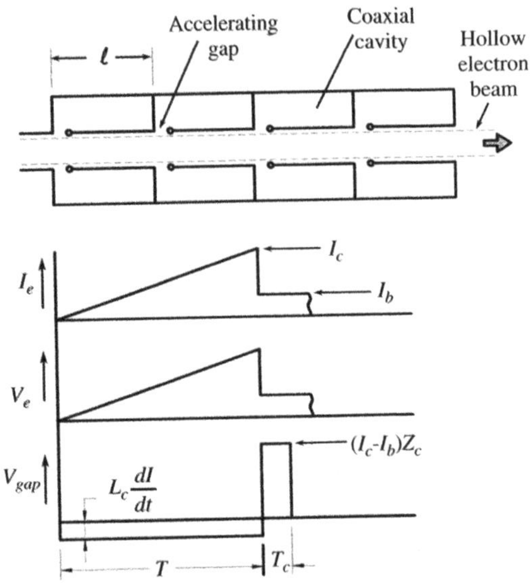

Figure 6.25 The autoaccelerator cavity structure and waveforms showing the injected current, the voltage developed across each gap during the current rise and the accelerating voltage during the current drop. [19]

6.5.3 Autoaccelerator Induction Linac

In the scheme shown in Figure 6.25, a hollow electron beam with current I_e, rising as a function of time, is formed by applying a voltage V_e of similar nature to a beam diode. This beam is injected into a vacuum-accelerating structure containing a series of coaxial cavities with each cavity having an associated gap. The electron beam excites the cavity. The stored energy density in the cavity continues to increase in the form of magnetic flux density during the slow-rising portion of e-beam current.

At the time T, when the e-beam current reaches a value I_c, the current is suddenly chopped to a very low value of I_b. At this time the magnetic energy stored in the cavity is returned to the electron beam via the induced voltage across the gap. The electrons near the tail of the pulse are now accelerated to a voltage $(I_c - I_b)Z_c$ where Z_c is the cavity impedance. The acceleration pulse lasts for a period $T_c = 2\ell/v_{pp}$ where ℓ is the length of the cavity and $v_{pp}(=c)$ is the wave propagation velocity. The advantages of an autoaccelerator are its simplicity because of absence of spark gap switches and its oil–vacuum interface insulator. The accelerating voltage across the gaps is automatically phase-matched to the velocity of electron beam. An autoaccelerator has been used to increase the energy of a 70 kA, 4.2 MeV electron beam to 7.4 MeV.

6.6 Design Examples

Example 6.1

An inductive switch, used in a magnetic pulse compression system, has N turns wound around a toroidal core of Metglas having the following parameters: (i) inner diameter D_1, (ii) outer diameter D_2, and (iii) total flux swing of ΔB ($2B_s$).

Calculate the time T required for saturation when the following voltage sources are connected across the switch: (a) $V(t) = V_0$, (b) $V(t) = V_0 \sin\omega t$, (c) $V(t) = V_0 \cdot t$.

Solution

The conditions to be satisfied for the saturation of flux in the core in an inductive switch can be written in the form of a general expression as

$$V(t) = N\frac{d\Phi_B}{dt}$$

$$\int V(t)dt = \int N d\Phi_B = \int N d(B \cdot A)$$

The area of the core is fixed. Integrating the above expression over the range from $-B_s$ to $+B_s$,

$$\int_0^T V(t) \cdot dt = \int_{-B_s}^{B_s} N \cdot A \cdot dB = 2 \cdot N \cdot A \cdot B_s$$

where N is the number of turns on the toroid, A is the core cross section, and B_s is the saturation flux density.

Case (a): $V(t) = V_0$

Inserting the expression $V(t) = V_0$ into the above expression and integrating, we can solve for the time to saturation for the specific voltage waveform:

$$\int_0^T V(t) \cdot dt = \int_0^T V_0 \cdot dt = V_0 \cdot T$$

$$V_0 \cdot T = 2 \cdot N \cdot A \cdot B_s$$

Solving for T, $T = \dfrac{2 \cdot N \cdot A \cdot B_s}{V_0}$

(continued)

(continued)

Case (b): $V(t) = V_0 \sin(\omega t)$

$$\int_0^T V(t)dt = \int_0^T V_0 \sin(\omega t) \cdot dt = \frac{V_0}{\omega} \cdot [1 - \cos(\omega T)]$$

$$\frac{V_0}{\omega} \cdot [1 - \cos(\omega T)] = 2 \cdot N \cdot A \cdot B_s$$

Solving for T, $T = \frac{1}{\omega} \cdot \cos^{-1}\left[1 - \left(\frac{2\omega NAB_s}{V_0}\right)\right]$

Case (c): $V(t) = V_0 \cdot t$

$$\int_0^T V(t) \cdot dt = \int_0^T V_0 \cdot t \cdot dt = \frac{1}{2} V_0 \cdot T^2$$

$$\frac{1}{2} V_0 \cdot T^2 = 2 \cdot N \cdot A \cdot B_s$$

Solving for T, $T = \sqrt{\frac{4 \cdot N \cdot A \cdot B_s}{V_0}}$

Example 6.2

The toroidal core of an inductive switch has 60 turns, an inner core diameter of 160 mm, an outer core diameter of 240 mm, and a height of 25 mm. The core has a saturated relative permeability $\mu_s = 1$, and a saturated flux $B_s = 1.5$ T. Calculate the time T_S for saturation for the three cases of waveforms in the example in Section 6.3.2. Assume that the peak voltage V_0 is 10 kV. In case of the sinusoidal waveform in case (b), assume a frequency of 2.5 kHz.

Solution

The area A of the core cross section is given by

$$A = \frac{1}{2}(D_2 - D_1) \cdot h$$
$$= \frac{1}{2}(240 - 160) \cdot 25 \text{ mm}^2 = 10^{-3} \text{ m}^2$$

Case (a): $V(t) = V_0$

Using the expression derived for this waveform in Example 2 for the time to saturation,

$$T = \frac{2 \cdot N \cdot A \cdot B_s}{V_0}$$

$$T = \frac{2 \times 60 \times 10^{-3} \times (1.5)}{10^4} = 18\ \mu\text{s}$$

Case (b): $V(t) = V_0 \sin(\omega t)$

$$T = \frac{1}{\omega} \cdot \cos^{-1}\left[1 - \left(\frac{2\omega NAB_s}{V_0}\right)\right]$$

$$T = \frac{1}{2\pi(2.5 \times 10^3)} \cdot \cos^{-1}\left[1 - \left(\frac{2 \times 2\pi(2.5 \times 10^3) \cdot (60 \times 10^{-3}) \cdot (1.5)}{10^4}\right)\right] = 64\ \mu\text{s}$$

Case (c): $V(t) = V_0 \cdot t$

$$T = \sqrt{\frac{4 \cdot N \cdot A \cdot B_s}{V_0}}$$

$$T = \sqrt{\frac{4 \cdot (60 \times 10^{-3}) \cdot (1.5)}{10^4}} = 6\ \text{ms}$$

Example 6.3

The two-stage MPC shown in Figure 6.13 has the following parameters: $C_0 = C_1 = C_2 = C_3 = 10$ nF, and $L_0 = 0.5$ mH. Each inductive switch employs six stacked toroidal Metglas cores. Each individual toroidal core has an inner diameter of 160 mm, an outer diameter of 240 mm, and a height of 25 mm. The saturation flux density of Metglas is 1.5 T. The inductive switches L_1 and L_2 are wound with 37 and 7 turns, respectively. Calculate the pulse time compression produced by each stage of the MPC.

Solution

To find the pulse time compression, the charging times of C_1, C_2, and C_3 must be determined. Pulse time compression for the first stage and second stage are given by

$$PTC_1 = \frac{T_2}{T_1}; \quad PTC_2 = \frac{T_3}{T_2}$$

(continued)

(continued)
where T_1 is the charging time for C_1, T_2 is the charging time for C_2, and T_3 is the charging time for C_3.

T_1, T_2, and T_3 are evaluated as follows:

Referring to Figure 6.15, C_1 (=C) is charged by C_0 (=C) through L_0. Hence, we can write

$$T_1 = \frac{1}{2f} = \pi\sqrt{\frac{L_0 C}{2}}$$

$$T_1 = \frac{1}{2f} = \pi\sqrt{\frac{(0.5\times 10^{-3})\cdot(10\times 10^{-9})}{2}} = 4.96\ \mu\text{s}$$

For the evaluation of T_2, it is seen from Figure 6.15 that C_2 (=C) is charged by C_1 (=C) through saturated inductance L_{S1} of inductive switch L_1:

$$T_2 = \pi\sqrt{\frac{L_{S1} C}{2}}$$

where

$$L_{S1} = \frac{\mu_r \mu_0 N^2 A}{\ell}$$

and $\mu_0 = 4\pi\times 10^{-7}$ H/m, $\mu_r = 1$, $N = 37$, and the area of the core cross section, A, is

$$A = \frac{6\cdot(D_2 - D_1)\cdot h}{2} = 60\times 10^{-4}\ \text{m}^2$$

The mean magnetic path length ℓ is calculated from

$$\ell = \frac{\pi\cdot(D_2 + D_1)}{2} = 0.628\ \text{m}$$

Substitution of the above values in L_{S1} gives

$$L_{S1} = 16\ \mu\text{H}$$

Substitution of L_{S1} in the above equation for T_2 gives its value as

$$T_2 = 880\ \text{ns}$$

Similarly, T_3 is calculated:

$$T_3 = \pi\sqrt{\frac{L_{S2} C}{2}}$$

$$L_{s2} = \frac{\mu_r \mu_0 N^2 A}{\ell}$$

$$L_{s2} = \frac{\left(4\pi \times 10^{-7}\right) \cdot 1 \times 7^2 \cdot \left(60 \times 10^{-4}\right)}{0.628} = 0.588\ \mu\text{H}$$

Substitution of L_{S2} in the above equation for T_3 gives its value as

$$T_3 = \pi\sqrt{\left(0.588 \times 10^{-6}\right) \cdot \left(5 \times 10^{-9}\right)} = 170\ \text{ns}$$

The pulse time compression for each stage is as follows:

$$\text{First stage}: \quad PTC_1 = \frac{T_2}{T_1} = \frac{0.88}{4.96} = 0.18$$

$$\text{Second stage}: \quad PTC_2 = \frac{T_3}{T_2} = \frac{170}{880} = 0.19$$

Example 6.4

An induction accelerator is designed to produce an electron beam of 30 MeV at 10 kA with an injected beam of 2 MeV. The individual pulsed power modules are rated at 500 kV and 50 ns. Determine the number of inductive cells and the cross-sectional areas for ferrite and Metglas. Assume that the saturation flux values are 5 kG for ferrite and 15 kG for Metglas.

Solution

The number of stages is determined by the energy gain required and the rating of the individual modules:

$$\text{Energy gain required} = 30 - 2\ \text{MeV} = 28\ \text{MeV}$$

$$\text{Number of stages} = \frac{28 \times 10^6}{500 \times 10^3} = 56$$

The cross-sectional area of ferrite and Metglas is calculated as follows:

$$V = N \cdot \frac{d\Phi_B}{dt}.$$

Since $N = 1$, $V \cdot dt = B \cdot A$.

(continued)

(*continued*)

Using the conversion factor 1 T = 10,000 Gauss, and substituting for $V = 500$ kV and $dt = 50$ ns and using the appropriate saturation flux values, the cross-sectional areas are

$A = 500\ \text{cm}^2$ for ferrite, and
$A = 167\ \text{cm}^2$ for Metglas.

The use of Metglas results in significant advantages over a ferrite material.

References

1 H. Knoepfel, *Pulsed High Magnetic Fields*, North Holland Publishers, 1970.
2 P.H. Ron, Configurations of Intense Pulse Power Systems for Generation of Intense Electromagnetic Pulses, *International Conference on Electromagnetic Interference and Compatibility (INCEMIC'99)*, New Delhi, India, p. 471, 1999.
3 P.H. Ron and R.P. Gupta, *Opening Switches*, National Research Council, NRC Technical Report No. TR-GD-007, NRC No. 25437, Canada, pp. 1–51, 1985.
4 P.H. Ron, V.K. Rohatgi, and R.S.N. Rau, Rise Time of a Vacuum Gap Triggered by an Exploding Wire. *IEEE Trans. Plasma Sci.*, PS-11, No. 4, p. 274, 1983.
5 P.H. Ron, V.K. Rohatgi, and R.S.N. Rau, Delay Time of a Vacuum Gap Triggered by an Exploding Wire. *J. Phys. D, Appl. Phys*, Vol. 17, p. 1369, 1984.
6 S. Nakajima, S. Arakawa, Y. Yamashita, and M. Shiho, Fe-Based Nanocrystalline FINEMET Cores for Induction Accelerators. *Nucl. Instrum. Method Phys. Res.*, A-331, pp. 318–322, 1993.
7 S.L. Birx, E.J. Lauer, L.L. Reginato, J. Schmidt, and M. Smith, Basic Principles Governing the Design of Magnetic Switches, Lawrence Livermore Laboratory, Report No. UCID-18831, 1980.
8 H. Deguchi, T. Hatakeyama, E. Murata, Y. Izawa, and C. Yamanaka, Efficient Design of Multi-Stage Magnetic Pulse Compression. *IEEE J. Quantum Electron.*, Vol. 30, No. 12, pp. 2934–2938, 1994.
9 P.H. Ron, Design and Construction of an Induction Linac with Ratings of 200 kV, 5 kA, 50 ns, 10–100 pps, Internal Report, BARC, Mumbai, India, 1998.
10 T. Akiba, et al., Development of an Inductive Voltage Accumulating System for a Free Electron Laser. *Nucl. Instrum. Methods Phys. Res. A* 259, p. 115, 1987.
11 I. Smith, P. Corcoran, H. Nishimoto, and D. Wake, Conceptual Design of a 30 MV Driver for LIBRA, *Proceedings of the 7th International Conference on High Power Particle Beams*, p. 127, 1988.
12 N.C. Christofilos, et al., High Current Induction Accelerator for Electrons. *Rev. Sci. Instrum.*, Vol. 35, No. 7, p. 886, 1964.

13 J.A. Nation, High Power Electron and Ion Beam Generation, in *Particle Accelerators*, Gardon and Breach Publishing, Vol. 10, pp. 1–30, 1979.
14 C.A. Kapetanakos and P. Sprangle, Ultra High Current Electron Induction Accelerators. *Phys. Today*, Vol. 38, No. 2, p. 58, 1985.
15 K.R. Prestwich, Electron and Ion Beam Accelerators. *AIP Conf. Proc.*, Vol. 249, No. 2, p. 1725, 1992.
16 D. Birx, Induction Linear Accelerators. *AIP Conf. Proc.*, Vol. 249, No. 2, p. 1554, 1992.
17 A.I. Pavlovskii, V.S. Bosamykin, G.D. Kuleshov, A.I. Gerasimov, V.A. Tananakin, and A.P. Klementev, Multielement Accelerators Based on Radial Lines. *Sov. Phys. Dokl.*, Vol. 20, p. 441, 1975.
18 D. Eccleshall, JK Temperley and CE Hollandsworth, “Charged, Internally Switched Transmission Line Configurations for Electron Acceleration,” IEEE Trans. on Nuclear Science, Vol. NS-26, No. 3, June 1979.
19 D. Keefe, *Research on High Current Accelerators*, Lawrence Berkeley Laboratory and Physical Dynamics, Inc., Report No. LBL-12210, April 1981.
20 I.D. Smith, A Novel Voltage Multiplication Scheme Using Transmission Lines. *IEEE Conference Record of the 15th Power Modulator Symposium*, p. 223, 1982.
21 M. Friedman, Autoacceleration of High Power Electron Beams. *Appl. Phys. Lett.* 41, p. 419, 1982.
22 C.A. Kapetanakos, et al. Compact, High Current Accelerators and Their Prospective Applications. *Phys. Fluids*, Vol. B3, No. 8, p. 2396, 1991.
23 D.V. Giri, *High Power Electromagnetic Radiators Nonlethal Weapons and Other Applications*, Harvard University Press, Chapters 4 and 5, 2004.

7

Energy Storage in Capacitor Banks

Energy storage capacitor banks are widely used in pulsed power for high-current applications, including exploding wire phenomena, shockless compression, and the generation, heating, and confinement of high-temperature, high-density plasmas, and their many uses. Advances in components used in large capacitor banks, including precise switches and high energy density, fast discharge capacitors have resulted in reliable operation.

This chapter covers various aspects involved in the design and construction of energy storage capacitor banks. Methods are described for reducing a complex capacitor bank system into a simple equivalent circuit made up of *L*, *C*, and *R* elements. Typical configurations and constructional aspects of capacitor banks are presented.

7.1 Basic Equations

The version of the *RLC* circuit that is common in pulsed power applications is shown in Figure 7.1. A capacitor is initially charged to a voltage V_0 by a power supply when the closing switch S_C is open and begins its discharge when S_C is closed at $t = 0$. The circuit elements *R* and *L* may be either physical elements or intrinsic equivalent values from physical properties such as conductor losses and the inductance associated with electrical connections.

Using *Kirchoff's voltage law* on the circuit of Figure 7.1 yields

$$V_C + V_L + V_R = 0 \tag{7.1}$$

Using the constitutive relations

$$V_R = R\,i(t)$$

$$V_L = L\frac{di(t)}{dt}$$

Foundations of Pulsed Power Technology, First Edition. Jane Lehr and Pralhad Ron.

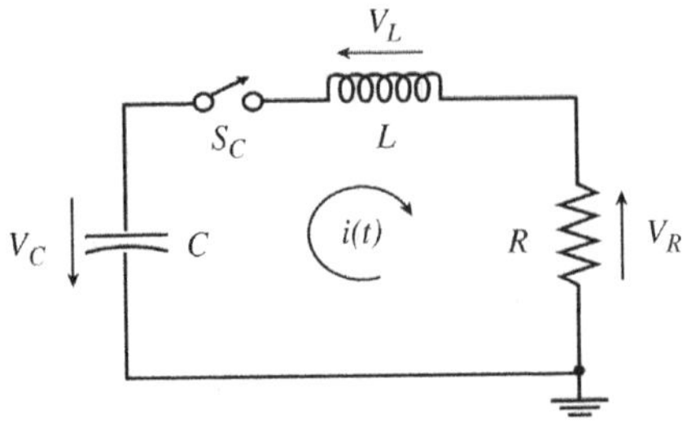

Figure 7.1 A capacitor or bank of capacitors with capacitance C is initially charged to a voltage V_0 and discharged through an inductor and resistance can represent many pulsed power circuits. The inductance and resistance may be the intrinsic values of the nonideal circuit.

Equation 7.1 is expressed as

$$V_c + L\frac{di(t)}{dt} + R\,i(t) = 0 \tag{7.2}$$

Differentiating (7.2) with respect to time yields

$$\frac{dV_C}{dt} + L\frac{d^2i(t)}{dt^2} + R\frac{di(t)}{dt} = 0 \tag{7.3}$$

And the constitutive relation for the capacitor provides a relationship between the voltage drop and the current

$$i(t) = C\frac{dV_C(t)}{dt} \tag{7.4}$$

Inserting (7.4) into (7.3) and rearranging terms,

$$\frac{d^2i(t)}{dt^2} + \frac{R}{L}\frac{di(t)}{dt} + \frac{1}{LC}i(t) = 0 \tag{7.5}$$

This can be expressed in a convenient form by defining

$$\alpha = \frac{R}{2L} \tag{7.6}$$

and

$$\omega_0 = \frac{1}{\sqrt{LC}} \tag{7.7}$$

so that (7.5) may be expressed as

$$\frac{d^2i(t)}{dt^2} + 2\alpha\frac{di(t)}{dt} + \omega_0^2 i(t) = 0 \tag{7.8}$$

The quantity α is called the attenuation factor and, in traditional circuits, is a measure of how fast the transient response dies away when the source is removed. The natural frequency of the circuit is ω_0. The solutions for (7.8) are a superposition of exponentials

$$i(t) = A_1e^{s_1t} + A_2e^{s_2t} \tag{7.9}$$

Solving (7.8) in the Laplace domain, the solutions of s_1 and s_2 are

$$s_1, s_2 = -\alpha \pm \sqrt{\alpha^2 - \omega_0^2} \tag{7.10}$$

so that the specific forms of the solutions depend on the relative values of α and ω_0 and the constants A_1 and A_2 are determined by the initial conditions. It is useful to define the quantity ξ

$$\xi = \frac{\alpha}{\omega_0} \tag{7.11}$$

which can be related back to the parameters of the *RLC* circuit through (7.6) and (7.7),

$$\xi = \frac{R}{2}\sqrt{\frac{C}{L}} \tag{7.12}$$

The solutions to (7.9) given in (7.10) can be expressed in terms of ξ as

$$s_1, s_2 = -\alpha \pm \omega_0\sqrt{\xi^2 - 1} = \omega_0\left(-\xi \pm \sqrt{\xi^2 - 1}\right) \tag{7.13}$$

The value of the quantity ξ determines the transient behavior of the circuit. Four cases are relevant:

Ideal (lossless) case $\xi = 0$
Overdamped response $\xi > 1$
Underdamped response $\xi < 1$
Critically damped case $\xi = 1$

7.1.1 Case 1: Lossless, Undamped Circuit $\xi = 0$

In this case, $\xi = 0$, $\alpha = 0$, and $R = 0$. From (7.8), the differential equation describing the circuit operation becomes

$$\frac{d^2 i(t)}{dt^2} + \omega_0^2 i(t) = 0 \tag{7.14}$$

The solutions are in the form

$$i(t) = A_1 \sin(\omega_0 t) + A_2 \cos(\omega_0 t) \tag{7.15}$$

When the switch closes at $t = 0$, the current is zero. Applying the initial condition that the current through an inductor cannot change instantaneously,

$$i_L(0_-) = i_L(0_+) = 0 \tag{7.16}$$

$$A_1 \sin(0) + A_2 \cos(0) = A_1(0) + A_2(1) = 0$$

and $A_2 = 0$

The other constant A_1 is determined by the initial condition on the capacitor:

$$V_c(0_-) = V_c(0_+) \tag{7.17}$$

Since there is no current flowing in the circuit at $t = 0_+$,

$$-V_c(0_+) = V_L(0_+) = V_0 \tag{7.18}$$

The negative sign in (7.18) reflects the sign convention adopted in Figure 7.1 that the voltage drop opposes the direction of current. Applying the initial condition yields

$$V_L(0) = L\frac{di}{dt}\bigg|_{t=0} = L[A_1\omega_0 \cos(0)] = V_0$$

yielding $A_1 = \frac{V_0}{\omega_0 L}$. The solution for the lossless case is

$$i(t) = \frac{V_0}{\omega_0 L}\sin(\omega_0 t) \tag{7.19}$$

The time to peak current T_{peak} can be calculated by setting the argument of the sine wave in (7.19) equal to $\pi/2$ giving

$$T_{peak} = \frac{\pi}{2}\sqrt{LC} \tag{7.20}$$

Although lossless circuits cannot be fully realized in practice, it is an important case for circuits requiring high peak currents. Large peak currents may be produced with a high-capacitance, low-inductance bank where the capacitors are arranged in parallel, which has the dual benefit of increasing the equivalent capacitance and reducing the equivalent inductance of each individual capacitor. The peak current of such a bank can be estimated from

$$I_{peak} = \frac{V_0}{\omega_0 L} = V_0\sqrt{\frac{C}{L}} \tag{7.21}$$

This result may also be obtained by considering that for a lossless, undamped circuit, the stored energy remains constant and alternates between capacitive energy ($1/2CV^2$) and inductive energy ($1/2LI^2$).

7.1.2 Case 2: Overdamped Circuit $\xi > 1$

In this case, from (7.13), the roots of the solution are real and the solution has the form

$$i(t) = A_1 \exp\left(\omega_0\left(-\xi + \sqrt{\xi^2 - 1}\right)t\right) + A_2 \exp\left(\omega_0\left(-\xi - \sqrt{\xi^2 - 1}\right)t\right) \tag{7.22}$$

Defining

$$\omega_D = \omega_0\sqrt{|\xi^2 - 1|} \tag{7.23}$$

and recognizing that

$$\omega_0\xi = \alpha \tag{7.24}$$

the solution (7.22) can be expressed as

$$i(t) = A_1 e^{-\alpha t} e^{\omega_D t} + A_2 e^{-\alpha t} e^{-\omega_D t}$$

Applying the initial condition (7.16) on the current through the inductor yields

$$A_2 = -A_1$$

$$i(t) = A_1 e^{-\alpha t}\left(e^{\omega_D t} - e^{-\omega_D t}\right) \tag{7.25}$$

Applying the initial condition (7.17) on the voltage across the capacitor to (7.25) yields

$$A_1 = \frac{V_0}{2\omega_D L}$$

And the solution for the underdamped case is

$$i(t) = \frac{V_0}{2\omega_D L} e^{-\alpha t}\left(e^{\omega_D t} - e^{-\omega_D t}\right) \tag{7.26}$$

The overdamped solution may be written as

$$i(t) = \frac{V_0}{\omega_D L}\sinh(\omega_D t)\cdot e^{-\alpha t} \tag{7.27}$$

The damping frequency ω_D can be expressed in terms of the circuit parameters as

$$\omega_D = \omega_0\sqrt{\frac{R^2}{4}\cdot\frac{L}{C} - 1} = \sqrt{\frac{R^2}{4L} - \frac{1}{LC}} \tag{7.28}$$

In terms of ω_0 and ξ, the current is

$$i(t) = \frac{V_0}{2\omega_0 L\sqrt{\xi^2 - 1}} \left\{\exp\left(\omega_0\left(-\xi + \sqrt{\xi^2 - 1}\right)t\right) - \exp\left(\omega_0\left(-\xi - \sqrt{\xi^2 - 1}\right)t\right)\right\} \tag{7.29}$$

The overdamped case is the discharge without oscillation.

7.1.3 Case 3: Underdamped Circuit $\xi < 1$

The solution to the underdamped case occurs when the term $\omega_0\sqrt{\xi^2 - 1}$ is imaginary. Using the defined relationships of (7.22) and (7.33), the solution to the underdamped circuit response is

$$i(t) = A_1 e^{-\alpha t} e^{j\omega_D t} + A_2 e^{-\alpha t} e^{-j\omega_D t} \tag{7.30}$$

and $j = \sqrt{-1}$

The solution (7.30) can be written as

$$i(t) = e^{-\alpha t}(A_1 \sin \omega_D t + A_2 \cos \omega_D t) \tag{7.31}$$

Again, applying the initial condition across the inductor,

$$A_2 = 0$$

Again, applying the initial condition on the capacitive voltage,

$$A_1 = \frac{V_0}{\omega_D L}$$

And the solution for the underdamped case is

$$i(t) = \frac{V_0}{\omega_D L} e^{-\alpha t} \sin \omega_D t \tag{7.32}$$

The underdamped case is an oscillatory wave with frequency ω_D in a decaying envelope determined by the attenuation constant α. In high-energy circuits, the underdamped response is usually avoided since a significant voltage reversal can occur across the capacitors leading to their failure.

7.1.4 Case 4: Critically Damped Circuit $\xi = 1$

In the critically damped case, $\sqrt{\xi^2 - 1} = 0$, and the roots to (7.13) are

$$s_1, s_2 = -\omega_0 \xi = -\alpha$$

Taking the inverse Laplace transform, the solution takes the form

$$i(t) = A_1 t\, e^{-\alpha t} + A_2 e^{-\alpha t}$$

Invoking the initial condition on the current through the inductor yields $A_2 = 0$, and applying the initial condition on the voltage across the inductor at $t = 0$,

$$i(t) = \frac{V_0}{L} t\, e^{-\alpha t} \tag{7.33}$$

The critically damped response represents the circuit response that decays the fastest without oscillation. There is generally too much uncertainty in a pulsed power driver to achieve critical damping.

7.1.5 Comparison of Circuit Responses

The current in the *RLC* circuit representing a capacitor bank is shown in Figure 7.2. The representative circuit has a capacitance of 10 μF, charged to an initial voltage of 100 kV with an equivalent inductance of 100 nF. The response of the circuit depends on the value of the resistance and is encompassed in the parameter ξ.

The current in the critically damped circuit may be used for a reference and is represented in Figure 7.2 with $\xi = 1$. The resistance to critically damp the circuit can be calculated from (7.12) and for the circuit parameters given, it is 200 mΩ. The underdamped responses, where $\xi < 1$, have larger peak currents than the critically damped response and oscillatory behavior. Capacitor banks may be designed to achieve these large peak currents and then use a crowbar switch to protect the circuit against the large voltage swings from the oscillations. The overdamped responses are shown in Figure 7.2 with $\xi >$ 1 with no oscillations and lower peak currents due to the higher loss in the circuit. It may be noted that the peak current in the lossless (undamped) case is 100 kA.

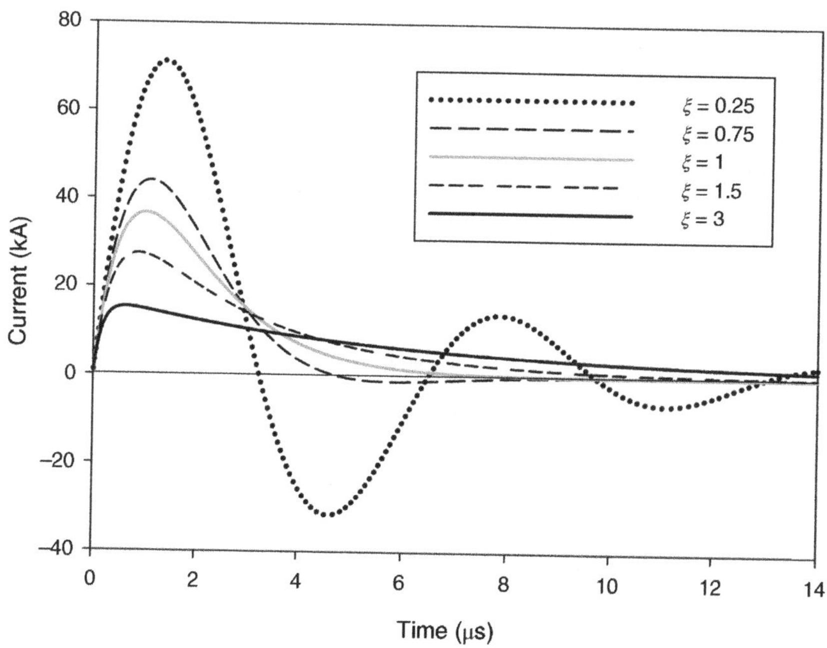

Figure 7.2 The time-dependent current $i(t)$ through the circuit of Figure 7.1 when $C = 10\ \mu$F, $L = 100$ nH, and $V_0 = 100$ kV for a range of values for ξ.

7.2 Capacitor Bank Circuit Topology

A typical system layout for a high-energy storage capacitor bank is shown in Figure 7.3. The energy storage capacitor is charged to a voltage V_0 by a power supply. The discharge sequence is started by a command trigger sent to the start switch, initiating the discharge of the capacitor bank. The discharge current is carried by transmission line and delivered to the load through a power feed.

A crowbar switch is often included in the capacitor bank circuit topology as a protection element. The crowbar switch is arranged so that it closes at a time after the main discharge has passed and provides a short circuit across the element to be protected. The crowbar is often used in series with a resistor to dissipate the excess energy.

The placement of a crowbar switch is determined by its purpose. For instance, high-energy density capacitors are susceptible to failure from excessive voltage reversal. In this case, the crowbar switch is placed as close to the capacitor as possible, often between the capacitor and the start switch. This allows only the desired current to go through the start switch, prolonging its life by minimizing erosion. Other applications, such as plasma confinement or high magnetic field generation, demand unidirectional current in an underdamped inductive load so that the crowbar is incorporated across the load. The crowbar switch may be self-breaking or externally triggered. Some applications, such as flux compression generators, may design the load to short itself during operation, essentially acting as a crowbar.

Crowbar switches can be difficult to implement in many instances because they must stay open while the main discharge is occurring and then close upon the command trigger when the voltage on the crowbar switch may be much lower. In many applications, the crowbar switch must trigger reliably at low percentages of its self-breakdown voltage.

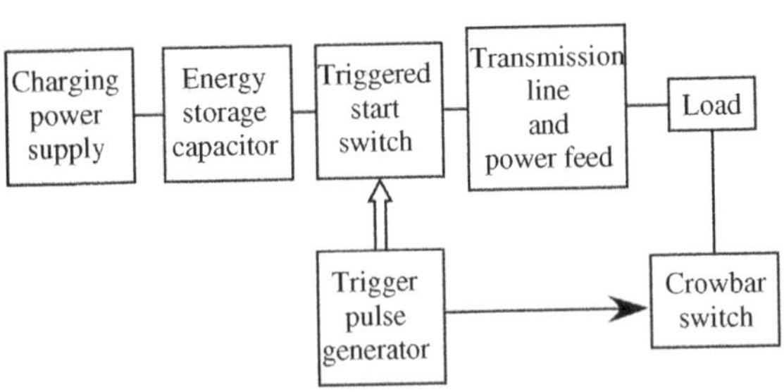

Figure 7.3 The circuit topology of an energy storage capacitor bank with a crowbar switch across the load. A crowbar switch protects the capacitor from excessive voltage reversal.

7.2.1 Equivalent Circuit of a Low-Energy Capacitor Bank

A circuit schematic of a low-energy capacitor bank is shown in Figure 7.4. It is comprised of energy storage capacitor C_s, start-charged through a resistor R to a voltage V_0, and discharged through the start switch S_s. A transmission line TL delivers the discharge current to the load L_{Load}, which is protected by a crowbar switch S_{cb} and energy dissipating resistor R_d.

The various symbols used in the equivalent circuit are as follows:

Capacitor:	L_c—equivalent series inductance of C_s
	R_c—equivalent series resistance of C_s
Start switch:	L_{SW}—self-inductance of the switch S_S
	R_{SW}—internal resistance of the switch S_S
Transmission line:	L_{TL}—inductance of transmission line of length ℓ
	R_{TL}—resistance of transmission line of length ℓ
Crowbar switch:	L_{cb}—self-inductance of crowbar switch
	R_{cb}—internal resistance of crowbar switch
	R_d—damping resistance in the crowbar circuit
Load:	L_{Load}—inductance of the load
	R_{Load}—resistance of the load

Prior to triggering the crowbar switch, the operation of the equivalent circuit in Figure 7.5 can be simplified to apply the equations of Section 7.1 with the following modifications:

$$L = L_c + L_s + L_{TL} + L_{Load}$$
$$R = R_c + R_s + R_{TL} + R_{Load}$$

The resultant modified equivalent circuit is so simple that the current waveform, the peak current, and the time to peak current can be calculated.

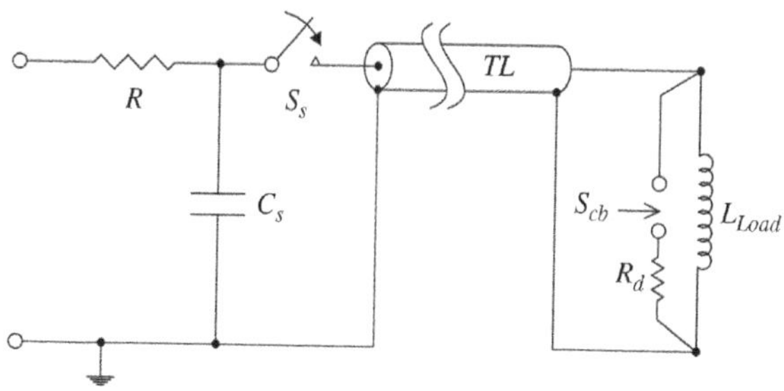

Figure 7.4 Capacitor discharge circuit with start switch (S_S) and crowbar switch (S_{cb}). The corresponding equivalent circuit is shown in Figure 7.5.

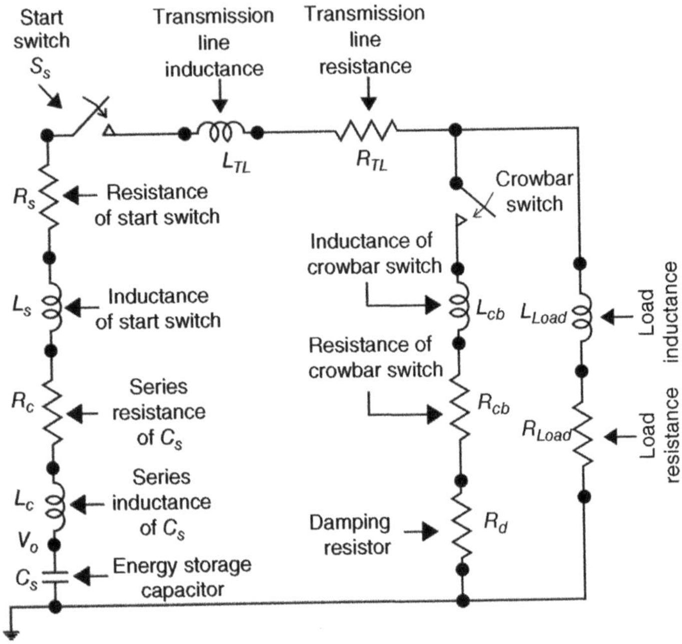

Figure 7.5 Equivalent circuit of a low-energy capacitor bank.

7.2.2 Equivalent Circuit of a High-Energy Capacitor Bank

A typical schematic for a large capacitor bank with stored energy in the range of 100 kJ to 10 MJ is shown in Figure 7.6 and its corresponding equivalent circuit in Figure 7.7. The capacitor and switch assembly consists of a large number N of subcapacitor banks (SB_1, SB_2, . . . , SB_N), each comprised of n energy storage capacitors (C_{s1}, C_{s2}, . . . , C_{sn}) with individual switches (S_{S1}, S_{S2}, . . . , S_{SN}). When the energy storage capacitors are charged to the desired voltage (V_0) by the charging supply, the trigger pulse generator supplies a fast- rising, high-voltage pulse to the individual switches, resulting in the simultaneous firing of all the switches. The current provided by the capacitor bank is carried by a group of transmission lines, which may be strip line or coaxial cables, connected in parallel. The power feed is typically a low-inductance configuration designed to receive the current from the transmission lines on one end and deliver it to the load on the other end.

The equivalent circuit of a large capacitor bank with N subbanks (each subbank containing n energy storage capacitors) discharged by N triggered start switches is illustrated in Figure 7.7. The discharge current of each subbank is carried by n' transmission lines, each of length ℓ, connected in parallel and

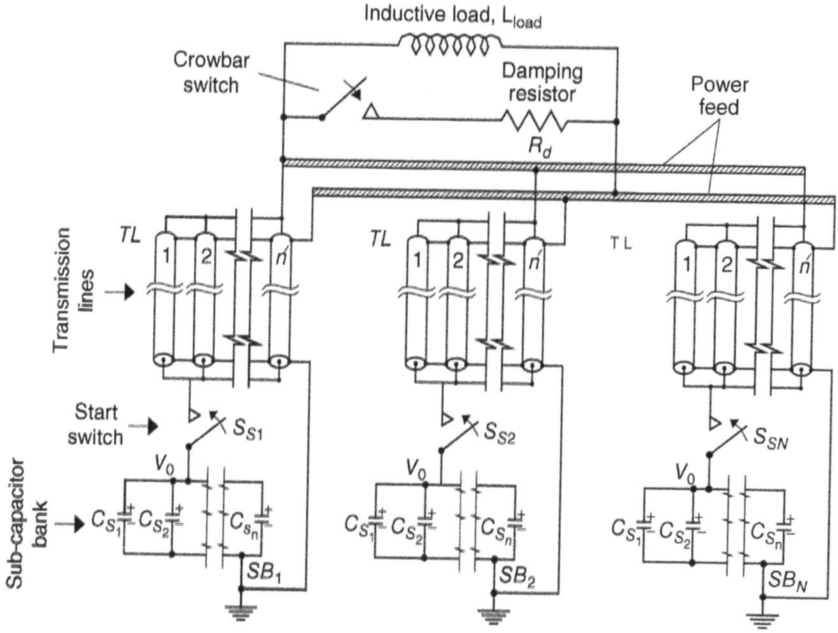

Figure 7.6 Circuit schematic of a large capacitor bank.

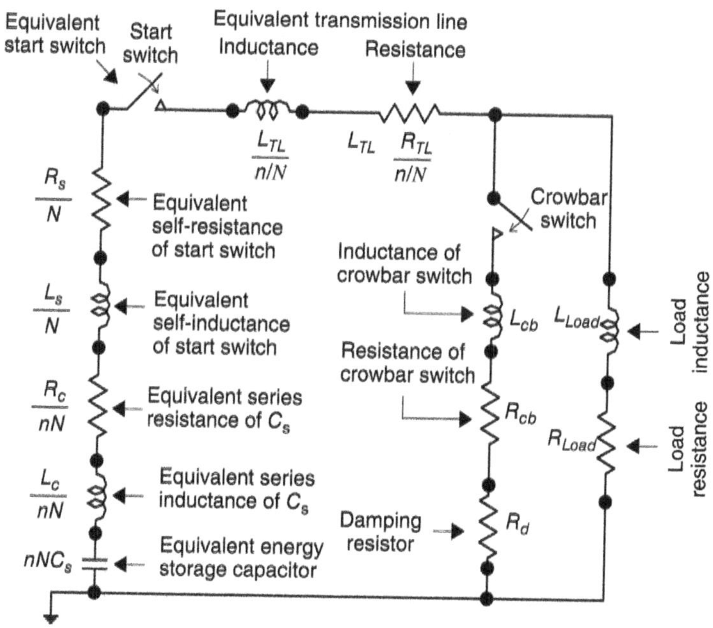

Figure 7.7 Equivalent circuit of a large capacitor bank.

terminated in the power feed that is connected to the load. The load is shunted by a crowbar switch in series with a damping resistor.

Before triggering the crowbar switch and combing terms, the parameters of the equivalent circuit can be reduced to just three parameters, L'', C'', and R'', where

$$L'' = \frac{L_c}{nN} + \frac{L_s}{N} + \frac{L_{TL}}{n'N} + L_{Load}$$

$$C'' = nNC$$

and $$R'' = \frac{R_c}{nN} + \frac{R_s}{N} + \frac{R_{TL}}{n'N} + R_{Load}$$

The circuit is then amenable to the evaluation of Section 7.1.

The equivalent circuit of Figure 7.7 is valid for the ideal condition where all the start switches fire simultaneously. In realistic circumstances, where there is an acceptable variation in the firing time of the various switches, the peak current will be less than the value predicted by the equivalent circuit, along with distortion in the waveshape. Moreover, depending on the circuit topology, the firing of adjacent switches can be problematic. When one switch fires, it may couple electromagnetically, effectively lowering the operating voltage across its closest neighbors. This serves to make those affected switches more difficult to trigger.

To obtain high-peak currents and good efficiency, the overall inductance of the circuit should be minimized, including choosing fast discharge capacitors with low internal inductance and resistance, as well as low-jitter start switches and a low-inductance transmission line [1,2].

7.3 Charging Supply

Power supplies can be an important design parameter because of the large energies involved. The power supply is generally available commercially in one of these charging modes: constant voltage, constant current, constant power, or resonant charging. The charging time of the bank depends on its overall capacitance, but shorter times are generally favored because of the danger of an untriggered discharge—known as a prefire.

7.3.1 Constant Voltage (Resistive) Charging

In constant voltage charging, the power supply has a power rating P_S and a variable voltage setting with a maximum voltage V_{max}. This implies that there is a maximum current I_{max} that may be drawn for the power supply, determined by

$$I_{max} = \frac{P_s}{V_{max}} \tag{7.34}$$

To avoid drawing excessive current, a current limiting resistor is inserted between the power supply and the charging load. The minimum resistor value R_{min} is determined by

$$R_{min} = \frac{V_{max}}{I_{max}} \tag{7.35}$$

the capacitor C is charged to a voltage V_0 through a series, current-limiting resistor R_{CL}. The voltage is constant, and the time to charge the capacitor is determined by the RC time constant. The voltage across the capacitor is given by

$$V_C = V_0\left(1 - e^{-(t/RC)}\right) \tag{7.36}$$

which is plotted as a function of time in Figure 7.8. The time to charge the capacitor T_{ch} depends on the time constant of the charging circuit R_cC. The capacitor is charged to 95% of its full voltage in three time constants and to 99% in five,

$$T_{ch} = 5R_cC \tag{7.37}$$

Rearranging terms,

$$T_{ch} = 5\left(\frac{V_0}{I_0}\right)C = 5 \times \frac{CV_0^2}{P_S}$$

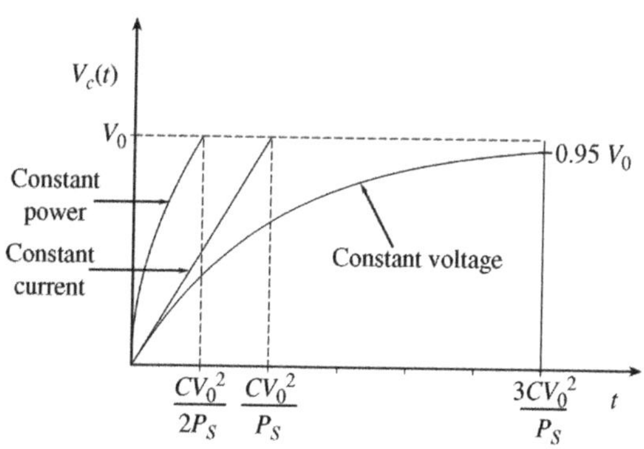

Figure 7.8 A constant current power charging supply is linear with time, while a constant power supply is faster with a $\sqrt{t}$ dependence. Both charge a capacitor more quickly than a constant voltage supply.

The energy dissipated in the resistor during charging is given by

$$E_R = \int_0^\infty \frac{V^2}{R_c} \cdot dt = \int_0^\infty \frac{\left(V_0 \cdot e^{-t/R_cC}\right)^2}{R_c} \cdot dt = \frac{1}{2} C V_0^2 \tag{7.38}$$

The energy supplied by the power supply E_{PS} to charge the capacitor C to a voltage V_0 through the current limiting resistor R_c is

$$E_{PS} = E_R + E_C = \frac{1}{2} C V_0^2 + \frac{1}{2} C V_0^2 \tag{7.39}$$

The efficiency of resistive charging, defined as the ratio of energy stored in capacitor to energy supplied by the power supply, is

$$\eta = \frac{E_C}{E_{PS}} = \frac{1/2 C V_0^2}{C V_0^2} = 50\% \tag{7.40}$$

Note from (7.33) that the energy lost in the resistor in resistive charging is independent of the value of R_c.

7.3.2 Constant Current Charging

Constant current is a popular choice for charging small capacitor banks, especially for high repetition rate applications. For a power supply with a power rating of P_S and a constant current of I_0, the charging rate of the capacitor can be calculated by

$$I_0 = C \frac{dV}{dt} \tag{7.41}$$

Since the constant is current,

$$\int_0^{V_c(t)} dV = \int_0^t \frac{I_0}{C} dt$$

the voltage on the capacitor $V_c(t)$ at any time t is

$$V_c(t) = \frac{I_0}{C} t \tag{7.42}$$

The charge is linear with time so that the time T_{ch} to charge the capacitor to a voltage V_0 is

$$T_{ch} = \frac{C V_0}{I_0} = \frac{C V_0^2}{P_S} \tag{7.43}$$

The growth of V_c as a function of time is shown in Figure 7.8, from where it can be seen that, for a given power P_S of the charging supply, charging a capacitor with a constant current is five times faster than resistive charging.

A constant current mode of charging can be implemented by using active electronic components. Constant current power supplies are often specified in units of energy per time (kJ/s) and are commercially available.

7.3.3 Constant Power Charging

As in the case of a constant current power supply, a constant power charging mode does not need a current-limiting resistor, so high efficiencies can be achieved [3]. For a constant power unit, the relation between the capacitive voltage V_C and current I_C,

$$P_S = I_C(t) \cdot V_C(t) \tag{7.44}$$

is a constant. The rate of voltage growth for a capacitor charged with a constant power unit follows from the basic charge balance relations for the capacitor:

$$I_C = C\frac{dV_c(t)}{dt}$$

Rearranging terms, multiplying both sides by $V_c(t)$, and integrating leads to

$$C \cdot V_c(t) \cdot dV_c = I_C(t) \cdot V_c(t) \cdot dt = P_s \cdot dt$$

Recognizing that P_S is a constant, $C \int_0^{V_c(t)} V_c(t) = P_S \int_0^t dt$yielding

$$\frac{1}{2}CV_C^2(t) = P_S \cdot t \tag{7.45}$$

For a constant power charging supply, the voltage on the capacitor rises with a square root time dependence:

$$V_c(t) = \sqrt{\frac{2P_S t}{C}} \tag{7.46}$$

The time required to charge to a voltage V_0 is given by

$$T_{ch} = \frac{CV_0^2}{2P_S} \tag{7.47}$$

The voltage growth curve for a constant power charging is shown in Figure 7.8. It is seen that, for a given power P_S of the power supply, the constant power charging scheme is 2 times faster than constant current charging and 10 times faster than resistive charging.

7.4 Components of a Capacitor Bank

The two most common implementations of capacitor/switch assemblies are common. One is to have a module make up of one or two capacitors with the switch mounted directly over the capacitor terminals so that each module has its individual switch. Another is a topology in which a group of capacitors are connected in parallel by a bus bar or parallel plate transmission line and share a start switch placed nearby.

7.4.1 Energy Storage Capacitor

The capacitors employed in large banks are of an energy storage type possessing the characteristics of low inductance, high peak current, large *di/dt*, good fault tolerance, long life, and excellent reliability. The capacitor elements, made up of alternate layers of metallic foil and dielectric film, are impregnated with dielectric fluid and assembled in a metal or plastic can. A number of these elements are internally connected in series to achieve the "voltage rating" and "capacitance rating." The capacitor manufacturing involves elaborate cycles of (i) capacitor winding in clean environment, (ii) hot air drying to remove bulk of moisture, (iii) vacuum drying, (iv) vacuum impregnation with chemically treated dielectric fluid, and finally (v) sealing in cans. The properties desired of a dielectric film are low dissipation factor for good thermal stability, high dielectric strength/dielectric constant for high-energy density, and good wettability to the impregnant to reduce the occurrence of voids and resultant localized weak spots for initiation of breakdown [1,4]. The dielectric fluid used for impregnation should have (i) low dielectric loss over a wide frequency range, (ii) high dielectric strength, (iii) low viscosity, (iv) low surface tension for good wettability of dielectric film surface, (v) ability to absorb gaseous products evolved during premature breakdowns, and (vi) nontoxicity to environment. In practice, the capacitors usually employ metallic foils of aluminum in the thickness range of 5–8 μm. The solid dielectric materials used are paper, polyester, polypropylene, polyethylene terepthalate (PET), polyvinylidene fluoride (PVDF), or mixed dielectric formed by combination of paper and polymer. Usually, operating electrical stresses larger than 1500 V/mil are used [2]. The dielectric fluids employed as impregnants are mineral oil, castor oil, and synthetic fluids such as dioctylphthalate (DOP), polybutene, and so on. The use of paper/polypropylene as mixed dielectric results in higher stored energy density in comparison to paper dielectric. This arises mainly due to smaller thickness, freedom from defects, lower dielectric losses, and higher dielectric strength of polypropylene. For use in Marx generators, plastic-cased capacitors are commercially available, which, of course, have to be mounted on stand-off insulators. In energy storage capacitors, the output terminals are brought out on single or double bushings made of porcelain, polyethylene, epoxy, or polyurethane. The terminal configurations may be coaxial vertical, coaxial radial, or parallel plate transmission line. When only the high-voltage terminal is brought out, the other terminal is connected to the metal casing.

7.4.1.1 Capacitor Parameters

The important parameters of energy storage capacitor are equivalent series resistance (*ESR*), equivalent series inductance (*ESL*), peak current, shot life (*N*), and DC life.

7.4.1.1.1 Equivalent Series Resistance

The *ESR* is representative of the dielectric and resistive losses inside the capacitor. The dielectric losses take place in the dielectric film and dielectric impregnating fluid and it is proportional to ($V^2C\ f\ tan\delta$), where f and $tan\delta$ represent the ringing frequency and loss angle, respectively. The dielectric losses are reduced by employment of dielectric film and dielectric fluid possessing low loss factor over a wide frequency and temperature range. The resistive losses may be reduced with good soldering and welding techniques for the leads and bushing. Extended foil construction has the advantage of reduced resistive losses in comparison to tab construction. The resistive losses increase due to skin effect at high ringing frequencies.

7.4.1.1.2 Equivalent Series Inductance

The *ESL* is representative of the energy stored in the inductance of the capacitor, which is not delivered to the load during the discharge. The *ESL* also reduces the *di/dt* capability of the capacitor. The *ESL* comprises the inductances contributed by the metallic foils, leads, and bushing. To reduce the *ESL*, the techniques employed are use of metal foil having high width-to-length ratio, paralleling of pads, extended foil construction, and use of radial bushings instead of conventional vertical bushings.

7.4.1.1.3 Peak Current Capability

The peak current capability of a capacitor is inversely proportional to its *ESL* and *ESR*. A capacitor carrying high peak current is subject to electromagnetic forces and hence greater care is required for maintaining the mechanical integrity of the interior structure. Extended foil construction allows much higher current-carrying capacity than tab construction, because of larger contact area between capacitor electrode and leads.

7.4.1.1.4 Shot Life

Shot life represents the number of charge–discharge cycles that the capacitor can withstand before failure. Because of the statistical nature of electrical breakdown, the shot life is usually expressed in terms of the percentage survival probability. If the capacitor is having a life of 10^5 shots for 90% survival, there is always a possibility of 10% capacitors failing up to 10^5 shots but with a certainty of 90% capacitors surviving to more than 10^5 shots. Usually the capacitor manufacturers establish the curves of "survival probability versus number of shots." The life of a capacitor is dependent on a number of factors such as the percentage voltage reversal, ringing frequency, temperature, and operating voltage. The typical life data curves for capacitors are shown in Figure 7.9.

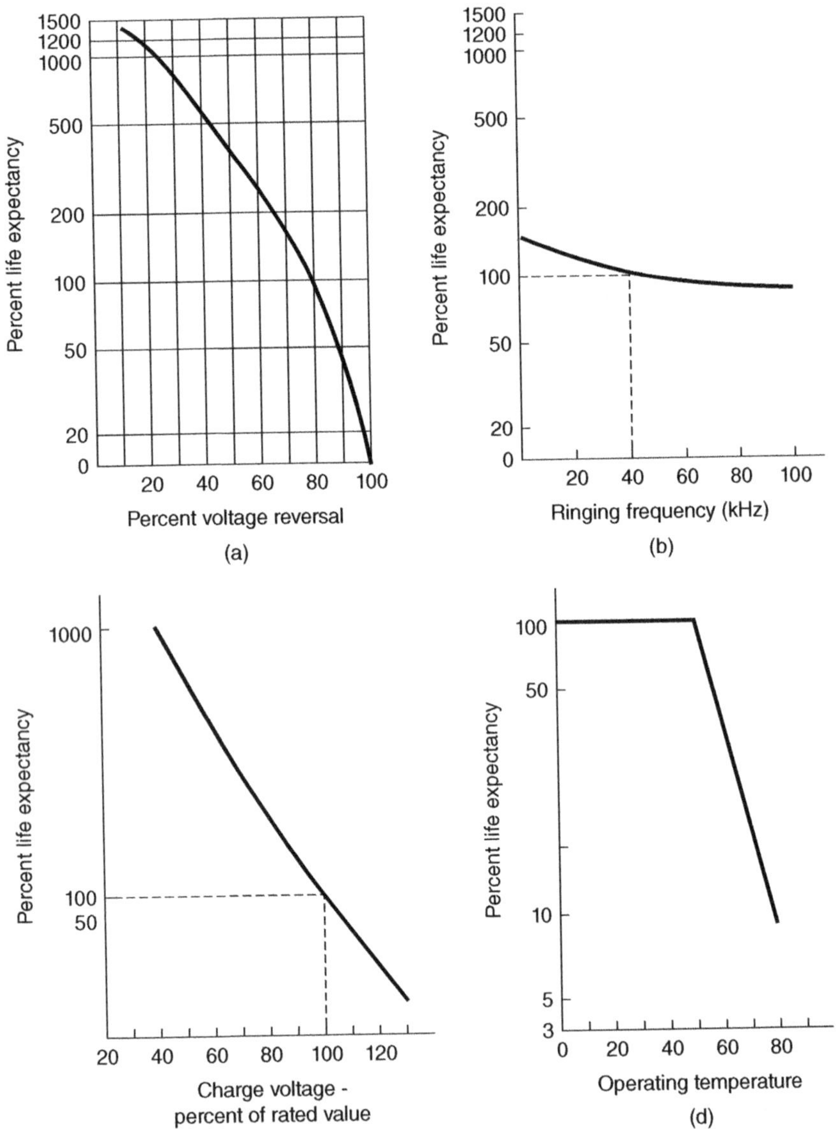

Figure 7.9 Curves of capacitor life expectancy versus (a) voltage reversal, (b) ringing frequency, (c) charge voltage, and (d) operating temperature.

7.4.1.1.5 DC Life

A capacitor also has a DC life, which is the accumulated time for which the DC voltage is held on the capacitor. The DC life is usually larger than 1000 h. In order to utilize the full shot life, it is advisable to employ fast charging of the capacitor. Large capacitor banks are generally charged to the operating voltage in 30–50 s.

7.4.1.2 Test Methods

The experimental methods available for the determination of *ESR* and *ESL* are (i) charging the capacitor to a low voltage and then observing the current waveforms after shorting the capacitor [1], (ii) variable inductance method [5], (iii) using a radio frequency (RF) oscillator to drive the capacitor and observing voltage and current waveforms at the resonant frequency [1], and (iv) the differential method, where measurements are carried out in series with an additional low-inductance capacitor of known capacitance [6].

The life testing of a capacitor is usually done under accelerated conditions by applying a voltage larger than the rated voltage. For paper/oil capacitor, if N_a is the shot life obtained at an elevated voltage V_a, the predicted life N_0 at the rated voltage V_0 is obtained by the following relation [7]:

$$\frac{N_0}{N_a} = \left(\frac{V_a}{V_0}\right)^{7.3} \tag{7.48}$$

7.4.1.3 Pulse Repetition Frequency

In order to get reasonable shot life for the capacitor under high repetition rates, specialized methods are necessary to remove the heat and mechanically reinforce the interior. Additionally, materials with high dielectric strength and low dielectric loss must be chosen.

7.4.1.4 Recent Advances

Pulsed power capacitors are used in numerous applications where high power is required. The requirements for pulsed power capacitors vary over a wide range of pulse widths. The time in which a capacitor can discharge can vary from less than a microsecond to tens of milliseconds. The primary measure of progress in pulsed power capacitors is volumetric energy density, often measured in Joules per cubic centimeter (J/cm^3). The wide ranges of requirements for the capacitor often result in confusion about this primary measurement of progress.

The greatest progress in capacitor technology has occurred in millisecond discharge capacitors. Initial drivers were commercial applications such as automatic external defibrillators (AEDs), which have expanded into other markets. Advanced thin film deposition technology has made available metalized polypropylene (MPP), which possesses the characteristics of uniform coating thickness across the width and length, freedom from defects, and extremely thin coatings (≈0.0003 μm). The MPP capacitors generally make use of coated aluminum or zinc electrodes. Because of extremely small thickness of coated electrode, the number of multilayer turns in a given cross section is manyfold compared to a foil/paper capacitor. This results in large surface area and hence higher energy density. Today's MPP capacitors possess energy density of ≈ 3 J/cm^3 at 1000 shots [8] in 2009, which is significant progress from the 1.3 J/cm^3 reported in 2003 [9] and the 0.6 J/cm^3 reported in 1986 [10]. The capability for stored energy [11,12] in a single container has increased from 20 kJ in 1975 to 50 kJ in 1980 to 100 kJ in 2000 for capacitors to drive flash lamps for the National Ignition Facility [13].

An additional outstanding advantage [14] of the MPP capacitor is its capability for fault tolerance, arising because of its self-healing against faults. Whenever a fault occurs, the high current density at the local fault site, combined with small amount of metal mass available there, results in the vaporization of a small fraction of the electrode area. The ultimate result of the fault is, therefore, not a puncture through the solid dielectric, but a fractional loss in the capacitance. This is called a soft failure mode. For an MPP capacitor, a capacitance reduction of 5% is considered to be the end of its life. If the capacitor continues to be used, the accumulation of gases increases rapidly and mechanical failure may occur by bulging of the capacitor body. In contrast, the dominant failure mode in foil/paper capacitors are dielectric punctures that result in a permanent short. The remaining capacitor elements, then, are subjected to higher voltage, and this enhanced stress results in drastic reduction in the life of the capacitor.

One of the primary limitations of capacitors is their discharge rate. In the microsecond discharge regime, General Atomics has achieved 200 V/μs at 10 kV in high-energy density, self-healing capacitors. Higher values can be achieved at higher energy densities by optimizing the electrodes for current carrying rather than self-healing. In contrast, the limitations on discharge rate in foil capacitors are due to parasitic inductance and not current density. Discharge rates of 0.1 MV/μs have been achieved in 10 kJ, large metal case Marx capacitors. New, 100 J plastic case capacitors, developed for energy storage in Linear Transformer Drivers, are capable of 0.2 MV/μs [14]. The operational parameters on these capacitors are an energy density of 0.1 J/cm^3 at 100 kVDC operating voltage and a 0.04 μF capacitance and a lifetime of 50,000 charge/discharge cycles [15].

Advances in several capacitor technologies are summarized in the Ragone plot in Figure 7.10 (Private communication, Joel Ennis, General Atomics, San Diego, CA.).

7.4.2 Trigger Pulse Generator

A large capacitor bank needs to supply peak current of 10–15 MA at more than 10^{12} A/s. This requirement cannot be met by a single spark gap. Depending on

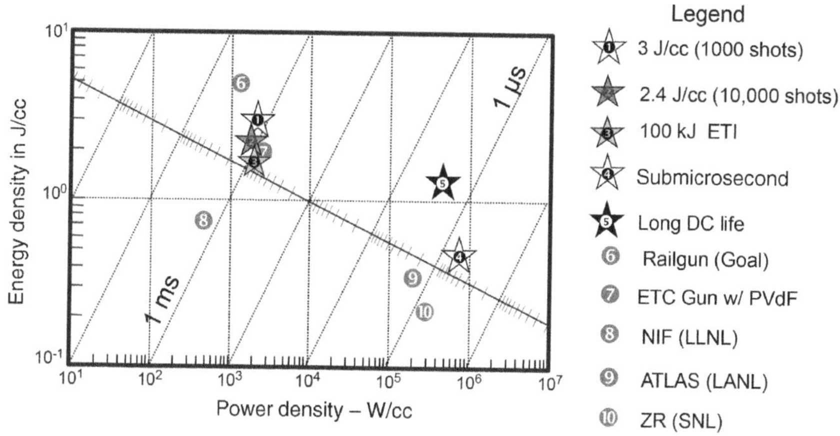

Figure 7.10 Power density of several capacitor technologies. (Courtesy of General Atomics Electronics Systems, Inc., San Diego, CA)

the design, many spark gaps would have to be operated in parallel with each spark gap usually mounted on its own subbank. Therefore, a need arises to synchronize [16–21] the firing of the spark gaps, so as to result in breakdown times within an allowable spread. When the spark gaps are connected to the load by means of low-inductance cables of length ℓ m, the allowable jitter in spark gap breakdown time is limited to the transit time of wave propagation through the cables, which is approximately $5 \times (2\ell)$ ns. If the actual spread is more than this, the slower spark gaps would be inhibited from firing, because of lowered stress due to reflections from the load end. It is thus seen that the jitter requirements in the spark gap breakdown work out to be in the range of 10–100 ns, depending upon the transmission line design. The jitter in a spark gap can be reduced by high overvolting factor and high rate of rise of trigger voltage. To achieve high performance from the spark gaps, the trigger voltage should be more than the bank voltage and dV/dt on the order of 1–10 kV/ns.

A typical layout of a pulse generator to trigger a 1 MJ capacitor bank is shown in Figure 7.11. A Blumlein generates 80 kV pulses to trigger two Marx generators, designated A and B, which in turn charge the eight cables of the master pulse generator and the (8 × 56) cables of the submaster pulse generators [21]. When the cables are charged to maximum voltage in 2 μs, another 80 kV pulse from the Blumlein, delayed suitably, triggers the master gap into conduction. The master pulse generator then delivers eight fast-rising, high-voltage pulses at the eight cables A–H. These pulses then trigger the eight submaster spark gaps, thereby generating 448 high-voltage, fast-rising pulses to trigger the 448 spark gaps of the main capacitor bank. The use of pulse charging by Marx generators in comparison to DC charging enables charging of the cables to a voltage much higher than their DC rating.

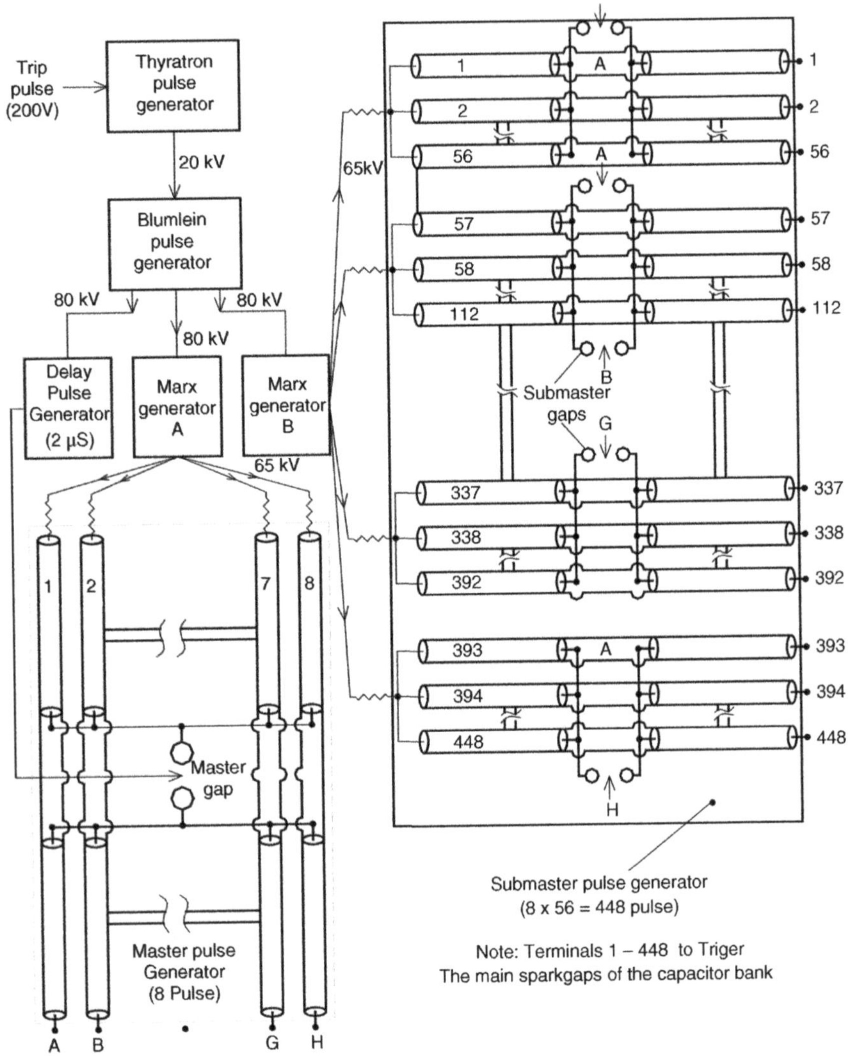

Figure 7.11 The arrangement of a trigger pulse generator for a 1 MJ capacitor bank.

7.4.3 Transmission Lines

The figure of merit [22] for a capacitor bank depends upon its ability to deliver high peak current and high rate of rise of current. The peak current is maximized for a given charge voltage V_0 by minimizing the total circuit inductance L. Each of the subsystems such as the capacitor–switch assembly, connecting leads, power feed, and load, contribute to the total circuit

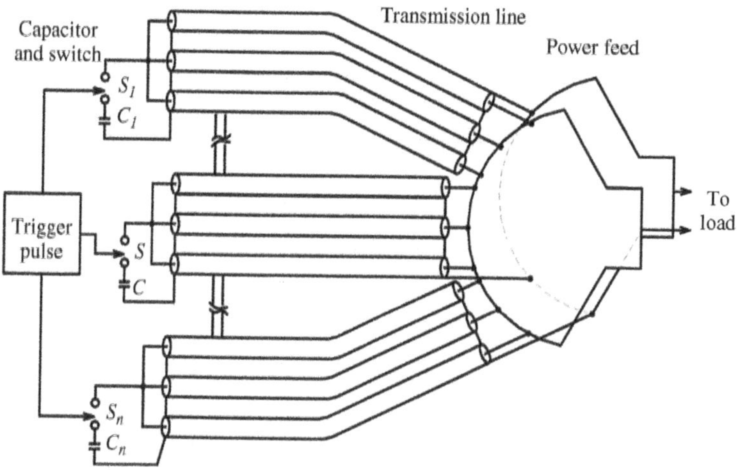

Figure 7.12 Configuration of a capacitor switch, transmission line, and power feed.

inductance. The inductance of the capacitor–switch assembly can be reduced to a negligible value by connecting a large number of capacitor–switch assemblies in parallel. Therefore, it becomes necessary that the inductance of the connecting leads, which join the capacitor bank switches on one end and the power feed on the other end, should be minimized. This may be achieved by using transmission lines in the form of coaxial cables or sandwich lines. A typical configuration of a capacitor bank–switch assembly, coaxial cables, and load is shown in Figure 7.12. Because of the complex nature of load, it is often difficult to match the impedance of the load to that of the transmission line. This gives rise to superposed oscillations of different harmonic frequencies, caused by reflections and re-reflections between the capacitor and load. These oscillations die out in a few transit times, before a steady state is reached. From this point of view, it is advisable to use as short a length of the transmission line as possible.

7.4.3.1 Coaxial Cables

The coaxial cable system is based on connecting in parallel a large number of coaxial cables, either of conventional type or of low-inductance type. The inductance per unit length of a coaxial cable is given by the following relation:

$$\tilde{L} = \frac{\mu_0}{2\pi} \times \ln\left(\frac{R_o}{R_i}\right) \tag{7.49}$$

where $\mu_0 = 4\pi \times 10^{-7}$ H/m = 1.2566×10^{-7} H/m and R_o/R_i is the ratio of inner diameter of outer conductor to outer diameter of inner conductor.

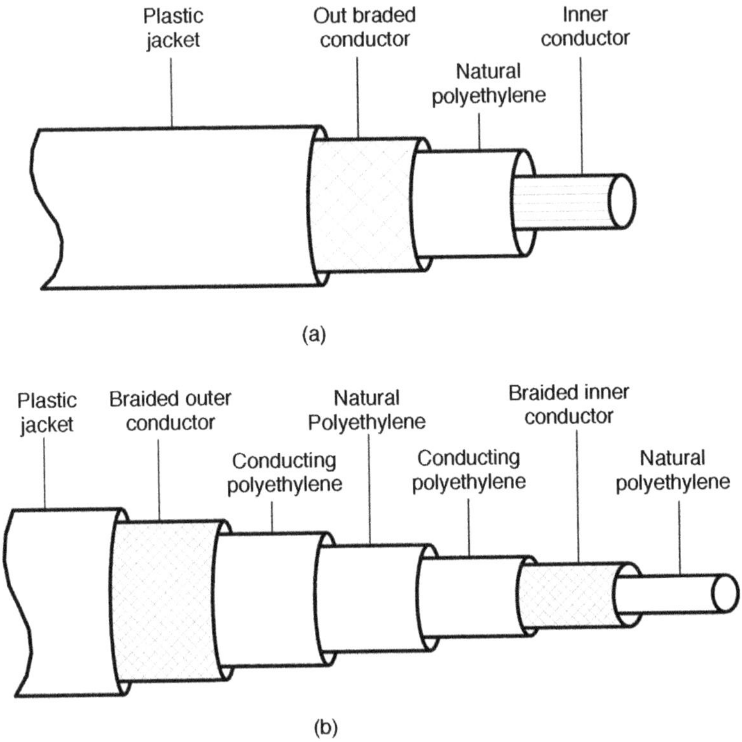

Figure 7.13 Structures of (a) conventional and (b) low-inductance coaxial cables.

Typical structures of conventional cable and low-inductance cables are shown in Figure 7.13. The conducting polyethylene in the low-inductance cable does the function of stress grading and hence enables closer spacing between the conductors. This results in a lower inductance in the range of 50–100 nH/m, compared to 200–300 nH/m, for a conventional cable. Low inductance and high capacitance per unit length of a low-inductance cable also results in a lower impedance of 10–20 Ω compared to 50–75 Ω of a conventional cable. For pulsed operation, a coaxial cable can be used at a voltage considerably above its root-mean-square (RMS) rating [23,24]. For example, RG 8/U with an RMS rating of 4 kV has been used at DC voltage of 60 kV continuously and at more than 200 kV for pulses of 20 μs. A 17/14 cable, a low-inductance cable having its outer conductor according to outer conductor of RG 17 and inner conductor according to outer conductor of RG 14, has been operated at pulsed current [11] of 10–40 kA. A low-inductance coaxial cable configuration preferred over a conventional cable, since it results in a more efficient arrangement.

7.4.3.2 Sandwich Lines

A sandwich line is made up of two or three metallic plates separated by a dielectric as shown in Figure 7.14. The inductance per unit length of this configuration is given by [25]

$$\tilde{L} = \mu_0(d/w)K \tag{7.50}$$

where K is a correction factor depending on the ratio of d/w. The commonly used materials for the metallic plates are aluminum or stainless steel. The dielectric is made up of many layers of Mylar. Depending on the voltage rating of the capacitor bank, a potential difference of 20–100 kV could exist between the two plates. To prevent the breakdown in the dielectric, certain precautions should be taken: the metallic surfaces facing the dielectric should be smooth; the edges of the plates should be rounded; there should be sufficient extension of Mylar sheets beyond the metal edges so as to prevent surface flashover; and air gaps are to be avoided at metal–dielectric interfaces. A gel-like spreadable dielectric may be used to fill in these voids. Tinning the electrode surfaces and applying a strong and uniform clamping force between the two metal plates help in securing an intimate contact.

The pressure generated by the magnetic field on the plates p_m is given by [2]

$$p_m = 2\pi \times \left(\frac{I}{w}\right)^2 \times 10^{-12}\,\text{atm} \tag{7.51}$$

where (i/w) is the current per unit width expressed in amperes per meter.

The external clamping pressure applied between the plates should be equal to or greater than p_m; otherwise, there will be distortion in the plates. Because of the pulsed nature of the discharge, the current will be conducted in a thin layer, with its skin depth being less than a millimeter. But a higher thickness of the plates will help in the mechanical strengthening of the structure. In comparison to coaxial cables, sandwich lines are more expensive, but they can provide smaller inductance, lower impedance, and higher peak current. They are generally used in situations where the experiment is located very near to the

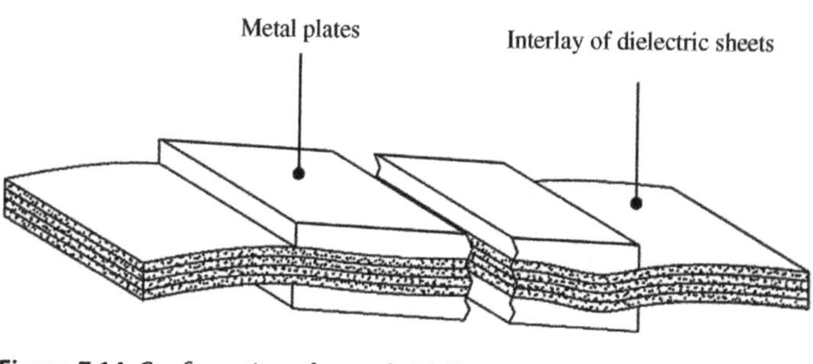

Figure 7.14 Configuration of a sandwich line.

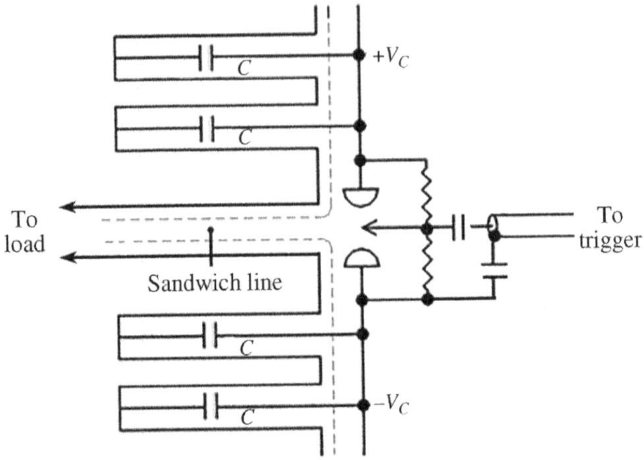

Figure 7.15 Configuration of a capacitor bank using a sandwich line.

capacitor bank. A typical configuration [26] of a capacitor bank using sandwich lines is shown in Figure 7.15.

7.4.4 Power Feed

The purpose of the power feed is to receive the discharge current from a large number of coaxial cables at one end and deliver it to the load on the other end. It should be mechanically configured for clamping and tightly securing a large number of cables. It should be designed for low inductance to enable efficient delivery of energy received from the capacitor bank. A collector is usually designed in a parallel plate configuration with metallic plates sandwiched by layers of dielectric sheets.

The electrical design at the cable-receiving end should ensure that no breakdown takes place across the surface of the cable, where the insulation is stripped and there is no occurrence of stress enhancement on the dielectric sheets [27]. Both the above factors are achieved by providing stress grading, by specially profiled electrodes, and by immersion in an insulating liquid of high dielectric constant. A typical electrical design for the insulation integrity of the cable–collector junction is shown in Figure 7.16 [27].

The current carried by a large number of coaxial cables has to be collected by a single collector plate that gives rise to a large current density, usually greater than 10 MA/m in large banks. At this current density, the pressure exerted on the plates [8] as calculated by (7.51) yields a value of 9000 psi. This emphasizes the need for strong clamping by an external uniform force. In a megajoule bank design [21], the collector plates are formed from thick aluminum plates. The

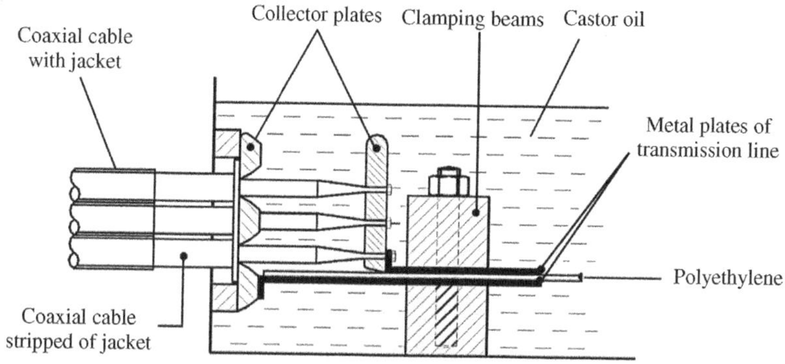

Figure 7.16 Electrical design for the cable–collector junction.

plates are backed by rubber sheets and clamped down by concrete blocks, held down by steel tie bolts.

7.5 Safety

An energy storage capacitor with a maximum stored energy of E_s^2 has a specification for fault energy E_f. If the external fault energy dumped into the capacitor exceeds E_f, catastrophic failure of the capacitor may take place by the bursting of the container with an explosive-like force, resulting in chunks of material flying at high velocities. External fault energy can be introduced into a capacitor in the following ways: (i) If a capacitor becomes faulty during the charging process, the healthy capacitors connected in parallel in a subbank discharge their stored energy into it, and (ii) if a capacitor develops a fault during the discharge of a capacitor bank, the maximum energy dumped into it will be an addition of the energy stored in that particular subbank containing the faulty capacitor and the energy stored in the rest of the subbanks. For a relatively low-energy capacitor bank (<100 kJ), the following usual safety precautions will suffice, namely, enclosing the capacitor bank in a container is typically sufficient.

For a large-energy (>100 kJ) capacitor bank, safeguards should be included to prevent/restrict the damage to human beings and facility. A few such features are (i) isolation of capacitors and subbanks by inductors, (ii) inclusion of fast-opening fuses in individual capacitors and subbanks, (iii) provision for emergency dump of the stored energy into a dummy load on command trigger, (iv) enclosing the subcapacitor bank in properly secured metal cabinets, and (v) locating the capacitor bank in a concrete bunker.

The metallic enclosure, in addition to providing safety against accidental explosion of the capacitor container, also provides protection against electromagnetic interference (EMI) effects arising from high magnitudes of *di/dt* and *dV/dt* in capacitor bank. For good shielding effectiveness of EMI, a capacitor bank cabinet, comprised of two enclosures, is recommended. One enclosure should be made of good electrical conductors such as aluminum or copper, so that the interfering electric fields are effectively attenuated. The other enclosure should be made of good magnetic metals such as iron or magnetic steel, so that it efficiently attenuates the magnetic fields. Use of galvanized iron is a good and simple choice for EMI suppression, since it contains iron (good magnetically) and zinc (good electrically). Conducting RF gaskets at the joints between panels ensures minimal RF leakage. For reliable and EMI-free performance of the capacitor bank, optimum use should be made of optical fibers for signal transmission and electrooptical/optoelectronic techniques for pulsed voltage and current measurement.

In a large capacitor bank, fuses [28] are included to limit the fault current due to internal or external faults, including overloads and failure of the capacitor itself. A fuse with low self-inductance should open rapidly when a critical current–time condition is exceeded, so as to limit the current and energy dumped into the faulty component, and isolate the source of the fault by continuing to provide high hold-off voltage without allowing restrike.

Generally, fuses are comprised of exploding wire or foil conductors in a low-inductance configuration. The conductor elements are embedded in a plasma-quenching medium made of a combination of materials including sand. The assembly is contained in a strong, reinforced mechanical structure to allow silent operation without the container bursting.

Figures 7.17 and 7.18 illustrate the salient features of an 85 kJ capacitor bank [29,30], wherein the safety features described above are implemented. The capacitor bank consists of eight subcapacitor banks (SB_1–SB_8), each having independent double-mode trigatron switches (TS_1–TS_8). On the application of a command trigger pulse V_T, the multiple-trigger pulse generator simultaneously generates 16 high-voltage trigger pulses comprised of 8 positive pulses (V_{AT1}–V_{AT8}) and 8 negative trigger pulses (V_{CT1}–V_{CT8}), which trigger the switches (TS_1–TS_8) into conduction. The energy stored in the capacitor bank is then discharged into the high magnetic field coil (FC) in series with a tuning coil (TC), so as to produce current in the form of a damped sinusoid. Each subbank has a capacitance of 54 μF and is made up of a parallel combination of three energy storage capacitors, each rated at 18 μF, 20 kV. The subbanks are charged to the desired voltage (V_c) by means of a variable HVDC supply rated up to 20 kV maximum. A vacuum contactor is used as an emergency dump switch (EDS) to deposit the energy stored in the subbanks into charging/dump resistors (R_1–R_8). The paralleling of individual capacitors is implemented through series inductors of 1.8 μH to protect healthy capacitors against damage due to excessive peak current in the event of failure of one of the capacitors.

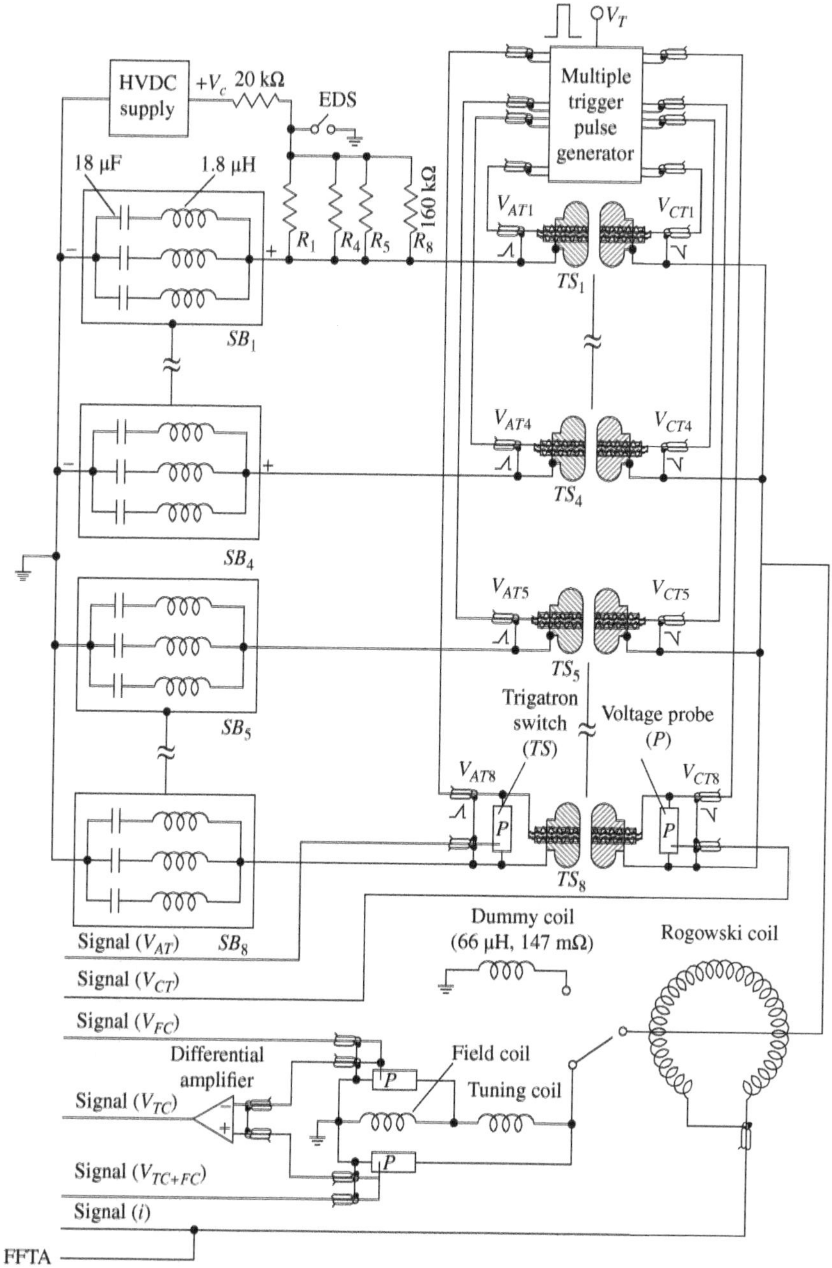

Figure 7.17 Schematic and measurement setup for a 85 kJ/20 kV capacitor bank. (Reprinted with permission from Ref. [29]. Copyright 1992, AIP Publishing LLC.)

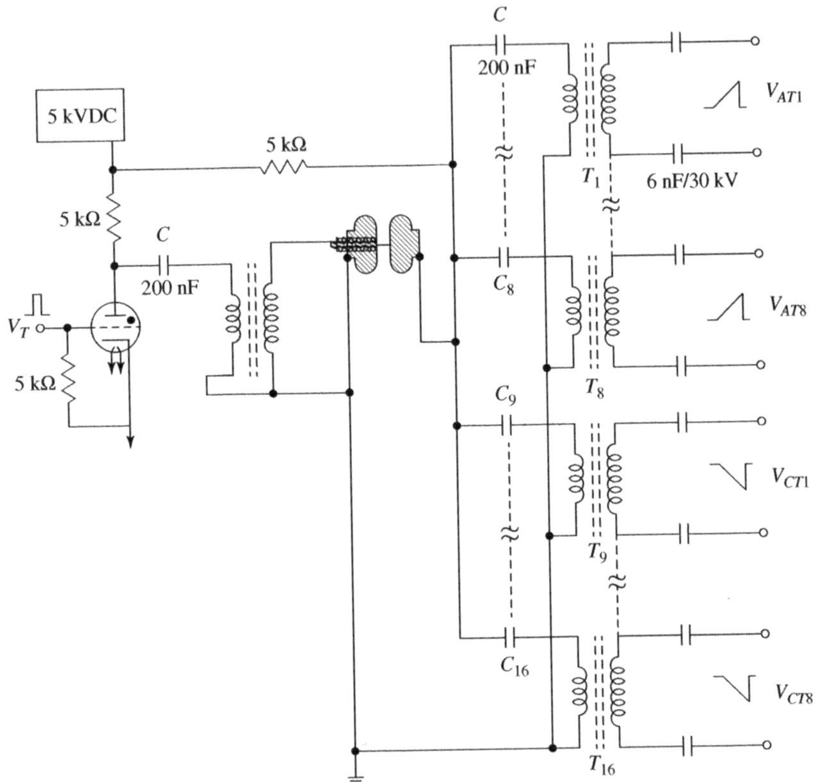

Figure 7.18 Circuit schematic of a 16 pulse multiple-trigger pulse generator. (Reprinted with permission from Ref. [29]. Copyright 1992, AIP Publishing LLC.)

The layout of a multiple-trigger pulse generator is shown in Figure 7.18. The capacitors (C_1–C_{16}) are charged to 5 kVDC. Application of a command trigger pulse V_T of 150 V, 10 μs to the hydrogen thyratron grid causes the discharge of charged capacitor C into the primary of pulse transformer T, thereby producing an auxiliary discharge of the cathode trigger gap of the single-mode master trigatron S, thus initiating it into conduction. Capacitors (C_1–C_{16}) then discharge into the primary windings of pulse transformers (T_1–T_{16}), respectively, through the master trigatron, thereby producing simultaneously 16 high-voltage pulses on their secondary windings. These 16 pulses are configured in the form of 8 positive pulses (V_{AT1}–V_{AT8}) and 8 negative pulses (V_{CT1}–V_{CT8}). They are transmitted to the trigger gaps of double-mode trigatrons (T_{S1}–T_{S8}) by coaxial cables via coupling capacitors rated at 6000 pF, 30 kV. The pulse transformers (T_1–T_{16}) are identical in construction and employ ferrite rods as high-permeability core, mineral oil for insulation, and spiral windings of alternate layers of copper and Mylar sheets.

The trigatron assembly consists of two main electrodes E_A and E_C with respective trigger electrodes T_A and T_C. The electrodes are made of stainless steel with main electrode gap spacing of 9.7 mm. The insulating medium is air at atmospheric pressure giving self-breakdown voltage of 25 kV. The trigatron switches are assembled in a double-enclosure box, the walls of which are made of plywood with an aluminum inner liner. These boxes serve multiple purposes: elimination of fire hazard due to hot flying particles originating from hot spot regions of electrodes, reduction of sound intensity produced by shock waves during sparking, and EMI shielding of the radiated interference from spark channels.

The field coil (FC) for generation of pulsed high-magnetic field (1.2×10^7 A/m) is in the form of a single-layer helix of 21 turns of hard copper busbars with cross section of 10 mm × 3 mm, wound on edge. The field coil carries a peak current of 56 kA at a ringing frequency of 1.1 kHz at the full rated energy of the capacitor bank. The high magnetic flux density (15 T) results in enormous mechanical stresses on the coil and the necessary strength is achieved by a combination of radial banding of polyglass and axial clamping by nonmagnetic stainless steel bolts.

The capacitor bank discharge associated with high values of V, i, dV/dt, and di/dt is a source of EMI and generates transients in functional and measurement circuits by various interference modes such as capacitive coupling, inductive coupling, common impedance coupling, and radiative coupling. The immunity against EMI is achieved by assembling the components in three electromagnetic shielded enclosures. The components illustrated in Figures 7.17 and 7.18 are divided into three sections: (a) subbanks SB_1–SB_4, trigatrons TS_1–TS_4, capacitors C_1–C_4 and C_9–C_{12}, and pulse transformers T_1–T_4 and T_9–T_{12}; (b) subbanks SB_5–SB_8, trigatrons TS_5–TS_8, capacitors C_5–C_8 and C_{13}–C_{16}, and pulse transformers T_5–T_8 and T_{13}–T_{16}; and (c) tuning and dummy coils. The aforementioned sections (a), (b), and (c) are contained in separate shielded enclosures. The shielded enclosures also serve as safety cages for protection against mechanical bursts of capacitors and other energy-handling components. As an additional precaution, before the capacitor bank is fired into the FC at full energy, the integrity of the system is checked by measurement of various signals V_{AT}, V_{CT}, V_{FC}, V_{TC}, and by test firing of the capacitor bank at lower energy into a dummy coil.

7.6 Typical Capacitor Bank Configurations

Table 7.1 lists some capacitor banks. A few details regarding their ratings, location, switches, transmission line, and trigger pulse generator are summarized.

Table 7.1 Capacitor bank configurations.

Location	Parameters	Switches	Transmission lines	Trigger Pulse generator
Culham Lab [21] UKAEA, Abingdon, Berks, UK	1250 μF, 40 kV, 1 MJ, 12 MA/12 μs	448 cascade gaps of 4 electrode type, atmospheric air insulation, breakdown jitter <35 ns	448 × 4 numbers of coaxial cables each 28 ft. long to carry discharge current	Single master gap and 8 submaster gaps generate 448 trigger pulses of 65 kV. Marx bank charges cables 20 ft. long
Culham Lab [31] UKAEA, Abingdon, Berks, UK	34,048 μF, 8 kV, 1 MJ, 1.6 MA/180 μs	112 trigatrons for main gaps. 28 ignitrons for clamp switches	112 × 3 numbers of coaxial cables each 12 m long to carry discharge current	Single master gap and 2 submaster gaps generate 112 trigger pulses of 35 kV
Ionized Gas Lab of CNEN [32], Frascati, Italy	1248 μF, 40 kV, 1 MJ, 6.7 MA	96 cascade gaps made of W–Cu alloy. Breakdown jitter < 30 ns	96 × 3 numbers of coaxial cables to carry the discharge current	Single master gap and 8 submaster gaps generated 8 × 12 (=96) trigger pulses of 48 kV
General Electric Research Laboratory [20], New York, USA	213 μF, 60 kV, 384 KJ, 4 MA/14 μs	4 cascade gaps insulated with atmospheric air; electrodes are hot-rolled steel and trigger is steel rod; jitter < 25 ns	4 sandwich lines in parallel plate; 3 ft. wide copper plates insulated by 6 polyethylene sheets	4 trigger pulses of 73 kV rising in 30 ns generated by pulse-charging 4 parallel plate lines to 90 kV and short circuiting at one end by a master gap
Max Planck Inst. [18], Munich, Germany	356 μF, 30 kV, 160 KJ, 1.75 MA/10 μs	40 triggered rail gaps at atmospheric air for main gaps; clamp switches also similar	40 × 3 numbers of low-inductance-type coaxial cables	40 trigger pulses generated by one master gap and 4 submaster gaps.
Sandia National Laboratories [11], USA	400 μF, 120 kV, 3 MJ, 2 MA, 2 numbers of 800 μF charged to ±60 kV	One solid dielectric switch triggered by exploding wire-initiated detonator	48 coaxial cables of 17/14 type connect the capacitor bank to the switch. Current carried from switch to load by similar 48 cables	
Bhabha Atomic Research Centre [29], Mumbai, India	432 μF, 20 kV,85 kJ, 56 kA/1.1 kHz	8 double-mode trigatrons; useful life 10,000 C; one switch per module	20 m long cables, 4 in parallel	One master gap and 16 ferrite cored transformers generate 16 simultaneous pulses
French-German Research Institute [33], St. Louis, France	865 μF, 10 kV, 50 kJ 90 kA at $di/dt = 600$ A/μs	4 thyristors with amplifying gates and interdigited gate structure in series		
Los Alamos National Laboratory [34], USA	36 MJ, 40–50 MA, 5–10 μs; 152 parallel Marx banks	2 rail gaps per Marx module; total number of switches = 300; useful switch life ≈200–300 C of accumulated charge	76 Mylar insulated transmission line; one transmission line per pair of modules	

7.7 Example Problems

Example 7.1

A capacitor bank, consisting of 24 capacitors in parallel, is charged by a power supply in 30 s. The capacitors have a capacitance value of 7.1 μF and a voltage rating of 40 kV. Calculate the following: (i) current rating of the power supply unit in "constant current mode" and (ii) power rating of the power supply unit in "constant power mode."

Solution

i. Substituting the values of $V_0 = 40 \cdot 10^3$ V, $C = 7.1 \times 10^{-6}$ F, and $t = 30$ s in Equation 7.34, we obtain the current rating I_0 of the "constant current mode as

$$I_0 = \frac{V_0 C}{t}$$

which reduces to $I_0 = \dfrac{(40 \times 10^3)(24)(7.1 \times 10^{-6})}{30} = 227$ mA.

ii. Substituting the values of $V_0 = 40 \times 10^3$ V, $C = 7.1 \times 10^{-6}$ F, and $t = 30$ s in Equation 7.36, we obtain the power rating P_0 of the power supply unit in "constant power mode" as

$$P_0 = \frac{CV_0^2}{2t}$$

which reduces to $P_0 = \dfrac{(7.1 \times 10^{-6}) \times (24) \times (40 \times 10^3)^2}{(2) \times (30)} = 4.54$ kW.

Example 7.2

A capacitor bank consists of five capacitors of 200 μF, each connected in parallel. Each capacitor has an inductance of 100 nH. The capacitor bank, charged to a voltage of 5 kV, is switched into an inductive load of 1.9 μH through a spark gap having an inductance of 80 nH. Calculate the peak discharge current.

Solution

In the equivalent circuit given in Figure 7.6, the following values apply in the present example:

(continued)

(continued)

$$C_s = 5 \times (200\ \mu\text{F}) = 1000\ \mu\text{F}$$

$$L_c = \frac{100\ \text{nH}}{2} = 20\ \text{nH}$$

$$L_{sw} = 80\ \text{nH}$$

$$L_\ell = 1.9\ \mu\text{H}$$

Hence, the overall capacitor bank is reduced to a single capacitor C_s discharging into an inductor of value L_0, where the values of C_s and L_0 are given by $C_s = 1000\ \mu\text{F}$ and $L_0 = L_c + L_{sw} + L_\ell$, which reduces to

$$L_0 = 20\ \text{nH} + 80\ \text{nH} + 1.9\ \mu\text{H} = 2000\ \text{nH}$$

The peak discharge current I_m is therefore given by

$$I_m = V_0\sqrt{\frac{C_s}{L_0}}$$

$$I_m = \left(5 \times 10^3\right)\sqrt{\frac{1000 \times 10^{-6}}{2000 \times 10^{-9}}} = 111.8\ \text{kA}$$

Example 7.3

A capacitor bank consists of four subbanks discharging in parallel into a common load. Each subbank has a capacitance of 7 μF charged to a voltage of 50 kV. The load current in each subbank is carried by five cables in parallel, each of 5 m in length. The individual values of inductances for capacitor, spark gap, and cable are 70, 30, and 140 nH/m, respectively. Calculate the following parameters: (i) total energy stored in the capacitor bank, (ii) peak current on a short circuit load, and (iii) ringing frequency on a short circuit load.

Solution

In the equivalent circuit given in Figure 7.8, the following values apply in the present example:

$$C_s' = 4 \times (7\ \mu\text{F}) = 28\ \mu\text{F}$$

$$L'_c = \frac{70\text{ nH}}{4} = 17.5\text{ nH}$$

$$L'_{sw} = \frac{30\text{ nH}}{4} = 7.5\text{ nH}$$

$$L'_{t\ell} = \frac{(5\text{ m}) \times (140)\text{ nH}/m}{(4) \times (5)} = 35\text{ nH}$$

Therefore, the total circuit inductance L'' of the equivalent circuit on a short circuited load is given by

$$L'' = L'_c + L'_{sw} + L'_{t\ell},$$

which reduces to $L'' = 17.5\text{ nH} + 7.5\text{ nH} + 35\text{ nH} = 60\text{ nH}$.

The required parameters are as follows:

a)

Energy stored in the capacitor bank E_t is given by

$$E_t = \frac{1}{2} C'_s V_0^2$$

$$E_t = \frac{1}{2}\left(28 \times 10^{-6}\right) \times \left(50 \times 10^3\right)^2 = 35\text{ kJ}$$

b) Peak current on a short-circuited load I_p is given by

$$I_p = V_0 \sqrt{\frac{C'_s}{L''}}$$

$$I_p = \left(50 \times 10^3\right) \sqrt{\frac{\left(28 \times 10^{-6}\right)}{\left(60 \times 10^{-9}\right)}} = 1.08\text{ MA}$$

c) Ringing frequency on a short-circuited load f_r is given by

$$f_r = \frac{1}{2\pi \cdot \sqrt{L'' C'_s}}$$

$$f_r = \frac{1}{2\pi \times \sqrt{\left(60 \times 10^{-9}\right) \times \left(28 \times 10^{-6}\right)}} = 130\text{ kHz}.$$

References

1 V. Valencia et al., High Repetition Rate, Long Life Capacitors Developed for Laser Isotope Separation Modulators. *Conference Record of the 15th Power Modulator Symposium*, p. 181, 1982.

2 E.L. Kemp, *Principal Features in Large Capacitor Banks*, Pulsed Power Lecture Series, Texas Tech University, Lubbock, TX.

3 H.K. Jennings, Charging Large Capacitor Banks in Thermonuclear Research. *Electr. Eng.*, Vol. 80, p. 419, 1961.

4 B.R. Hayworth and D. Warrilow, Constant Power Charging Supplies for High Voltage Energy Transfer, CSI Technical Note # 109 A, 1977.

5 B.R. Hayworth, How to Tell a Nanohenry from a Microfarad. *Electron. Instrum.*, Vol. 8, No. 4, p. 36, 1972.

6 W.C. Nunnally, M. Kristansen, and M.O. Hagler, Differential Measurement of Fast Energy Discharge Capacitor Inductance and Resistance. *IEEE Trans. Instrum. Meas.*, Vol. 24, No. 2, p. 112, 1975.

7 R. Litte and R. Limpaecher, *Test of High Performance Capacitors*. Proceedings of the 4th International Pulsed Power Conference, p. 387, 1983.

8 W. MacDougall, J.B. Ennis, Y.H. Yang, R.A. Cooper, J.E. Gilbert, J.F. Bates, C. Naruo, M. Schneider, N. Keller, S. Joshi, T.R. Jow, J. Ho, C.J. Scozzie, and S.P.S. Yen, High Energy Density Capacitors for Pulsed Power Applications. *Proceedings of the IEEE Pulsed Power Conference*, 2009.

9 F.W. MacDougall, J.B. Ennis, R.A. Cooper, J. Bates, and K. Seal, High Energy Density Pulsed Power Capacitors. *Proceedings of the IEEE International Pulsed Power Conference*, pp. 513–517, 2003.

10 F.W. MacDougall, J.B. Ennis, Y.H. Yang, K. Seal, S. Phatak, B. Spinks, N. Heller, C. Naruo, and T.R. Jow, Large, High Energy Density Pulse Discharge Capacitors Characterization. *Proceedings of the IEEE Pulsed Power Conference*, pp. 1215–1218, 2005.

11 E.C. Cnare and C.B. Dobbie, A 3 MJ Capacitor Bank for the SNL Electro-Explosive Facility. *Proceedings of the 4th International Pulsed Power Conference*, p. 193, 1983.

12 W.J. Sarjeant, F.W. MacDougall, D.W. Larson, and I. Kohlberg, Energy Storage Capacitors: Aging and Diagnostic Approaches for Life Validation. *IEEE Trans. Magnet.*, Vol. 33, No. 1, p. 501, 1997.

13 T. Scholz, P. Windsor, and M. Hudis, Development of a High Energy Density Storage Capacitor for NIF. *Proceedings of the International Pulsed Power Conference*, pp. 114–117, 1999.

14 W.J. Sarjeant, J. Zirnheld, and F.W. MacDougall, Capacitors. *IEEE Trans. Plasma Sci.*, 26, No. 5, p. 1368, 1998.

15 J.B. Ennis, F.W. MacDougall, X.H. Yang, R.A. Cooper, K. Seal, C. Naruo, B. Spinks, P. Kroessler, and J. Bates, Recent Advances in High Voltage, High

Energy Capacitor Technology. *Proceedings of the IEEE International Pulsed Power Conference*, pp. 282–285, 2007.

16 B. Augsburger, B. Smith, I.R. McNab, Y.G. Chen, D. Hewkin, K. Vance, and D. Disley, Royal Ordnance 2.4 MJ Multi-Module Capacitor Bank. *IEEE Trans. Magnet.*, Vol. 31, No. 1, p. 16, 1995.

17 R.F. Fitch and N.R. McCormick, Low Inductance Switching Using Parallel Spark Gaps. *Proc. IEE*, 106 A, No. Suppl. 2, p. 117, 1959.

18 R.H. Suess and G. Mueller, Operating Characteristics of Spark Gaps and Bank Assembly for an Electron Ring Accelerator Experiment. *7th Symposium on Fusion Technology*, p. 423, 1972.

19 T.E. James, Fast High Current Switching Systems for Megajoule Capacitor Bank, Report No. CLM-L 23, Culham Laboratory. Abingdon, Berkshire, UK, 1973.

20 L.M. Goldman, H.C. Pollock, J.A. Reynolds, and W.F. Westendorp, Spark Gap Switching of a 384 kJ Low Inductance Capacitor Bank. *Rev. Sci. Instrum.*, Vol. 33, No. 10, p. 1041, 1962.

21 N.R. McCormick, A One Megajoule, Low Inductance Capacitor Bank for Nuclear Fusion Research, Report No. CLM-P171, UKAEA Research Group, Culham Lab, Abingdon, Berkshire, UK.

22 F.B.A. Frungel, *High Speed Pulse Technology*, Vol. 1, Academic Press, 1965.

23 R.A. Gross and B. Miller, Plasma Heating by Strong Shockwaves, in *Methods of Experimental Physics: Volume 9 – Part A: Plasma Physics*, L. Marton, ed., Academic Press, 1970.

24 D.P. Biocourt, Electrical Characteristics of Coaxial Cable, Report No. LA3318, Los Alamos Laboratory, November 1965.

25 F.W. Grover, *Inductance Calculations*: Working Formulas and Tables, 1946 and 1973, Dover Phoenix Edition, 2004.

26 G. Boult, M. M. Kekez, G. D. Lougheed, W. C. Michie, and P. Savic, 1 MJ Bank Facility, Laboratory Technical Report, No. LTR-GD-81, National Research Council, Canada.

27 T.E. James, Field Sketching as Applied to Electric Field Design Problems, in *High Voltage Pulse Technology*, L.L. Alston, ed., Oxford University Press, 1968.

28 Maxwell Catalog *High Voltage Capacitor Protection Fuses: Fuse Selection Guide*, Maxwell Laboratories Inc.

29 P.H. Ron, K. Nanu, S.T. Iyengar, K.V. Nagesh, R.K. Rajawat, and V.R. Jujaray, An 85 kJ High Performance Capacitor Bank with Double Mode Trigatrons. *Rev. Sci. Instrum.*, Vol. 63, No. 1, p. 37, 1992.

30 P.H. Ron, K. Nanu, S.T. Iyengar, and V.K. Rohatgi, Single and Double Mode Gas Trigatrons: Main Gap and Trigger Gap Interactions. *J. Phys. D Appl. Phys.*, Vol. 21, p. 1738, 1988.

31 G.C.H. Heywood et al., High Current Capacitor Banks for a 2 MJ High Beta Toroidal Experiment (HBTX). *6th International Symposium on Fusion Technology, Aachen, Germany*, p. 219, 1970.

32 A. Marconcini, Constructive Technologies of an Energy Storage Capacitor Bank 1 MJ-40 kV, Report of Passoni and Villa, Milano, Italy, 1982.

33 E. Spahn, G. Burderer, and F. Hatterer, Compact 50 kJ Pulse Forming Unit, Switched by Semiconductors. *IEEE Trans. Magnet.*, Vol. 31, No. 1, p. 78, 1995.

34 W.M. Parsons, The Atlas Project: A New Pulsed Power Facility for High Energy Density Physics Experiments. *IEEE Trans. Plasma Sci.*, Vol. 25, No. 2, p. 205, 1997.

lectrical Breakdown in Gases

ases are insulators, but in the presence of an electric field of sufficiently high nplitude, a current can flow by generating free charges from within the gas self by ionization. Under right conditions, a low-impedance channel is formed id the gas experiences a breakdown.

Electrical breakdown is a complex, multistage phenomenon beginning with e kinetics of charged particles that occurs when a threshold is exceeded. iarge carriers can be generated by photoionization from an external light urce or by collisions in the gas by a sufficiently high electric field. Depending the magnitude of the applied voltage and the specifics of the geometry, any imber of complex physical phenomena may occur. A comprehensive survey gas discharge phenomena is beyond the scope of this book, but a number of xts can be found [1,2].

In this chapter, a short treatise on atomic collisions in gases is followed by a scription of Townsend's experiments in the early twentieth century. The schen curve is presented and followed by a description of the predominant eories of spark formation. Other breakdown phenomena, such as coronas and e hollow electrode carrier generation used in pseudosparks, are also dis-ssed. Finally, the efficient use of gaseous insulation with intershields is scussed.

1 Kinetic Theory of Gases

ctrical breakdown occurs in a gas when a high-conductivity channel is med between cathode and anode. A key issue in gas discharge physics is how s high conductivity is built up by the electronic atom and ionic collision ocesses in the gas and the quantitative treatment is reduced to the study of ticle behavior and the conditions of motion. However, the nature and mode motion of the neutral molecules is essential knowledge. Therefore, before a dy is made of the behavior of gases under the influence of an electric field, it is propriate to review the basic principles of the kinetic theory of gases pertinent

idations of Pulsed Power Technology, First Edition. Jane Lehr and Pralhad Ron.

to gaseous ionization and breakdown. From there, the various concepts of ionic and electronic can be inferred.

8.1.1 The Kinetic Theory of Neutral Gases

The gas theory discussed here assumes that all molecules are electrically neutral. The relationship between the macroscopic quantities of pressure p, temperature T, and volume V_{Vol} of a gas is described by what we now call the ideal gas law:

$$pV_{Vol} = nR_{univ}T \tag{8.1}$$

where

n is the number of moles of gas

and

R_{univ} is the universal gas constant, with a value = 8314 J/K.

The ideal gas law is based on empirical observations but provides a bridge to microscopic quantities. It is sometimes convenient to express Equation 8.1 in terms of the gas density N—the number of molecules per unit volume:

$$N = \left(\frac{nN_A}{V_{Vol}}\right) \tag{8.2}$$

where N_A (=6.02×10^{23} molecules/mol) is Avogadro's number. Equation 8.1 is expressed as

$$p = \frac{nN_A}{V_{Vol}} \cdot \frac{R_{univ}}{N_A} \cdot T = N \cdot \frac{R_{univ}}{N_A} \cdot T \tag{8.3}$$

Equation 8.3 is more commonly written

$$p = Nk_BT \tag{8.4}$$

where

k_B is the Boltzmann constant ($= R_{univ}/N_A = 1.3806 \times 10^{23}$ (J/K-molecule))

Equations 8.3 and 8.4 relate the macroscopic quantity pressure to the microscopic number density N, the number of molecules per unit volume. The number density N has the advantage of being an absolute (invariant) quantity, whereas the gas pressure must be corrected for temperature. The number density N_2 of the gas at any pressure p_2 and temperature T_2 (in Kelvin) can be obtained from the known number density N_1 of the gas corresponding to a pressure p_1 and temperature T_1:

$$N_2 = N_1 \cdot \frac{p_2}{p_1} \cdot \frac{T_1}{T_2} \tag{8.5}$$

where the temperatures T_1 and T_2 are in Kelvin.

8.1.1.1 Maxwell–Boltzmann Distribution of Velocities

Building on the work of Robert Boyle, Swiss mathematician Daniel Bernoulli (1700–1782) formulated a quantitative theory in his book on hydrodynamics [3]. He derived Boyle's law by computing the force exerted on a movable piston by the impacts of a number ($= n \cdot N_A$) of molecules with mass m moving with a speed v in a closed space of volume V_{Vol}. That is, if the volume is lower, the pressure is proportionally higher as Boyle stated, $pV_{Vol} = constant$.

Bernoulli also showed that the pressure is proportional to the kinetic energy of the particles ($1/2mv^2$) since the frequency of impacts is proportional to the speed and the force of each impact is proportional to the momentum mv. This explained the observed fact that increases of pressure arising from equal increases in temperature are proportional to the density. Bernoulli's theory introduced the idea that heat or temperature could be identified with the kinetic energy of particles in an ideal gas. His insight, however, was not widely accepted.

In the 1820s, an Englishman named John Herapath derived the relationship between pressure and molecular speed [4]. The paper, rejected by the Royal Society but published later in an 1836 issue of *Railway Magazine*, was the first published calculation of the average speed of a molecule.

The relationship between pressure and the average kinetic energy of the molecules in the gas are shown in Figure 8.1. A box with sides of length a, b, and c contains a gas whose molecules have mass m and all travel with the same speed v. The pressure and temperature are assumed to be uniform throughout the gas volume, meaning that 1/3 of the total number of molecules are traveling along each of the x-, y-, and z-axes.

The total number of molecules contained within a gaseous volume V_{Vol} is ($n N_A$) and the number of molecules traveling in each of the x-, y-, and z-directions is $1/3n N_A$. When molecules traveling with velocity v in the x-direction collide with the wall, the change in momentum is $2mv$. The average time for a molecule to travel the length of the box is b/v and the rate of collisions on each of the six walls of the box is $v/2b$ per second. The force F exerted on the wall of the box is equal to the rate of change of momentum per molecule times

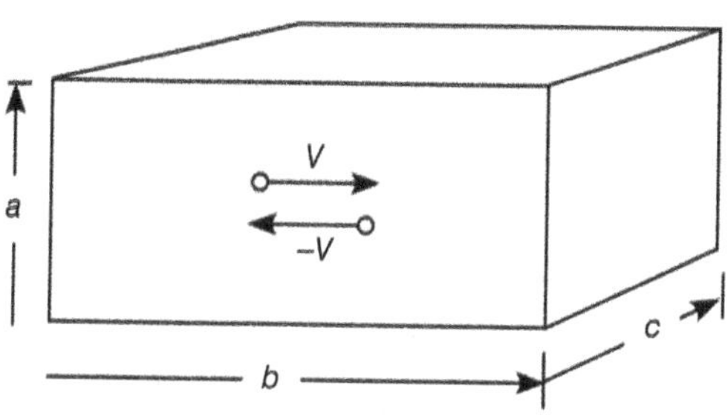

Figure 8.1 A box with dimensions a, b, and c contains n moles of a gas at a pressure p.

the average number of molecules hitting the wall:

$$F = \left\{ \left(\frac{v}{2b} \right) \cdot (2mv) \right\} \times \left(\frac{nN_A}{3} \right) \tag{8.6}$$

The force per unit area exerted by the molecules hitting the wall is the pressure p, given by

$$p = \frac{F}{ac} = \left(\frac{mv^2 \times nN_A}{3abc} \right) = \left(\frac{2}{3} \right) \times \frac{nN_A}{V_{Vol}} \times \frac{1}{2} mv^2 \tag{8.7}$$

Invoking Equation 8.2, pressure is given by

$$p = N \cdot \left(\frac{2}{3} \right) \cdot \left(\frac{1}{2} mv^2 \right) \tag{8.8}$$

Comparing (8.8) with (8.4),

$$\left(\frac{1}{2} mv^2 \right) = \frac{3}{2} \cdot k_B T \tag{8.9}$$

The surprisingly simple results of Equations 8.8 and 8.9 relate the macroscopic quantities of pressure and absolute temperature to the average kinetic energy per molecule. Unfortunately, Herapath's derivation went unnoticed. In the 1850s, various difficulties with the prevalent theories of heat were noted, but little progress occurred until James Clerk Maxwell read the work of the German physicist Rudolph Clausius, who built upon the basic atomic theory that a gas consisted of flying molecules with velocities that were dependent on pressure. Clausius explained why gases with particles that should have traveled at a speed of thousands of meters per second traveled far slower in reality. He concluded that molecules collide with each other and developed the idea of the mean free path as a measurement of the average distance traveled by a molecule between collisions. In 1859, Clausius published a paper giving a calculation for the mean free path in terms of the average distance between molecules and the distance between the centers of colliding molecules at impact. The concept of the mean free path, still relevant today, is discussed in the next section.

At that time, it was widely accepted that the molecules all traveled at the same speed, but Maxwell noticed that collisions would result in particles having different speeds. He realized the molecular velocities are actually distributed over a wide range and calculated their distribution, shown in Figure 8.2, to be

$$f(v_r) = \frac{4}{\sqrt{\pi}} v_r^2 e^{-v_r^2} \tag{8.10}$$

where $v_r = v/v_p$ and $v_p = \sqrt{2k_B T/m}$ is the most probable value of molecular velocity.

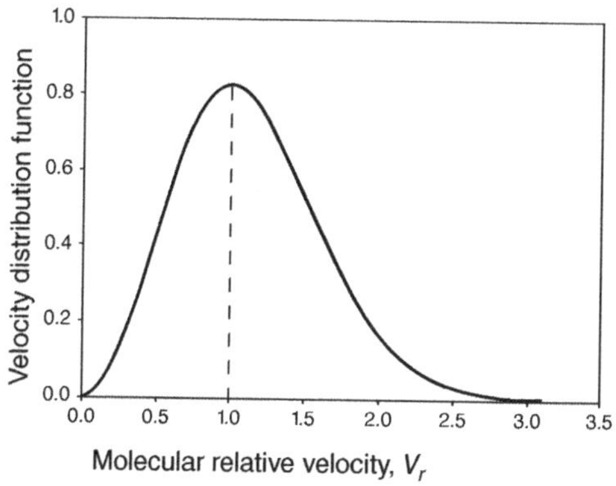

Figure 8.2 Maxwell's distribution of molecular velocities is not symmetric about its most probable value v_p—an indication that the most probable value ($v_r = 1$) is different from its average value. The energy corresponding to the most probable molecular velocity is k_BT while the molecular energy corresponding to the average velocity is $3/2k_BT$.

This was the first time that the problem was considered probabilistic. In 1868, Austrian physicist Ludwig Boltzmann modified Maxwell's distribution to explain heat conduction and proved that at a given temperature, the energy distribution is unique. The resulting expression is known as the Maxwell–Boltzmann velocity distribution.

Equation 8.10 was derived under the specific conditions of a neutral gas in thermal equilibrium, without diffusion or acceleration in the presence of an electric field. However, the Maxwell–Boltzmann velocity distribution indicates a small fraction of molecules within a volume will have a significant velocity. This kinetic energy is available to be exchanged with other molecules through collisions and may result in ionizing collisions. The "gas" then is composed mostly of neutral molecules but with some small number of electrons and ions at the same temperature T. This is known as a thermal distribution. In this case, the average particle velocity is

$$\frac{1}{2}mv_e^2 = \frac{1}{2}m_i v_{ie}^2 = \frac{1}{2}m_e v_{ee}^2 = \frac{3}{2}\cdot k_B T \tag{8.11}$$

where m, m_i, and m_e are the masses of the neutral gas molecules, ions, and electrons respectively, and v_e, v_{ie}, and v_{ee} are their average velocities, respectively.

8.1.1.2 Mean Free Path

In his first paper on kinetic theory, citing earlier work by Joule as well as contemporary work by Krönig, Clausius showed that molecules can move with great speed (~460 m/s)—speeds much greater than the magnitude of the bulk fluid velocity. Although Clausius was a well-respected scientist, this new work was met with skepticism. A key objection to Clausius' work was the observation that gases take a long time to intermingle. Dutch meteorologist H.D. Buys-Ballot posed the question [4]: If the molecules are moving so fast, then "how does it happen that tobacco smoke, in rooms, remains so long extended in immoveable layers?" In response, Clausius rethought the theory and realized that if the kinetic theory was right, then so was the calculated molecular speed. This led to the key insight that the molecules must not be traveling in straight paths for long but must be colliding with other molecules. Until this point, it was commonly believed that molecules collided only with vessel walls and not with other molecules, because the molecules were so tiny. It is interesting to note that gases with molecules that interact and those that do not interact give the same macroscopic explanation of pressure. It is because the pressure on the wall depends only on the density of molecules close to the wall and their velocity distribution. That is, pressure is the result of the average kinetic energy of the molecules. Clausius concluded that the molecules generally went in a straight line and were influenced by other molecules in the gas only if they were close and termed the average distance traveled by a molecule between collisions the "mean free path." Even at a low pressure (low densities), molecules collide with each other frequently.

The kinetic theory of gases was advanced tremendously by the concept of a mean free path. To illustrate the concept of a mean free path λ, a gas may be modeled as a collection of stationary solid spheres—"billiard balls"—with radius r_1 and number density N that are bombarded with a beam of particles with radius r_2, as shown in Figure 8.3. A collision occurs whenever the distance between the molecule and the particle is less than (r_1+r_2). This effective collision radius is known generally as the impact parameter b. The particle beam has an area A and is traveling much faster than the gas molecules resulting

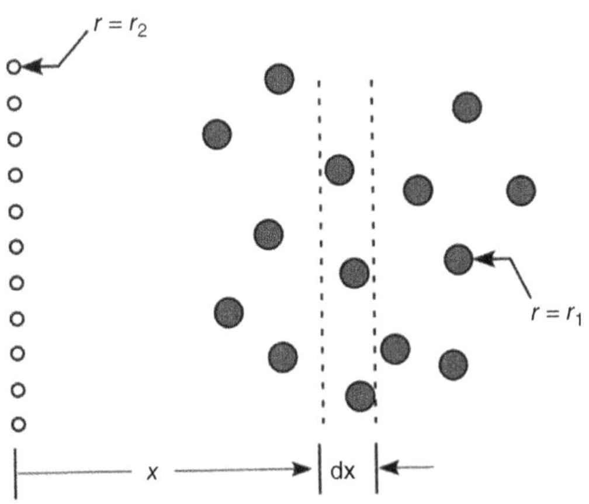

Figure 8.3 A beam of particles with radius r_2 is aimed at a gas of molecules with radius r_1.

in a flux Γ. As the particle beam travels through the gas, the flux decreases as particles are scattered out of the beam because of collisions with the gas molecules. At a distance x, the flux is decreased from its initial value of Γ_0 at $x=0$ to the value $\Gamma(x)$.

Using the effective collision radius, if the particle encounters a molecule, a collision occurs. The probability of collision, then, is related to the area the molecules occupy with the time it takes the beam to travel the distance dx. The particle beam covers an area A and sees a gas of uniform number density N before it, as shown in Figure 8.4. The probability that the fraction of the beam flux that reaches a distance x and then undergoes a collision as it travels from x to $(x+dx)$ is proportional to the fraction of the total beam area A, which is occupied by a molecule with an effective collision radius b, known as the impact parameter.

For a particle beam with an area A passing through a gas with number density N,

- the area occupied by one molecule is πb^2,
- the number of molecules contained in area A and width dx is $NAdx$, and
- the total area covered by $NAdx$ molecules is $\pi b^2 \cdot NAdx$.

As the beam of molecules moves through the gas, some collisions will occur before reaching a distance x. If $\Gamma(x)$ is the flux of molecules that reach the distance x without collision, then $d\Gamma$ is the number of molecules that reach a distance x without colliding and then collide within the distance dx. The fraction of the particle beam that will collide between x and $x+dx$ is proportional to the fraction of the area that the gas molecules occupy, yielding

$$d\Gamma = -\Gamma(x)\frac{\pi b^2 \cdot NAdx}{A} = -\Gamma(x) \cdot N \cdot \pi b^2 \cdot dx \tag{8.12}$$

Figure 8.4 The beam sees *NAdx* molecules within its flux area *A* as it travels a distance *dx*.

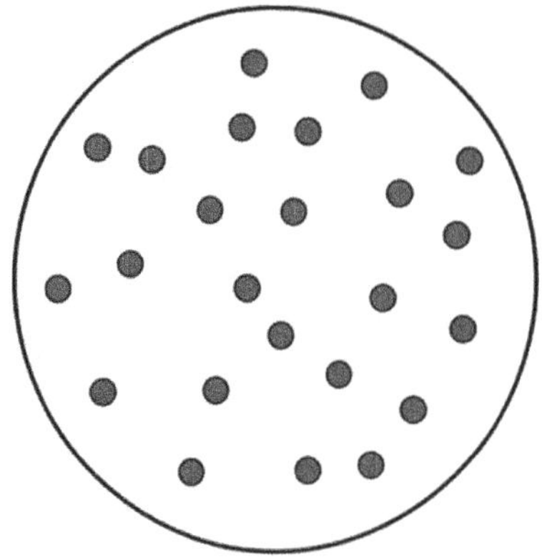

Integrating (8.12)

$$\int_0^x \frac{d\Gamma}{\Gamma(x)} = -\int_0^x N \cdot \pi b^2 \cdot dx$$

yields

$$\Gamma(x) = \Gamma_0 e^{-N \cdot \pi b^2 \cdot x} = \Gamma_0 e^{-(x/\bar{\lambda})} \tag{8.13}$$

These equations have a very limited validity. Nevertheless, these results using the "billiard ball" model are satisfactory even though the physical mechanism of collision is, in reality, very different. For instance, when two particles approach each other, it is difficult to determine if a collision has occurred because molecules do not have a definite radius. To account for these more realistic conditions, it is customary to use the collision cross section $\sigma(v)$ as the "effective area." It is related to the impact parameter b by

$$\sigma(v) = \pi b^2 \tag{8.14}$$

The cross section depends strongly on the relative velocities (kinetic energies) of the collision partners and is written $\sigma(v)$. The collision cross section is related to the mean free path $\bar{\lambda}$ by the relation

$$\bar{\lambda} = \frac{1}{\sigma N} \tag{8.15}$$

In fact, the effective radius—or the cross section—of a molecule varies with the conditions. The quantity collision cross section σ is the total target area of the collision partners' one molecule and is a strong function of the relative velocity:

$$\Gamma(x) = \Gamma_0 e^{-(N\sigma)x} = \Gamma_0 e^{-(x/\bar{\lambda})} \tag{8.16}$$

where $\bar{\lambda}$ is the mean free path.

Equation 8.10 suggests the free path has a distribution function. As the beam of molecules is moving in a certain direction, some will collide before reaching a certain distance x. If $\Gamma(x)$ is the flux of molecules that reach the distance x without collision, then $d\Gamma$ is the number of molecules that reach a distance x without colliding and then collide within the distance dx.

The length of the free path is statistical and clearly depends on the gas density. Some molecules in the gas can travel a considerable distance between collisions while others impact frequently. The free paths, then, will be distributed around a mean value $\bar{\lambda}$. Equation 8.16 gives the distribution of free paths as depicted in Figure 8.5 by an exponential curve. This distribution has no maximum value. At $x = \bar{\lambda}$, 37% of the molecules have a mean free path greater than or equal to $\bar{\lambda}$.

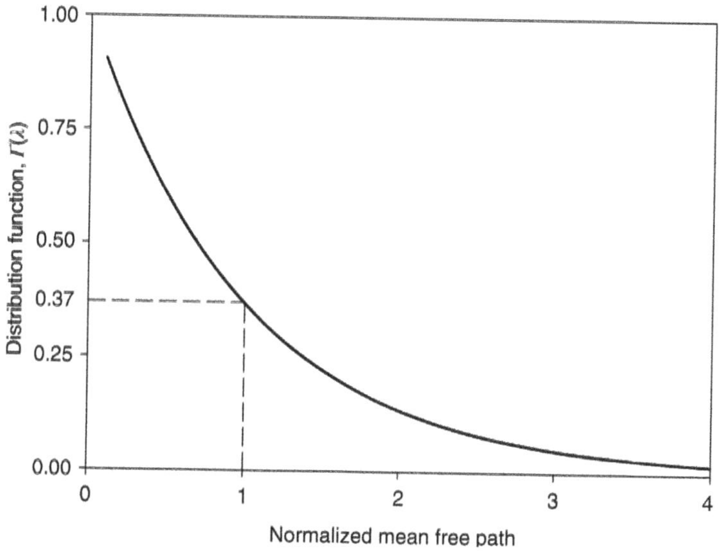

Figure 8.5 The distribution function $\Gamma(\lambda)$ is the fraction of molecules having a mean free path equal to or exceeding $\bar{\lambda}$. At its average value, $\lambda/\bar{\lambda} = 1$, and 37% of the molecules have a mean free path greater than the average $\bar{\lambda}$.

8.1.2 The Kinetic Theory of Ionized Gases

In Section 8.1, the physical processes of collisions and the distribution of molecular energies in gases have been described. The collisions were only between molecules that were treated as rigid solid bodies (billiard balls). The highly constrained but simple treatment illustrates the important principles of collisions.

In general, collisions can be categorized into elastic and inelastic collisions. In elastic collisions, energy is not exchanged between the colliding partners—but the momentum is redistributed. For instance, Clausius' insight that the molecules must be colliding with each other was the description of an elastic collision. Inelastic collisions—where energy is exchanged between the collision partners—are a complex phenomenon that can result in the production of molecules and particles with a net charge in the gas.

In the presence of charge carriers, a gas that has been a perfect insulator starts to show the properties of a conductor. A gas exhibits conductivity in proportion to its concentration of free charges and is a complex function of the number of free electrons and ions and their mobility and velocity. It is instructive to be familiar with the production of charged particles in gases.

8.1.2.1 Energy Gained from the Electric Field

The likelihood of charged particles and molecules colliding increases greatly when an electric field is present because the charged particles move under the influence of an electric field and gain energy. In the presence of an electric field $\vec{E}$, electrons and other charged particles and molecules move under the influence of the force $\vec{F}$:

$$\vec{F} = q\vec{E} = m\vec{a} \tag{8.17}$$

where q is the net charge on the particle and equals e for an electron, m is the mass of the particle, and $\vec{a}$ is the acceleration vector. The neutral molecules in the gas are unaffected by the electric field. It is important to note that the charged particles travel along the vector electric field line. The charged particle is accelerated by the force exerted on it by the electric field. The acceleration increases the velocity of the charged particle and hence its kinetic energy. The net gain in energy from the electric field is then available to be lost in inelastic collisions that may result in the generation of more free particles.

8.1.2.2 Elastic Collisions

As the charged particle moves, it encounters a gas molecule. It is first attracted, but as it moves closer it is repulsed, resulting in a change of direction (momentum), as shown in Figure 8.6. For instance, it was assumed that energy was not exchanged in the collision, only momentum. In fact, when two molecules collide, the fields of their electrons in the outer shells start to interact. These fields become distorted during the interval and energy is stored during the collision interval. A more accurate treatment of the complex phenomena involves quantum mechanics.

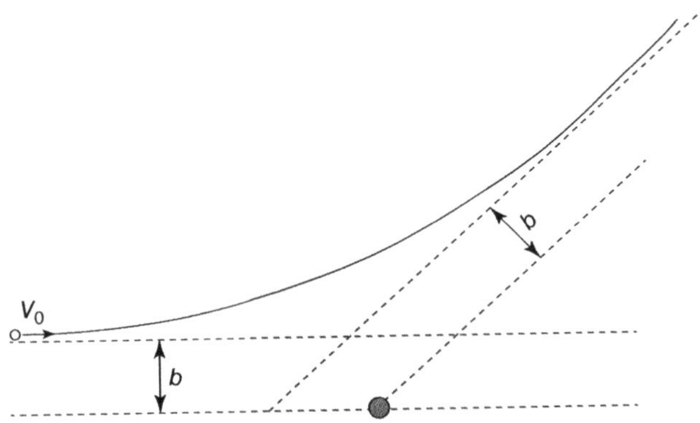

Figure 8.6 The collision of a particle and a molecule may be elastic or inelastic.

8.1.2.3 Inelastic Collisions

In an inelastic collision, energy is exchanged between the collision partners; a portion of the kinetic energy before collision is converted into the potential energy of one of the particles in the system. This increase in the potential energy manifests as an ionization or excitation of the particle. In an atom or a molecule, the lowest energy state is the ground state and the highest is when the molecule loses the electron in its valance band—the ionized state. The difference in energy between the ionized and the ground states of the molecule is the ionization energy. If the added potential energy is less than required for ionization but more than the difference to the next highest energy state, the outermost electron jumps to that higher energy state and is said to be excited. Excited molecules are also called metastable molecules because of their relatively long lifetimes.

Generally speaking, whenever a conversion from kinetic to potential energy occurs, the collision is inelastic. Inelastic collisions—those where energy is exchanged between the colliding partners—are important to the development of gas discharges because inelastic collisions result in the generation of free charges.

8.1.2.3.1 Ionization

The process of liberating an electron from a gas particle is known as ionization and is given by

$$\text{Ionization reaction}: \quad A + B \rightarrow A + B^{+} + e$$

In the ionization reaction, two neutral molecules collide with an exchange of energy in the amount of the ionization potential of one of the partners. The result of this exchange is a positive ion and a free electron. The number of charged particles resulting from the above ionization reaction between two neutral molecules in thermal distribution is small. The neutral molecules A and B have energies described by Maxwell–Boltzmann statistics and the likelihood of an energy exchange sufficient to cause ionization at room temperature is small (but nonzero). In fact, free electrons are available in neutral gases with an energy distribution determined by their temperature and are called thermal electrons.

The process of ionization is dominated by electron acceleration in an electric field and is greatly aided by the appearance of initiatory electrons. Electrons may originate in the gas by ionization. Electrons may also originate from the cathode (negative electrode) by electron emission—liberating an electron from a solid. Ionization is the process of liberating an electron from a molecule with either the simultaneous production of a positive ion or with the increase of positive charge. Both processes are of paramount importance in the gaseous conductivity.

In a gas subject to an electric field, charge production is dominated by electron impact ionization. Electrons gain energy from the electric field rapidly because of their small size and that energy is then available to be lost in a collision. The process is summarized as

$$A + e \rightarrow A^{+} + e + e \qquad \text{(electron impact ionization)}$$

where A is the gas molecule, A^{+} is the positive ion of the gas molecule, and e is an electron. The ionizing collision results in the production of a positive ion, a secondary electron, and the primary electron. The secondary electron is the one produced by the ionizing collision. Note that the colliding (primary) electron is not consumed in the interaction, but exits with kinetic energy lower by at least the ionization energy of the gas molecule. Impact ionization is perhaps the most important process in the breakdown of gases, but is not sufficient alone to result in breakdown. This simple model offers a reasonable model of electron scattering in gases. In general, the scattering cross section is a function of the electron energy and collision type.

Ionizing collisions occur when a primary electron has energy greater than the ionizing potential V_i of the gas and gives up that energy to the gas molecule during a collision. The electron gains energy by accelerating under the force exerted on it by the electric field. The electron's mean free path is defined as average distance an electron travels before undergoing a collision.

While the net motion of the electrons is in the direction of the applied electric field, the electrons do not travel in a straight path because of collisions, as shown in Figure 8.7. This implies then that the energy to collisionally ionize a molecule is gained only in the distance traveled in the direction of the electric field. Denote the mean free path in the direction of the electric field as λ_{Ei}. The energy gained by the electron from the electric field as it travels a distance λ_{Ei} must equal the ionization energy of the molecule:

$$q_e E \cdot \lambda_{Ei} \geq q_e V_i \tag{8.18}$$

where E is the electric field and q_e is the electronic charge.

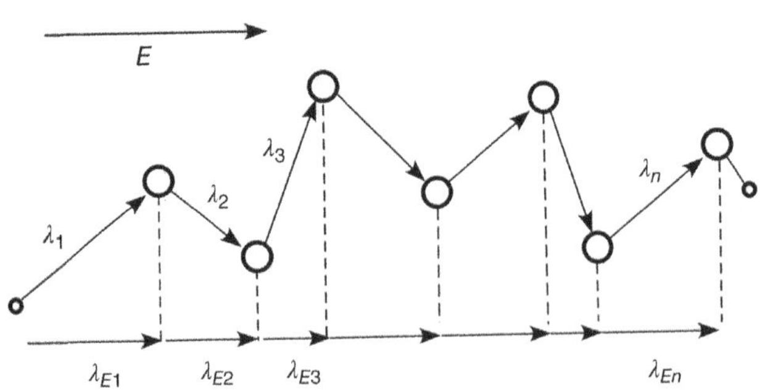

Figure 8.7 The electron travels a distance λ_i between collisions. but only gains energy for the length λ_{Ei} traveled in the direction of the electric field E.

8.1.2.3.2 Photoionization and Collisions with Excited Molecules

A molecule may also increase its potential energy by absorbing a quantum of energy less than that required for ionization, but that is sufficient to allow the valance electron to jump to a higher energy (excited) state. The excited molecule is also called a metastable molecule because of its long lifetime (1–10 ms). The generation of a metastable molecule is

$$A + B \rightarrow A^* + B \qquad \text{(metastable production)}$$

where A^* is an excited molecule.

Molecules can also be excited by absorbing energy in a collision with an electron in the reaction:

$$A + e \rightarrow A^* + e \qquad \text{(electron impact excitation)}$$

Metastable ionization may also occur when an excited molecule and an electron collide in a reaction known as step ionization:

$$A^* + e \rightarrow A^+ + e \qquad \text{(step ionization)}$$

The long lifetime increases the likelihood that an energy-absorbing collision will occur and lead to ionization. The metastable molecule is important because the injection of a relatively small amount of energy can cause the electron to fall back to its ground state with the simultaneous emission of a photon by the de-excitation reaction:

$$A^* + e \rightarrow A + hf + e \qquad \text{(de-excitation)}$$

where h is Planck's constant and f is the frequency of the emitted light. A metastable molecule may lose its potential energy not only as light emission but by imparting kinetic energy to the colliding electron, which then acquires a higher velocity and is able to ionize. Photons are also generated by the recombination of an ion and an electron by the reaction [4]:

$$A^+ + e \rightarrow A + hf \qquad \text{(radiative recombination)}$$

Other important collision processes involving electrons are as follows:

$$A^* + e \rightarrow A^{**} \quad \text{(dielectronic excitation)}$$

$$A^{**} \rightarrow A^+ + e \quad \text{(autoionization)}$$

$$A^{**} \rightarrow A + hf \quad \text{(dielectronic recombination)}$$

These photons are then available to ionize other molecules. Photoionization—molecular ionization by the absorption of photons—is a critical process in the development of gas discharge.

$$A^* + hf \rightarrow A^+ + e \quad \text{(step photoionization)}$$

$$A + hf \rightarrow A^+ + e \quad \text{(photoionization)}$$

The direct photoionization reaction above implies that for photoionization to occur, the photon energy must exceed the ionization energy. In reality, photoionization is a very complex process because it involves the interaction of radiation with matter. To illustrate, photons with a wavelength of 125 nm (UV) have an energy of 9.9 eV and are known to ionize almost all gases despite the fact that almost all molecules and atoms have ionization energies greater than 9.9 eV. Moreover, the presence of dust or water vapor is known to emit electrons through photon absorption. In monoatomic gases, absorption is easily interpreted as the removal of one electron by the absorption of a photon. In polyatomic gases, the process appears more as a band of acceptable energies. All photoionization occurs between 6 and 50 eV.

8.1.2.3.3 Penning Ionization and Other Complex Collisions

Penning ionization is a collision between a neutral and an excited molecule that produces ionization:

$$A^* + B^* \rightarrow A^+ + B + e \quad \text{(Penning ionization)}$$

The excited molecule may be an impurity or an engineered mixture known as a Penning mixture. A Penning mixture is a mixture of an inert gas with a small amount of a quench gas. The gases are chosen so that the quench gas has a lower ionization potential than the first excited state of the inert gas. The inert gas becomes excited from any number of collision processes and can ionize the quench gas through the Penning effect. The Penning effect may occur from trace amounts of impurities in the gas or by an engineered mixture. The most common example of an engineered Penning mixture is the neon lamp that is not filled with pure neon but with a Penning mixture of neon and less than 2% of argon. The mixture is easier to ionize than either gas alone, resulting in a lower applied voltage. The Penning mixture used in plasma display panels is usually helium or neon with a small percentage of xenon at pressures of a few hundred Torr. Gas ionization detectors also use Penning mixtures comprised of argon/xenon, neon/argon, and argon/acetylene.

More complex reactions can involve multiple collision partners as well as ions and neutrals:

$$A^+ + e + e \rightarrow A^* + e \quad \text{(three body collision)}$$

$$A^+ + B \rightarrow A^+ + B^* + e \quad \text{(ion impact excitation)}$$

$$A^+ + B + e \rightarrow A^* + B \quad \text{(three-body collision)}$$

$$A^+ + B \rightarrow A^+ + B^+ + e \quad \text{(ion impact ionization)}$$

It should be noted that the collision process is inherently statistical in nature and must be deduced from macroscopic measurements. The most important macroscopic processes of charged particles in gases are drift in the electric field, self-diffusion, ionization and excitation of the gas molecules, scattering, formation of ion clusters, attachment and detachment of electrons and molecules, and recombination of electrons and ions and charge exchange.

8.1.2.4 Total Collisional Cross Section

In Section 8.1.1.2, the concept of the collision cross section σ was introduced using the billiard ball model of atoms. The concept of the collision cross section is useful in describing the impact process and its products. Each collision reaction has its own collision cross section and is generally a function of the relative velocity (energy) of the colliding pair.

If the particle undergoes σN collisions per unit path, some of them will be elastic, some will lead to excitation—a jump in the valence electron of the target molecule, and others will result in the ionization of the target molecule. The total cross section is made up of individual cross sections. For instance, the total collisional cross section $\sigma(v)$ may consist of three dominant processes such as electron impact with cross section $\sigma_{el}(v)$, excitation with cross section $\sigma_{ex}(v)$, and excitation with cross section $\sigma_{ion}(v)$. The total collisional cross section is the sum of the cross sections of the individual processes, given by

$$\sigma(v) = \sigma_{el}(v) + \sigma_{ex}(v) + \sigma_{ion}(v) \tag{8.19}$$

Each collision has its own mean free path λ_i so that a collision path in a gas might look like that shown in Figure 8.8.

In almost all discharge work, it is of vital interest to know the exact numerical value of the number of ionizations an electron produces by collision with neutral, unexcited molecules as it transverses the gap. The number of electrons produced per unit length in the field direction, α, is known as the ionization coefficient, the primary ionization coefficient, or Townsend's first ionization coefficient. The primary ionization coefficient is related to the ionization cross section.

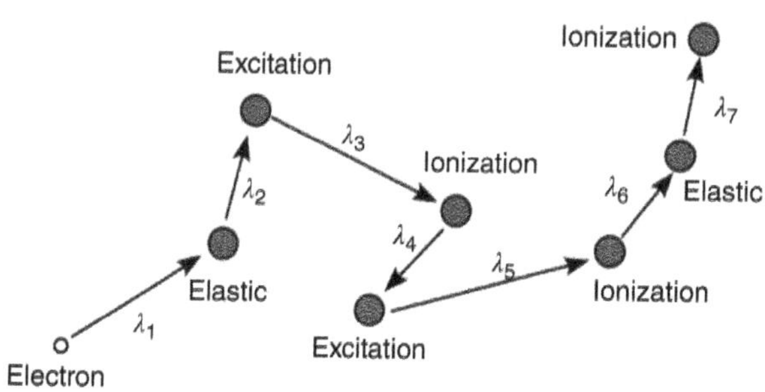

Figure 8.8 An electron moves through a gas under the influence of an electric field.

8.2 Early Experiments in Electrical Breakdown

8.2.1 Paschen's Law

The first significant studies of electrical breakdown were conducted by de la Rue and Muller (1880) [5] and by Paschen (1889) [6] using large parallel plate electrodes, spaced a distance d apart, immersed in gas at a constant pressure p and temperature, and with a steady voltage V applied. The electrodes were assumed to be clean and smooth. It was found that for any given values of p and d, the gas remained almost a perfect insulator for all values of V up to a certain threshold value V_b at which electrical breakdown occurred and a high current could pass, limited only by the impedance of the external circuit. Moreover, it was found at the voltage scaled as the product of the pressure and distance (pd), the similarity parameter. This important result was accompanied by the realization that electrical breakdown was initiated by electrons either supplied by irradiating the cathode or by naturally occurring chance electrons supplied by radioactivity or cosmic rays.

Paschen's law states that the breakdown voltage V_b (or sparking potential) is a unique function of the parameter (pd) for a given gas and electrode material under uniform field conditions that may be succinctly expressed as

$$V_b = \Psi(pd) \tag{8.20}$$

The Paschen curve is unique to each gas but has a characteristic shape, shown in Figure 8.9, where the breakdown voltage has a minimum value $(pd)_{min}$ and

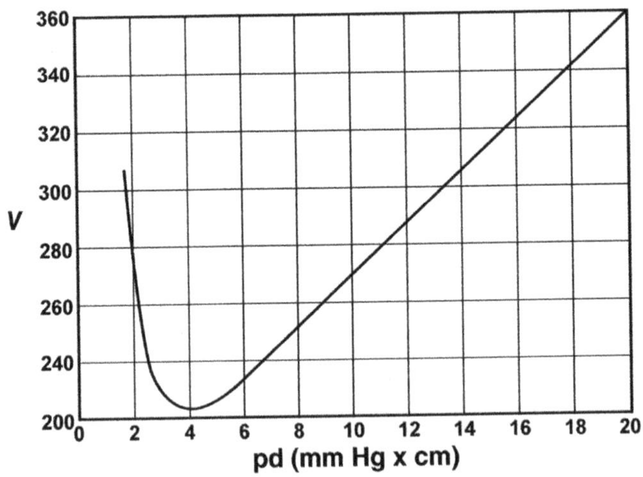

Figure 8.9 The breakdown curve for helium, measured by Townsend and McCallum, shows the characteristic shape of a Paschen curve [7]. The portion of the curve to the left of the Paschen minimum rises rapidly in contrast to the right side. (With permission of Oxford University Press).

increases rapidly on either side. At $(pd)_{min}$, the energy required to produce ionization is generally a few times the ionization potential of the gas. Typical values of the extrema on the Paschen curve are breakdown voltages on the order of 200–300 V and $(pd)_{min} \sim 5$ Torr mm.

A discharge is characterized in relation to the minimum in the Paschen curve. The region of the Paschen curve when $pd < (pd)_{min}$ is known as the "left side of the Paschen curve." Here, V_b increases rapidly as the pressure decreases for a fixed gap d because of the decreasing number of molecules available for collisions. For breakdown to occur, ionization must be provided by high electric fields, resulting in a steep rise in the left side of the Paschen curve with decreasing values of pd. On the right side of the Paschen minimum, there is sufficient number of scatterers to provide electron multiplication so that ionization occurs at moderate electric fields and the curve rises much more slowly with increasing pd. In principle, the breakdown voltage could become infinite, but in practice this does not happen because additional electrons are available as the voltage gets large because field emission occurs.

8.2.2 Townsend's Experiments

In 1901, John Sealy Townsend initiated studies of current growth in gases at pressures below 100 Torr. The system, consisting of an anode (A) and a cathode (K) separated by a distance d and connected to an ultrastable variable voltage source V, is shown in Figure 8.10. The voltage source was a battery bank that provided voltages up to ~2 kV and established a uniform electric field ($E_0 = V_0/d$) between planar electrodes. The cathode surface was irradiated with ultraviolet emission from a spark gap. A small initial current i_0, on the order of 10^{-13} A, of photoelectrons (electrons produced with light) was collected at the anode at low applied voltages [8,9].

Townsend found that with a fixed gap, an increase in the voltage produced an increase of current $i(V)$ until a spark was produced. However, he also found that

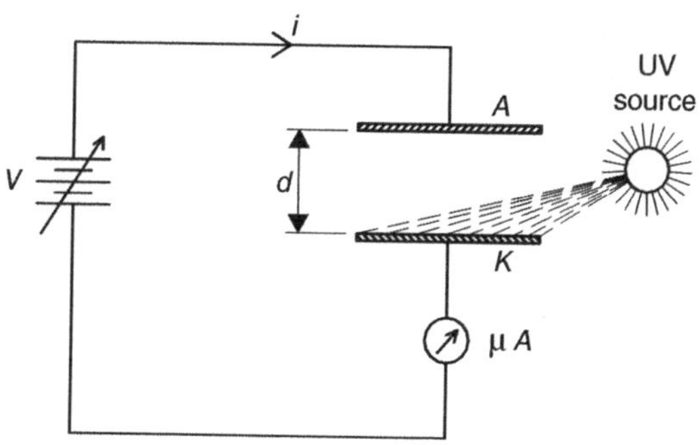

Figure 8.10 Experimental setup for the prebreakdown current growth.

when the voltage was held constant, the current depended on the gap distance *d*. Clearly, the current was a function of both the applied voltage *V* and gap distance *d*.

Townsend considered that electrons could generate other electrons by collisions with neutral atoms and postulated that it took a certain critical energy, characteristic of each molecule, to liberate an electron. The initial electrons therefore had to acquire an energy that exceeded the critical energy, which is now called the ionization energy. It should be noted that in Townsend's experiments, the discharge does not necessarily result in light emission, and it is sometimes called the Townsend dark discharge.

A typical current growth characteristic (i–V curve) is shown in Figure 8.11, where three regions (behavior I, II, and III) have been identified.

8.2.2.1 Region I: The Ionization-Free Region

As the voltage is gradually increased, the current increases until it reaches the saturation value i_0, because of charge carriers generated by photoemission from the UV illumination of the cathode. At very low voltages, $V < V_0$, only a fraction of the saturation current flows in the circuit because the electrons emitted by the cathode are scattered in random directions because of their collisions with the gas molecules. As the applied voltage is increased, the number of electrons collected by the anode is also increased because the directed average drift velocity of the electrons because of the electric field exceeds the average random velocity component.

8.2.2.2 Region II: The Townsend First Ionization Region

When the voltage is increased beyond V_1, the current is increased to a level greater than i_0. This increased current is caused by the flow of additional

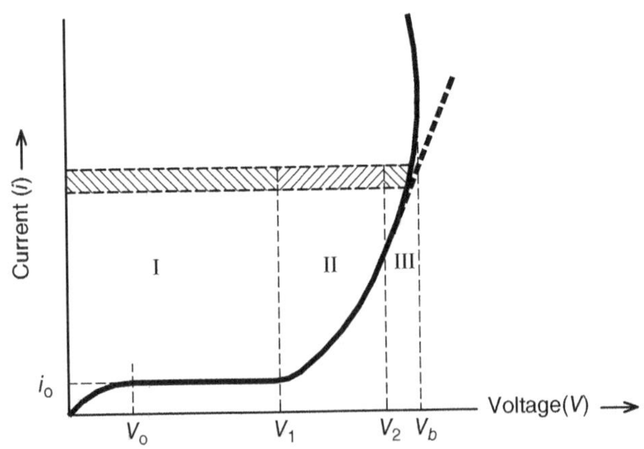

Figure 8.11 Current growth characteristic in a gas discharge.

electrons created by the ionizing collisions between gas molecules and primary electrons. Townsend found that in region II, the normalized current measured at the anode was directly proportional to the gap distance d when plotted on a semi-log scale. That is,

$$\ln\left(\frac{i(d)}{i_0}\right) = \alpha d \quad \text{or} \quad i(d) = i_0 e^{\alpha d} \tag{8.21}$$

In other words, in region II, the current increases exponentially with gap distance. The proportionality constant α is known as the Townsend primary ionization coefficient and is defined as the number of ionizing events performed by an electron in a 1 cm path along the field. Its value depends on the gas species and is a strong function of the applied electric field. The Townsend first ionization region is identified with the electron avalanche, discussed in Section 8.2.4.

8.2.2.3 Region III: Townsend Second Ionization Region

When the voltage is further increased beyond V_2, the current is found to increase much faster than in region II, where the discharge is not self-sustaining. Electrical breakdown is said to occur when the discharge is self-sustaining—when the discharge proceeds even when the initial electron source is removed. That is, a single-electron avalanche is not sufficient to carry the circuit-limited current and Townsend breakdown requires a mechanism for the generation of successor avalanches. Townsend postulated that the successor avalanches were formed by positive ions, resulting from ionizing collisions with primary electrons, bombarding the cathode, and liberating electrons. These electrons are referred to as secondary electrons and act as the seeds for successor avalanches. An expression for the current growth is derived in Section 8.3.1.

8.2.3 Paschen's Law Revisited

The Paschen curve is intricately linked to the Townsend discharge and can be obtained from Townsend's first coefficient α, also known as the ionization coefficient of the gas, and the γ coefficient quantifying the number of secondary electrons produced at the cathode per ion of the primary avalanche. The ionization coefficient α has units of number of ionizations per centimeter and is a strong function of the electric field. It is typically expressed as

$$\frac{\alpha}{p} = f\left(\frac{E}{p}\right) \tag{8.22}$$

Townsend proposed the following semiempirical form for pressures higher than that corresponding to the Paschen minimum:

$$\frac{\alpha}{p} = A \exp\left(\frac{-B}{E/p}\right) \tag{8.23}$$

Table 8.1 Values of the gas constants A and B and their range of validity [10].

Gas	$A(\text{Torr cm})^{-1}$	$B\left(\frac{V}{\text{Torr cm}}\right)$	$\left(\frac{E}{p}\right)\left(\frac{V}{\text{Torr–cm}}\right)$
Air	14.6	365	100–800
N_2	12	342	100–600
H_2	5.1	138.8	20–600
He	3	34	20–150
Ar	13.6	235	100–600
CO_2	20	466	500–1000

where A and B are constants (only) of the gas. Typical values of these constants for common gases are given in Table 8.1. A note of caution is warranted, however, because many of the values given in Table 8.1 were obtained with gases of unknown purity.

The values given for the constants in Table 8.1 may be used for estimating purposes only within the indicated range of E/p. In his derivation of this relationship, Townsend assumes the electrons do not gain energy from collisions, that the electric field is so strong that electrons always move in the direction of the field, and that the probability of ionization is zero for energies less than the ionization energy and unity for energy greater than the ionization energy. In actuality, the relation between Townsend's primary ionization coefficient α, the applied electric field E, and the gas pressure p is very complex and the subject of continued scrutiny.

There are also other functional forms of the relation between α and E that are a better fit. However, Equation 8.23 is often quoted, not only for historical reasons, but also because it has yielded some analytical treatments that have helped explain observed phenomena, such as a functional form of the Paschen curve.

The Paschen curve can be fit to the equation

$$V_b = \frac{B(pd)}{C + \ln(pd)} \tag{8.24}$$

where

$$C = \ln A - \ln\left\{\ln\left(1 + \frac{1}{\gamma}\right)\right\} \tag{8.25}$$

Experimental investigations [11] confirm Equation 8.24 for pd values below 200 Torr cm. The constant C takes into account the weak (double-logarithmic) influence of γ, where γ is the probability that a secondary electron will be ejected

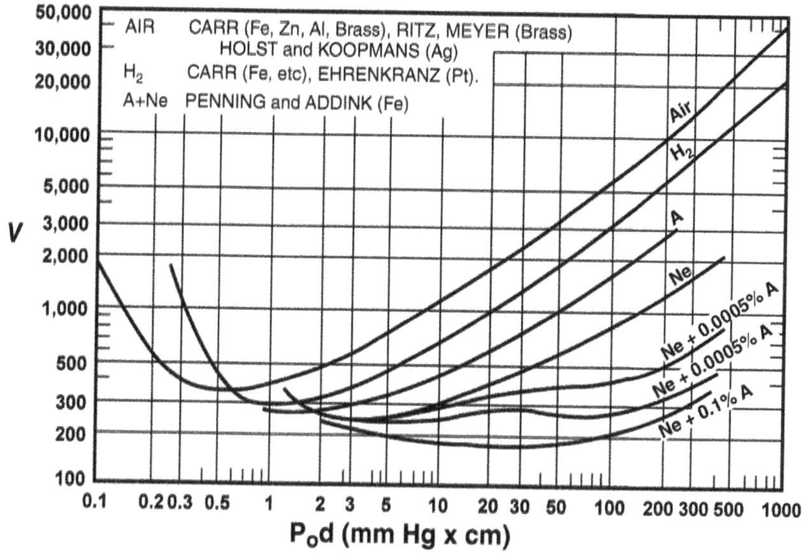

Figure 8.12 Paschen curves for a number of common gases [12]. (Reproduced with permission of Oxford University Press.)

from the cathode upon impact by an ion. The constants A and B are parameters of the gas alone and vary with the type of gas, as shown in Table 8.1. Thus, although Paschen curves have a characteristic shape, the specific shape of the curve varies significantly with the gas, as shown in Figure 8.12.

Because the breakdown voltage is a mathematical function of the parameter pd, the extrema can be calculated from Equations 8.14 and 8.15. The value of the minimum breakdown voltage $(V_b)_{min}$ and the location of the Paschen minimum $(pd)_{min}$ are [13]

$$(V_b)_{min} = \frac{(2.718) \cdot B}{A} \cdot \ln\left(1 + \frac{1}{\gamma}\right) \tag{8.26}$$

$$(pd)_{min} = \frac{(2.718)}{A} \cdot \ln\left(1 + \frac{1}{\gamma}\right) \tag{8.27}$$

If the voltage on the gap is lower than $(V_b)_{min}$, breakdown will not occur in the uniform field gap regardless of the pressure or gap spacing.

Equations 8.26 and 8.27 show that the location of the Paschen minimum and its associated breakdown voltage depends on the condition of the cathode and the cathode material through the electron feedback parameter γ. The effect can be quite dramatic, as illustrated in Figure 8.13, where the Paschen curve for hydrogen is measured near the minimum for a variety of cathode materials. Note that both the magnitude of the breakdown voltage and the location of the

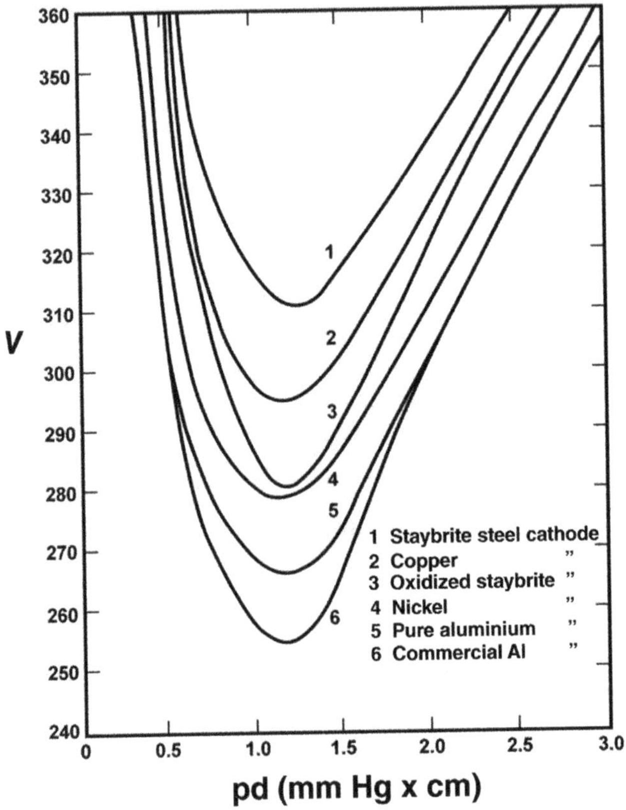

Figure 8.13 Paschen curves in hydrogen, in the vicinity of the Paschen minimum, for different cathode materials [15]. (Reproduced with permission of Oxford University Press.)

minimum are affected. Typical values of the breakdown voltage at the Paschen minimum are on the order of 300 V [14].

Air is perhaps the most important gas for engineering applications and has been extensively studied. Numerous scaling relations have been proposed. At room temperature—even at reduced pressures, gaps larger than 100 μm (0.1 mm) have *pd* values well to the right of the Paschen minimum where the increase in breakdown voltage is linear.

In air, the breakdown voltage is given by [16]

$$V_b = (24.22)\frac{pd}{760} \cdot \frac{293}{T} + 6.08\sqrt{\left(\frac{pd}{760} \cdot \frac{293}{T}\right)} \tag{8.28}$$

where

p is the pressure in Torr,

T is the gas temperature in K,
d is the gap spacing in cm, and
V is the breakdown voltage in kV.

From Equation 8.28, it can be surmised that at a constant pressure, the breakdown electric field is not a constant value but varies with gap distance. This is demonstrated in Example 8.2 that plots (8.28) along with experimental data.

Paschen curves are typically generated for uniform electric fields. The breakdown voltage, practically observed at very long gaps (or high pressures) and at very short gaps (or low pressures), are less than the values predicted from Paschen's law. For long gaps, nonuniformity of the electric field arises from space–charge effects, whereas for short gaps, the nonuniformity arises from asperities on the electrode surfaces. Consequently, field emission plays an important role at high- and low-pressure regimes because of high electric fields. There is a renewed interest in Paschen curves for applications involving very small gaps—microgaps—where the gap dimension is measured in micrometers [17,18].

Paschen-like curves can be generated for nonuniform field geometries but lack the generality of those generated in uniform electric fields because of the very strong dependence of the primary ionization coefficient α on the local electric field. In practice, in nonuniform field conditions, like those found in spark gap switches, the breakdown voltage is measured as a function of pressure for a fixed gap distance and is known as self-breakdown or V–p curves.

8.2.4 The Electron Avalanche

An electron avalanche forms from initiatory electrons that may be present naturally, or emitted from a metal by a photon, for instance. Under the action of an applied electric field, the free electron is accelerated and collides with the background neutral gas molecule, ionizes the molecules, and creates new electrons. These new electrons are also accelerated by the applied electric field and undergo collisions that lead to further ionization of the gas and more electrons. This successive generation of additional electrons through successive collisions is called cascade ionization and is depicted in Figure 8.14. In this fashion, the number of electrons in the gas grows exponentially.

The electron avalanche occurs through a primary ionization process described in Townsend's region II. It was first described by Townsend, who later also developed the mathematic description. Townsend surmised that the initial electrons generated by the UV source generated additional electrons in the course of crossing the gap under the influence of the electric field. The current is related to the electron number density by the relation $I = q_e n v_d$, where q_e is the electronic charge, n is the number density of electrons, and v_d is the electron drift velocity. Therefore, by direct analogy with Equation 8.19, the

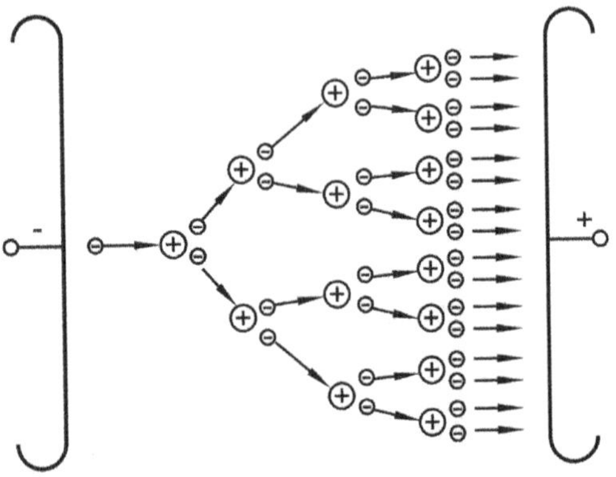

Figure 8.14 A cascade ionization initiated by a single electron in a uniform electric field.

number of electrons reaching the anode a distance d away is

$$n(d) = n_0 e^{\alpha d} \tag{8.29}$$

Townsend's first ionization coefficient α describes the complex physical process, which includes several types of collisions. More generally, an expression for the increase in electron number density $dn(x)$ for an electron density $n(x)$ traveling a distance $x + \Delta x$ from the cathode is

$$dn(x) = n(x) \cdot \alpha \cdot dx \tag{8.30}$$

Integration of Equation 8.30 with the initial condition $n(x=0) = n_0$, leads to

$$n(x) = n_0 e^{\alpha x} \tag{8.31}$$

Thus, an initiating electron, starting from the cathode on its drift toward the anode, causes multiplication of charge carriers because of ionization of the gas molecules. The additional electrons cause further multiplication of charge carriers, resulting in an exponential current growth.

The electrons are accelerated in the electric field and travel through the gas at their drift velocity ($\sim 10^7$ cm/s). The advancing electron cloud leaves behind a cluster of positive ions, which drift slowly toward the cathode. The positive ions are also accelerated in the applied field but in the opposite direction to the electrons and travel much slower ($\sim 10^5$ cm/s) because of their mass. This difference in velocity between the charged particles leads to separation of charges. Additionally, the electrons in the avalanche head diffuse because of the repulsion of like charges. As a result of the charge motion, the electrons are concentrated at the head of the avalanche, while the positive ions are distributed throughout the tail, with the maximum density of positive ions occurring just

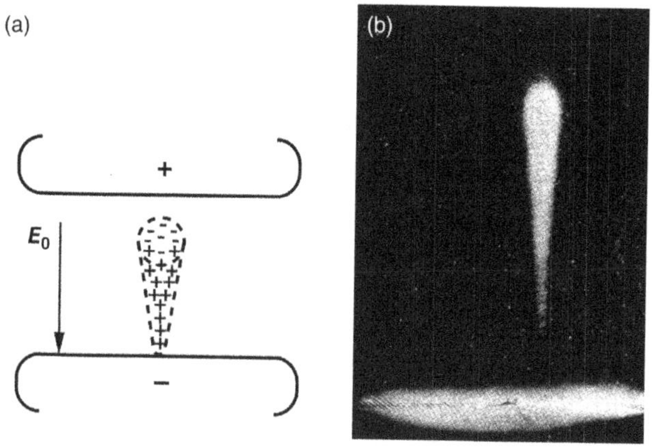

Figure 8.15 The distribution of charged particles in a single avalanche (a) and Raether's picture of an electron avalanche taken in a cloud chamber [19]. (Reproduced with permission of Oxford University Press.)

behind the head. This process of multiplication of charge carriers by collisions and the resulting charge distribution is termed a Townsend avalanche or an electron avalanche.

The charge carrier distribution in a single-electron avalanche, when the avalanche has advanced to a distance x from the cathode, is shown in Figure 8.15a along with a reproduction of one of Raether's photographs [19] of an electron avalanche (Figure 8.15b) taken in a Wilson cloud chamber. A Wilson cloud chamber is a closed vessel containing a supersaturated vapor, used extensively for particle detection until the advent of the bubble chamber in the 1950s. The trajectories of charged particles can be recorded because the ions act as condensation nuclei.

The avalanche, then, moves at a speed determined by the electron drift velocity. The process continues as long as the external illumination is present. In region II, an important characteristic of the current growth is that if the UV source is removed, it results in the condition $I_0 = 0$ in Equation 8.3, and the current decays to zero. It should be noted that although current growth is recorded, the discharge is not self-sustaining and is not electrical breakdown.

8.3 Mechanisms of Spark Formation

The theory of electrical breakdown in gases is dominated by two models: the Townsend discharge and the streamer mechanism. It is interesting to note that both mechanisms develop from an initial electron avalanche and scale with the similarity parameters (pd) and (E/p) where p is the gas pressure, d is the gap distance, and E is the magnitude of the applied electric field. An excellent treatise on the theories and experiments devoted to the understanding of the development of the electrical breakdown of a gas-filled gap can be found in Ref. [20].

The Townsend mechanism is electrical breakdown by the development of successive avalanches in uniform electric fields. Hence, long formative time lag is needed because avalanches travel at the electron drift velocity and the contribution of slow-moving ions is critical to the development of successor avalanches. With the development of high-voltage impulse generators, experimental observations regarding breakdown that could not be explained with the Townsend theory began to emerge. Rogowski measured time lag shorter than an electron transit time [21]. Other experiments indicated that for high pressures and large values of d, the breakdown voltage was largely independent of the cathode material. Additionally, breakdown was filamentary in character, which was inconsistent with the Townsend theory of breakdown. Efforts to resolve these inconsistencies led to the development of the streamer theory.

Townsend breakdown dominates at low pd values in uniform electric fields and the streamer mechanism dominates high pressures, long distances, large overvoltages, and nonuniform electric field geometries.

8.3.1 The Townsend Discharge

The Townsend mechanism is electrical breakdown by the development of successive avalanches. The formation of secondary avalanches is necessary because a single-electron avalanche is not sufficient to carry the full-circuit-limited current. Townsend breakdown requires a mechanism for the generation of additional current by successor avalanches, which in turn requires the generation of secondary electrons. Townsend postulated that a secondary mechanism would occur from the positive ions generated in the primary avalanche bombarding the cathode and emitting electrons, termed secondary electrons.

Let n_0 continue to represent the number of primary electrons emitted from the cathode and n_s represent the total number of secondary electrons leaving the cathode. The total number of electrons leaving the cathode n_T is

$$n_T = n_0 + n_s \tag{8.32}$$

If γ is the number of secondary electrons produced at the cathode per ionizing collision in the gap, then the number of secondary electrons produced is

$$n_s = \gamma \cdot n_T \cdot \left(e^{\alpha d} - 1\right) \tag{8.33}$$

because each electron leaving the cathode produces $\left(e^{\alpha d}-1\right)$ additional electrons, on average, through collisions with neutral atoms. If n_T electrons are in the gap, then an average of $n_T \cdot \left(e^{\alpha d}-1\right)$ ionizing collisions occur.

Combining Equations 8.23 and 8.24 yields

$$n_T = n_0 + \gamma \cdot n_T \cdot \left(e^{\alpha d}-1\right) \tag{8.34}$$

rearranging terms,

$$n_T = \frac{n_0}{1-\gamma \cdot (e^{\alpha d}-1)} \tag{8.35}$$

and the number of electrons arriving at the anode is

$$n(d) = n_T e^{\alpha d} = \frac{n_0\, e^{\alpha d}}{1-\gamma \cdot (e^{\alpha d}-1)} \tag{8.36}$$

The steady-state current measured at the anode is

$$i(d) = \frac{i_0\, e^{\alpha d}}{1-\gamma \cdot (e^{\alpha d}-1)} \tag{8.37}$$

Mathematically, the Townsend breakdown criterion can be derived from Equation 8.37 by setting the denominator equal to zero:

$$1-\gamma \cdot \left(e^{\alpha d}-1\right) = 0 \tag{8.38}$$

Since the term $e^{\alpha d} \gg 1$, the above relation may be written as

$$e^{\alpha d} = \frac{1}{\gamma} \tag{8.39}$$

Equation 8.39 illustrates that the generation of secondary electrons and successor avalanches is critical for electrical breakdown to occur by the Townsend mechanism.

Raether [19] provided an excellent depiction of the evolution of a Townsend discharge from a primary avalanche, shown in Figure 8.16. In part (a), the primary avalanche initiates secondary avalanches from the cathode surface, which launch a number of successor avalanches illustrated in part (b). After a number of generations, positive space charge builds up near the anode surface, shown in part (c). The successor avalanches continue to build until the current is limited by the circuit.

Townsend breakdown occurs in uniform electric fields at short gap lengths. Short gaps are necessary to ensure a good source of secondary avalanches from cathode processes. The primary ionization coefficient α has such a strong dependence on the electric field that even perturbations from uniformity on the order of 5% produce sufficient space charge that the applied electric field is distorted and breakdown does not proceed from successor avalanches. The long formative time lag is needed because the avalanche travels at the electron drift

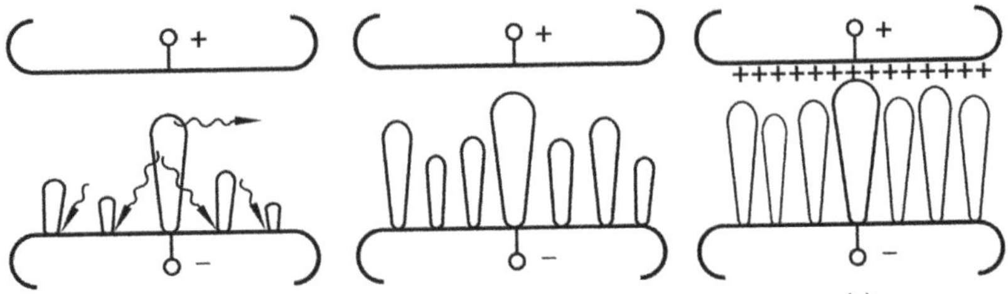

Figure 8.16 Raether's illustrations of the Townsend mechanism of electrical breakdown by successive avalanches [19]. The first avalanche in part (a) produces photons that impact the cathode, resulting in successor avalanches in part (b) and also in circuit-limited electrical breakdown (c). (Reproduced with permission of Oxford University Press.)

velocity and several avalanche "generations" may be required to build up sufficient charged particle to carry the full circuit current. Moreover, the contribution of slow-moving ions—ionic drift velocities are over 100 times slower than electron drift velocities—is critical in the development of successor avalanches. Successor avalanches develop from secondary electron generation and are strongly dependent on the cathode material. The Townsend theory of breakdown is well understood, but suffers because predictions require detailed knowledge of the condition of the cathode, including material, surface roughness, and impurities.

8.3.1.1 Multiple Secondary Mechanisms

Townsend originally postulated that secondary avalanches were initiated by electrons produced by bombardment of the cathode by ions produced in the primary avalanche. Later, other mechanisms were identified. Secondary electrons are also produced by the photoelectric effect when a photon of sufficient energy is absorbed by the cathode and emits an electron.

The secondary emission coefficient from positive ions, γ_i, is defined as the number of electrons emitted at the cathode per incident positive ion. The secondary emission coefficient from photons is designated δ and is defined as the number of electrons emitted at the cathode per incident photon. When the electron swarm reaches the anode, the positive ions left behind drift toward the cathode. Maximum emission takes place when these positive ions, just behind the head of the avalanche, strike the cathode.

The time for this to develop is given by the positive ion transit time τ_+, given by

$$\tau_+ = \frac{x}{v_d^+} \tag{8.40}$$

where

v_d^+ is positive ion drift velocity.

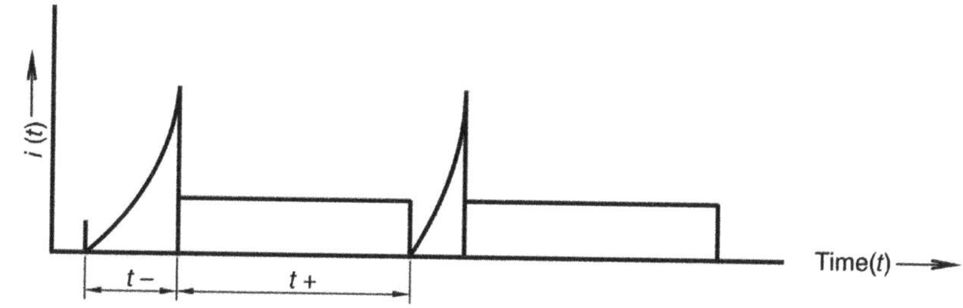

Figure 8.17 Current growth from the γ_i process for $\gamma_i \cdot \left(e^{\alpha d} - 1\right) = 1$.

At this time, corresponding to $(\tau_{-} + \tau_{+})$, the next electron avalanche starts from the cathode, which gives rise to another progeny, and this continuing process results in multiple avalanches. The current growth from multiple avalanches caused by the γ_i process is shown in Figure 8.17 for the steady-state condition of $\gamma_i \cdot \left(e^{\alpha d} - 1\right) = 1$. Figure 8.18 shows the current growth due to the δ process for the condition of $\delta \cdot \left(e^{\alpha d} - 1\right) = 1$. It may be observed from Figures 8.17 and 8.18 that the avalanche development for the γ_i process is considerably slower than the δ process because of the difference in the velocity of the positive ions ($\sim 10^5$ cm/s) and photons ($\sim 10^{10}$ cm/s). The avalanche development for the condition of $\delta \cdot \left(e^{\alpha d} - 1\right) = 1$ is shown in Figure 8.19. The

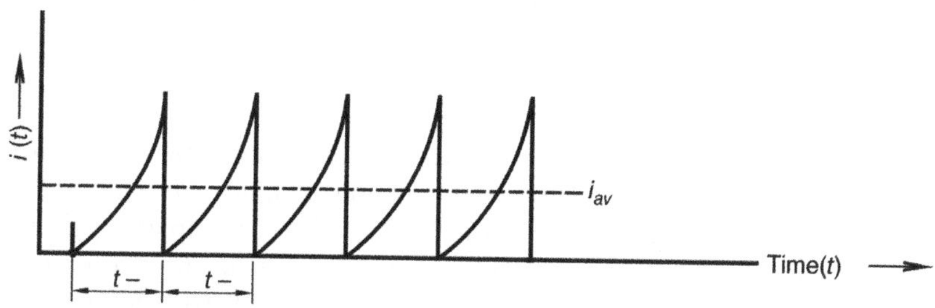

Figure 8.18 Current growth from the δ process for $\delta \cdot \left(e^{\alpha d} - 1\right) = 1$.

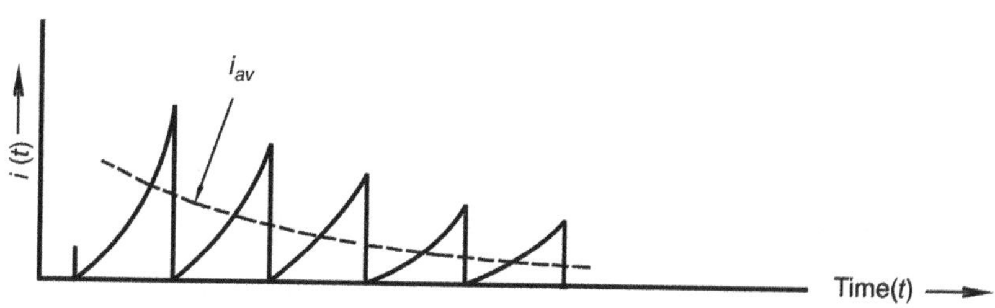

Figure 8.19 Current growth from the δ process for $\delta \cdot \left(e^{\alpha d} - 1\right) < 1$.

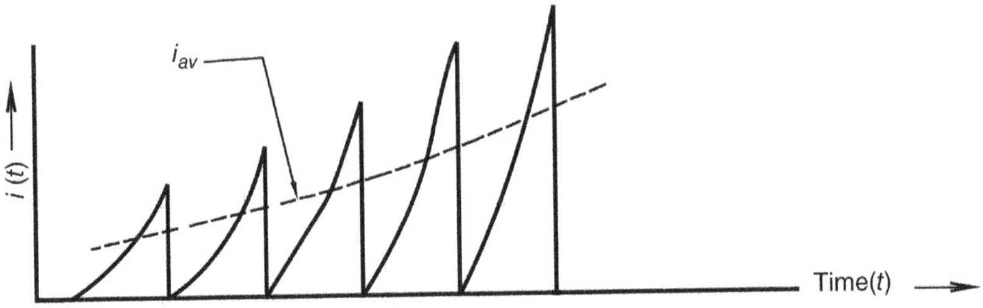

Figure 8.20 Current growth from the δ process for $\delta \cdot \left(e^{\alpha d} - 1\right) > 1$.

daughter avalanche being smaller in size than its parent leads to a gradual decay of current. Figure 8.20 shows the avalanche development for $\delta \cdot \left(e^{\alpha d} - 1\right) > 1$, leading to larger succeeding avalanches, fast current growth, and rapid breakdown of the gap. The supply of electrons to initiate successor avalanches leading to breakdown can come from any number of collision processes.

The Townsend discharge travels at a speed determined by the electron drift velocity in the applied electric field, ~10^7 cm/s. It is distinguished by relatively long formative times to breakdown and the space charge in the gap is insufficient to significantly distort the applied electric fields. The threshold field for Townsend discharge onset is known as the critical field, which is 26 kV/(bar cm) for air and 89 kV/(bar cm) for sulfur hexafluoride (SF_6). In a Townsend breakdown, light emission may be present or the discharge could remain dark.

8.3.1.2 Generalized Townsend Breakdown Criterion

The effect of numerous secondary processes may be lumped together into a single secondary coefficient ω that is a function of both E/p and x [22]. Schumann noted that the structure of the Townsend theory implies a polarity effect that is not observed in experiments and proposed a generalized secondary ionization coefficient $\omega(x)$ that is proportional to the Townsend primary ionization coefficient $\alpha(x)$, which eliminates that difficulty [23]. The Townsend current is

$$i(d) = \frac{i_0 \cdot e^{\alpha d}}{\left[1 - (\omega/\alpha) \cdot (e^{\alpha d} - 1)\right]} \tag{8.41}$$

where

$\omega = \beta + \alpha\gamma_i + \delta + \varepsilon_m + \varepsilon_{ex} + \eta$,

$\beta\, n_x dx$ is the number of electrons produced by collision by n_x positive ions traveling a distance dx in the electric field,

$\gamma_i\, n_x$ is the number of electrons produced when n_x ions are incident on the cathode,

$\delta\, n_x dx$ is the number of photoelectrons emitted from the cathode on incidence of the fraction of photons generated when n_x electrons travel a distance dx in the electric field,

$\varepsilon_m, \varepsilon_{ex}$ are coefficients that describe the effect of metastable and excited molecules on the cathode, and

η is a coefficient that accounts for the electrons produced by photoionization.

The individual coefficients are known as Townsend secondary ionization coefficients and ω/α is the generalized Townsend secondary coefficient representing the combined action of all the individual processes [6,24]. The positive feedback processes, other than emission at the cathode from positive ions and photons, which contribute to the current growth, are collisional ionizations from positive ions drifting toward the cathode, photoionization of the gas, and space charge effects. In general, the generalized Townsend secondary ionization coefficients are a function of (E/p) and of the geometry.

An important characteristic of current growth in region III is that, at a certain voltage V_b, the positive feedback (or regeneration) because of secondary emission from the cathode becomes strong enough such that

$$\frac{\omega}{\alpha}(e^{\alpha d} - 1) = 1 \tag{8.42}$$

Mathematically, when the condition of Equation 8.42 is met, an infinite current flows. In fact, this condition indicates the current is "self-sustained." At V_b, even if the irradiation of the cathode by the source is removed, the current continues to flow, and this corresponds to the Townsend breakdown criterion. The voltage V_b then is known as the sparking voltage or the breakdown voltage and the distance d_s is termed the sparking distance.

It is to be noted that the Townsend breakdown criterion is the threshold condition for breakdown for obtaining a steady-state current and does not result in a low-impedance arc. A slight overvoltage condition, that is, $\Delta V > (V - V_b)$, leads to rapid breakdown of the gap with a sufficiently low spark impedance that the current in the circuit is determined by the external resistance.

8.3.1.3 Townsend Criterion in Nonuniform Geometries

Although Townsend discharge is operative in uniform fields, the Townsend criterion can be converted to a form appropriate to nonuniform electric fields:

$$\int_0^{x_0} \omega(x) \exp\left\{\int_0^{x} \alpha(x')dx'\right\} dx = 1 \tag{8.43}$$

where α is the primary ionization coefficient, ω is the generalized secondary coefficient of ionization, x is the coordinate along the avalanche path, and x_0 defines the total distance along the breakdown path from the electrode surface to the avalanche head. This form of the Townsend Criterion is most useful for determining when electrical breakdown will not occur. Nonuniform electric fields tend to drive the discharge into a regime where space charge is important (the streamer regime, described in Section 8.3.2) because the primary ionization coefficient α is a very strong function of the applied electric field, that is, $\alpha(E)$.

8.3.1.4 Modifications for Electronegative Gases

In electronegative gases, collisions may remove electrons in a process known as "electron capture" or "electron attachment," represented by the reaction

$$A + e \rightarrow A^- \qquad \text{(electron attachment)}$$

where A is the molecule of the attaching gas.

Geballe and Reeves [25] modified the primary ionization coefficient for electron attachment. When attachment is present, the current growth equation has been shown [26] to be

$$i(d) = \frac{i_0\left[(\alpha/(\alpha-\eta))\cdot e^{(\alpha-\eta)d} - (\eta/(\alpha-\eta))\right]}{1-\gamma(\alpha/(\alpha-\eta))\{e^{(\alpha-\eta)d}-1\}} \tag{8.44}$$

where

η—the attachment coefficient—is the number of attachments per electron per centimeter.

It is common to use the notation $\overline{\alpha} = \alpha - \eta$. Sulfur hexafluoride ($SF_6$) and air are two important, and well-studied, electronegative gases that find widespread use in pulsed power applications.

8.3.2 Theory of the Streamer Mechanism

The Townsend theory successfully accounted for the observed dependence of the breakdown voltage on gas density and electrode spacing. However, during later investigations of electrical breakdown of long, atmospheric air gaps, difficulties arose in explaining experimental observations with Townsend's theory. For instance, in spark gaps at atmospheric pressure, with electrode spacing ~1 cm, very short delay times (less than 1 μs) were obtained. Short formative time lag occurs when the gap is subject to impulse voltages just a few percent above the DC breakdown voltage. Townsend's theory is breakdown by successive avalanches that develop on a timescale determined by the ion drift velocity. Because positive ion transit times across the gap are on the order of 10 μs, the observed delay time to breakdown in the long gaps was too short to have involved a series of successive avalanches produced by ions impinging on the cathode. Furthermore, the breakdown in long gaps appeared to be independent of cathode material. The branched and irregular growth of these sparks is also difficult to reconcile with the Townsend theory. A new theory of the mechanism of electrical breakdown was developed to account for these and other observations [27].

Photographs of these spark discharges in long gaps revealed the existence of narrow, luminous discharges originating from either the anode or from the middle of the gap rather than the broad development from the cathode expected for a Townsend discharge. These discharges became known as "streamers" —evocative of a thin band of bright light, attached at one end to an electrode and floating toward the other—or "kanals" from the German word for channel. In modern times, the word "streamer" is almost exclusively used, but occasionally one comes across a description of the "kanal mechanism."

Following a large amount of experimental work, the streamer mechanism of a spark channel was proposed in 1940 by Meek [28] for cathode-directed (positive) streamers and independently by Raether [29] for anode-directed (negative) streamers. The streamer mechanism predicts a transition to a self-propagating conducting channel from a single-electron avalanche. Streamers develop when the charge density at the head of the avalanche becomes so large that it distorts the applied electric field. That is, the space charge in the avalanche head generates a self-electric field that is on the order of the applied electric field. Under these conditions, the avalanche is said to become critical and transforms itself into a streamer, which progresses much more rapidly. In a uniform field, and only in a uniform field, it has been found that the avalanche-to-streamer transition necessarily leads to a total breakdown of the gap [26].

An essential principle of breakdown in gases by the streamer mechanism is space charge distortion of the local electric field at the tip of the streamer. Although the fundamentals of streamers are well accepted, how a streamer evolves into a spark is still an active research topic. The streamer theory has generated controversy because of the difficulty in quantitatively describing the ionization process [30]. The role of photoionization in streamer development has been postulated to play a critical role, but the ability of an avalanche to produce photons energetic enough to ionize in a pure gas has been questioned [31–33]. Other ionization processes, associative ionization of excited atoms and runaway electrons, have been advanced to account for streamer propagation [32–35].

8.3.2.1 Criterion for Streamer Onset

The streamer theory of the spark was developed as a consequence of many experimental observations, including critical insights gained from cloud chamber results from Raether as well as others. In 1928, Slepian drew attention to the space charge fields produced in the track of an individual electron avalanche and its effect on the growth of the avalanche [36]. Loeb proposed that a series of electron avalanches formed a line across the gap [37]. von Hippel and Franck [38] calculated the general redistribution of the field in a uniform gap by the positive space charge developed by successive electron avalanches and showed an enhanced ionization in the later avalanches, which allowed the breakdown of the gap to occur in a time consistent with electron drift speeds. As a consequence, the streamer mechanism was proposed independently by Meek [28] and Raether [29] in 1940.

The streamer onset criterion, the condition under which an electron avalanche transitions to a streamer, has two well-known forms that are essentially equivalent. The first is associated with Meek, who hypothesized that an electron avalanche would transform into a streamer when the electric field generated by the space charge at the head of the avalanche E_r was on the order of the applied external field E_0:

$$E_r \sim E_0 \tag{8.45}$$

The other form is associated with Raether, who did extensive experiments on electron avalanches and streamers in a cloud chamber where he measured the critical length of the avalanche until it transitioned into a streamer, x_{cr}. From his measurements, he deduced the critical gain factor (αx_{cr}), which is related to the number of electrons in the avalanche head as it becomes a streamer, N_{cr}:

$$N_{cr} \sim 10^8 \tag{8.46}$$

These two conditions for streamer onset are essentially equivalent. Both Meek and Raether validated their postulates by comparing their predictions of the breakdown voltage of atmospheric air in uniform fields with experimental data. The fact that the streamer criterion enables the uniform field breakdown voltage in atmospheric air to be calculated in rough agreement with measurements should not be considered to justify the theory. Greater emphasis should be placed on the general explanation of the physical processes leading to spark breakdown and the fact that it is based on gas mechanisms only, exclusive of cathode effects.

8.3.2.1.1 Meek's Analysis

Meek proposed the following mechanism for breakdown: In a uniform electric field, breakdown by the streamer mechanism occurs by the transition of an electron avalanche—launched from the cathode and traveling toward the anode—into a self-propagating streamer that develops from the anode to the cathode to form a conduction filament between the electrodes.

Meek considered the case observed experimentally by Raether [19,29] where a single-electron avalanche travels from cathode to anode and becomes critical at the anode, resulting in a streamer propagating from anode to cathode and then breakdown. Meek proposed the following criterion for the transition of an electron avalanche to a streamer:

> A streamer will develop when the radial field about the positive space charge in an electron avalanche attains a value of the order of the external applied field. [28]

Meek chose this case and argued that such a streamer will form when the radial field E_r about the positive space charge in an electron avalanche is sufficiently large to draw into the main avalanche channel some of the electrons that are liberated just outside of it (but closer to the cathode) by photo-ionization. Meek postulated that this happens if $E_r \sim E_0$. Previously, a rough qualitative understanding of streamer development had been proposed by Loeb where principal consideration had been given to the electric field distortion produced by the positive space charge but in the direction of the external applied field [40]. While the breakdown process was pictured there in a qualitative fashion, no quantitative criterion was given for streamer formation and breakdown. Loeb did demonstrate that under sparking conditions, the positive space charge in an electron avalanche is on the same order of magnitude as the applied field [40]. Loeb also noted the axial field distortion produced by the positive space inhibits the advance of the electron avalanche toward the anode but favors the development of a positive streamer toward the cathode [41]. Meek solved this conundrum by quantifying the criterion using the radial electric field [28]. That is, the space charge also produces a distortion perpendicular to the applied electric field—in the radial direction.

To calculate the electric field near the head of a streamer, assume the electrons are in a spherical volume located at the head of the avalanche and the ion density producing the distortion in the electric field is contained in a sphere of radius r located just behind it, as shown in Figure 8.21. This model is justified because the space charge distribution process is rapidly cumulative. In fact, about five times as many ions are produced in the final 10% of a 1 cm gap as in the preceding 9 mm.

According to Meek, the electric field E_r at the surface of a spherical volume containing a volume density of positive ions N_i with radius r is

$$E_r = \frac{Q_{tot}}{4\pi\, \varepsilon_0\, r^2} \tag{8.47}$$

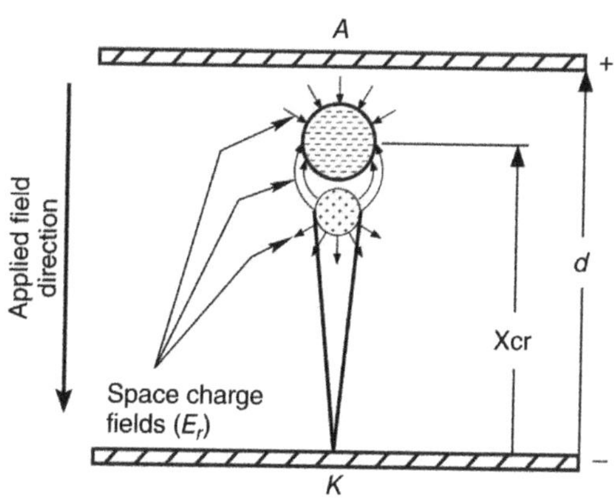

Figure 8.21 The distribution of charges in an electron avalanche used to calculate the radial electric field produced by an ion distribution.

The total charge contained in the spherical volume is given by

$$Q_{tot} = \left(\frac{4}{3}\pi r^3\right) \times (q_e N_i) \tag{8.48}$$

where q_e is the electronic charge. Combining Equations 8.47 and 8.48 yields

$$E_r = \frac{r}{3} q_e N_i \tag{8.49}$$

An estimate of the number density of positive ions just behind the electron cloud in the avalanche head is needed. It would not be accurate to assume all the positive ions in the avalanche are located just behind the head. The positive ions are distributed throughout the conical volume of the avalanche because of their reduced velocity compared to electrons. However, the volumetric ion density N_i needed to calculate the distortion of the electric field can be related to the number of ions n_i created in the avalanche as it develops from x to $x + dx$.

The number of positive ions n_i in an avalanche with an avalanche length x is $e^{\alpha \cdot x}$. As the avalanche travels an additional distance dx, the number of additional ions created is $dn_i(x) = \alpha \cdot e^{\alpha \cdot x} \cdot dx$ and they are assumed to reside in a cylindrical volume of radius r and length dx so that the volume of the cylinder is $(\pi r^2) \cdot dx$. The volumetric ion density located just behind the electron cloud in the avalanche head then is given by

$$N_i = \frac{dn_i(x)}{\pi r^2 \cdot dx} = \frac{\alpha \cdot e^{\alpha \cdot x}}{\pi r^2} \tag{8.50}$$

Raether related the avalanche head size to the distance the avalanche traveled in the electric field through the diffusion equation [39]. As the avalanche travels across the gap, the avalanche head, comprised mostly of electrons, grows through diffusion, according to

$$r^2 = 2Dt \tag{8.51}$$

with a time t

$$t = \frac{x}{v_e} \tag{8.52}$$

The velocity of the avalanche is determined by the rate at which the electrons travel in the gas and is related to the applied electric field E by the electron mobility μ_e:

$$v_e = \mu_e E \tag{8.53}$$

Equation 8.51 can be expressed as

$$r^2 = \left(\frac{2D \cdot x}{\mu_e \cdot E_0}\right) \tag{8.54}$$

Inserting Equations 8.50 and 8.51 into Equation 8.49 reduces to

$$E_r = \frac{1}{3}\frac{q_e \cdot \alpha \cdot e^{\alpha \cdot x}}{r} = \frac{1}{3}\frac{q_e \cdot \alpha \cdot e^{\alpha \cdot x}}{\sqrt{2D \cdot x/(\mu_e \cdot E)}} \tag{8.55}$$

The ratio of the diffusion coefficient D to the electron mobility μ_e can be found from

$$\frac{D}{\mu_e} = \frac{k_B T_e}{q_e} \tag{8.56}$$

where T_e is the electron temperature and k_B is the Boltzmann constant. The electron temperature is related to the electron energy gained from the electric field by

$$\frac{3}{2} k_B T_e = q_e V \tag{8.57}$$

Thus, by estimating material properties and setting $x = d$ for the threshold condition of the cathode-directed (positive) streamer, Meek derived the condition for air at a pressure p as

$$E_r = 5.28 \times 10^{-7} \frac{\alpha e^{\alpha d}}{\sqrt{d/p}} \frac{\text{V}}{\text{cm}} \tag{8.58}$$

which could be compared with experiment [41,42]. For air, this equation is satisfied for $\alpha = 18.6$. It should also be noted that Meek and Loeb used parameters experimentally determined by Raether in their estimates in deriving Equation 8.58.

8.3.2.1.2 Raether's Analysis

Raether's approach was similar to Meek's approach, but focused on interpreting his measurements and observations on electron avalanches and their transition to streamers. Raether hypothesized that the avalanche transformed into a streamer became critical when the number density of electrons in the avalanche head reached a certain size and treated the critical avalanche length x_{cr}—the length of the avalanche when it transformed into a streamer—as the parameter. Instead of requiring Meek's imposed condition that the primary avalanche reach the anode for a streamer to form, he allowed for critical avalanche lengths less that the gap length ($x_{cr} < d$), which occurs in an anode-directed (negative) streamer.

To obtain an approximate value for the field at the head of the avalanche, Raether assumes the positive ions to be contained in a sphere of radius that is calculated from the diffusion process along with an empirical value of the thermal energy of the electrons. In atmospheric air, the thermal energy of the electrons was found to be 1.5 eV. Raether's criterion for the electron avalanche to streamer condition in atmospheric air is

$$\alpha x_{cr} = 17.7 + \ln(x_{cr}) \tag{8.59}$$

where x_{cr} is measured in centimeter. The threshold case of Raether's criterion is when the electron avalanche transverses the gap and only becomes critical at the anode so that the streamer onset criterion occurs by setting $x_{cr} = d$ (the gap length) and in Equation 8.59 becomes

$$\alpha d = 17.7 + \ln(d) \tag{8.60}$$

Because the natural logarithm is a slowly varying function and the range of gap distances are relatively small, Equation 8.60 can be estimated by

$$\alpha d \approx 18 \text{ or } 20 \tag{8.61}$$

Thus, according to Raether [43], the streamer breakdown takes place when the number of electrons in the avalanche head reaches a critical number, given by

$$N_{cr} = e^{\alpha\, x_{cr}} = 10^8 \text{ or } 10^9 \tag{8.62}$$

Equations 8.61 and 8.62 both are forms of what is known as Raether's criterion. The streamer breakdown time, obtained by substituting $\alpha \cdot x_{cr} = 20$ into Equation 8.52, is given by

$$\tau_f = \frac{20}{\alpha\, v_d} \tag{8.63}$$

Equation 8.63 is a statement that the streamer formation time is dominated by the time required for the avalanche to become critical.

8.3.2.2 The Electric Field Along the Avalanche

Regardless of the polarity of the applied voltage, streamers start with an electron avalanche traveling from cathode to anode. A single electron accelerates in the electric field and causes further ionization. When the original electron has traveled a distance x in the direction of the applied field, the number of electrons created is $e^{\alpha x}$. The quantity $e^{\alpha x}$ is sometimes referred to as the "gain factor."

The electric field at the head of the avalanche characterized the space charge electric field in terms of the radial electric field E_r to differentiate it from the field along the axis, E_M, ahead of the advancing avalanche, which is roughly

$$E_M = E_0 + E_r \tag{8.64}$$

However, the criterion $E_r \sim E_0$ is not a sharp boundary. The space charge in the avalanche head begins to distort the electric field in the gap when the radial field E_r is a fraction of the applied field E_0. This is expressed as

$$E_r = k' \cdot E_0 \tag{8.65}$$

Further work showed the ratio of radial space charge field to the applied field, $\acute{\kappa}$, could be quite small. Using an avalanche head radius of 130 μm—a value determined by Raether in his cloud chamber experiments, Meek and Loeb calculated the radial field E_r to be 6 kV/cm, a mere 20% of the applied electric field [41,42]. On the other hand, Raether determined that value of $\acute{\kappa}$ for anode-directed streamers is very close to 1 [43]. In general, value of $\acute{\kappa}$ is considered to be in the range of 0.1–1, as opposed to Meek's original hypothesis that $\acute{\kappa} = 1$. It is important to bear in mind that many of the assumptions made in the derivation of the avalanche field—that the positive space charge is concentrated in a sphere, for example—are crude approximations that preclude too great attention being paid to the exact value of $\acute{\kappa}$.

In front of the avalanche head, the space charge field increases the magnitude of the electric field along the axis to

$$|E_M| = (1 + \acute{\kappa}) \cdot E_0 \tag{8.66}$$

The electric fields also vary along the length of the avalanche because of the separation of charges. The electrons travel at a speed of ~10^7 cm/s, while the positive ions have a speed of ~10^5 cm/s because of their larger mass and the avalanche develops as a cloud of electrons with a trail of positive space charge across the gap, as shown in Figure 8.15. Raether depicted the fields along the avalanche, as shown in Figure 8.22.

8.3.2.3 A Qualitative Description of Streamer Formation

The steamer onset criterion describes the conditions under which an electron avalanche transforms into a streamer. Although the concept of a streamer has been the subject of much investigation, the breakdown criterion has changed

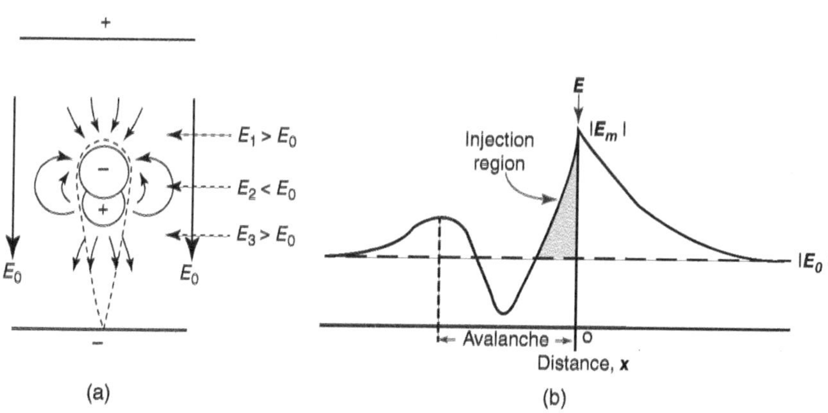

Figure 8.22 The electric fields within the avalanche are affected by the distribution of space charges [19]. (Reproduced with *permission from Oxford University Press)* In front of the head and within the tail of the avalanche, the local electric fields are greater than the applied field [20].

little since its introduction in 1940. The streamer onset criterion neglects all physical processes that must take place between the avalanche reaching the critical size and the collapse of the voltage. It is assumed that when the avalanche reaches critical size, an unstable condition persists and the current becomes self-sustained. The precise nature of this unstable condition is still an active research area.

From the discussion of the electric fields in the vicinity of the avalanche, we know that avalanches initiate with electron motion in an electric field and travel in the direction from cathode to anode. The mechanism of secondary charge carrier generation, however, is not clear. Because electrical breakdown occurs in either polarity, we know that streamers can travel in either direction. In the next sections, the physical processes that occur and lead to a streamer and breakdown is described.

Because the physical processes are dependent on polarity, the characteristics of the streamers are different. The cathode-directed streamer is closely linked to photoionization and results in a very fast discharge traveling at a velocity of $\sim 10^8$ cm/s. The anode-directed streamer travels more slowly than the positive streamer with a velocity up to an order of magnitude less. The cathode-directed streamer has a lower threshold for onset so that, in cases most relevant to pulsed power applications, positive and negative streamers propagate simultaneously. A qualitative description of positive and negative streamers in uniform fields is given as well as the more common mechanism where streamers are launched in both directions.

8.3.2.3.1 The Cathode-Directed (Positive) Streamer

Suppose that a primary electron avalanche develops from the cathode, travels in the applied electric field, and reaches its critical avalanche length just as the avalanche reaches the anode. When the avalanche has crossed the gap, the electrons are swept into the anode, the positive ions remaining in a cone-shaped volume extending across the gap. A highly localized space charge field near the anode results in the launching of a streamer in the direction of the cathode. This cathode-directed streamer, also known as a positive streamer, grows with the help of photoionization. Electrons are produced by photons emitted from the densely ionized gas near the anode. These photoelectrons act as initiators of auxiliary avalanches of the streamer that are attracted to the strong electric fields generated by the positive space charge in the head of the advancing streamer. The greatest multiplication in the auxiliary avalanches will occur along the axis of the main avalanche where the space charge field supplements the external field. The streamer absorbs the new avalanches, neutralizing some charge and advancing the streamer toward the cathode, as illustrated in Figure 8.23. The process develops as a self-propagating streamer that effectively extends the anode toward the cathode and results in a conducting filament of

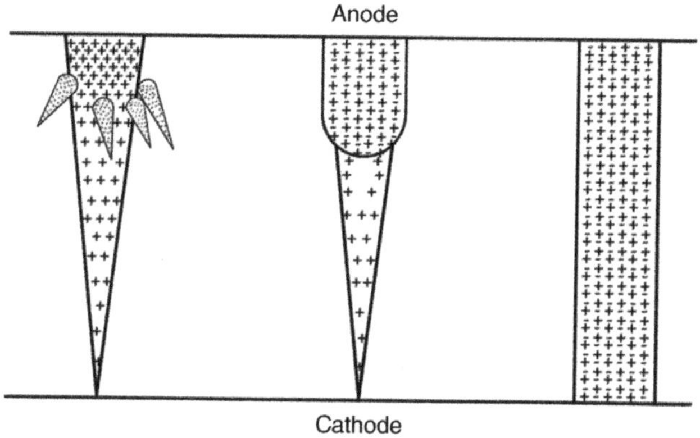

Figure 8.23 The progression of the cathode-directed streamer envisioned by Meek [44]. An electron avalanche first reaches the anode, leaving a weakly ionized cone shaped trail; the positive streamer launched from the anode progresses to the cathode, fed by auxiliary electron avalanches that are absorbed into the advancing streamer. (Reproduced with permission from Oxford University Press.)

highly ionized gas between the electrodes. Raether's depiction of the cathode-directed streamer is shown in Figure 8.24.

8.3.2.3.2 The Anode-Directed (Negative) Streamer

Raether proposed the formation mechanism for anode-directed streamers. Anode-directed streamers occur when the primary avalanche becomes sufficiently strong before reaching the anode. Raether's criterion states that negative streamers develop when the initiating avalanche mechanism produces a sufficient number of electrons ($\sim 10^8$) to give an electric field due to space charge comparable to the applied electric field. The total enhanced field promotes

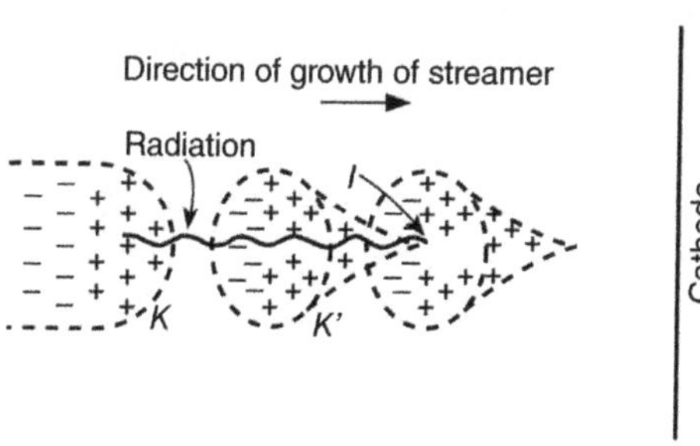

Figure 8.24 Raether believed photoionization was crucial to the formation of streamers by providing seeds for additional electron avalanches to feed the growth of the streamer [45]. (Reproduced with permission of Oxford University Press.)

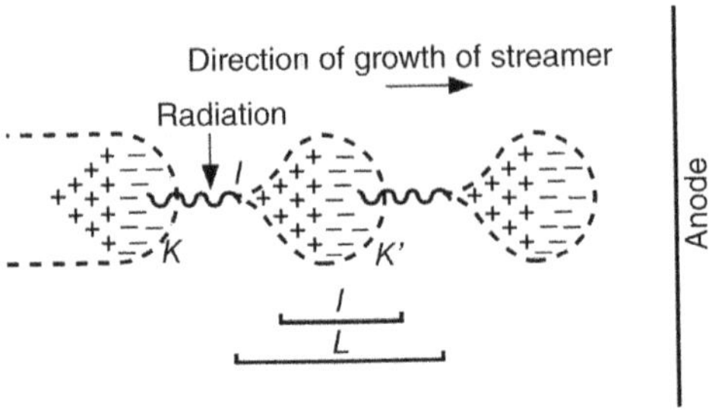

Figure 8.25 The negative (anode-directed) streamer as depicted by Raether advances by absorbing auxiliary electron avalanches formed ahead of the streamer [45]. (Reproduced with permission of Oxford University Press.)

auxiliary anode-directed electron avalanches ahead of the negative (anode-directed) streamer, as depicted in Figure 8.25. The auxiliary avalanches are initiated by electrons formed by photoionization and electrons in the gap space ahead of the streamer.

8.3.2.4 Streamer Criterion in Nonuniform Electric Fields

Both Meek and Raether used uniform field geometries in their development of the criterion for the avalanche-to-streamer transition. Most applications, however, incorporate both nonuniform geometries and transient electric fields. Both factors make the likelihood of electrical breakdown by the Townsend mechanism very unlikely. Even a deviation from uniformity of only 5% leads to breakdown by the streamer mechanism. Nonuniform geometries can introduce local field enhancements, which enhances the rate of local space charge generation, driven by the increase in the primary ionization coefficient α, which is a very strong function of the applied electric field—that is, $\alpha(E)$. Moreover, the breakdown by the Townsend mechanism takes significantly longer time than a streamer to develop into a current-carrying arc.

The criterion for streamer onset in nonuniform electric fields is estimated using the gain factor $e^{\alpha x}$ and Raether's criterion of critical number of electrons (Equation 8.62). In a uniform electric field, the number of electrons generated from a single electron is $e^{\alpha x_{cr}} \sim 10^8$ or $\alpha \cdot x_{cr} \sim 20$. In nonuniform geometries, the number of electrons generated per initial electron is obtained by integrating along an electric field line:

$$\int_0^{\ell} \alpha(E) \cdot dx \sim 20 \tag{8.67}$$

The integral along the field line in Equation 8.67 is typically evaluated numerically.

8.3.2.5 The Overvolted Streamer

An anode-directed streamer generally takes place close to the cathode, especially when the overvoltage or the gap size is large. In this case, the primary avalanche becomes critical close to the cathode and streamers grow toward both electrodes. The successive stages in the development of streamer breakdown are shown in Figure 8.26.

Stage 1: The critical space charge field E_r, causing substantial field distortion in the gap, is produced from an increasing charge carrier density in the head of the avalanche. The space charge field in the region between the avalanche head and the anode as well as between the avalanche tail and the cathode adds to the applied field and the resultant field is increased substantially. The tail containing the positive ion carriers decelerates the advancement of the avalanche head and the ionization efficiency of the front electrons.

Stage 2: The photoelectrons produced in the gas medium initiate auxiliary avalanches along the radial field lines and feed into the tail of the main avalanche. Simultaneously, the auxiliary avalanches also develop toward the anode because of photoelectrons produced in the region between the avalanche head and the anode.

Stage 3: The tail of the main avalanche continues to elongate and propagates rapidly toward the cathode as a result of the feeding in of charge carriers by the subsidiary avalanches.

Stage 4: Ultimately, the stem of the main avalanche bridges the gap between the cathode and the anode. Electrons are then drawn into the stem, resulting in the formation of conducting channel between the

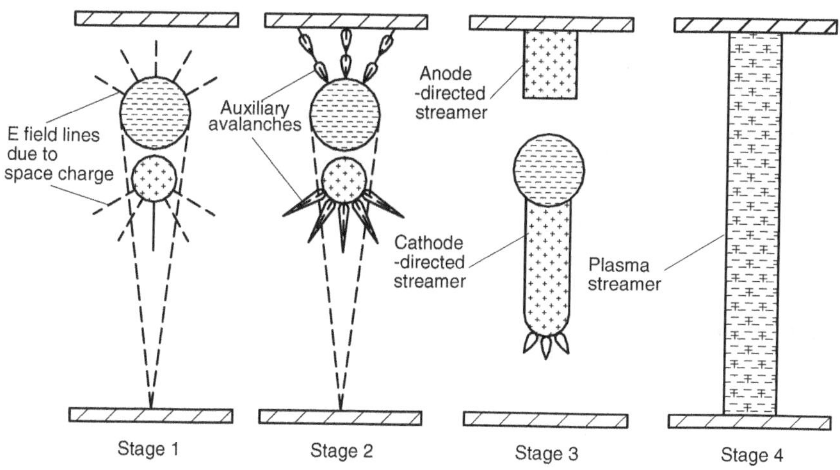

Figure 8.26 Successive stages of streamer breakdown.

electrodes. The impedance of this channel remains high until the channel is fully thermalized and capable of carrying full-circuit current.

Evidence for the overvolted streamer mechanism is shown in Figure 8.27, where the velocity of propagation of light emitted from a discharge in atmospheric laboratory air is measured with a streak camera [46]. A streak image is obtained by imaging through a slit and recording its evolution in time. In Figure 8.27, the streamer initiates at the cathode (bottom electrode) and travels toward the anode. Approximately 20 ns into the evolution of the anode-directed streamer, a streamer initiates from the anode and travels toward the cathode.

8.3.2.6 Pedersen's Criterion

Pedersen developed a methodology for establishing the breakdown criterion in compressed gases. According to Pedersen, streamer breakdown criterion for air [47,48] can be written as

$$\alpha_x \exp\left\{\int_0^x \alpha dx\right\} = H(x, \rho) \tag{8.68}$$

where α_x is the numerical value of α at the avalanche head and $H(x,p)$ is an unknown function dependent on critical avalanche distance and air density.

The semiempirical relation of Equation 8.68 is a modified version of the equation derived by Meek, which takes into consideration the nonuniformity of electric field. For air at atmospheric pressure, Equation 8.68 simplifies to

$$\ln \alpha_x + \int_0^x \alpha dx = H(x) \tag{8.69}$$

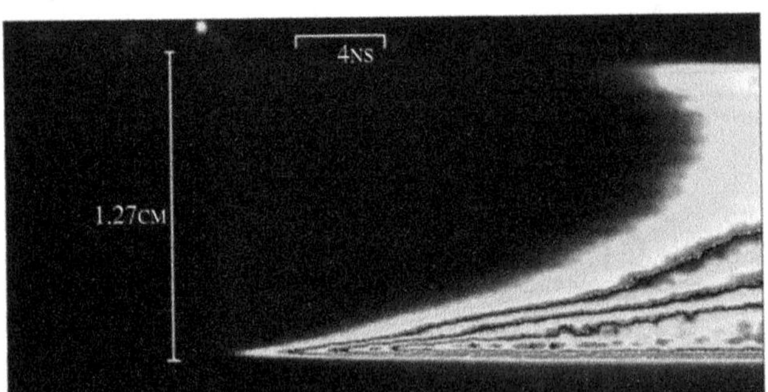

Figure 8.27 The streak image of a discharge in laboratory atmospheric air [46]. The horizontal axis corresponds to gap distance and the vertical axis to time. The scale is shown in the upper quadrant of the streak image. The discharge initiates at the cathode (at the bottom of the streak data) with a constant velocity. Before the anode-directed streamer is across the gap, a cathode-directed streamer starts and results in acceleration of the discharge.

The steps in the evaluation of streamer breakdown voltage for a given nonuniform gap are as follows:

Step 1: $H(x)$ is evaluated from the following relation assuming a uniform field for the test gap:

$$H(x) = \ln \alpha + \alpha d \tag{8.70}$$

where d is the test gap spacing and α is the Townsend first ionization coefficient obtained from previously established data at the measured breakdown field of the uniform field gap.

Step 2: For the test gap, E versus x, and hence α versus x, is determined at an assumed breakdown voltage V_b and the left side of Equation 8.70 is evaluated.

Step 3: If the evaluated left-side value in step 2 agrees with the $H(x)$ value obtained in step 1, the assumed breakdown voltage is the correct steamer breakdown voltage. If the left-side value in step 2 is not equal to $H(x)$, another value for V_b is assumed and the iterations of step 2 are repeated until the assumed breakdown voltage satisfies Equation 8.70.

Pedersen did the above calculations for air at atmospheric pressure for sphere–sphere gap geometry of various ratios of diameter to gap distance, D/d, and found agreement within 2% of the established values. The Pedersen criterion has been used to estimate the breakdown voltage of compressed gases such as air, nitrogen, and SF_6 and has been found to agree well with experimental results [49].

8.4 The Corona Discharge

Corona is a luminous, audible discharge that occurs when an excessive localized electric field gradient causes ionization in the surrounding gas. It manifests easily in highly nonuniform electric field geometries, such as point-to-plane electrodes or cylindrical geometries with inner conductors made of wire. Corona is characterized by a colored glow frequently visible in a darkened environment. The audible discharge, usually a subtle hissing sound, increases in intensity with increasing output voltage. Ozone, an odorous unstable form of oxygen, is frequently generated during this process. Rubber is destroyed by ozone, and nitric acid can be created if sufficient moisture is present. These items have detrimental effects on materials, inclusive of electrical insulators.

A high electric field gradient exists only in the small region near the highly stressed region (the point electrode or the inner wire in a cylindrical geometry) with a negligible field gradient in the remaining gap spacing. Thus, the Townsend breakdown criterion is satisfied only for a small region, where the gas is ionized. The corona is accompanied by a luminous glow around the high-field electrode and is usually accompanied by a hissing noise. If the corona current is time resolved, it is found to consist of recurring current pulses occurring either regularly or randomly.

For a negative point corona (when the high-field electrode is negative), these pulses are known as Trichel pulses [50]. The electron avalanches develop from the point electrode and propagate toward the anode. The multiplication of charge carriers takes place only in a small region around the point and the electron swarm drifts in the low-field region without causing further ionizations. These electrons are slowed down because of the effect of the positive ions left behind by the avalanche and attach to gas molecules forming negative ions. The presence of the negative ions reduces the electric field at the point electrode and the discharge extinguishes [51]. When the positive and negative ions have drifted away to their respective electrodes, the original high-field conditions are re-established, giving rise to another Trichel current pulse. This process continues, giving rise to a steady-state corona current. Trichel pulses are regularly placed and only appear in gases exhibiting some electron attachment, notably air and sulfur hexafluoride.

For a positive point corona [52,53], the process of current growth is complex and seems to consist of the following stages: (i) An electron avalanche, initiated by electrons available in the gas near the high-field region, propagates into the point anode. (ii) Streamers develop in the high-field region. (iii) Lateral avalanches feed into the streamer core, giving rise to a corona burst evidenced by pulses observed on the current waveform. (iv) A negative ion cloud is formed between the point and the positive ions, inhibiting the growth of the streamer (see Figure 8.28).

Corona is of practical importance because it results in power loss in transmission line and other components, results in radio frequency

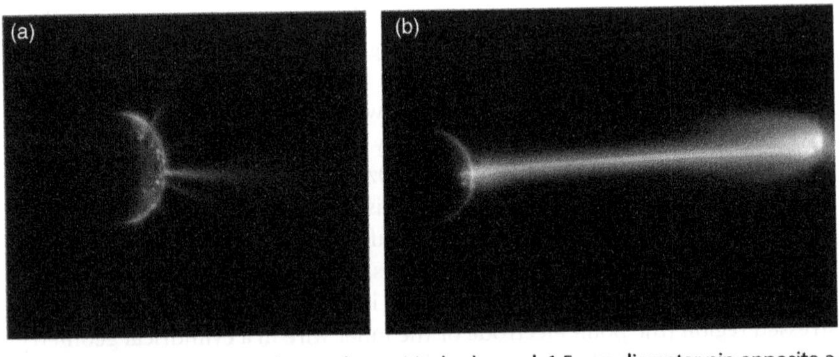

Figure 8.28 Corona is formed around a positively charged, 1.5 mm diameter pin opposite a grounded plane. In part (a) the corona is stable, and in part (b) the corona transitions into a high-current discharge.

interference, and can reduce the service life of solid and liquid insulation by initiating partial discharges and chemical decomposition. A good high-voltage design takes corona generation into account and provides design countermeasures to limit the possibility of problems developing. It can also be used as an advantage, as evidenced by the variety of corona-stabilized switches discussed in Chapter 4.

8.5 Pseudospark Discharges

Pseudospark breakdown takes place at pressure–distance product (*pd*) values lying to the left side of the Paschen minimum from about 10 to 10^{-4} mbar-mm for a specific electrode geometry, shown in Figure 8.29. The mechanism of pseudospark discharge is explained in terms of prebreakdown regime and breakdown regime [54–65].

8.5.1 The Prebreakdown Regime

In the prebreakdown regime, as in the case of Townsend mechanism, the electrons are accelerated toward the anode by an electric field, causing ionizing collisions resulting in the formation of electron–ion pairs. The electrons are collected by the anode. The ions drift toward the cathode, causing ionizing collisions and subsequent multiplication of positive ions, resulting in an ion avalanche. The gas pressure is low where positive ions are more efficient ionizers than electrons. The positive ion swarms pass through the cathode hole and form an ion space charge. A virtual anode is thus formed behind the cathode with a potential of several hundred volts.

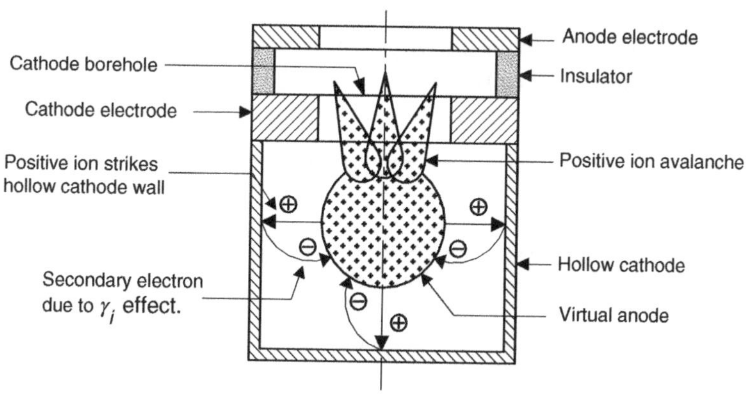

Figure 8.29 Electrode geometry and mechanics in a psuedospark discharge.

8.5.2 Breakdown Regime

In this regime, most of the charged particle generation takes place behind the cathode, between the virtual anode and back surface of the cathode. The plasma in the virtual anode is surrounded by a Debye layer of thickness λ_D with a high electric field through which the ions transport to the cathode. The Debye layer thickness λ_D and the potential drop V_D across the Debye layer are given by

$$\lambda_D = \left(\frac{\varepsilon_0 k_b T_e}{q_e^2 n_e}\right) \tag{8.71}$$

$$V_D = \frac{k_b T_e}{2q_e} \tag{8.72}$$

where

ε_0 is the dielectric constant,
k_B is the Boltzmann's constant,
T_e is the electron temperature,
N_e is the electron density, and
q_e is the electronic charge.

The electrons generated at the back surface of the cathode are accelerated to the virtual anode, where they cause a further multiplication of charge carriers. When the plasma density inside the virtual anode reaches a particular critical value, a fast breakdown takes place with the current being carried by a well-defined discharge formed on the central axis of the cathode–anode geometry. An interesting feature of pseudospark is that the virtual anode gives rise to an intense electron beam, with current densities exceeding 10^6 A/cm^2, in the direction of the anode and an ion beam in the opposite direction.

The exact mechanisms leading to pseudospark breakdown are still not well understood. However, experimental evidence exists regarding the sequence of events during breakdown, as determined by the measurements of voltage, current, and light emissions in the discharge. The various stages in the discharge development are shown in Figure 8.30.

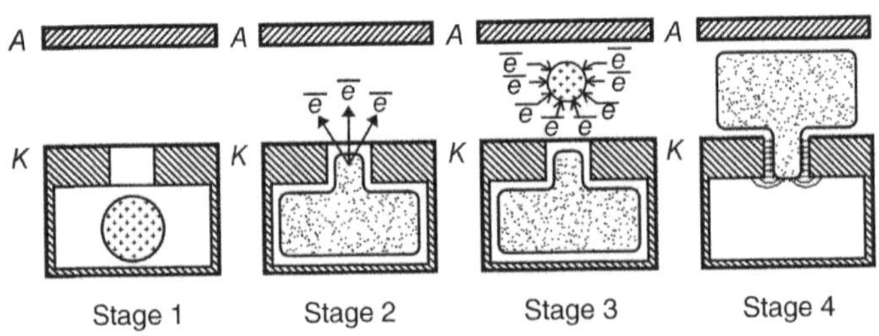

Figure 8.30 Sequence of events in a pseudospark discharge.

Stage 1— *virtual anode formation:* A virtual anode of critical charge density is formed behind the cathode electrode. In the case of self-breakdown, it is formed by the injection of positive ions from Townsend discharge in the main gap. In the case of a triggered pseudospark switch, it is formed by plasma injection in the hollow cathode volume by an external trigger discharge.

Stage 2— *hollow cathode discharge formation and electron injection into the main gap*: During this stage, a hollow cathode discharge forms and grows in intensity and volume. The mechanisms active in the hollow cathode volume are photoionization of gas, photoemission from the cathode, and secondary emission from the cathode by γ_i process.

Stage 3— *positive ion space charge and anode plasma formation:* The hollow cathode discharge injects an intense electron beam into the main gap. This gives rise to an intense positive ion space charge near the anode, caused mainly by the lower mobility of ions. Charge carriers are accelerated in the space charge-intensified electric field and multiplication takes place. This results in the formation of anode plasma and the plasma stem grows in intensity and volume.

Stage 4— *anode plasma spreads into the borehole:* The plasma discharge spreads and enters into the cathode borehole, where it becomes super dense. At this stage, the hollow cathode discharge disappears, cathode spots are formed, and the current is carried by the plasma. The mechanisms of sputtering and ionization are active at the borehole.

8.6 Breakdown Behavior of Gaseous SF_6

As an insulator for high-current components, sulfur hexafluoride (SF_6) is not as commonly used as it once was because of concerns about greenhouse gases as well as its prohibitive price. Historically, SF_6 at high pressure found widespread use in pulse power applications and modern electrical equipment because of its exceptional dielectric strength, generally attributed to its strong affinity for electrons, limiting the growth of the electron avalanche.

The dielectric strength of SF_6 at high pressure deteriorates because of a number of factors including the electrode material—its type, surface area and surface finish, the gap spacing, gas contamination by conducting particles, and the presence of an insulating support spacers [66]. To successfully exploit the high dielectric strength of high-pressure SF_6, the gas must be maintained at high purity and low moisture content and care must be taken in the choice of components and their assembly. The important aspects of breakdown behavior in SF_6 are briefly discussed in the following sections.

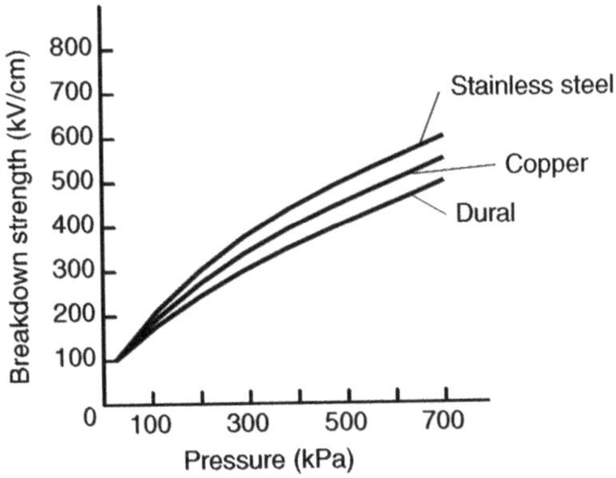

Figure 8.31 Effect of electrode material on the breakdown strength of compressed SF_6 [67].

8.6.1 Electrode Material

A typical curve [67] showing the dependence of breakdown strength of SF_6 at various pressures is shown in Figure 8.31 for the electrode materials stainless steel, copper, and Dural. The data were taken with parallel–plane electrodes of 70 mm diameter and a gap spacing of 2 mm. The electrode material has a pronounced effect at high pressures with stainless steel producing the highest breakdown strength and Dural the lowest. The properties of mechanical strength, melting temperature, and work function also decrease in the same order, implying that processes involving the heating and mechanical deformation of microscopic inhomogeneities and detachment of metal particles from the electrode surfaces become involved in the breakdown process along with field emission when the electric field at the electrodes is strong enough (250–300 kV/cm). Brass is a commonly used electrode material that has shown superior performance in high-pressure SF_6 switches.

8.6.2 Surface Area and Surface Finish

The curves [68] depicting the behavior of breakdown strength of SF_6 as a function of surface area and surface finish are shown in Figure 8.32. The data were taken by applying voltage at 60 Hz AC for the electrode configurations of (i) parallel–plane gaps of 10 cm electrode diameter and 5 mm gap spacing with a resulting surface area of 80 cm^2, (ii) coaxial cylinders with a13.2 cm inner diameter, a 15.2 cm outer diameter, and 11.6 cm length with an annular gap spacing of 10 mm and resulting surface area of 480 cm^2, and (iii) coaxial cylinders with a 23.75 cm inner diameter, a 25.75 cm outer diameter, and 40.2 cm length with an annular gap of 10 mm and resulting surface area of 3000 cm^2. The data indicate the breakdown strength of SF_6 reduces with

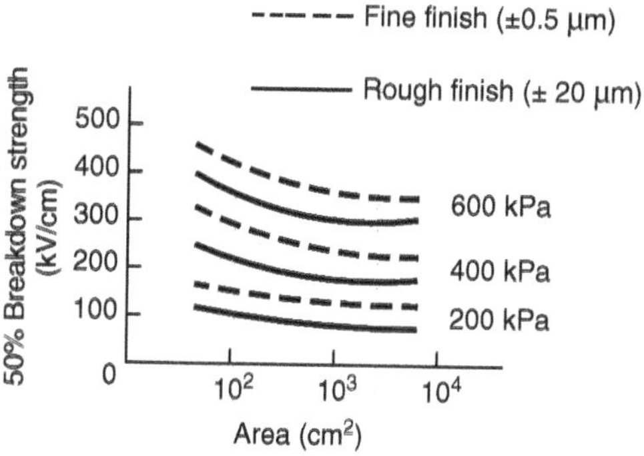

Figure 8.32 Effect of electrode finish and surface area [68].

increasing surface area and increasing electrode surface roughness. Rougher metallic surfaces have more surface asperities, increasing local field enhancements, and electron emission.

8.6.3 Gap Spacing and High Pressures

The behavior of the breakdown voltage in SF_6 as a function of pressure–distance product (*pd*) shows anomalies, as shown in Figure 8.33 [69]. A Rogowski-profile parallel–plane electrode configuration with a 43 cm diameter is subject to a 50 Hz AC voltage at various gap settings. Although the Paschen's law predicts the breakdown voltage to be dependent only on the product *pd* and not on individual values of *p* or *d*, a departure from Paschen's law is seen at high SF_6 gas pressures, where the breakdown voltage becomes dependent on both

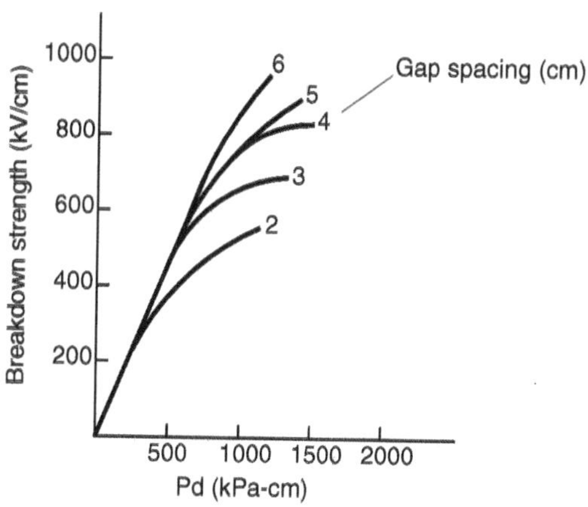

Figure 8.33 The effect of gap spacing on the breakdown strength of compressed SF_6 [69].

the gap spacing and pressure. This effect is seen in the self-break characteristics of high-pressure SF_6 switches, where the V–p curve loses its linearity with increasing pressure. These deviations from Paschen's law are from high-field effects that are known to strongly affect breakdown in SF_6.

8.6.4 Insulating Spacer

The breakdown voltage of SF_6 reduces substantially when the electrodes are bridged by an insulating spacer [71,72], as shown by the characteristics in Figure 8.34. The percentage reduction in the breakdown voltage with the introduction of a spacer is much higher at higher pressures. The data in Figure 8.34 were taken with a DC voltage applied to parallel–plane electrodes with 10 mm diameter cylindrical epoxy spacers with lengths of 5, 7.5, and 10 mm. The reduction in breakdown voltage is attributed to microdischarges at stress concentrations either on the insulator surface or at the spacer–electrode contact, and charging of the insulator surface. Techniques exist to maximize the electrical breakdown strength in the vicinity of a spacer [73,74].

8.6.5 Contamination by Conducting Particles

When SF_6 is contaminated with conducting particles, the breakdown voltage is found to be lowered significantly in comparison to the pure gas [74]. Figure 8.35 shows the voltage–pressure characteristics of SF_6 contaminated by copper wire particles [75]. The experimental arrangement consists of various lengths of a 0.4 mm diameter wire particle in a coaxial geometry having radii of 125 and

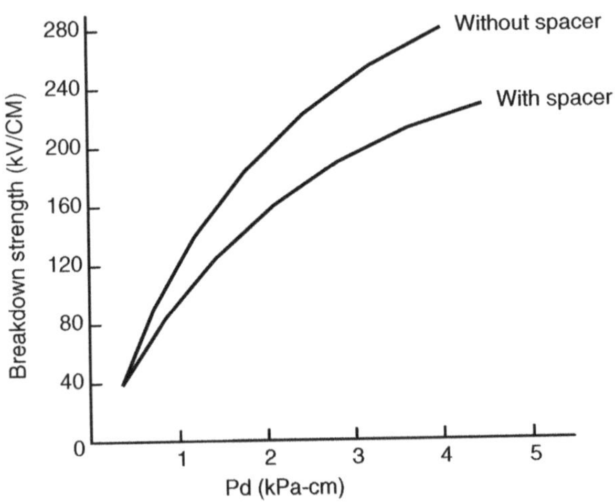

Figure 8.34 Effect of the inclusion of a spacer on the electrical breakdown of SF_6 [70].

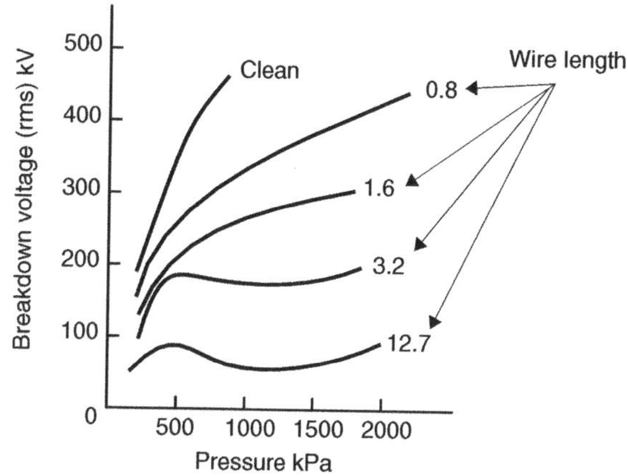

Figure 8.35 Effect of conducting particles on the electrical breakdown strength of SF_6 [75].

75 mm. The breakdown voltage reduces with larger wire lengths. One mechanism proposed for the reduction in breakdown voltage is the occurrence of microdischarges when a particle is sufficiently close to an electrode. Non-conducting particles have been found to have much less effect than conducting particles. The purity of the gas can be maintained by using particle traps and filters.

8.7 Intershields for Optimal Use of Insulation

The nonuniform electric field distribution that exists between the electrodes in a spherical or cylindrical geometry creates inefficiencies in the use of insulation. In standard cylindrical or spherical configurations, the insulation is heavily stressed near the inner electrode and is underutilized near the outer electrode because of low electrical stress. A uniform stress throughout the volume makes maximum use of the dielectric. The use of one or more intershields can attain a good approximation of uniform fields in these geometries.

The design of intershields has been developed by means of a three-dimensional electrolytic tank by Kiss et al. [76] as well as theoretically in the form of formulas by Boag [77]. The formulas derived by Boag have been used in the design of an electrostatic particle generator by Chick and Petrie [78] and a Cockcroft–Walton multiplier by Ron [79].

8.7.1 Cylindrical Geometry

Cylindrical geometries are extensively used in pulsed power technologies for coaxial capacitors and transmission lines. A main feature is that no edges, with their associated field enhancements, are necessary, but the nonuniformity of the electric field in the interelectrode gap results in a lower stored energy density than could be achieved with more uniform fields. The introduction of an additional cylinder—an intershield—results in a more uniform electric field distribution and an increased energy density.

In this section, the electric field in the standard cylindrical geometry is used to introduce the equivalent gap length, which is then used as the optimization parameter to determine the radial location of the intershield and increased efficiency of the insulation.

8.7.1.1 Two-Electrode Concentric Cylinders

The electric field at any radius r in the cylindrical geometry shown in Figure 8.36a is given by

$$E(r) = \frac{V}{r \cdot \ln(R_2/R_1)}, \quad \text{for} \quad R_1 \leq r \leq R_2 \tag{8.73}$$

where

V is the voltage applied to the inner conductor,
R_2 is the outer radius of the cylinder, and
R_1 is the inner radius of the cylinder.

The outer conductor is a fixed dimension R_2 and is held at ground potential. The field between the cylinders is azimuthally symmetric. The maximum electric field E_M occurs at the inner conductor and varies with r, as shown in Figure 8.36b.

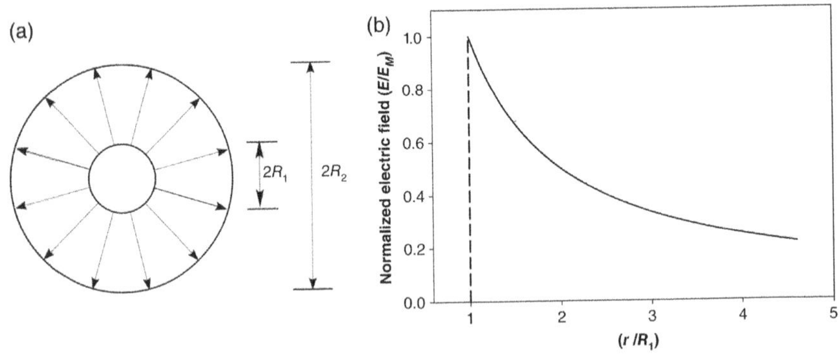

Figure 8.36 The electric field between two concentric cylinders is uniformly distributed as shown in part (a) and varies as $1/r$ as shown in part (b). The peak electric field occurs at $r = R_1$.

8.7.1.1.1 *The Equivalent Gap Length*

The peak electric field occurs at the surface of the inner conductor. Defining the quantity G as the *equivalent gap length*,

$$G = \frac{V}{E(R_1)} \tag{8.74}$$

For two concentric cylinders,

$$G = \frac{V}{E(R_1)} = R_1 \ln\left(\frac{R_2}{R_1}\right) \tag{8.75}$$

Assuming a fixed value for R_2, the optimum value of R_1 to maximize G can be found by setting $dG/dR_1 = 0$ and solving for R_1,

$$\ln\left(\frac{R_2}{R_1}\right) = 1 \quad \text{or} \quad \frac{R_1}{R_2} = \frac{1}{e} = 0.3679 \tag{8.76}$$

This is the result obtained in Chapter 3. It is convenient to normalize G to the fixed value R_2, and from Equation 8.75,

$$\frac{G}{R_2} = \frac{R_1}{R_2}\ln\left(\frac{R_2}{R_1}\right) \tag{8.77}$$

The maximum value of G/R_2 can be calculated and found to be

$$\left.\frac{G}{R_2}\right|_{Max} = \frac{V}{R_2 \cdot E(R_1 = (R_2/e))} = \frac{R_1}{R_2} = 0.3679 \tag{8.78}$$

A curve of G/R_2 versus R_1/R_2 is plotted in Figure 8.37.

8.7.1.2 Cylindrical Geometry with an Intershield

In order to utilize the insulation effectively, another conducting cylinder, or intershield, can be inserted between the inner and outer radii of the concentric cylinders, at a position to be determined, r', as indicated in Figure 8.38. The intershield is charged capacitively to a voltage V', by the voltage V applied to the inner conductor R_1. The intershield alters the electric field distribution in the geometry, increasing the energy storage density.

The electric fields in the two regions must be treated separately:

$$E(r) = \frac{(V - V')}{r \cdot \ln(r'/R_1)}, \quad \text{for} \quad R_1 \le r \le r' \tag{8.79}$$

$$E(r) = \frac{V'}{r \cdot \ln(R_2/)}, \quad \text{for} \quad r' \le r \le R_2 \tag{8.80}$$

where

$E(r)$ is the electric field at a radius r,
V is the voltage on the inner cylinder,

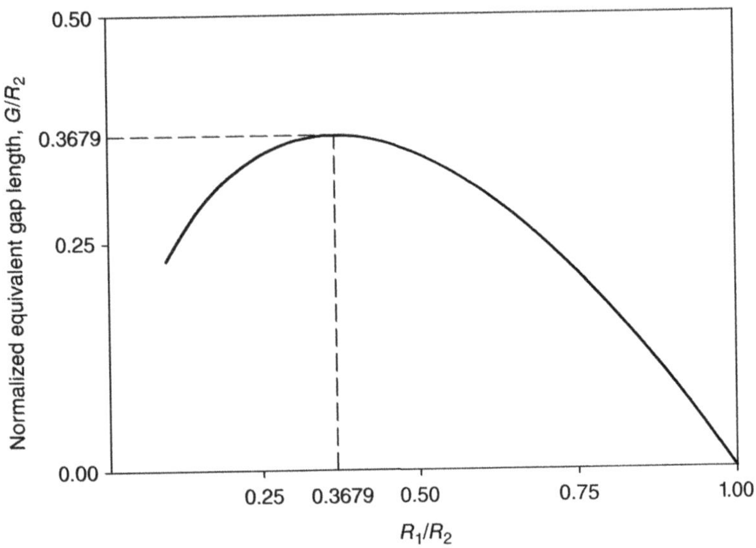

Figure 8.37 The variation of the normalized equivalent gap length with the normalized radius shows the maximum occurs at $R_2 = eR_1$.

V' is the voltage on the shield,
r' is the radius of shield, and
R_1 is the radius of inner conductor.

Again, the maximum electric fields are located at the innermost conductor in each region. The maximum electric fields are as follows:

$$\text{Region I}: \quad E(R_1) = \frac{(V - V')}{R_1 \cdot \ln(r'/R_1)} \tag{8.81}$$

$$\text{Region II}: \quad E(r') = \frac{V'}{r' \cdot \ln(R_2/r')} \tag{8.82}$$

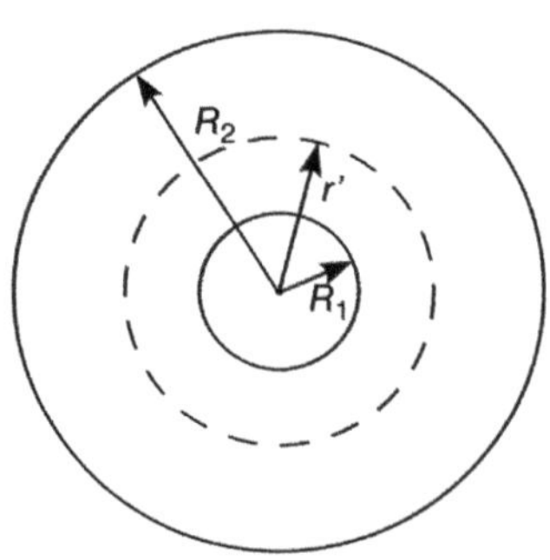

Figure 8.38 Concentric cylinders with an intershield located at $r = r'$. The intershield charges to a voltage V' when a voltage V is applied to the center conductor.

The maximum voltage V' is determined by setting the maximum electric fields to the electrical breakdown fields for the insulating medium.

$$E(R_1) = E(r') = E_{BD}$$

The voltage at the intershield V' can be solved in terms of the applied voltage and dimensions and is found to be

$$V' = V \cdot \frac{r' \ln (R_2/r')}{R_1 \ln (r'/R_1) + r' \ln (R_2/r')} \tag{8.83}$$

Substituting the expression for V' from Equation 8.83 into Equation 8.82 and rearranging terms, the electric field at R_1 can be expressed solely as a function of V,

$$E(R_1) = \frac{V}{R_1 \ln (r'/R_1) + r' \ln (R_2/r')} \tag{8.84}$$

and the equivalent gap length G can be written

$$G = \frac{V}{E(R_1)} = R_1 \ln\left(\frac{r'}{R_1}\right) + r' \ln\left(\frac{R_2}{r'}\right) \tag{8.85}$$

The insulation is optimized for a given outer radius R_2, when

$$\frac{dG}{dR_1} = \frac{dG}{dr'} = 0$$

The condition $dG/dR_1 = 0$ yields

$$\frac{R_1}{r'} = \frac{1}{e} = 0.3679 \tag{8.86}$$

The condition $dG/dr' = 0$ yields

$$\frac{R_1}{r'} + \ln\left(\frac{R_2}{r'}\right) = 1 \tag{8.87}$$

and must be satisfied simultaneously with Equation 8.86, so that

$$\frac{R_2}{r'} = e^{1-(1/e)} = 1.882 \tag{8.88}$$

From Equations 8.86 and 8.88, we can write

$$\frac{R_1}{R_2} = \frac{R_1}{r'} \cdot \frac{r'}{R_2} = \frac{(0.3679)}{(1.882)} = 0.1955 \tag{8.89}$$

Normalizing Equation 8.85 with respect to R_2,

$$\frac{G}{R_2} = \frac{R_1}{R_2} \ln\left(\frac{r'}{R_1}\right) + \frac{r'}{R_2} \ln\left(\frac{R_2}{r'}\right) \tag{8.90}$$

yields a maximum normalized equivalent gap length:

$$\left.\frac{G}{R_2}\right|_{Max} = 0.5314 \tag{8.91}$$

8.7.1.3 Intershield Effectiveness

The inclusion of one intershield results in better utilization of insulation by

$$\eta = \frac{G/R_2|_{\text{With shield}} - G/R_2|_{\text{Without shield}}}{G/R_2|_{\text{Without shield}}} \times 100$$
$$\eta = \frac{(0.5315) - (0.3679)}{(0.3679)} \times 100 = 44\% \tag{8.92}$$

8.7.2 Spherical Geometry

Spherical geometries are rarely used in practice, but are included here to illustrate the generality of the method.

8.7.2.1 Two Concentric Spheres

The field on the surface of the inner sphere is given by

$$E(R_1) = \frac{V}{R_1(1 - (R_1/R_2))} \tag{8.93}$$

where

R_1 is the radius of the inner sphere;
R_2 is the radius of the outer sphere.

The equivalent gap length G can be written as

$$G = \frac{V}{E(R_1)} = R_1\left(1 - \frac{R_1}{R_2}\right) \tag{8.94}$$

For optimum utilization of insulation, $dG/dR_1 = 0$, yielding an optimum ratio $R_1/R_2 = 0.5$.

8.7.2.2 A Spherical Geometry with an Intershield

Using a methodology similar to the above for concentric cylinders, the voltage at the intershield is found to be

$$V' = V \cdot \frac{r'(1 - (r'/R_2))}{R_1(1 - (R_1/r')) + r'(1 - (r'/R_2))} \tag{8.95}$$

The electric field at the inner conductor can be expressed:

$$E(R_1) = \frac{V}{R_1(1 - (R_1/r')) + r'(1 - (r'/R_2))} \tag{8.96}$$

Equivalent gap length G can be written as

$$G = R_1\left[1 - \frac{R_1}{r'}\right] + r'\left[1 - \frac{r'}{R_2}\right] \tag{8.97}$$

For optimum utilization of insulation, $dG/dR_1 = dG/dr' = 0$ and we find the optimal ratios to be $R_1/r = 0.5$, $r/R_2 = 0.625$, and $R_1/R_2 = 0.3125$. The normalized equivalent gap length is

$$\frac{G}{R_2} = \frac{R_1}{R_2}\left[1 - \frac{R_1}{r'}\right] + \frac{r'}{R_2}\left[1 - \frac{r'}{R_2}\right] \tag{8.98}$$

and yields a maximum normalized equivalent gap length, $G/R_2|_{Max} = 0.3906$.

8.8 Design Examples

Example 8.1

Calculate the number density of sulfur dioxide gas at a pressure of 1 bar and a temperature of 20 °C.

Solution

Using Equation 8.4 and rearranging terms, $N = p/k_BT$.

The Boltzmann constant is $k_B = 1.3806 \times 10^{23}$ J/(K-molecule). Converting the temperature into units of Kelvin, $T = 20\,°\text{C} + 273 = 293$ K. Using the conversion factor 1 bar $= 10^5$ (N/m^2) and remembering that 1 J = 1(N × m),

$$N = \frac{p}{k_BT} = \frac{10^5\ \text{N/m}^2}{1.3806 \times 10^{23}\ \text{J/(K-molecule)} \times (293\ \text{K})} = 2.47 \times 10^{25}\,\frac{\text{molecules}}{m^3}$$

Note that the number density at a certain pressure and temperature is independent of the gas type or molecular mass, and the above calculation is a general result.

Example 8.2

Show that the breakdown electric field in air under uniform field conditions is not a constant value but varies with gap distance in atmospheric air at room temperature.

Solution

Using Equation 8.28, we can apply the stated conditions of atmospheric air ($p = 760$ Torr) and room temperature ($T = 293$) so that the relation reduces to

$$V_{BD} = (24.22) \cdot d + 6.08\sqrt{d}$$

(cotinued)

Applying the uniform field conditions,

$$E_{BD} = \frac{V_{BD}}{d} = \left(24.22 + \frac{6.08}{\sqrt{d}}\right)\frac{\text{kV}}{\text{cm}} \tag{8.99}$$

The electric field can be plotted as a function of gaps distance, shown in Figure 8.39. Experimental data [57] are shown alongside the curve computed from Equation 8.76 showing good agreement.

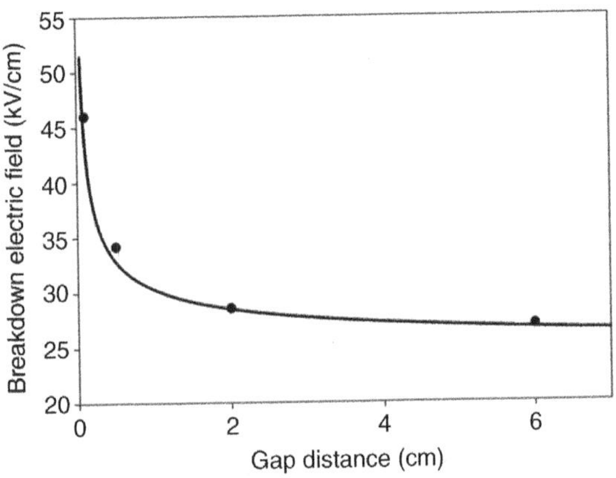

Figure 8.39 The breakdown electric field of atmospheric air as a function of gap distance agrees well with measured data.

For long gaps, the electrical breakdown field approaches 24 kV/cm at standard temperature and pressure (STP), which is the critical value in atmospheric air. The often quoted value of 30 kV/cm is valid for a gap distance of 1 cm at a pressure of 760 Torr and a temperature of 293 K.

Example 8.3

Calculate the minimum diameter of an isolated sphere that can be charged to 6 MV in atmosphere at sea level.

Solution

The electric field on the outer surface of the inner sphere is given by

$$E = \frac{V}{R_1(1 - (R_1/R_2))}$$

For an isolated sphere, $R_2 \rightarrow \infty$, yielding

$$E = \frac{V}{R_1}$$

substituting $V = 6$ MV and $E \sim 30$ kV/cm for air breakdown at atmospheric pressure at sea level, the minimum sphere radius is 2 m.

Example 8.4

An optimally designed cylindrical gap with outer cylinder radius of 0.5 m and insulated with air at 10^5 Pa (STP) has one intershield. Calculate (i) the maximum voltage that the inner cylinder can sustain without breakdown, (ii) the voltage on the inner shield, and (iii) the efficiency of insulation utilization compared to a cylindrical gap without a shield.

Solution

An optimum design results in $R_2/r' = 1.882$, $R_1/r' = 0.3679$, and $R_2/R_1 = 5.117$.

$$R_2 = 0.5 \text{ m}$$

$$R_1 = \frac{R_2}{(5.117)} = 0.0977 \text{ m}$$

and

$$r' = \frac{R_2}{(1.882)} = 0.2657 \text{ m}$$

According to Equation 8.56, the equivalent gap length can be written as

$$G = \frac{V}{E(R_1)} = R_1 \ln\left(\frac{r'}{R_1}\right) + r' \ln\left(\frac{R_2}{r'}\right)$$

$$G = (0.0977) \ln(e) + (0.2657) \ln(1.882) = 0.2657 \text{ m}$$

i. The maximum holdoff voltage is found using the breakdown field of air (3 MV/m) at STP.

$$V_{Max} = G \cdot E_{BD} = (0.2657 \text{ m})\left(3 \frac{\text{MV}}{\text{m}}\right) = 800 \text{ kV}$$

(*cotinued*)

ii. For an optimized geometry, the potential on the intershield can be written:

$$\frac{V'}{V} = \frac{(0.632)r'/R_1}{1+(0.632)r'/R_1} = \frac{(0.632)(2.718)}{1+(0.632)(2.718)} = (0.632)$$

For an optimized geometry,

$$V' = (0.632)\,V$$
$$V' = (800\text{ kV}) \times (0.632) = 505\text{ kV}$$

iii. The efficiency of insulation utilization compared to the utilization of the cylinder without an intershield can be written from Equations 8.56 and 8.63 as

$$\eta = \frac{G/R_2|_{\text{With shield}} - G/R_2|_{\text{Without shield}}}{G/R_2|_{\text{Without shield}}} \times 100$$

$$\eta = \frac{(0.2657/0.5) - 0.3679}{0.3679} \times 100 = 44\%$$

Example 8.5

Calculate and graph the electric fields within concentric cylinders with an intershield charged to a voltage of 100 kV in an optimal geometry with an outer radius of 0.5 m.

Solution

From Example 4, the optimum ratios and values for $R_2 = 0.5$ have been calculated to be $R_2/r' = 1.882$, $R_1/r' = 0.3679$, and $R_2/R_1 = 5.117$, $R_2 = 0.5$ m, $R_1 = 0.0977$ m, and $r' = 0.2657$ m. The fields are given by Equations 8.81 and 8.82 as

$$E(r) = \frac{(V - V')}{r \cdot \ln(r'/R_1)}, \quad \text{for} \quad R_1 \leq r \leq r'$$
$$E(r) = \frac{V'}{r \cdot \ln(R_2/r')}, \quad \text{for} \quad r' \leq r \leq R_2$$

The expression for the electric field can be simplified by invoking the result derived in Example 2 that V' is 63% of V at the optimum values. Figure 8.40 is a graphic representation.

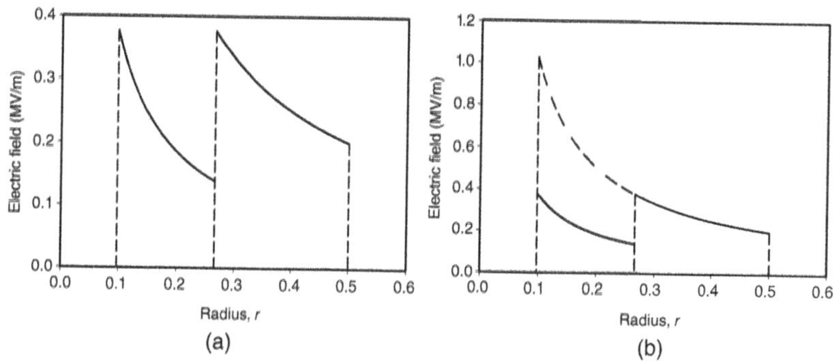

Figure 8.40 The electric fields (a) in concentric cylinders with an intershield. In part (b), the dotted curve is the radial electric field for an optimized cylindrical case without an intershield. The inclusion of an intershield reduces the electric field on the inner conductor by more than a factor of 2.

$$E(r) = \frac{(1-0.632)V}{r\cdot\ln(e)} = \frac{0.368V}{r}, \quad \text{for} \quad R_1 \le r \le r'$$
$$E(r) = \frac{(0.632)V}{r\cdot\ln(1.882)} = \frac{V}{r}, \quad \text{for} \quad r' \le r \le R_2$$

Example 8.6

An optimally designed spherical gap between two spherical electrodes with the radius of the earthed outer surface being 0.5 m and employing air at 10^5 Pa has been provided with one intershield. Calculate (i) maximum voltage that the inner sphere can sustain without breakdown, (ii) voltage on the intershield, and (iii) efficiency of insulation utilization vis-à-vis plane spherical gap without shield.

Solution

Optimum design results in $R_1/r' = 0.5$, $r'/R_2 = 0.625$ and $R_1/R_2 = 0.3125$.

$$R_2 = 0.5\ \text{m}$$

$$r' = (0.625) \times R_2 = 0.3125\ \text{m}$$

and

$$R_1 = (0.5) \times r' = 0.1563\ \text{m}$$

(cotinued)

The equivalent gap length can therefore be written from Equation 8.68 as

$$G = \frac{V}{E(R_1)} = r'\left[1 - \frac{r'}{R_2}\right] + R_1\left[1 - \frac{R_1}{r'}\right]$$

$$G = (0.3125\ \mathrm{m})[1 - (0.625)] + (.1563)[1 - (0.5)] = 0.1953\ \mathrm{m}$$

i. Maximum sustainable voltage on the inner sphere without breakdown is given by

$$V_{Max} = G \cdot E_{BD} = (0.1953\ \mathrm{m}) \times \left(3\frac{\mathrm{MV}}{\mathrm{m}}\right) = 586\ \mathrm{kV}$$

ii. Voltage on the intershield is given by Equation 8.66.

$$V' = V \cdot \frac{r'(1 - (r'/R_2))}{R_1(1 - (R_1/r')) + r'(1 - (r'/R_2))}$$

Plugging in the optimal ratio for values for R_1/r' and r'/R_2, the voltage at the intershield is

$$V' = 60\%V$$

$$V' = (0.60)(586\ \mathrm{kV}) = 352\ \mathrm{kV}$$

iii. Efficiency of insulation utilization is given by

$$\eta = \frac{G/R_2|_{\text{With shield}} - (G/R_2)|_{\text{Without shield}}}{G/R_2|_{\text{Without shield}}} \times 100$$

The optimal equivalent gap length without a shield has not been calculated.

$$\frac{G}{R_2} = \frac{R_1}{R_2}\left(1 - \frac{R_1}{R_2}\right)$$

$$\left.\frac{G}{R_2}\right|_{Max} = (0.5)(1 - (0.5)) = 0.25$$

$$\eta = \frac{(0.1953/0.5) - 0.25}{0.25} \times 100 = 56\%$$

References

1 Yu. Raizer, *Gas Discharge Physics*, Springer, 1991.
2 A. Fridman and L. Kennedy, *Plasma Physics and Engineering*, Taylor & Francis, 2004.
3 S.G. Brush, *History of the Kinetic Theory of Gases. Available from* http://punsterproductions.co/~sciencehistory.pdf/italenc.pdf.
4 M. Fowler, *Molecular Collisions*, 2008. Available from http://galileo.phys.virginia.edsu/classes/152.mf1i.spring 02/Molecular Collisions.htm (accessed January 15, 2012).
5 W. De La Rue and H.W. Muller, Experimental Researchers on the Electric Discharge with the Cloride of a Silver Battery. *Philos. Trans. R. Soc. Lond.*, Vol. 171, pp. 65–116, 1880.
6 F. Paschen, Uber die zum funkenubergang in Luft, Wasserstoff und Kohlensaure bei verschiedenen Druckeneforderliche Potentialdifferenz. *Ann. Phys. Chem.*, Vol. 37, p. 69, 1889.
7 J.S. Townsend and S.P. McCallum, *Philos. Mag.*, Vol. 17, p. 678, 1934; J.M. Meek and J.D. Craggs, eds., *Electrical Breakdown in Gases*, Oxford University Press, p. 89, 1953.
8 J.S. Townsend, *The Theory of Ionization of Gases by Collisions*, Constable and Co., 1910.
9 F. Llewellyn-Jones, The Development of Theories of the Electrical Breakdown of Gases, in E.E. Kunhardt and L.H. Luessen, eds., *Electrical Breakdown and Discharges in Gases*, Plenum Press, New York, 1983.
10 A. von Engel and M. Steenbeck, Elektrische Basenbladungen, ihre Phsik u. Technik, in J.D. Cobine, ed., *Gaseous Conductors: Theory and Engineering Applications*, Vol. 1, Dover Publications, New York, pp. 98 and 149, 1958.
11 F. Llewellyn Jones and J.P. Henderson, The Influence of the Cathode on the Sparking Potential of Hydrogen. *Philos. Mag.*, Vol. 28, pp. 185–191, 1939.
12 J.M. Meek and J.D. Craggs, *Electrical Breakdown in Gases*, Oxford University Press, p. 84, 1953.
13 A.H. von Engel, *Ionized Gases*, Oxford Press, London, 1955.
14 C.M. Cooke and A.H. Cookson, The Nature and Practice of Gases as Electrical Insulators. *IEEE Trans. Electr. Insul.*, Vol. 13, p. 239, 1978.
15 F.L. Jones, *Philos. Mag.*, Vol. 28, p. 192, 1939; J.M. Meek and J.D. Craggs, eds., *Electrical Breakdown in Gases*, Oxford University Press, London, p. 91, 1953.
16 F.M. Bruce, Calibration of Uniform-Field Spark-Gaps for High-Voltage Measurement at Power Frequencies. *J. Inst. Electr. Eng. II: Power Eng.*, Vol. 94, No. 38, pp. 138–149, 1947.
17 D.B. Go and D.A. Pohlman, A Mathematical Model of the Modified Paschen's Curve for Breakdown in Microscale Gaps. *J. Appl. Phys.*, Vol. 107, No. 10, p. 103303, 2009.

18 P.G. Slade and E.D. Taylor, Electrical breakdown in atmospheric air between closely spaced electrical contacts, *IEEE Trans. Compon. Packaging Manuf. Technol.*, Vol. 25, pp. 390–396, 2002.

19 J.M. Meek and J.D. Craggs, eds., *Electrical Breakdown in Gases*, Oxford University Press, 1978.

20 E.E. Kunhardt, Electrical Breakdown of Gases: The Prebreakdown Stage. *IEEE Trans. Plasma. Sci.*, Vol. 8, No. 3, pp. 130–138, 1980.

21 W. Rogowski, Stossspannung und Durchschlag bei Gasen, *Electrical Engineering (Archiv fur Elektrotechnik)* Vol. 20.1, pp. 99–106, 1928.

22 F. Llewellyn Jones and A.B. Parker, Mechanism of Electric Spark. *Nature*, Vol. 165, p. 960, 1950.

23 W.O. Schumann, *Elektriche Druchbruchfeldstarke von Gasen: Theretische Grundlagen und Anwenddung*, Springer, 1923.

24 J. Dutton, Spark Breakdown in Uniform Fields, in J.M. Meek and J.D. Craggs, eds., *Electrical Breakdown of Gases*, Oxford Press, p. 209, 1978.

25 R. Geballe and M.L. Reeves, A Condition on Uniform Field Breakdown in Electron Attaching Gases. *Phys. Rev.*, Vol. 92, p. 867, 1953.

26 E. Nasser, *Fundamentals of Gaseous Ionization and Plasma Electronics*, John Wiley & Sons, Inc., New York, 1971.

27 L.B. Loeb, *Fundamental Processes of Electrical Discharge in Gases*, John Wiley & Sons, Inc., New York, 1939.

28 J.M. Meek, A Theory of the Spark Discharge. *Phys. Rev.*, Vol. 57, pp. 722–730, 1940.

29 H. Raether, *Arch. Elektrotech.*, Vol. 34, p. 49, 1940.

30 R.V. Hodges, R.N. Varney, and J.F. Riley, Probability of Electrical Breakdown: Evidence for a Transition Between the Townsend and Streamer Breakdown Mechanisms. *Phys. Rev. A*, Vol. 31, No. 4, p. 2610, 1985.

31 E.D. Lozanskii, Development of Electron Avalanches and Streamers, *Usp. Fiz. Nauk*, Vol. 117, p. 493, 1975; *Sov. Phys. Usp.*, Vol. 18, p. 893, 1975.

32 S. Nijdam, F.M. van de Wetering, R. Blanc, E.M. van Veldhuizen, and U. Ebert, Probing Photo-Ionization: Experiments on Positive Streamers in Pure Gases and Mixtures. *J. Phys. D Appl. Phys.*, Vol. 43, C p. 145204, 2010.

33 E.E. Kunhardt and W.W. Byszewski, Development of Overvoltage Breakdown at High Gas Pressure, *Phys. Rev. A*, Vol. 21, p. 2069, 1980.

34 L.P. Babich, *High-Energy Phenomena in Electric Discharges in Dense Gases: Theory, Experiment and Natural Phenomena*, Futurepast Inc., 2003.

35 H. Krompholz, L.L. Hatfield, A.A. Neuber, K.P. Kohl, J.E. Chaparro, and H.Y. Ryu, Phenomenology of subnanosecond gas discharges at pressures below one atmosphere. *Trans. Plasma. Sci.*, Vol. 34, No. 3, pp. 927–936, 2006.

36 J.H. Cox and J. Slepian, Effect of Ground Wires on Traveling Waves, *Electrical World*, Vol. 91, p. 768, 1928.

37 L.B. Loeb, Fundamental Processes of Electrical Discharge in Gases (John Wiley and Sons, Inc., New York, 1939), Chap. IX.

38 A. von Hippel and J. Franck, Electrical Penetration and Townsend Theory, *Z. Phys.*, Vol. 57, p. 696, 1929.
39 H. Raether, Untersuchung der Elektronenlawine mit der Nebelkammer, *Zeitschrift für Physik*, Vol. 107, pp. 91–110, 1937.
40 L.B. Loeb and A.F. Kip, Electrical Discharges in Air at Atmospheric Pressure The Nature of the Positive and Negative Point-to-Plane Coronas and the Mechanism of Spark Propagation, *Journal of Applied Physics* Vol. 10.3, pp. 142–160, 1939.
41 L.B. Loeb and J.M. Meeks, The Mechanism of Spark Discharge in Air at Atmospheric Pressure. *J. Appl. Phys.*, Vol. 11, No. 6, pp. 438–447, 1940.
42 L.B. Loeb and J.M. Meek, *The Mechanism of the Electric Spark*, Stanford University Press, 1941.
43 H. Raether, Über den Aufbau von Gasentladungen. I., *Zeitschrift für Physik A Hadrons and Nuclei* Vol. 117.5, pp. 375–398, 1941.
44 J.M. Meek and J.D. Craggs, *Electrical Breakdown of Gases*, Oxford Press, p. 255, 1953.
45 J.M. Meek and J.D. Craggs, *Electrical Breakdown of Gases*, Oxford Press, p. 266, 1953.
46 J.M. Lehr, L.K. Warne, R.E. Jorgenson, Z.R. Wallace, K.C. Hodge, and M. Caldwell, Streamer Initiation in Volume and Surface Discharges in Atmospheric Gases. *Proceedings of the IEEE Power Modulator Symposium*, 2008.
47 A. Pedersen, Calculation of Spark Discharge on Corona Starting Voltages in Non-Uniform Fields. *IEEE Trans. Power Apparatus Syst.*, C Vol. 86, p. 200, 1967.
48 A. Pedersen, On the Electrical Breakdown of Gaseous Dielectrics: An Engineering Approach. *IEEE Trans. Dielectr. Electr. Insul.*, Vol. 24, No. 5, pp. 721–739, 1989.
49 N.H. Malik, Streamer Breakdown Criterion for Compressed Gases. *IEEE Trans. Electr. Insul.*, Vol. 16, p. 463, 1981.
50 G.W. Trichel, The Mechanism of the Negative Point to Plane Corona Near Onset. *Phys. Rev.*, Vol. 54, p. 1078, 1938.
51 E. Kuffel and M. Abdullah, *High Voltage Engineering*, Pergamon Press, 1970.
52 W. Hermstein, Die Entwicklung der Positiven Vorentladugengen in Luft Zaum Durchschlag. *Archiv Elektr.*, Vol. 45, No. 4, pp. 279–288, 1960.
53 M. Goldman, and R.S. Sigmond, Corona and Insulation. *IEEE Trans. Electr. Insul.*, Vol. 17, p. 90, 1982.
54 J. Christiansen and Ch. Schultheiss, Production of High Current Particle Beams by Low Pressure Spark Discharges. *Z. Phys.*, Vol. 290, No. 1, pp. 35–41, 1979.
55 D. Bloess, I. Kamber, H. Riege, G. Bittner, V. Bruckner, J. Christiansen, K. Frank, W. Hartmann, N. Lieser, Ch. Shultheiss, R. Seebock, and W. Steudtner,

The Triggered Pseudo-Spark Chamber as a Fast Switch and as a High Intensity Beam Source. *Nucl. Instrum. Methods*, Vol. 205, p. 173, 1983.

56 J. Christiansen et al., The Triggered Pseudo-Spark, *XVIth International Conference on Phenomena in Ionized Gases, Dusseldorf*, 1983.

57 A.V. Kozyrev, Yu. D. Korolev, V.G. Rabotkin, and I.A. Shemyakin, Processes in the Breakdown Stage of a Low Pressure Discharge and the Mechanism of Discharge Initiation in Pseudospark Switches. *J. Appl. Phys.*, Vol. 74, No. 9, p. 5366, 1993.

58 M. Stetter, P. Felsner, J. Christiansen, K. Frank, A. Gortler, G. Hintz, T. Mehr, R. Stark, and R. Tkotz, Investigation of the Different Discharge Mechanisms in Pseudospark Discharges. *IEEE Trans. Plasma Sci.*, Vol. 23, No. 3, p. 283, 1995.

59 R. Stark, O. Almen, J. Christiansen, K. Frank, W. Hartmann, and M. Stetter, An Investigation of the Temporal Development of the Pseudospark Discharge. *IEEE Trans. Plasma Sci.*, Vol. 23, No. 3, p. 294, 1995.

60 L.C. Pitchford, N. Quadoudi, J.P. Boeuf, M. Legentil, V. Puech, J.C. Thomas, Jr., and M.A. Gundersen, Triggered Breakdown in Low Pressure Hollow Cathode Discharges. *J. Appl. Phys.*, Vol. 78, No. 11, p. 77, 1995.

61 M. Stetter, P. Felsner, J. Christiansen, K. Frank, T. Mehr, and R. Tkotz, First Experimental Observation of the Ignition of a Superdense Glow Before the Glow-to-Arc Transition in a Pseudospark Discharge. *J. Appl. Phys.*, 79, No. 2, p. 631, 1996.

62 K. Frank, E. Dewald, C. Bickes, U. Ernst, M. Iberler, J. Meier, U. Prucker, A. Rainer, M. Schlaug, J. Schwab, J. Urban, W. Weisser, and D.H.H. Hoffmann, Scientific and Technological Progress of Pseudospark Devices. *IEEE Trans. Plasma. Sci.*, Vol. 27, No. 4, p. 1008, 1999.

63 Y. Korolev and K. Frank, Discharge Formation Processes and Glow-to-Arc Transition in Pseudospark Switch. *IEEE Trans. Plasma. Sci.*, Vol. 27, No. 5, p. 1525, 1999.

64 A. Anders, S. Anders, and M.A. Gundersen, Model for Explosive Electron Emission in a Pseudospark Superdense Glow. *Phys. Rev. Lett.*, Vol. 71, No. 3, p. 364, 1993.

65 W. Hartmann and M.A. Gundersen, Origin of Anomalous Emission in a Superdense Glow Discharge. *Phys. Rev. Lett.*, Vol. 60, No. 23, p. 2371, 1988.

66 N.H. Malik and A.H. Qureshi, Breakdown Mechanisms in Sulphur Hexafluoride. *IEEE Trans. Electr. Insul.*, Vol. 13, No. 3, p. 135, 1978.

67 B.A. Goryunov, Dielectric Strength of Compressed SF_6 and the Electrode Material and Surface Structure. *Sov. Phys. Tech. Phys.*, Vol. 20, No. 1, p. 66, 1975.

68 T. Nitta, N. Yamada, and Y. Fujiwara, Area Effect of Electrical Breakdown in Compressed SF_6. *IEEE Trans. Power Apparatus Syst.*, Vol. 93, No. 2, p. 623, 1974.

69 Y. Kawaguchi, K. Sakata, and S. Menju, Dielectric Breakdown of SF_6 in Nearly Uniform Fields. *IEEE Trans. Power Apparatus Syst.*, Vol. 90, No. 3, p. 1072, 1971.

70 J.R. Laghari, Spacer Flashover in Compressed Gases. *IEEE Trans. Electr. Insul.*, Vol. 20, No. 1, p. 83, 1985.

71 J.H. Mason, Discharges, *IEEE Trans. Electr. Insul.*, Vol. 13, No. 4, p. 211, 1978.

72 C.M. Cooke and C.M. Trump, Post Type Support Spacer for Compressed Gas Insulated Cables. *IEEE Trans. Power Apparatus Syst.*, Vol. 92, No. 5, p. 1441, 1973.

73 T. Takuma and T. Watanabe, Optimum Profiles of Disc Type Spacers for Gas Insulation. *Proc. IEE*, Vol. 122, No. 2, p. 183, 1975.

74 J.R. Laghari and A.H. Qureshi, A Review of Particle Contaminated Gas Breakdown. *IEEE Trans. Electr. Insul.*, Vol. 16, No. 5, p. 388, 1981.

75 A.H Cookson and O. Farish, Particle Initiated Breakdown Between Coaxial Electrodes in Compressed SF_6. *IEEE Trans. Power Apparatus Syst.*, Vol. 92, p. 871, 1973.

76 A. Kiss, E. Koltay, and A. Szalay, Electrode System of Improved Stress Uniformity for Pressurized Van de Graaff Generators. *Nucl. Instrum. Methods*, Vol. 46, p. 130, 1967.

77 J.W. Boag, The Design of the Electric Field in a Van de Graaff Generator. *Proc. IEE*, Vol. 100, IV, p. 63, 1953.

78 D.R. Chick and D.P.R. Petrie, An Electrostatic Particle Accelerator. *Proc. IEE*, Vol. 103B,, p. 132, 1956.

79 P.H. Ron, An Appraisal of the Insulation Requirements of Cockcroft-Walton-Type High Voltage D.C. Equipment, M.Sc. thesis, Institute of Science and Technology, University of Manchester, UK, October 1969.

9

Electrical Breakdown in Solids, Liquids, and Vacuum

Electrical insulation is the key to reliability in pulsed power systems. The choice of insulation is among gases, solids, liquids, and vacuum. The gases as well as liquids at high pressure in flow mode are made use of in high-repetition-rate spark gaps, where the ionized matter and decomposed products need to be rapidly removed from the sparking region. Solid insulators function as mechanical supports, enclosures, and feed through. Thin films of solid insulation are used in energy storage capacitors and pulse-forming lines (PFLs) for high energy density storage, and advances in metalized films with their self-healing properties are revolutionary. Liquid insulators are used in Marx generators, Tesla transformers, and PFLs as well as spark gaps. New insulating oils have come into widespread use for high-voltage applications with advances in compact and portable pulsed power systems. Deionized water, with its exceptionally high dielectric constant, is popular for low-impedance PFLs. Magnetically insulated transmission lines (MITLs) may be used to reduce the inductance of power feeds to vacuum loads. Vacuum is also widely used for insulating diodes for generation of electron and ion beams.

The important topic of electrical breakdown in gases has been discussed in Chapter 8. This chapter brings out the essential features of prebreakdown and breakdown in solids, liquids, and vacuum. The various aspects of interaction among the insulators and with electrodes, especially from the viewpoint of insulation coordination, are discussed.

9.1 Solids

Solid insulators are used in pulsed power equipment for many reasons and under a variety of conditions. Solid insulators have an unmatched electrical breakdown strength and are often used in conjunction with oils or gels to mitigate breakdown in air pockets or crevices. This pairing must be done carefully to divide the electric field uniformly across the resultant insulator. This is a common technique for energy storage capacitors, Tesla transformers and

Foundations of Pulsed Power Technology, First Edition. Jane Lehr and Pralhad Ron.

helical lines that are wound with insulating films and metal foils are immersed in a liquid dielectric. Common solid film insulators are paper, polypropylene, and Mylar. Solids are used extensively for mechanical support. High-voltage leads may be fed through solid insulators such as cylindrical bushings or radial disks, where the feedthroughs must provide hermetic seals to isolate the dielectrics on either side. For outdoor installations, the feedthrough insulator must operate in humid and polluted conditions. In the case of spark gaps filled with a liquid dielectric, handling high voltage and current, the arci will produce a shockwave. For repetitive pulsed power systems, thermal considerations such as effective cooling become important for satisfactory performance of solid dielectrics at high temperatures. The choice of insulators for use in pulsed power systems requires knowledge of breakdown mechanisms and performance under varying conditions as well as commercial availability.

9.1.1 Breakdown Mechanisms in Solids

Studies of electrical breakdown in solid dielectrics are very important because of their wide use. When breakdown occurs, solids are usually permanently damaged. The mechanisms leading to the electrical breakdown in solids is a complex phenomenon and varies according to the time duration of voltage application. The mechanisms may be categorized as intrinsic, thermal, or electromechanical breakdown, partial discharges, and electrical trees.

9.1.1.1 Intrinsic Breakdown

The highest values of breakdown strength are realized when other sources of imperfections in the material and testing are eliminated. The timescale for intrinsic breakdown is on the order of 10 ns and is assumed to be electronic in nature. Electrons in the valence band gain sufficient energy from the application of a high electric field to cross the energy gap and enter the conduction band. With sufficient electrons in the conduction band, intrinsic breakdown occurs. Intrinsic breakdown [1–4] strengths are high and may be in the range of 5–10 MV/cm. In laboratory tests, measurements of intrinsic breakdown stress is obtained by the elimination of imperfections such as (i) nonuniformity in the fields, (ii) internal discharges originating from imperfections such as foreign particles or voids, (iii) external discharges due to weak ambient surrounding the solid dielectric, (iv) mechanical damage, or (v) field-induced chemical attacks. The intrinsic breakdown strength may only be attainable in laboratory setting and not in practical systems.

Typical arrangements of electrode/sample geometry for the measurement of intrinsic breakdown strength are shown in Figure 9.1. Such an arrangement allows good field uniformity for the small region of thickness of the specimen under test. Since the breakdown strengths are high, very thin specimens of solid dielectric are used. This allows the use of reasonable values for the applied

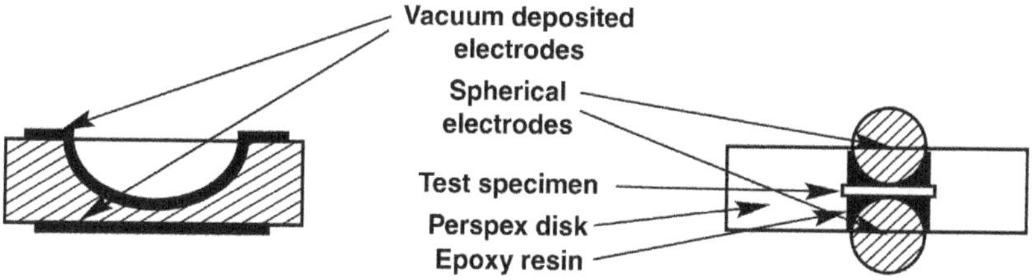

Figure 9.1 Typical setup for measurement of intrinsic field strength in solids.

voltage, as well as reduces the probability of occurrence of inclusions or voids within the specimen thickness. However, provision should be made for proper mechanical support of the specimen, failing which the thin specimen may be subjected to further thinning by creep under electromechanical force and erroneous results. Short duration pulses with high rate of rise of voltage are employed, which also helps in reducing the probability of occurrence of other types of extraneous unintended breakdowns such as thermal breakdown, which involves Joule heating. Intrinsic breakdown remains an ideal concept as it is the highest value obtainable after all other mechanisms have been eliminated.

9.1.1.1.1 Frohlich Criterion

Frohlich considered the case of amorphous solids where the concentration of conduction electrons is sufficiently high to make electron–electron collisions dominant [4]. Under the influence of an electric field, the conduction electrons in a solid dielectric are accelerated and may gain energy from the field. The energy gain depends on the magnitude of the electric field, electron energy, and temperature. Energy is also lost to the lattice in electron–lattice collisions and in collisions with other electrons. A typical figure for energy exchange is shown in Figure 9.2.

The energy gained by an electron in an external electric field is proportional to $W^{3/2}$, whereas energy lost to the lattice is proportional to $W^{-1/2}$. The Frohlich criterion postulates if the net energy gained by an electron from the electric field is greater than the energy lost to the lattice, the electron is continuously accelerated, resulting in a state of instability, and intrinsic breakdown is said to occur.

From Figure 9.2, it appears an electron accelerated by an electric field $F_1 < F_C$ gains energy until it reaches an energy W' at field F_1. However, this is not possible since $W' > I_0$, and I_0 is the energy required for ionization. Such an electron would rapidly lose its energy to an ionizing collision and its energy would come down below I_0. Due to this mechanism, the available conduction electrons can have all energies but only up to I_0. Thus, the Frohlich criterion would be satisfied for applied fields greater than a critical field F_c. For

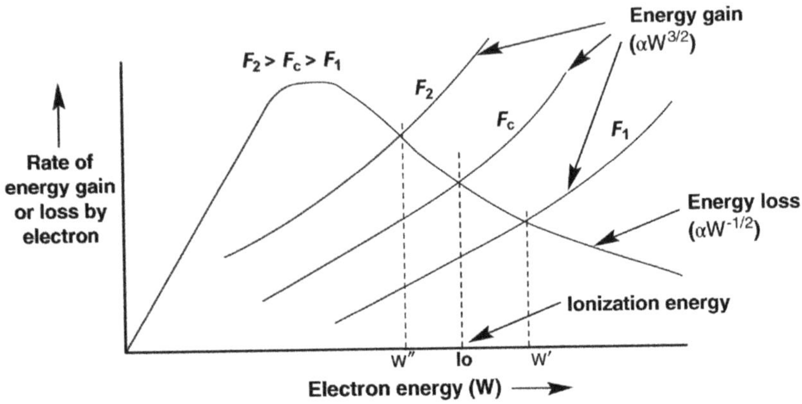

Figure 9.2 The exchange of energy between the electron and lattice relative to the ionization energy determines the concentration of free electrons.

example, this could happen to an electron possessing energy W'' at an applied field $F_2 \geq F_C$. According to the Frohlich theory, the intrinsic breakdown strength does not depend on the specimen thickness or waveform or duration of applied field, provided the duration exceeds the formative time lag.

9.1.1.1.2 Avalanche Criterion

In this theory, conduction electrons gain sufficient energy from the applied field to release further electrons from the lattice, similar to impact ionization in gas. The breakdown takes place when the electron avalanche reaches a critical size. According to this theory, the breakdown strength would be dependent upon the specimen thickness and electrode geometry. The 'time to breakdown' would be dependent upon the overvoltage applied to the specimen.

The avalanche criterion is also known as a "low energy criterion" because of the critical role played by low-energy electrons, in contrast to the "high-energy criterion" of Frohlich, where the electrons possessing high energy near to I_0 become important.

9.1.1.2 Thermal Breakdown

If the rate at which heat is generated in the solid dielectric is greater than the rate at which the heat is dissipated, the temperature of the solid rises (at least locally) in the dielectric, and may result in thermal breakdown [5–7]. The phenomenon is illustrated in Figure 9.3. An applied electric field results in the generation of heat in the solid insulator due to conduction or dielectric losses and depends on the magnitude of the applied field. During this process, heat is also lost through diffusion to the surroundings, raising the temperature of the dielectric volume. If the heat generated is greater than the heat lost, the thermal equilibrium is unstable, and thermal runaway may occur.

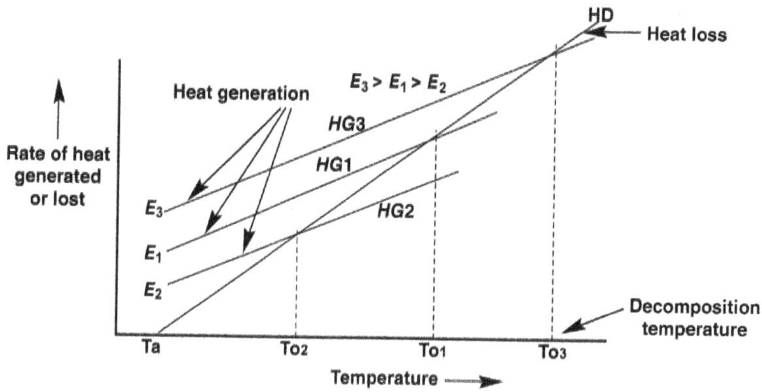

Figure 9.3 Thermal breakdown considerations for fixed cooling system HD and variable applied fields E_1, E_2, and E_3.

The heat generated in the dielectric under pulsed conditions is the volumetric power density lost to conduction in the insulator by Joule heating, HG_{DC}:

$$HG_{\mathrm{DC}} = \sigma E^2$$

where

E	=	applied pulsed electric field
σ	=	conductivity of the solid material

Conservation of energy requires that the heat input to the sample must be equal to the heat conducted away from the sample plus the heat used to raise the temperature of the solid. The MKS unit of HG_{DC} is $\mathrm{W/m^3}$. The energy required to raise the temperature of a dielectric related to its volumetric heat capacity c_v, an intrinsic material property with units of $\mathrm{J/m^3K}$.

Rate of accumulation in the dielectric:

$$c_v \frac{dT}{dt}$$

Heat lost to the surroundings through surface with area A:

$$\frac{\partial}{\partial x}\left(k_{th}\frac{\partial T}{\partial x}\right)$$

The energy lost through dielectric losses raises the internal temperature of the dielectric and is lost to the surroundings through thermal diffusion.

An energy balance yields

$$HG_{\mathrm{DC}} = \sigma E^2 = c_v \frac{dT}{dt} + \frac{\partial}{\partial x}\left(k_{th}\frac{\partial T}{\partial x}\right) \tag{9.1}$$

In the event of a rapid buildup of heat, no heat is lost to the surroundings and $(\partial/\partial x)(k_{th}(\partial T/\partial x)) = 0$ and if the heat generated is lost to the surroundings, $c_v(dT/dt) = 0$.

In the case of AC, heat is generated by dielectric losses from dipole rotation as well as Joule heating. The rate of heat generated per unit volume under an AC electric field, HG_{AC}, is

$$HG_{\text{AC}} = E^2 \cdot 2\pi f \cdot \varepsilon_0\varepsilon_r \tan\delta$$

where

E	=	value of the AC electric field
f	=	frequency in hertz
ε_0	=	permittivity of free space
ε_r	=	relative permittivity of the solid
$\tan\delta$	=	dielectric loss tangent of the solid

The heat conduction equation for heating of an insulator subject to an AC electric field is

$$HG_{\text{AC}} = E^2 \cdot 2\pi f \cdot \varepsilon_0\varepsilon_r \tan\delta = c_v\frac{dT}{dt} + \frac{\partial}{\partial x}\left(k_{th}\frac{\partial T}{\partial x}\right) \tag{9.2}$$

The heat generated from the application of the electric field is proportional to E^2 and the heat conducted away depends on the thermal conductivity and surface temperature. Although not directly evident from the above formulas, the conductivity σ and the dielectric losses ($\tan\delta$) increase with temperature. Generally, the thermal breakdown need not be considered for DC applications because of the low conductivity of good insulators, but it may be important in pulsed applications where high electric fields may be used. For example, the thermal breakdown strength of commercial insulators at room temperature is very high—on the order of 10 MV/cm—making thermal breakdown very unlikely. Thermal breakdown mechanisms become important for varnished cellulose materials with high dielectric losses, where the thermal breakdown strength can become as low as 100 kV/cm at power frequencies. Even for low loss materials, thermal breakdown considerations become important for operation at high temperatures or high frequencies. Thermal breakdown is a well-documented phenomenon.

9.1.1.3 Electromechanical Breakdown

When a voltage is applied to a solid dielectric of initial thickness d_0 between two electrodes, a mechanical force will be exerted on the dielectric by the force of attraction between the surface charges. This compressive force at a voltage V

decreases the dielectric thickness to d and increases the mechanical stress on the material. The compressive force p_c is given by

$$p_c = \frac{1}{2}\varepsilon_0\varepsilon_r E^2 = \frac{1}{2}\varepsilon_0\varepsilon_r\left(\frac{V}{d}\right)^2 \tag{9.3}$$

If the dielectric has a Young's modulus Y, then the mechanical compressive strength is given by Hooke's law:

$$p_c = Y \ln\left(\frac{d_0}{d}\right)$$

Following Stark and Garton [8], the electric stress must be in equilibrium with the mechanical strength, and above two expressions are equal. Rearranging terms,

$$V^2 = \frac{2Y}{\varepsilon_0\varepsilon_r} d^2 \ln\left(\frac{d_0}{d}\right) \tag{9.4}$$

Differentiating with respect to d, it may be shown that an extremum occurs at $d = 0.6 \times d_0$. As the voltage across the sample increases, its thickness decreases. When $d < 0.6 \times d_0$, the compressive force exceeds the strength of the material and causes mechanical damage to the dielectric. Electromechanical breakdown has to be carefully avoided in case of testing for intrinsic breakdown.

9.1.1.4 Partial Discharges

A partial discharge (PD) occurs inside voids embedded in solid dielectrics [9,10]. Typically, these voids are imperfections in the solid material. The partial discharge phenomenon is illustrated in Figure 9.4, where Figure 9.4a shows a void of thickness t' embedded in a dielectric of thickness t. Figure 9.4b shows the following waveforms:

a) V_1 is the voltage applied across the electrodes.
b) V_2 is the prospective voltage across the void in the absence of breakdown in the void.
c) V_3 is the actual voltage across the void, in the event of breakdown taking place in the void.

When the prospective voltage across the void reaches a value V_b, corresponding to the breakdown voltage across the void, a spark forms in the void and the voltage across the void collapses to a low value. When the spark extinguishes, the voltage in the void V_3 starts rising again, until V_3 again reaches the value of V_b, another breakdown takes place inside the void. The above action results in a series of discharge pulses across the void, which continues until the prospective voltage V_2 falls below the value of V_b. Similar formation of discharge pulses happens on the negative cycle as well.

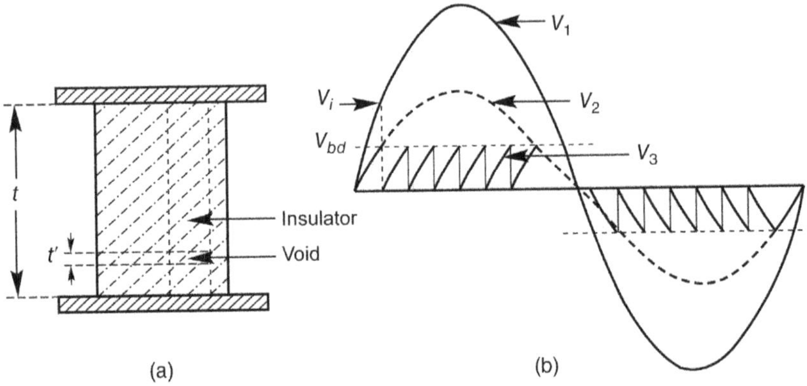

Figure 9.4 Partial discharge mechanism and waveforms. V_1: applied voltage; V_2: prospective voltage across void; V_3: actual voltage across void.

The value of V_b is governed by the Paschen curve and the *pd* value of the void, where the quantity d is the thickness t' of the void and p is its gas pressure. The magnitude of the voltage on the applied voltage waveform V_1 at the time the void voltage V_3 reaches V_b is known as the inception voltage, V_{inc}. The magnitude of the voltage on the applied waveform when the sparking inside the void extinguishes is known as the extinction voltage V_{ext}.

The following expressions can be written for the PD phenomenon:

a) Inception voltage:

$$V_{inc} = E_b t' + \frac{E_b(t - t')}{\varepsilon_r} = E_b\left[t' + \frac{(t - t')}{\varepsilon_r}\right] \tag{9.5}$$

where

E_b = breakdown field $(=V_b/t')$

ε_r = relative permittivity of the solid dielectric

b) The charge ΔQ dissipated in the void per partial discharge pulse is given by

$$\Delta Q = C_{void} \cdot \Delta V \tag{9.6}$$

where

C_{void} = void capacitance

ΔV = voltage drop across void $\approx V_b$

The charge dissipated in the void per cycle of applied voltage is given by

$$Q/cycle = 2 \times n_{PD} \times \Delta Q \tag{9.7}$$

where

n_{PD} = number of partial discharges per half-cycle of applied voltage.

If the positive and negative breakdown voltages in the void are different, then the number of PD pulses will not be the same in each half cycle. In such a case, the above formula needs to be modified.

c) The total charge (Q) dissipated per second is given by

$$Q/\mathrm{s} = 2 \times n_{PD} \times f \times \Delta Q \tag{9.8}$$

where

f = frequency of the applied voltage waveform.

The energy dissipated in the void causes erosion, tracking, treeing, and electrochemical deterioration. The damage is gradual and breakdown of the insulator at operating voltages may take a time period of years, depending on the design conditions. Deterioration of the insulator by PD may be studied in the laboratory, under accelerated conditions, by the application of high overvoltages or high-frequency voltage.

The measurement of PD parameters—discharge magnitude and number of discharges per unit time—can be carried out with a PD detector. The PD detector measures the apparent charge appearing at the terminals of the insulator and not the actual charge dissipated at the void. Typical units of the discharge magnitude are picocoulombs. PD measurements are sensitive to external disturbances and their accuracy requires a high-resolution detection system.

9.1.1.5 Electrical Trees

Electrical trees [11–16] can be classified into (a) dry trees and (b) water trees. The formation of these individual trees depends upon the properties of dielectric and the environment in which they operate. Over a period of time, which may extend to few years at the operating voltage, the trees cause the total breakdown of the insulator.

9.1.1.5.1 Dry Trees

Electrical dry trees are hollow tubes, resembling the branches of trees, which are formed inside a dielectric due to electrical stress. These trees can be generated in a laboratory, under accelerated conditions, by using electrode geometrics in the form of point–plane or point–point as shown in Figure 9.5.

The voltage waveform for the formation of electrical dry trees may be AC or impulse. The diameters of the channels lie in the range of 10–500 μm and they are filled with mixtures of gases evolved from the decomposition of dielectric material. The nucleation sites for the initiation of trees are regions in the insulator where localized field enhancements take place, such as asperities on the electrodes, embedded foreign particles, or voids. The initiation of the trees may be due to many factors, but the most probable seems to be electromechanical. For example, electromechanical force at an asperity may cause intense local pressure to cause fissures or microscopic cracks in the dielectric.

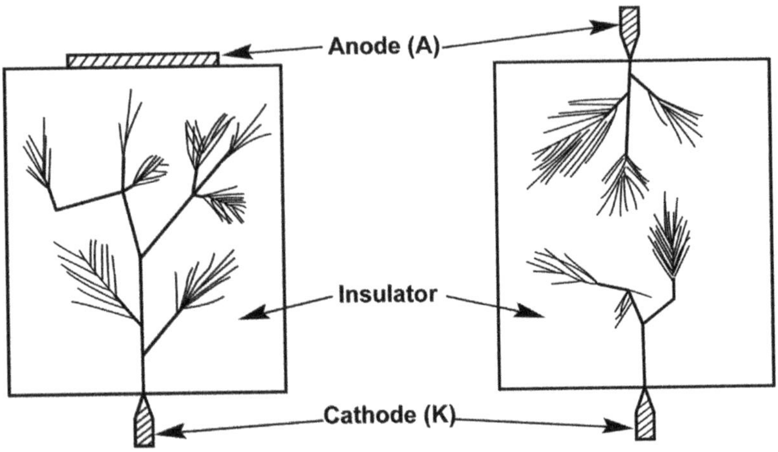

Figure 9.5 Electrical trees in an insulator under two different electrode geometries.

The spark channels resulting from electrical discharges in these cracks deposit their thermal energy, leading to erosion, tracking, gas evolution, and decomposed products. The accelerated charged particles impact the walls of the cavities with high velocities, leading to their growth.

In day-to-day working systems, the operating voltage stresses used in the design of insulation are very low. Hence, the tree growth proceeds at a very slow rate. However, in a laboratory, on the application of high-voltage stresses, significant tree growth can occur even in minutes. On the application of an impulse of extremely high field strength of short duration, sufficiently long trees can develop even in a single impulse. When the trees are localized to a small region, the discharges may not greatly affect the overall performance of the insulator. However, when the trees grow to occupy a major length of the insulator, the remaining unbridged portion of the insulator will be subjected to extremely high stresses, leading to a disruptive breakdown.

9.1.1.5.2 Water Trees

If the dielectric samples are hydrophilic and the experiments of Figure 9.5 are carried out with the insulator immersed in water, it is found that the tree channels that are formed are filled with water. Such electrical trees are called "water trees." An interesting property of water trees is that when the electric stress is removed, the channels become hollow and dry, the water being reabsorbed in the solid dielectric. On the application of a voltage stress, the water reappears in the tree channels. Since many chemicals and conducting salts dissolve easily in the water, the electrical conductivity of water trees compared to dry trees is high, leading to rapid growth. In the early days, the failure mode in underground buried cables was due to the formation of water trees, but advances in cable manufacturing has resulted in a high immunity toward water trees. It has been found that incorporation of a metallic screen near the outer braid results in the suppression of tree growth in underground cables.

9.1.2 Methods of Improving Solid Insulator Performance

In the examples illustrated below, the insulation performance and its operating life are enhanced by using the concepts of (i) employing layers of insulating films instead of a single layer of equivalent thickness, (ii) improving the contact area at the interface between electrode and dielectric by metallization and oil impregnation, (iii) controlling a nonuniform field with corona guards and equipotential rings, and (iv) modifying insulator shapes and surface profiles to reduce the interaction of charge carriers at the surface.

9.1.2.1 Insulation in Energy Storage Capacitors

Energy storage capacitors are important components in pulsed power systems. The insulation system in a capacitor consists of alternate layers of insulating films and metallic foils. The insulating films usually employed are paper, polymer, or a mixed dielectric comprised of paper and polymer in combination. The insulating films are used in multiple layers of 5 or 7 to account for faults in the form of impurities or voids in any of the layers. For example, in a helical line [17] storage element, as many as 35 layers of Mylar sheets are wound over one another. Popular polymer films are polyethylene terephthalate (PET), also known as Mylar, polypropylene, and polyvinylidene fluoride (PVDF). The liquid insulators such as mineral oil, castor oil, dioctylphthalate, and polybutene are used.

The recent notable advances [18] have been in the use of metalized polymer films, in which electrodes such as aluminum or zinc are vapor deposited onto a polymer film to a thickness of approximately 0.0003 μm. This results in tremendous space saving, thus allowing packing of many more multilayers, in comparison to discrete foil capacitors that employ aluminum foils of 5 μm thick. Hence, the major gain is toward high energy density and availability of large energy in a single capacitor unit of reasonable dimensions. The metallization also reduces field intensification at electrode–dielectric junction, thus enabling operation at extremely high field strengths. The extreme thinness [19] of the electrode gives other advantages such as fault tolerance with no catastrophic failure and self-clearing of the fault by vaporization or oxidation of electrode areas in the region of the fault, although the deterioration does occur in the form of gradual reduction of capacitance.

9.1.2.2 Surge Voltage Distribution in a Tesla Transformer

A high-voltage surge appears on the secondary winding of a Tesla transformer whenever a high voltage-to-earth fault occurs. Typical aspects regarding construction and voltage stresses are shown in Figure 9.6.

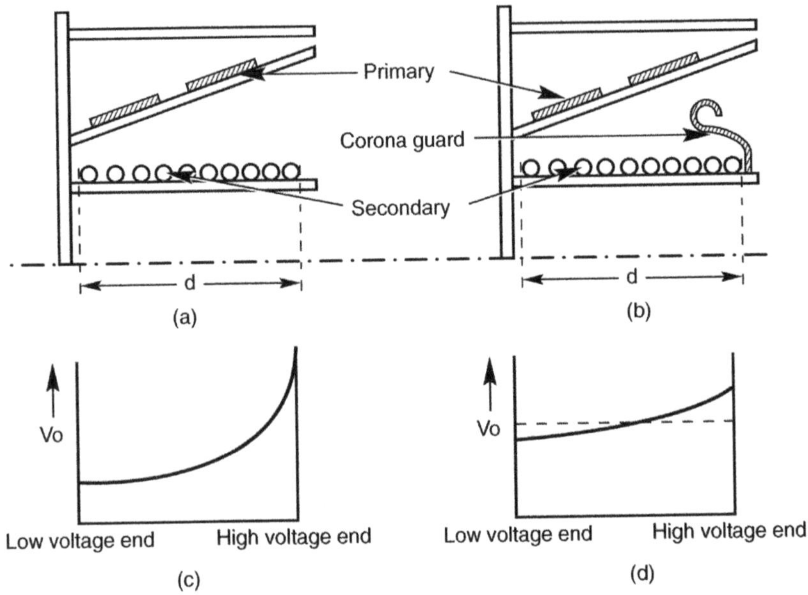

Figure 9.6 Constructional aspects and voltage stresses in a Tesla transformer. (a) Construction without corona guard. (b) Construction with corona guard. (c) Surge voltage distribution without corona guard. (d) Surge voltage distribution with corona guard [20].

Detailed descriptions of the constructional aspects of Tesla transformer are available in Chapter 2. The surge voltage distribution along the high-voltage winding is dictated mainly by the values of leakage and self-capacitance between (i) turn to turn, (ii) secondary turns to primary turns, (iii) primary turns to earth, and (iv) secondary turns to earth. It is seen from Figure 9.7a and c that in the absence of a corona guard, the surge voltage distribution is nonuniform. The turn-to-turn insulation on the high-voltage end is stressed much more than turn-to-turn insulation at the low-voltage end, thus increasing the probability of breakdown on the former. However, this situation is corrected and the surge voltage distribution is made more uniform [20] by the addition of a corona guard on the high-voltage end, as shown in Figure 9.7b and d.

9.1.2.3 Surface Flashover in Standoff Insulators

The shape and surface profile of the standoff insulator for achieving maximum surface flashover voltage depends upon the operating environment. If the insulator is to operate in vacuum, it should be shaped in the form of a frustum of a cone. The surface should be rough and the insulator be mounted with the broad side on the negative electrode, as shown in Figure 9.7d. This type of arrangement will enhance the surface flashover value by decreasing the role of triple junction and secondary electron emission from the insulator surface.

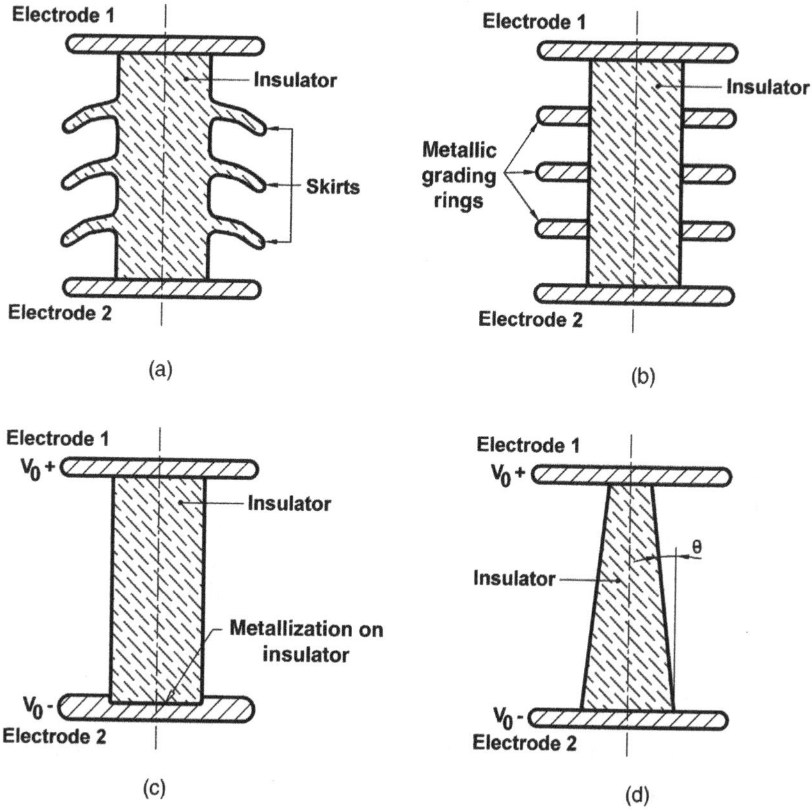

Figure 9.7 Techniques of improving surface flashover on insulators. (a) Skirts on insulator. (b) Grading rings on insulator. (c) Metallization of insulator ends on negative electrode. (d) Providing an appropriate flare angle for insulator on negative electrode.

Metallization of the insulator surface at the area of contact with the electrode, as shown in Figure 9.7c, will enhance the breakdown strength by eliminating the voids and inclusions at the interface. The provision of equipotential rings as shown in Figure 9.7b will help in all the operating environments. If the solid insulator is to operate outdoors under humid and polluted environments, the inclusion of skirts on the insulator as shown in Figure 9.7a is beneficial. The performance in this case will be improved because of increase in the effective length of the insulator, resulting in reduced leakage current. For outdoor insulators near the sea (high saline and humid environment), the leakage current along the surface is high because of increased conductivity. In these cases, there is a likelihood of dry band formation and arcing at the neck of the insulator and good care and maintenance in the form of frequent cleaning of the insulator surface is required.

9.1.2.4 General Care for Fabrication and Assembly

The components in a pulsed power system need to have reliable insulation for operation at high voltage levels. Hence, precautions need to be taken to carry out the fabrication and assembly in a clean room. The components need careful handling so as to avoid contamination by grease, oil, dust, and fingerprints. Metal parts should be finished with a good polished surface and rounded edges. During the winding of electrodes and insulating sheets, as in the case of Tesla transformers, appropriate tension should be used to achieve good uniformity and good interface contact without introducing inclusions, voids, or gaps. The insulator sheets should have enough overlap to prevent surface flashover between adjacent electrodes.

9.2 Liquids

Liquid dielectrics play a very important role in the components of pulsed power systems, such as Marx generators, Tesla transformers, PFLs, and spark gaps. Even an all-liquid pulsed power system is feasible, where every component of the system uses a liquid dielectric. Such an example would be an oil-insulated Marx generator, which charges an intermediate oil/water-filled PFL and finally is switched by an oil or water spark gap. However, the properties required of a liquid dielectric for every component are different. The dielectric in a Marx generator needs a high dielectric strength, the dielectric in a PFL needs a high dielectric constant and low conductivity as well as high dielectric strength, and that in a spark gap needs high dielectric strength combined with high thermal conductivity and minimum decomposition products as well as self-healing properties. In addition to good electrical properties, the other desirable properties for liquid dielectric are (i) good thermal properties, (ii) low viscosity, (iii) low flammability, (iv) good chemical and thermal stability, (v) all properties remaining good even at low temperatures, (vi) environmental considerations, and (vii) low cost.

9.2.1 Breakdown Mechanisms in Liquids

Based on the vast amount of work [21–47] done in the field of prebreakdown and breakdown in liquid dielectrics, the breakdown mechanisms can be classified into particle alignment, electronic breakdown, and streamers in bubbles.

9.2.1.1 Particle Alignment

Solid impurities always exist in a liquid either as fibers or as dispersed particles. Even the best of the particle filters will not be able to remove the particles of extremely small sizes. It is a common observation that a lot of convection currents are set up in a liquid dielectric due to particle movements even at low applied voltage. If the relative permittivity of the solid ε_1 is different from that of the liquid ε_2, in the presence of an electric field a particle of radius R will experience a force given by

$$\vec{F}_E = \frac{\varepsilon_1 - \varepsilon_1}{\varepsilon_1 + 2\varepsilon_2} \cdot R^3 \left(\vec{E} \cdot \nabla \vec{E} \right) \tag{9.9}$$

The above force tends to concentrate the solid impurities to the region of the center of the electrodes where the field is fairly uniform. There is also a diffusion force F_D that tends to take the particles to the area of low impurity concentration. If $F_E > F_D$, the alignment of the particles takes place along the center of the electrodes and breakdown in the liquid takes place along the aligned particles.

9.2.1.2 Electronic Breakdown

An electron injected into the liquid from an asperity on the electrode or released anywhere in the liquid medium starts gaining energy from the applied field. At the same time, the electron will be continuously losing energy to the molecules of the liquid with number density N_ℓ in the form of nonionizing collisions, such as elastic, vibration, and excitation processes. Consider a hydrocarbon liquid and predominantly vibration processes. The corresponding losses will be proportional to $N_\ell \cdot \sum n_i \cdot \sigma_i$, where n_i is the number of component chemical groups in the molecule and σ_i is the collision cross section for that group. An equivalent mean free path λ can be written in the form of $\lambda^{-1} = N_\ell \cdot \sum n_i \cdot \sigma_i$. Under the normal circumstances, the λ will be too small for the energy $eE \cdot \lambda$ to be close to the ionization energy V_i. However, impact ionization could take place under the circumstances of low values of N_ℓ and $N_\ell \cdot \sum n_i \cdot \sigma_i$, which may happen at elevated temperatures at high field strength near an asperity of an electrode. The energy losses of an electron then become small. Some electrons will gain more energy from the field than they lose to the molecules. Such electrons will get continuously accelerated and can generate additional electrons by impact ionization of the molecule. Breakdown of the liquid is supposed to take place when the avalanche of the electrons reaches a critical size at very high fields [21–23].

9.2.1.3 Streamers in Bubbles

Many researchers believe that when breakdown takes place in a liquid dielectric, it is invariably preceded by propagation of streamers in the low-density vapor or bubbles but with no agreement regarding the formation of the bubble in the liquid. Most present-day research seem to search for the source of the bubble [24].

The dynamic processes that take place in low-vapor-density regions of the bubbles are similar to charge carrier multiplication by electron impact ionization and formation of electron avalanches in gases. Thus, the streamer mechanism of liquid breakdown is similar to the growth of electrical trees in a solid due to discharges in a void. Once sufficient space charge is formed, the remaining liquid dielectric in front of the space charge will experience high-field intensification. The shockwaves in front of the streamer head combined with thermal dissipation will lead to more low-vapor-density areas, increased ionization processes, and rapid streamer propagation. When the streamer bridges the electrodes, full-scale breakdown of the liquid takes place.

9.2.1.3.1 *Streamer Structures*

Similar to the characterization of electrical trees in a solid such as stems, branches, and subbranches, the streamer channel in liquids can be identified as primary, secondary, and tertiary. The primary channels are comparatively thick with a radius of 60–90 μm and they propagate in the paraxial direction. The secondary streamers sprout from the primary channels in various directions. The tertiary streamers (~5 μm) branch off from the secondary streamers. The secondary and tertiary streamers are the result of space charge distortion due to the high-density streamer front of the primary. It seems that while the primary streamer is generated at an electrode asperity, the secondary and tertiary streamers are initiated at the other regions of the liquids. Light emission is observed from the primary and secondary as well as tertiary streamers.

9.2.1.3.2 *Diagnostics*

The understanding of streamers in liquids has advanced appreciably because of advances in diagnostic techniques such as electro-optics, optoelectronics, fast digital oscilloscopes, Kerr cells, photomultiplier tubes, image converters, optical and Schlieren photography, and optical spectroscopy. These techniques have allowed the recording of current pulses, light pulses, voltage gradients inside the streamers, and time-resolved photography of streamers. These in turn have resulted in the estimation of streamer velocity and plasma temperature/density inside the streamers as well as material composition of the streamers.

9.2.1.3.3 *Effect of Hydrostatic Pressure*

Hydrostatic pressure has a large influence on streamer and the breakdown voltage increases with pressure. The streamers are found to initiate and grow at higher fields, when the hydrostatic pressure is high. The higher the applied field, the higher the pressure required to stop the current and light pulses. An increase of breakdown strength by a factor of 3 or 4 has been observed for the transformer oil when the hydrostatic pressure is increased to 4 MPa. It has been shown that electrical breakdown can occur in liquids even at pressures of 100 atm.

9.2.2 Mechanisms of Bubble Formation

The mechanisms stipulated for bubble formation inside a liquid dielectric are based on the interaction of electric field with (i) foreign particles, thereby exerting a force on them, (ii) asperities on electrode causing field emission, (iii) chemical interaction with molecules causing their dissociation, and (iv) release of the already existing gas dissolved in the liquid.

9.2.2.1 Krasucki's Hypothesis

Krasucki's microscopic observations [25] of liquids under breakdown indicated that a vapor bubble grew continuously and when a critical size was reached, breakdown took place. If the field was removed when the bubble had grown to a critical size, it collapsed much faster than a corresponding air bubble, thus indicating that the pressure inside the bubble must be zero. The essential condition for the bubble to start is the creation of zero pressure somewhere inside the liquid. He observed that in the presence of impurity particles, the bubble grew preferentially on the particles. In pure liquids, the bubble always grew near the electrode surface.

Krasucki considered the balance between two forces acting on a particle: electrostatic force trying to lift the particle off the liquid–electrode interface and hydrostatic pressure and surface tension trying to retain the particle on the liquid–electrode interface. When these two forces are equal, zero pressure is created. This condition is expressed by

$$E = 358\sqrt{\frac{1}{\varepsilon}\left(p + \frac{2\gamma_s}{r_b}\right)}\ \frac{\text{V}}{\text{cm}} \tag{9.10}$$

where

ε	=	permittivity of the liquid
p	=	hydrostatic pressure
γ_s	=	surface tension
r_b	=	radius of the particle

According to the above relation, the breakdown voltage increases with (i) decreasing radius r_b of the particle, (ii) increasing surface tension γ_s, and (iii) increasing hydrostatic pressure p.

9.2.2.2 Kao's Hypothesis

According to Kao [26,27], a bubble once created starts elongating in the direction of the field, keeping its volume constant. When it elongates to a particular length d, analogous to the pd value at the minimum on the Paschen curve for gas breakdown, electrical breakdown takes place inside the bubble. The condition for breakdown at a field strength E_0 is expressed as

$$E_0 = \frac{1}{\varepsilon_1 \varepsilon_2}\left[\left(\frac{2\pi \cdot \gamma_s \cdot [2\varepsilon_1 + \varepsilon_2]}{r_b}\right)\left(\frac{\pi}{4}\sqrt{\frac{(V_b)_{\min}}{2r_b E_0}}\right) - 1\right]^{1/2} \tag{9.11}$$

where

ε_1	=	permittivity of the liquid
ε_2	=	permittivity of the bubble
r_b	=	initial radius of the bubble
γ_s	=	surface tension of the liquid
$(V_b)_{\min}$	=	breakdown voltage at the Paschen minimum

Although the hydrostatic pressure does not appear in the formula, the dependence of breakdown strength on hydrostatic pressure is implied through the dependence of r_b on pressure. When the pressure is high, the initial radius of the bubble will be small and hence require a high breakdown field.

9.2.2.3 Sharbaugh and Watson Hypothesis

According to this hypothesis, an intensified local electric field E_m, appearing at an asperity of a cathode electrode, gives rise to field emission at high current density j_e. Because of a very short range of electron in a liquid, the power is dissipated to a small liquid mass m. As a result, the liquid gets converted to a low-density vapor in a localized region. The breakdown then ensues in this bubble. The following expressions are written to prove the feasibility of the process:

$$\Delta W = E_m \cdot j_e \tag{9.12}$$

$$\Delta H = m\left[C_p(T_b - T_a) + \ell_b\right] \tag{9.13}$$

where

ΔW	=	rate of electrical energy input into the liquid
ΔH	=	thermal energy
C_p	=	specific heat
T_b	=	boiling temperature
T_a	=	ambient temperature
ℓ_b	=	latent heat of vaporization

Sharbaugh and Watson [21,29] have substituted practical values for E_m of few MV/cm, j_e of few A/cm^2, and calculated the rate of electrical energy input ΔW. They have also calculated ΔH for n-hexane by substituting the known constants for C_p, T_b, T_a, and ℓ_b. They have shown that, with a pulse duration of a few microseconds, enough energy is available from field emission to vaporize a small mass of liquid ahead of an asperity into a bubble.

This thermal model exhibits a marked pressure dependence of breakdown strength. As the pressure of the liquid is increased, its boiling point T_b increases and hence more field is required to form the bubble. The model also shows the dependence of breakdown field on molecular properties C_p and ℓ_b.

9.2.3 Breakdown Features of Water

Water as a liquid dielectric finds a special status in pulsed power systems because of (i) high energy density in energy storage, (ii) low impedance capability in PFLs, (iii) self-healing properties post-breakdown, (iv) easy maintenance, (v) low cost, and (vi) ease of disposal, even in large quantities. It is widely used in low-impedance PFLs and intermediate storage capacitors. As a result, a large number of studies have been conducted to investigate the pre-breakdown phenomena and breakdown in water [30–38]. Many of the breakdown characteristics of water are similar to those of other dielectric liquids. The distinguishing features of water dielectric arise because of the wide disparity in the dielectric constants of water and other insulating liquids.

9.2.3.1 Dependence of Breakdown Strength on Pulse Duration

Water shows pronounced dependence of breakdown voltage on pulse duration [35,36]. For example, for short-duration (7–30 ns) submegavolt pulses, the electric field strength is nearly twice as large in comparison to longer duration pulses (50 ns to 1 ms). The relevant formulas are given in Chapter 3.

9.2.3.2 Dependence of Breakdown Voltage on Polarity

For pulse duration in the range of 50 ns to 1 μs, the breakdown field strength for negative polarity on the high-field electrode is nearly twice that for positive polarity.

9.2.3.3 Electric Field Intensification

It is believed that in water dielectric, there is enhanced field intensification at the asperities due to collective orientation of the dipolar water molecules at the liquid–electrode interface [30,32]. In general, this implies that for any liquid dielectric, higher field intensification and lower breakdown strength will result with a larger dielectric constant. Thus, the electrical breakdown strength of water (ε_r=80) is lower than propylene carbonate (ε_r=65), which is lower than many most oils with relative dielectric constants between one and five.

9.2.4 Methods of Improving Liquid Dielectric Performance

The various factors responsible for lowering the breakdown strength of liquid dielectrics are discussed above. This section presents a few examples of practical systems, where the causes responsible for lowering the insulation performance have been removed. The examples illustrated pertain to components that use either oil or water.

9.2.4.1 New Compositions

Some PFLs use vegetable oils such as castor oils (ε_r=4.7) for their dielectric because of its higher dielectric constant compared to that of mineral oil (ε_r=2.4). However, most vegetable oils are hygroscopic [39]; hence, the units employing vegetable oils should be sealed well to prevent the entry of air and moisture. As an additional precaution, antioxidants may be added even in sealed units to increase the service life.

Synthetic oils like PAO (poly-alpha-olefin), a silicone oil, have shown good performance [40] in closing switches operating at high powers and high repetition rates because of their favorable features, such as (i) resistance to oxidation, thus enabling longer service life, (ii) lower viscosity, enabling continued operation even at lower temperatures, and (iii) good lubrication properties and absence of frothing, thus showing good compatibility for use with hydraulic pumps. In closing switches handling high-power transmission, the liquid dielectric should be used in the forced flow mode at high pressures and velocity to rapidly remove the gases evolved by molecular dissociation and erosion products from the electrodes.

9.2.4.2 Addition of Electron Scavengers

The addition of electron scavengers to liquid dielectrics has been found to increase the breakdown voltage. However, when breakdown takes place at higher voltage, the streamer characteristics are altered. For example, in chlorocyclohexane, which contains chlorine atom as an electron scavenger, the streamer velocity is 10 times [24] that of cyclohexane. Adding electron scavengers like carbon tetrachloride (CCl_4) increases the number and amplitude of current pulses for negative streamers. This may be due to detrapping of electrons at high fields, thus making available a large number of charge carriers.

9.2.4.3 Liquid Mixtures

The superior performance of mixed dielectrics in gases and solids is well known. Examples are SF_6 and N_2 in gases and paper and polypropylene in solids. In liquid dielectrics used for PFLs, mixtures of water (ε_r=80) and ethylene glycol (ε_r=44) have shown advantages [34] toward increased intrinsic time constant, which makes possible the use of longer charging times, and *in situ* adjustment of PFL impedance to that of the load, simply by varying the weight percentage of the two dielectrics.

9.2.4.4 Impregnation

When using liquid dielectrics for impregnation of insulating films and metallic foils in energy storage capacitors or PFLs, it is important to carry out the process at high temperatures and vacuum environment. This enables efficient removal of air trapped at the electrode–liquid interface and also for complete impregnation of the total interface area.

9.2.4.5 Purification

When water is used as a PFL dielectric, the water should be freed of foreign particles and free ions, by passing through particle filters and deionizer columns. Cooling the water to lower temperatures [41] near zero increases the breakdown voltage as well as resistivity. Thus, an addition of a chiller unit in the water conditioning system will give a lot of advantages toward high energy storage density and longer charging times without the accompanying ohmic losses.

9.3 Vacuum

For an ideal vacuum, where there is total absence of medium between the electrodes, there should be no electrical breakdown at all. However, at high applied voltage, breakdown does take place because of charge carriers being injected into the gap by electrodes and medium being made available in the form of availability of adsorbed gases, absorbed gases, and metal vapors released from the electrodes. In addition, for use of vacuum as insulation, it becomes essential to use solid insulators as mechanical supports for the electrodes. The electrical breakdown could then take place as a surface flashover across the solid insulator, the insulator surface being an electrically weaker medium than vacuum.

A few examples of components in a pulsed power system that utilize vacuum for insulation are (i) spark gaps for switching, (ii) diodes for production of particle beams, flash X-rays, high-power microwaves, and (iii) transmission lines for feeding pulsed power into the load.

In order to achieve high-voltage holdoff, minimization of the component size, and long-time reliable performance, it is important to understand the breakdown mechanisms and the methods of improving the breakdown strength. The important aspects of performance of vacuum insulation and breakdown mechanisms are discussed in this section.

9.3.1 Vacuum Breakdown Mechanisms

Vacuum is a very good insulator. If $pd<10^{-3}$ Torr cm, electrons crossing the gap do so essentially without colliding. This does not mean, however, that electrical breakdown does not occur in vacuum. In this section, the proposed mechanisms leading to electrical breakdown in vacuum are discussed.

9.3.1.1 ABCD Mechanism

In this mechanism [48], an electron emitted from the cathode upon striking the anode produces C photons and A positive ions. Each photon and positive ion upon striking the cathode produces D and B electrons, respectively. These additional particles in turn produce more charged particles. The production of charged particles in the gap becomes cumulative when the applied field exceeds a particular value, which results in

$$AB + CD \geq 1 \tag{9.14}$$

This is when the conductivity of the gap between the electrodes increases to a sufficiently high value to result in electrical breakdown of vacuum. The photons can be in the form of soft X-rays or hard X-rays produced by the impact of energetic electrons on the anode. The photons can also be in the visible and ultraviolet regions produced by recombination of electrons and positive ions. The positive ions can be produced by electron impact ionization of adsorbed or absorbed gases released from the electrodes.

The values of A, B, C, and D of Equation 9.14 were measured [49] in an experiment using highly degassed electrodes and were not found to be suitable to lead to electrical breakdown.

The probability of breakdown due to the ABCD mechanism is high at large impulse field intensity and in cases where there is (i) high gas evolution from electrodes due to desorption, (ii) metal vapor formation, or (iii) unfavorable micro-projection geometry on the electrode surfaces.

9.3.1.2 Field Emission-Initiated Breakdown

The current density at a microprotrusion on the cathode electrode is described by the Fowler–Nordheim (FN) field emission equation:

$$j_c = C_1 E_p^2 e^{-C_2/E_p} \frac{A}{\text{cm}^2} \tag{9.15}$$

where

$$C_1 = \frac{1.54 \times 10^{-6}}{\varphi t^2(y)}$$

$$C_2 = \left(6.83 \times 10^7\right) \cdot \varphi^{3/2} v(y)$$

$$y = \left(3.79 \times 10^{-4}\right) \cdot \frac{\sqrt{E}}{\varphi}$$

where

φ	=	work function
E_p	=	field at the microprotrusion tip

Since $t(y)$ and $v(y)$ are slowly varying functions, for most practical purposes Equation 9.15 can be approximated to

$$j_c = \frac{1.54 \times 10^{-6}}{\varphi} E_p^2 e^{-(6.83\times10^7)\varphi^{3/2}/E_p} \frac{A}{\text{cm}^2} \tag{9.16}$$

$$E_p = \beta E_{avg} = \beta \frac{V}{d} \approx \left(2 + \frac{h}{r}\right) \cdot \frac{V}{d} \tag{9.17}$$

where

V	=	applied voltage
h	=	height of microprojection
r	=	tip radius
d	=	gap spacing
β	=	field intensification factor ~10 to 1000

The geometrical parameters referred to in Equations 9.15–9.17 are illustrated in Figure 9.8.

The following observations can be made on the consequences of Fowler–Nordheim field emission:

1) Depending on the geometrical parameters of a microprotrusion and the applied field intensity, the field intensity E_p and current density j_c at a microprojection would be on the order of 10^6–10^8 V/cm and 10^8–10^{10} A/cm^2, respectively. Such a large current density would mean the emission of a large number of electrons n_e from the cathode. This could lead to electrical breakdown of vacuum by continuous multiplication of charge carriers by the *ABCD* and other possible mechanisms.
2) The large field emission current density would lead to Joule heating of the microprojection, resulting in melting, vaporization, and plasma formation. The metal vapor with varying degrees of ionization, introduced into the gap, could lead to vacuum breakdown.
3) The metal vapor would also be produced by the heating of an anode spot by the impact of a high-energy electron beam.

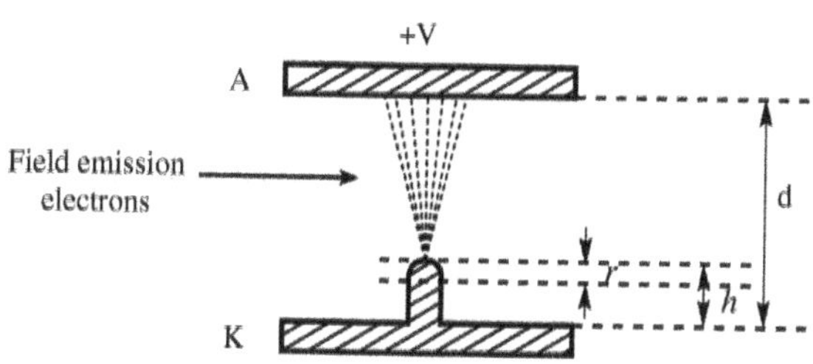

Figure 9.8 Geometrical parameters of a microprotusion.

4) Impurities on the cathode electrode having low work function would be the sites for high patch fields and hence high field emission current density.

9.3.1.3 Microparticle-Initiated Breakdown

Microparticles or clumps that traverse the gap are caused by the following:

1) Loosely adhering material that are detached from an electrode by electrostatic force.
2) Electrostatic force that pulls out microprojections, especially made soft due to Joule heating by field emission current.
3) Vaporization of the anode material by pulsed heating $\left(\int_0^t n_e eV dt\right)$ by accelerated electron beam caused by field emission at the microprotrusion.
4) Vaporization of cathode material by Joule heating $\left(\left(\int_0^t i_c^2 R dt\right)\right)$ of the microprotrusion by field emission current.

The breakdown mechanism [50] by microparticle can be written

$$Q_p = A\varepsilon_0 E \tag{9.18}$$

$$W_i = Q_p V = A\varepsilon_0 \frac{V^2}{d} \tag{9.19}$$

where

Q_p	=	charge acquired by the particle
A	=	surface area of the microparticle
E	=	electric field
V	=	applied voltage
d	=	gap spacing
W_i	=	energy of the microparticle at the impact point on the anode

The energy of the microparticle at the point of impact on the anode, W_i, can also be written as

$$W_i = \frac{1}{2} m_p v_p^2 \tag{9.20}$$

where

m_p	=	mass of the microparticle
v_p	=	impact velocity of the microparticle

According to Cranberg [50], the breakdown will occur if the kinetic energy given by Equation 9.20 exceeds some critical value $W_i = W_c$. Setting $W_i = W_c$ and $V = V_b$, and solving Equation 9.19 for V_b,

$$V_b = \sqrt{\frac{W_c d}{A\varepsilon_0}} = C\sqrt{d} \tag{9.21}$$

According to Slivkov [51], the breakdown voltage V_b is given by

$$V_b = \sqrt{\frac{W_c d}{A\varepsilon_0}} = Cd^{0.625} \quad (9.22)$$

Equations 9.21 and 9.22 give the relationship between breakdown voltage and gap spacing for the breakdown mechanisms based on metallic clumps. The basic assumption in the breakdown criteria by clump mechanism is that when Equation 9.21 or 9.22 is satisfied, the impact velocity of the microparticle would be so large that it may cause additional microprojections when it gets embedded on the impacting electrode.

9.3.1.4 Plasma Flare-Initiated Breakdown

This type of breakdown takes place in a vacuum gap d on the application of a high-voltage V short-duration Δt pulse. The high field intensity $E_p = \beta(V/d)$ at a microprotrusion causes a large field emission current density (j_c), which results in a large field emission current (i_c) at the tip. The mass of the tip, m, having a resistance R explodes into a plasma flare if the Joule energy input $E_i = i_c^2 \cdot R \cdot \Delta t$ exceeds the energy given by a critical value E_c given by

$$E_c = m\left[C_p(T_m - T_0) + L_v\right] \quad (9.23)$$

where

C_p = specific heat
L_v = latent heat of vaporization
T_m = melting point of electrode material
T_0 = initial temperature

The higher the degree of ionization in the plasma flare, the higher the energy difference $E_i - E_c$. This plasma flare starts expanding toward the anode and the breakdown takes place when it bridges the gap.

A plasma flare can also start from the anode when the electron emission from the cathode-initiated plasma front is high enough to explode the anode material at the impact point. The gap will be bridged by the plasma when the two plasma fronts meet. The plasma flare-initiated breakdown mechanism has been studied in great detail [52,53] in a cylindrical vacuum gap by injecting a plasma formed by an exploding wire located along the axis of a coaxial vacuum gap. In this setup, the following mechanisms postulated for vacuum breakdown were experimentally studied: (i) field emission, (ii) metallic clumps, (iii) atomized vapor, (iv) partially ionized vapor, and (v) plasmas with varying degrees of ionization. The various mechanisms were characterized by the delay time and rise time of the breakdown current.

9.3.2 Improving Vacuum Insulation Performance

There are a number of practices that can be employed to improve the insulating properties of vacuum. This section elucidates those common practices, many of which are common sense. Evidence of others, like the work hardening of electrode materials, is summarized.

9.3.2.1 Conditioning

When an electrode gap in vacuum breaks down, the initial breakdown voltages are low. With successive breakdown events, the breakdown voltage steadily increases until, after a large number of breakdowns, it attains almost a steady value. This breakdown voltage is known as the "conditioned breakdown voltage." The improvement in breakdown voltage by conditioning has been attributed to the smoothing of microprotrusions and partial removal of surface layers of adsorbed gas from the electrodes.

9.3.2.1.1 Current Conditioning

In this method, the current in the breakdown event is restricted to hundreds of microamperes by including a current limiting resistor in the high-voltage power supply. A breakdown pulse removes a microprojection and the following pulse shifts to another microprojection site. After a large number of such current-limited breakdowns, the entire surface area of the electrode is smooth. Typical time of conditioning may vary from 30 min to a few hours depending upon the initial roughness condition of electrode surfaces. DC and AC voltages are employed, starting from about 50% of the expected breakdown voltage. However, AC voltage has the advantage of conditioning both the electrodes. The next step of incremental voltage is applied when the discharge pulses in the previous step cease.

A typical diagram for V–I characteristics as a function of time for current conditioning is shown in Figure 9.9. The initial voltage at which the breakdown sets in is V_1 as can be seen from the discharge current pulses. V_4 is the conditioned breakdown voltage achieved after the current conditioning period. V_2 and V_3 are the intermediate step voltages at which the conditioning is carried out.

9.3.2.1.2 Spark Conditioning

In this method, impulse voltages with durations of hundreds of nanoseconds are used. The spark currents during breakdown are limited to a few amperes. Breakdown conditioning currents up to 2 kA using a 1.2/50 lightning-like impulses (rise time t_r=1.2 μs, fall time t_f=50 μs) are also used [54] for large gaps. After the spark conditioning, the samples have been subjected [55]

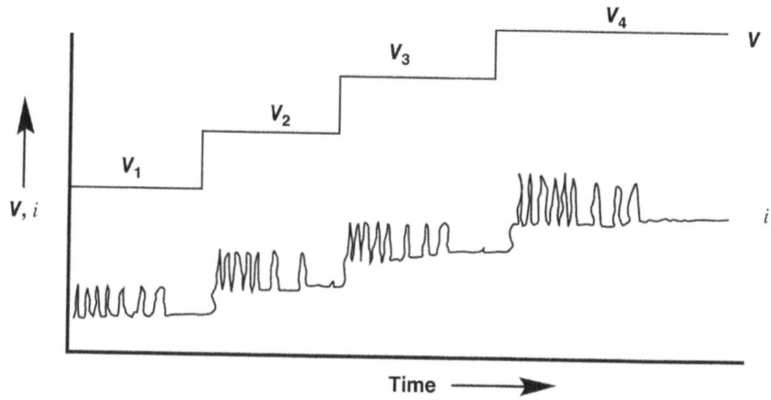

Figure 9.9 Current conditioning characteristics: *V–i* versus time.

to X-ray diffraction for analyzing residual stresses on the electrodes and X-ray photoelectric spectroscopy for analyzing changes in chemical composition and energy of the valence band electrons. The X-ray photoelectric spectroscopy analysis results have shown that chemical cleaning of the electrodes takes place, reducing participation by impurities in the breakdown process, and the valence band electron energy is changed, changing the field emission characteristics. The X-ray diffraction analysis indicated the residual stresses are changed, and the electrode surfaces are work-hardened, thereby reducing the participation of clumps in the breakdown process.

9.3.2.1.3 Glow Discharge Conditioning

Glow discharges in gases like hydrogen, helium, argon, nitrogen, SF_6, or dry air can be easily established between electrodes for conditioning. Anderson [56] used glow discharge for cleaning of vacuum electrodes as far back as 1935. The gas pressure p is chosen to make the pd product occur at the Paschen minimum for the gas in question, allowing the lowest value of voltage to be used. The conditioning of the electrodes takes place by sputter-cleaning due to the bombardment of positive ions from the plasma discharge.

Cuneo successfully used the following plasma discharge technique [57,58] for the cleaning of an ion diode: A 13.5 MHz, 100 W radio frequency (RF) source was used to produce a glow discharge in a gas mixture of 90% argon and 10% oxygen at a pressure of 5–10 mTorr. A potential of 300 V developed across the electrodes. The argon component of the discharge provides physical sputtering, whereas oxygen provides chemical combustion of heavy hydrocarbons into easily removable components such as CO, CO_2, and H_2O. A continuous flow of gas allowed the removal of impurities. The cleaning procedure was carried out for 30–60 min and another discharge was carried out with pure argon to remove any unreacted oxygen from the system.

9.3.2.1.4 Outgassing and Annealing

Outgassing is minimized by heating the electrodes in a high-temperature vacuum furnace at temperatures ranging from 250 to 1500 °C with time periods up to several hours. The temperature is selected based on the temperature capability of the solid insulator supports, with ceramic, alumina, and glass being able to withstand higher temperatures than polymers. The electrodes can be annealed separately at the highest temperatures to achieve the full advantages of annealing such as the removal of residual stresses. This results in a work-hardened electrode with superior electrical performance.

Studies have been conducted [59] on the effects of annealing mirror-finished oxygen-free copper (OFC) electrodes at 400 and 700 °C for 1 h for a vacuum spark gap. The mirror finish was accomplished with electrochemical buffing (ECB) and diamond turning (DT). The gaps were spark-conditioned with 100–500 pulses of 100 kV impulse having rise and fall times of 64 and 700 μs, respectively. The effect on the breakdown strength before and after conditioning for the gaps employing ECB- and DT-finished OFC copper electrodes with and without annealing are shown in Table 9.1.

The importance of the role of residual stress on electrical breakdown can be assessed from Table 9.1. Higher annealing temperatures produce higher breakdown strengths due to the more complete removal of residual stresses from Table 9.1.

9.3.2.1.5 Comparison of Conditioning Techniques

The characteristics for the relative performance by the different methods of conditioning are shown qualitatively in Figure 9.10. The best performance is obtained by employing a combination of conditioning techniques.

Table 9.1 The first and conditioned breakdown strength for OFC copper electrodes.

Electrode preparation	Initial breakdown field (MV/m)	Conditioned breakdown field (MV/m)
ECB, nonannealed	15.6	80
ECB, annealed at 400 °C	39.1	120
ECB, annealed at 700 °C	16.0	115
DT, nonannealed	71.4	160
DT, annealed at 400 °C	49.4	230
DT, annealed at 700 °C	14.9	250

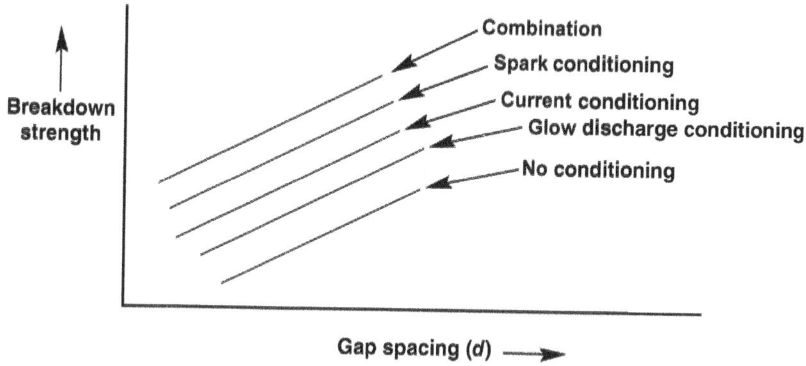

Figure 9.10 Relative performance by various conditioning techniques [54].

9.3.2.2 Surface Treatment and Coatings

Significant increases in the breakdown voltage have been obtained with electrolytic deposition of 1 μm-thick coatings on copper electrodes. A coating of cobalt–molybdenum alloy resulted in a 25% increase, and a 35% increase with cobalt–tungsten alloy [60]. Sputter-coated, 3 μm-thick gold films are used [57] for magnetically insulated transmission lines (MITLs), where the voltage stresses are quite large. Substantial increases in breakdown strengths have been obtained by carrying out pulsed electron beam irradiation and ion implantation on the electrode surfaces. This may result in work hardening of the surface and the breakdown due to clump mechanism will shift to higher voltage.

9.3.3 Triple-Point Junction Modifications

A triple-point junction [61] is formed at the contact area where a metal and insulator meet in the presence of another insulator such as vacuum. A common location for a triple point is where an electrode and insulating spacer meet in a vacuum, as shown in Figure 9.11.

If the contact is imperfect, a void or a gap results and the electric field at the triple junction will be enhanced to ($\varepsilon_r \beta E_{av}$), where ε_r is the relative dielectric constant of the insulator. This gives rise to enhanced field emission at the microprotrusions at the triple junction, thus lowering the breakdown strength. This situation can be improved by metalizing the insulator surface at the contact area and securing a firm contact in a recess formed in the cathode electrode, as shown in Figure 9.11.

The breakdown stress for the modified triple junction will be higher due to the elimination of the void and the shielding of the emitting area by the cathode. Improving the triple junction at the anode does not help much. However, if the gap is subject to AC voltage, the triple junction at the anode will need to be improved as well.

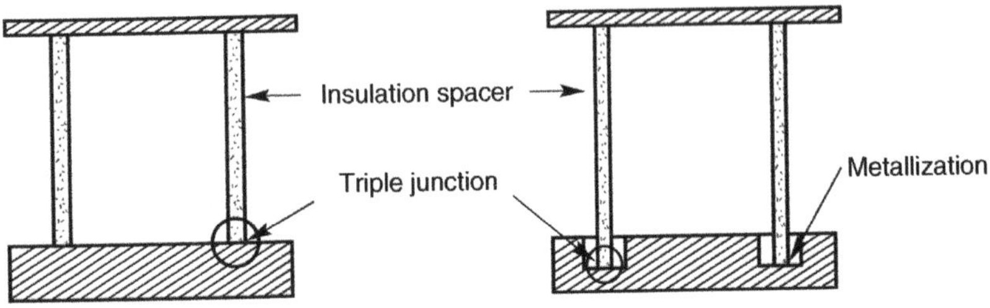

Figure 9.11 Triple junction in a vacuum gap [61].

9.3.4 Vacuum Magnetic Insulation

The presence of a magnetic field greatly enhances the insulating properties of vacuum gaps. The physics is rather straightforward to understand but difficult to predict. The electron trajectories with and without crossed magnetic field for a planar vacuum gap are shown in Figure 9.12. With no magnetic field, the electrons emitted from the cathode move along the field lines and reach the anode in the shortest possible path, as shown in Figure 9.12a. Under the influence of crossed magnetic field B_1, the electrons are subjected to a force $F_x = j_y x B_z$ and will acquire a velocity component in the *x*-direction. As a result, the electron will reach the anode in a longer path, as shown in Figure 9.12b. When the magnetic field is increased to a critical value $B_2 > B_1$ as shown in Figure 9.12c, the electron just grazes the anode surface but does not get collected by the anode. At the very high magnetic field $B_3 \gg B_2$, the electron is trapped and orbits near the cathode. A large number of electrons emitted by the cathode thus form a cloud on the cathode. Under the conditions of Figure 9.12d, the breakdown strength of the vacuum gap will be very high, on the order of many MV/cm, since the electrons are deprived of interaction with the anode.

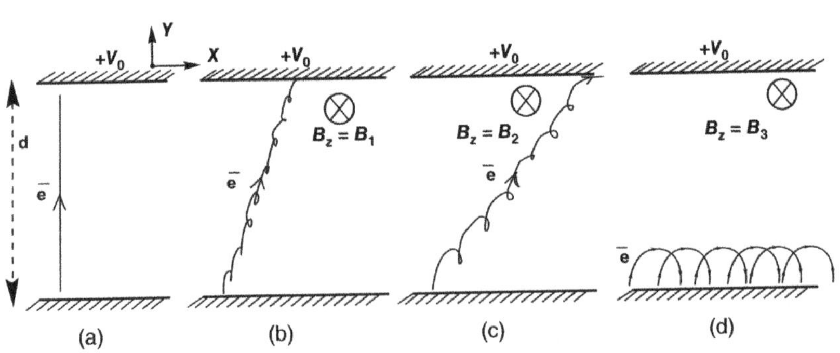

Figure 9.12 Electron trajectories in a vacuum gap with and without crossed magnetic field. (a) No Mag field. (b) Mag field = B_1. (c) Mag field = $B_2 > B_1$. (d) Mag field = $B_3 > B_2$.

The exact prediction of critical magnetic field is difficult, considering the three-dimensional variations in electric field, magnetic field, and electron velocity. However, under simplified assumptions, various models give the values of critical magnetic field as follows [62–66]:

Model 1 (Plane–Parallel Gap) [64]:

$$B_z^{crit} = \frac{1}{d}\sqrt{\frac{2eV}{r_e}} \times \sqrt{1 + \frac{eV}{2m_e c^2}} \tag{9.24}$$

Model 2 (Plane–Parallel Gap) [62]:

$$B_z^{crit} = \frac{(m_e c)\sqrt{\gamma^2 - 1}}{ed} \tag{9.25}$$

$$\gamma = 1 + \frac{eV_0}{m_e c^2}$$

where

r_e	=	classical electron radius
V_0	=	applied voltage
d	=	gap spacing
m_e	=	electron mass
c	=	velocity of light

For coaxial cylindrical gap, where the inner conductor is a cathode of radius R_C and the outer conductor an anode of radius R_A, the following expressions give the critical magnetic field.

Model 3 (Coaxial Gap) [65]:

$$B_z^{crit} = (m_e c^2) \times \left(\frac{2R_A}{R_A^2 - R_C^2}\right)\sqrt{U_0^2 + 2U_0} \tag{9.26}$$

$$B_\Theta^{crit} = \left(\frac{m_e c^2}{e}\right) \times \left(\frac{0.2\sqrt{U_0^2 + 2U_0}}{R_c^2 \cdot \ln(R_A/R_A)}\right) \tag{9.27}$$

where

$$U_0 = V_0 \frac{e}{m_e c^2}$$

In the above discussions, the crossed magnetic field was assumed to be applied by an external source. However, in pulsed power systems with currents in the range of hundreds of kiloamperes to megaamperes, the required critical magnetic fields are self-generated by the systems themselves. Two important applications that make use of self-generated high magnetic fields for providing magnetic insulation are MITLs supplying pulsed power to vacuum loads, ion diodes, and a magnetically insulated line oscillator (MILO), a relativistic high-power microwave source.

An external magnetic field whose magnitude is sufficient to provide magnetic insulation for electrons will not prevent ions from crossing the gap, since $m_i > m_e$. Ion diodes use this phenomenon for extracting the ion beam while suppressing the electron beam. A typical arrangement of an ion diode is shown in Figure 9.13.

For the production of light ion beams (LIBs), the anode electrode is covered by a polymer sheet. By the application of an intense pulsed high voltage in the megavolt region, the light ion beams of hydrogen and carbon are drawn from the plasma. The intense ion beam can be made to impact the cathode electrode for material studies or other applications or can be extracted through the cathode using a perforated cathode electrode.

Magnetic insulation may also use the self-magnetic field of the current flowing in a low-impedance diode $Z_{\text{diode}} \approx 1\,\Omega$ to limit current flow. For diodes with self-generated magnetic fields, the critical electron current for magnetically insulation is given by [66]

$$I_e(\text{kA}) = (8.5) \cdot \beta_e \cdot \gamma \cdot \left(\frac{R}{d}\right) \tag{9.28}$$

where

$$\beta_e = \sqrt{1 - \frac{1}{\gamma^2}} \tag{9.29}$$

$$\gamma = 1 + \frac{eV_0}{m_e c^2}$$

and

R = diode–cathode radius
d = anode–cathode gap spacing

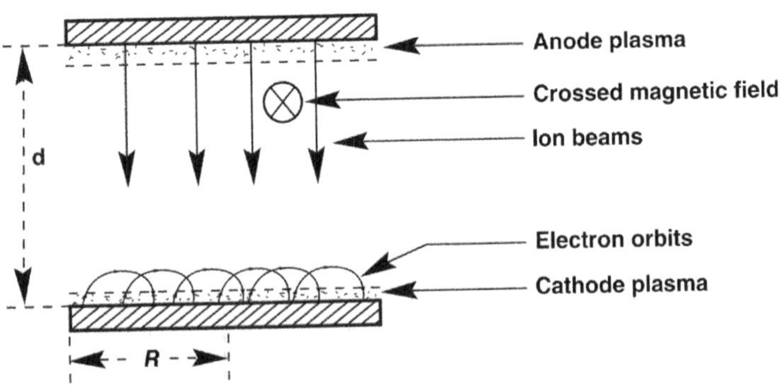

Figure 9.13 Configuration of a planar ion diode using vacuum magnetic insulation.

and

V_0 = applied voltage

The ion current is given by

$$I_i = F \cdot I_e = 8.5 \cdot \left(\frac{F}{1+F}\right) \cdot \gamma \cdot \frac{R}{d} \cdot \ln\left(\gamma + \sqrt{(\gamma^2 - 1)}\right) \tag{9.30}$$

where

$$F = \frac{1}{2}\left(\frac{\beta_i}{\beta_e} \cdot \frac{R}{d}\right) \tag{9.31}$$

$$\beta_i = \sqrt{\frac{ZeV}{M_i C^2}} \tag{9.32}$$

9.3.5 Surface Flashover Across Solids in Vacuum

Until ultrahard levels are achieved, vacuum is an excellent insulator. The breakdown voltage rises rapidly as lower pressures (higher vacuum levels) are achieved because the growth of the electron avalanche is inhibited by the lack of ionizing collision partners. Instead, electrical breakdown tends to occur along a dielectric surface in a phenomenon known as surface flashover. These insulators are used in vacuum devices as standoff posts, windows, or barriers.

Electrons emitted from the cathode gain energy in the electric field. Upon impact with the insulator, one or more secondary electrons are produced, which then gain energy from the electric field and produce their own secondary electrons. Thus, the dielectric surface is the source of electrons to feed the developing avalanche by a process known as secondary electron emission. As this process is underway, another mechanism, known as electron-stimulated desorption – where the electrons impacting the dielectric liberates gas trapped or adsorbed by the surface, comes into play. Electrical breakdown can develop in this growing gas density above the dielectric that ultimately leads to surface flashover in vacuum. Usually, the dielectric is damaged by the flashover and loses its insulating properties because of a decrease in the surface resistivity.

9.3.5.1 Secondary Electron Emission from Dielectric Surfaces

When energetic particles impinge on a solid, they can impart their energy, exciting electrons within the material. If this energy is sufficient to overcome surface energy barriers, such as the work function, electron affinity, or surface charge potential, electrons can escape from the material. A typical SEE curve from an insulator is shown in Figure 9.14. The extent of electron emission from the material can be quantified as the ratio of incident particle flux to emitted particle flux, and is termed the electron yield or secondary electron emission (SEE) yield. The yield is traditionally denoted by the coefficient δ and is either

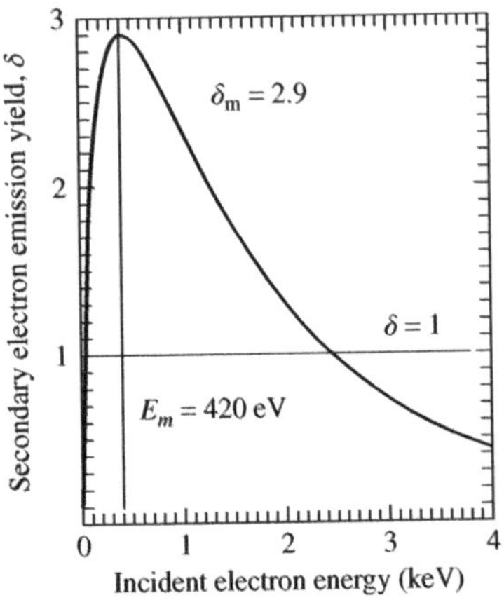

Figure 9.14 A secondary electron emission yield curve. The maximum yield is 2.9 electrons per electron, which occurs at an incident electron energy of 420 eV.

dimensionless or expressed in (*electron/electron*). The yield depends strongly on the energy of the impacting electron.

Secondary electron emission yields are relevant to many technical applications, including the development of advanced electron multiplier detectors [67], where high-yield materials are desired for more sensitive particle detection; scanning electron microscopy [68,69], where low-energy electron emission provides a means for material surface imaging; Auger electron spectroscopy [70], where core-level electron emission provides a signature of surface elemental composition; as well as space vehicle charging and plasma display panels.

Electron yield measurements on dielectrics are more difficult to make than on conductors, since any charge that is deposited in the material cannot be easily dissipated. During measurements of electron emission properties of insulators, charge builds up near the surface because of low bulk and surface conductivity. The resulting sample potentials that develop can affect incident electron energies as well as shifts of the emitted electron spectrum, and consequently lead to significant alterations in the measured electron yields. To minimize insulator charging, pulsed-electron beams coupled with neutralization sources are used.

The magnitude of the secondary electron emissions coefficient and insulator charging can depend on a number of parameters such as incident beam angle [71,72], substance purity, crystalline structure [73], temperature [74], insulator thickness [75], surface cleanliness [76], surface topography [72], sample potential [77], and previous electron irradiation exposure leading to increased defect density and trapped charge [67,78], Numerous conductor electron spectra, yield curves, and yield parameters are available on the NASA SEE Charge Collector Knowledge Base [79].

9.3.5.2 Saturated Secondary Electron Emission Avalanche

A model that has proven to be quite successful in explaining experimental results is the saturated secondary electron emission avalanche (SSEEA) model first proposed by Boersch et al. [80] and advanced by Milton [81] and Bergeron [82]. Anderson and Brainard [83] added the mechanism of electron-stimulated desorption of gases at the surface as a critical element in surface flashover initiation. Pillai and Hackam [84] formulated these processes into an initiation criterion for electrical breakdown showing the square root dependence on the insulator length.

The mechanism of surface flashover across a solid insulator in a vacuum gap is illustrated in Figure 9.15. Figure 9.15a shows an electron emitted from the triple-point junction at the cathode *K*, hopping along the insulator surface under the influence of an electric field. At the impact point on the insulator, the secondary emission of electrons from the surface takes place. These secondary electrons also get accelerated toward the anode and join the original primary band of impacting electrons. This process of multiplication of primary and secondary electrons continues along the surface in an expanding avalanche. The impacting electrons also liberate gas from the surface in a process known as electron-stimulated desorption. The advancing electron charges in the avalanche head also increases the net electric field intensity across the insulator, further driving the flashover toward the anode *A* shown in Figure 9.15b. If the insulator shape is modified to a frustum of a cone with the base at the cathode as shown in Figure 9.15(c), an electron emitted from the triple junction will reach the anode, traveling under the force exerted by the electric field, without hitting the insulator surface.

9.3.5.2.1 The SSEEA Saturation Condition

It is thought that surface flashover is the result of a negative feedback process where the impacting electron energy is always driven back to the

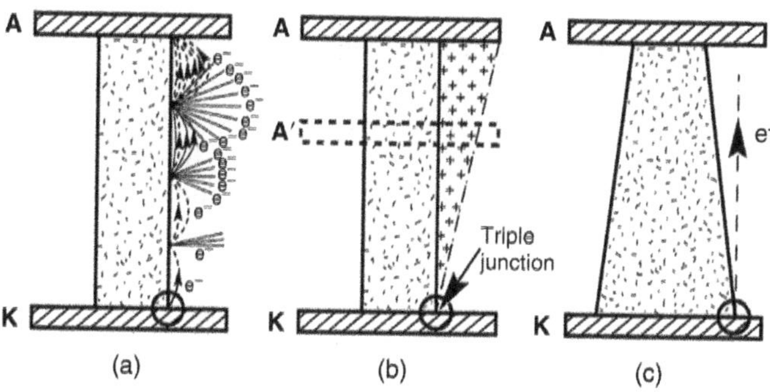

Figure 9.15 Mechanism of surface flashover in vacuum. (a) Electron emission from triple junction and secondary electron emission from insulator surface. (b) Surface charging due to secondary emission and field intensification by equivalent reduced gap *A′K*. (c) the effective breakdown strength increases when electrons do not impact the insulator.

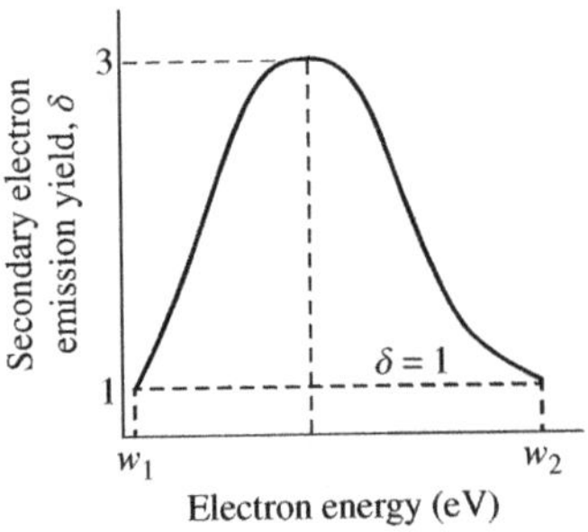

Figure 9.16 Generic secondary electron emission (SEE) yield characteristics for dielectrics.

condition at the lower incident electron energy where δ=1. This condition where impacting electrons produce one secondary is known as the *saturation condition.*

A generic, exaggerated SEE yield curve for a dielectric material is shown in Figure 9.16, which will be used to illustrate the saturated secondary electron emission avalanche condition that arises from the energies of the impacting electrons being driven to energy w_1, the lower energy where the yield equals 1.

To illustrate the negative feedback process, consider an electron emitted from the triple point under the influence of an electric field applied parallel to the insulator surface that impacts the surface with an energy greater than w_1. Since $\delta > 1$ for this case, the negative charge leaving the point of impact (in the form of secondary electrons) will be greater than the negative charge arriving at the point of impact (the impacting electron), and thus the insulator surface will attain a *net positive charge.* A successive electron leaving the cathode will be attracted by this positive charge and spends less time in the applied electric field, thus gaining less energy from the field. The result is to drive the energy of the incident electron back toward w_1. Now consider that an electron strikes the surface with an energy less than w_1. This produces a net negative charge on the surface that repels a successive electron. This electron then gains more energy from the applied field, and the operating point will again tend toward w_1. After a period of time (on the order of nanoseconds), the energies of all of the impacting electrons will converge to w_1. This convergence, corresponding to the condition $\delta = 1$, is referred to as the *saturated secondary electron avalanche condition.* This saturation will occur for all initial energies at impact of less than w_2. Impact energies greater than w_2 constitute, in principle, a runaway condition but, due to the large magnitude of w_2 (1–2 keV), such a runaway has not yet been observed experimentally in surface flashover.

9.3.5.2.2 *Electron Trajectories*

For the geometry in Figure 9.17, the insulator is inserted between the anode and the cathode in a planar geometry in vacuum, establishing an applied, uniform electric field E_p between the electrodes and parallel (tangential) to the surface. Electrons are emitted from the cathode in the high local electric field region

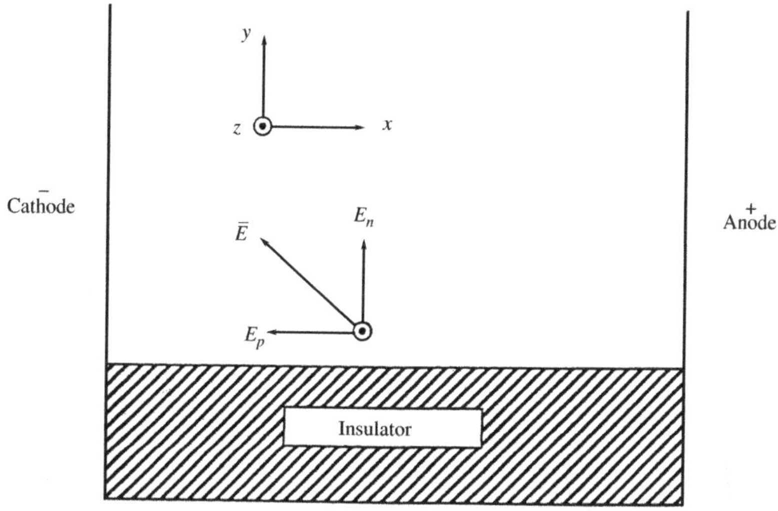

Figure 9.17 Illustration of the spatial coordinates and the relationship between the insulator surface and the electric fields used to formulate the equations of motion of electrons.

where the cathode, insulator, and vacuum meet known as the triple point. Under an electric field, the force on the electron is

$$\vec{F} = q\vec{E} \tag{9.33}$$

where $\vec{F}$ is the force, q is the charge of the particle, and $\vec{E}$ is the electric field. If the electric field is entirely parallel to the surface, the electrons will also travel parallel. However, very near the surface local changes caused by polarization, surface roughness and imperfections cause some electrons to impact the surface. When the electron impacts the surface with sufficient energy that another electron is emitted, it leaves behind a *net positive charge*, which, in turn, sets up an electric field normal to the surface. The resultant electric field acting on the subsequent electrons near the surface is given by

$$\vec{E} = E_p\hat{x} + E_n\hat{y} \tag{9.34}$$

where E_n is the electric field component normal to the insulator surface due to surface charges and E_p is the applied electric field.

The equations of motion to describe the electron accelerating in the electric field are

$$\frac{dv_x}{dt} = \frac{-e}{m}(-E_p) \tag{9.35}$$

and

$$\frac{dv_y}{dt} = \frac{-e}{m}(E_n) \tag{9.36}$$

where e is the electronic charge, m is the electronic mass, and v_x and v_y are the $\hat{x}$ and $\hat{y}$ components of the velocity. The Pillai–Hackam model assumes that a secondary electron is emitted normal to the surface with an initial velocity v_0. The initial conditions are

$$v_x(0) = 0, \quad v_y(0) = v_0, \quad x(0) = 0, \quad y(0) = 0 \tag{9.37}$$

$$v_x(t) = \frac{e}{m} E_p t \tag{9.38}$$

$$v_y(t) = v_0 - \frac{e}{m} E_n\, t \tag{9.39}$$

Integration of (9.38) and (9.39) gives the trajectories

$$x(t) = \frac{e}{m} E_p \frac{t^2}{2} \tag{9.40}$$

$$y(t) = v_0 \cdot t - \frac{e}{m} E_n \frac{t^2}{2} \tag{9.41}$$

The trajectory of an electron emitted from the surface shows that the electrons are driven back into the surface, resulting in additional secondary electrons. The time t_0 at which the electron impacts the surface is obtained by setting $y(t_0) = 0$ into (9.41) and solving for t_0:

$$t_0 = \frac{2mv_0}{eE_n} \tag{9.42}$$

Solving for $v_x(t)$ at impact when $t = t_0$ gives

$$v_x(t_0) = 2v_0 \frac{E_p}{E_n} \tag{9.43}$$

which along with $v_y(t_0) = v_0$ gives the electron energy at impact w_i:

$$w_i = \frac{m}{2}\left(v_x^2 + v_y^2\right) = \frac{m}{2} v_0^2 \left(1 + 4\left(\frac{E_p}{E_n}\right)^2\right) \tag{9.44}$$

After the secondary electron avalanche has saturated, the impact energy w_i will be equal to w_1 – the lower energy at which the secondary electron emission yield δ equals 1. Equation 9.44 can then be solved to give the normal field component E_n at saturation in terms of the applied electric field E_p and the material properties, giving

$$E_n = E_p \sqrt{\frac{2w_0}{w_1 - w_0}} \tag{9.45}$$

where

$$w_0 = (1/2) m v_0^2 \tag{9.46}$$

and w_1 is the emission energy. From (9.45), the angle φ between the electric field and the insulator surface after saturation is

$$\tan\varphi = \frac{E_n}{E_p} = \sqrt{\frac{2w_0}{w_1 - w_0}} \tag{9.47}$$

9.3.5.2.3 SSEEA Breakdown Criterion: The Pillai–Hackam Model

Pillai and Hackam [84] formulated a breakdown criterion for surface flashover in vacuum by incorporating electron-stimulated desorption of gas from the insulator surface as described by Anderson and Brainard [83] along with the SSEEA model as its basis. After saturation of the avalanche, the electrons impacting the surface constitute a current density perpendicular to the insulator surface, with the amount of gas desorbed assumed to be proportional to this current density. The electrons in the avalanche undergo ionizing collisions with the desorbed gas, and Paschen breakdown then occurs when the density of the desorbed gas reaches a critical value. Pillai and Hackam determined the critical applied electric field for breakdown E_p in terms of the insulator length, the parameters describing the gas desorption process, and the SEE characteristics of the dielectric.

From the normal electric field component at saturation, an electron current density normal to the surface, J_n, is given by

$$J_n = \frac{2\varepsilon_0 v_1 E_p^2 \tan\varphi}{w_1} \tag{9.48}$$

where v_1 is the velocity associated with the energy w_1 and ε_0 is the permittivity of free space. This current density lead to an electron-stimulated gas desorption rate D_0 given by

$$D_0 = \frac{\gamma}{e} J_n \tag{9.49}$$

where γ is the gas desorption coefficient defined as the number of molecules desorbed per unit impacting electron. Combining (9.48) and (9.49),

$$D_0 = \frac{2\varepsilon_0 v_1 E_p^2 \tan\varphi}{e w_1} \tag{9.50}$$

The density of desorbed gas N_d is

$$N_d = \frac{D_0}{v_d} \tag{9.51}$$

where v_d is the velocity with which the desorbed gas molecules leave the insulator surface. The amount of desorbed gas per unit area of the insulator M is then given by

$$M = N_d \ell = \frac{D_0}{v_d} \ell \tag{9.52}$$

where ℓ is the length of the insulator. The amount of gas desorbed can be recast in terms of the applied electric field via (9.50) giving

$$M = \frac{2\varepsilon_0 \nu_1 \gamma E_p^2 \tan\varphi}{v_d e w_1} \tag{9.53}$$

As the quantity of gas above the insulator surface increases, the local gas density increases that lead to Paschen breakdown in the desorbed gas. Defining the crucial amount of gas required for breakdown as M_{cr}, (9.53) gives the applied electric field at breakdown $E_{p_{bd}}$ as

$$E_{p_{bd}} = \sqrt{\frac{M_{cr} v_d e w_1}{2\varepsilon_0 \nu_1 \gamma \ell \tan\varphi}} \tag{9.54}$$

The breakdown voltage under these uniform field conditions is then given by

$$V_{bd} = \sqrt{\frac{M_{cr} v_d e w_1 \ell}{2\varepsilon_0 \nu_1 \gamma \tan\varphi}} \tag{9.55}$$

which has a $V_{bd} \propto \sqrt{\ell}$, the length of the insulator. The Pillai–Hackam model incorporates the key features of the initial charging of the insulating surface, the strong influence of desorbed gas from the insulator, and the dependence on the insulator length.

9.3.5.2.4 Mitigation Techniques

For a given geometry and dielectric properties, a material with a higher value of w_1 should result in greater pulsed surface flashover holdoff as follows: The primary electron must spend more time in the interelectrode electric field to gain the higher energy associated with the higher value of w_1. This results in fewer secondary electron-producing impacts with the dielectric surface and a lower rate of electron-stimulated gas desorption from the surface. It is this growing gas density above the insulator surface that ultimately leads to surface flashover in vacuum.

If the insulator shape is modified to a frustum of a cone with the base at the cathode as shown in Figure 9.15c, it is seen that an electron emitted from the triple junction will reach the anode traversing along the field line, without hitting the insulator surface. Such an arrangement will prevent the charging of the insulator surface and hence results in high breakdown strength of the gap.

There is a wide body of literature on this subject [85–101] and an excellent review of techniques to inhibit surface flashover in vacuum can be found in Ref. [93]. Several techniques have been shown to increase the surface flashover potential of insulators in vacuum, including the use of surface coatings [85–87] and magnetic insulation [85,89–92].

9.4 Composite Dielectrics

Most insulating systems are comprised of multiple dielectrics arranged in series or in parallel. The simplest composite dielectric system is one where two dielectrics are the same materials. This takes advantage that two layers have a higher dielectric strength than a single sheet with the equivalent thickness. In other cases, the composite dielectric may exist as part of the design – such was when a paper insulator is impregnated with oil – or may arise due to a fabrication difficulty – such as air trapped between solid insulators. When composite insulation has components with different dielectric constants, utilization of the material may be impaired.

Examine the composite dielectric shown in Figure 9.18 comprised of two rectangular insulating slabs with an overall thickness d. One slab has a dielectric constant ε_1 and a thickness d_1 and the other has a dielectric constant ε_2 and a thickness d_2. The overall thickness $d = d_1 + d_2$. When a voltage V_0 is applied, an electric field is established in each slab whose magnitude is determined by the dielectric constant and thickness of each slab. The two dielectrics may be considered to be capacitors in parallel. However, here the electric fields will be considered to highlight the role the dielectric constant plays in the field distribution.

The boundary condition at the interface of the two dielectrics requires the normal component of the displacement vector $\vec{D} = \varepsilon\vec{E}$ to be continuous in the absence of free surface charge. Because the geometry is planar, the electric field is normal to the interface and the boundary condition becomes

$$\varepsilon_1 E_1 = \varepsilon_2 E_2 \tag{9.56}$$

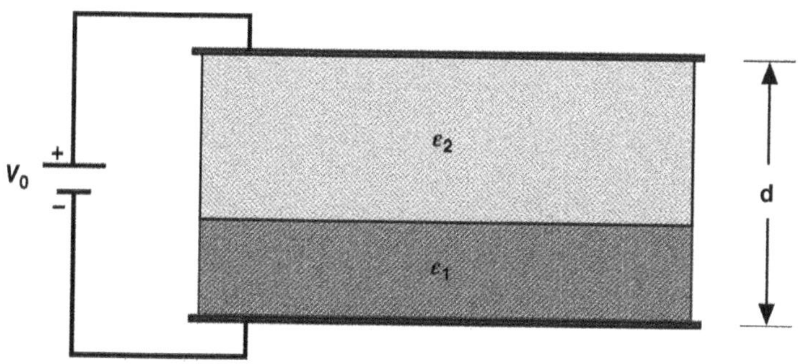

Figure 9.18 A dielectric comprised of two rectangular slabs.

This partition of the electric fields according to the dielectric constants establishes a voltage across each slab. A voltage V_1 is established across slab 1 and is related to the electric field E_1 by

$$E_1 = \frac{V_1}{d_1} \tag{9.57}$$

Likewise, in slab 2,

$$E_2 = \frac{V_2}{d_2} \tag{9.58}$$

Combining Equations 9.56, 9.57, and 9.58,

$$\varepsilon_1 \frac{V_1}{d_1} = \varepsilon_2 \frac{V_2}{d_2} \tag{9.59}$$

Voltage is a scalar quantity and therefore its value must be equal at the dielectric interface:

$$V_0 = V_1 + V_2 \tag{9.60}$$

Using (9.37) in (9.38)

$$V_0 = \frac{\varepsilon_2\, d_1}{\varepsilon_1\, d_2} V_2 + V_2$$

Rearranging terms and solving for V_2,

$$V_2 = \frac{\varepsilon_1 d_2}{\varepsilon_1 d_2 + \varepsilon_2 d_1} V_0 \tag{9.61}$$

For electrical breakdown considerations, the electric field partition is a main concern and

$$E_2 = \frac{\varepsilon_1}{\varepsilon_1 d_2 + \varepsilon_2 d_1} V_0 \tag{9.62}$$

Again using (9.34),

$$E_1 = \frac{\varepsilon_2}{\varepsilon_1}, \quad E_2 = \frac{\varepsilon_2}{\varepsilon_1 d_2 + \varepsilon_2 d_1} V_0 \tag{9.63}$$

This is an important consideration for solid dielectrics impregnated with liquids such as the combination of pressboard and transformer oil commonly used in transformers where the lower dielectric constant and lower breakdown strength of the transformer oil must be considered. It is also an important consideration for air gaps between insulators where the lower dielectric constant of the air creates a higher electric field.

9.5 Design Examples

Example 9.1

In Section 9.1.1.3, an expression for the electromechanical breakdown is given. Derive the location of the damage threshold given as $d = 0.6 \cdot d_0$.

Solution

The relationship between applied voltage and the reduced thickness is given by

$$V^2 = \frac{2Y}{\varepsilon_0 \varepsilon_r} d^2 \ln\left(\frac{d_0}{d}\right)$$

Let $x = d/d_0$, and the above equation can be represented as

$$V^2 = \left(\frac{2Y}{\varepsilon_0 \varepsilon_r}\right) \cdot x^2 \cdot d_0^2 \ln\left(\frac{1}{x}\right)$$

To find the location of the maximum value, set $dV/dx = 0$:

$$2V\frac{dV}{dx} = 0 = \frac{2Y}{\varepsilon_0 \varepsilon_r} \cdot d_0^2 \cdot \left[2x \ln\left(\frac{1}{x}\right) - x\right]$$

which reduces to

$$0 = x \cdot \left[2 \ln\left(\frac{1}{x}\right) - 1\right]$$

Throwing out the trivial solution, $x = 0$, the solution for the maximum is

$$\ln\left(\frac{1}{x}\right) = \frac{1}{2}$$

and

$$x = 0.6 \quad \text{or} \quad d = 0.6 \times d_0$$

The relationship between the applied voltage the normalized thickness x is given by

$$V = \sqrt{\frac{2Y}{\varepsilon_0 \varepsilon_r}} \cdot x \cdot d_0 \sqrt{\ln\left(\frac{1}{x}\right)}$$

To plot, the undistorted average electric field will be normalized with a quantity known as the apparent stress E_a. The apparent stress is the average electric field in the compressed solid. That is, the value of the apparent stress, $E_a = [V(x = 0.6)]/d_0$ at which the collapse occurs is obtained by inserting the value x=0.6 into

$$E_a = \frac{V(x = 0.6)}{d_0} = \sqrt{\frac{2Y}{\varepsilon_0 \varepsilon_r}} \times (0.6) \times \sqrt{\left(\frac{1}{2}\right)}$$

The graph of this function is shown in Figure 9.19.

(continued)

(continued)

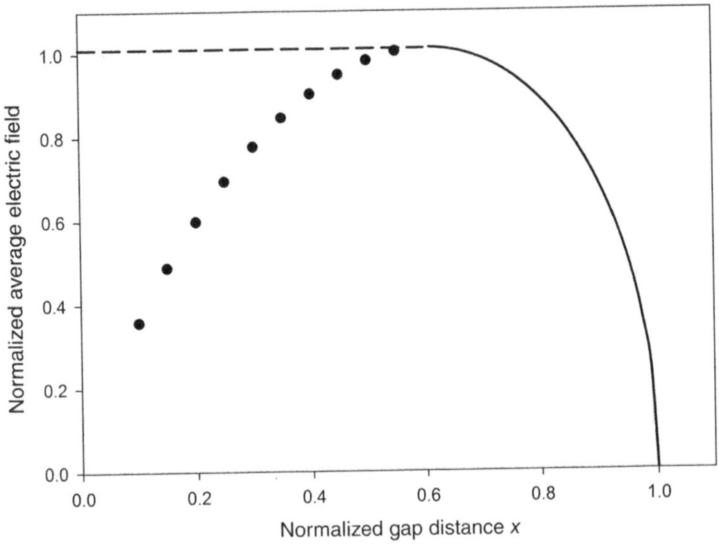

Figure 9.19 The graph of the applied electric field normalized to the apparent stress E_a has a maximum at a distance $d=0.6d_0$. Only the values of x greater than 0.6 are valid because the material mechanically breaks due to the compressive stress induced by the electric field.

Note that the apparent field – which causes the dielectric to collapse – depends on the material properties of the material; the strength related through the Young's modulus Y, and the measure of the electric field in the material related through the relative permittivity.

Example 9.2

A compact pulsed power system is being developed and the design creates a small gap of 8 mm that must be insulated. The system uses an oil dielectric with a dielectric constant of 2.2 and a dielectric strength of 250 kV/cm. To increase the voltage holdoff, a composite dielectric is being considered. A solid dielectric with a dielectric constant of 4.4 and a dielectric strength of 500 kV/cm has been identified as a candidate material. Sheets with a thickness of 3 mm are commercially available. Will this increase the voltage holdoff?

Solution

If only the oil is used to fill the crevice, the maximum field is the value at breakdown:

$$E_{BD}^{\text{oil}} = \frac{V_0}{d}$$

For $d = 8$ mm and $E_{BD}^{oil} = 25$ kV/mm, the maximum voltage is 200 kV.

If a composite dielectric is to be used, the allowable electric field is the breakdown value in each medium. The crevice is 8 mm, and the solid is 3 mm thick ($= d_1$), indicating the thickness of the oil $d_2 = 5$ mm. The parameters are

$$\begin{array}{llll} \text{Solid}: & E_{BD}^{solid} = E_1 = 50\ \text{kV/mm} & \varepsilon_1 = 4.4\,\varepsilon_0, & d_1 = 3\ \text{mm} \\ \text{Oil}: & E_{BD}^{oil} = E_2 = 25\ \text{kV/mm} & \varepsilon_2 = 2.2\,\varepsilon_0, & d_2 = 5\ \text{mm} \end{array}$$

$$E_1 = E_{BD}^{solid} = 50 = \frac{2.2}{(4.4)(5) + (2.2)(3)} \times V_0$$

Solving for V_0, the maximum allowable charging voltage in the solid is $V_0^{solid} = 650$ kV.

$$E_2 = E_{BD}^{oil} = 25 = \frac{4.4}{(4.4)(5) + (2.2)(3)} \times V_0$$

Solving for V_0, the maximum allowable charging voltage in the oil is $V_0^{oil} = 162.5$ kV. Since a voltage of 162 kV is sufficient to breakdown the oil, it is the maximum voltage. However, it may be noted that the maximum voltage when only the oils is used in the crevice is 200 kV, and composite dielectric is not helpful.

This is not unexpected since the higher dielectric constant tends to push the electric field into the insulator with the lower dielectric constant. In this case, that insulator also has the weaker breakdown strength.

Example 9.3

An electric field applied to an insulator dissipates a constant power uniformly over its length *d*. Derive the condition for the maximum temperature in the insulator when the heat capacity of the material is neglected.

Solution

The heat diffused to the surroundings given an input $HG = HD$:

$$\frac{P_G}{V_{Vol}} = \frac{\partial}{\partial x}\left(k_{th}\frac{\partial T}{\partial x}\right)$$

$$\int_0^x P_G \cdot dx = \left(k_{th}\frac{\partial T}{\partial x}\right) \times V_{Vol}$$

$$P_G x = \left(k_{th}\frac{\partial T}{\partial x}\right) \times V_{Vol}$$

(continued)

(*continued*)

$$\int_0^d P_G x \cdot dx = \left(k_{th} \int_{T_a}^{T} dT\right) \times V_{Vol}$$

$$\int_0^d P_G x \cdot dx = \left(k_{th} \int_{T_a}^{T} dT\right) \times V_{Vol}$$

$$P_G \cdot d^2 = k_{th}(T - T_a) \times V_{Vol}$$

$$V_{Vol} = d \cdot A$$

$$\frac{P_G \cdot d}{k_{th} A} + T_a = T$$

Example 9.4

An oscillatory voltage at 1 MHz is applied across a rectangular insulating block of dimensions 100 mm long×100 mm wide×10 mm thick. The insulator has a relative dielectric constant ε_r=3.5; a loss tangent tanδ=0.05; thermal conductivity k_c=0.11 W/mK; and a decomposition temperature T_d=100 °C. Calculate the following: (i) the minimum cooling required on the broad side of the insulator to sustain a voltage gradient of 238 kV/m across the thickness, and (ii) minimum thermal breakdown voltage for a cooling system, which maintains an ambient surface temperature of 25 °C on the broad area side of the insulator.

Solution

The solution of the problem involves two steps: (i) calculation of the rate of heat generation within the insulator due to the applied high-frequency field and (ii) calculation of the heat conducted away from the high-temperature region of the block to the low-temperature ambient region.

Since the cooling is from the broad side, it is obvious that the central plane of the insulator will have a maximum temperature (T_m) and heat will be conducted away symmetrically on either side toward the broad surfaces with ambient temperature (T_a).

Case (i): Rate of heat generation *HG* inside the dielectric can be written as

$$\begin{aligned} HG &= V \cdot \frac{V}{1/\omega C} \cdot \tan\delta \\ &= V^2 \cdot \omega C \cdot \tan\delta \\ &= (E \cdot d)^2 \cdot 2\pi f \cdot \tan\delta \cdot \frac{\varepsilon_r \varepsilon_0 A}{d} \\ &= (E \cdot d)^2 \cdot 2\pi f \cdot \tan\delta \cdot \frac{\varepsilon_r \varepsilon_0 (\ell \cdot b)}{d} \\ &= 2\pi \varepsilon_r \varepsilon_0 E^2 f \tan\delta \cdot (\ell \cdot b \cdot d) \end{aligned}$$

where

ℓ = length of the insulator
b = width of the insulator
d = thickness of the insulator
E = applied field strength
f = frequency
$\tan\delta$ = loss tangent of the insulator

Substituting ε_0=8.85×10^{-12}, ε_r=3.5, f=10^6, tanδ=0.05, ℓ=b=100×10^{-3}, and d=10×10^{-3}, the heat generated is found to be 55 W.

The heat HD conducted away is given by

$$HD = \frac{(k_c \cdot A)}{d} \cdot (T_m - T_a) = \frac{(k_c \cdot \ell \cdot b)}{d} \cdot (T_m - T_a)$$

Considering symmetrical heat conduction in two directions from the central plane with T=T_m toward broad sides with temperature T=T_a, we can rewrite the formula for HD:

$$HD = 2\frac{(k_c \cdot \ell \cdot b)}{d/2} \cdot (T_m - T_a)$$

Substituting k_c=0.11, ℓ=100.10^{-3}, b=100×10^{-3}, d=5×10^{-3}, and T_m=T_d=100, we get

$$HD = (0.44) \times (100 - T_a)\,\mathrm{W}$$

Under the conditions of equilibrium, the heat generated is dissipated away. Setting HG=HD, the relation is

$$55 = (0.44) \times (100 - T_a)$$

$$T_a = -25°\mathrm{C}$$

This calculation indicates that when the applied field gradient is 238 kV/m or more, forced external cooling is necessary to maintain the broadside insulator surfaces at (−25 °C) or better. Otherwise, since the heat generated exceeds the heat dissipated, the maximum temperature inside the insulator will rise to more than the decomposition temperature leading to thermal breakdown.

(continued)

(*continued*)

(ii) The heat generated in the bulk of the insulator due to the applied voltage is given by

$$HG = \left(9.73 \times 10^{-10}\right) \cdot E^2 \text{ W}$$

The heat conducted away from the high-temperature region, T_m, to the low-temperature region, T_a, is given by

$$HD = (0.44) \times (T_m - T_a) \text{ W}$$

For equilibrium conditions, equate *HG* and *HD*:

$$(0.44) \times (100 - 25) = \left(9.73 \times 10^{-10}\right) \cdot E^2$$

Solving for the electric field E=184 kV/m or a voltage of 1.84 kV. If no special cooling system is employed, at a temperature T_a=25 °C, an electrical field gradient of 184 kV/m can be sustained. If the gradient increases beyond this, the insulation failure will occur by thermal breakdown.

References

1 J.J. McKeown, Proc. *Inst. Elect. Eng.*, Vol. 112, Iss. 4, p. 824–828, April, 1965.
2 D.B. Watson, W. Heyes, K.C. Kao, and J.H. Calderwood, Some Aspects of Dielectric Breakdown of Solids. *IEEE Trans. Electr. Insul.*, Vol. 1, No. 2, p. 30, 1965.
3 R. Stratton, Theory of Dielectric Breakdown in Solids, in *Progress in Dielectrics*, Vol. 3, John Wiley & Sons, Inc., New York, 1961.
4 H. Frohlich, On the Theory of Dielectric Breakdown in Solids. *Proc. R. Soc. A*, Vol. 188, No. 1015, 1947. doi:10.1098/rspa.1947.0023.
5 S. Whitehead, *Dielectric Breakdown in Solids*, Oxford University Press, 1951.
6 H.F. Church, Thermal Breakdown, Chemical and Electrochemical Deterioration, in *High Voltage Technology*, L.L. Alston, ed., Oxford University Press, 1968.
7 E. Kuffel and M. Abdullah, *High Voltage Engineering*, Pergamon Press, 1970.
8 K.H. Stark and G.C. Garton, *Nature*, Vol. 176, p. 1225, 1955.
9 R. Bartnikas, Partial Discharges: Their Mechanism, Detection and Measurements. *IEEE Trans. Dielectr. Electr. Insul.*, Vol. 9, No. 5, p. 763, 2002.
10 P.H.F. Morshuis, Degradation of Solid Dielectrics Due to Internal Partial Discharges: Some Thoughts on Progress Made and Where to Go Now. *IEEE Trans. Dielectr. Electr. Insul.*, Vol. 12, No. 5, p. 905, 2005.

11 P.P. Budenstein, On the Mechanism of Dielectric Breakdown in Solids. *IEEE Trans. Electr. Insul.*, Vol. 15, No. 3, p. 225, 1980.
12 E.J. McMohan, A Tutorial on Treeing. *IEEE Trans. Electr. Insul.*, Vol. 13, No. 4, p. 277, 1978.
13 A. Bulinski et al., Water Treeing in Binary Linear Polyethylene Blends. *IEEE Trans. Dielectr. Electr. Insul.*, Vol. 1, No. 6, p. 949, 1994.
14 P. Ross, Inception and Propagation Mechanisms of Water Treeing. *IEEE Trans. Dielectr. Electr. Insul.*, Vol. 5, No. 5, p. 66, 1998.
15 M. Ieda, M. Nagao, and M. Hihita, High Field Conduction and Breakdown in Insulating Polymers: Present Situation and Future Prospects. *IEEE Trans. Dielectr. Electr. Insul.*, Vol. 1, No. 5, p. 34, 1994.
16 K.D. Wolter, J.F. Johnson, and S. Tanaka, Degradation Product Analysis for Polymeric Dielectric Materials Exposed to Partial Discharges. *IEEE Trans. Electr. Insul.*, Vol. 13, No. 5, p. 327, 1978.
17 V.P. Singhal, B.S. Narayan, K. Nanu, and P.H. Ron, Development of Blumlein Based on Helical Line Storage Elements. *Rev. Sci. Instrum.*, Vol. 72, No. 3, p. 1862, 2001.
18 W.J. Sarjeant, J. Zirnheld, and F.W. MacDougall, Capacitors. *IEEE Trans. Plasma Sci.*, Vol. 26, No. 5, p. 1368, 1998.
19 W.J. Sarjeant, F.W. Macdougall, D.W. Larson, and I. Kohlberg, Energy Storage Capacitors: Aging, and Diagnostic Approaches for Life Validation. *IEEE Trans. Magn.*, Vol. 33, No. l, p. 501, 1997.
20 E.A. Abramyan, Transformer Type Accelerators for Intense Electron Beams. *IEEE Trans. Nucl. Sci.*, Vol. 18, No. 3, p. 447, 1971.
21 A.H. Sharbaugh, T.C. Devins, and S.J. Rzad, Progress in the Field of Electrical Breakdown in Dielectric Liquid. *IEEE Trans. Electr. Insul.*, Vol. 13, No. 4, p. 249, 1979.
22 T.J. Lewis, Molecular Structure and the Electrical Strength of Liquified Gases. *J. Electrochem. Soc.*, Vol. 107, p. 185, 1960.
23 A. Adamczewski, *Ionization, Conductivity and Breakdown in Dielectric Liquids*, Taylor & Francis, London, 1969.
24 A. Beroual, M. Zahn, A. Badent, K. Kist, A.J. Schwabe, H. Yamashita, K. Yamazawa, M. Danikas, W.G. Chadband, and Y. Torshin, Propagation and Structure of Streamers in Liquid Dielectrics. *IEEE Electr. Insul. Mag.*, p. 6, 1998.
25 Z. Krasucki, Breakdown of liquid dielectrics, In *Proceedings of the Royal Society of London A: Mathematical, Physical and Engineering Sciences*, Vol. 294, no. 1438, pp. 393–404. The Royal Society, 1966.
26 K.C. Kao and J.B. Higham, The Effects of Hydrostatic Pressure, Temperature, and Voltage Duration on the Electric Strength of Hydrocarbon Liquids. *J. Electrochem. Soc.*, Vol. 108, p. 522, 1961.
27 K.C. Kao, Deformation of Gas Bubbles in Liquid Drops in an Electrically Stressed Insulating Liquid. *Nature*, Vol. 208, pp. 279–280, 1965.

28 M. Khalifa, *High Voltage Engineering: Theory and Practice*, Marcel Dekker, 1990.
29 P.K. Watson and A.H. Sharbaugh, High Field Conduction Currents in Liquid *n*-Hexane Under Microsecond Pulses. *J. Electrochem. Soc.*, Vol. 108, p. 522, 1956.
30 D.A. Wetz, J. Mankowski, J.C. Dickens, and M. Kristiansen, The Impact of Field Enhancement and Charge Injection on the Pulsed Breakdown Strength of Water. *IEEE Trans. Plasma Sci.*, Vol. 34, No. 5, p. 16, 2006.
31 H.M Jones and E.E. Kunhardt, Pulsed Dielectric Breakdown of Pressurized Water and Salt Solutions. *J. Appl. Phys.*, Vol. 77, No. 2, p. 795, 1995.
32 J. Qian, R.P. Joshi, K.H. Schoenbach, J.R. Woodworth, and G.S. Sarkisov, Model Analysis of Self and Laser Triggered Electrical Breakdown of Liquid Water for Pulsed Power Applications. *IEEE Trans. Plasma Sci.*, Vol. 34, No. 5, p. 1680, 2006.
33 W.F. Schmidt, *Liquid State Electronics of Insulating Liquids*, CRC Press, 1997.
34 M.T. Buttram, Area Effects in the Breakdown of Water Subjected to Long Term (~100 μs) Stress. *IEEE Conference Record of 15th Power Modulator Symposium*, p. 168, 1982.
35 J.P. Vandevender and T.H. Martin, Untriggered Water Switching. *IEEE Trans. Nucl. Sci.*, Vol. 22, p. 979, 1975.
36 J.C. Martin and I. Smith, AWRE Note SSWA/JCM/704/49, AWRE, 1970.
37 M. Zahn, Y. Ohki, D.B. Fenneman, R.J. Gripshover, and V.H. Gehman, Jr., Dielectric Properties of Water and Water/Ethylene Glycol Mixtures for Use in Pulsed Power System Design. *Proc. IEEE*, Vol. 74, No. 9, pp. 1182–1221, 1986.
38 J.D. Shipman, The Electrical Design of the NRL Gamble II, 100 Kilojoule, 50 Nanosecond, Water Dielectric Pulse Generator Used in Electron Beam Experiments. *IEEE Trans. Nucl. Sci.*, Vol. 18, p. 243, 1971.
39 T.V. Oommen, Vegetable Oils for Liquid-Filled Transformers. *IEEE Electr. Insul. Mag.*, p. 6, 2002.
40 P. Norgard, R.D. Curry, and R. Sears, Poly-α Olefin Synthetic Oil: A New Paradigm in Repetitive High Pressure Oil Switches. *IEEE Trans. Plasma Sci.*, Vol. 34, No. 5, p. 1662, 2006.
41 V.Y. Ushakov, *Breakdown in Liquids*, Springer, 2008.
42 S. Katsuki, H. Akiyama, A. Ghazala, and K.H. Schoenbach, Parallel Streamer Discharges Between Wire and Plane Electrodes in Water. *IEEE Trans. Dielectr. Electr. Insul.*, Vol. 9, No. 4, p. 499, 2002.
43 A. Denat, High Field Conduction and Prebreakdown Phenomenon in Dielectric Liquids. *IEEE Trans. Dielectr. Electr. Insul.*, Vol. 13, No. 3, p. 518, 2006.
44 A.H. Sharbaugh, J.C. Devins, and S.J. Rzad, Review of Past Work Done on Liquid Breakdown. *IEEE Trans. Electr. Insul.*, Vol. 15, No. 3, p. 167, 1980.

45 T.J. Lewis, Breakdown Initiating Mechanisms at Electrode Interface in Liquids. *IEEE Trans. Dielectr. Electr. Insul.*, Vol. 10, No. 6, p. 948, 2003.
46 R. Tobazeon, Prebreakdown Phenomenon in Dielectric Liquid. *IEEE Trans. Dielectr. Electr. Insul.*, Vol. 1, No. 6, p. 1132, 1994.
47 H. Bluhm, *Pulsed Power Systems: Principles and Applications*, Springer, 2006.
48 R. Hawley and A.A. Zaky, in *Progress in Dielectrics*, J.B. Birks, ed., Vol. 7, Haywood, London, p. 115, 1967.
49 J.G. Trump and R.J. van de Graaff, The Insulation of High Voltages in Vacuum. *J. Appl. Phys.*, Vol. 18, p. 327, 1947.
50 L. Cranberg, The Initiation of Electrical Breakdown in Vacuum. *J. Appl. Phys.*, Vol. 23, p. 518, 1952.
51 I.N. Slivkov, Mechanism for Electrical Discharge in Vacuum. *Sov. Phys. Tech. Phys.*, Vol. 2, p. 1928, 1957.
52 P.H. Ron, V.K. Rohatgi, and R.S.N. Rau, Rise Time of a Vacuum Gap Triggered by an Exploding Wire. *IEEE Trans. Plasma Sci.*, Vol. 11, No. 4, p. 274, 1983.
53 P.H. Ron, V.K. Rohatgi, and R.S.N. Rau, Delay Time of a Vacuum Gap Triggered by an Exploding Wire. *J. Phys. D. Appl. Phys.*, Vol. 17, p. 1369, 1984.
54 H.G. Bender and H.C. Kärner, Breakdown of Large Vacuum Gaps Under Lightning Impulse Stress. *IEEE Trans. Dielectr. Electr. Insul.*, Vol. 23, p. 37, 1988.
55 K. Ohira, R. Iwai, S. Kobayashi, and Y. Saito, Parameters Influencing Breakdown Characteristics of Vacuum Gaps During Spark Conditioning. *IEEE Trans. Dielectr. Electr. Insul.*, Vol. 6, No. 4, p. 457, 1999.
56 H.W. Anderson, *Electr. Eng.*, Vol. 54, p. 1315, 1935.
57 M.E. Cuneo, The Effect of Electrode Contamination, Cleaning, and Conditioning on High Energy Pulsed Power Device Performance. *IEEE Trans. Dielectr. Electr. Insul.*, Vol. 6, No. 4, p. 469, 1999.
58 P.R. Menge and M.E. Cuneo, Quantitative Cleaning Characterization of Lithium-Fluoride Ion Diode. *IEEE Trans. Plasma Sci.*, Vol. 25, No. 2, p. 252, 1997.
59 S. Kobayashi, Recent Experiments on Vacuum Breakdown of Oxygen Free Copper Electrodes. *IEEE Trans. Dielectr. Electr. Insul.*, Vol. 4, No. 6, p. 841, 1997.
60 W. Opydo, J. Mila, R. Batura, and J. Opydo, Electric Strength of Vacuum Systems with Co–Mo Alloy Coated on Electrodes. *IEEE Trans. Dielectr. Electr. Insul.*, Vol. 2, No. 2, p. 271, 1995.
61 M.J. Kofoid, Effect of Metal–Dielectric-Junction Phenomena on High-Voltage Breakdown over Insulators in Vacuum. *IEEE Trans. Power Apparatus Syst.*, Vol. 79, No. 3, p. 999, 1960.
62 S. Shope, J.W. Poukey, K.D. Bergeron, D.H. McDaniel, A.J. Toepfer, and J.P. Vandevender, Self-Magnetic Insulation in Vacuum for Coaxial Geometry. *J. Appl. Phys.*, Vol. 49, No. 7, p. 3675, 1978.

63 K. Yatsui, A. Tokuchi, H. Tanaka, H. Ishizuka, A. Kawai, E. Sai, K. Masugata, M. Ito, and M. Matsui, Geometric Focusing of Intense Pulsed Ion Beams from Racetrack Type Magnetically Insulated Diodes. *Laser Part. Beams*, Vol. 3, Part 2, p. 119, 1985.
64 S. Humphries Jr. Self-Magnetic Insulation of Pulsed Ion Diodes. *Plasma Phys.*, Vol. 19, p. 399, 1977.
65 R.J. Barker and P.F. Ottinger, Steady State Numerical Solution of Magnetically Insulated Charge Flow in Coaxial Geometry, NRL Memorandum Report 4654, Naval Research Laboratory, Washington, DC, December 8, 1981.
66 J.J. Ramirez, A.J. Toepfer, and M.J. Clauser, Pulsed Power Applications to Intense Neutron Source Development. *Nucl. Instrum. Methods*, Vol. 145, p. 179, 1977.
67 A. Shih, J. Yater, P. Pehrsson, J. Butler, C. Hor, and R. Abrams, Secondary Electron Emission from Diamond Surfaces. *J. Appl. Phys.*, Vol. 82, No. 4, pp. 1860–1867, 1997.
68 L. Reimer and H. Drescher, Secondary Electron Emission of 10–100 keV Electrons from Transparent Films of Al and Au. *J. Phys. D.*, Vol. 10, pp. 805–815, 1977.
69 H. Seiler, Secondary Electron Emission in the Scanning Electron Microscope. *J. Appl. Phys.*, Vol. 54, No. 11, pp. R1–R18, 1983.
70 M. Belhaj, S. Odof, K. Msellak, and O. Jbara, Time-Dependent Measurement of the Trapped Charge in Electron Irradiated Insulators: Applications to Al_2O_3-sapphire. *J. Appl. Phys.*, Vol. 88, No. 5, pp. 2289–2294, 2000.
71 R.E. Davies, Measurement of Angle-Resolved Secondary Electron Spectra, PhD Dissertation, Utah State University, 1999.
72 Y.C. Yong, J.T.L. Thong, and J.C.H. Phang, Determination of Secondary Electron Yield from Insulators Due to a Low-kV Electron Beam. *J. Appl. Phys.*, Vol. 84, No. 8, pp. 4543–4548, 1998.
73 N.R. Whetten, and A.B. Laponsky, Secondary Electron Emission from MgO Thin Films. *J. Appl. Phys.*, Vol. 30, No. 3, pp. 432–435, 1959. S. Yu, T. Jeong, W. Yi, J. Lee, S. Jin, J. Heo, and D. Jeon, Double-to Single-Hump Shape Change of Secondary Electron Emission Curve for Thermal SiO_2 Layers. *J. Appl. Phys.*, Vol. 79, No. 20, pp. 3281–3283, 2001.
74 J.B. Johnson and K.G. McKay, Secondary Electron Emission of Crystalline MgO. *Phys. Rev.*, Vol. 91, No. 3, pp. 582–587, 1953.
75 S. Yu, T. Jeong, W. Yi, J. Lee, S. Jin, J. Heo, and D. Jeon, Double-to Single-Hump Shape Change of Secondary Electron Emission Curve for Thermal SiO_2 Layers. *J. Appl. Phys.*, Vol. 79, No. 20, pp. 3281–3283, 2001.
76 R.E. Davies and J.R. Dennison, Evolution of Secondary Electron Emission Characteristics of Spacecraft Surfaces. *J. Spacecr. Rockets*, Vol. 34, pp. 571–574, 1997.

77 W. Yi, S. Yu, W. Lee, I.T. Han, T. Jeong, Y. Woo, J. Lee, S. Jin, W. Choi, J. Heo, D. Jeon, and J.M. Kim, Secondary Electron Emission Yields of MgO Deposited on Carbon Nanotubes. *J. Appl. Phys.*, Vol. 89, No. 7, pp. 4091–4095, 2001.
78 J. Cazaux, Electron-Induced Secondary Electron Emission Yield of Insulators and Charging Effects. *Nucl. Instrum. Methods Phys. Res. B*, Vol. 244, No. 2, pp. 307–322, 2006.
79 J.R. Dennison, W.Y. Chang, N. Nickles, J. Kite, C.D. Thomson, J. Corbridge, and C. Ellsworth, Final Report Part III: Materials Reports, NASA Space Environments and Effects Program Grant, Electronic Properties of Materials with Application to Spacecraft Charging (available in electronic format through NASA SEE as part of the SEE Charge Collector Knowledgebase) 2002.
80 H. Boersch, H. Hamisch and W. Ehrlich, Surface Discharges over Insulators in Vacuum. *Z. Angew. Phys.*, Vol. 15, pp. 518–525, 1963.
81 O. Milton, Pulsed Flashover of Insulator in Vacuum. *IEEE Trans. Electr. Insul.*, Vol. 7, No. 1, pp. 9–15, 1972.
82 K.D. Bergeron, Theory of the Secondary Electron Avalanche at Electrically Stressed Insulator: Vacuum Interfaces. *J. Appl. Phys.*, Vol. 48, No. 7. pp. 3073–3080, 1977.
83 R.A. Anderson and J.P. Brainard, Mechanism of Pulsed Surface Flashover Involving Electron Stimulated Desorption. *J. Appl. Phys.*, Vol. 51, No. 3, pp. 1414–1421, 1980.
84 A.S. Pillai and R. Hackam, Surface Flashover of Solid Dielectrics in Vacuum. *J. Appl. Phys.*, Vol. 53, No. 4, pp. 2983–2987, 1982.
85 L.L. Hatfield, G.R. Leiker, E.R. Boerwinkle, H. Krompholz, R. Korzekwa, M. Lehr, and M. Kristiansen, Methods of Increasing the Surface Flashover Potential in Vacuum. *Proceedings of the XIII International Symposium on Discharges and Electrical Insulation in Vacuum*, Vol. 1, pp. 241–245, 1988.
86 H.C. Miller and E.J. Furno, The Effect of Mn/Ti Surface Treatment on Voltage Holdoff Performance of Alumina Insulators in Vacuum. *J. Appl. Phys.*, Vol. 49, pp. 5416–5420, 1978.
87 T.S. Sudarshan and J.D. Cross, The Effect of Chromium Oxide Coatings on Surface Flashover of Alumina Spacers in Vacuum. *IEEE Trans. Electr. Insul.*, Vol. 11, pp. 32–35, 1976.
88 E.W. Gray, Vacuum Surface Flashover: A High Pressure Phenomenon. *J. Appl. Phys.*, Vol. 58, No. 1, pp. 132–141, 1985.
89 A.A. Avdienko, Surface Breakdown of Solid Dielectrics in Vacuum I: Characteristics for Breakdown of Insulators Along the Vacuum Surface. *Sov. Phys. Tech. Phys.*, Vol. 22, No. 8, pp. 982–985.
90 J.P. VanDevender, D.H. McDaniel, E.L. Neau, R.E. Mattis and K.D. Bergeron, Magnetic Inhibition of Insulator Flashover. *J. Appl. Phys.*, Vol. 53, No. 6, pp. 4441–4447, 1982.

91 J. Golden and C.A. Kapetanakos, Flashover Breakdown of an Insulator in Vacuum by a Voltage Impulse in the Presence of a Magnetic Field. *J. Appl. Phys.*, Vol. 48, No. 4, pp. 3073–3038, 1977.

92 C.W. Mendel, D.B. Siedel, and S.E. Rosenthal, A Simple Theory of Magnetic Insulation from Basic Physical Considerations. *Laser Part. Beams*, Vol. 1, Part 3, pp. 311–320, 1983.

93 H.C. Miller, Flashover of Insulators in Vacuum. *IEEE Trans. Electr. Insul.*, Vol. 28, No. 4, pp. 512–527, 1993.

94 O. Yamamoto, T. Takuma, M. Fakuda, S. Nagata, and T. Sonoda, Improving Withstand Voltage by Roughening the Surface of an Insulator Spacer Used in Vacuum. *IEEE Trans. Dielectr. Electr. Insul.*, Vol. 10, No. 4, p. 550, 2003.

95 J.M. Wetzer and P.A.F. Wouters, High Voltage Design of Vacuum Components. *IEEE Trans. Dielectr. Electr. Insul.*, Vol. 2, No. 2, p. 203, 1995.

96 G.A. Mesyats and D.I. Proskurovsky, *Pulsed Electrical Discharge in Vacuum*, Springer, 1989.

97 G.A. Mesyats, *Explosive Electron Emission*, URO-Press, Ekaterinburg, 1998.

98 G.A. Mesyats, *Pulsed Power*, Kluwer Academic/Plenum Publishers, 2004.

99 R.V. Latham, *High Voltage Vacuum Insulation*, Academic Press, London, 1981.

100 R.V. Latham, *High Voltage Vacuum Insulation: A New Perspective*, Authorhouse, 2006.

101 V.Y. Ushakov, *Insulation of High Voltage Equipments*, Springer, 2004.

10

Pulsed Voltage and Current Measurements

In many pulsed power systems, the magnitudes of the resultant electrical pulses lie in the range of 100 kV to 10 MV and 10 kA to 10 MA and pulse durations ranging from nanoseconds to microseconds. Conventional measurement systems for waveform display can handle signals of a few volts. The fundamental problem, therefore, lies in transforming extremely high-level signals to low-enough magnitudes without affecting the original waveform. Pulse voltage measurement techniques include crest voltmeters, voltage dividers, capacitive probes, electro-optical transducers, and reflection attenuators. Pulsed currents can be monitored with resistive current shunts, Rogowski coils, *B*-dot probes, current transformers, and magneto-optic devices.

10.1 Pulsed Voltage Measurement

10.1.1 Spark Gaps

Spark gaps can be used for the measurement of crest values of voltage pulses from 1 kV to 2.5 MV, but do not give information about the waveshape [1–4]. The spark gap electrodes are in the form of spheres of diameter *D*, made from aluminum or copper sheets having a smooth surface and uniform curvature. Proper clearances between the spheres and the surrounding grounded objects should be observed, according to international standards, so that the electric field conditions in the gap region are not affected. A resistor of 100 kΩ to 1 MΩ can be inserted in series with the high-voltage sphere to reduce the pitting on the spheres by spark, but this resistor is usually omitted for fast rise time waveforms. The gap spacing required for causing four to six flashovers in 10 applied pulses determines the 50% breakdown value, and this represents the peak voltage value.

Table 10.1 The air density correction factor.

Relative air Density (δ)	0.70	0.75	0.80	0.85	0.90	0.95	1.0	1.05	1.10	1.15
Correction Factor (K)	0.72	0.77	0.82	0.86	0.91	0.95	1.0	1.05	1.09	1.13

10.1.1.1 Peak Voltage of Pulses (>1 μs)

Tables 10.1 and 10.2 list the breakdown voltages for a standard lightning pulse of 1.2/50 μs for various gap spacing S and sphere diameters D [5]. The values given in Table 10.2 are valid for 50 Hz AC/DC and impulse with durations of 1/50 μs and longer. Over the gap spacing $S < 0.5\ D$, the accuracy of the measurement is within ±3%, but accuracy decreases at S/D values greater than 0.5. The gap breakdown voltage increases with relative air density, which is a function of both the ambient air pressure and temperature. The relative air density δ is given by

$$\delta = \frac{P}{760} \cdot \frac{273 + 20}{273 + T} = \frac{0.386P}{273 + T} \tag{10.1}$$

where P is the barometric pressure in millimeter of mercury (mmHg) and T is the temperature in degrees centigrade. For changes in δ, suitable correction

Table 10.2 Peak breakdown values for positive lightning impulse at 20°C, 760 mmHg for sphere gaps with one sphere grounded.

	Voltage kV peak							
	Sphere diameter (cm)							
Gap spacing (mm)	**6.25**	**12.5**	**25**	**50**	**75**	**100**	**150**	**200**
5	17.2	16.8						
10	31.9	31.7						
20	59	59						
50		134	138	138	138			
100			254	263	265	266	266	266
500						1040	1150	1190
1000								1840

Table 10.3 Peak breakdown values for negative lightning impulse AC and DC at 20°C, 760 mmHg for sphere gaps with one sphere grounded.

	Voltage kV peak							
	Sphere diameter (cm)							
Gap spacing (mm)	6.25	12.5	25	50	75	100	150	200
5	17.2	16.8						
10	31.9	31.7						
20	58.5	59.0						
50		129	137	138	138			
100			244	263	265	266	266	266
500							1130	1180
1000								1840

factor K can be obtained from Table 10.1. The actual breakdown voltage V_a can be obtained from Equation 10.2:

$$V_a = K \cdot V_s \tag{10.2}$$

where V_s is the breakdown voltage at STP, given in Tables 10.2 and 10.3.

10.1.1.2 Peak Voltage of Pulses (<1 μs)

The delay time of the spark gap is defined as the time from first application of the voltage to when the gaps current becomes circuit-limited. The delay time is considered to have two components: the statistical time lag and the formative time lag [6]. The statistical time lag is the time required for a free electron to become available in the gap. The formative time lag is the further time required for the acceleration of this electron to cause ionization growth and feedback mechanism to lead to the ultimate breakdown. In non-irradiated gaps, the electron availability depends on the thermal distribution of electrons in the atmosphere or natural radiation, and cannot be guaranteed during the short time when the pulse is available. Reliable and consistent breakdown for short duration pulses is possible by providing free electrons to the gap, which may be achieved in several ways. A radioactive source can be incorporated into the high-voltage electrode. The radiation sources usually employed are radium, cobalt, or polonium having an activity of around 0.5 mCi. Free electrons can also be provided by ultraviolet radiation provided by another spark gap or a quartz mercury arc lamp with a minimum rating of 35 W. Large gap spacings, with their longer breakdown times, are unsuitable for producing consistent breakdown for fast pulses. It may be summarized, in brief, that good irradiation and

Table 10.4 Breakdown voltages for small sphere gaps for one sphere grounded at 20°C and 760 mmHg.

	Kilovolts peak					
	Sphere diameter (cm)					
Gap spacing (mm)	**0.5**	**1.0**	**1.5**	**2.5**	**3.0**	**4.0**
0.01	1.06	1.03	1.01	0.995	0.988	0.980
0.05	2.89	2.82	2.80	2.77	2.77	2.77
0.1	4.85	4.75	4.70	4.64	4.64	4.65
0.5	15.37	17.80	17.75	17.44	17.45	17.42
1.0	20.15	26.50	29.75	31.40	31.70	31.78

short gap spacing are the key requirements of spark gaps for measurement of crest voltages of nanosecond pulses. Because spark gaps with small gap spacing can handle only small voltages of not more than tens of kilovolts, they have to be used in conjunction with calibrated voltage dividers for measurement of very high voltages. Table 10.4 gives the breakdown voltages for small sphere gaps [3].

10.1.2 Crest Voltmeters

Crest voltmeters are based on measuring the average rectified current passing through an accurately known capacitor, as in the case of Chubb-Fortescue method [7] shown in Figure 10.1 or measuring the DC voltage across a storage capacitor charged to near peak value by a rectifying diode [8,9], as shown in Figure 10.2.

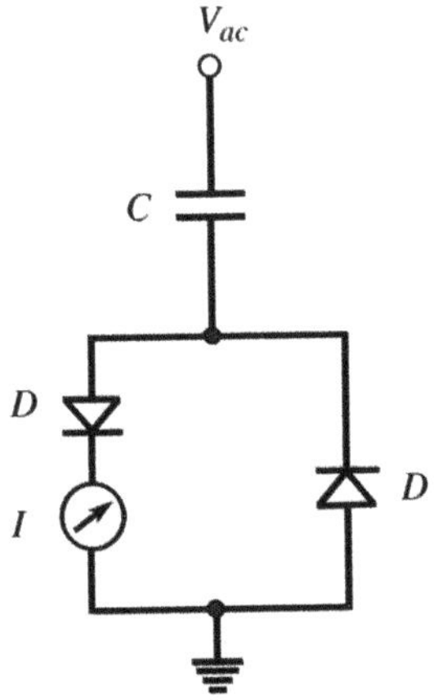

Figure 10.1 The Chubb–Fortescue method of peak voltage measurement.

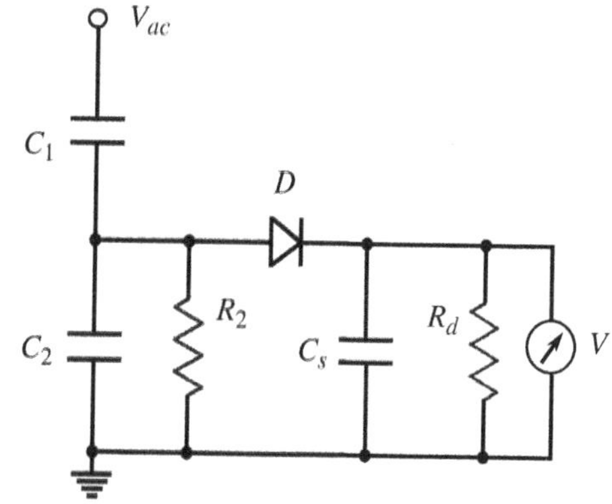

Figure 10.2 Peak voltage measurement using diode rectifiers.

In the Chubb-Fortescue method, the peak voltage is given by

$$V_p = \frac{I_{av}}{2f\,C} \tag{10.3}$$

where f is the frequency of the applied voltage.

In the rectified voltage method, the peak voltage is given by

$$V_p = K \cdot V_s \tag{10.4}$$

where K is the ratio of the capacitor divider and V_s is the DC voltage to which C_s is charged.

The above methods, used for periodic AC voltage, are not directly useful for measuring the peak voltage of unidirectional single pulses. The circuit of Figure 10.2 can be modified for unidirectional pulses by eliminating the resistors R_2 and R_d, and is shown in Figure 10.3.

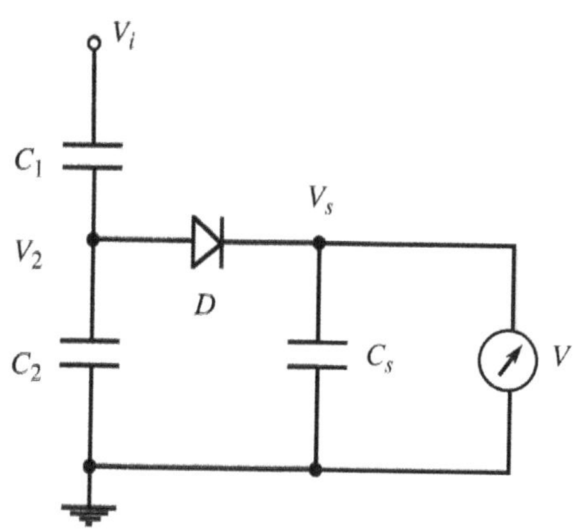

Figure 10.3 Peak voltmeter circuit for impulse waveform.

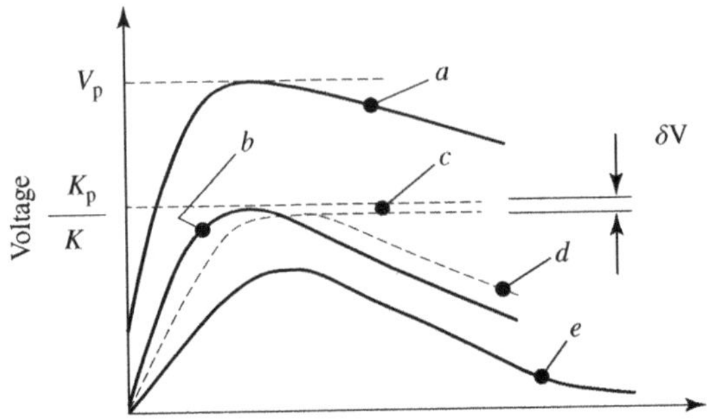

Figure 10.4 Waveform in a peak voltmeter circuit. (a) Input voltage V_i. (b) Capacitive divider output voltage V_2. (c) Voltage across C_s with diode with finite T_c and $T_d = h$. (d) Waveform across C_s with diode with finite T_c and T_d. (e) Waveform across C_s with diode with discrete resistor in place of diode.

In this circuit, the storage capacitor C_s gets charged to V_p/K through the rectifier D. When the input voltage decreases to a value less than V_p, the diode is reverse-biased and the voltage V_p/K is maintained on C_s. The voltage on C_s is measured by an electrostatic voltmeter V. The waveforms of input voltage V_i, voltage divider output V_2, and output voltage across C_s, V_s, are shown in Figure 10.4. Figure 10.4 indicates that the error δV in the peak voltage measurement is caused by (a) a finite charging time-constant $T_c = C_s R_f$ caused by the forward resistance R_f of the diode and (b) a finite discharge time constant $T_d = C_s R_\ell$ caused by the equivalent leakage resistance R_ℓ of the reverse biased diode, electrostatic voltmeter, and insulators. Commercially available peak voltmeters use improved active circuitry to considerably reduce the error.

10.1.3 Voltage Dividers

Voltage dividers accept the full voltage at the input and produce an output voltage, suitably scaled down by a factor known as division ratio, which can be carried to a digitizer for waveform display. For pulsed voltages, the frequency response of the probe must be faster than its highest frequency component. In many applications, voltage dividers must be calibrated *in situ* to account for stray reactance effects. Voltage dividers can be classified into resistive dividers, capacitive dividers, and mixed dividers.

10.1.3.1 Resistive Divider

A resistive divider is a conventional method for measuring the potential difference V_L across a load R_L. The equivalent measurement circuit is shown in Figure 10.5 where the resistive divider is represented by the dotted box and

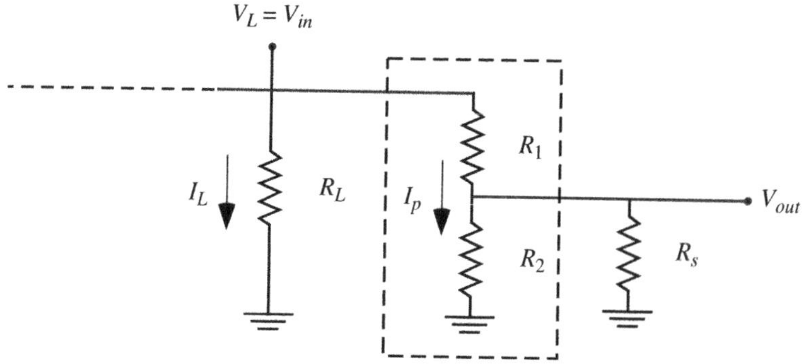

Figure 10.5 A resistive voltage divider, shown in the dotted box, is used to measure the voltage across a load resistor R_L and recorded on a scope with internal shunt impedance R_S.

the resistor R_S is the impedance of the scope or digitizer recording the output voltage V_{out}.

The resistive divider ideally consists of pure resistive components R_1 and R_2 connected to each other in series and which is in turn connected in parallel to the load R_L. The voltage across the load V_L is the input voltage V_{in} to the resistive divider. The output voltage V_{out} is delivered with a cable to the scope, which has a shunt resistance R_S.

10.1.3.1.1 *Ideal Resistive Divider*

For proper operation of the resistive voltage divider, the relative value of these resistances must be considered:

1) The internal resistance of the scope R_S is connected in parallel to the output section of the probe R_2. If the value of R_2 is chosen so that $R_2 \ll R_S$, then the equivalent value of the parallel combination

$$\left(R_S \| R_2\right) = \frac{R_S R_2}{R_S + R_2} \sim R_2, \quad \text{for} \quad R_2 \ll R_S \tag{10.5}$$

Under these conditions, the majority of the probe current travels through resistor R_2. Resistive voltage dividers are generally terminated in the high-impedance shunt impedance option of the scope. If $R_2 \sim R_S$ and the condition in Equation 10.5 is not met, then the equivalent resistance of the parallel combination $\left(R_S \| R_2\right)$ must be used for accurate measurements.

2) Applying *Kirchoff's Voltage Law* to the voltage divider circuit in Figure 10.5.

$$V_{in} = \left(\left(R_S \| R_2\right) + R\right) \times I_P \tag{10.6}$$

$$V_{out} = \left(R_S \| R_2\right) \times I_P \tag{10.7}$$

Combining (10.6) and (10.7) by solving for the current through the probe yields

$$\frac{V_{in}}{((R_S \| R_2) + R_1)} = \frac{V_{out}}{(R_S \| R_2)} \tag{10.8}$$

When expressed as

$$\frac{V_{out}}{V_{in}} = \frac{(R_S \| R_2)}{((R_S \| R_2) + R_1)} \tag{10.9}$$

it is known as the *voltage division ratio*. Imposing the condition $V_{out} \ll V_{in}$ leads to the relative value $(R_S \| R_2) \ll R_1$

3) In order to measure the voltage across the load while not interfering with the operation of the circuit, the probe current I_p must be negligible compared to the load current I_L.

$$I_p \ll I_L \tag{10.10}$$

From (10.6),

$$\frac{V_{in}}{R_1 + (R_S \| R_2)} \ll \frac{V_L}{R_L} \tag{10.11}$$

Since

$$V_L = V_{in} \tag{10.12}$$

$$R_L \ll R_1 + (R_S \| R_2) = R_p \tag{10.13}$$

where R_p is the input impedance of the resistive voltage divider.
If the probe draws too much current, the voltage divider is said to "load the circuit" by drawing too much current. This can not only result in an erroneous measurement but can also lead to a potentially dangerous situation.

Typically, a voltage divider must withstand a large input voltage so the resistor column is in the form of a large number of discrete resistors or a single column of a liquid resistor with a tap-off for the output voltage. In the absence of stray capacitance, a resistive voltage divider produces an ideal output voltage that is frequency independent. In practice, because stray capacitance is inevitable, resistive voltage dividers are used for DC or for very low frequencies.

10.1.3.1.2 Effect of Stray Capacitance

Whereas in theory the resistive divider has an ideal frequency response, in practice, the frequency response is limited by stray reactance [10–12]. For very high voltages, the resistive divider may have a large number of stages so that the voltage across any single stage is limited. This, however, results in a significant

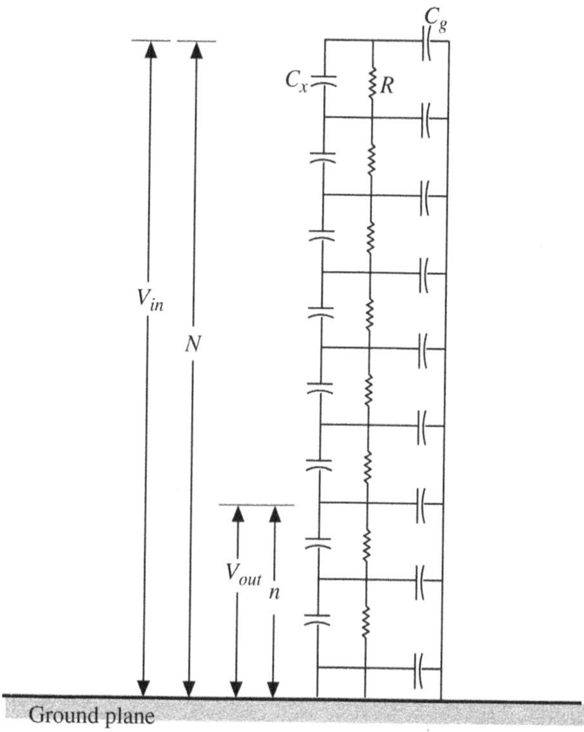

Figure 10.6 The equivalent circuit of a resistive divider with its stray capacitances.

stray capacitance that limits the frequency response. While stray inductances are also present, their effect on the response of the resistive voltage divider is much smaller, though efforts to minimize it are prudent.

Figure 10.6 represents the equivalent circuit of a resistive voltage divider including stray capacitance for pulsed applications. The voltage divider consists of N identical elements in series with resistance R. Each element also has a capacitance to ground C_g and a stray capacitance across each element C_x.

In practice, the stray capacitance that occurs in the physical implementation of the voltage probe can severely limit the high-frequency performance of a resistive divider and limit its ability to accurately measure the rate of rise of a fast pulse. Although the resistive elements also have an associated inductance, the value can be made small with good design practice and thus has a minimal effect on the frequency response. To illustrate, the circuit of the resistive divider with stray capacitance, shown in Figure 10.6, has been analyzed for a step function input voltage. The resistive divider is made of N discrete elements, each with a resistance, and a stray capacitance to the next element C_x as well as a stray capacitance to ground C_g. The output voltage V_{out} is taken at a fraction n of the total number of elements N so that the steady-state voltage divider ratio is the fraction n/N of the input voltage V_{in}.

Defining

V_{in} is the rectangular input voltage to be measured applied at $t = 0$
V_{out} is the output of the voltage divider
N is the total number of resistive elements
n is the number of resistance elements from the ground end where V_{out} is measured
C_g is the capacitance of each element to ground
C_x is the capacitance across each element

For convenience, define

$$C_G = NC_g$$

$$C_C = \frac{C_x}{N}$$

and

$$R_o = NR$$

Bellaschi calculated the divider's time-dependent response to a step function input as [12]

$$V_{out} = \frac{n}{N} V_{in} + \frac{V_i}{\pi} \sum_{k=1}^{\infty} \frac{2 \sin(k\pi(n/N))}{k \cos(k\pi)} \times \frac{e^{-\beta \cdot t}}{1 + k^2\pi^2(C_C/C_G)} \tag{10.14}$$

where

$$\beta = \frac{k^2\pi^2}{R_0(C_G + k^2\pi^2 C_C)}$$

The effects of stray capacitance on the performance of a resistive divider are summarized in Figure 10.7. The transient response of the resistive divider is dominated by the stray capacitance before settling to the steady-state value of $(n/N)V_{in}$.

Increasing the ground capacitance C_g in particular has a detrimental effect on the frequency response of the voltage probe. Physically, this can be explained on the basis of the time required to charge C_g through R. In order to get the best rise time performance, the values of R and C_g/C_x should be as small as possible. However, the input resistance of the probe R_o is limited by concerns regarding loading the probe. For resistive dividers to respond to fast rising voltage pulses, the resistance should be kept relatively low and have small ground capacitance. In practice, the undesirable effects of the stray capacitance to ground can be minimized by ensuring the voltage divider has a uniform electric field distribution throughout, as discussed in the next section.

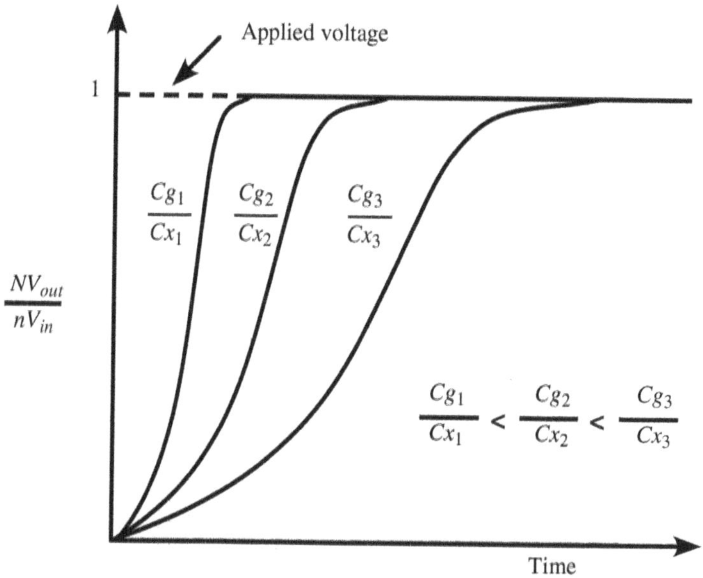

Figure 10.7 The step response of a resistive divider.

10.1.3.1.3 Effect of Stray Inductance

In addition to the stray capacitance, the inductance of the voltage divider may contribute to poor high-frequency performance. With no stray capacitance, the rise time of the divider is limited by L/R, the inductive time constant. Lower value of L in the voltage divider can be obtained by constructing the divider from solid carbon resistors instead of the more inductive variety of commercially available high-voltage resistors. The inductance of the low-voltage arm plays a significant role and should be reduced by connecting many resistors in parallel in a low-inductance geometry [13,14]. The low-voltage arm of the resistive divider should be shielded in a metallic enclosure to reduce electromagnetic interference.

10.1.3.1.4 Grading Rings in Resistive Dividers

The rise time of the resistive divider can be improved by supplying the charging current for C_g by a source other than the elements of the resistive divider. This can be achieved by incorporating grading rings or corona guards on the resistive divider as shown in Figure 10.8. In Figure 10.8a–c, the charging current for C_g is supplied capacitively by the grading rings and this current does not flow through the resistor column. The grading rings also result in increased values of the stray capacitance across the element C_x, which also contributes to reducing the rise time of the divider.

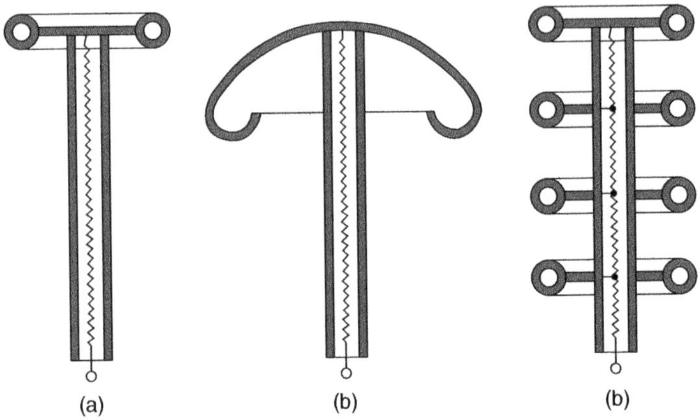

Figure 10.8 Different types of grading electrodes for a resistive divider. (a) Grading ring on top. (b) Corona ring on top. (c) Multiple grading rings.

The grading rings also affect the voltage distribution along the divider length by making it more uniform, as shown in Figure 10.9. The uniform voltage distribution improves the frequency response and reduces the probability of electrical breakdown near the top end of the divider.

10.1.3.1.5 Shielded Dividers

A shielded resistive divider consists of two separate columns, as shown in Figure 10.10 [15,16]. The inner column is comprised of resistors R_1 and R_2, and the outer one is a cylindrical resistor R. The inner column is the main voltage

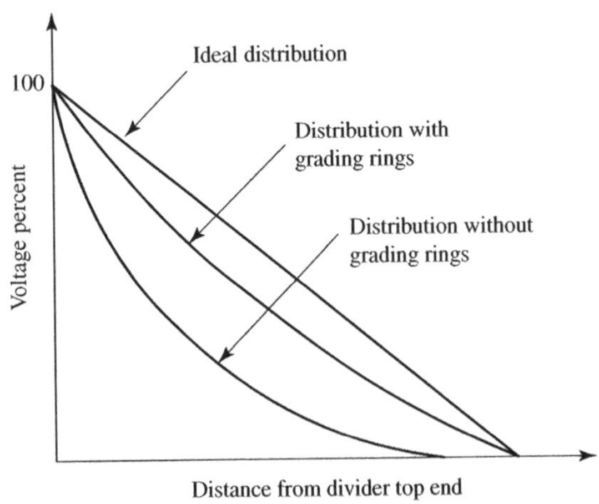

Figure 10.9 Voltage distribution along the resistor column with and without grading rings.

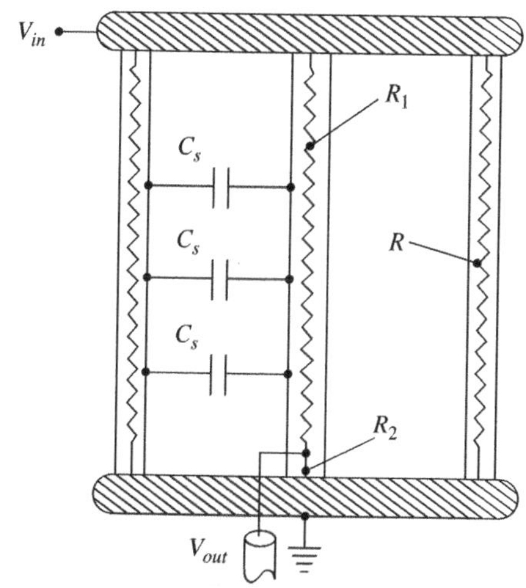

Figure 10.10 A shielded resistive divider.

divider and the outer resistive column serves to supply the charging currents for the stray ground capacitances and reduce the electromagnetic pickup. The charging current through the stray capacitance between the columns C_s is negligible, since the voltage across it is nearly zero. The shielding resistor R may use plastic film resistors, resistive cloth, resistive metal, or a liquid resistive column.

10.1.3.1.6 Dividers with Liquid Resistors

A resistive voltage divider using a uniform column of liquid resistor in place of discrete resistors has several advantages. The overall inductance is lower because of the absence of connecting leads between adjacent resistor elements, which also minimizes corona effects. A liquid resistive divider is capable of handling very large power densities and is self-healing in the event of accidental electrical breakdown. The divider ratio is unaffected, in spite of changes in the resistivity caused by variations in temperature, field gradient, or frequency, which may be important in large facilities with little environmental control.

Copper sulfate liquid dividers are very popular in nanosecond pulsed power technology for measuring pulse waveforms with amplitudes of millions of volts (Figure 10.11) [16–21]. The divider resistance may vary from tens of ohms to hundreds of ohms and currents of a few kiloamperes are drawn through the divider. Employment of such a low-value resistance makes the effects of stray capacitance negligible and results in low rise time. However, when the timescale is reduced to the nanosecond regime, the high dielectric constant of the water may cause the RC time constant of the resistive divider to become too large to accurately measure fast rise times. Copper electrodes and copper sulfate

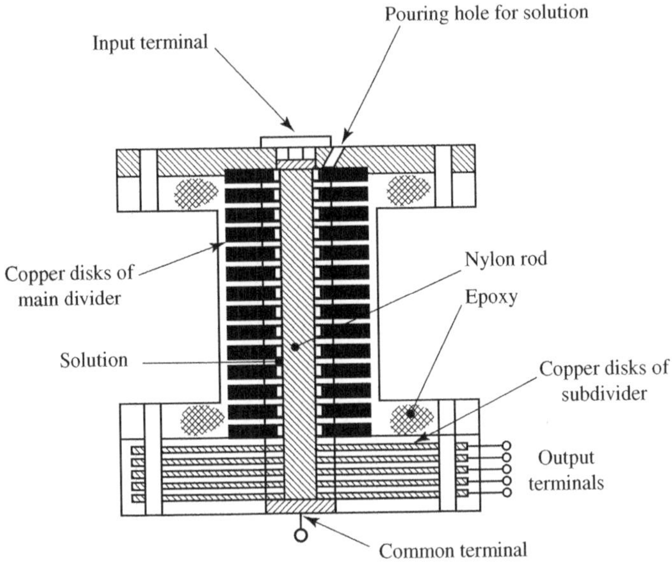

Figure 10.11 A $CuSO_4$ divider with a variable divider ratio [22].

($CuSO_4$) solution prepared with deionized water must be employed in the divider. $CuSo_4$ solution can be used at voltage gradients up to 50 kV/cm. Other solutions may also be used. Thiosulfate mixed with deionized water may be used with aluminum electrodes.

Liquid resistors can be made in any number of ways. A $CuSO_4$ divider with variable divider ratios [22] of 75, 37.5, 25, and 18.75 is shown in Figure 10.12. Dividers with liquid resistors may also be used in conjunction with a discrete resistor subdivider after careful analysis of the additional shunt resistance.

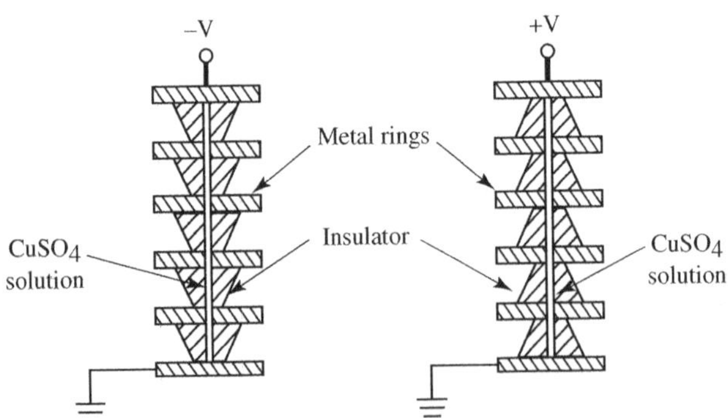

Figure 10.12 Insulator profiles for positive and negative voltages in vacuum.

When $CuSO_4$ dividers are used for measurement of pulsed voltages in vacuum, special care needs to be taken to prevent surface flashover across the insulator. The insulator surface exposed to vacuum should be profiled in such a way that the tangential component of the electric along the surface is reduced. This ensures that the electron multiplication, due to electrons emitted by secondary emission on the insulator surface, is inhibited, thereby resulting in high surface breakdown value. In vacuum, the insulator angle used depends on the polarity of the measured voltage and is shown in Figure 10.12.

10.1.3.2 Capacitive Dividers

A voltage divider can also be made from capacitive elements in series. Like the resistive voltage divider, the condition for the low-voltage arm is that the capacitor has to withstand the voltage. In a capacitor divider, $C_1 \ll C_2$, so that the majority of the voltage drop is across C_1 and the voltage output across C_2 is low enough to be taken to the oscilloscope for waveform display. The equations describing the operation of the capacitive voltage divider may be derived by the equivalent circuit of Figure 10.13.

From Kirchoff's voltage law:

$$V_{in} = V_{C_1} + V_{out} \tag{10.15}$$

The current flowing through the capacitor C_1 is the probe current I_p

$$I_p = C_2 \frac{dV_{C_2}}{dt} \tag{10.16}$$

The probe current I_p divides by *Kirchoff's current law* through the parallel combination of C_1 and the shunt resistance of the scope R_S.

$$I_p = I_{C_2} + I_{R_s} \tag{10.17}$$

where

$$I_{C_2} = C_2 \frac{dV_{out}}{dt} \tag{10.18}$$

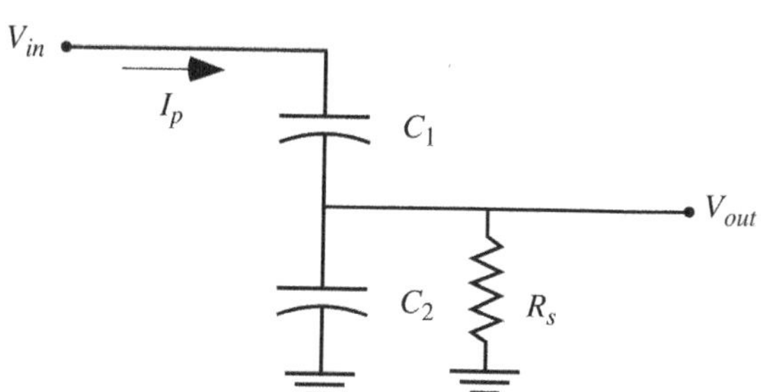

Figure 10.13 The equivalent circuit of a capacitive voltage divider.

and

$$I_{R_S} = \frac{V_{out}}{R_s} \tag{10.19}$$

Combining Equation (10.15) through (10.19) yields

$$C_1 \frac{dV_{C_1}}{dt} = C_2 \frac{dV_{out}}{dt} + \frac{V_{out}}{R_s} \tag{10.20}$$

Rearranging terms

$$\frac{dV_{C_1}}{dt} = \frac{C_2}{C_1} \frac{dV_{out}}{dt} + \frac{V_{out}}{C_1 R_s} \tag{10.21}$$

Differentiating Equation 10.15

$$\frac{dV_{in}}{dt} = \frac{dV_{C_1}}{dt} + \frac{dV_{out}}{dt} \tag{10.22}$$

Substituting (10.22) into (10.21) and combining terms

$$\frac{dV_{in}}{dt} = \left(\frac{C_1 + C_2}{C_1}\right) \frac{dV_{out}}{dt} + \frac{V_{out}}{C_1 R_s} \tag{10.23}$$

Equation 10.23 is the general solution to for a capacitive divider and can be solved by Laplace transform. Since V_{in} is the signal to be reconstructed from the measured signal V_{out}

$$\frac{V_{in}}{V_{out}} = \left(\frac{C_1 + C_2}{C_1}\right) + \frac{1}{sR_s C_1} \tag{10.24}$$

This gives the bandwidth of the capacitive voltage divider as

$$\omega_{3dB} = \frac{1}{R_S(C_1 + C_2)} \tag{10.25}$$

Capacitive dividers are used in many applications with numerous implementations [23–28]. Capacitive dividers have two important limiting cases: the low- and high-frequency responses.

10.1.3.2.1 Low-Frequency Response

For slow rising pulses – those where the high-frequency components are much less than the 3 dB bandwidth – the capacitive voltage divider gives an output voltage of

$$V_{out} = \frac{C_1}{C_1 + C_2} V_{in} \tag{10.26}$$

For slow pulses, the capacitive divider has a response similar to the resistive divider.

10.1.3.2.2 High-Frequency Response: The V-dot Probe

For fast rising pulses of limited duration, a capacitive probe can be constructed where the vast majority of the frequency components are much greater than its bandwidth given in (10.25). In this case, the response of the capacitive probe is dominated by the derivative term in (10.23) and the voltage measured on the scope is

$$V_{out} = R_S C_1 \frac{dV_{in}}{dt} \tag{10.27}$$

The input waveform is reconstructed by carefully integrating the measured waveform V_{out}. Under these conditions, the capacitive probe is usually referred to as a *V*-dot probe, with the "dot" referring to the time derivative. The *V*-dot probe is also known, perhaps more precisely, as a *D*-dot because of the dependence on the dielectric constant of the insulator. However, these terms are used interchangeably in practice. The *V*-dot probe acts like a high-pass filter with a corner frequency given by ω_{3dB}.

10.1.3.2.3 Configurations of Capacitive Dividers

Capacitor dividers can be designed using a number of discrete capacitors analogous to the resistive divider. However, the inductance introduced by the interconnecting leads limits the high-frequency response. The most common design of a capacitive voltage divider is to use the natural capacitance existing between the electrodes and creating a low-voltage leg of the capacitive divider by installing a small capacitance created with a dielectric and a floating electrode near the ground plane. A capacitive divider using commercial high-voltage cable is shown in Figure 10.14.

The floating electrode on the left is similar to a patch antenna. An auxiliary dielectric between the ground and floating electrode creates the capacitance of the low-voltage leg of the divider. The dimensions of the floating electrode as well as the dielectric properties of the auxiliary dielectric determine the value of the low-voltage capacitor and the resulting calibration factor. The floating electrode is connected to the inner conductor of the BNC cable output for extraction of the output signal. The voltage divider on the right has a larger capacitance because the entire inner surface is utilized in the formation of the low-voltage capacitance, which will also limit the effect of stray capacitance to that of the direction of wave propagation. This type of capacitive divider may be used for any frequency range and is typically used in coaxial configurations

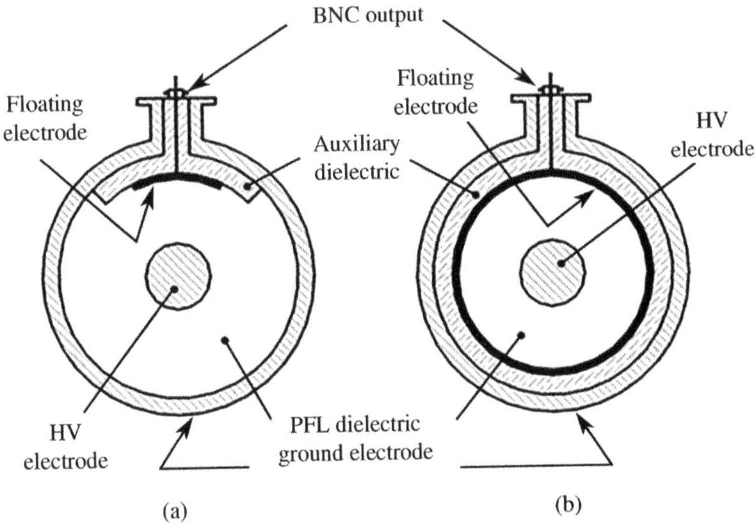

Figure 10.14 Two configurations of a capacitive voltage divider installed into a high-voltage cable. In part (a) the floating electrode is a patch that covers only a fraction of the azimuthal, whereas in part (b) the coaxial cable with the floating electrode makes a segment of triax.

because the current is uniformly distributed. The floating electrode divider may be an integral part of a high-voltage pulse generator [29–34].

Another common configuration is to create a capacitive voltage divider in structures where the electric fields are confined by inserting a coaxial capacitor in its ground plane of as the low-voltage leg, as shown in Figure 10.15. This is generally a short segment of coaxial cable where the ground braid of the diagnostic cable can be connected to the ground plane directly shown in Figure 10.15a. This coaxial capacitor, along with the capacitance associated with the fringe electric fields, forms the capacitor C_2. The high-voltage leg C_1 is formed by the capacitance from the high-voltage electrode to exposed node of C_2. The equivalent circuit is shown in Figure 10.15b and R_S is the low-impedance (50 Ω) termination of the scope.

In many cases, $C_2 \gg C_1$, which gives an effective bandwidth of

$$\omega_{3dB} = \frac{1}{R_S C_2} \tag{10.28}$$

A factor in the design is to have the strength of V_{out} large enough to overcome the signal to noise ratio in the harsh electromagnetic environment while keeping it low enough to be below the electrical breakdown strength of the connectors.

V-dot probes are compatible for pulse-forming lines such as triplate lines as well as coaxial geometry and are extensively used in for fast-rising, high-voltage

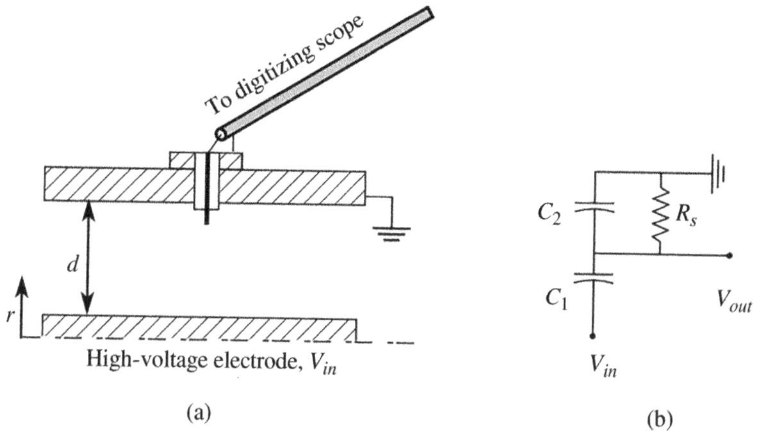

Figure 10.15 The (a) implementation of a *V*-dot probe located in the ground plane of a coaxial device and (b) the equivalent circuit.

pulses of short duration where the all the frequency components are much higher than the bandwidth of the probe. In the event that this is not the case, and the frequency contents of the pulse are near the 3 dB bandwidth, a more extensive calibration can be performed in the frequency domain [35–42].

10.1.4 Electro-optical Techniques

Electro-optical techniques for measuring voltage involve the electric field-induced birefringence of some insulating materials. A monochromatic probing light beam is modulated by the unknown electric field to be measured and the voltage waveform can be recovered by unfolding the modulation information. Compared to other methods of pulsed voltage measurement, elctro-optical techniques have the unique feature of freedom from electromagnetic interference. The refractive index of the electro-optical transducer is given by $n = n_o + aE + bE^2 + \ldots$, where n_o is the refractive index without an applied voltage, E is the applied electric field, and $(a, b, \ldots)$ are the electro-optic coefficients of the crystal [42,43]. The Kerr cell behavior corresponds to the E^2 dependence, that is, the third term of the expression, whereas the Pockels cell corresponds to the linear dependence on E, that is, the second term of the expression.

10.1.4.1 The Kerr Cell

In 1875, John Kerr discovered that a glass plate to which a strong electric field was applied became birefringent. Birefringence is the optical property of a material where the refractive index n depends on the polarization and direction

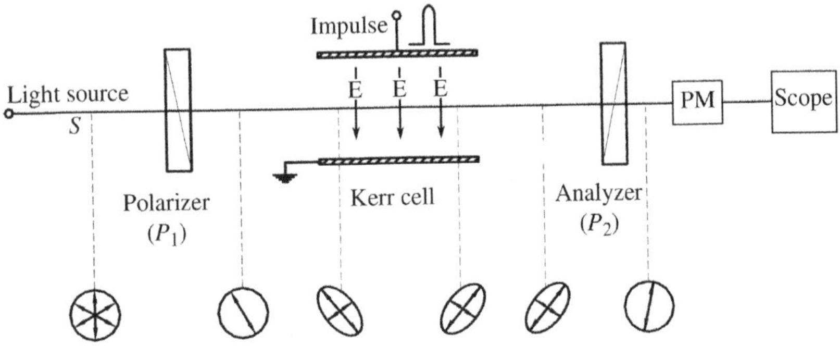

Figure 10.16 Electro-optical Kerr cell system and states of polarization along light path.

of propagation of light. The Kerr effect is the induced birefringence of a material in a response to a strong electric field. The Kerr effect is also called the quadratic electro-optic effect because the induced change in refractive index is proportional to the square of the electric field. All materials show a Kerr effect, but certain liquids and solids have a sufficiently strong response that it may be used as a diagnostic.

In a Kerr measurement, monochromatic, vertically polarized light impinges on a Kerr cell that is set at a 45° angle relative to the vertical. A voltage applied to the Kerr cell with dimension d establishes an electric field E in the Kerr medium, causing it to become birefractive (birefringent). The modulated beam is then passed through another polarizer aligned at a 90° angle from the vertical into a photomultiplier or other device to convert light intensity into a voltage that may be measured on a scope. The experimental layout for a Kerr measurement is shown in Figure 10.16.

The laser probe S is vertically polarized by passing through polarizer P_1, and oriented 45° with respect to the electric field inside the Kerr cell. With this relative orientation, the incoming beam of light is comprised of two beams: one having its electric field vector parallel to the applied field (the extraordinary beam) and the other perpendicular (the normal beam) [44]. The applied electric field induces birefringence in the Kerr medium, resulting in the two component beams travelling at different propagation speeds ($v = c/n$) as they pass through the Kerr medium, yielding a phase difference Δ upon exiting the cell, given by

$$\Delta = 2\pi \frac{\ell}{\lambda}(n_e - n_n) \tag{10.29}$$

where n_e and n_n are the refractive indices in the extraordinary and normal directions and λ is the wavelength of the probing laser light in a vacuum. It can be seen from (10.29) that the phase displacement is proportional to the polarizability of the Kerr medium and can be rewritten as

$$\Delta = 2\pi K \ell E^2 \tag{10.30}$$

where the proportionality constant K is known as the Kerr constant, ℓ is the length of the cell, and E is the applied electric field. The beam is elliptically polarized as it comes out of the Kerr cell. The analyzer P_2 is cross-polarized (90° orientation) with respect to P_1 so that only the component of light parallel to P_2 is transmitted. The phase shift between the normal and extraordinary light beam bending the Kerr cell is recorded by measuring the intensity $I$of the laser probe using detectors such as photodiodes, photomultipliers, or optical photography, allowing thereby the reconstruction of the signal. The relation between the intensity and the phase difference is

$$I = I_0 \sin^2\left(\frac{\Delta}{2}\right) \tag{10.31}$$

where I_0 is the luminosity measured when the polarizer and analyzer are aligned in the same direction and the Kerr cell electric field is zero. For a phase difference between the two waves of $\lambda/2$, a linearly polarized wave results from the superposition of the extraordinary and the normal beams after passing through the Kerr cell. The voltage at which this occurs is called the "half wave voltage" and shows as a maximum in the intensity measurement.

The Kerr cell method of voltage measurement can be used to measure time-dependent electric fields [45–53]. From Equation 10.30, an increasing electric field produces a continuously increasing phase change $\Delta(t)$, which is equivalent to the rotation of the plane of polarization. The intensity of the transmitted light is given by

$$I(t) = I_0 \sin^2\left(\frac{\Delta(t)}{2}\right) \tag{10.32}$$

where I_0 is the initial calibration as used in (10.31).

Combining (10.32) with (10.30)

$$I(t) = I_0 \sin^2\left(\pi K \ell E^2(t)\right) \tag{10.33}$$

Equation 10.33 predicts an alternating series of maxima and minima as E is continuously increased. The first maximum occurs at the value corresponding to the half wave voltage, when the argument in both (10.32) and (10.33) equals $\pi/2$, and the value of the electric field is E_M. Setting

$$\frac{\Delta}{2} = \pi K \ell E^2(t) = \frac{\pi}{2}$$

The electric field at the first maximum E_M is given by

$$E_M = \frac{1}{\sqrt{2K\ell}} \tag{10.34}$$

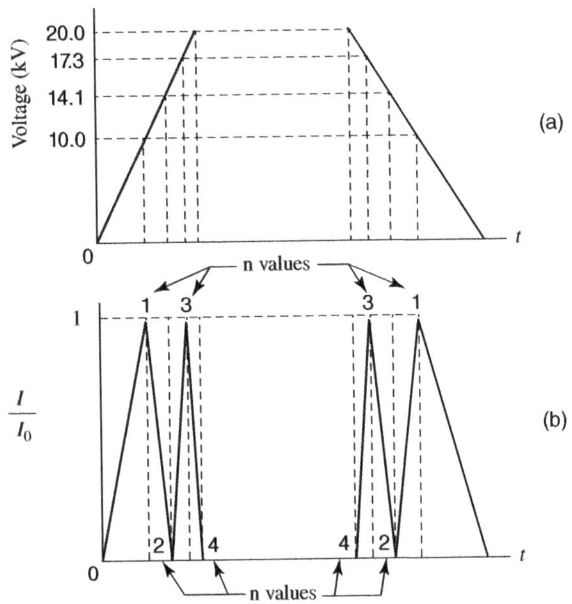

Figure 10.17 The input signal to a Kerr cell in part (a) produces the signal in part (b).

Equations 10.33 and 10.34 can be combined to yield

$$\frac{I(t)}{I_0} = \sin^2\left(\frac{\pi}{2}\frac{E^2(t)}{E_M^2}\right) \tag{10.35}$$

The transmission maxima occur when the argument of Equation 10.35 equals odd multiples of $\pi/2$ so that $(E/E_M)^2 = 1, 3, 5, 7 \ldots$ and transmission minima occur at multiples of π, when $(E/E_M)^2$ values are 2, 4, 6, 8 . . . . This forms the basis for recovering the applied impulse waveform. A typical pattern of Kerr cell-produced signal using a photodetector for a trapezoidal input waveform is shown in Figure 10.17. In this example, a cell having $d = 1$ cm and $E_M \times d =$ 10 kV has been assumed. The maxima occur at 10 kV and 17.3 kV, whereas the minima appear at 14.1 and 20 kV.

Experimentally, $I(t)$ at any instant is obtained by measuring the time-dependent photodetector current. The value of $E_M \times d$, the minimum voltage required to be applied to the Kerr cell for obtaining the first transmission maximum, can be determined experimentally [49,50]. Direct optical photography of time-resolved streak-fringe pattern has also been used in place of photodetector techniques [51]. The dimensions, shape, and spacing of the cell electrodes should be designed for uniform field and high breakdown voltage. The Kerr cell as well as the detector system should be screened against the stray light.

The Kerr cell medium must be a liquid exhibiting a large Kerr constant K. Some polar liquids, such as nitrotoluene ($C_7H_7NO_2$) and nitrobenzene ($C_6H_5NO_2$), exhibit very large Kerr constants. Water also has a significant Kerr constant of $9.4 \times 10^{-14}\,mV^{-2}$, compared to $4.4 \times 10^{-12}\,mV^{-2}$ for nitrobenzene. The cell electrodes are usually made of nickel or aluminum. A typical Kerr cell might require voltages on the order of 30 kV because the Kerr effect is relatively weak.

10.1.4.2 The Pockels Cell

When a linearly polarized light beam passes through a Pockels cell subjected to an electric field, the phase retardation Φ, induced between the two orthogonal components of light, is given by [43]

$$\Phi = \frac{2\pi}{\lambda} \cdot \Delta n \cdot \ell \tag{10.36}$$

where λ is the wavelength of light, Δn the induced birefringence, and ℓ the optical length of the Pockels cell. The induced birefringence Δn is given by

$$\Delta n = n_0^3 \cdot \gamma \cdot \frac{V_i}{d} \tag{10.37}$$

where γ is the electro-optical coefficient of the Pockels cell crystal, V_i is the amplitude of the applied voltage, and d the distance between the electrodes of the Pockels cell. Therefore, Equation 10.36 can be rewritten as

$$\Phi = \frac{2\pi}{\lambda} \cdot n_0^3 \cdot \gamma \cdot \frac{V_i}{d} \cdot \ell \tag{10.38}$$

The Pockels cell can be configured for longitudinal modulation, where the direction of the incident light beam is parallel to the electric field, as shown in Figure 10.18a, or transverse modulation, where the direction of the incident beam is perpendicular to the direction of the electric field, as shown in Figure 10.18b.

In case of longitudinal modulation, because $\ell = d$, Equation 10.32 simplifies to

$$\Phi = \frac{2\pi}{\lambda} \cdot n_0^3 \cdot \gamma \cdot V_i \tag{10.39}$$

which is independent of the dimensions of the Pockels cell.

It is seen from (10.38) and (10.39) that the applied voltage waveform can be recovered by the determination of the phase retardation Φ, between the incident and modulated beam. The various materials [43,53,54] used for the Pockels cell are single crystals of BSO($Bi_{12}SiO_{20}$), BTO ($Bi_{12}TiO_{20}$), BGO ($Bi_4Ge_3O_{12}$), $LiNiO_3$, $LiTaO_3$, and so on. The electro-optic effect of the Pockels cell should have negligible dependence on temperature and vibrations. The

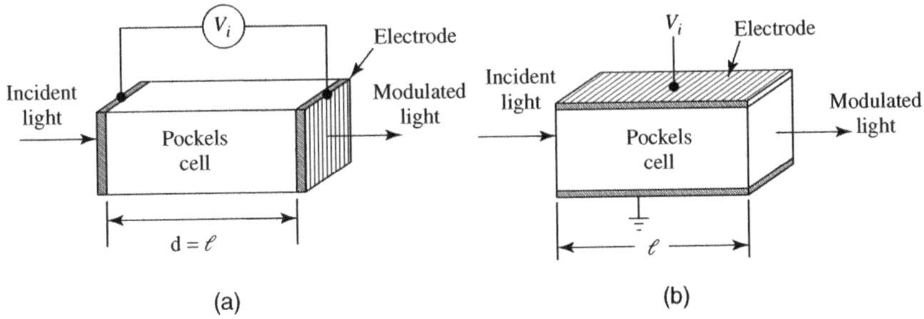

Figure 10.18 Techniques of longitudinal and transverse modulation. (a) Longitudinal modulation. (b) Transverse modulation.

sensor crystal should have high volume and surface resistivity to minimize leakage current.

10.1.4.2.1 Transverse Modulation

A Pockels cell voltage measurement system based on transverse modulation is shown in Figure 10.19. The light beam from the LED is carried up to the optical sensor element by an optical fiber. The light beam passes through the components of the transducer comprised of polarizer, quarter wave plate, and the Pockels cell. At the entrance to the Pockels cell, the beam is circularly polarized, but is converted to an elliptically polarized beam after being modulated by the applied voltage on the Pockels cell. The analyzer splits the incident beam, of intensity J_o, into two mutually perpendicular beams of intensities J_1 and J_2. These light beams are then transmitted through optical fibers and converted to electrical signals by PIN diodes. The unknown input voltage V_i can then be computed from the output signal voltage V_0 with signal processing [45].

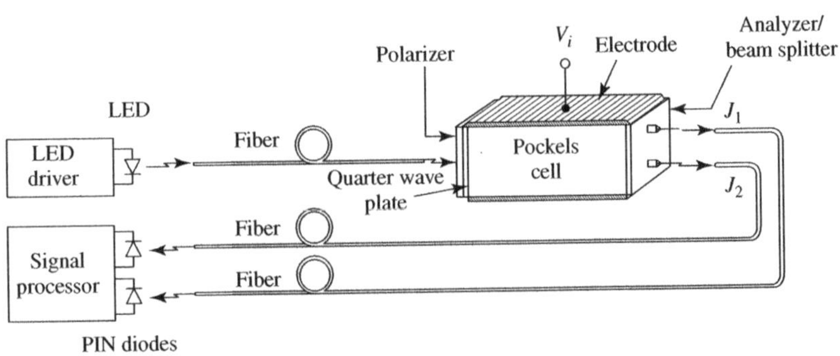

Figure 10.19 Schematic of Pockels cell voltage measurement based on transverse modulation.

In the system described above, voltage up to 3 kV (rms) can be applied to the Pockels cell. For voltage values higher than 3 kV, it is necessary to attenuate the input voltage by means of a voltage divider.

10.1.4.2.2 Longitudinal Modulation

A Pockels cell voltage measurement system based on longitudinal modulation is shown in Figure 10.20 [43]. As shown in Figure 10.20a, the overall system is comprised of laser source, optical fibers, polarizer, $\lambda/4$ waveplate, multisegmented Pockels sensor system, analyzer, and detector. The main advantage of this system is its capability to measure hundreds of kilovolts directly without the need for an external voltage divider. A series combination of N crystals (ε_1, d_1) and $(N-1)$ dielectric spacers (ε_2, d_2) results in a built-in capacitive divider, as shown in Figure 10.20b. The electric field appearing across individual crystals can be written as

$$E = \frac{V_{C_1}}{Nd_1} = \frac{1}{Nd_1}\frac{C_2}{C_1 + C_2}V_i = \frac{V_i}{Nd_1 + (\varepsilon_1/\varepsilon_2)(N-1)d_2} \tag{10.40}$$

The maximum voltage that can be measured is limited by the intrinsic electrical breakdown of the Pockels crystal, the surface flashover across the external crystal surface, and the overall linearity in the transmitted light intensity with applied voltage. The flashover voltage can be increased by the use of sulfur hexafluoride (SF_6) gas as the surrounding medium. The system performance is

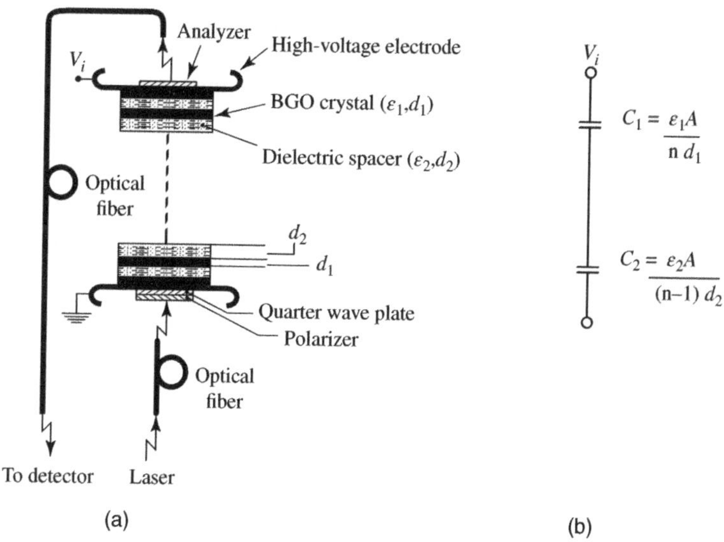

Figure 10.20 The voltage sensing system (a) and equivalent circuit (b) of a multisegmented Pockels cell voltage measurement based on longitudinal modulation.

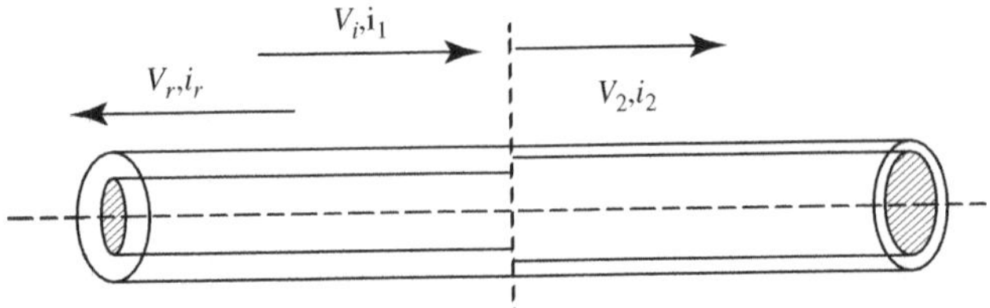

Figure 10.21 A simple reflection attenuator.

also dependent upon (a) maintenance of the state of polarization by the fiber, (b) piezoelectric effect in the crystal, giving rise to mechanical vibrations, and (c) electro-optic effect in the dielectric spacer. The system may be used for the measurement of DC, AC, and pulses.

10.1.5 Reflection Attenuator

A reflection attenuator is based on the reflection of the incident pulse at a mismatched boundary on the transmission line. The simplest implementation of a reflection attenuator is a transmission line of impedance Z_1 connected to an output line of impedance Z_2, shown in Figure 10.21. The incident voltage and current are represented by V_i and i_i. The reflected voltage and current on *line* 1 is V_r and i_r, and V_2 and i_2 are the parameters transmitted into the output *line* 2 with impedance Z_2.

An expression for the attenuation can be derived from the transmission line theory as follows.

$$V_2 = V_i + V_r \tag{10.41}$$

$$i_2 = i_i + i_r \tag{10.42}$$

$$i_2 = V_2/Z_2 \tag{10.43}$$

$$i_i = V_i/Z_1 \tag{10.44}$$

$$i_r = V_r/Z_1 \tag{10.45}$$

Substitution of Equations 10.43 through 10.45 into Equation 10.42 results in

$$\frac{V_2}{Z_2} = \frac{V_i}{Z_1} - \frac{V_r}{Z_1} \tag{10.46}$$

Substituting for V_2 from Equation 10.41 and simplifying yields

$$V_r = V_i \frac{Z_2 - Z_1}{Z_2 + Z_1} \tag{10.47}$$

$$V_2 = V_i \frac{2Z_2}{Z_2 + Z_1} \tag{10.48}$$

And

$$\text{Attenuation} = \frac{V_i}{V_2} = \frac{1}{2}\left(1 + \frac{Z_1}{Z_2}\right) \tag{10.49}$$

Equation 10.49 shows that the attenuation can be increased by increasing the mismatch between the two lines so that the input line has a high Z_1 and an output line having small Z_2. If a single, low-impedance line is unattainable, the output line can be made up of several lines in parallel.

10.2 Pulsed Current Measurement

10.2.1 Current Viewing Resistor

A current-viewing resistor (CVR) is a shunt with a highly accurate resistance R_{CVR} inserted in the current path. The voltage waveform developed across R_{CVR} is transmitted by a coaxial cable connected to an oscilloscope. In order for this signal to truly represent the current, the CVR should have negligible inductance and the resistive value should be low enough as not to disturb the original current flow. In addition to the proper design of the current shunt itself, the design of the measuring circuit comprising of coaxial cable and its connections to the current shunt and the oscilloscope should be such that no additional noise signals are picked up due to electromagnetic interference and reflections from terminations. This can be achieved through careful attention to the electromagnetic topology [56–59].

The important design parameters for a current shunt are the resistance R; the tolerance in resistance ΔR, the maximum energy dissipation; and the physical dimensions.

10.2.1.1 Energy Capacity

A CVR is rated for its energy capacity or maximum energy absorption E_{max}. This criterion is related to the heating of the resistive element, which affects the accuracy of the measuring resistance value R_{CVR}. For pulsed applications, the energy dissipated in the shunt-measuring resistance of the CVR, E_{CVR}, is

$$E_{CVR} = R_{CVR} \int i^2(t)dt \tag{10.50}$$

where the integral is calculated over the pulse duration. For square pulses, (10.50) is given by

$$E_{CVR} = R_{CVR} \times I_p^2 \times T_p \tag{10.51}$$

where I_p is the peak current and T_p is the pulse duration of the square pulse. The CVR should be chosen such that $E_{CVR} < E_{max}$. If E_{CVR} exceeds E_{max}, then changes in the calibrated value of R_{CVR} can occur and affect the accuracy of the measurement. The pulse energy capacity is the maximum energy dissipated in the CVR over a duration sufficiently short that losses are negligible.

Current viewing resistors in capacitor banks merit special consideration since a major fraction of the stored energy is dissipated external to the CVR. For a capacitor bank having a stored energy E_{Stored}

$$\frac{E_{CVR}}{E_{Stored}} = \frac{R_{CVR}}{R_{ext} + R_{CVR}} \tag{10.52}$$

where R_{ext} is the external resistance of the capacitor bank. Unfortunately, prior to measurement, R_{ext} is typically not known. However, from consideration of the peak current, practical CVR parameters such as resistance and output voltage,

$$\frac{E_{CVR}}{E_{Stored}} \sim \frac{1}{10} \tag{10.53}$$

for an underdamped capacitor bank [63]. Commercially, CVRs are available with measurement uncertainties of 0.2%.

10.2.1.2 Configurations

A current shunt is a four-terminal device, two input terminals for the main current and two terminals for the output voltage. The various configurations of the current shunt are shown in Figure 10.22. In Figure 10.22a are shown the folded strip shunt and folded twisted wire shunt, which are based on bifilar geometry [60–62]. The features of minimum enclosed volume and opposing current flow in the bifilar resistive element result in a minimum inductance configuration.

The tubular current shunts shown in Figure 10.23a are very popular [63–66]. Self-supporting tubular shunts can be mechanically strengthened by plotting

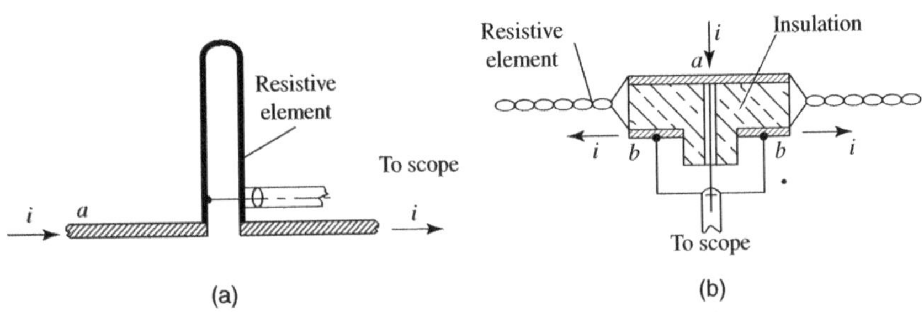

Figure 10.22 Two small volume, folded bifiliar shunt implementations for measuring currents. (a) Folded strip shunt. (b) Parallel twisted shunt.

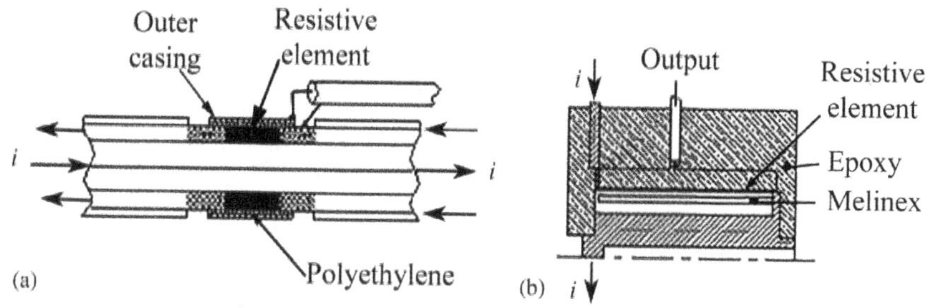

Figure 10.23 Self-supporting tubular current viewing resistors. (a) Series *F* current viewing resistor. (b) Coaxial cable shunt.

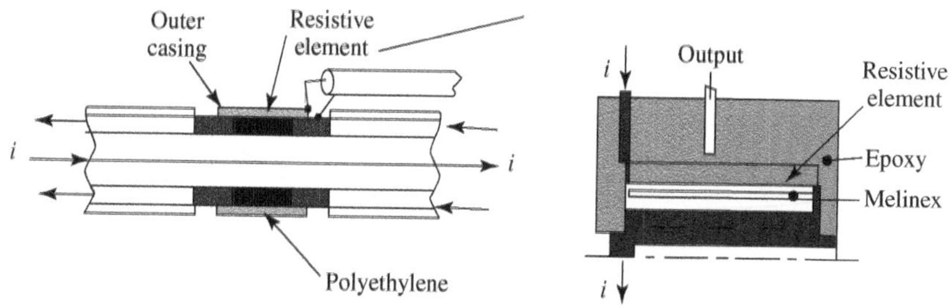

Figure 10.24 Externally reinforced tubular current viewing resistors.

the resistive element in an epoxy or by having a backup shell, as shown in Figure 10.23b. Tubular current shunts are commercially available with a variety of end connectors. The end connectors often determine the frequency response. The resistive elements may be made of materials such as nichrome, manganin, constantin, or stainless steel (Figure 10.24).

A radial graphite disc shunt, shown in Figure 10.25a, possesses a rise time of 350 ps [67]. Current shunts are also based on carbon composition discrete resistors or $CuSO_4$ solution resistors [68], as shown in Figure 10.25b and c. The variation of resistivity with current density for $CuSO_4$ solution resistors is shown in Figure 10.25c [68]. The behavior of resistivity versus current density for $CuSO_4$ and carbon composition resistors is shown in Figure 10.26 [68].

10.2.1.3 Tolerance in Resistance

Due to the energy dissipated E_{CVR} in the resistor and as a result of temperature rise T, the resistance changes from its initial value R_{CVR} to a new value R'. The tolerance ΔR_{CVR} is governed by

$$E_{CVR} = m \cdot C_p \cdot T = V \cdot \rho' \cdot C_p \cdot T \tag{10.54}$$

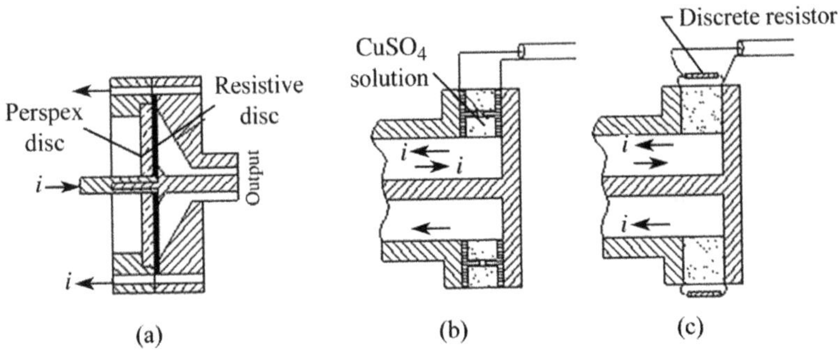

Figure 10.25 Disk current shunt configurations can use high-quality discrete resistors or those constituting liquids. (a) Radial disk current shunt. (b) Copper sulfate shunt. (c) Discrete resistor shunt.

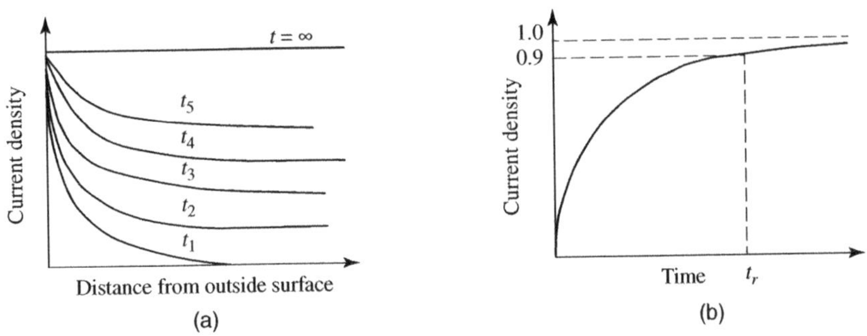

Figure 10.26 Current density distribution due to the skin effect.

where m is the mass of the resistor material, C_p is the specific heat of the resistor material, and p' is the mass density. The temperature coefficient of resistance α gives rise to

$$R' = R_{CVR}(1 + \alpha T) \tag{10.55}$$

from which

$$T = \frac{\Delta R_{CVR}}{\alpha R_{CVR}} \tag{10.56}$$

Substituting for T from 10.54 into 10.56 results in

$$\frac{\Delta R_{CVR}}{R_{CVR}} = \frac{\alpha E}{V \cdot \rho' \cdot C_p} \tag{10.57}$$

For a given current shunt, the tolerance in resistance can be evaluated, once the value of E_{CVR} is known from Equation 10.57.

10.2.1.4 Physical Dimensions

The length and area of cross section of the resistive element can be determined from the following relations:

$$R = \frac{\rho \cdot \ell}{A} \tag{10.58}$$

$$V = \ell \cdot A \tag{10.59}$$

From 10.58 and 10.59, we get

$$A = \sqrt{\frac{\rho \cdot V}{R}} \tag{10.60}$$

$$\ell = \sqrt{\frac{R \cdot V}{\rho}} \tag{10.61}$$

10.2.1.5 Frequency Response

Bandpass of a well-designed CVR can be very large with an upper limit determined by the skin effect in the resistive element, though efforts should be made to minimize the inductance of the CVR. As a result of the skin effect, the current takes a finite time to diffuse into the interior of the resistor volume. The distribution of current density as a function of distance from the outer surface is shown in Figure 10.26a. Figure 10.26b shows the effective current density averaged over the entire cross section as a function of time.

The rise time t_r contributed by the current diffusion time [64,69,70] is given by

$$t_r(\text{skin effect}) = \frac{\mu d^2}{4\rho} \tag{10.62}$$

where ρ is the resistivity of the material, μ is the permeability of the current shunt material, and d is the wall thickness of the tubular current shunt. Inspection of (10.60) shows that choosing a high-resistivity material with a thin wall will minimize the rise time due to the skin effect.

10.2.2 Rogowski Coil

In a Rogowski coil, shown in Figure 10.27, a conducting wire is wound around the minor axis of a toroid, which surrounds a time-varying current.

The rate of change of the magnetic flux density produced by the time-varying current induces a voltage in the conductor. This voltage can be quantitatively related to the current being measured. Rogowski coils can be constructed to either measure the current or its time derivative depending on the specific choices of components. However, both operate on the same principle: a

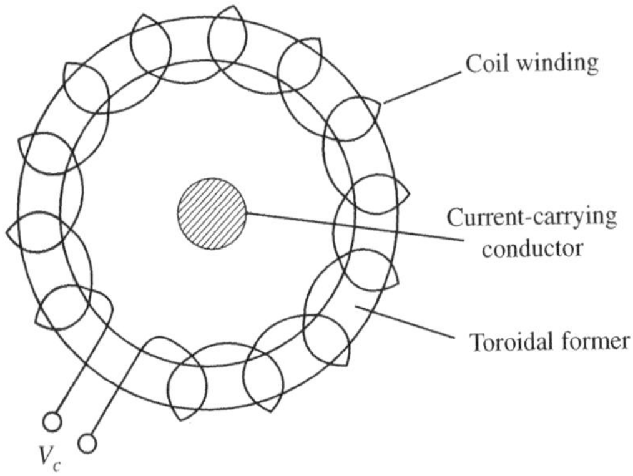

Figure 10.27 Principle of a Rogowski coil.

time-varying current through a wire produces an associate magnetic field and a time-varying magnetic field through a loop produces a voltage.

10.2.2.1 Voltage Induced in the Rogowski Coil

An analytical expression for the voltage induced in the Rogowski coil can be derived under the assumptions that an air core is employed. Additionally, it is assumed that the flux density over the entire cross section of the coil, the cross section of the coil over the entire length of the winding, and the winding density are constant.

To calculate the voltage induced in the Rogowski coil, the magnetic field from the current-carrying conductor is first calculated from Ampere's law,

$$\oint_C \vec{H} \cdot d\vec{\ell} = i \quad \text{(Ampere's law)}$$

where $\vec{H}$ is the magnetic field vector and i is the current enclosed by the contour C.

It can be shown that the magnetic field for a long wire is inversely proportional to the radial distance from the wire, directly proportional to the current i and in the polar $(\hat{\theta})$ direction. The magnetic field impinging on the Rogowski coil from the current travelling through the wire is

$$H(r,t) = \frac{i(t)}{2\pi r} \tag{10.63}$$

The magnetic field intensity $\vec{B}$ is related to the magnetic field by

$$\vec{B} = \mu\vec{H} \tag{10.64}$$

If the mean radius of the coil winding is r_0, then the value of the magnetic intensity from (10.63) and (10.64) is

$$B(r_0, t) = \frac{\mu\, i(t)}{2\pi\, r_0} \tag{10.65}$$

The magnetic flux density Φ_B through the N turns of area A of the Rogowski coil is given by

$$\Phi_B = N \oint \vec{B} \cdot d\vec{A}$$

which reduces to

$$\Phi_B(t) = \frac{\mu N A}{2\pi r_0} i(t) \tag{10.66}$$

Using Faraday's law

$$V_c = -\frac{d\Phi_B}{dt}$$

$$V_c = \frac{\mu N A}{2\pi r_0} \frac{di(t)}{dt} \tag{10.67}$$

Equation 10.65 implies that the voltage induced in the coil V_c is proportional to the derivative of the current being measured. Recognizing that $2\pi r_0 = \ell_{RC}$ is the mean circumference of the coil, (10.67) can be rewritten:

$$V_c = \mu A \times \left(\frac{N}{\ell_{RC}}\right) \times \frac{di(t)}{dt} \tag{10.68}$$

where N/ℓ_{RC} is the winding density of the coil.

It is important to note that provided the pulse length of the measured current is longer than the transit time of the Rogowski coil, the voltage induced in the coil is independent of the position of the current-carrying conductor within the window. Moreover, flux linkages are associated only with the minor turns, not with the major turns, and the results of Equation 10.68 are valid for any shape and cross section of the toroid provided the coil completely encloses the current. The case where the pulse width of the current being measured is shorter than the transit time of the Rogowski coil is addressed by Krompholtz [71].

10.2.2.2 Compensated Rogowski Coil

The simple Rogowski coil of Figure 10.26 is prone to errors caused by flux linkages ϕ' within the window parallel to the major axis. This situation arises when the current-carrying conductor is not perpendicular to the toroidal plane, and an additional voltage $Vc' = (d\Phi'/dt)$ is added to V_c. In practical Rogowski

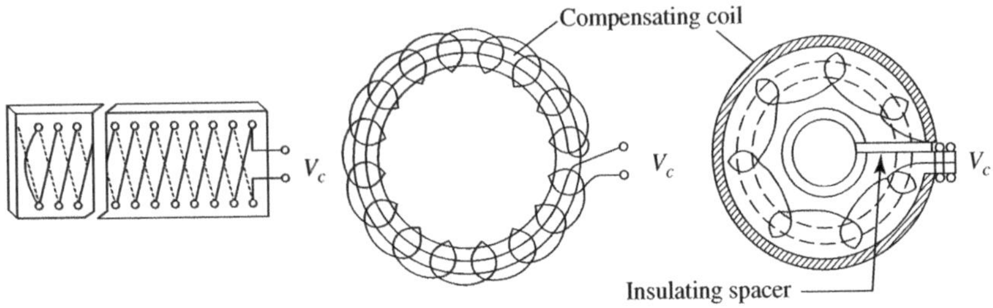

Figure 10.28 Compensated Rogowski coils may be constructed in a number of ways. (a) Bifiliar coil on a belt. (b) One-turn compensation within the coil. (c) One-turn compensation outside the coil.

coils, the voltage V'_c is cancelled by using bifilar winding [4,72,73] or by introducing an additional turn [74–77], as shown in Figure 10.28. However, the error associated with stray flux linkages to the minor cross section by extraneous current sources outside the window remains uncorrected but can be minimized by choosing small area.

10.2.2.2.1 Sensitivity

The sensitivity K of a differentiating Rogowski coil, defined as volts per amp, can be written as [77]

$$K = \frac{V_0}{i(t)} = \left(\frac{\mu_0 N a^2}{2r_0}\right) \cdot \frac{1}{RC} = \left(\frac{\mu_0 NA}{\ell}\right) \cdot \frac{1}{RC} \tag{10.69}$$

Equation (10.67) indicates that the overall sensitivity of the differentiating Rogowski depends on the coil constant ($\mu_o NA/\ell$) and integrator time constant (RC). A high sensitivity can be obtained by choosing a large minor cross section, a large winding density, and a low integrator time constant.

10.2.2.2.2 High-Frequency Response

The probe cannot record rise times faster than its inductive time constant, $L_c/(Z_o + R_p)$, where R_p corresponds to the coil resistance. As L_c is proportional to (N^2A), a reduction in N improves the rise time but reduces the sensitivity [78]. It is advisable to keep low N for good high-frequency response and then increase A for increased sensitivity. However, as the area increases, the sensitivity may be limited by noise from extraneous flux linkage with the coil's cross section.

10.2.2.2.3 Low-Frequency Response

The low-frequency response or the pulse width performance of a differentiating Rogowski depends on the integrator time constant (RC). To produce a minimum droop on the pulse width, RC should be chosen to be higher than 10 times the pulse width [10].

10.2.2.2.4 Advantages and Limitations

The differentiating Rogowski coil has good immunity from electromagnetic interference because of the high-voltage amplitudes propagated to the oscilloscope. It should be noted that the high-frequency components generated by the differentiating Rogowski coil are susceptible to frequency-dependent attenuation in the coaxial cable. Moreover, transmitted high (di/dt) values can induce sufficient voltage at the coil and cable termination to cause electrical breakdown and limit the highest measurable current or the shortest duration.

10.2.2.3 Self-Integrating Rogowski Coil

The equivalent circuit of a self-integrating Rogowski coil is shown in Figure 10.29, where no external integrator is employed. Instead, the coil inductance L_c in series with a low-value resistor R forms the L_cR integrating circuit. Under the conditions $\omega L_c \gg R$ and $L_c/R \gg \tau$, the entire coil voltage V_c is dropped across L_c.

The coil current (i_c) is given by

$$V_c = L_c \frac{di_c}{dt} \tag{10.70}$$

Substituting the value of V_c from 10.65 yields

$$\frac{\mu_0 NA}{\ell_{RC}} \frac{di}{dt} = L_c \frac{di_c}{dt}$$

which on simplification leads to

$$i_c = \frac{\mu_0 NA}{\ell_{RC} \cdot L_c} \cdot i \tag{10.71}$$

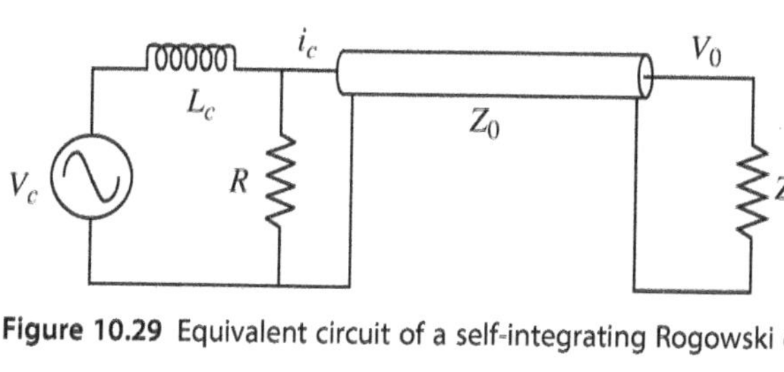

Figure 10.29 Equivalent circuit of a self-integrating Rogowski coil.

$$V_0 = i_c R = \frac{\mu_0 NA \cdot R}{\ell_{RC} \cdot L_c} \cdot i \tag{10.72}$$

It is seen from Equation 10.72 that the output voltage of a self-integrating Rogowski is directly proportional to the current being measured.

10.2.2.3.1 Sensitivity

The sensitivity of a self-integrating Rogowski is given by

$$\frac{V_0}{i} = \frac{\mu_0 NA}{\ell_{RC}} \cdot \frac{R}{L_c} \tag{10.73}$$

Since L is proportional to N^2A, the sensitivity is inversely proportional to N, which is in contrast to a differentiating Rogowski.

10.2.2.3.2 High-Frequency Response

The high-frequency response of a self-integrating Rogowski coil is not dependent on L_c/R as in the case of a differentiating Rogowski. Instead, it is limited by the self-resonant frequency (f_r) of the coil, values of the *LCR* circuit, and the transmission line properties of the helical [73,76,79–82] line formed between the coil conductor and the shield-compensating conductor. The value of f_r, determined by L_c and coil stray capacitance C_p should be higher than the highest-frequency component encountered. The best rise time performance is obtained by setting the *LCR* circuit values for critical damping [77,84] with R much smaller than the characteristic impedance of the helical line.

10.2.2.3.3 Low-Frequency Response

For a self-integrating Rogowski coil, the low-frequency response depends on the integrator time constant $L_c/(R+Z_o)$, with a high value resulting in a good pulse width response. The value of L_c can be substantially increased by employing a low-loss ferrite core [83]. For very fast rising pulse, with nanosecond rise time, the ferrite core is unable to respond and behaves like an air core, resulting in a good frequency response.

10.2.2.3.4 Advantages and Limitations

The advantages of self-integrating Rogowski coil are that it can respond to very fast rising pulses that are not limited by (L_c/R) value and extraneous voltages generated by stray flux linkages are rapidly damped [84] due to the very small value of R. It is particularly suitable for measuring currents in electron beam accelerators, where there is a likelihood of charge collection by electron impact. The major limitation of a self-integrating Rogowski is its low sensitivity.

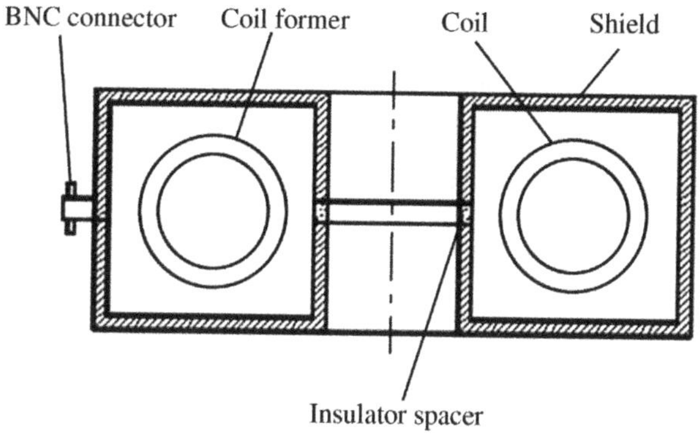

Figure 10.30 A shielded Rogowski coil.

10.2.2.4 Construction

For mechanical integrity and stability in geometry, the Rogowski coil is generally encapsulated in epoxy, which in addition improves the electrical insulation. Generally, the terminals are brought out on a BNC connector mounted on the coil. To reduce the extraneous noise due to electromagnetic interference, a metallic housing is used to serve as a shield. A circumferential slot provided on the shield, as shown in Figure 10.30, prevents short circuit while allowing the flux to link with the coil cross section.

10.2.3 Inductive (B-dot) Probe

A *B*-dot or inductive probe, shown in Figure 10.29, is a particular case of a Rogowski coil, where the coil does not form a closed loop. Although basically meant for spatial flux mapping, it can also be employed for pulsed current measurement [39,42,85,86]. The voltage induced in an inductive probe can be approximated to

$$V_c = \frac{\mu_0 NA}{2\pi r_0}\frac{di}{dt} \tag{10.74}$$

where N is the number of turns, A is the area of the coil, and r_o is the distance between current-carrying conductor and inductive probe.

Inductive or *B*-dot probes can be designed for external integration or self-integration (Figure 10.31). The disadvantages of an inductive probe are that it is very sensitive to external flux linkage and mounting the probe such that the flux lines pass perpendicular to the coil cross section is difficult. Care must be taken since a *B*-dot probe needs fresh *in situ* calibration whenever the distance r_o is charged. A *B*-dot probe can easily be made by connecting one or more turns to the inner and outer conductors of a commercial cable.

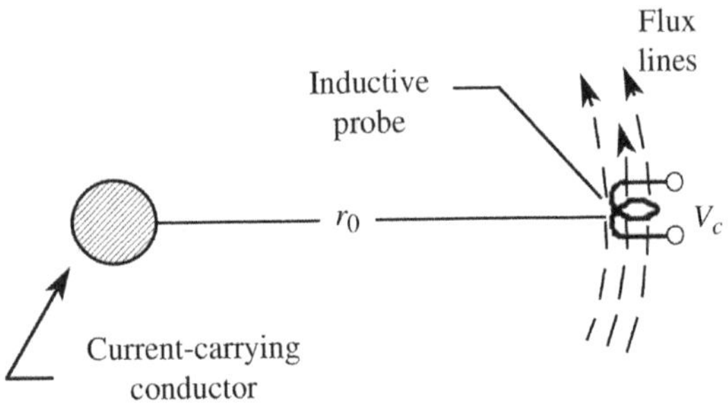

Figure 10.31 Pulsed current measurement by an inductive probe.

10.2.4 Current Transformer

A current transformer is a particular case of a Rogowski coil, employing iron core or ferrite core instead of air core, which results in a coupling coefficient of near unity. Under the conditions that $\omega L_s \gg R_\ell$, where R_ℓ is the resistance in the secondary circuit where the output voltage V_o is developed, V_o can be written as [83]

$$V_0 = i_p(t)\frac{N_p}{N_s}R_\ell \tag{10.75}$$

where i_p is the primary current, N_p is the primary turns, and N_s is the secondary turns.

For a single-turn primary, Equation 10.69 simplifies to

$$V_0 = \frac{i_p(t) \cdot R_\ell}{N_s} \tag{10.76}$$

As a result of including the ferromagnetic core, the value of L_s, and therefore (L_s/R), is increased, improving the low-frequency response. The increase in the magnetic permeability improves the sensitivity and hence extends its usefulness for the measurement of low currents.

On the other hand, the ferromagnetic core limits the high-frequency response due to a lowered secondary resonant frequency $[f_r = 1/2\pi\sqrt{(L_s C_s)}]$ and the core saturation sets a limit on the product of current and time. Current transformers, used to measure alternating currents on high-voltage power transmission lines, need a careful insulation design.

10.2.5 Magneto-optic Current Transformer

The advantages of optical current transformers over the conventional measurements arise from the use of optical fibers instead of electrically conducting wires, allowing the optical current transformer to be used in unique experimental

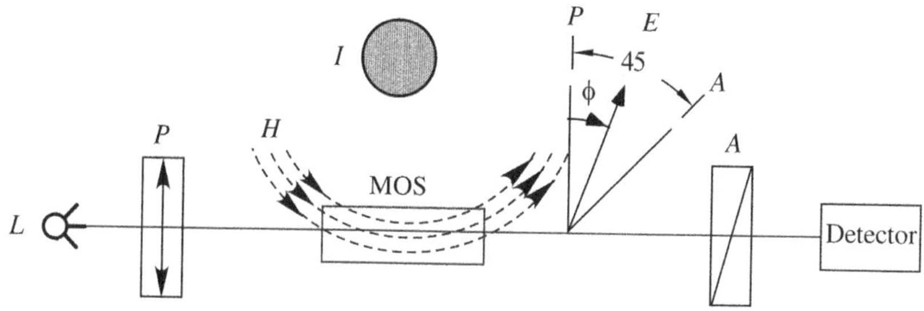

Figure 10.32 Scheme of a magneto-optic current transformer where *L* is the light source, *P* is the polarizer, *MOS* is the magneto-optic sensor, *I* is the current-carrying conductor, *H* is the magnetic field produced by the conductor, and *A* is the analyzer.

conditions. Advances and miniaturization in monochromatic light sources, optical components, and optoelectronic detectors will only increase the compactness of magneto-optic current transformers.

10.2.5.1 Basic Principles

A magneto-optic current transformer configured to measure a time-varying current is shown in Figure 10.32 [87,88]. It is based on Faraday effect, where the plane of polarization of a linearly polarized light, passing through a magneto-optic material subjected to a longitudinal magnetic field H is rotated by an angle ϕ [73,89]. The Faraday rotation ϕ is given by

$$\phi = V \int_0^{\ell} H \cdot d\ell \tag{10.77}$$

where V is the Verdet's constant and ℓ is the length of magneto-optic material.

If H is constant along the length, Equation 10.71 may be written as

$$\phi = V \cdot H \cdot \ell \tag{10.78}$$

Thus, Faraday rotation consists of experimentally determining the rotation angle ϕ from which H can be derived. From H, the current i can be obtained. The methods of determining ϕ in the optical mode are quite complicated, so optoelectronic [90–92] techniques are usually employed. The two-step process for obtaining the rotation angle ϕ are: (1) the analyzer converts the rotation angle into intensity modulation of the light beam and (2) the detector, consisting of fast photodiodes and associated electronics, converts the optical signal transmitted by the analyzer into an electrical signal whose voltage is proportional to the intensity of light. Phase modulation and interferometric fringe techniques have also been used to convert the Faraday rotation angle into a proportionate electrical signal [93,94].

10.2.5.2 Intensity Relations for Single-Beam Detector

The analyzer is usually oriented at 45° to the plane of polarization set by the polarizer. If E_o is the electric field vector of the incident light, the component of E_o along the polarization plane of the analyzer is given by

$$E_a = E_0 \cos(45 - \phi) = \frac{E_0}{\sqrt{2}}(\sin\phi + \cos\phi) \tag{10.79}$$

Because the intensity of light is proportional to E^2, the light intensity J_a at the output of the analyzer is given by

$$J_a = \frac{J_0}{2}(1 + \sin 2\phi) \tag{10.80}$$

Equation 10.74 shows that the relationship between J_a and ϕ is nonlinear. However, for a restricted range in ϕ, the error from linearity is very small. The drawback of the single-beam analyzer scheme is that the output signal is dependent on the intensity of the input light, the variation of which causes a variation in the output signal. However, this error can be eliminated by using a differential split-beam arrangement.

10.2.5.3 Intensity Relations for Differential Split-Beam Detector

A schematic of a differential split-beam detector is shown in Figure 10.33. In this arrangement, the analyzer A in the form of a Wollaston prism splits the incoming linearly polarized light into two mutually orthogonal beams J_1 and J_2, each of them at 45° to the polarizer plane fixed by P.

The intensities J_1 and J_2 are now given by

$$J_1 = \frac{J_0}{2}(1 + \sin 2\phi) \tag{10.81}$$

$$J_2 = \frac{J_0}{2}(1 - \sin 2\phi) \tag{10.82}$$

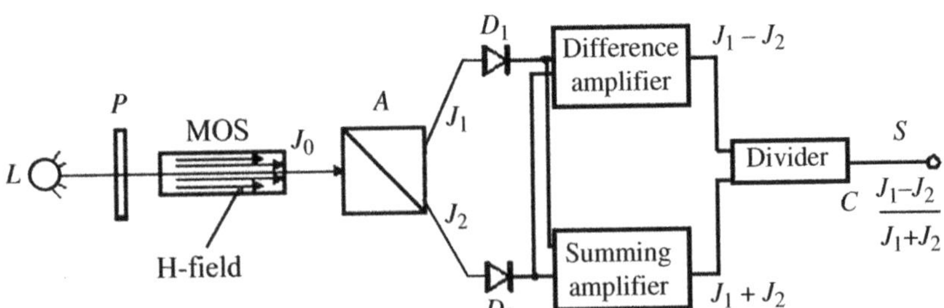

Figure 10.33 Differential split-beam detector where L is the light source, P is the polarizer, MOS is the magneto-optic sensor, A is the analyzer (a Wollaston prism), and D_1 and D_2 are photodetectors.

The output electrical signal S after detector electronics is given by

$$S = C\frac{J_1 - J_2}{J_1 + J_2} = C \cdot \sin 2\phi \tag{10.83}$$

It is seen from Equation 10.83 that the output signal S is now independent of J_o, thus making the system immune to input intensity variations.

10.2.5.4 Light Source

The requirements to be satisfied by the light source are long life, high radiance, good coherence, small divergence angle, and monochromaticy. Common light sources are spectrum lamps with monochromatic filters, lasers, laser diodes, and LEDs.

10.2.5.5 Magneto-optic Sensor

The magneto-optic [73,95–101] sensor should have low loss and little or no birefringence that would produce a phase retardation superimposed on Faraday rotation angle and transparency to light at the probe wavelength. A large Verdet's constant avoids the need for multiple passes of light in the magneto-optic sensor [102]. Magneto-optic sensors are usually made of diamagnetic- or paramagnetic-doped glasses such as lead glass, quartz glass, crown glass, flint glass, or rare earth garnets like YIG, TAG, or TGG. Single-mode optical fibers [103–110] are frequently used in magneto-optic current transformer applications.

10.2.5.6 Frequency Response

The relaxation time [106] of Faraday effect in glasses is about 10^{-10} s, which means that for most practical purposes the upper frequency limit is set by the photomultiplier/photodiode and other electronics.

10.2.5.7 Device Configurations

Magneto-optical transformers can take a variety of configurations [43,53–55,107,108,111–114]. There are three examples. In the first, the Faraday cell completely surrounds the current-carrying conductor, as shown in Figure 10.34a. In this case, the effective optical length corresponding to ℓ in Equation 10.72 is $2(a + b)$. The sensitivity of the device is increased by increased path length of the light beam. In such devices, employing bulk material for the Faraday cell, where there are multiple reflections in the interior, care needs to be taken to minimize the "reflection-induced phase difference." In the second example, the Faraday cell is placed in the air gap of a ferromagnetic core, as shown in Figure 10.34b. The value of H is thus increased substantially due to the high permeability of the core; therefore, the sensitivity in this case is increased

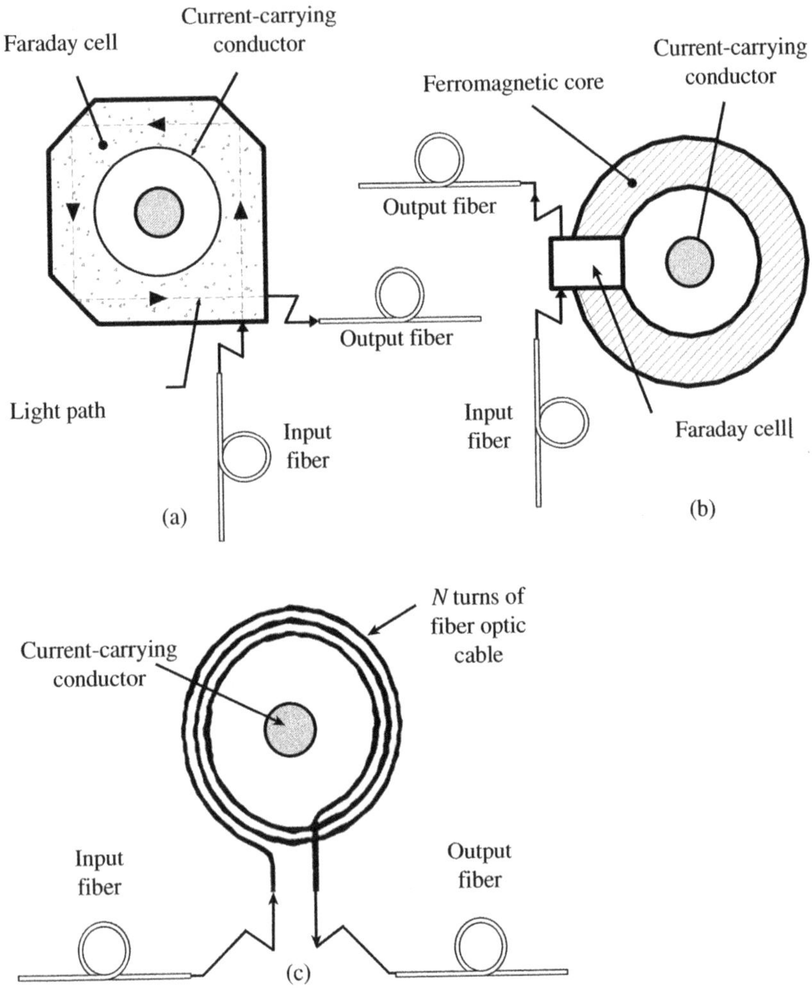

Figure 10.34 Faraday sensors can take a variety of forms. (a) Faraday cell surrounding the current-carrying conductor. (b) Faraday cell in an air gap of a ferromagnetic core. (c) Fiber optic cable wound around the current-carrying conductor as a Faraday sensor.

by increasing H rather than ℓ. In the third example, the Faraday cell is formed by winding N turns of optical fiber around the current-carrying conductor, as shown in Figure 10.34c. The effective path length of the light beam through the fiber is given by (πDN), where D is the diameter of the bobbin on which the fiber is wound. For such a configuration, care has to be taken to minimize the intrinsic birefringence and bend-induced linear birefringence in the fibers; otherwise the sensitivity will be affected.

10.3 Design Examples

Example 10.1

A resistive voltage divider shown in Figure 10.18a has 10 resistors, each of 10 MΩ in its high-voltage arm. The low-voltage arm has a resistor of 100 kΩ in parallel with a capacitor of 10 nF. The divider is used to measure the input voltage waveform with a maximum voltage of 400 kV. Calculate the following: (i) attenuation of the divider, (ii) voltage and wattage rating of the resistors in the high-voltage arm, and (iii) capacitances to be connected in parallel with each resistor in the high-voltage arm.

Solution

Attenuation of the divider is given by

$$A = \frac{V_o}{V_i} = \frac{R_2}{R_1 + R_2}$$

$$A = \frac{(100\ \mathrm{k\Omega})}{(10\ \mathrm{M\Omega}) + (100\ \mathrm{k\Omega})} = \frac{1}{1000}$$

Voltage and wattage rating of each resistor are given by

$$\text{Voltage rating} = \frac{(400\ \mathrm{kV})}{10} = 40\ \mathrm{kV}$$

$$\text{Power rating} = \frac{V^2}{R} = \frac{\left(40 \times 10^3\right)^2}{10 \times 10^6} = 160\ \mathrm{W}$$

The condition for frequency independence as per Equation 10.16 is given by

$$C_1R_1 = C_2R_2 = \left(10 \times 10^{-9}\right) \cdot \left(10 \times 10^3\right)s = 1\ \mathrm{ms}$$

Hence,

$$C_1 = \frac{\left(10^{-3}\right)}{\left(10 \times 10 \times 10^6\right)} = 10\ \mathrm{pF}$$

Hence, capacitance (C_x') to be connected across each resistor in the high-voltage arm is given by

$$C_x' = 10 \times 10\ \mathrm{pF} = 100\ \mathrm{pF}$$

Example 10.2

A coaxial floating electrode capacitor divider, employing water dielectric ($\varepsilon_r = 80$) ($\varepsilon_r = 80$), shown in Figure 10.16, has the following dimensions: diameter of the high-voltage electrode: $D_1 = 150$ mm; inner diameter of the grounded electrode: $D_2 = 250$ mm; diameter of the floating electrode: $D_f = 240$ mm; and the thickness = 100 mm.

Calculate (i) attenuation of the divider and (ii) errors in voltage measurement when the low-voltage arm is terminated into impedance of 50 Ω and 1 MΩ, respectively.

Solution

i) The attenuation is given by

$$K = \frac{V_0}{V_i} = \frac{C_1}{C_1 + C_2}$$

where C_1 as per Equation 3.2 is given by

$$C_1 = \frac{2\pi\varepsilon_0\varepsilon_r}{\ln(D_f/D_1)} \times \ell$$

$$= 2\pi \frac{(8.85 \times 10^{-12})(80)}{\ln(240/150)} \times 10 \times 10^{-3} = 946 \text{ pF}$$

Similarly,

$$C_2 = \frac{2\pi\varepsilon_0\varepsilon_r}{\ln(D_2/D_f)} \times \ell$$

$$= \left(\frac{2\pi(8.85 \times 10^{-12})(80)}{\ln(250/240)}\right) \times 100 \times 10^{-3} = 11 \text{ nF}$$

Hence,

$$\frac{V_0}{V_i} = \frac{C_1}{(C_1 + C_2)} = \frac{946 \times 10^{-12}}{(946 \times 10 - 12 + 11 \times 10^{-9})} = 0.085$$

ii) The error in voltage measurement is given by

$$\delta V = V_i\left(\frac{C_1}{C_1 + C_2}\right) \cdot \left(1 - e^{-(t/T)}\right)$$

a) For 50 Ω termination, the time constant (T) for the low-voltage arm is given by

$$T = C_2R_2 = \left(11 \times 10^{-9}\right) \times (50) = 550\ \text{ns}$$

Hence, at $(t) = T = 550$ ns

$$\delta V = V_i \times 0.085 \times \left(1 - e^{-1}\right) = 0.058V_i$$
$$= 5.8\%\ \text{of the input voltage}$$

b) For 1 MΩ termination, the time constant T for the low-voltage arm is given by

$$T = C_2R_2 = \left(11 \times 10^{-9}\right) \times \left(1 \times 10^{6}\right) = 11\ \text{ms}$$

Hence, the error δV in the voltage will be negligible at 550 ns, but it will be 5.8% for a long-duration pulse at $(t) = 11$ ms. The error numbers in the above calculations indicate the reasons why a capacitor divider is generally not terminated into a 50 Ω impedance, but instead into a 1 MΩ impedance.

Example 10.3

A Rogowski coil has 100 turns wound around an air-cored toroid having a mean major radius of 30 mm and a minor radius of 5 mm. The peak input current to be measured is 1.0 A at 1 MHz. The Rogowski output is connected to a passive RC integrator with a time constant of $100T$, where T is the oscillatory period of the input current. Calculate the following: (i) inductance of the coil, (ii) coil output voltage, and (iii) sensitivity of the Rogowski coil.

Solution

i) The input current can be represented by

$$i(t) = i_0 \sin \omega t$$

where

$$i_0 = 1\ \text{A}, \quad \omega = 2\pi f = 2\pi \times 10^6, \quad T = \frac{1}{f} = 10^{-6} = 1\ \mu\text{s}$$

Inductance L is given by

$$L = \frac{\mu_0 A \cdot N^2}{\ell}$$

(continued)

(continued)

where

$$\mu_0 = 4\pi \times 10^{-7} \frac{H}{m}$$

A = minor cross section of toroid

$$= \pi a^2 = \pi\left(5 \times 10^{-3}\right)^2 = 0.785 \times 10^{-4}\ \mathrm{m}^2$$

ℓ = mean magnetic path length in the core

$$= 2\pi r_0 = 2\pi\left(30 \times 10^{-3}\right) = 0.1885\ \mathrm{m}$$

$N = 100$

and

$$L = 5.2\ \mu\mathrm{H}$$

ii) The coil output as per Equation 10.61 is given by

$$V_c = \left(\frac{\mu_0 NA}{\ell}\right) \cdot \left(\frac{di}{dt}\right) = \left(\frac{\mu_0 NA}{\ell}\right) i_0 \omega \cos(\omega t)$$

Substituting the values of μ_0, N, A, ℓ, i_0, and ω in the expression for V_c, we get

$$V_c = 0.329 \cos(\omega t)$$

iii) Integrator output as per Equation 10.62 is given by

$$V_0 = \left(\frac{\mu_0 NA}{\ell}\right) \cdot \left(\frac{1}{RC}\right) \cdot i \cdot (t)$$

$$= \left(\frac{\left(4\pi \times 10^{-7}\right) \times (100) \times \left(0.785 \times 10^{-4}\right)}{0.1885}\right) \times \left(\frac{1}{100 \times 10^{-6}}\right) i_0 \sin(\omega t)$$

$$= 5.2 \times 10^{-3} \sin \omega t$$

iv) The sensitivity of the Rogowski is given by

$$K = \frac{V_0}{i(t)} = \frac{5.2 \times 10^{-3} \sin \omega t}{1 \sin \omega t} = 5.2\ \frac{\mathrm{mV}}{\mathrm{A}}$$

References

1 J.D. Craggs and J.M. Meek, *High Voltage Laboratory Technique*, Butterworths, London, 1954.

2 W.G. Hawlay, *Impulse Voltage Testing*, Chapman & Hall Ltd., London, 1959.

3 F.B.A. Frungel, *High Speed Pulse Technology*, Vol. II, Academic Press, New York, 1965.

4 A.J. Schwab, *High Voltage Measurement Techniques*, MIT, UK, 1972.
5 IEEE Inc., 4–1978: IEEE Standard Techniques for High Voltages Testing, 1978.
6 J.M. Meek and J.D. Craggs, *Electrical Breakdown of Gases*, Oxford University Press, London, 1953.
7 L.W. Chubb and C. Fortescue, Calibrating of the Sphere Gap Voltmeter. *Trans. AIEE*, Vol. 32, p. 739, 1913.
8 R. Davis, G.W. Bowdler, and W.G. Standring, The Measurement of High Voltages with Special Reference to the Measurement of Peak Voltages. *Proc. IEE*, Vol. 68, p. 1222, 1930.
9 W.P. Baker, A Novel High Voltage Peak Voltmeter. *Proc. IEE*, Vol. A103, p. 519, 1956.
10 R.J. Thomas, High Impulse Current and Voltage Measurement. *IEEE Trans. Instrum. Meas.*, Vol. 19, No. 2, p. 102, 1970.
11 P.R. Howard, Errors in Recording Surge Voltages. *Proc. IEE, Part I*, Vol. 99, p. 371, 1952.
12 P.L. Bellaschi, The Measurement of High Surge Voltages. *Trans. AIEE*, Vol. 52, p. 544, 1933.
13 H.D. Sutphin, Subnanosecond High Voltage Attenuator. *Rev. Sci. Instrum.*, Vol. 43, No. 10, p. 1535, 1972.
14 G.R. Mitchel and C. Melancon, Subnanosecond Protection Circuits for Oscilloscope Inputs. *Rev. Sci. Instrum.*, Vol. 56, No. 9, p. 1804, 1985.
15 R.E. Dollinger and D.L. Smith, Shielded High Voltage Probes. *IEEE Trans. Electron Devices*, Vol. 26, No. 10, p. 1553, 1979.
16 Y. Kubota et al., A Quick Response High Voltage Divider. *Jpn. J. Appl. Phys.*, Vol. 15, No. 10, p. 2037, 1976.
17 D.G. Pellinen and S. Heurlin, A Nanosecond Risetime Megavolt Divider. *Rev. Sci. Instrum.*, Vol. 42, No. 6, p. 824, 1971.
18 D.G. Pellinen and I. Smith, A Reliable Multimegavolt Voltage Divider. *Rev. Sci. Instrum.*, Vol. 43, No. 2, p. 299, 1972.
19 D.G. Pellinen et al., A Picosecond Risetime High Voltage Divider. *Rev. Sci. Instrum.*, Vol. 45, No. 7, p. 944, 1974.
20 H. Matsuzawa and T. Akitsu, Output Voltage Improvement of the Coaxial Marx-Type High Voltage Generator. *Rev. Sci. Instrum.*, Vol. 56, No. 12, p. 2287, 1985.
21 Z.Y. Lee, Subnanosecond High Voltage Two Stage Resistive Divider. *Rev. Sci. Instrum.*, Vol. 54, No. 8, p. 1060, 1983.
22 P.H. Ron, Rise and Delay Time of a Vacuum Gap Triggered by an Exploding Wire, PhD thesis, Indian Institute of Science, Bangalore, India, 1984.
23 R.J. Thomas, Response of Capacitive Voltage Dividers. *Microwaves*, Vol. 6, pp. 50–53, 1967.
24 E. Slamecka and W. Waterscheck, Transient Response of Impulse Voltage Dividers. *Siemens Z. (in German)*, Vol. 41, p. 63, 1967.

25 How to Make Accurate High Voltage AC Measurements, Application Notes, Model 13200 Voltage Divider, IIT, Jennings.
26 M.M. Brady and K.G. Dedrick, High Voltage Pulse Measurement with a Precision Capacitive Voltage Divider. *Rev. Sci. Instrum.*, Vol. 33, No. 12, p. 1421, 1962.
27 W.R. Fowkes and R.M. Rowe, Refinements in Precision Kilovolt Pulse Measurement. *IEEE Trans. Instrum. Meas.*, Vol. 15, No. 4, p. 284, 1966.
28 D.M. Barrett, S.R. Byron, E.A. Crawford, D.H. Ford, W.D. Kimura, and M.J. Kushner, Low Inductance Capacitive Probe for Spark Gap Voltage Measurements. *Rev. Sci. Instrum.*, Vol. 56, No. 11, p. 211, 1985.
29 R.D. Genuario and J.C. Blackburn, 300 kV Rectangular Pulse Generator with Nanosecond Rise Time. *Rev. Sci. Instrum.*, Vol. 45, No. 12, p. 1546, 1974.
30 N.W. Harris, High Voltage Probe for Liquid Immersion. *Rev. Sci. Instrum.*, Vol. 45, No. 7, p. 961, 1974.
31 E.A. Edson and G.N. Detzal, Capacitive Voltage Divider for High Voltage Pulse Measurement. *Rev. Sci. Instrum.*, Vol. 52, No. 4, p. 604, 1981.
32 C.A. Ekdahl, Voltage and Current Sensors for a High Density Z-Pinch Experiment. *Rev. Sci. Instrum.*, Vol. 51, No. 12, p. 1645, 1980.
33 W.Z. Fam, A Novel Transducer to Replace Current and Voltage Transformers in High Voltage Measurements. *IEEE Trans. Instrum. Meas.*, Vol. 45, No. 1, p. 190, 1996.
34 P. Osmokrovic, D. Petkovic, and O. Markovic, Measuring Probe for Fast Transients Monitoring in Gas Insulated Substations. *IEEE Trans. Instrum. Meas.*, Vol. 46, No. 1, p. 36, 1997.
35 C.S. Wong, Simple Nanosecond Capacitive Divider. *Rev. Sci. Instrum.*, Vol. 56, No. 5, p. 767, 1985.
36 F.P. Burch, On Potential Dividers for Cathode Ray Oscillographs. *Philos. Mag.*, Vol. 13, p. 760, 1932.
37 R. Keller, Wideband High Voltage Probe. *Rev. Sci. Instrum.*, Vol. 35, No. 8, p. 1057, 1964.
38 G.G. Wolzak et al., Capacitive Measurement of High DC Voltages. *Rev. Sci. Instrum.*, Vol. 52, No. 10, p. 1572, 1981.
39 C.E. Baum, Electromagnetic Sensors and Measurement Techniques, in *Fast Electrical and Optical Measurements*, NATO ASI Series E: Applied Sciences, J.E. Thompson and L.H. Luessen, eds., Vol. 1, Martinus Nijhoff Publishers, pp. 73–144, 1986.
40 L.K. Warne, L.I. Basilio, W.A. Johnson, M.E. Morris, M.B. Higgins, and J.M. Lehr, Capacitance and Effective Area of Flush Monopole Probes, Sandia National Laboratories Internal Report, SAND2004-3994, 2004.
41 W. Pfeiffer, Ultrafast Electrical Voltage and Current Monitors, in *Fast Electrical and Optical Measurements*, NATO ASI Series E: Applied Sciences, J.E. Thompson and L.H. Luessen, eds., Vol. 1, Martinus Nijhoff Publishers, pp. 145–174, 1986.

42 M.S. DiCapua, High Speed Electric Field and Voltage Measurements, in *Fast Electrical and Optical Measurements*, NATO ASI Series E: Applied Sciences, J.E. Thompson and L.H. Luessen, eds., Vol. 1, Martinus Nijhoff Publishers, pp. 175–222, 1986.

43 C.E. Baum, E.L. Breen, J.C. Giles, J.P. O'Neill, and G.D. Sower, Sensors for Electromagnetic Pulse Measurement Both Inside and Away from Nuclear Source Regions. *IEEE Trans. Antennas Propag.*, Vol. 26, No. 1, pp. 22–35, 1978.

44 J.C. Santos, M.C. Taplamacioglu, and K. Hidaka, Pockels High Voltage Measurement System. *IEEE Trans. Power Deliv.*, Vol. 15, No. 1, p. 8, 2000.

45 D.C. Wunsch and A. Erteze, Kerr Cell Measuring System for High Voltage Pulses. *Rev. Sci. Instrum.*, Vol. 35, No. 7, p. 816, 1964.

46 W. Botticher et al., Electro-Optical Measurement of Current and Voltage in Fast High Pressure Glow Discharges. *J. Phys. E: Sci. Instrum.*, Vol. 11, p. 248, 1978.

47 E.C. Cassidy and H.N. Cones, A Kerr Electro-Optical Technique for Observation and Analysis of High Intensity Electric Fields. *J. Res. Natl. Bur. Stand.*, Vol. 73, Nos. 1 and 2, p. 5, 1969.

48 E.E. Bergmann and G.P. Kolleogy, Measurement of Nanosecond HV Transients with the Kerr Effect. *Rev. Sci. Instrum.*, Vol. 48, No. 12, p. 1641, 1977.

49 D.F. Nelson, The Modulation of Laser Light. *Sci. Am.*, Vol. 218, No. 6, p. 17, 1968.

50 E.C. Cassidy et al., Calibration of a Kerr Cell System for High Voltage Pulse Measurements. *IEEE Trans. Instrum. Meas.*, Vol. 17, No. 4, p. 313, 1968.

51 E.C. Cassidy, Development and Evaluation of Electrooptical High Voltage Pulse Measurement Techniques. *IEEE Trans. Instrum. Meas.*, Vol. 19, No. 4, p. 395, 1970.

52 E.C. Cassidy, Pulsed Laser Kerr System Polarimeter for Electrooptical Fringe Pattern Measurement of Transient Electrical Parameters. *Rev. Sci. Instrum.*, Vol. 43, No. 6, p. 886, 1972.

53 S.Y. Ettinger and A.C. Venezia, High Voltage Pulse Measuring System Based on Kerr Effect. *Rev. Sci. Instrum.*, Vol. 34, No. 3, p. 221, 1963.

54 M. Kanoi, G. Takahashi, T. Sato, M. Higaki, E. Mori, and K. Okumara, Optical Voltage and Current Measuring System for Electric Power Systems. *IEEE Trans. Power Deliv.*, Vol. 1, No. 1, p. 91, 1986.

55 Cruden, A. Richardson, Z.J. Macdonald, J.R. Andonovic, I. Laycock, W., and Bennett, A., Compact 132 kV Combined Optical Voltage and Current Measurement System. *IEEE Trans. Instrum. Meas.*, Vol. 47, No. 1, p. 219, 1998.

56 T. Mitsui, K. Hosoe, H. Usami, and S. Miyamoto, Development of Fiberoptic Voltage Sensors and Magnetic Field Sensors. *IEEE Trans. Power Deliv.*, Vol. 2, No. 1, p. 87, 1987.

57 S. Kobayashi, A. Horide, I. Takagi, M. Higati, G. Takahashi, E. Mori, and Y. Yamagowa, Development and Field Test Evaluation of Optical Voltage and Current Transformers for Gas Insulated Switchgear. *IEEE Trans. Power Deliv.*, Vol. 7, No. 2, p. 815, 1992.

58 C.E. Baum, Electromagnetic Topology for the Analysis and Design of Complex Electromagnetic Systems, in *Fast Electrical and Optical Measurements*, NATO ASI Series E: Applied Sciences, J.E. Thompson and L.H. Luessen, eds., Vol. 1, Martinus Nijhoff Publishers, pp. 467–548, 1986.

59 C.E. Baum, E.L. Breen, F.L. Pitts, G.D. Sower, and M.E. Thomas, The Measurement of Lightening Environmental Parameters Related to Interaction with Electronic Systems. *IEEE Trans. Electromagn. Compat.*, pp. 123–137, 1982.

60 W. Graf, System Design: Practical Shielding and Grounding Techniques Based on Electromagnetic Topology, in *Fast Electrical and Optical Measurements*, NATO ASI Series E: Applied Sciences, J.E. Thompson and L.H. Luessen, eds., Vol. 1, Martinus Nijhoff Publishers, pp. 567–583, 1986.

61 E.F. Vance, Electromagnetic-Interference Control. *IEEE Trans. Electromagn. Compat.*, Vol. 22, No. 4, pp. 319–328, 1980.

62 R.J. Thomas, High Voltage Pulse Reflection-Type Attenuator with Subnanosecond Response. *IEEE Trans. Instrum. Meas.*, Vol. 16, No. 2, p. 146, 1967.

63 J.H. Park, Shunts and Inductors for Surge Current Measurement. *J. Res. Natl. Bur. Stand.*, Vol. 39, p. 191, 1947.

64 F.B. Silsbee, Notes on the Design of Four Terminal Resistance Standards for AC. *J. Res. Natl. Bur. Stand.*, Vol. 4, p. 73, 1930.

65 P.L. Bellaschi, Heavy Surge Currents, Generating and Measurement. *Electr. Eng.*, Vol. 53, p. 86, 1934.

66 Catalog on "Coaxial Shunts for Surge Current and High Frequency Measurement: Nanoseconds to Megamps," T & M Research Products, USA.

67 E. Thornton, A Metal-Foil Shunt for Measuring Submicrosecond Duration High Current Pulses. *J. Phys. E: Sci. Instrum.*, Vol. 3, p. 962, 1970.

68 E. Thornton, Subnanosecond Rise Time in Metal-Foil Coaxial Shunts. *J. Phys. E: Sci. Instrum.*, Vol. 8, p. 1052, 1975.

69 T&M Research, http://www.tandmresearch.com/index.php?page=products; last accessed 9/30/2016.

70 L. Hogberg, 1 GC/s Bandwidth Shunts for Measurement of Current Transients in the 10 A–10 kA Range. *J. Sci. Instrum.*, Vol. 42, p. 273, 1965.

71 P.H. White and B.R. Withey, Carbon Composition and Copper Sulphate Shunts for the Measurement of Submicrosecond High Current Pulses. *J. Phys. E: Sci. Instrum.*, Vol. 3, p. 757, 1970.

72 F.D. Bennett and J.W. Marvin, Current Measurement and Transient Skin Effect in Exploding Wire Circuits. *Rev. Sci. Instrum.*, Vol. 33, No. 11, p. 1218, 1962.

73 H. Krompholz et al., Nanosecond Current Probe for High Voltage Experiments. *Rev. Sci. Instrum.*, Vol. 55, No. 1, p. 127, 1984.
74 R. Malewski, New Device for Current Measurement in Exploding Wire Circuits. *Rev. Sci. Instrum.*, Vol. 39, No. 1, p. 90, 1968.
75 E.S. Wright and R.G. John, Miniature Rogowski Coil Probes for Direct Measurement of Current Density Distribution in Transient Plasma. *Rev. Sci. Instrum.*, Vol. 36, p. 1891, 1965.
76 H. Knoepfel, *Pulsed High Magnetic Fields*, North Holland Pub. Co., 1970.
77 B. McCormack and J.P. Thomas, Design, Fabrication and Testing of the TFTR Rogowski Coil and Diamagnetic Loop. *9th Symposium on Engineering Problems in Thermonuclear Fusion*, Vol. II, p. 1130, 1981.
78 W.J. Sarjieant and E. Branner, A Current Transformer for Microsecond Pulses. *Proc. IEEE*, p. 359, 1968.
79 J. Cooper, On the High Frequency Response of a Rogowski Coil. *J. Nucl. Energy, Part C: Plasma Phys.*, Vol. 5, p. 285, 1963.
80 D. Pellinen, A Segmented Faraday Cup to Measure kA/cm^2 Electron Beam Distributions. *Rev. Sci. Instrum.*, Vol. 43, No. 11, p. 1654, 1972.
81 G. Dworschak et al., Production of Pulsed Magnetic Fields with a Flat Pulse Top of 440 kOe and 1 msec Duration. *Rev. Sci. Instrum.*, Vol. 45, No. 2, p. 243, 1974.
82 G. Decker and D.L. Honea, Magnetic Probes with Nanosecond Response Time for Plasma Experiments. *J. Phys. E: Sci. Instrum.*, Vol. 5, p. 481, 1975.
83 M.M. Brady, Simple Pulse Current Transformer to Check Magnetron Pulse Current. *J. Sci. Instrum.*, Vol. 44, p. 71, 1967.
84 W. Stygar and G. Gerdin, High Frequency Rogowski Coil Response Characteristics. *IEEE Trans. Plasma Sci.*, Vol. 10, No. 1, p. 40, 1982.
85 V. Nassisi and A. Luches, Rogowski Coils: Theory and Experimental Results. *Rev. Sci. Instrum.*, Vol. 50, No. 7, p. 900, 1979.
86 D.G. Pellinen and P.W. Spence, A Nanosecond Rise Time Megampere Current Monitor. *Rev. Sci. Instrum.*, Vol. 42, No. 11, p. 1699, 1971.
87 A.M. Stefanovskii, Rogowski Loop for Measurement of Currents of Nanosecond Duration. *Instrum. Exp. Tech.*, p. 375, 1967.
88 J.M. Anderson, Wide Frequency Range Current Transformers. *Rev. Sci. Instrum.*, Vol. 42, No. 7, p. 915, 1971.
89 D.G. Pellinen et al., Rogowski Coil for Measuring Fast, High Level Pulsed Currents. *Rev., Sci., Instrum.*, Vol. 51, No. 11, p. 1535, 1980.
90 M.S. DiCapua, High Speed Magnetic Field and Current Measurements, in *Fast Electrical and Optical Measurements*, NATO ASI Series E: Applied Sciences, J.E. Thompson and L.H. Luessen, eds., Vol. 1, Martinus Nijhoff Publ., pp. 223–262, 1986.
91 V. Nassisi and A. Luches, Diagnostic Probe for Intense Electron Beams. *Rev. Sci. Instrum.*, Vol. 48, No. 11, p. 1400, 1977.

92 A.J. Rogers, Optical Technique for Measurement of Current at High Voltage. *Proc. IEE*, Vol. 120, No. 2, p. 261, 1973.

93 C.E. Baum, Sensors for Measurement of Intense Electromagnetic Pulses. *Proceedings of the 3rd International Pulsed Power Conference*, pp. 179–185, 1981.

94 A.J. Rogers, Optical Methods for Measurement of Voltage and Current on Power Systems. *Opt. Laser Technol.*, Vol. 9, No. I 6, p. 273, 1977.

95 Papp, A. and Harms, H., Magnetooptical Current Transformer. 1: Principles. *Appl. Opt.*, Vol. 19, No. 22, p. 3729, 1980.

96 H. Aulich, Magnetooptical Current Transformer. 2: Components. *Appl. Opt.*, Vol. 19, No. 22, p. 3735, 1980.

97 H. Harms and A. Papp, Magnetooptical Current Transformer. 3: Measurements. *Appl. Opt.*, Vol. 19, No. 22, p. 3741, 1980.

98 P.J. Wild, A Phasemeter for Photoelectric Measurement of Magnetic Fields. *Rev. Sci. Instrum.*, Vol. 41, No. 8, p. 1163, 1980.

99 J.E. Thompson et al., Optical Measurement of High Electric Field and Magnetic Field. *IEEE Trans. Instrum. Meas.*, Vol. 25, No. 1, p. 1, 1976.

100 F.B.A. Frungel, *High Speed Pulse Technology: Capacitor Discharge Engineering*, Vol. III, Academic Press, 1976.

101 N. George et al., Faraday Effect at Optical Frequencies in Strong Magnetic Fields. *Appl. Opt.*, Vol. 4, No. 2, p. 253, 1965.

102 Y. Toshihiko, Compact and Highly Efficient Faraday Rotators Using Relatively Low Verdet Constant Faraday Materials. *Jpn. J. Appl. Phys.*, Vol. 9, No. 4, p. 745, 1980.

103 G.A. Massey et al., Electromagnetic Field Components, Their Measurement Using Linear Electro-Optic and Magneto-Optic Effects. *Appl. Opt.*, Vol. 14, No. 1, p. 2712, 1975.

104 F.J. Sansalone, Compact Optical Isolator. *Appl. Opt.*, Vol. 10, No. 10, p. 2329, 1971.

105 K. Kyuma et al., Fiber Optic Current and Voltage Sensors Using Bi_{12} Ge O_{20} Single Crystals. *J. Lightw. Technol.*, Vol. 1, No. 1, p. 93, 1983.

106 H. Schneider et al., Low Birefringence Single Mode Optical Fibers: Preparation and Polarization Characteristics. *Appl. Opt.*, Vol. 17, No. 19, p. 3035, 1978.

107 A.M. Smith, Polarization and Magneto-Optic Properties of Single Mode Fiber. *Appl. Opt.*, Vol. 17, No. 1, p. 52, 1978.

108 K. Kyuma et al., Fiber Optic Measuring System for Electric Current by Using a Magnetooptic Sensor. *IEEE J. Quantum Electron.*, Vol. 18, No. 10, p. 1619, 1982.

109 T.G. Gilallorenzi et al., Optical Fiber Sensor Technology. *IEEE J. Quantum Electron.*, Vol. 18, No. 4, p. 626, 1982.

110 G.L. Chandler and F.C. Jahoda, Current Measurements by Faraday Rotation in Single Mode Optical Fibers. *Rev. Sci. Instrum.*, Vol. 56, No. 5, p. 852, 1985.

111 J. McCartan and M.R. Barrault, An Optical Magnetic Probe. *J. Sci. Instrum.*, Vol. 44, p. 265, 1967.

112 Emerging Technologies Working Group , Optical Current Transducers for Power Systems: A Review. *IEEE Trans. Power Deliv.*, Vol. 9, No. 4, p. 1778, 1994.

113 Y.N. Ning, Z.P. Wang, A.W. Palmer, K.T.V. Grattan, and D.A. Jackson, Recent Progress in Optical Current Sensing Techniques. *Rev. Sci. Instrum.*, Vol. 66, No. 5, p. 3097, 1995.

114 R. Shukla, A. Shyam, S. Chatuvedi, R. Kumar, D. Lathi, V. Chaudhary, R. Verma, K. Debnath, L. Sharma, J. Sonara, K. Shah, and B. Adhikary, Nanosecond Rise Time Air-Core Current Transformer for Long Pulse Current Measurement in Pulsed Power Systems. *Rev. Sci. Instrum.*, Vol. 76, No. 12, 2005. doi: http://dx.doi.org/10.1063/1.2149011

11

Electromagnetic Interference and Noise Suppression

The measurement of signals in the presence of electromagnetic interference is subject to errors because the displayed signal is a resultant of the true signal and a noise signal. Proper suppression techniques may be adapted to minimize the noise signal. Noise suppression techniques include shielded cables and enclosures, proper grounding and optimal bundling, and routing of cables and electrical isolation obtained through optics, power line filters, and isolation transformers. After discussing the principles of interference modes and noise suppression, techniques for achieving electromagnetic compatibility at the source and noise-free measurements for the monitor are presented. This chapter conveys the general principles of electromagnetic compatibility, and a specialized and rigorous treatment of electromagnetic shielding is given in Chapter 12.

11.1 Interference Coupling Modes

Interference is any electromagnetic disturbance that interrupts, obstructs, or otherwise degrades or limits the effective performance of electronics and electrical equipment. It can be induced intentionally, as in electronic warfare, or unintentionally, as the result of spurious emissions and responses. In the context of this chapter, electromagnetic interference is unintentional and concerns an induced signal on a victim circuit caused by the emission of electromagnetic radiation by a source circuit. Often, the terms "electromagnetic interference coupling" and "cross talk" are used interchangeably in discussions about electromagnetic interference. However, cross talk inherently refers to circuits and conductors located in proximity to one another where the coupling path is characterized by their mutual reactance. Electromagnetic coupling is a more general term, which includes the linkage of electric, magnetic, or electromagnetic fields from one conductor to another, regardless of their relative location. Interference can occur with both AC and transient excitations and is highly dependent on the signal frequency. The broad categories of interference

Foundations of Pulsed Power Technology, First Edition. Jane Lehr and Pralhad Ron.

coupling modes that introduce noise into a measurement system are capacitive, inductive, radiative, and common impedance.

In general, interference between circuits is due to a combination of electric and magnetic coupling [1–7]. The interference is referred to as *capacitive coupling* when the circuits are coupled through an electric field – such as at high frequencies. In contrast, *inductive coupling* – interference caused by magnetic fields – occurs either at low frequencies or where low circuit impedances are encountered. Inductive coupling dominates when either (or both) of the lines are shielded and their lengths are significantly shorter than a half-wavelength, then the impedance from any point on the shield to ground will be low and the electric coupling will be typically small compared to magnetic coupling. In the case of open wires, if the lengths are short compared to the half-wavelength, if the terminating resistances are significantly smaller than the characteristic impedances of open wires, then the net coupling will be dominated by magnetic coupling. In cases where the impedance restriction for open wires is violated, electric coupling should also be considered.

11.1.1 Coupling in Long Transmission Lines

11.1.1.1 Capacitive Coupling

The capacitive coupling [1] between adjacent transmission lines is illustrated in Figure 11.1. The circuit contains open wires of length ℓ, separated by a distance D, and supported at a height h above the ground. The interfering circuit *Line 1* is driven by a voltage generator V_1 through its internal resistance R_1 and terminated in a resistor $R_{\ell 1}$.

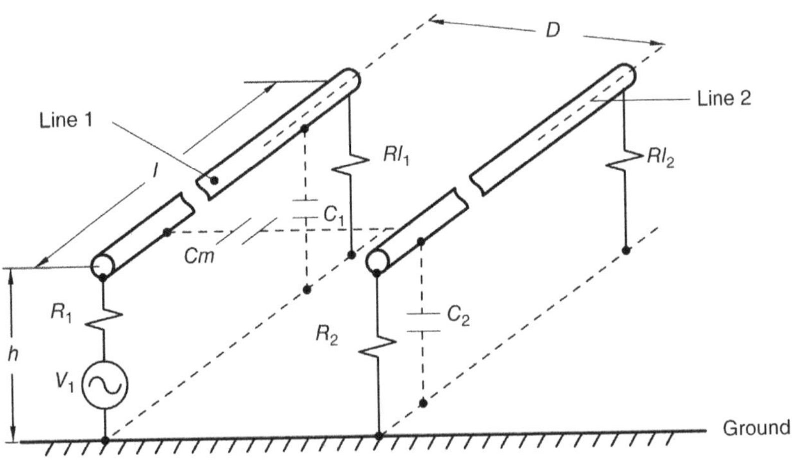

Figure 11.1 Two open parallel wires above a ground plane may be capacitively coupled if their lengths are long compared to their signal wavelength. The signal in *Line 1* induces a voltage in victim *Line* 2.

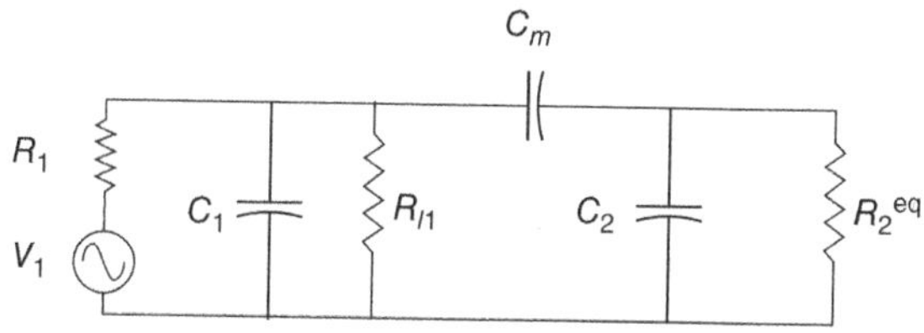

Figure 11.2 High-frequency circuit equivalent for capacitive coupling between two open wires [2]. The resistance R_2^{eq} is the parallel combination of R_2 and R_{ℓ_2} and $C_m = \tilde{C}_m \cdot \ell$, $C_1 = \tilde{C}_1 \cdot \ell$, and $C_2 = \tilde{C}_2 \cdot \ell$.

The victim circuit *Line 2*, which runs parallel to *Line 1*, does not contain any source and is terminated in resistors R_2 at one end and $R_{\ell 2}$ at the other. *Line 1* is assumed to have distributed inductance $\tilde{L}_1$ and capacitance $\tilde{C}_1$ and *Line* 2, $\tilde{L}_2$, and $\tilde{C}_2$, respectively. The cables are coupled by their mutual capacitance per unit length $\tilde{C}_m$ so that when a high-frequency signal propagates along *Line 1*, a voltage V_2 appears on *Line 2*. The magnitude of the induced voltage V_2 may be estimated from the high-frequency lumped equivalent circuit [2] shown in Figure 11.2.

The voltage induced on the victim *Line 2* is given by

$$V_2 = V_1 \cdot \frac{\left(X^{/}_{c2}\middle|\middle|R^{/}_2\right)}{\left(X^{/}_{c2}\middle|\middle|R^{/}_2\right) + X^{/}_{cm}} \cdot \frac{Z_{EQ}}{R_1 + Z_{EQ}} \tag{11.1}$$

where $Z_{EQ} = \left(X^{/}_{C_1}||R_{l_1}\right)\middle|\middle|\left(X^{/}_{cm} + R^{/}_2||X^{/}_{c_2}\right)$ and $X\|R$ is the equivalent impedance when X and R are connected in parallel.

By considering the two lines as cables with distributed parameters, the voltage capacitively coupled into a fractional length $d\ell$ of *Line 1* at any point P in the time domain can be written as

$$V_{d\ell} = \left[\left(d\ell \cdot \tilde{C}_m\right) \cdot \frac{d\xi_p}{dt}\right] Z_2 \tag{11.2}$$

where Z_2 is the characteristic impedance of *Line 2* and ξ_p is the voltage at point P on the *Line 1*.

The capacitive coupling between the two circuits can be reduced to zero by having a perfectly conducting sheet introduced between the lines and connecting it to ground. However, if the sheet possesses finite resistivity, a potential difference would develop between two points on the sheet, which would again lead to capacitive coupling. A coaxial cable with a braided shield, although it considerably reduces the capacitive coupling to adjacent circuits, cannot entirely prevent it, because of the apertures in the shield and its finite conductivity. By installing the monitoring instruments in a shielded metallic enclosure, made with highly conductivity materials, the external electric field lines terminate on the enclosure, thus preventing capacitive coupling to the interior circuits.

11.1.1.2 Radiative Coupling

In addition to the continuous electromagnetic radiation from transmitters or other ambient sources, pulse power systems which involve the fast switching of high currents and high voltages give rise to transient electromagnetic (EM) radiation. This unintentional electromagnetic radiation can couple into signal transmission lines and manifest as a voltage pulse evident in measured signals. The induced voltage can be determined analytically [3] or experimentally [4]. The magnitude of this radiatively coupled noise voltage could be much larger than the true signal, thus making accurate signal measurement impossible. Spurious voltage signals can also couple through the power connection causing a power frequency harmonic on the measurement [5]. Enclosures shielded from electromagnetic radiation can be used to protect against this induced voltage [6]. Radiated coupling can be minimized by either placing an shielded enclosure (1) around a source so that the electromagnetic radiation is attenuated to the outside or (2) around the measurement equipment for protection from an impinging source. Often both the methods are jointly employed. By a proper arrangement of shielded cables and enclosures, and by efficient grounding technique, accurate measurement of low level signals even in the presence of intense electromagnetic fields is possible.

11.1.1.3 Inductive Coupling

If ξ_p is the voltage at any point P on the transmission line, the inductive voltage induced in a length $d\ell$ of an adjacent transmission line in the time domain can be written as

$$V_{dl_2} = (M \cdot dl)\frac{d}{dt}\left(\frac{\xi_p}{Z_1}\right) \tag{11.3}$$

From Equations 11.2 and 11.3, the voltage induced in the time domain due to the combined effects of capacitive and inductive coupling can be written as

$$V_{dl_2} = \left(Z_2 \cdot C_m - \frac{M}{Z_1}\right) dl \cdot \frac{d\xi_p}{dt} \tag{11.4}$$

11.1.2 Common Impedance Coupling

Coupling of signals from one circuit into another can occur if two or more circuits share a common impedance. This is illustrated in Figure 11.3, where a conductor of impedance Z_C is common to a generator circuit (V_g, Z_{0g}, $Z_{\ell g}$) of low impedance and a victim circuit (Z_{0R}, $Z_{\ell R}$) of high impedance [7]. Assuming $Z_{0g} + Z_{\ell g} \gg Z_C$, the current I_g through Z_C is given by

$$I_g = \frac{V_g}{Z_{0g} + V_{\ell g}} \tag{11.5}$$

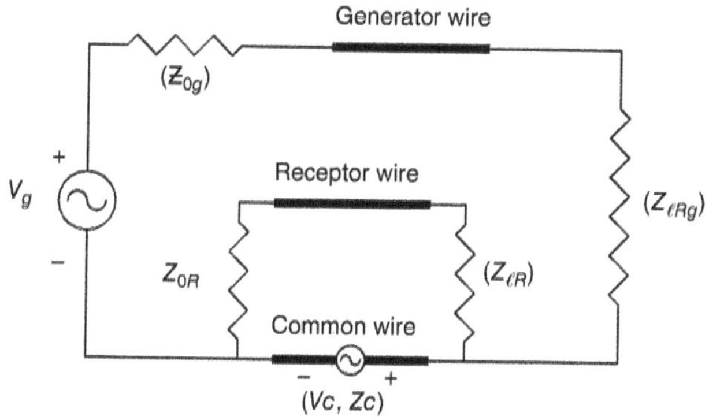

Figure 11.3 Common impedance coupling.

and the voltage V_C developed across the conductor with impedance Z_C is given by

$$V_C = \frac{V_g}{Z_{0g} + Z_{\ell g}} \cdot Z_C \tag{11.6}$$

The voltage $V_{\ell R}$ coupled across the load $Z_{\ell R}$ of the victim circuit can be written as

$$V_{\ell R} = \frac{Z_{\ell R}}{Z_{0R} + Z_{\ell R}} \cdot \frac{V_g}{Z_{0g} + Z_{\ell g}} \cdot Z_C \tag{11.7}$$

If this noise voltage $V_{\ell R}$ is comparable to or greater than the legitimate signal in that circuit, malfunction or false measurement will take place, depending on whether the victim circuit is a control circuit or a monitor circuit. Equation 11.7 indicates that the voltage transferred to the victim circuit due to common impedance coupling can be reduced by reducing the value of the common impedance.

Examples of common impedance coupling are ground loop coupling and conducted interference through the power lines. The ground loop coupling can be reduced by the techniques of single-point grounding and bypassing the signal shield current to another low-impedance conductor connected in parallel. Conducted interference introduced from the power supply mains can be reduced by incorporating power line filters and isolation transformers.

11.1.3 Coupling of Short Transmission Lines over a Ground Plane

Inductive coupling is pervasive in pulsed power circuits because of the extensive use of ground planes paired with the relatively short electrical lengths of PFLs used in pulse compression. Most conventional treatments of interference phenomena use long cable lengths and a point ground that is not easily adaptable to cables over a ground plane – which alters the inductance.

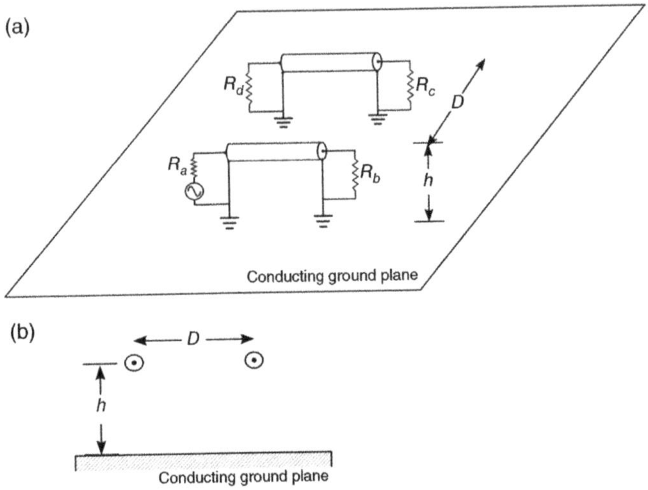

Figure 11.4 The current flowing in *Line 1* induces a voltage on the victim circuit *Line 2* and induces a voltage across the resistor R_d. The lines may be either open or shielded cables located a distance h over a ground plane and separated by a distance D.

The generic circuit for analysis, shown in Figure 11.4, is comprised of two transmission lines of length ℓ separated by a distance D and located at a height h over a ground plane. The two transmission lines may be either open wires or shielded coaxial cables. Inductive coupling by associated magnetic fields dominate the interference phenomena when the length ℓ of the transmission lines are significantly shorter than the half-wavelength of the excitation pulse,

$$\ell \ll \frac{\lambda}{2} \quad \text{or} \quad \ell \ll \frac{c}{2f} \tag{11.8}$$

When the condition (11.8) is met, and either of the lines are shielded cables, the impedance from any point of the shield to the ground will be low and the electric coupling will be small relative to the magnetic coupling. For open wires, in addition to the condition (11.8), inductive coupling will dominate provided the terminating resistances are significantly smaller than the characteristic impedances of the open wire. In cases where the impedance restrictions on open wires are violated, electric coupling should also be considered.

The interfering cable, *Line 1*, is driven by a voltage generator V_1 through its internal resistance R_a, yielding a current I_1 and terminated in a resistor R_b. If the cables are inductively coupled, the magnetic field associated with the current I_1 in *Line 1* induces a voltage in the victim circuit *Line 2*. For this analysis, the victim circuit *Line 2* has no source and the induced voltage can be measured across one of the terminating resistors R_c or R_d. In general, *Line 2* would also be carrying a signal and the induced voltage would add to it as noise.

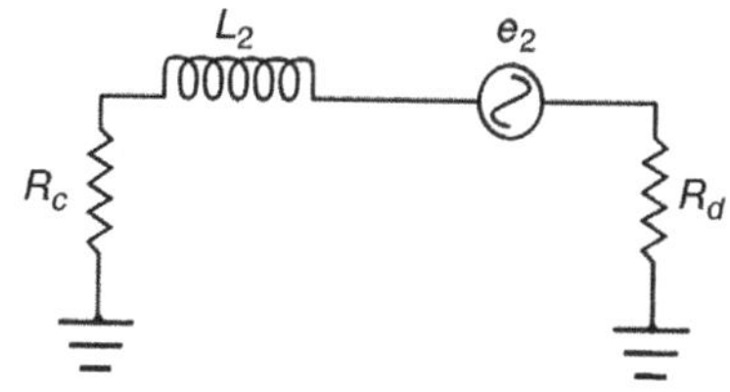

Figure 11.5 The victim circuit may be replaced by a series circuit and a voltage generator representing the noise induced on *Line 2* by the current in *Line 1*.

With inductive coupling, the interference depends only on the current of the source line and is independent of its other characteristics, such as voltage or characteristic impedance. The victim circuit may be replaced by a series circuit consisting of the terminating resistance and inductance of the victim line and voltage generator. Thus, the victim circuit may be replaced by a series circuit consisting of the terminating resistance and inductance of the victim line and a voltage generator, as shown in Figure 11.5.

11.1.3.1 Voltages Induced by Transients

Mohr's analysis [8] for two transmission lines over a ground plane consists of determining the open-circuit voltage of the equivalent generator of the victim circuit $e_2(t)$ in Figure 11.5. For interference by transient signals, the voltage induced in the victim circuit has the form

$$e_2(t) = \frac{MI_1}{\tau}\alpha \qquad (11.9)$$

where

- M is the mutual inductance between circuits,
- I_1 is the peak value of the source transient current,
- τ is the time constant for the source current, and
- α is the factors that accounts for shield or shields used.

Once the generator's open-circuit voltage $e_2(t)$ in the equivalent victim circuit is found, circuit analysis can be used to obtain the interfering voltage e_d induced on the terminating resistor R_d. Here, the solutions are presented without derivation, but Mohr shows that the analysis is accurate for even the fastest transients.

Four cases of lines over a ground plane are analyzed where the source and victim lines may be either open wires or shielded cables. The results for the important case of a source current with an exponential rise is given by

$$i_1(t) = I_1\left(1 - e^{-t/\tau}\right) \qquad (11.10)$$

11.1.3.1.1 Two Open Wires

The simplest case of inductive inference is where two open wires are linked by the magnetic field generated by the current flowing in the source line and characterized by their mutual inductance. The equivalent open-circuit voltage of the victim circuit is

$$e_2(t) = \frac{MI_1}{\tau} \cdot e^{-t/\tau} \tag{11.11}$$

The voltage drop across the inductor L_2 must be accounted for and the interference voltage $e_d(t)$ measured at the resistor R_d is given by

$$e_d(t) = \frac{MI_1}{\tau} \cdot \frac{R_d}{R_c + R_d} \cdot \left\{ \frac{1}{1 - \left[L_2/((R_c + R_d) \cdot \tau)\right]} \cdot \left(e^{-t/\tau} - e^{-((R_c+R_d)t)/L_2}\right) \right\} \tag{11.12}$$

The induced voltage may be considered to be divided by R_c and R_d and then attenuated by the expression in brackets. This attenuation factor is obtained by comparing (11.12) with (11.9). The peak interference voltage $V_d^{\max}$ occurs at a time t_m corresponding to the mathematical extrema of the attenuation factor and is given by

$$t_m = \frac{\tau}{L_2} \frac{\ln\left[(R_c + R_d) \cdot \tau\right]}{\left[((R_c + R_d) \cdot \tau)/L_2\right] - 1} \tag{11.13}$$

The peak interference voltage V_d^{max} is obtained from (11.11) by inserting the value of t_m:

$$V_d^{max} = e_d(t_m) = \frac{MI_1}{\tau} \cdot \frac{R_d}{R_c + R_d} \cdot \left\{ \frac{1}{1 - \left[(L_2)/(R_c + R_d)\tau\right]} \cdot \left(e^{-t_m/\tau} - e^{-((R_c+R_d)t_m)/L_2}\right) \right\} \tag{11.14}$$

11.1.3.1.2 Shielded Wire-to-Open Wire

The induced voltage in the victim loop is

$$e_2(t) = \frac{MI_1}{\tau} \cdot \frac{1}{(L_{s1}/R_{s1}\tau) - 1} \cdot \left(e^{-(R_{s1}/L_{s1}) \cdot t} - e^{-t/\tau}\right) \tag{11.15}$$

If the time constant $L_2/(R_c + R_d)$ of the victim circuit is small compared to both τ and (L_{s1}/R_{s1}) – as is usually the case – the voltage drop across L_2 may be neglected. The voltage induced across the resistance R_d is

$$e_d(t) = e_2(t)\frac{R_d}{R_c + R_d} \tag{11.16}$$

$$e_d(t) = \frac{MI_1}{\tau} \cdot \frac{1}{(L_{s1}/R_{s1}\tau) - 1} \cdot \frac{R_d}{R_c + R_d} \cdot \left(e^{-(R_{s1}/L_{s1})\cdot t} - e^{-t/\tau}\right) \tag{11.17}$$

The time corresponding to the peak in the induced interference is

$$t_m = \frac{\tau \ln(\tau \cdot R_{s1}/L_{s1})}{[(\tau \cdot R_{s1}/L_{s1}) - 1]} \tag{11.18}$$

The peak voltage induced in the victim circuit is

$$V_d^{max} = \frac{MI_1}{\tau} \cdot \frac{R_d}{R_c + R_d} \cdot \frac{1}{(L_{s1}/R_{s1}\tau) - 1} \cdot \left(e^{-(R_{s1}/L_{s1})\cdot t_m} - e^{-t_m/\tau}\right) \tag{11.19}$$

11.1.3.1.3 Open Wire-to-Shielded Wire Coupling

The analysis is similar to that above but the open wire is excited and the victim circuit is a shielded cable. Peak output voltage across R_d is given by

$$V_d^{max} = \frac{MI_1}{\tau} \cdot \frac{R_d}{R_c + R_d} \cdot \frac{1}{(L_{s2}/R_{s2}\tau) - 1} \cdot \left(e^{-(R_{s2}/L_{s2})\cdot t_m} - e^{-t_m/\tau}\right) \tag{11.20}$$

Equation 11.20 is equivalent to (11.19) with the exception that the shield is on cable 2.

11.1.3.1.4 Shielded Wire-to-Shielded Wire

The shields complicate the analysis because the ground plane forms another transmission line with the cable's shield. Mohr transforms the shielded cable into an equivalent open wire. The equivalent open-circuit voltage in the victim line is

$$\frac{e_2(t)}{MI_1} \cdot \left(\frac{L_{s1}}{R_{s1}} - \frac{L_{s2}}{R_{s2}}\right) = \frac{1}{1 - (R_{s1}/L_{s1})} \cdot \left(e^{-(R_{s1}/L_{s1})\cdot t} - e^{-t/\tau}\right) - \frac{1}{1 - (R_{s2}/L_{s2})} \cdot \left(e^{-(R_{s2}/L_{s2})\cdot t} - e^{-t/\tau}\right) \tag{11.21}$$

If both cables are identical, $L_{s1} = L_{s2} = L_s$ and $R_{s1} = R_{s2} = R_s$, then (11.16) reduces to the much simpler form,

$$e_2(t) = \frac{MI_1}{\tau}\left(\frac{\tau(R_s/L_s)}{1 - \tau(R_s/L_s)}\right)^2 \cdot \left(e^{-t/\tau} - e^{-(R_s/L_s)\cdot t}\right) + \frac{t \cdot (R_s/L_s)}{1 - \tau(R_s/L_s)} \cdot e^{-(R_s/L_s)\cdot t} \tag{11.22}$$

Assuming the time constant of the victim circuit is short compared to τ, the induced voltage $e_d(t)$ is effectively divided only between R_c and R_d, so that

$$e_d(t) = e_2(t)\frac{R_d}{R_c + R_d} \tag{11.23}$$

The time corresponding to maximum output is the transcendental equation

$$t_m\left(\frac{R_s}{L_s} - \frac{1}{\tau}\right) - \ln\left[\frac{R_s}{L_s}t_m \cdot \left(\frac{R_s}{L_s}\tau - 1\right) + 1\right] = 0 \tag{11.24}$$

which can be solved by an iterative process. Once t_m is known, the peak-induced voltage can be found by inserting this value into v:

$$V_d^{max} = e_d(t_m) = e_2(t_m) \cdot \frac{R_d}{R_c + R_d} \tag{11.25}$$

11.1.3.2 Modification of Inductances by the Ground Plane

Mohr takes an interesting approach of first analyzing the alteration of the cable inductance by the ground plane and then reducing the equivalent parameters of the shielded cable to that of an equivalent open wire. Thus, all cases can be treated as open wires over ground plane. Both the self-inductance of the cable as well as the mutual inductance between the cable pairs are altered by the presence of the ground plane.

11.1.3.2.1 *Open Wire over a Ground Plane*

For an open wire of length ℓ and diameter d, held at a height h over a ground plane, the inductance per unit length $\tilde{L}_{ow}$ is

$$\begin{aligned} \tilde{L}_{ow} &= 0.14 \times \log_{10}\left(\frac{4h}{d}\right)\frac{\mu\text{H}}{\text{ft}} \\ &= 0.46 \times \log_{10}\left(\frac{4h}{d}\right)\frac{\mu\text{H}}{\text{m}} \end{aligned} \tag{11.26}$$

and plotted in Figure 11.6.

The mutual inductance per unit length $\tilde{M}_{ow}$ of a pair of open wires separated by a distance D and held at a height h above a ground plane is

$$\begin{aligned} \tilde{M}_{ow} &= 0.07\log_{10}\left[1 + \left(\frac{2h}{D}\right)^2\right]\frac{\mu\text{H}}{\text{ft}} \\ &= 0.23\log_{10}\left[1 + \left(\frac{2h}{D}\right)^2\right]\frac{\mu\text{H}}{\text{m}} \end{aligned} \tag{11.27}$$

and plotted in Figure 11.7.

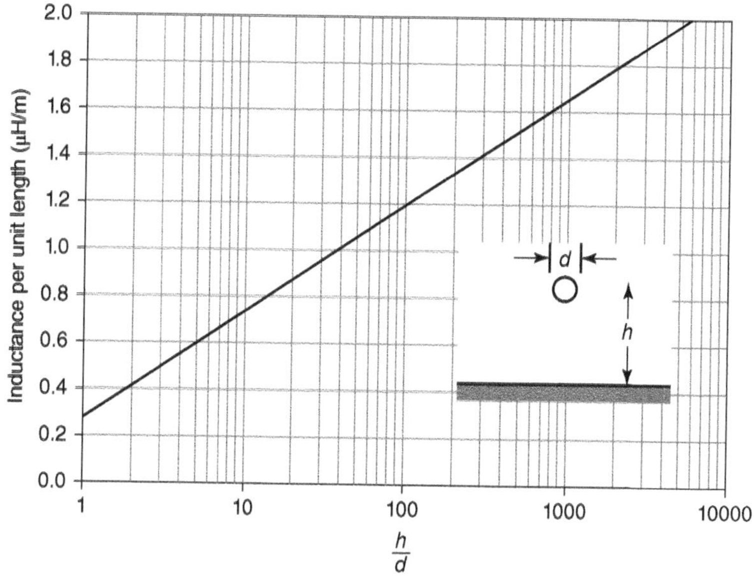

Figure 11.6 The inductance per unit length of an open wire over a ground plane, Equation 11.26, is a function of the wire diameter and the height above the ground plane.

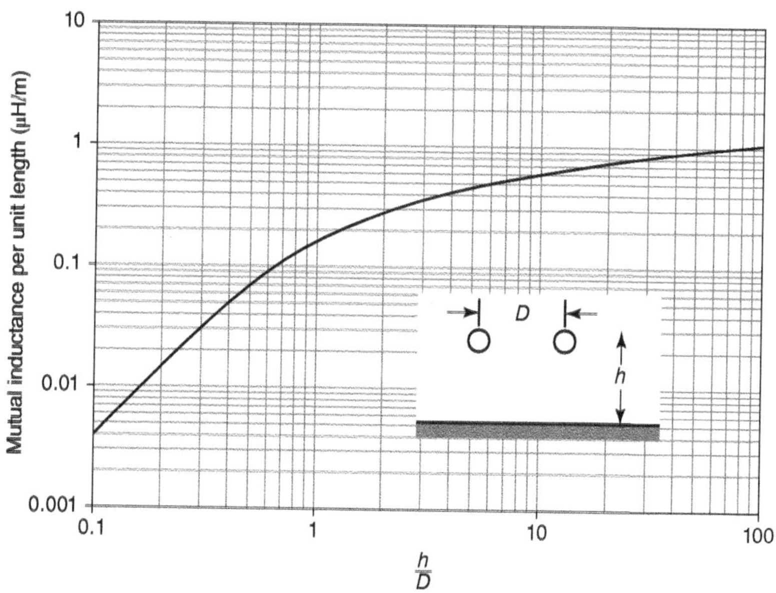

Figure 11.7 The mutual inductance per unit length of two open wires spaced a distance D apart and held a distance h above the ground plane (Equation 11.27).

11.1.3.2.2 Shielded Wire over a Ground Plane

Interference problems involving shielded wires may be handled similarly to open wire-to-open wire problems by reducing the shielded lines to equivalent open wire lines as shown in Figure 11.8. The shielded line is considered to be made up of two isolated wires – representing the inner and outer conductors – over a ground plane, coupled by their mutual inductance L_m.

The coaxial cable has an inductance $\tilde{L}_c$ given by

$$\tilde{L}_c = 0.46\log_{10}\left(\frac{d_s}{d_i}\right)\frac{\mu\text{H}}{\text{m}} \tag{11.28}$$

where d_s is the diameter of the shield and d_i the diameter of the inner conductor.

The transformation of the cable inductance to a representation by two open wires requires expressions for each conductor's inductance as well as their mutual inductance. The inductance of the inner (L_i) and shield (L_s) conductors can be obtained from (11.26) for each conductor as [8]

$$\tilde{L}_i = 0.14\log_{10}\left(\frac{4h}{d_i}\right)\frac{\mu\text{H}}{\text{ft}} = 0.46\log_{10}\left(\frac{4h}{d_i}\right)\frac{\mu\text{H}}{\text{m}} \tag{11.29}$$

$$\tilde{L}_s = 0.46\log_{10}\left(\frac{4h}{d_s}\right)\frac{\mu\text{H}}{\text{ft}} = 0.46\log_{10}\left(\frac{4h}{d_s}\right)\frac{\mu\text{H}}{\text{m}} \tag{11.30}$$

The inductance of the cable is

$$L_c = L_i + L_s - 2L_m \tag{11.31}$$

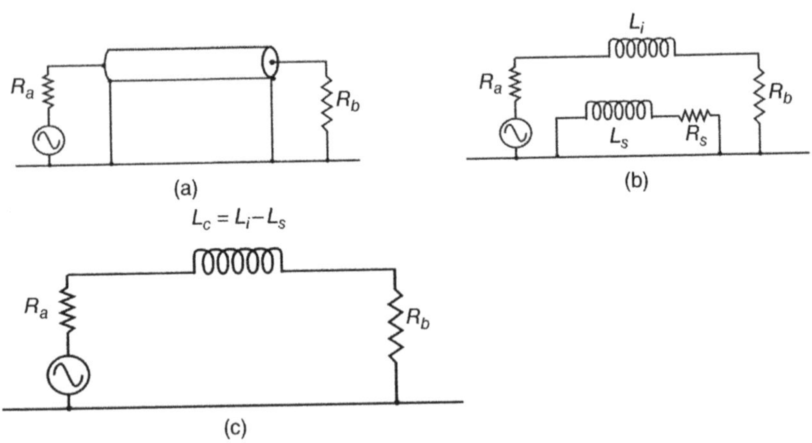

Figure 11.8 The coaxial cable held above a ground plane in part (a) can be represented by the inductances of its conductors, shown in part (b) and their mutual inductances. Analysis shows that the shielded cable above a ground plane can be represented by the open-wire equivalent circuit in part (c) with an equivalent inductance L_c.

where

$$\tilde{L}_m = 0.46 \log_{10}\left(\frac{4h}{d_s}\right) \frac{\mu\text{H}}{\text{m}} \tag{11.32}$$

Noting that $L_m = L_s$, (11.31) reduces to

$$L_c = L_i - L_s \tag{11.33}$$

The return current is considered to be conducted partly by the ground plane. Additionally, the shield is imperfect, and has a resistance R_s, which is assumed to be independent of frequency (negligible skin effect). Thus, the shielded cable over a ground plane can be treated as an open wire over a ground plane with an equivalent inductance of L_c.

11.2 Noise Suppression Techniques

Noise can be introduced into a signal circuit by either intersystem interference or intrasystem interference [9]. Intersystem interference originates from outside the operating system by components such as broadcast transmitters, communication systems, or automobile ignition systems, or may be conducted through the power line. Intrasystem interference originates within the subsystems by rapid changes in current and voltage levels, corona at high voltage levels, spark gaps and switching transients, or by conducted interference from one subsystem to others, either from the power line or via common ground loops. Both the types of interference can be suppressed by employing a combination of techniques such as the use of shielded cables and shielded enclosures, efficient grounding and avoidance of ground loops, and installation of power line filters and isolation transformers.

11.2.1 Shielded Enclosure

A shielded enclosure is essentially a room made of connected conducting walls. It attenuates the EM radiation across the boundary defining the shielded enclosure. The shielded enclosures can be classified [10,11] into demountable types and permanent types. The latter, constructed *in situ* by welding thick steel plates, are generally more expensive and used in special applications where either a very large size or a high degree of shielding, especially at low frequencies, is required. The demountable enclosures, shown in Figure 11.9, are assembled from standard panels. The individual panels are made of plywood sheets laminated on both sides by zinc-plated steel sheets. Specialized types of radio frequency (RF)-tight mechanical joints are used for joining each panel to its neighboring panels. The enclosures are provided with RF doors, fitted with copper–beryllium spring

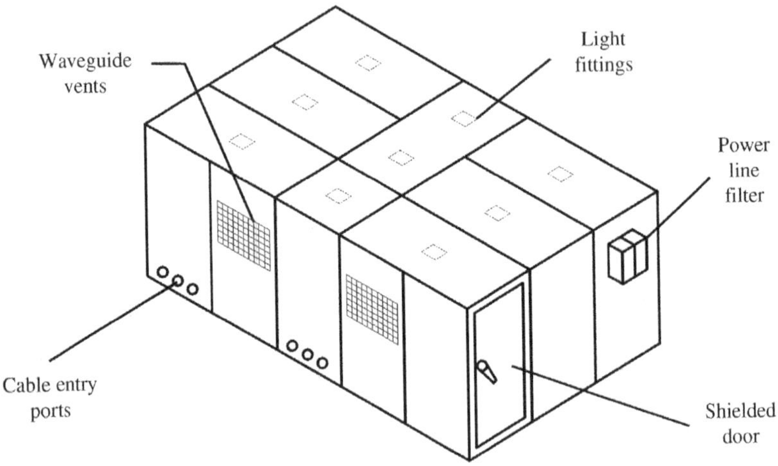

Figure 11.9 A typical shielded enclosure.

fingers, for entry or exit. The power input to the equipment inside the enclosure enters through isolation transformers and power line filters.

The required degree of shielding provided by the enclosure is determined by the frequency and strength of the interfering signals, the frequency and amplitude of the signals to be measured, and the required accuracy of the measurement. Where there is a need to eliminate the RF reflection from the interior walls of the enclosure, anechoic chambers are used [12]. In anechoic chambers, the walls are fitted with assembly of conical pieces fabricated from carbon-impregnated foamed plastic that absorbs RF energy.

The shielding effectiveness *SE* of an enclosure expressed in decibels can be defined as

$$SE = 10\log\frac{P_2}{P_1} = 20\log\frac{E_2}{E_1} = 20\log\frac{H_2}{H_1} \tag{11.34}$$

where P_2, E_2, H_2 are the power, electric field, and magnetic field, respectively, of the EM wave at a point, without the enclosure, and P_1, E_1, H_1 are the power, electric field, and magnetic field, respectively, of the EM wave at the same point, with the enclosure.

A generalized expression [13–17] for shielding effectiveness can be written as

$$SE = A + R + B \tag{11.35}$$

where A is the absorption loss within the barrier, R is the reflection loss from the air–barrier interface, and B is the correction factor to account for the barrier–air interface and multiple reflections within the barrier.

The components A, R, and B are illustrated in Figure 11.10.

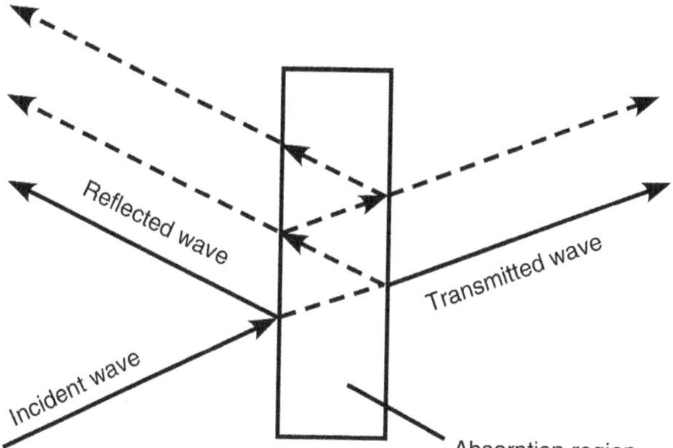

Figure 11.10 An electromagnetic wave incident on a barrier gets partially reflected and transmitted and has multiple internal reflections due to the thickness.

11.2.1.1 Absorption Loss (A)

The absorption loss A in decibel units is given by

$$A = 8.686 \cdot t \cdot \sqrt{\pi \mu \sigma f} = 131 \cdot t \cdot \sqrt{\mu_r \sigma_r f} \tag{11.36}$$

where t is the material thickness in meters, μ ($=\mu_o\mu_r$) is the magnetic permeability of the material, μ_o is the permeability of free space ($4\pi \times 10^{-7}$H/m), μ_r is the relative permeability, σ ($\sigma_{Cu}\sigma_r$) is the conductivity of the material mho/m, and f is the frequency in Hz. The conductivity may be expressed as the relative conductivity σ_r, that is, the conductivity relative to copper:

$$\sigma_r = \frac{\sigma}{\sigma_{Cu}}$$

The conductivity of copper, σ_{Cu}, is ~0.59×10^8 mho/m at 20°C.

11.2.1.2 Reflection Loss (R)

The reflection loss R in decibel is given by

$$R = 20 \log \frac{|1+K|^2}{4|K|} \tag{11.37}$$

and

$$K = \frac{Z_\omega}{Z_s} \tag{11.38}$$

where Z_ω is the wave impedance in ohms and Z_s is the intrinsic shield impedance in ohms.

11.2.1.2.1 Plane Wave Impedance (η_0)

The EM wave is considered to be a plane wave at a distance from the radiating source on the order of or greater than 6λ, where λ is the wavelength. The wave impedance η_0, for plane waves, in free space is given by 120π or 377 Ω.

11.2.1.2.2 High-Impedance Electric Field

The electromagnetic wave is considered to be a high-impedance electric field [18], when the E field is predominant. Under these circumstances, the wave impedance is much larger than 377 Ω. The wave impedance for a high-impedance electric field is given by

$$\left|\frac{Z_\omega}{\eta_0}\right| = \frac{1}{(1+\beta r)^2} \cdot \left[\frac{(\beta r)^6 + 1}{(\beta r)^2}\right]^{1/2} \approx \beta r \tag{11.39}$$

when $(\beta r) \ll 1$.

11.2.1.2.3 Low-Impedance Magnetic Field

The electromagnetic wave is considered to be a low-impedance magnetic field when the H field is predominant. Under these circumstances, the wave impedance is much smaller than 377 Ω. The wave impedance for a low-impedance magnetic field is given by

$$\left|\frac{Z_\omega}{\eta_0}\right| = \left[1 + (\beta r)^2\right] \cdot \left[\frac{(\beta r)^2}{(\beta r)^6 + 1}\right]^{1/2} \approx \beta r \tag{11.40}$$

when $\beta r \ll 1$.

11.2.1.2.4 Wave Impedance Versus βr

Typical characteristics [14] of "wave impedance versus βr" for electric and magnetic dipoles are shown in Figure 11.11, from where it is seen that at large distances from the radiating source the wave impedance approaches the plane wave impedance of 377 Ω.

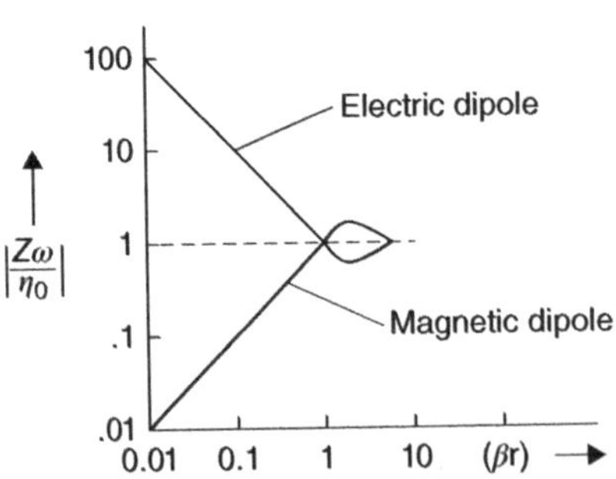

Figure 11.11 Wave impedance versus (βr) for electric and magnetic dipoles [14].

11.2.1.2.5 Intrinsic Shield Impedance (Z_s)

The shield impedance Z_s is given [10] by

$$Z_S = (1 + j) \cdot \sqrt{\frac{\pi \mu f}{\sigma}}$$

Or

$$|Z_S| = \sqrt{\frac{2\pi \mu f}{\sigma}} = 3.69 \times 10^{-7} \sqrt{\frac{\mu_r f}{\sigma_r}} \quad (11.41)$$

11.2.1.3 Correction Factor (β)

The correction factor β in decibel is given [10] by

$$\beta = 20 \cdot \log \left| 1 - \frac{(1-K)^2}{(1+K)^2} \cdot e^{-2\gamma t} \right| \quad (11.42)$$

where $\gamma = (1 + j)\sqrt{\pi \mu \sigma f}$.

In practice, when A is greater than 10 dB, the correction factor β is usually neglected.

11.2.1.4 Shielding Effectiveness for Plane Waves

For an enclosure made of 1mm-thick copper sheet, typical far-field values of absorption and reflection losses at various frequencies are shown in Table 11.1.

The important observations made from Table 11.1 are (a) the absorption loss is small at low frequencies but increases at high frequencies, due to its $\sqrt{f}$ dependence; (b) the reflection loss is high at low frequencies but decreases with increasing frequency. At increasing frequency, Z_s increases, resulting in a

Table 11.1 Shielding effectiveness for copper (1 mm thick) for plane waves assuming $Z_\omega = 377\ \Omega$ (far-field values).

Frequency f (Hz)	Intrinsic shield impedance ($Z_s \Omega$)	K factor (Z_ω/Z_s)	Absorption A (dB)	Reflection R (dB)	Shielding effectiveness SE (dB) SE = A+R
10	11.6×10^{-7}	32.5×10^7	0.4	158	158
100	36.9×10^{-7}	10×10^7	1.3	148	149
10^3	116×10^{-7}	3.25×10^7	4	138	142
10^4	369×10^{-7}	1.02×10^7	13	128	141
10^5	116.7×10^{-6}	0.3×10^7	41	118	159
10^6	369×10^{-6}	0.1×10^7	131	110	241
10^7	116.7×10^{-5}	0.03×10^7	414	98	512

reduced mismatch between Z_s and Z_w; (c) at low frequencies, the shielding is due primarily to reflection losses, whereas at high frequencies, it is due to absorption losses. The absorption loss is proportional to $(\mu_r\sigma_r)^{1/2}$ and hence by the use of commercial iron ($\mu_r = 200$, $\sigma_r = 0.17$) in comparison to copper ($\mu_r = 1$, $\sigma_r = 1$), a nearly sixfold improvement in shielding may be achieved; and (d) the absorption loss A is proportional to the wall thickness t of the shielded enclosure. Hence, considerable improvement in shielding can be obtained by increasing the thickness of the shield.

The previous expressions are approximate. More exact models are available, including the finite size of the enclosure modeled as sphere or a cylinder [19–21].

11.2.1.5 Shielding Effectiveness for High-Impedance *E* and Low-Impedance *H* Fields

Typical near-field values of reflection losses at $f = 10$ kHz for high-impedance E fields and low-impedance H fields, as calculated from the Equations 11.37 through 11.41, are shown in Table 11.2. Two important points to be noted are as follows: (a) At a given frequency, as the distance is reduced, the E field reflection loss is increased due to increased mismatch between wave impedance and shield impedance. (b) At a given frequency, as the distance is reduced, the H field reflection loss is reduced due to improved matching between the wave and shield impedances.

Another important feature that is not explicitly seen from Table 11.2 but obvious from Figure 11.11 is the similarity principle between distance and frequency, wherein decreasing the distance is equivalent to decreasing the frequency. Due to this similarity principle, observations of Table 11.2 can also be interpreted as follows: (i) At a given distance, the high-impedance E field reflection loss increases with decreasing frequency. (ii) At a given distance, the low-impedance H field reflection loss decreases with decreasing frequency.

Table 11.2 Reflection losses for high-impedance *E*-fields and low-impedance *H*-fields at 10 kHz for copper (near-field values).

Distance r (m)	Near field parameter (β)	Wave impedance (Z_ω/η_0)		Reflection losses	
		E-field ($\approx 1/\beta r$)	H-field ($\approx \beta r$)	E	H
0.1	0.21×10^{-4}	4.7×10^4	0.21×10^{-4}	222	34
1.0	2.09×10^{-4}	4.7×10^3	2.09×10^{-4}	201	42
10	20.9×10^{-4}	4.7×10^2	20.9×10^{-4}	181	74
100	209×10^{-4}	4.7×10	2.09×10^{-2}	61	94
1000	2090×10^{-4}	4.7	2.09×10^{-1}	141	114

In summary, it may be concluded that while it is easy to shield the high-impedance E fields over the entire frequency range, it is very difficult to shield low-impedance H fields at the low frequencies. Effective shielding for low-impedance H fields at low frequencies can be attained with a combination of steel and copper sheets. The thick steel plates increase the absorption loss with its high $\mu_r\sigma_r$, as governed by Equation 11.36. Reflection loss is increased by using copper, with its low relative permeability μ_r and high conductivity σ_r, resulting in low Z_S.

11.2.1.6 Typical Shielding Effectiveness of a Simple Practical Enclosure

The variation of attenuation with frequency of a simple demountable shielded enclosure employing 0.5mm-thick, zinc-plated, steel sheets is shown in Figure 11.12. As discussed earlier, the attenuation values for low-frequency magnetic fields are low, whereas a flat attenuation of better than 100 dB can be obtained for electric fields over a broad frequency range. Theoretically, the shielding effectiveness should continue to increase indefinitely with frequency; however, in practice, the shielding effectiveness drops at high frequency due to leaks, gaps, and defects at the joints.

11.2.1.7 Twisted Shielded Pair

The twisted shielded pair is a commonly used technique where two wires are twisted to cancel the magnetic coupling and enclosed in a conducting shield,

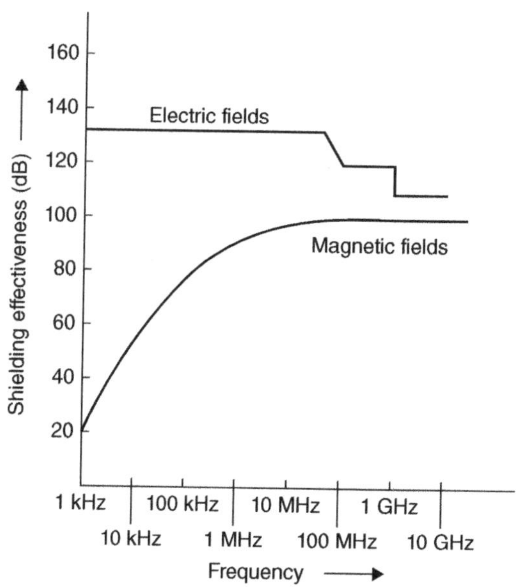

Figure 11.12 The typical attenuation characteristics of a simple shielded enclosure.

often braided, to minimize the impinging electromagnetic fields from coupling to the wires. It is important that shield be terminated circumferentially at both the source and load junctions to minimize noise introduced at these points.

11.2.2 Grounding and Ground Loops

An ideal measurement system with minimum interference is one in which there is complete absence of a ground plane. Such a system, as shown in Figure 11.13a, is, of course, subjected to capacitive and inductive interference from nearby conductors, but is free from the common impedance coupling from the ground loops. In a practical system, shown in Figure 11.13b, a ground plane with a finite conductivity is always present; grounding the pulsed power system chassis on the left to the monitor on the right side is necessary for the safety of the operator from electrical shock. The part *EF* of the ground loop has a voltage drop of $i_{g\ell}\, z_{g\ell}$, where $i_{g\ell}$ is the ground current due to many external circuits and $z_{g\ell}$ is the ground loop impedance. The value of $i_{g\ell}$ could go very high in case of a system fault, giving rise to false alarms or malfunctions in control circuits, and can overshadow the low-level signals in diagnostic circuits.

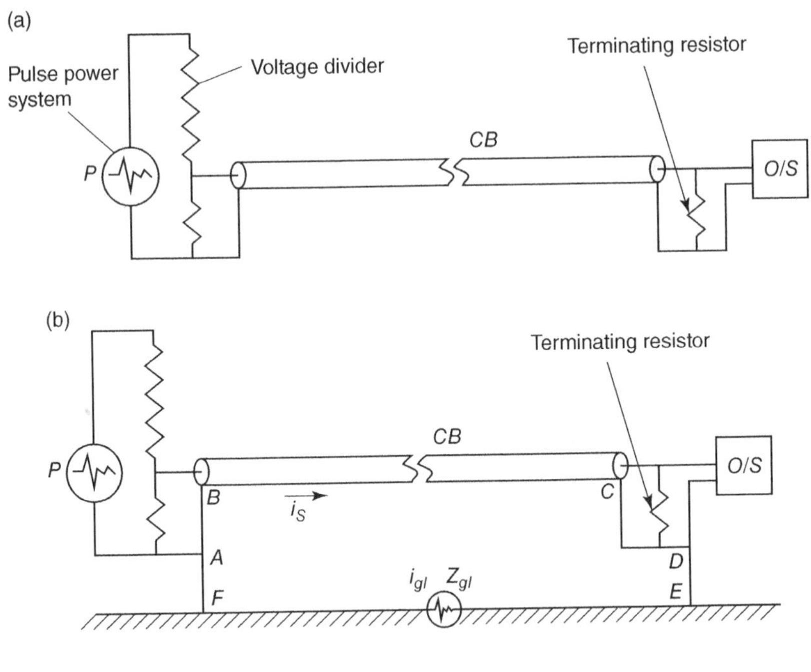

Figure 11.13 A measuring system (a) idealized and (b) in practice.

The voltage drop across the ground loop drives a current i_s through the cable shield, giving rise to a voltage $i_s Z_c \ell$ in the signal circuit, where Z_c is the surface transfer impedance of the cable and ℓ is the length of the cable [22,23]. It should be appreciated that the common impedance coupling due to ground loop would be present, even if there is no physical electrical connection of points A and D to ground; this connection is provided by the stray capacitances C_{AF} and C_{DE}.

It is thus seen that the result of the ground loops is to cause a current flow in the shield of the signal cable, which in turn is responsible for inducing the noise in the signal circuit. The techniques of suppression of ground loop coupled noise are based on (i) reducing the shield current by providing a lower impedance bypass path, (ii) elimination of or reducing the length of the ground loop by single-point grounding, and (iii) breaking the ground loop by optical isolation.

11.2.2.1 Low-Impedance Bypass Path

A low-impedance bypass path, in parallel to the signal shield, can be provided by a copper strip [25] running parallel to the signal cable or by the outer shield of a double-shielded cable or by a copper [26,27] mat embedded in the floor concrete. Typical bypass through a copper strip is shown in Figure 11.14. The figure also shows ferrite cores [28] on the signal cable and inductances L_1 and L_2 in the paths AF and DE for further improvement in the reduction of shield current. However, L_1 and L_2 should not be so large as to compromise safety against shock hazard to operating personnel. When a large number of signal cables are involved, they can run together as a single bundle and their shields can be joined together at the ends and connected to the bypass path. The cables should run as close as possible [29] to the bypass conductor, by minimizing the loop area enclosed by them to reduce the inductive pickup.

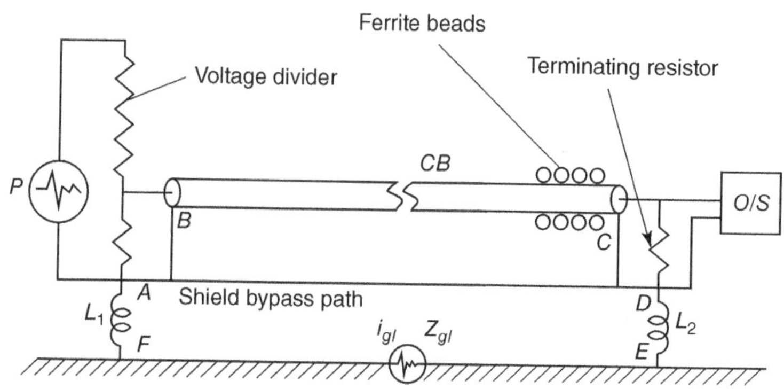

Figure 11.14 Shield currents reduction with a bypass conductor.

11.2.2.2 Single-Point Grounding

The noise introduced into the signal circuit by a ground loop can be minimized by reducing the ground loop impedance $Z_{g\ell}$. One way to achieve this is to reduce the length of the ground loop *EF*. In the limit, this leads to single-point grounding when the points *E* and *F* coincide, as shown in Figure 11.15. The grounding connections of the various circuits are now done at the ground tree [30,31], which is the only ground for the entire system. The ground tree should be made of thick conductors possessing very low impedance and buried to a suitable depth; its branches should be welded to the trunk to reduce its impedance as well as to maintain electrical integrity. The ground resistance can be decreased by using multiple rods and chemical treatment of the soil by using chemicals such as sodium chloride, magnesium sulfate, copper sulfate, and calcium chloride [32,33].

11.2.2.3 Breaking Ground Loops with Optical Isolation

Single-point grounding, shown in Figure 11.15, does not entirely eliminate the problem of ground loop coupling, since the capacitance to ground C_{DE} present at the monitor provides a closed path to ground at high frequencies. The ground loop can be broken by using an insulator for signal transmission rather than an electrically conducting conventional cable. This can be achieved by converting the low-level electrical signal on the pulsed power side into an optical signal carried by an optical fiber. An overall system of optical transmission and reception [34,35] consists of (a) converting the electrical signal into an optical signal with a light-emitting diode (LED) or a laser diode, (b) carrying the optical signal with an optical fiber, and (c) reconverting the optical signal into electrical signal with a photodiode or photo transistor. The optical isolation technique, in addition to providing immunity to ground loop coupling, also

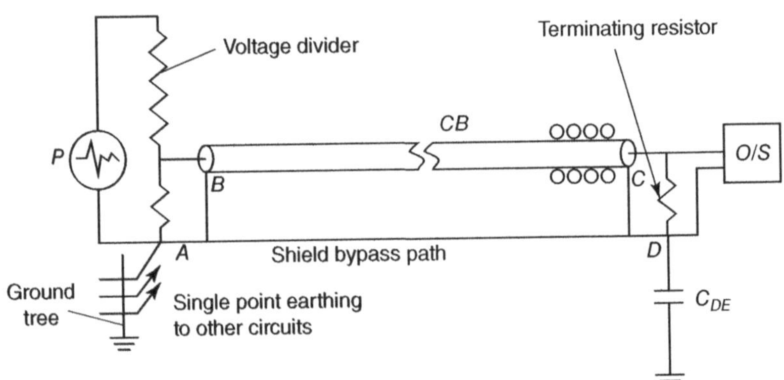

Figure 11.15 A single-point grounding scheme.

eliminates interference due to capacitive coupling, inductive coupling, and radiated coupling. The optical isolation technique is more costly, and is frequently restricted to a few signals.

11.2.3 Power Line Filters

Electrical interference, generated by equipment, can be fed back into the AC mains, which in turn couples this noise to other equipment connected to the same supply via common impedance coupling. The interference can also enter into the equipment through the primary of the distribution transformer [36]. The interference signal is mixed with the genuine signal, giving rise to faulty measurements. A power line filter, introduced between the mains and the equipment, is the most efficient tool for suppression of electromagnetic interference (EMI), which occurs in the power supply network. Passive *LC* mains filters suppress electromagnetic interference in two basic ways: The capacitor elements shunt the interference to ground, and the series inductor elements raise the impedance of the line making the shunt capacitor elements even more effective.

11.2.3.1 Types of Filters

Power line filters can be classified, as shown in Figure 11.16, into (a) reactive π or T type, employing lumped elements of capacitors and inductors, (b) absorptive filters using lossy ferrites and dielectrics, (c) mixed filters employing combination of reactive and absorptive types, and (d) reactive feed through capacitors. For high insertion loss, the filters can be connected in series. An absorptive lossy filter is electrically equivalent to a frequency-dependent resistor in series with the power line [38]. The materials chosen for an absorptive filter are either magnetic ferrites containing high iron and zinc content or a dielectric of semiconducting and ferroelectric type [39]. These materials are incorporated in the insulation in cable-type structure or in the

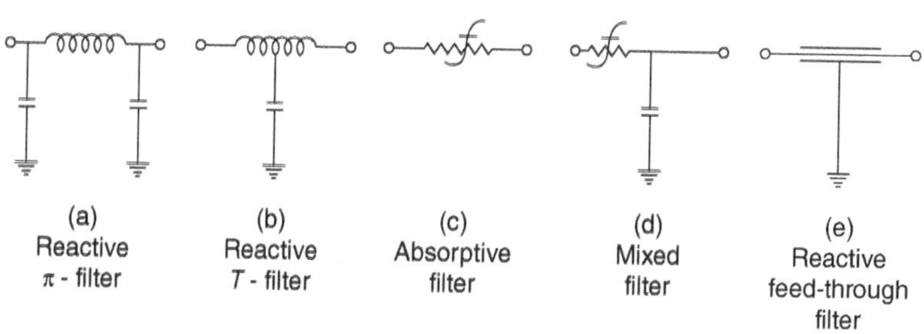

Figure 11.16 Types of power line filters.

core in helical structures. The materials provide low insertion loss at power line frequency but high insertion loss in the stop band of the interfering frequencies. The main advantage of the absorptive filter over the reactive filter is the absence of spurious response caused by *LC* resonance exhibited by the latter [40,41].

11.2.3.2 Insertion Loss

The suppression effectiveness of the power line filters is characterized by insertion loss (*IL*) as a function of frequency and expressed in decibels. The insertion loss is given by [37]

$$IL = 10 \log \frac{P_0}{P_n} = 20 \log \frac{V_0}{V_n} \text{ dB} \tag{11.43}$$

where P_0 and V_0 are, respectively, the power and voltage at the load without the power line filter and P_n and V_n are the power and voltage at the load with the power line filter.

The insertion loss as a function of frequency is shown in Figure 11.17. The insertion loss should be negligible at power line frequencies (0, 50, 60, and 400 Hz), allowing these frequencies to "pass" and should be very high at the interfering frequency – the "stop band."

In practice, a filter's insertion loss as a function of frequency is quite complex and depends on the impedance of the power line source and the load, which are generally not well known. The stop band should be very wide to protect against a wide variety of interfering signals. High-performance filters can have insertion losses over 100 dB from frequencies of 14 kHz to 10 GHz [42].

A typical setup for measuring the insertion loss characteristics is shown in Figure 11.18. The insertion loss is obtained from the ratio of the signal output voltage without and with the filter in the circuit. The DC source enables passing the rated bias current through the filter. The buffer network prevents the high-

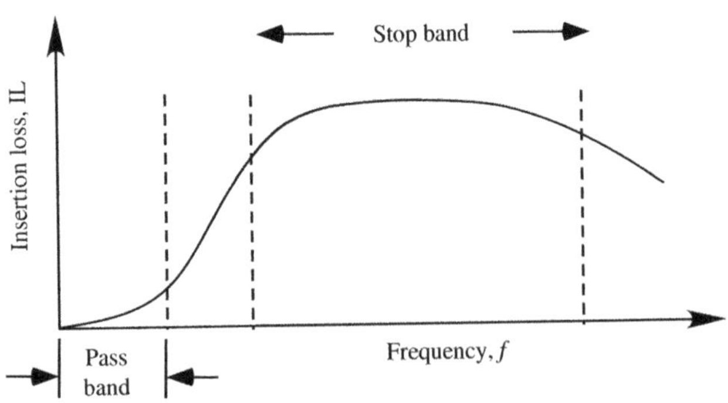

Figure 11.17 Insertion loss as a function of frequency showing the pass band and the stop band.

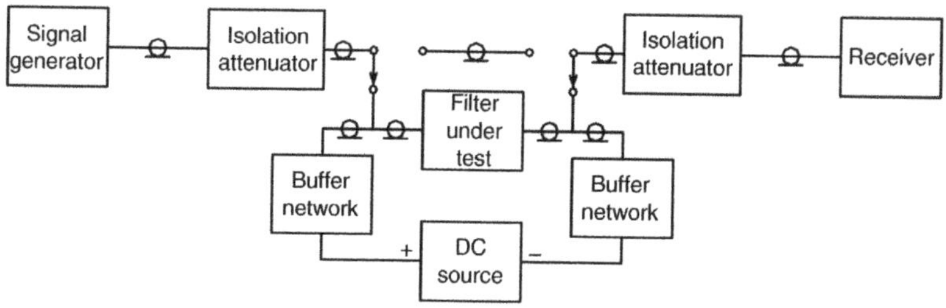

Figure 11.18 Typical setup for an insertion loss measurement.

frequency signal from passing into the DC source. The isolation attenuators provide impedance matching between the signal generator and the filter as well as between the filter and the receiver [42,43].

11.2.4 Isolation Transformer

In addition to the power line filter, an isolation transformer [40] can be used to further improve the degree of suppression for interference coupled from mains to equipment and from equipment to mains. The isolation transformer shown in Figure 11.19 is a conventional iron-core transformer with a Faraday shield inserted between the primary and secondary windings. The Faraday shield, in the form of a metallic foil, is connected to the transformer case and the power line ground. The interference signal entering from the mains or the equipment is now bypassed to ground via the capacitances C_{pF} and C_{Fs} instead of coupling into one another via C_{ps}. The shielding performance of the isolation transformer can be improved by adding multiple shields and box shields.

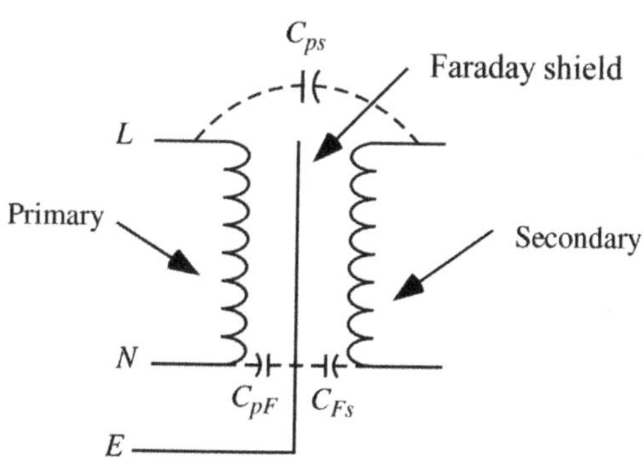

Figure 11.19 An isolation transformer.

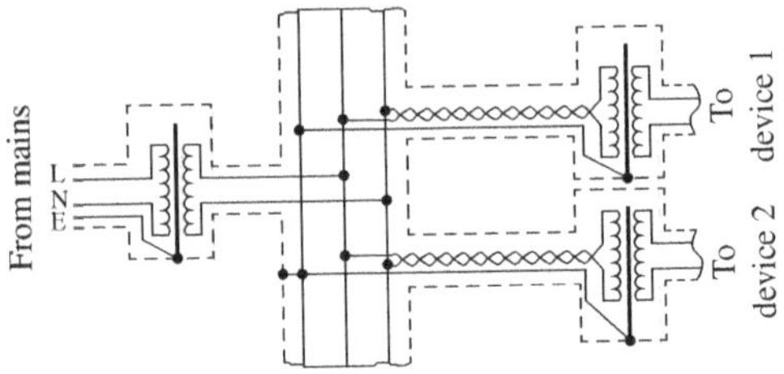

Figure 11.20 Distribution of main supply power with isolation transformers.

The diagnostic equipment, employing active amplifiers, needs to be supplied with mains power and can be isolated from each other [30] by incorporating individual isolation transformers, as shown in Figure 11.20. The isolation transformer, in addition to providing suppression to interfering signals from the mains, also provides decoupling of the power line ground from the instrument chassis. The isolation transformer may or may not suppress common mode interference depending on the specific design. In many pulsed power applications, common mode is a dominant source of signal interference.

11.3 Well-Shielded Equipment Topology

Figure 11.21 illustrates a schematic of an electronic system, consisting of three subsystems with interconnected power and signal cables. The time-varying currents in the circuits of the subsystems and cables give rise to radiated emissions and conducted (into wires) emissions [45]. The radiated emissions

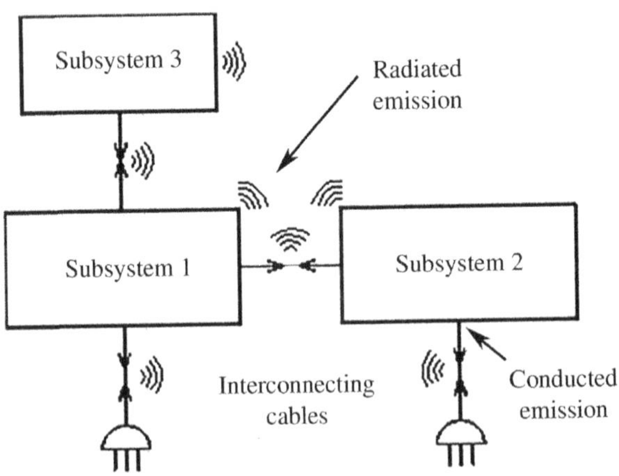

Figure 11.21 Conducted and radiated emissions between subsystems.

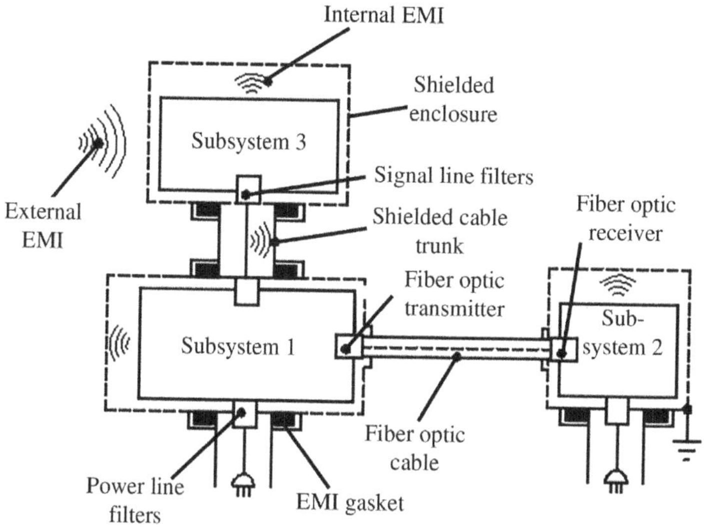

Figure 11.22 EMI suppression techniques.

are due to various current-carrying conductors acting as antennas, which may or may not be resonant [46]. The conducted emissions arise from coupling modes from one circuit to another: capacitive, inductive, radiative, or through common impedance.

Figure 11.22 illustrates intrasystem interference, where the unwanted signals are injected into one subsystem from another subsystem within the system. Intersystem interference due to the coupling of radiated or conducted emissions from an external system also affects the equipment. The overall result of the EMI, whether intrasystem or intersystem, is the introduction of unwanted signals (noise) into a victim system. When the EMI-induced noise level exceeds an acceptable threshold level, the susceptibility level, the consequences are erroneous measurements, malfunction, false alarms, frequent interruptions, and damage.

In order that a given electronic system should not affect the other external systems, it is necessary to reduce its "emission level." To make a given system to continue to operate reliably in an outside EMI environment, it is necessary to increase its "susceptibility level." Figure 11.22 illustrates the same system of Figure 11.21, modified into an EMI immune system by incorporating the EMI suppression techniques of a shielded enclosure, shielded cables, cable-shielded trunking system [47], power and signal line filters [48,49], isolation transformers [50,51], EMI gaskets, and electro-optical/optoelectronic converters. It is thus seen from Figure 11.22 that the EMI emission arising from within the system does not leak to the outside and vice versa.

11.3.1 High-Interference Immunity Measurement System

A typical measurement system possessing high EMI immunity, derived by incorporating the various noise suppression techniques discussed above, is shown in Figure 11.23. The low-level signal transmission system, described here, employs a double-shielded cable CB that is grounded at a single point *A* at the transmitter end by joining the two shields together and connecting it to the low-voltage end of the measuring resistor of the divider [28,52]. If a double-shielded cable is not available, a single-shielded cable encased in a copper tube can be utilized. The signal monitor, an oscilloscope in this case, is located inside a shielded enclosure. Inserting ferrite cores or beads over the inner shield of the signal cable improves the signal-to-noise ratio. Because the oscilloscope chassis should not be connected to the power supply ground, it should be supplied with AC mains through an isolation transformer. A power line filter can be added after the isolation transformer for more effective screening of mains-borne interference. The shielding effectiveness of the measurement system will not be affected if the single-point earth is shifted to the shielded enclosure instead of at the transmitter end. When a large number of diagnostic cables are involved, the cables can be bundled together and carried in a diagnostic conduit made of galvanized iron. Good RF joints between conduit interconnections should be obtained by the use of conductive elastomer gaskets [53]. While bundling the cables together, it is advisable to separate out the power cables from the low-level signal cables. The conduit serves as an additional shield, in addition to its purpose of routing the cables to the various diagnostic sensors. The ends of the conduit should be joined to the cable shields at one end and to the shielded enclosure at the other end.

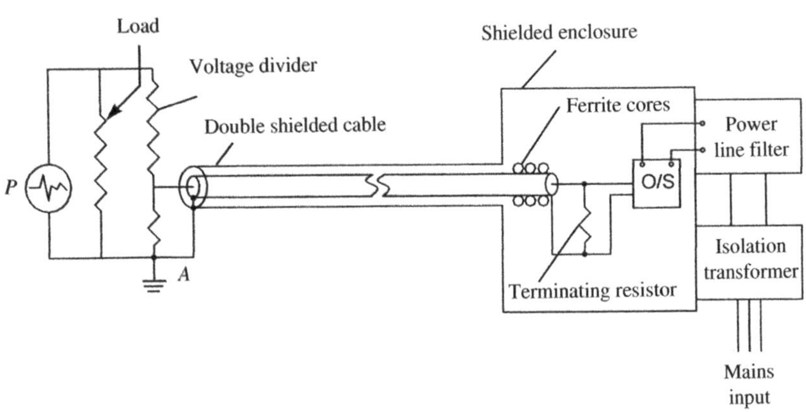

Figure 11.23 A measurement system with high EMI immunity.

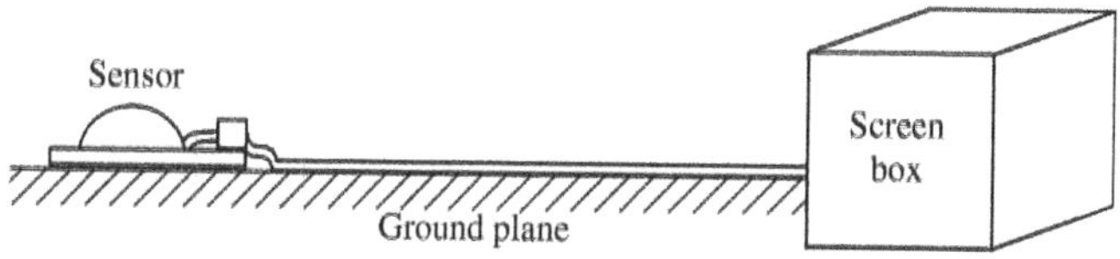

Figure 11.24 Configuration of sensor, signal cable, and screen box containing a recorder. The measurement cable is placed in contact with the ground plane.

11.3.2 Immunity Technique for Free Field Measurements

In measuring EM field parameters, a sensor as well as a recorder is used. Even when good measurements are taken, the signal can be corrupted while being transmitted to the recording device if sufficient care is not taken to fit these cables into the experimental configuration [54–56] The requirements to be met are as follows: (i) The cables should not disturb the EM quantities to be measured. (ii) The current and charge-per-unit-length magnitudes on the signal-carrying coaxial cables should be minimized to reduce the noise pickup on the recorded signal. These requirements are met by making the instrumentation cabling part of the conductor topology, as shown in Figure 11.24, where the cable runs along the conductor and the cable shield makes a continuous electrical contact with it. Because the cable leaving the sensor does not protrude into the upper region, it does not affect the field parameters present there. In addition, the sensor ground plane makes a perfect electrical contact with the conductor surface, thus using it as a local ground plane, which behaves as a part of the sensor. The recorder is placed in a screen box, which should make a good electrical contact with the experimental conductor.

Sometimes, when the size of the screen box is large, it could cause significant EM scattering to the sensor introducing noise on the sensor output. Such effects can be reduced by locating the screen box far away from the sensor or/and positioning the sensor and screen box such that the experimental conductor shadows the sensor. A good example of this would be positioning the sensor and screen box on the opposite sides of an aircraft fuselage or wing.

11.4 Design Examples

Example 11.1

An open wire is carrying a transient pulse with a peak current of 7 kA and has a 10–90% rise time of 3 μs. The lines are located 4 in. over a ground plane. The RG58C/U cable is terminated by resistors $R_c = 1\ \text{k}\Omega$ and $R_d = 3\ \text{k}\Omega$

(continued)

(continued)

Solution

In this case, the open wire is the source and the shielded wire is the victim circuit. The commercial cable is RG 58 C/U, a single-braid cable with a characteristic impedance of 50 Ω. The parameters are available from the manufacturer's product literature as follows:

Shield diameter:	$d_s = 0.116$ in.
Shield resistance	$R_s = 4.7\ \text{m}\Omega/\text{ft}$
Inner conductor diameter	$d_i = 0.035$ in.
Cable inductance per unit length	$\tilde{L}_c = 0.07\ \mu\text{H/ft}$

The time constant τ is calculated from the 10–90% rise time,

The time constant	$\tau = \frac{t_r}{2.2} = \frac{3}{2.2}\ \mu\text{s} = 1.35\ \mu\text{s}$
The cable inductance	$L_c = \tilde{L}_c \cdot \ell = 0.073 \frac{\mu\text{H}}{\text{ft}} \times 10\,\text{ft} = 0.73\ \mu\text{H}$
Time constant of victim circuit	$\frac{L_c}{R_c + R_d} = \frac{0.73\ \mu\text{H}}{1\ \text{k}\Omega + 3\ \text{k}\Omega} = 0.18\ \text{ns}$

The time constant of the victim circuit is significantly smaller than the time constant of the source circuit τ and the voltage effectively divides between R_c and R_d. The maximum induced voltage is given by (11.20) as

$$V_d^{max} = \frac{MI_1}{\tau} \cdot \frac{R_d}{R_c + R_d} \cdot \frac{1}{(L_{s2}/R_{s2}\tau) - 1} \cdot \left(e^{-(R_{s2}/L_{s2}) \cdot t_m} - e^{-t_m/\tau}\right)$$

The mutual inductance is calculated from (11.27) or read from Figure 11.7 with $h/D = 4$,

$$M = 0.07 \log_{10}\left[1 + \left(\frac{2 \times 4}{1}\right)^2\right] \times 10 = 1.3\ \mu\text{H}$$

The voltage division ratio is $\frac{R_d}{R_d + R_d} = \frac{3\ \text{k}\Omega}{1\ \text{k}\Omega + 3\ \text{k}\Omega} = \frac{3}{4}$

The dimensionless factor $\frac{R_{s2} \cdot \tau}{L_{s2}} = 0.021$

The time t_m at which the peak of the induced voltage occurs may be calculated from (11.13):

$$\tau_m = (1.35)\frac{\ln(0.021)}{(0.021 - 1)} = 5.3\ \mu\text{s}$$

The effective shielding factor $\frac{1}{(L_{s2}/R_{s2}\tau)-1}\cdot\left(e^{-(R_{s2}/L_{s2})\cdot t_m}-e^{-t_m/\tau}\right)=0.019$

The induced voltage

$$V_d^{max}=\frac{MI_1}{\tau}\cdot\frac{R_d}{R_c+R_d}\cdot\frac{1}{(L_{s2}/R_{s2}\tau)-1}\cdot\left(e^{-(R_{s2}/L_{s2})\cdot t_m}-e^{-t_m/\tau}\right)$$

$$=\left\{\frac{(1.3)}{1.35}\times\frac{3}{4}\times(0.019)\right\}\times I_1=96\,\text{V}$$

The open wire carrying a signal with a 3 μs rise time and a peak current of 7 kA induces a peak voltage of 96 V on the shielded cable.

Example 11.2

For commercial iron sheet ($\mu_r=200$, $\sigma_r=0.17$) of 1mm thickness, calculate the shielding effectiveness for plane waves at various frequencies from 10 Hz to 10 MHz, for every increase of one order.

Solution

As per Equation 11.35, the overall shielding effectiveness *SE* for plane waves is given by

$$SE=A+R+B\approx A+R$$

where *A* is the absorption loss and *B* is the reflection loss. The formula for *A* as given by Equation 11.36 is

$$A=8.686\cdot t\sqrt{\pi\mu\sigma f}=131\cdot t\sqrt{\pi\mu_r\sigma_r f}$$

Substituting the values of $t=10^{-3}$ m, $\mu_r=200$, $\sigma_r=0.17$, the absorption loss is calculated to be

$$A=0.764\sqrt{f} \tag{11.21}$$

Similarly, as per Equation 11.37, the reflection loss *R* is given by

$$R=20\log\frac{|1+K|^2}{4|K|}\approx 20\log\frac{K}{4}=20\log\frac{Z_w}{4Z_s}$$

Substituting the values of $Z_w=120\pi$ and $Z_s=3.69\times10^{-7}\cdot\sqrt{\sigma_r/(\mu_r f)}$, the equation for *R* works out to

$$R=20\log\cdot\left(\frac{74\times10^5}{\sqrt{f}}\right) \tag{11.22}$$

(continued)

(continued)

Table 11.3 Shielding effectiveness for plane waves for iron of 1mm thickness.

Freq (*f*)	Z_s (Ω)	$K = Z_w/Z_s$	*A* (dB)	*R* (dB)	*SE*=*A*+*R*
10	4×10^{-5}	9.4×10^{6}	2.4	127.4	129.8
100	1.26×10^{-4}	3×10^{6}	7.64	117.5	125.1
10^3	4×10^{-4}	9.4×10^{5}	24.0	107.4	131.4
10^4	1.26×10^{-3}	3×10^{5}	76.4	97.5	174.0
10^5	4×10^{-3}	9.4×10^{4}	241.0	87.4	328.4
10^6	1.26×10^{-2}	3×10^{4}	764.0	77.5	841.5
10^7	4×10^{-2}	9.4×10^{3}	2416.0	67.4	2483.4

The values of shielding effectiveness *SE*, the sum of absorption loss *A*, and reflection loss *R* for various frequencies are listed in Table 11.3.

From Table 11.3, the important points to note are as follows: (i) The absorption loss is low at low frequencies and reflection loss is high at low frequencies for iron. (ii) The shielding at high frequencies is mainly due to absorption loss in case of iron. (iii) Comparison between copper in Table 11.1 and iron in Table 11.3 indicates that their properties for shielding effectiveness are complementary, copper being good at high frequencies due to its high electrical conductivity and iron being good at low frequencies due to its high magnetic permeability. This is the reason why galvanized iron – which has components of iron and zinc possessing high μ_r and high σ_r respectively – finds wide usage as shielding material in shielded enclosures. (iv) The conductivity term σ_r appears in the numerator of both expressions for *A* and *R*, hence indicating that superconductors are ideal for 100% electromagnetic shielding

Example 11.3

A circuit consists of an open wire of radius 1mm and length of 1m running parallel to a perfectly conducting ground plane at a height of 100mm. It is energized by a signal $V_1 = 1000 \sin \omega t$. The source and load impedances are resistors with values of 0.1 and 100 Ω, respectively. Another low-signal circuit runs nearby, which again has an open wire of radius of 1 mm and a length of 1 m running parallel to the ground plane at a height of 100 mm with resistors of 0.1 and 100 Ω on the source side and load side, respectively. The two open wires are separated by a distance of 10 mm. Calculate the voltage induced in the second circuit due to capacitive coupling.

Solution

The system of circuits is similar to Figure 11.1. The equivalent circuit can be approximated by Figure 11.2. Hence, the capacitively coupled voltage V_2 appearing in Circuit 2 due to the energizing signal V_1 in Circuit 1, as given by Equation 11.1, is written as

$$V_2 = V_1 \cdot \frac{(X'_{c2}||R'_2)}{(X'_{c2}||R'_2) + X'_{cm}} \cdot \frac{Z}{R_1 + Z}$$

where $R'_2 = (R_2||R_{l\ell2})$

$$Z = \left(X'_{C1}\left|\left|R_{\ell1}\right)\right.\right||(X'_{Cm} + (R'_2||X'_{C2}))$$

We can calculate the various circuit parameters as follows:

$$C'_1 = \ell \cdot C_1$$

where

C_1' = lumped capacitance between wire 1 and infinite ground plane

C_1 = capacitance per unit length between wire 1 and infinite ground plane for wire of radius a and height h above ground plane

$$C_1 = \frac{2\pi\varepsilon_0}{\ln(2h/a)}$$

and

$$C'_1 = \ell \cdot \frac{2\pi\varepsilon_0}{\ln(2h/a)} = \frac{2\pi \cdot (8.85 \times 10^{-12})}{\ln((2 \times 100)/1)} = 10.5\ \text{pF}$$

$$X'_{C1} = \frac{1}{\omega C'_1} = \frac{1}{2\pi(100 \times 10^3) \cdot (10.5 \times 10^{-12})} = 150\ \text{k}\Omega$$

Similarly,

$$C'_2 = \ell \cdot C_2$$

where

C'_2 = lumped capacitance between wire 2 and ground plane

C_2 = capacitance per unit length between wire 2 and the infinite ground plane for a wire 2 of radius a and height h above ground plane.

(continued)

(*continued*)

$$C_2 = \frac{2\pi\varepsilon_0}{\ln(2h/a)}$$

$$C_2' = \ell \cdot \frac{2\pi\varepsilon_0}{\ln(2h/a)} = \frac{2\pi \cdot (8.85 \times 10^{-12})}{\ln((2 \times 100)/1)} = 10.5 \text{ pF}$$

$$X'_{C2} = \frac{1}{\omega C_1'} = \frac{1}{2\pi(100 \times 10^3) \cdot (10.5 \times 10^{-12})} = 150 \text{ k}\Omega$$

Similarly,

$$C_m' = \ell \cdot \tilde{C}_m$$

where

C_m' = lumped capacitance between wire 1 and wire 2

and $\tilde{C}_m$ = mutual capacitance per unit length between wire 1 and wire 2 separated by distance D, which is given by

$$\tilde{C}_m = \frac{\pi\varepsilon_0}{\ln(D/a)}$$

$$C_m' = \ell \cdot \frac{\pi\varepsilon_0}{\ln(D/a)} = \frac{\pi \cdot (8.85 \times 10^{-12})}{\ln(10/1)} = 12 \text{ pF}$$

$$X'_{Cm} = \frac{1}{\omega C_m'} = \frac{1}{2\pi(100 \times 10^3) \times (12 \times 10^{-12})} = 130 \text{ k}\Omega$$

Similarly,

$$R_2' = (R_2 \| R_{\ell 2}) = \frac{1}{(1/R_2) + (1/R_{\ell 2})} \approx 0.1 \ \Omega$$

$$(X'_{C1} \| R_{\ell 1}) = \frac{1}{\sqrt{(1/X'_{C1})^2 + (1/R_{\ell 1})^2}} = \frac{1}{\sqrt{(1/10^4) + (1/(225 \times 10^4))}} \approx 100 \ \Omega$$

$$(X'_{C2} || R_2') = \frac{1}{\sqrt{(1/X'_{C2})^2 + (1/R_2')^2}} = \frac{1}{\sqrt{(1/0.1) + (1/(150 \times 10^3))}} \approx 0.1 \ \Omega$$

$$Z = \left(X'_{C_1} || R_{l_1}\right) \Big|\Big| \left(X'_{cm} + R_2' || X'_{C_2}\right) = 100 \ \Omega \| (130 \ k\Omega + 0.1 \ \Omega) \approx 100 \ \Omega$$

Substituting the above calculated parameters into the equation for V_2,

$$V_2 = V_1 \cdot \frac{(X'_{c2}||R'_2)}{(X'_{c2}||R'_2) + X'_{cm}} \cdot \frac{Z}{R_1 + Z}$$

$$= (1000) \cdot \left[\frac{100\,\Omega}{(0.1\,\Omega + 100\,\Omega)} \times \frac{0.1\,\Omega}{(0.1\,\Omega + 130\,k\Omega)}\right] = 0.77\,\text{mV}$$

This example illustrates that a circuit carrying a fast-rising, high-voltage impulse would capacitively couple substantial voltage into a nearby, small-signal circuit, warranting the use of EMI shielding to the source circuit as well as victim circuit.

References

1 L.D. Jambor et al. Parallel Wire Susceptibility Testing for Signal Lines. *IEEE Trans. Electromagn. Compat.*, 8, No. 2, p. 111, 1966.

2 D.E. Merewether and T.F. Ezell, The Effect of Mutual Inductance and Mutual Capacitance on the Transient Response of Braided Shield Coaxial Cables. *IEEE Trans. Electromagn. Compat.*, 18, No. 1, p. 15, 1976.

3 C.W. Harrison, Jr. and C.D. Taylor, Response of a Terminated Transmission Line Excited by a Plane Wave Field for Arbitrary Angles of Incidence. *IEEE Trans. Electromagn. Compat.*, Vol. 15, No. 3, p. 118, 1973.

4 H. Naito and A.V. Shah, Time Domain Measurements of Voltage Induction by Transient Electromagnetic Fields. *IEEE Trans. Instrum. Meas.*, 27, No. 1, p. 38, 1978.

5 V.M. Turesin, Electromagnetic Compatibility Guide for Design Engineers. *IEEE Trans. Electromagn. Compat.*, 9, No. 3, p. 139, 1967.

6 D.A. Miller and J.E. Bridges, Review of Circuit Approach to Calculate Shielding Effectiveness. *IEEE Trans. Electromagn. Compat.*, 10, No. 1, 1968.

7 R.J. Mohr, Coupling Between Open and Shielded Wire Lines over a Ground Plane. *IEEE Trans. Electromagn. Compat.*, 9, No. 2, p. 34, 1967.

8 C.R. Paul, Prediction of Cross Talk in Ribbon Cables: Comparison of Model Predictions and Experimental Results. *IEEE Trans. Electromagn. Compat.*, 20, No. 3, p. 394, 1978.

9 J.H. Pluck, *Electromagnetic Shielding and Suppression Techniques*, RFI Industries Pty. Ltd., Australia.

10 E.S. Kesney, Shielded Enclosures for EMC and Tempest Testing, *Guide for Design of Shielded Facilities*, Keene Corporation, Ray Proof Division, 1975.

11 E.S. Kesney, Construction Methods and Evaluation of RFI/EMC Shielded Construction Systems, *Guide for Design of Shielded Facilities*, Keene Corporation, Ray Proof Division, 1975.

12 P. Zais, *Use an RF Shield to Prevent EMI from Altering Your Results*, Industrial Research and Development, March 1983.

13 P.R. Bannister, New Theoretical Expressions for Predicting Shielding Effectiveness for the Plane Shielded Cases. *IEEE Trans. Electromagn. Compat.*, 10, No. 1, p. 2, 1968.

14 R.B. Schulz et al., Low Frequency Shielding Resonance. *IEEE Trans. Electromagn. Compat.*, 10, No. 1, p. 7, 1968.

15 R.B. Schulz, ELF and VLF Shielding Effectiveness of High Permeability Materials. *IEEE Trans. Electromagn. Compat.*, 10, No. 1, p. 95, 1968.

16 L.F. Babcock, Shielding Circuits from EMP. *IEEE Trans. Electromagn. Compat.*, 9, No. 2, p. 45, 1967.

17 R.B. Cowdell, New Dimensions in Shielding. *IEEE Trans. Electromagn. Compat.*, 10, No. 1, p. 158, 1968.

18 S.L. O'Yound et al., Survey of Techniques for Measuring R.F. Shielding Enclosures. *IEEE Trans. Electromagn. Compat.*, 10, No. 1, p. 72, 1968.

19 E.F. Vance, Electromagnetic-Interference Control. *IEEE Trans. Electromagn. Compat.*, Vol. 22, No. 4, pp. 319–328, 1980.

20 E.L. Breen, *The Application of B-dot and D-dot Sensors to Aircraft Skin Surfaces*, Instrumentation Development Memo 3, July 1974.

21 C.E. Baum, Electromagnetic Sensors and Measurement Techniques, in *Fast Electrical and Optical Measurements*, J.E. Thompson and L.H. Luessen, eds., Vol. 1, NATO ASI Series E: Applied Sciences, p. 73, 1986.

22 A.J. Schwab, Electromagnetic Interference in Impulse Measuring Systems. *IEEE Trans. Power Apparatus Syst.*, 93, No. 1, p. 333, 1974.

23 L.J. Greenstein and H.G. Tobin, Analysis of Cable Coupled Interference. *IEEE Trans. Radio Freq. Interference*, Vol. 5, No. 1, p. 43, 1963.

24 G.I. Chandler, II, Design of the CTX Diagnostics Screen Room. *8th Symposium on Engineering Problems in Fusion Research, Chicago, IL*, October 26–29, 1981.

25 B.A. Gregory, *An Introduction to Electrical Instrumentation and Measurement System*, 2nd edn, Macmillan Press Ltd., 1981.

26 J.G. Sverak et al., Safe Substation Grounding: Part I. *IEEE Trans. Power Apparatus Syst.*, 100, No. 9, p. 4281, 1981.

27 J.G. Sverak et al., Safe Substation Grounding: Part II. *IEEE Trans. Power Apparatus Syst.*, 101, No. 10, p. 4006, 1982.

28 G.I. Chandler, II, *Grounding, Shielding and Transient Suppression Techniques Developed for the Control and Data Acquisition System*, Engineering Problems in Fusion Research, Chicago, IL, 1981.

29 R.A. Fitch, Salving Diagnostics in the Pulse Power Environment. *Proceedings of the Pulsed Power Conference*, Paper IIID5, 1976.

30 M.E. Norris, TFTR Diagnostics Grounding. *Proceedings of the 9th Symposium on Engineering Problems in Fusion Research*, p. 1134, 1981.

31 F. Faulkner et al., TFTR Grounding System, *Proceedings of the 9th Symposium on Engineering Problems in Fusion Research*, p. 1438, 1981.
32 C. Jensen, Grounding Principles and Practice II: Establishing Grounds. *Electr. Eng.*, Vol. 4, No. 2, p. 68, 1945.
33 R. Rudenberg, Grounding Principles and Practice I: Fundamental Considerations on Ground Currents. *Electr. Eng.*, Vol. 64, No. 1, p. 1, 1945.
34 G. Chandler et al., Use of Fiber Optics to Eliminate Ground Loops and Maintain Shielding Integrity on a Controlled Fusion Experiment. *Proceedings of the Society of Photo Optical Instrumentation Engineers*, Vol. 190, LASL Optical Conference, May 23–25, 1979.
35 H.I. Bassen and R.J. Hoss, An Optically Linked Telemetry System for Use with Electromagnetic Field Measurement Probes. *IEEE Trans. Electromagn. Compat.*, 20, No. 4, p. 483, 1978.
36 R.M. Vines et al., Noise on Residential Power Distribution Circuits. *IEEE Trans. Electromagn. Compat.*, 26, No. 4, p. 161, 1984.
37 D.B. Clark et al., Power Line Insertion Loss Evaluated in Operational Type Circuits. *IEEE Trans. Electromagn. Compat.*, 10, No. 2, p. 243, 1968.
38 H.M. Hoffart, Electromagnetic Interference Reduction Filters. *IEEE Trans. Electromagn. Compat.*, 10, No. 2, p. 225, 1968.
39 F. Mayer, Absorptive Lines as RFI Filters. *IEEE Trans. Electromagn. Compat.*, 10, No. 2, p. 224, 1968.
40 H. Weidmann and W.J. McMartan, Two Worst-Case Insertion Loss Test Methods for Passive Power Line Interference Filter. *IEEE Trans. Electromagn. Compat.*, 10, No. 2, p. 257, 1968.
41 H.W. Denny and W.B. Warren, Lossy Transmission Line Filters. *IEEE Trans. Electromagn. Compat.*, 10, No. 4, p. 363, 1968.
42 Lindgren, *Power Line Filters*. Available at www.ets-lindgren.com/pdf/N192XSeries.pdf accessed December 17, 2015.
43 S.H. Eisbruck and F.A. Giordano, A Survey of Power Line Filter Measurement Techniques. *IEEE Trans. Electromagn. Compat.*, 10, No. 2, p. 238, 1968.
44 P.C. Laughlin, Ultra-Isolation Transformers. *Instruments and Control Systems*, Vol. 34, p. 2250, 1961.
45 P.H. Ron, Concepts in Electromagnetic Interference and Protection, in *Electromagn. Compat.*, G.K. Deb, ed., Institution of Electrical and Telecommunication Engineers, Delhi, India, 1995.
46 E.K. Miller and J.A. Landt, Direct and Time Domain Techniques for Transient Radiation and Scattering from Wires. *Proc. IEEE*, Vol. 68, No. 11, p. 1396, 1980.
47 Falcon Trunking Systems, Cable Management Systems, http://www.falcontrunking.co.uk/pdf/falcon_electrical_cat.pdf, Issue 12, 2016 accessed 4/23/201
48 LMI Lectromagnetics Inc., OT Lectroline Filters, Catalogue No. F 300 A, 1985.

49 J. Brown, Understanding how Ferrites can Prevent and Eliminate RF Interference to Audio Systems, http://audiosystemsgroup.com/SAC0305Ferrites.pdf; accessed 4/23/17

50 G. Chandler, *Grounding and Shielding, Computer Interfacing and Control of Pulsed Power Systems*, Pulsed Power Lecture Series, No. 37, Texas Tech. University, Lubbock, TX.

51 R.C. McLoughlin, Ultra Isolation Transformers. *Instrum. Control Syst.*, Vol. 34, p. 2250, 1961.

52 A.J. Schwab, *High Voltage Measurement Techniques*, MIT Press, 1972.

53 EMI Shielding Theory and Gasket Design Guide, Chomerics, https://www.sealingdevices.com/documents/EMI%20Shielding%20Theory%20Gasket%20Design%20Guide%20of%20Chomerics.pdf; last accessed 4/28/2017

54 K.S.H. Lee, ed., *EMP Interaction: Principles, Techniques, and Reference Data*, Hemisphere Publishing, 1986.

55 O. Hartel, *Electromagnetic Compatibility by Design*, 4th edn, R&B Enterprises, West Conshohocken, PA, 1995.

56 C.E. Baum, The Role of Scattering in Electromagnetic Interference Problems, in *Electromagnetic Scattering*, P.L.E. Uslenghi, ed., Academic Press, 1978.

2

M Topology for Interference Control

terference control for large systems is complicated because of the physical d electrical complexity of the systems. For instance, the interaction of lsed energy with a large aerospace or communication system is not nited to coupling with the system's external structure: Energy can pene- ate the system's outer surface and may propagate in the interior, couple to bles, and finally manifest as voltages and currents. Electromagnetic pology allows partitioning of complex systems into tractable subsystems. nce the complex system has been decomposed into simpler systems, the stem can be examined for compromising apertures and penetrating nductors. Ultimately, it is easier and cheaper to use the good practices ustrated here to maintain shielding integrity than to search for interfer- ce signals. Electromagnetic topology results in a systemic approach to erference control.

The fundamental problem of interference control is the protection of a tential victim from the source [1], as shown in Figure 12.1. In the figure, S_I presents the intense interfering external source, which may be nuclear- nerated electromagnetic pulses [2], high-power microwaves [3,4], ultra- deband radiating systems [5], or lightning [6,7] that has the potential to luce high voltage and high current transients in the victim electronics tem S_V. Since the magnitudes of the induced voltage and current transients ıld be many orders of magnitude higher than the signal amplitudes in it will lead to malfunction or damage of S_V [8–10]. A broadband, high- plitude source, such as a pulsed power system, needs equally broadband lation techniques for its control. At high peak amplitudes and high quencies, conventional shielding techniques are inadequate, and the nature electromagnetic interference (EMI) control becomes maintaining shield egrity [8–13]. The fundamental step in maintaining shield integrity is to ablish the topology of the electromagnetic shields within the facility or tem. The electromagnetic topology considers the "closedness" of the ducting surfaces and their nesting properties (both series and parallel),

dations of Pulsed Power Technology, First Edition. Jane Lehr and Pralhad Ron.

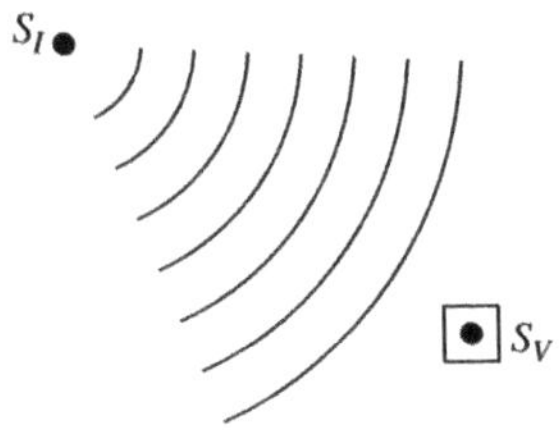

Figure 12.1 Schematic of an interfering source S_I and its victim electronics S_V.

which can attenuate or shield the incident electromagnetic energy as it propagates through the system.

The subject discussed in this chapter, namely, electromagnetic topology and its relation to interference control, differs from that discussed in Chapter 11, where infinite planar shields, plane waves, and low frequencies were considered. Electromagnetic topology results in an overall system approach that gives a high confidence in eliminating electromagnetic interference and meeting system performance requirements. This chapter covers the various aspects involved in partitioning of systems by nested enclosures and topological concepts to preserve shielding integrity. Electromagnetic energy can penetrate into an enclosure either through apertures and penetrating conductors in the cavity surface or by diffusion through the cavity walls. The concept of nested enclosures is introduced in Section 12.1; then the chapter is organized in order of relative importance for maintaining the shielding integrity of nested enclosures. Section 12.2 shows the proper techniques of passing grounds and other conductors through the shield. Section 12.3 discusses the effects of apertures on shielding effectiveness and Section 12.4 shows how shielding effectiveness can be degraded by fields diffusing through conductive surfaces. The chapter also includes a design example to calculate the voltage induced in a metallic loop placed on a spherical shield of different metals and thicknesses.

12.1 Topological Design

The essence of EMI control for pulsed power systems is to separate principal volumes into zones, establish shielding integrity, and maintain it by limiting interfering currents and fields that can breach it [14–17]. In practical systems, there are often several layers of shielding forming a set of nested shields, as illustrated in Figures 12.2–12.7. Each layer of the nested shields provides a barrier to EMI. In the actual systems, the outermost shield may represent a building, housing sophisticated equipment, or an aircraft, for example. The next shielding layer may be cabinets containing electronic systems or a diagnostic screen room. Additional shielding layers may be another metallic cabinet for additional protection for the enclosed electronic systems.

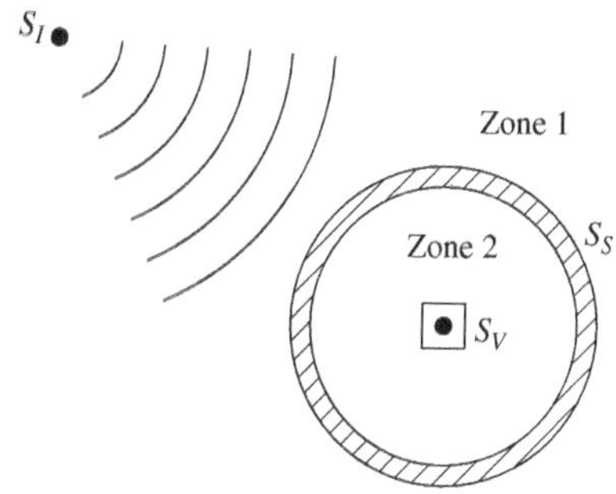

Figure 12.2 An interfering source S_I located in Zone 1 impinges on the victim electronics S_V located in Zone 2. Zone 1 and Zone 2 are defined by the metallic shield surface S_S.

12.1.1 Series Decomposition

The ideal solution for the protection from EMI would be to totally enclose S_V by a continuous metallic surface S_S possessing infinite conductivity σ and without any surface discontinuities, as shown in Figure 12.2. However, it is usually not possible to implement because in a practical system S_V would likely need many inputs, such as power, signal cables, cooling tubes, or ventilation windows. The practical protection approach, therefore, is shown in Figure 12.3, where S_S is provided with a penetration discontinuity to accommodate the input requirements for S_V. However, this results in leakage of electric fields E_0 magnetic fields H_0, or electromagnetic fields P_0, which would be incident on S_V and hence interfere with its functioning. Therefore, the practical considerations of providing an aperture in the shield have reduced the shielding efficiency of S_S.

To enhance shielding effectiveness, it may be necessary to use two nested shields, S_{S1} and S_{S2}, with both of the shields having penetration discontinuities, as shown in Figure 12.4. If powerful EMI sources are located in S_V itself, it may necessitate another nested shield S_{S3} in the subsystem of S_V, as shown in Figure 12.5, to contain the interference generated from S_V itself.

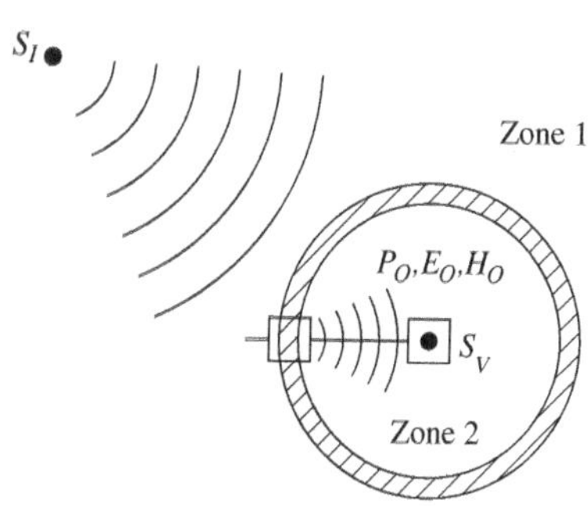

Figure 12.3 A model of an interfering source S_I and victim electronics S_V when separated by a metallic shield surface S_S with a penetration for system inputs. The penetrating discontinuity is done improperly and allows interfering signals to enter the enclosure.

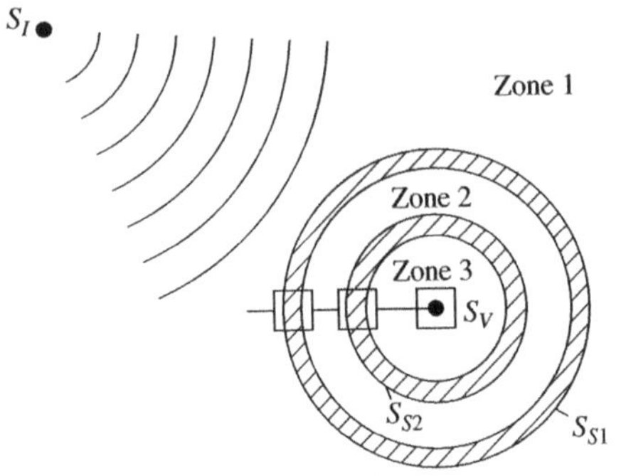

Figure 12.4 The nested shields S_{S1} and S_{S2} separating an interfering source S_I and its victim electronics S_V define Zone 1, Zone 2, and Zone 3.

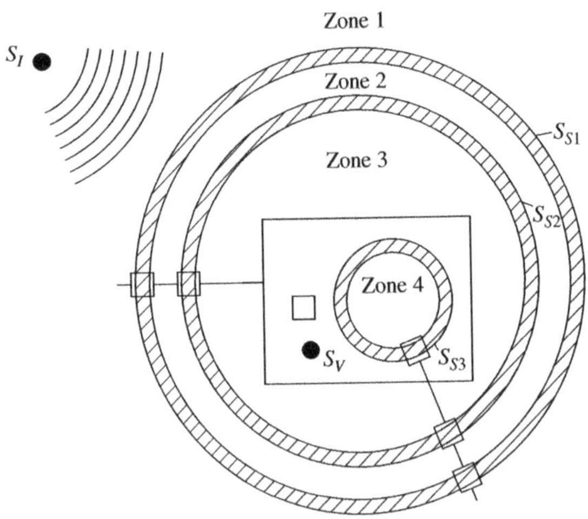

Figure 12.5 Schematic of nested shields including S_{S3} for containing radiating signals from S_V.

12.1.2 Parallel Decomposition

Parallel decomposition is another topology that can be used to achieve a greater degree of shielding effectiveness for a given system. In parallel decomposition, the nested shields partition the space around the electronic equipment, for example, inside the first level of shielding S_{S1}, as shown in Figure 12.6. As in series decomposition, proper grounding techniques must be used to maintain shielding integrity.

The subshields S_{S2}^1 and S_{S2}^2 may also be connected in such a way as to become a single zone, Zone 3, as shown in Figure 12.7. This can be easily implemented by using shielded cable to connect the two subshields, for instance. Parallel decomposition has particular appeal for some applications such as the

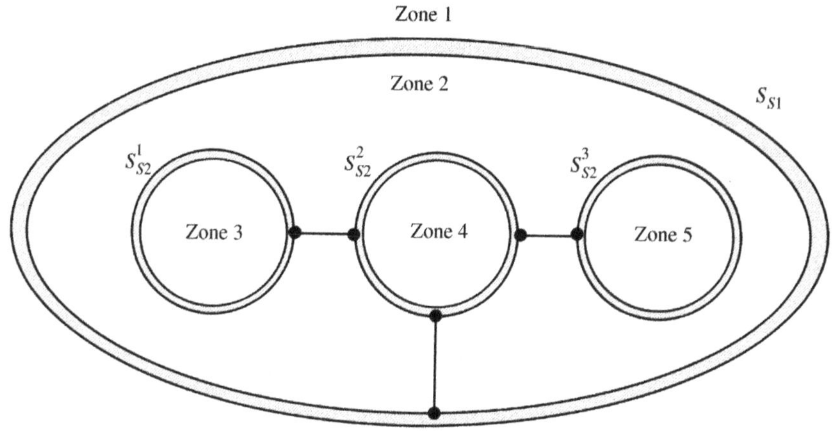

Figure 12.6 Parallel decomposition of an electromagnetic topology.

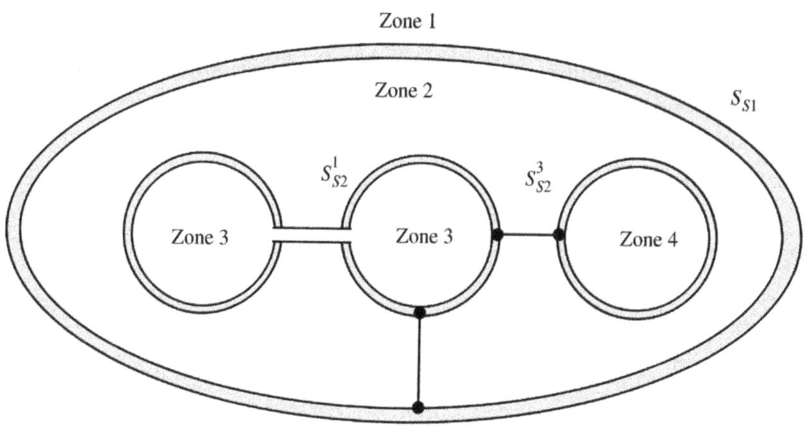

Figure 12.7 A shielded cable is used to connect two subshields together forming one topological zone, Zone 3.

aerospace industry, where the outer shield (the aircraft skin) is a given part of the electromagnetic topology. It has been shown that by putting the power source in Zone 2, the number of necessary filters is minimized to transport power to a subshield [18].

12.2 Shield Penetrations

As noted in the previous section, we frequently must penetrate the shield in order to supply power and communicate with equipment inside the shield. If the enclosure is a building, ventilation as well as plumbing, heat, or fuel will also need to be supplied. Each of these openings and penetrating conductors represent a compromise of the shield integrity. Inherent to the concepts of electromagnetic topology is the fact that currents on conductors attach to the shield flow predominantly on the surface to which the conductor is attached.

12.2.1 Necessity for Grounding

Whereas the shield is used to control internal potential differences of external origin, grounding is used to control potential differences of internal origin. If the isolated shields (S_{S1}, S_{S2}, S_{S3}), shown in Figures 12.5–12.7, are left ungrounded, they may assume high potentials due to static charges, which from the point of view of safety is not acceptable. The various shield surfaces, therefore, need to be grounded, as shown in Figure 12.8. The parameters L and R represent the inductance and resistance of the grounding conductor of Zone 0. The potential to which the shield surface will rise ($V = iR + L\,di/dt$) depends on the current i flowing through the ground conductor of Zone 0 and the complex impedance of the ground conductor.

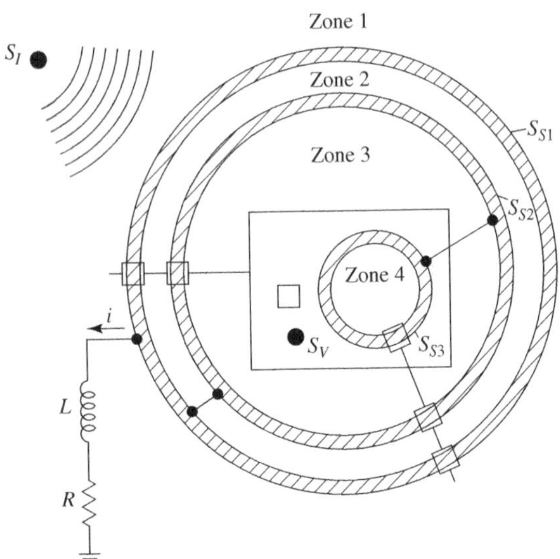

Figure 12.8 Proper grounding, discussed in the next section, preserves the shielding integrity of each nested enclosure. The connections inside and outside a shield are made at different locations to avoid corrosion leading to conductive feedthrough.

12.2.2 Grounding Conductors

Grounding, in general, is more of a safety consideration than an interference control technique. However, if done incorrectly, grounding can completely defeat a shield and distribute the interference signal. Good technique is simple: Topologically, ground conductors should never penetrate shield surfaces.

The various methods of connecting grounding conductors to the shield surface are shown in Figure 12.9. In Figure 12.9a, the external grounding conductor is connected to the outer Surface 1-1 in Zone 0. The transient

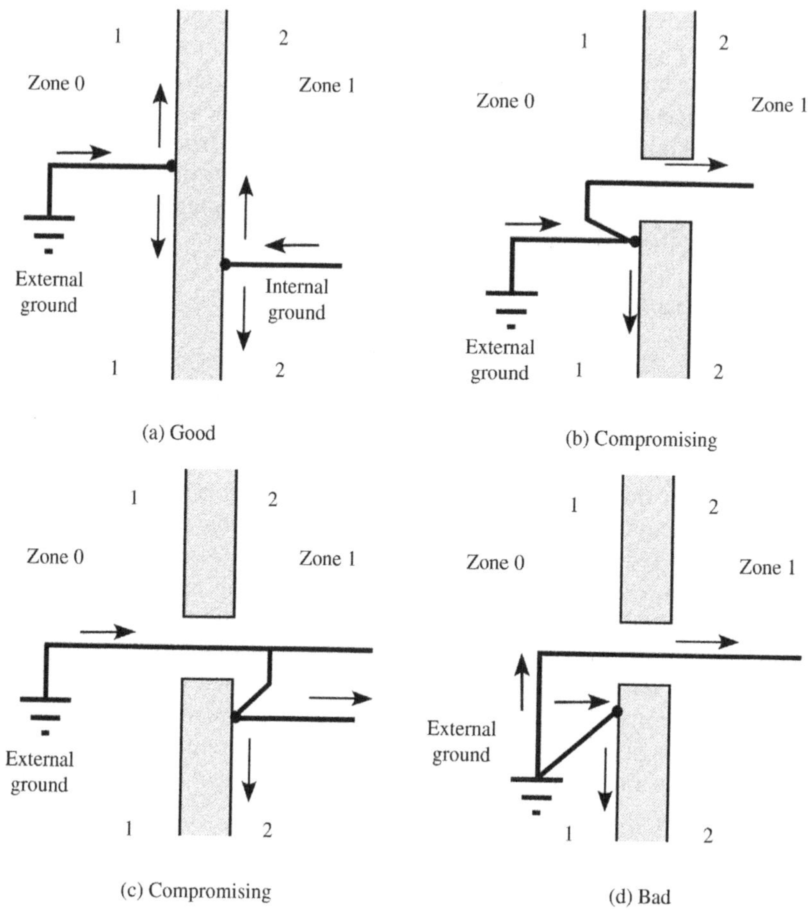

Figure 12.9 Methods of connecting grounding conductors to surface shields and their performance. For illustrations (a) through (d) Surface 1-1 is the outer shield and Surface 2-2 is the inner shield surface [1]. (Reproduced with permission from the Summa Foundation.)

currents carried by the grounding conductor are distributed to the outer Surface 1-1. Due to the skin effect, most of the current will flow on Surface 1-1, and if the shields are designed optimally, the current on Surface 2-2 in Zone 1 will be negligible. Considering that any current on Surface 2-2 would produce interference in Zone 1, this method of grounding connection is good. In Figure 12.9b, although the external ground is connected to Surface 1-1, the grounding connection required for the other nested shield enclosures for the inside zones is taken into the protected environment through a penetration discontinuity. Considering that the grounding conductor carries current, this would produce interference in the protected zone by capacitive, inductive, radiative, and ground loop coupling. Thus, this method of grounding connection is not advisable. The method of grounding connection shown in Figure 12.9c is worse, because it results in interference in the protected zone both by the high current on the inner Surface 2-2 as well as the interference current carried by the grounding conductor itself. The method of taking out the grounding conductor right from the external ground into the protected zone through the penetration discontinuity, as shown in Figure 12.9d, is the worst, because it carries very large interference currents from the ground.

12.2.3 Groundable Conductors

Groundable conductors may be transmission line cable shields, waveguide, or pipes to carry water or ventilating air that need to be grounded. Figure 12.10a shows the method where the penetrating tube is secured firmly to the penetration hole in the shield. The preferred method is welding rather than bolting. In this method, the interfering current carried by the tube is diverted to the outer Surface 1-1 of the shield and negligible current is carried into the protected Zone 1. This method, therefore, results in the best performance toward electromagnetic compatibility. The method shown in Figure 12.10b is not advisable, because the penetrating tube partially carries the interfering current along with it into the protected Zone 1. In addition, the current density on the inner Surface 2-2 is more than that on outer Surface 1-1, which again makes the protected zone vulnerable. The worst method is as shown in Figure 12.10c, where the entire interfering current is carried by the penetrating tube into the safe zone, without any diversion of this current onto the shield surface.

12.2.4 Insulated Conductors

Examples of insulated conductors are power input cables or signal cables, which may be AC, DC, or impulse and diagnostic cables for carrying voltage or current signals from sensors. These insulated conductors also carry interference

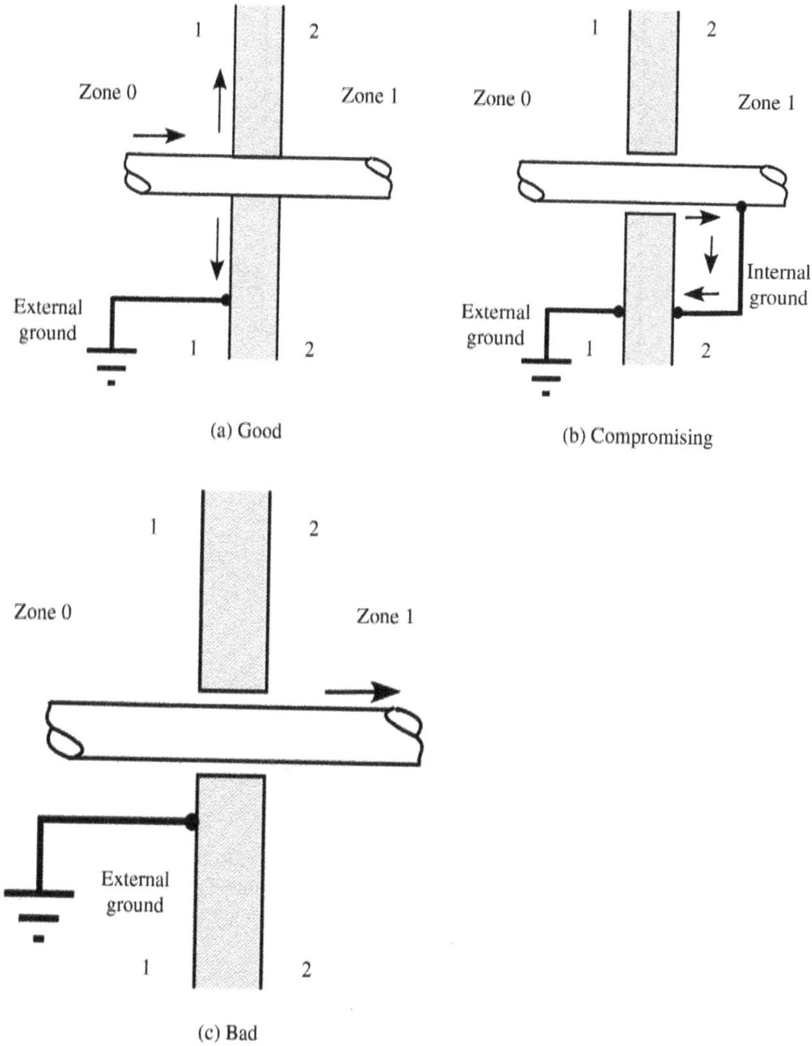

Figure 12.10 Methods of connecting groundable conductors to shield surfaces and their performance. Surface 1-1 is the outer shield surface and Surface 2-2 is the inner shield surface [1]. (Reproduced with permission from the Summa Foundation.)

currents and it is imperative that these interfering currents should not be carried from one zone into another. These currents must be diverted onto the corresponding shield surface by means of power line filters, signal line filters, and surge diverters, whose details have already been discussed in Chapter 10. The insulated conductors, along with their accompanying filters and diverters,

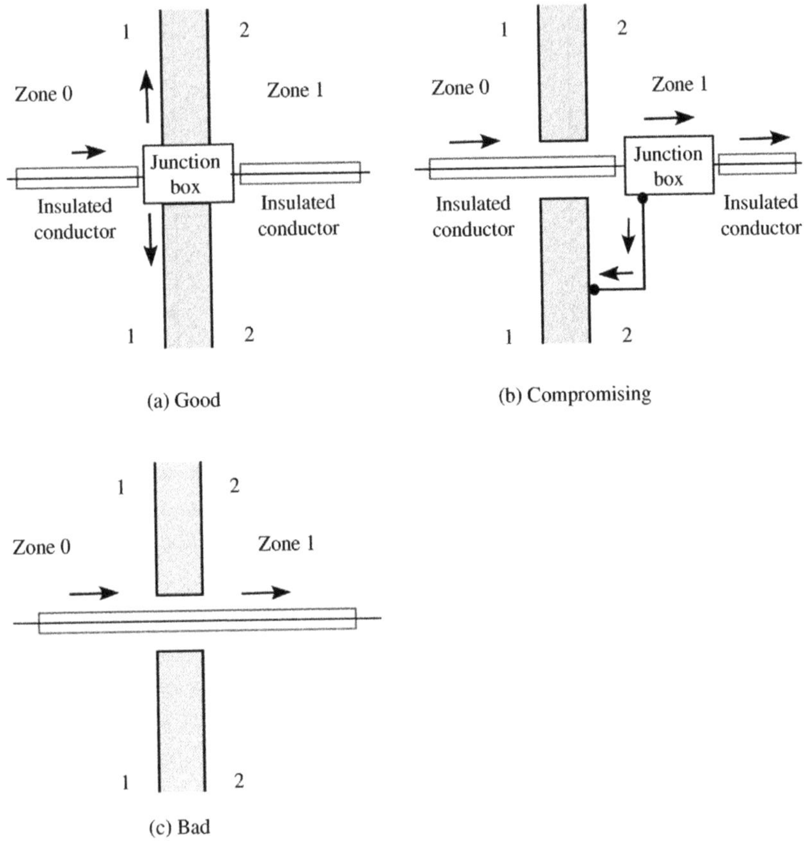

Figure 12.11 Methods of connecting junction boxes containing insulated conductors to shield surfaces and their performance. Surface 1-1 is the outer shield and Surface 2-2 is the inner shield surface [1]. (Reproduced with permission from the Summa Foundation.)

should be bunched together on the basis of voltage/current magnitudes. Such bunches should be enclosed in metallic shielding boxes and the interfering currents diverted to the box surface. These metal boxes should be mechanically secured and electrically connected to the corresponding shields separating the electromagnetic zones. The methods of connecting metallic junction boxes to the shield are shown in Figure 12.11. The condition when the junction box is secured to the shield with a good mechanical and electrical connection is shown in Figure 12.11a. An ideal method of securing the junction box to the shield is by welding. In this case, no interference current is carried by the insulated conductors into the protected zone and the entire interfering current is diverted to the outer shield Surface 1-1 via the junction box. The current density distribution is high on the outer shield surface and minimal on the inner shield surface. It is therefore seen that this method of connection produces good shielding integrity resulting in high electromagnetic compatibility. It should be noted that commercial filters exist that mount to the screen room wall, either inside or outside, but topologically it is the same as given in Figure 12.11a. Figure 12.11b shows the method where the junction box is located in protected Zone 1 and connection to the shield is done on the inner shield surface. This results in bad performance because the interfering current is carried by the insulated conductor partially into the protected zone, resulting in a high current density on the inner shield Surface 2-2. Both these factors contribute to poor performance with respect to electromagnetic compatibility. The scheme shown in Figure 12.11c results in the worst performance because of the absence of the junction box; the insulated conductors carry the entire interfering signal into the protected zones because it is a transmission line that supports TEM modes.

12.3 Shield Apertures

The best possible configuration is to have a completely closed surface with no apertures in the shield where both the magnetic and electric fields are confined within its origination zone. Apertures, however, are usually required for access such as windows and doors, ventilation, and a host of other facility requirements. These apertures must be designed so that access is provided without compromising the electromagnetic compatibility of the system.

The degree to which an aperture degrades the electromagnetic integrity of a shield depends on wavelength of the impinging field [9]. Figure 12.12 illustrates

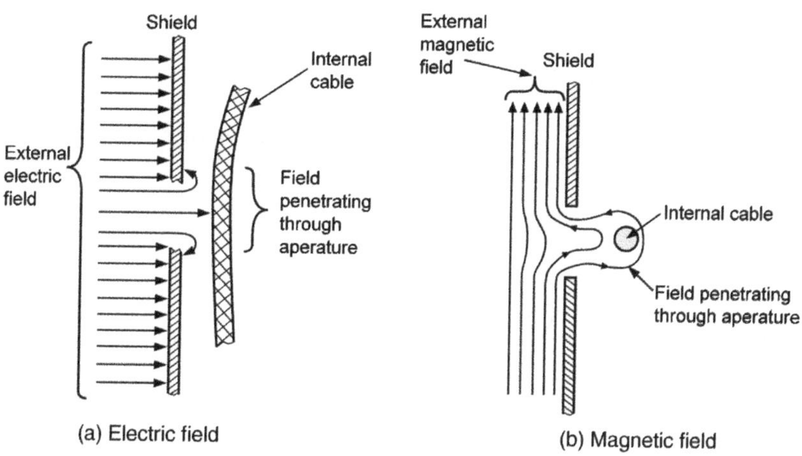

Figure 12.12 Penetration of small apertures by an electromagnetic wave [1]. (Reproduced with permission from the Summa Foundation.)

fields where the aperture is small compared to the wavelength of external source. In Figure 12.12a, a portion of the external electric field that would have terminated on the outside of the shield fringes through the aperture and induces charges on the inner shield and internal cables. Similarly, in Figure 12.12b, a portion of the magnetic field that would have been bound by surface currents in the shield fringes through the aperture and induces a voltage on an internal cable.

In Figure 12.12, the aperture is small compared to the wavelength of external field, and the penetration is greatly reduced from the impinging wave. On the other hand, if the aperture is large compared to the wavelength of the external field, the incident wave can "shine through" the aperture, as shown in Figure 12.13. The wave behaves optically, with the wave reflected from the conducting surface but the portion that is incident on the aperture transmitted into the shield. The "shine-through" wave is attenuated very little as it penetrates the aperture, and nearly the peak field strength is transmitted into the interior of the shield, compromising the integrity of the shield.

Large apertures are sometimes needed and the observation that the extent of field penetration depends on aperture size can be exploited. Figure 12.14a shows a magnetic field penetrating a single large aperture through which the magnetic field diffuses deep into the protected Zone 1 and induces transients due to *dB/dt* induced effects, which may be large. The penetration of the

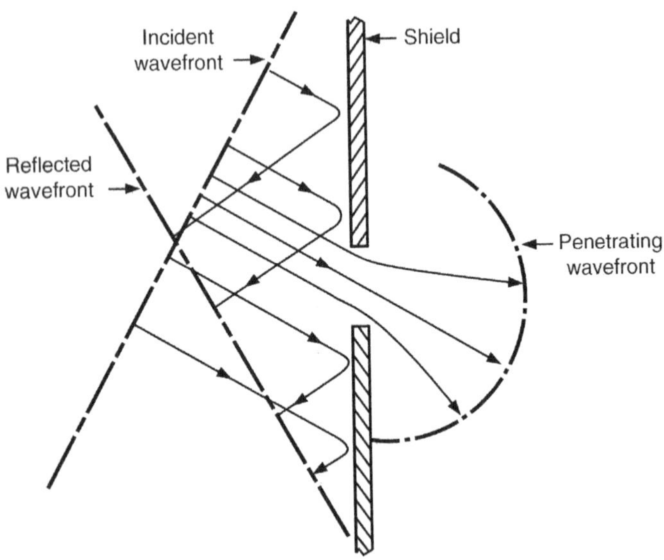

Figure 12.13 Penetration of a large aperture allows the impinging wave to shine through into the shielding enclosure [1]. (Reproduced with permission from the Summa Foundation.)

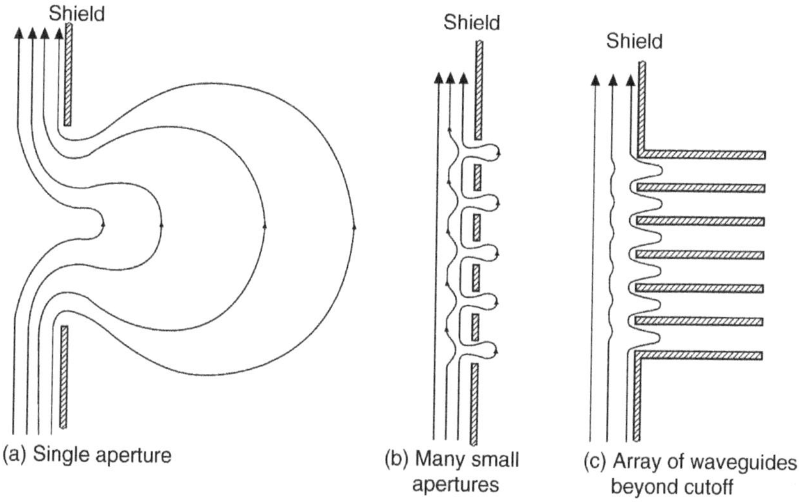

Figure 12.14 The penetration of apertures by a field [1]. (Reproduced with permission from the Summa Foundation.)

magnetic field through the large aperture can be reduced substantially by configuring the single large aperture into a number of smaller apertures, as shown in Figure 12.14b. One common technique to make these small apertures is to use a metallic screen that is electrically connected to the enclosure to cover a large aperture. The situation can be further improved by adding depth to the many small apertures. The length of the apertures is sized to be waveguides beyond cutoff, as shown in Figure 12.14c, where the penetration can be reduced drastically.

Optical fibers do not conduct electricity; thus, may be used to mitigate EMI problems. However, some optical fibers have metal covers or metal cladding. In this case, the metal must be discontinued as the shield is penetrated. It may be convenient to convert electrical signals into optical signals via electro-optical techniques and then transport the fibers through proper shield apertures, as discussed in this section, from one zone to another and then reconvert into electrical signals. The details for some electro-optical techniques are discussed in Chapters 10 and 11.

12.4 Diffusive Penetration

Continuously closed shields are, by far, the most effective electromagnetic shields because they severely limit the penetration of high frequencies and reflect propagating waves throughout the spectrum. This section applies only to continuously closed surfaces. The effects of apertures, penetrations, and grounding conductors are much more damaging to shielding integrity.

Diffusive penetration of electromagnetic waves into conductors is a result of its conductivity. Because materials have a finite conductivity, the shield is not perfectly conducting and the impinging waves are not totally reflected and the resultant current density penetrates into the metal of the enclosure. Diffusion is a function of frequency f and is characterized by the skin depth δ expressed as

$$\delta = \sqrt{\frac{1}{\pi f \sigma \mu}} \tag{12.1}$$

where the electrical conductivity σ and magnetic permeability μ are material properties of the enclosure wall.

The relation between the wall's skin depth and its thickness is illustrated in Figure 12.15. If the wall is highly conductive or very thick so that the skin depth is much less than the wall thickness, as shown in Figure 12.15b, the current density all resides on the outside of the wall and the enclosure is effectively shielded. The current density for the electrically thin case, where the wall thickness is less that the skin depth, is shown schematically in Figure 12.15c. A current density J_2 resides on the inside surface of the enclosure, resulting in a magnetic field inside the enclosure. In practice, the conductivity of the metallic walls is usually very high, and the true source of the penetrating fields is the normal component of the impinging magnetic field.

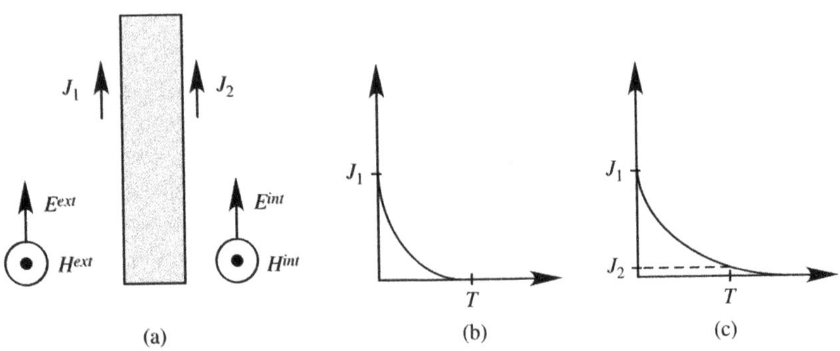

Figure 12.15 The relation between skin depth δ and the wall thickness T of the shielding enclosure determines the extent of diffusive penetration. (a) The external electromagnetic field impinges on a conductive wall of thickness T, inducing a current density J_1 on the outer surface of the wall. (b) When the wall thickness is much greater than the skin depth, the enclosure is effectively shielded. (c) If the skin depth is greater than the thickness of the wall, a current of magnitude J_2 will reside on the inner surface of the wall.

The effectiveness of the enclosure can be evaluated by comparing its wall thickness to the skin depth of the dominant or characteristic frequency of the incident pulse, $2\pi f_0$ (=$c\ \tau_0$).

$\delta \ll T$	The wall is electrically thick; the fields on the inside of the enclosure are much smaller than on the outside. Most of the incoming electromagnetic energy is shielded out by attenuation and reflection.
$\delta \gg T$	The wall is electrically thin; the diffusion mechanism allows for effective penetration by the normal component of magnetic field; diffusion by electric fields is negligible; the shape of the enclosure becomes important due to focusing effects.
$\delta \sim T$	Diffusive penetration becomes important; using typical values for wall materials and thicknesses corresponds to frequencies on the order of hundreds of kilohertz.

12.4.1 Cavity Fields

Cavity fields can be calculated for certain simple enclosure shapes for which working formulas can be derived. From Figure 12.16, let an electromagnetic pulse $H^{ext}(t)$ of width τ_0 be incident on a closed metallic shell of conductivity σ, permeability μ, and wall thickness T. Let V be the volume of the interior of the metallic shell, S be its surface area, and the penetrating pulse be denoted $H^{int}(t)$.

The nature of the internal pulse can be described by examining the important time parameters. These are the width of the external (incident) pulse τ_0; the fall time τ_f of the penetrant pulse; the transit time τ_a (= r_a/c) for

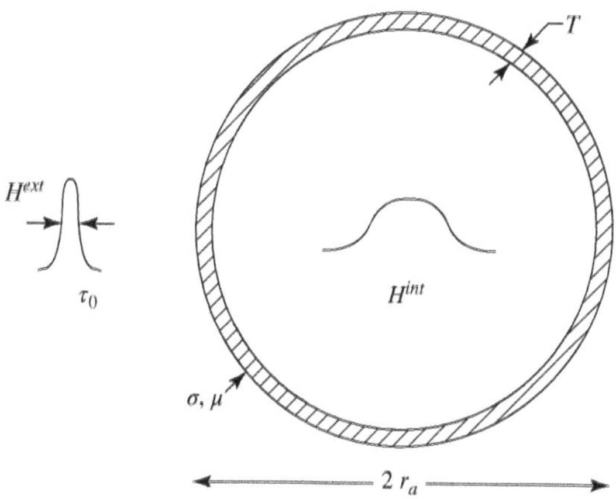

Figure 12.16 An external pulse impinging on a completely closed, conductive spherical shell of wall thickness T. The penetrant pulse is "smeared" by diffusing through the wall [1]. (Reproduced with permission from the Summa Foundation.)

a signal travelling through the enclosure; and the diffusion time τ_D of the enclosure wall given by

$$\tau_D = \sigma \mu T^2 \tag{12.2}$$

For an enclosure made from aluminum several millimeters thick and several meters in diameter, the transit time is on the order of tens of nanoseconds and the diffusion time is on the order of tens of microseconds. The pulse is "smeared" by the effective *L/R* decay time of the induced eddy current loops in the enclosure. Thus, $H^{ext}(t)$ may be considered an impulse function compared to H^{int} because $\tau_0 \ll \tau_D$. For incident pulse widths on the order of hundreds of nanoseconds, the fall time of the penetrating pulse is on the order of tens of milliseconds.

The time evolution of the pulse is as follows:

$t=0$	$H^{ext}(t)$ arrives at the outer surface of the enclosure.
$t<\tau_D$	$H^{int}(t)$ is insignificant since the diffusion time τ_D has not yet elapsed.
$t\sim\tau_D$	$H^{int}(t)$ reaches a peak value and $H^{ext}(t)$ has long passed the enclosure.
$t>\tau_D$	The enclosure de-energizes due to the decay of eddy currents [14].

12.4.1.1 Frequency Domain Solutions

There are three very useful shapes for interference control that have been extensively studied: parallel plates, cylindrical shells, and spherical shells. In this section, only the spherical shell will be examined because it can be used to represent a facility enclosure because the dimensions are all finite. An analysis for the other canonical shapes can be found in the excellent reference book by Lee [1]. It should be noted that the shielding effectiveness depends on the shape of the shield structure as well as the conductivity, permeability, and thickness of the shell. The shape is important because it concentrates the field within the enclosure.

The equations to be solved in the Laplace domain are

$$\begin{aligned} \nabla \times \vec{E}(\vec{r},s) &= -s\vec{B}(\vec{r},s) \\ \nabla \times \vec{H}(\vec{r},s) &= \begin{cases} \sigma \cdot \vec{E}(\vec{r},s) & \text{inside the shell} \\ 0 & \text{outside the shell} \end{cases} \end{aligned} \tag{12.3}$$

with the boundary conditions on the tangential components of $\vec{E}$ and $\vec{H}$ being continuous at the interface. It is assumed that both the interior and exterior of the enclosure have the permeability of free space, and that the wall itself has permeability μ.

Define the transfer function $T(s)$ as

$$T(s) = \frac{H^{int}(s)}{H^{ext}(s)} \tag{12.4}$$

Solving the basic Equation 12.4 yields the transfer function for the spherical shell $T_s(s)$ with radius r_a as

$$T_S(s) = \frac{3}{3\cosh\left(\sqrt{s\tau_d}\right) + \left(C\sqrt{s\tau_d} + \left(2/(C\sqrt{s\tau_d})\right)\right)\sinh\left(\sqrt{s\tau_d}\right)} \tag{12.5}$$

where

$$C = \frac{\mu_0 r_a}{\mu T} \quad \text{and} \quad s = j\omega = j2\pi f$$

The quantity C is a function of the size and material of an enclosure and makes a convenient parameter for calculations. The above expression for the transfer function $T_S(s)$ can be simplified for electrically thin shells:

$$T_S(s) = \frac{1}{1 + s\mu_0\sigma T(r_a/3)} \tag{12.6}$$

This formula is valid for physically thin shells and when $f \ll (c/2r_a)$ and the enclosure dimension r_a is large. Typically, these expressions are valid to within 5% [1].

In general, the shielding effectiveness SE can be calculated from the transfer function.

$$SE = 20\log\left|\frac{H^{ext}}{H^{int}}\right| = -20\log|T_S(s)|\ \text{dB} \tag{12.7}$$

The SE for a thin-walled sphere is graphed in Figure 12.17, using the transfer function $T_S(s)$ given in Equation 12.6.

12.4.1.2 Time Domain Solutions

For many pulsed power applications, the external field H^{ext} can be treated as an impulse function relative to the internal cavity field because $\tau_0 \ll \tau_D$. Defining an effective impulse strength as

$$H_0 = \int_{-\infty}^{\infty} H^{ext}(t)dt \tag{12.8}$$

The external field can be expressed as

$$H^{ext}(t,x) = H_0\delta\left(t - \frac{x}{c}\right)$$

For highly conductive shells, the parameter C is large, and the transfer function is

$$T_s(s) = \frac{1}{\cosh\left(\sqrt{s\tau_D}\right) + 3C\sqrt{s\tau_D}\sinh\left(\sqrt{s\tau_D}\right)} \tag{12.9}$$

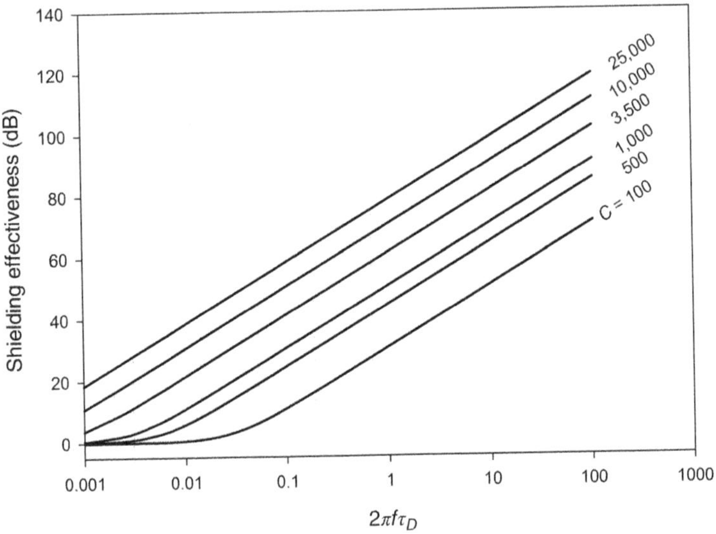

Figure 12.17 The shielding effectiveness *SE* for a thin spherical shell as a function of the normalized frequency $2\pi\tau_D f$ for various values of the "size and material" parameter *C*.

and the time history of the cavity field is then

$$H^{int}(t) = \frac{H_0}{2\pi \cdot j} \int_{-j\infty+\Omega}^{j\infty+\Omega} \frac{e^{st}\,ds}{\cosh\left(\sqrt{s\tau_D}\right) + 3C\sqrt{s\tau_D}\sinh\left(\sqrt{s\tau_D}\right)} \tag{12.10}$$

This integral can be computed by means of an infinite series. The early time approximation of the above is given as

$$H^{int}(t) \cong \frac{2H_0}{3C\tau_D\sqrt{\pi}}\sqrt{\frac{\tau_D}{\tau}}e^{-(\tau_D/4t)}, \quad \text{for} \quad \frac{t}{\tau_D} \le 0.1 \tag{12.11}$$

The intermediate and late time approximation is

$$H^{int}(t) \cong \frac{H_0}{3C\tau_D}\left[e^{-(t/3C\tau_D)} - 2e^{-(\pi^2 t/\tau_D)} + 2e^{-(4\pi^2 t/\tau_D)}\right], \quad \text{for} \quad \frac{t}{\tau_D} \ge 0.1 \tag{12.12}$$

These expressions describe the entire time history of the cavity for conditions where the shells are electrically thin and the diffusion time across the shell is much greater than the external pulse width. The normalized magnitude of the internal magnetic field H^{int} is plotted for various values of *C* in Figure 12.18. These two expressions describe the entire time history of the diffusive field in the cavity to within an accuracy of 0.1% [15].

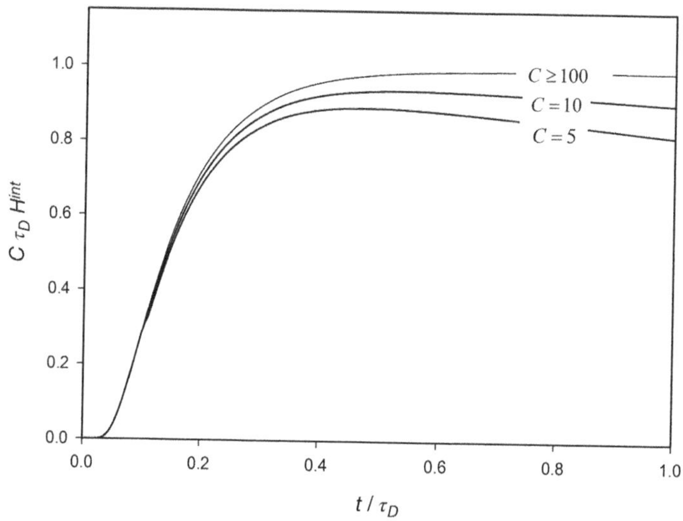

Figure 12.18 The time variations of the penetrant pulse subject to a unit impulse for various values of the "size and material" parameter C [1]. (Reproduced with permission from the Summa Foundation.)

12.4.2 Single Panel Entry

In the above section, the diffusion of electromagnetic fields through the walls of the enclosure was described. Impinging electromagnetic fields diffuse through the conductive walls of the shielding enclosure and induce surface currents on the inside surface. However, if the enclosure walls carry no current, then fields cannot diffuse into the cavity. Thus, the performance of the shield can be enhanced if large currents can be prevented from flowing over large areas of the shielding enclosure. This is particularly important if the shielding enclosure has many openings such as mesh walls or a shield with many openings or poorly bonded joints.

Recognizing the importance of limiting the area over which the induced surface currents flow leads to the concept of a single-entry panel. Figure 12.19a shows the implementation of the single-entry panel concept where *all penetrating conductors* and the *external grounding conductors* are located on one side of the enclosure. Moreover, there should be only *one* such panel anywhere on the shield. The current flowing over the shield surface is small because there is no exit path on other parts of the shield. The currents entering on one conductor must either be reflected back on the same conductor or leave through one of the other conductors on the single panel. In contrast is Figure 12.19b, where entry points are placed in various locations across the shielding surface, allowing surface currents to flow over large areas of the shield,

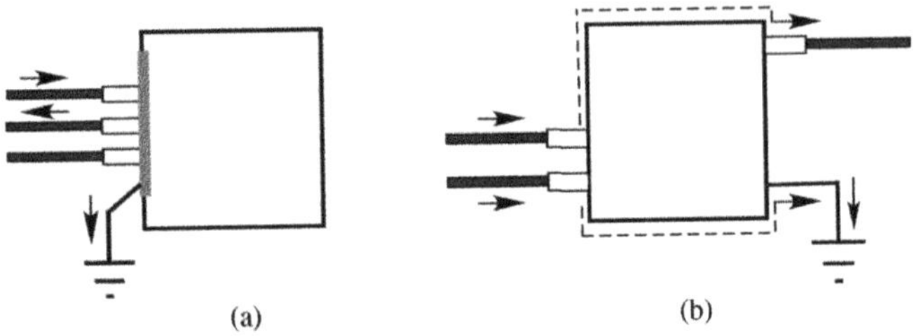

Figure 12.19 The placement of external penetrations can influence the shielding integrity of the enclosure. A single-entry panel, illustrated in part (a), confines the surface currents and minimizes the induced penetrating fields. The shielding integrity of the enclosure is maintained. When the external penetrations are randomly distributed over the enclosure, surface currents flow, resulting in a degradation of its shielding integrity [1]. (Reproduced with permission from the Summa Foundation.)

exciting any leaks in its path, and exiting through a path on the opposite wall. Whereas the random entry approach permits excitation of any flaws in the shielding surface, the single-entry panel approach concentrates the surface currents on the entry panel, thereby preserving the integrity of the shield. Conversely, the single-entry panel allows for poorer shielding quality on the remainder of the shield.

12.4.3 Voltages Induced by Diffusive Penetration

For a closed metal sphere, the voltage induced by diffusive penetration through a shielding enclosure can be estimated by the voltage induced by an impulse in an inductive loop the same size as the spherical enclosure [8]. The loop has an area of πr_a^2 with peak induced voltage V_{pk} of

$$V_{pk} \approx \pi r_a^2 \mu_0 H_0 \left(\frac{H_{pk}}{t_{pk}} \right) \tag{12.13}$$

where H_0 is the amplitude of the incident impulse, H_{pk} is the peak internal magnetic field, and t_{pk} is the time at which the peak occurs.

12.5 Design Examples

Example 12.1

A magnetic intensity impulse of strength $H_0 = 3.5 \times 10^{-5}$ As/m is incident on a copper enclosure whose diameter is 10 m. Calculate the voltage induced inside the cavity by diffusive penetration for the cases when the wall thickness T is 0.1,

0.5, 1.5, and 10 mm. The conductivity of copper is 5.8×10^7 $(\Omega\,m)^{-1}$ and the relative permeability is 1.

Solution

For a closed metal shield, the voltage induced on conductors inside the enclosure can be estimated by the peak voltage measured in a loop the same physical size as the enclosure. This induced voltage is given by Equation 12.13.

To calculate the peak internal magnetic field, we must first determine the diffusion time τ_D and the parameter C, which is related to the size and material of the enclosure. The expressions for τ_D and C are

$$C = \frac{r_a}{3\mu_r T} \quad \text{and} \quad \tau_D = \mu \sigma T^2$$

T (mm)	C	τ_D
0.1	33,333	0.72 μs
0.5	6,666	18.2 μs
1	3,333	72 μs
5	666	1.8 ms
10	333	7.3 ms

From Figure 12.16 we see that the normalized penetrating magnetic field at the center of a spherical enclosure saturates at a value of approximately 1 for values of the parameter C above 100 and the maximum occurs at a time $t/\tau_D \sim 0.4$. Thus,

$$C\,\tau_D H^{int} \cong 1 \quad \text{and} \quad \frac{t}{\tau_D} \cong 0.3$$

Equation 12.13 then becomes

$$V_{pk} \approx \frac{\pi\, r_a^2 \mu_0 H_0}{0.3 C \tau_D^2}$$

For a copper enclosure with a dimension of 10 m, the voltages induced by diffusion through the walls of various thicknesses are as follows:

(continued)

(*continued*)

T (mm)	V_{pk}
0.1	2.6 V
0.5	20.8 mV
1	2.6 mV
5	21 μV
10	2.6 μV

From the above table we can see that a very thin shield will suffice to reduce the induced voltages to well below the level of internally generated voltages.

Example 12.2

Given a shielding enclosure for a high-power microwave system operating at 1 GHz, what size aperture is possible in the shielding enclosure without shine-through? What about an EMP system operating at 300 MHz?

Solution

a) The wavelength associated with a 1 GHz HPM source is

$$\lambda = \frac{c}{\nu} = \frac{3 \times 10^8 \text{ m/s}}{1 \times 10^9 \text{ Hz}} = 0.3 \text{ m}$$

Therefore, any aperture on the order of 0.1 m, or 10 cm, would be considered a large aperture.

b) If the source was an EMP source with significant spectral components of around 300 MHz, the associated wavelength is

$$\lambda = \frac{c}{\nu} = \frac{3 \times 10^8 \text{ m/s}}{300 \times 10^6 \text{ Hz}} = 1 \text{ m}$$

and only large windows and doors would be susceptible to shine-through penetrations.

References

1 E.F. Vance, Electromagnetic Interference Control. *IEEE Trans. Electromagn. Compat.*, Vol. 22, No. 4, pp. 319–328, 1980.

2 K.S.H. Lee, *EMP Interaction: Principles, Techniques, and Reference Data*, Hemisphere Publishing, 1986.

3 C.D. Taylor and D.V. Giri, *High-Power Microwave Systems and Effects*, Taylor & Francis, 1994.
4 J. Benford, J. Swegle and E. Schamiloglu, High Power Microwaves, Third Edition, CRC press, New York, 2015 J. Swegle, *High-Power Microwaves*, Artech House, 1992.
5 D.V. Giri, *High Power Electromagnetic Radiators*, Harvard University Press, 2004.
6 M.A. Uman, *Lightning*, Dover Publications, 1982.
7 R.L. Gardner, *Lightning Electromagnetics*, Hemisphere Publishing, 1990.
8 F.M. Tesche, M.V. Ianoz, and T. Karlsson, *EMC Analysis Methods and Computational Models*, Wiley-IEEE, 1997.
9 C.E. Baum, Electromagnetic Topology for the Analysis and Design of Complex Electromagnetic Systems, in *Fast Electrical and Optical Measurements*, NATO ASI Series E: Applied Sciences, J.E. Thompson and L.H. Luessen, Vol. 1, Martinus Nijhoff Publishers, pp. 467–548, 1986.
10 C.E. Baum, Electromagnetic Sensors and Measurement Techniques, *Fast Electrical and Optical Measurements*, NATO ASI Series E: Applied Sciences, J.E. Thompson and L.H. Luessen, Vol. 1, Martinus Nijhoff Publishers, pp. 467–548, 1986.
11 C.E. Baum, The Theory of Electromagnetic Interference Control, in *Modern Radio Science 1990*, J. Bach Anderson, Oxford University Press, 1990.
12 F.M. Tesche, Topological Concepts for Internal EMP Interaction. *IEEE Trans. Electromagn. Compat.*, Vol. 20, No. 1, pp. 60–64, 1978.
13 E.F. Vance and W. Graf, The Role of Shielding in Interference Control. *IEEE Trans. Electromagn. Compat.*, Vol. 30, No. 3, 294–297, 1988.
14 W. Graf and E.F. Vance, Shielding Effectiveness and Electromagnetic Protection, *IEEE Trans. Elect. Comp.* Vol. 30, No. 3, pp 289–293, 1988.
15 G. Bedrosian and K.S.H. Lee, EMP Penetration Through Metal Skin Panels and into Aircraft Cavities, Interaction Notes, Note 314, August 1976.
16 C.W. Harrison, Jr., and C.H. Papas, On the Attenuation of Transient Field by Imperfectly Conducting Spherical Shells, Interaction Notes, Note 34, July 1964.
17 K.S.H. Lee and G. Bedrosian, Diffusive Electromagnetic Penetration into Metallic Enclosures, *IEEE Trans. Ant. Prop.* Vol. AP-27, No. 2, pp. 194–198, 1979.
18 C.E. Baum, Sublayer Sets and Relative Shielding Order in Electromagnetic Topology, Interaction Notes, Note 416, April 1982.

ıdex

dations of *Pulsed Power Technology*, First Edition. Jane Lehr and Pralhad Ron.

F

G

H

T